CRC HANDBOOK OF

THERMODYNAMIC DATA *of* POLYMER SOLUTIONS *at* ELEVATED PRESSURES

CRC HANDBOOK OF

THERMODYNAMIC DATA *of* POLYMER SOLUTIONS *at* ELEVATED PRESSURES

Christian Wohlfarth

CRC Press
Taylor & Francis Group
Boca Raton London New York

CRC Press is an imprint of the
Taylor & Francis Group, an **informa** business
A TAYLOR & FRANCIS BOOK

CRC Press
Taylor & Francis Group
6000 Broken Sound Parkway NW, Suite 300
Boca Raton, FL 33487-2742

First issued in paperback 2019

ISBN-13: 978-0-8493-3246-3 (hbk)
ISBN-13: 978-0-367-39330-4 (pbk)
Library of Congress Card Number 2006004732

Library of Congress Cataloging-in-Publication Data

Wohlfarth, C.
    CRC handbook of thermodynamics data of polymer solutions at elevated pressures / Christian Wohlfarth.
        p. cm.
        Includes bibliographical references and index.
        ISBN 0-8493-3246-X (alk. paper)
        1. Polymer solutions--Thermal properties--Handbooks, manuals, etc. 2. Thermodynamics--Handbooks, manuals, etc. I. Title.

    QD381.9.S65W636 2005
    547'.704569--dc22
                                                                        2004058530

PREFACE

Knowledge of thermodynamic data of polymer solutions is a necessity for industrial and laboratory processes. Such data serve as essential tools for understanding the physical behavior of polymer solutions, for studying intermolecular interactions, and for gaining insights into the molecular nature of mixtures. They also provide the necessary basis for any developments of theoretical thermodynamic models. Scientists and engineers in academic and industrial research need such data and will benefit from a careful collection of existing data. The *CRC Handbook of Thermodynamic Data of Polymer Solutions at Elevated Pressures* continues the two former ones, the *CRC Handbook of Thermodynamic Data of Copolymer Solutions* and the *CRC Handbook of Thermodynamic Data of Aqueous Polymer Solutions*, in providing a reliable collection of data for polymer solutions from the original literature.

The *Handbook* is again divided into seven chapters: (1) Introduction, (2) Vapor-Liquid Equilibrium (VLE) Data and Gas Solubilities at Elevated Pressures, (3) Liquid-Liquid Equilibrium (LLE) Data of Polymer Solutions at Elevated Pressures, (4) High-Pressure Fluid Phase Equilibrium (HPPE) Data of Polymer Solutions, (5) Enthalpy Changes in Polymer Solutions at Elevated Pressures, (6) PVT Data of Polymers and Solutions, and (7) Pressure Dependence of the Second Virial Coefficients (A_2) of Polymer Solutions. Finally, appendices quickly route the user to the desired data sets. Thus, the book covers all the necessary areas for researchers and engineers who work in this field.

In comparison with low-molecular systems, the amount of data for polymer solutions at elevated pressures is still rather small. About 650 literature sources were perused for the purpose of this *Handbook*, including some dissertations and diploma papers. About 1600 data sets, i.e., about 400 VLE/gas solubility isotherms, 250 LLE and 800 HPPE data sets, a number of volumetric and enthalpic data and some second virial coefficients, are reported. Additionally, tables of systems are provided where results were published only in graphical form in the original literature to lead the reader to further sources. Data are included only if numerical values were published or authors provided their numerical results by personal communication (and I wish to thank all those who did so). No digitized data have been included in this data collection. The *Handbook* is the first complete overview about this subject in the world's literature. The closing day for the data collection was June, 30, 2004. The *Handbook* results from parts of a more general database, *Thermodynamic Properties of Polymer Systems*, which is continuously updated by the author. Thus, the user who is in need of new additional data sets is kindly invited to ask for new information beyond this book via e-mail at wohlfarth@chemie.uni-halle.de. Additionally, the author will be grateful to users who call his attention to mistakes and make suggestions for improvements.

The *CRC Handbook of Thermodynamic Data of Polymer Solutions at Elevated Pressures* will be useful to researchers, specialists, and engineers working in the fields of polymer science, physical chemistry, chemical engineering, material science, biological science and technology, and those developing computerized predictive packages. The *Handbook* should also be of use as a data source to Ph.D. students and faculty in Chemistry, Physics, Chemical Engineering, Biotechnology, and Materials Science Departments at universities.

Christian Wohlfarth
Merseburg, August 2004

About the Author

Christian Wohlfarth is Associate Professor for Physical Chemistry at Martin Luther University Halle-Wittenberg, Germany. He earned his degree in chemistry in 1974 and wrote his Ph.D. thesis on investigations on the second dielectric virial coefficient and the intermolecular pair potential in 1977, both at Carl Schorlemmer Technical University Merseburg. In 1985, he wrote his habilitation thesis, *Phase Equilibria in Systems with Polymers and Copolymers*, at Technical University Merseburg.

Since then, Dr. Wohlfarth's main research has been related to polymer systems. Currently, his research topics are molecular thermodynamics, continuous thermodynamics, phase equilibria in polymer mixtures and solutions, polymers in supercritical fluids, PVT behavior and equations of state, and sorption properties of polymers, about which he has published approximately 100 original papers. He has written the books *Vapor-Liquid Equilibria of Binary Polymer Solutions*, *CRC Handbook of Thermodynamic Data of Copolymer Solutions*, and *CRC Handbook of Thermodynamic Data of Aqueous Polymer Solutions*.

He is working on the evaluation, correlation, and calculation of thermophysical properties of pure compounds and binary mixtures resulting in six volumes of the *Landolt-Börnstein New Series*. He is a respected contributor to the *CRC Handbook of Chemistry and Physics*.

CONTENTS

5. ENTHALPY CHANGES IN POLYMER SOLUTIONS AT ELEVATED PRESSURES

6. PVT DATA OF POLYMERS AND SOLUTIONS

7. PRESSURE DEPENDENCE OF THE SECOND VIRIAL COEFFICIENTS (A_2) OF POLYMER SOLUTIONS

APPENDICES

1. INTRODUCTION

1.1. Objectives of the handbook

Knowledge of thermodynamic data of polymer solutions is a necessity for industrial and laboratory processes. Furthermore, such data serve as essential tools for understanding the physical behavior of polymer solutions, for studying intermolecular interactions, and for gaining insights into the molecular nature of mixtures. They also provide the necessary basis for any developments of theoretical thermodynamic models. Scientists and engineers in academic and industrial research need such data and will benefit from a careful collection of existing data. However, the database for polymer solutions is still modest in comparison with the enormous amount of data for low-molecular mixtures, and the specialized database for polymer solutions at elevated pressures is even smaller. On the other hand, especially polymer solutions in supercritical fluids are gaining increasing interest (1994MCH, 1997KIR) because of their unique physical properties, and thermodynamic data at elevated pressures are needed for optimizing applications, e.g., separation operations of complex mixtures in the high-pressure synthesis of polymers, recovery of polymer wastes, precipitation, fractionation and purification of polymers, and polymers in green chemistry processes.

Basic information on polymers can be found in the *Polymer Handbook* (1999BRA). Some data books on polymer solutions appeared in the early 1990s (1990BAR, 1992WEN, 1993DAN, and 1994WOH), but most data for polymer solutions have been compiled during the last decade. A data book with information on copolymer solutions appeared in 2001 (2001WOH) and on aqueous polymer solutions in 2004 (2004WOH). No databooks or databases dedicated specially to polymer solutions at elevated pressures presently exist. Thus, the intention of the *Handbook* is to fill this gap and to provide scientists and engineers with an up-to-date compilation from the literature of the available thermodynamic data on polymer solutions at elevated pressures. The *Handbook* does not present theories and models for polymer solution thermodynamics. Other publications (1971YAM, 1990FUJ, 1990KAM, 1999KLE, 1999PRA, and 2001KON) can serve as starting points for investigating those issues.

The data within this book are divided into six chapters:

- Vapor-liquid equilibrium (VLE) data and gas solubilities for binary or ternary polymer solutions
- Liquid-liquid equilibrium (LLE) data of (quasi)binary, ternary, or quaternary polymer solutions
- High-pressure fluid phase equilibrium (HPPE) data of (quasi)binary, ternary, or quaternary polymer solutions
- Enthalpy changes for polymer solutions at elevated pressures
- PVT data of some selected polymers and their solutions as well as excess volumes and densities at elevated pressures
- Second virial coefficients (A_2) of polymer solutions at elevated pressures

Data from investigations applying to more than one chapter are divided and appear in the relevant chapters. Data are included only if numerical values were published or authors provided their results by personal communication (and I wish to thank all those who did so). No digitized data have been included in this data collection, but a number of tables include systems based on data published in graphical form.

1.2. Experimental methods involved

At higher pressures, the classification of experimental data into vapor-liquid, liquid-liquid, or fluid-fluid equilibrium data is not always simple. Also the expression "elevated pressure" is more or less relative. With respect to the latter, data are included here if at least some data points of a system were measured at a pressure above normal pressure. With respect to a classification of experimental data into the chapters 2, 3, or 4, practical reasons as well as theoretical considerations are applied. Practical reasons are often chosen for systems when gas solubility data are given, but for some systems, gas solubilities are given together with the cloud points in chapter 4. Theoretical considerations are taken into account for most polymer systems with subcritical solvents and supercritical fluids. Van Konynenburg and Scott (1980VAN) cited six classes of phase diagrams and showed that almost all known types of phase equilibria of binary mixtures can be classified within their scheme. More recently, Yelash and Kraska (1999YEL) developed global phase diagrams for mixtures of spherical molecules with non-spherical molecules also including polymers. So, most of the systems in this book could at least be categorized into the corresponding chapters in accordance to their rules. Nevertheless, sometimes problems remain for a number of systems with subcritical or supercritical fluid solvents. Therefore, in chapter 4, the type of equilibrium within one data set is sometimes stated individually for each data point.

Accordingly, a classical discussion of experimental methods for vapor-liquid or liquid-liquid equilibrium measurements does not really fit here. Information about experimental methods for polymer solutions at ordinary pressures can be found in (1975BON) and (2000WOH) or in the two former handbooks (2001WOH and 2004WOH). Here, the classification also used by Christov and Dohrn (2002CHR) is chosen where experimental methods for the investigation of high-pressure phase equilibria are divided into two main classes, depending on how the composition is determined: analytical methods (or direct sampling methods) and synthetic methods (or indirect methods).

Analytical methods

Analytical methods involve the determination of the composition of the coexisting phases. This can be done by taking samples from each phase and analyzing them outside the equilibrium cell at normal pressure or by using physicochemical methods of analysis inside the equilibrium cell under pressure. If one needs the determination of more information than the total polymer composition, i.e., if one wants to characterize the polymer with respect to molar mass (distribution) or chemical composition (distribution), the sampling technique is unavoidable.

Withdrawing a large sample from an autoclave causes a considerable pressure drop, which disturbs the phase equilibrium significantly. This pressure drop can be avoided by using a variable-volume cell, by using a buffer autoclave in combination with a syringe pump or by blocking off a large sampling volume from the equilibrium cell before pressure reduction. If only a small sample is withdrawn or if a relatively large equilibrium cell is used, the slight pressure drop does not affect the phase composition significantly. Small samples can be withdrawn using capillaries or special sampling valves. Often sampling valves are directly coupled to analytical equipment.

Analytical methods can be classified as isothermal methods, isobaric-isothermal methods, and isobaric methods.

Using the *isothermal mode*, an equilibrium cell is charged with the system of interest, the mixture is heated to the desired temperature, and this temperature is then kept constant. The pressure is adjusted *in the heterogeneous region* above or below the desired equilibrium value depending on how the equilibrium will change pressure. After intensive mixing over the time necessary for equilibrating the system, the pressure reaches a plateau value. The equilibration time is the most important point for polymer solutions. Due to their (high) viscosity and slow diffusion, equilibration will often need many hours or even days. The pressure can be readjusted by adding or withdrawing of material or by changing the volume of the cell if necessary. Before analyzing the compositions of the coexisting phases, the mixture is usually given some time for a clear phase separation. Sampling through capillaries can lead to differential vaporization (especially for mixtures containing gases and high-boiling solvents as well) when no precautions have been taken to prevent a pressure drop all along the capillary. This problem can be avoided with an experimental design that ensures that most of the pressure drop occurs at the end of the capillary. Sometimes, one or more phases will be recirculated to reduce sampling problems. The use of physicochemical methods of analysis inside the equilibrium cell, e.g., by a spectrometer, avoids the problems related to sampling. On the other hand, time-consuming calibrations can be necessary. At the end, isothermal methods need relatively simple and inexpensive laboratory equipment. If carried out carefully, they can produce reliable results.

Isobaric-isothermal methods are often also called dynamic methods. One or more fluid streams are pumped continuously into a thermostated equilibrium cell. The pressure is kept constant during the experiment by controlling an effluent stream, usually of the vapor phase. One can distinguish between continuous-flow methods and semi-flow methods. In continuous-flow methods, both phases flow through the equilibrium cell. They can be used only for systems where the time needed to attain phase equilibrium is sufficiently short. Therefore, such equipment is usually not applied to polymer solutions. In semi-flow methods, only one phase is flowing while the other stays in the equilibrium phase. They are sometimes called gas-saturation methods or pure-gas circulation methods and can be used to measure gas solubilities in liquids and melts or solubilities of liquid or solid substances in supercritical fluids.

Isobaric methods provide an alternative to direct measurements of pressure-temperature compositions in liquid phase compositions in vapor phase (P-T-w^L-w^V) data. This is the measurement of (P-T-w^L) data followed by a consistent thermodynamic analysis. Such isobaric analytical methods are usually made in dynamic mode, too.

The main advantages of the analytical methods are that systems with more than two components can be studied, several isotherms or isobars can be studied with one filling, and the coexistence data are determined directly. The main disadvantage is that the method is not suitable near critical states or for systems where the phases do not separate well. Furthermore, dynamic methods can be difficult in their application to highly viscous media like concentrated polymer solutions where foaming may cause further problems.

Synthetic methods

In synthetic methods, a mixture of known composition is prepared and the phase equilibrium is observed subsequently in an equilibrium cell (the problem of analyzing fluid mixtures is replaced by the problem of "synthesizing" them). After known amounts of the components have been placed into an equilibrium cell, pressure and temperature are adjusted so that *the mixture is homogeneous*. Then temperature or pressure is varied until formation of a new phase is observed. This is the common way to observe cloud points in demixing polymer systems. No sampling is necessary. Therefore, the experimental equipment is often relatively simple and inexpensive. For multicomponent systems, experiments with synthetic methods yield less information than with analytical methods, because the tie lines cannot be determined without additional experiments. This is specially true for polymer solutions where fractionation accompanies demixing.

The appearance of a new phase is either detected visually or by monitoring physical properties. Visually, the beginning of turbidity in the system or the meniscus in a view cell can be observed. Otherwise, light scattering is the common method to detect the formation of the new phase. Both visual and non-visual synthetic methods are widely used in investigations on polymer systems at elevated pressures. The problem of isorefractive systems (where the coexisting phases have approximately the same refractive index) does not belong to polymer solutions where the usually strong concentration dependence of their refractive index prevents such a behavior. If the total volume of a variable-volume cell can be measured accurately, the appearance of a new phase can be observed from the abrupt change in the slope of a pressure-volume plot. An example is given in chapter 4. The changes of other physical properties like viscosity, ultrasonic absorption, thermal expansion, dielectric constant, heat capacity, or UV and IR absorption are also applied as non-visual synthetic methods for polymer solutions.

A common synthetic method for polymer solutions is the $(P\text{-}T\text{-}w^L)$ experiment. An equilibrium cell is charged with a known amount of polymer, evacuated and thermostated to the measuring temperature. Then the second component (gas, fluid, solvent) is added and the pressure increases. The second component dissolves into the (amorphous or molten) polymer and the pressure in the equilibrium cell decreases. Therefore, this method is sometimes called pressure-decay method. Pressure and temperature are registered after equilibration. No samples are taken. The composition of the vapor phase is calculated using a phase equilibrium model if two or more gases or solvents are involved. The determination of the volume of gas or solvent vaporized in the unoccupied space of the apparatus is important as it can cause serious errors in the determination of their final concentrations. The composition of the liquid phase is often obtained by weighing and using the material balance. By repeating the addition of the second component into the cell, several points along the vapor/gas-liquid equilibrium line can be measured. This method is usually applied for all gas solubility/vapor sorption/vapor pressure investigations in systems with polymers.

The synthetic method is particularly suitable for measurements near critical states. Simultaneous determination of *PVT* data is possible.

Some problems related to systems with polymers

Details of experimental equipment can be found in the original papers compiled for this book and will not be presented here. Only some problems should be summarized that have to be obeyed and solved during the experiment.

The polymer solution is often of an amount of some cm^3 and may contain about 1g of polymer or even more. Therefore, the equilibration of prepared solutions can be difficult and equilibration is usually very time consuming (liquid oligomers do not need so much time, of course). Increasing viscosity makes the preparation of concentrated solutions more and more difficult with further increasing the amount of polymer. Solutions above 50-60 wt% can hardly be prepared (depending on the solvent/polymer pair under investigation). All impurities in the pure solvents have to be eliminated. Degassing of solvents (and sometimes of polymers too) is absolutely necessary. Polymers and solvents must keep dry. Sometimes, inhibitors and antioxidants are added to polymers. They may probably influence the position of the equilibrium. The thermal stability of polymers must be obeyed, otherwise, depolymerization or formation of networks by chemical processes can change the sample during the experiment.

Certain principles must be obeyed for experiments where liquid-liquid equilibrium is observed in polymer-solvent (or supercritical fluid) systems. To understand the results of LLE experiments in polymer solutions, one has to take into account the strong influence of polymer distribution functions on LLE, because fractionation occurs during demixing. Fractionation takes place with respect to molar mass distribution as well as to chemical distribution if copolymers are involved. Fractionation during demixing leads to some effects by which the LLE phase behavior differs from that of an ordinary, strictly binary mixture, because a common polymer solution is a multicomponent system. Cloud-point curves are measured instead of binodals; and per each individual feed concentration of the mixture, two parts of a coexistence curve occur below (for upper critical solution temperature, UCST, behavior) or above the cloud-point curve (for lower critical solution temperature, LCST, behavior), i.e., produce an infinite number of coexistence data.

Distribution functions of the feed polymer belong only to cloud-point data. On the other hand, each pair of coexistence points is characterized by two new and different distribution functions in each coexisting phase. The critical concentration is the only feed concentration where both parts of the coexistence curve meet each other on the cloud-point curve at the critical point that belongs to the feed polymer distribution function. The threshold point (maximum or minimum corresponding to either UCST or LCST behavior) temperature (or pressure) is not equal to the critical point, since the critical point is to be found at a shoulder of the cloud-point curve. Details were discussed by Koningsveld (1968KON, 1972KON). Thus, LLE data have to be specified in the tables as cloud-point or coexistence data, and coexistence data make sense only if the feed concentration is given. This is not always the case, however.

Special methods are necessary to measure the critical point. Only for solutions of monodisperse polymers, the critical point is the maximum (or minimum) of the binodal. Binodals of polymer solutions can be rather broad and flat. Then, the exact position of the critical point can be obtained by the method of the rectilinear diameter:

$$\frac{(\varphi_B{}^I - \varphi_B{}^{II})}{2} - \varphi_{B,crit} \propto (1 - \frac{T}{T_{crit}})^{1-\alpha} \tag{1}$$

where:

$\varphi_B{}^I$	volume fraction of the polymer B in coexisting phase I
$\varphi_B{}^{II}$	volume fraction of the polymer B in coexisting phase II
$\varphi_{B,crit}$	volume fraction of the polymer B at the critical point
T	(measuring) temperature
T_{crit}	critical temperature (LLE)
α	critical exponent

For solutions of polydisperse polymers, such a procedure cannot be used because the critical concentration must be known in advance to measure its corresponding coexistence curve. Two different methods were developed to solve this problem: the phase-volume-ratio method (1968KON) where one uses the fact that this ratio is exactly equal to one only at the critical point, and the coexistence concentration plot (1969WOL) where an isoplethal diagram of values of $\varphi_B{}^I$ and $\varphi_B{}^{II}$ vs. the feed concentration, φ_{0B}, gives the critical point as the intersection of cloud-point curve and shadow curve. Details will not be discussed here. Treating polymer solutions with distribution functions by continuous thermodynamics is reviewed in (1989RAE) and (1990RAE).

Measurement of enthalpy changes in polymer solutions

Up to now, there is only one paper on measuring the excess enthalpy of a polymer solution at elevated pressures (1999LOP) where a liquid oligomer was investigated, see chapter 5. Measurement of the enthalpy change caused by mixing a given amount of a liquid polymer with an amount of pure solvent in a mixing experiment was made there with a flow mixing cell and flow rates were selected to cover the whole composition range. To determine the heat effect, a modified Setaram calorimeter based on heat-flux principle was used.

The enthalpy of mixing, $\Delta_M h$, is given by

$$\Delta_M h = n_A H_A + n_B H_B - (n_A H_{0A} + n_B H_{0B}) \tag{2a}$$

and the excess enthalpy, H^E, is given by

$$H^E = \Delta_M h / (n_A + n_B) \tag{2b}$$

where:

H_A	partial molar enthalpy of solvent A
H_B	partial molar enthalpy of polymer B
H_{0A}	molar enthalpy of pure solvent A
H_{0B}	molar enthalpy of pure polymer B
n_A	amount of substance of solvent A
n_B	amount of substance of polymer B

PVT/density measurement for polymer melt and solution

There are two widely practiced methods for the *PVT* measurement of polymers and polymer solutions:

1. Piston-die technique
2. Confining fluid technique

which were described in detail by Zoller in papers and books (1986ZOL and 1995ZOL). Thus, a short summary is sufficient here.

In the piston-die technique, the material is confined in a rigid die or cylinder, which it has to fill completely. A pressure is applied to the sample as a load on a piston, and the movement of the piston with pressure and temperature changes is used to calculate the specific volume of the sample. Experimental problems concerning solid samples need not be discussed here, since only data for the liquid/molten (equilibrium) state are taken into consideration for this handbook. A typical practical complication is leakage around the piston when low-viscosity melts or solutions are tested. Seals cause an amount of friction leading to uncertainties in the real pressure applied. There are commercial devices as well as laboratory-built machines which have been used in the literature.

In the confining fluid technique, the material is surrounded at all times by a confining (inert) fluid, often mercury, and the combined volume changes of sample and fluid are measured by a suitable technique as a function of temperature and pressure. The volume change of the sample is determined by subtracting the volume change of the confining fluid. A problem with this technique lies in potential interactions between fluid and sample. Precise knowledge of the *PVT* properties of the confining fluid is additionally required. The above-mentioned problems for the piston-die technique can be avoided.

For both techniques, the absolute specific volume of the sample must be known at a single condition of pressure and temperature. Normally, these conditions are chosen to be 298.15 K and normal pressure (101.325 kPa). There are a number of methods to determine specific volumes (or densities) under these conditions. For polymeric samples, hydrostatic weighing or density gradient columns were often used.

The tables in chapter 6 do not provide specific volumes below the melting transition of semicrystalline materials or below the glass transition of amorphous samples, because *PVT* data of solid polymer samples are non-equilibrium data and depend on sample history and experimental procedure (which will not be discussed here).

PVT data of polymers have been measured by many authors and it is not the intention of this handbook to present all available data. Only a certain number of polymers and their data sets are selected here to provide the necessary information with respect to the pressure dependence of the thermodynamic properties for the main classes of polymer solutions in the preceding chapters. Further *PVT* data for many polymers are given in (1995ZOL), for copolymers in (2001WOH). Parameters of the Tait equation for a number of polymers are given in (1999BRA) and (2003WOH).

Measurement of densities for polymer solutions at elevated pressures can be made today by U-tube vibrating densimeters. Such instruments are commercially available. Calibration is often made with pure water. Otherwise, densities can also be measured by the above discussed *PVT* equipment or in the equilibrium cell of a synthetic method.

Excess volumes are determined by

$$V^E_{spez} = V_{spez} - (w_A V_{0A,\, spez} + w_B V_{0B,\, spez})$$
(3a)

or

$$V^E = (x_A M_A + x_B M_B)/\rho - (x_A M_A/\rho_A + x_B M_B/\rho_B)$$
(3b)

where:

V^E	molar excess volume at temperature T
V^E_{spez}	specific excess volume at temperature T
$V_{0A,\, spez}$	specific volume of pure solvent A at temperature T
$V_{0B,\, spez}$	specific volume of pure polymer B at temperature T
ρ	density of the mixture at temperature T
ρ_A	density of pure solvent A at temperature T
ρ_B	density of pure polymer B at temperature T

Determination of second virial coefficients A_2

The osmotic virial coefficients are defined via the concentration dependence of the *osmotic pressure*, π, of a polymer solution, e.g., (1974TOM):

$$\frac{\pi}{c_B} = RT\left[\frac{1}{M_n} + A_2 c_B + A_3 c_B^2 + ...\right] \tag{4}$$

where

A_2	second osmotic virial coefficient
A_3	third osmotic virial coefficient
c_B	(mass/volume) concentration of polymer B
M_n	number-average relative molar mass of the polymer
R	gas constant
T	(measuring) temperature

There are a couple of methods for the experimental determination of the second virial coefficient. Scattering methods (classical light scattering, X-ray scattering, neutron scattering) are usually applied to investigate its pressure dependence, e.g., (1972SCH).

Scattering methods enable the determination of A_2 via the common relation:

$$\frac{Kc_B}{R(q)} = \frac{1}{M_w P_z(q)} + 2A_2 Q(q)c_B + ... \tag{5}$$

with

$$q = \frac{4\pi}{\lambda}\sin\frac{\theta}{2} \tag{6}$$

where:

K	a constant that summarizes the optical parameters of a scattering experiment
M_w	mass-average relative molar mass of the polymer
$P_z(q)$	z-average of the scattering function
q	scattering vector
$Q(q)$	function for the q-dependence of A_2
$R(q)$	excess intensity of the scattered beam at the value q
λ	wavelength
θ	scattering angle

Depending on the chosen experiment (light, X-ray or neutron scattering), the constant K is to be calculated from different relations. For experimental and theoretical details please see the corresponding textbooks (1972HUG, 1975CAS, 1982GLA, 1986HIG, 1987BER, 1987KRA, 1987WIG, and 1991CHU).

1.3. Guide to the data tables

Characterization of the polymers

Polymers vary by a number of characterization variables (1991BAR, 1999PET). The molar mass and their distribution function are the most important variables. However, tacticity, sequence distribution, branching, and end groups determine their thermodynamic behavior in solution too. Unfortunately, much less information is provided with respect to the polymers that were applied in most of the thermodynamic investigations in the original literature. For copolymers, the chemical distribution and the average chemical composition are also to be given. But, in many cases, the samples are characterized only by one or two molar mass averages and some additional information (e.g., T_g, T_m, ρ_B, or how and where they were synthesized). Sometimes even this information is missed.

The molar mass averages are defined as follows:

number average M_n

$$M_n = \frac{\sum_i n_{B_i} M_{B_i}}{\sum_i n_{B_i}} = \frac{\sum_i w_{B_i}}{\sum_i w_{B_i} / M_{B_i}} \tag{7}$$

mass average M_w

$$M_w = \frac{\sum_i n_{B_i} M_{B_i}^2}{\sum_i n_{B_i} M_{B_i}} = \frac{\sum_i w_{B_i} M_{B_i}}{\sum_i w_{B_i}} \tag{8}$$

z-average M_z

$$M_z = \frac{\sum_i n_{B_i} M_{B_i}^3}{\sum_i n_{B_i} M_{B_i}^2} = \frac{\sum_i w_{B_i} M_{B_i}^2}{\sum_i w_{B_i} M_{B_i}} \tag{9}$$

viscosity average M_η

$$M_\eta = \left(\frac{\sum_i w_{B_i} M_{B_i}^a}{\sum_i w_{B_i}} \right)^{1/a} \tag{10}$$

where:

a	exponent in the viscosity-molar mass relationship
M_{Bi}	relative molar mass of the polymer species B_i
n_{Bi}	amount of substance of polymer species B_i
w_{Bi}	mass fraction of polymer species B_i

Measures for the polymer concentration

The following concentration measures are used in the tables of this handbook (where B always denotes the main polymer, A denotes the solvent, and in ternary systems C denotes the third component):

mass/volume concentration

$$c_A = m_A/V \qquad c_B = m_B/V \tag{11}$$

mass fraction

$$w_A = m_A/\Sigma\, m_i \qquad w_B = m_B/\Sigma\, m_i \tag{12}$$

mole fraction

$$x_A = n_A/\Sigma\, n_i \qquad x_B = n_B/\Sigma\, n_i \qquad \text{with} \qquad n_i = m_i/M_i \text{ and } M_B = M_n \tag{13}$$

volume fraction

$$\varphi_A = (m_A/\rho_A)/\Sigma\,(m_i/\rho_i) \qquad \varphi_B = (m_B/\rho_B)/\Sigma\,(m_i/\rho_i) \tag{14}$$

segment fraction

$$\psi_A = x_A r_A/\Sigma\, x_i r_i \quad \psi_B = x_B r_B/\Sigma\, x_i r_i \qquad \text{usually with } r_A = 1 \tag{15}$$

base mole fraction

$$z_A = x_A r_A/\Sigma\, x_i r_i \quad z_B = x_B r_B/\Sigma\, x_i r_i \qquad \text{with } r_B = M_B/M_0 \text{ and } r_A = 1 \tag{16}$$

where:

c_A	(mass/volume) concentration of solvent A
c_B	(mass/volume) concentration of polymer B
m_A	mass of solvent A
m_B	mass of polymer B
M_A	relative molar mass of the solvent A
M_B	relative molar mass of the polymer B
M_n	number-average relative molar mass
M_0	molar mass of a basic unit of the polymer B
n_A	amount of substance of solvent A
n_B	amount of substance of polymer B
r_A	segment number of the solvent A, usually $r_A = 1$
r_B	segment number of the polymer B
V	volume of the liquid solution at temperature T
w_A	mass fraction of solvent A
w_B	mass fraction of polymer B
x_A	mole fraction of solvent A
x_B	mole fraction of polymer B

z_A	base mole fraction of solvent A
z_B	base mole fraction of polymer B
φ_A	volume fraction of solvent A
φ_B	volume fraction of polymer B
ρ_A	density of solvent A
ρ_B	density of polymer B
ψ_A	segment fraction of solvent A
ψ_B	segment fraction of polymer B

For high-molecular polymers, a mole fraction is not an appropriate unit to characterize composition. However, for oligomeric products with rather low molar masses, mole fractions were sometimes used. In the common case of a distribution function for the molar mass, $M_B = M_n$ is to be chosen. Mass fraction and volume fraction can be considered as special cases of segment fractions depending on the way by which the segment size is actually determined: $r_i/r_A = M_i/M_A$ or $r_i/r_A = V_i/V_A = (M_i/\rho_i)/(M_A/\rho_A)$, respectively. Classical segment fractions are calculated by applying $r_i/r_A = V_i^{\text{vdW}}/V_A^{\text{vdW}}$ ratios where hard-core van der Waals volumes, V_i^{vdW}, are taken into account. Their special values depend on the chosen equation of state or simply some group contribution schemes, e.g., (1968BON, 1990KRE) and have to be specified.

Volume fractions imply a temperature dependence and, as they are defined in equation (14), neglect excess volumes of mixing and, very often, the densities of the polymer in the state of the solution are not known correctly. However, volume fractions can be calculated without the exact knowledge of the polymer molar mass (or its averages). Base mole fractions are sometimes applied for polymer systems in earlier literature. The value for M_0 is the molar mass of a basic unit of the polymer. Sometimes it is chosen arbitrarily, however, and has to be specified.

Tables of experimental data

The data tables in each chapter are provided there in order of the names of the polymers. In this data book, mostly source-based polymer names are applied. These names are more common in use, and they are usually given in the original sources, too. Structure-based names, for which details about their nomenclature can be found in the *Polymer Handbook* (1999BRA), are chosen in some single cases only. CAS index names for polymers are not applied here. Finally, the list of systems and properties in order of the polymers in Appendix 1 is made by using the names as given in the chapters of this book.

Within types of polymers the individual samples are ordered by their increasing average molar mass, and, when necessary, systems are ordered by increasing temperature. In ternary systems, ordering is additionally made subsequently according to the name of the third component in the system. Each data set begins with the lines for the solution components, e.g., in binary systems

Polymer (B):	**poly(dimethylsiloxane)**	**2002PFO**
Characterization:	$M_n/\text{kg.mol}^{-1} = 20$, $M_w/\text{kg.mol}^{-1} = 60$,	
	Baysilone M10000, GE Bayer Silicones, Germany	
Solvent (A):	**chlorodifluoromethane** $CHClF_2$	**75-45-6**

where the polymer sample is given in the first line together with the reference. The second line provides then the characterization available for the polymer sample. The following line gives the solvent's chemical name, molecular formula, and CAS registry number.

In ternary and quaternary systems, the following lines are either for a second solvent or a second polymer or a salt or another chemical compound, e.g., in ternary systems with two solvents

Polymer (B):	**polyethylene**	**1990KEN**
Characterization:	M_n/kg.mol^{-1} = 8, M_w/kg.mol^{-1} = 177, M_z/kg.mol^{-1} = 1000,	
	HDPE, DSM, Geleen, The Netherlands	
Solvent (A):	**n-hexane** **C$_6$H$_{14}$**	**110-54-3**
Solvent (C):	**nitrogen** **N$_2$**	**7727-37-9**

or, e.g., in ternary systems with a second polymer

Polymer (B):	**polystyrene**	**2000BEH**
Characterization:	M_n/kg.mol^{-1} = 93, M_w/kg.mol^{-1} = 101.4, M_z/kg.mol^{-1} = 111.9,	
	BASF AG, Germany	
Solvent (A):	**cyclohexane** **C$_6$H$_{12}$**	**110-82-7**
Polymer (C):	**polyethylene**	
Characterization:	M_n/kg.mol^{-1} = 13, M_w/kg.mol^{-1} = 89, M_z/kg.mol^{-1} = 600,	
	LDPE, Stamylan, DSM, Geleen, The Netherlands	

or, e.g., in quaternary (or higher) systems like

Polymer (B):	**ethylene/norbornene copolymer**		**2003LEE**
Characterization:	M_n/kg.mol^{-1} = 34.8, M_w/kg.mol^{-1} = 83.3, 50 mol% ethylene,		
	T_g/K = 408.15, Union Chemical Laboratories, Taiwan		
Solvent (A):	**ethene** **C$_2$H$_4$**		**74-85-1**
Solvent (C):	**toluene** **C$_7$H$_8$**		**108-88-3**
Solvent (D):	**bicyclo[2,2,1]-2-heptene (norbornene)**	**C$_7$H$_{10}$**	**498-66-8**

There are some exceptions from this type of presentation within the tables for A_2 data, the UCST and LCST data, or the *PVT* data of pure polymers. These tables are prepared in the forms as chosen in 2001WOH .

The originally measured data for each single system are sometimes listed together with some comment lines if necessary. The data are usually given as published, but temperatures are always given in K. Pressures are sometimes recalculated into kPa or MPa.

Because many investigations on liquid-liquid or fluid-fluid equilibrium in polymer solutions at elevated pressures are provided in the literature in figures only, chapters 3 and 4 contain additional tables referring to types of systems, included components, and references.

Final day for including data into this *Handbook* was June, 30, 2004.

1.4. List of symbols

a	exponent in the viscosity-molar mass relationship
a_A	activity of solvent A
A_2	second osmotic virial coefficient
A, B, C	parameters of the Tait equation
c_A	(mass/volume) concentration of solvent A
c_B	(mass/volume) concentration of polymer B
$\Delta_M H$	(integral) enthalpy of mixing
H^E	excess enthalpy
K	a constant that summarizes the optical parameters of a scattering experiment
K^∞	weight fraction-based partition coefficient at infinite dilution
m_A	mass of solvent A
m_B	mass of polymer B
M	relative molar mass
M_A	molar mass of the solvent A
M_B	molar mass of the polymer B
M_n	number-average relative molar mass
M_w	mass-average relative molar mass
M_η	viscosity-average relative molar mass
M_z	z-average relative molar mass
M_0	molar mass of a basic unit of the polymer B
MI	melting index
n_A	amount of substance of solvent A
n_B	amount of substance of polymer B
P	pressure
P_0	standard pressure (= 0.101325 MPa)
P_A	partial vapor pressure of the solvent A at temperature T
P_{0A}	vapor pressure of the pure liquid solvent A at temperature T
P_{crit}	critical pressure
$P_z(q)$	z-average of the scattering function
q	scattering vector
$Q(q)$	function for the q-dependence of A_2
R	gas constant
$R(q)$	excess intensity of the scattered beam at the value q
r_A	segment number of the solvent A, usually $r_A = 1$
r_B	segment number of the polymer B
T	(measuring) temperature
T_{crit}	critical temperature
T_g	glass transition temperature
T_m	melting transition temperature
V, V_{spez}	volume or specific volume at temperature T
V_0	reference volume
V^E	excess volume at temperature T
V^{vdW}	hard-core van der Waals volume
ΔV	volume difference $V - V_0$ (in swelling experiments)

w_A	mass fraction of solvent A
w_B	mass fraction of polymer B
$w_{B,crit}$	mass fraction of the polymer B at the critical point
x_A	mole fraction of solvent A
x_B	mole fraction of polymer B
z_A	base mole fraction of solvent A
z_B	base mole fraction of polymer B
α	critical exponent
γ_A	activity coefficient of the solvent A in the liquid phase with activity $a_A = x_A\gamma_A$
λ	wavelength
φ_A	volume fraction of solvent A
φ_B	volume fraction of polymer B
$\varphi_{B,crit}$	volume fraction of the polymer B at the critical point
ρ	density (of the mixture) at temperature T
ρ_A	density of solvent A
ρ_B	density of polymer B
π	osmotic pressure
θ	scattering angle
Ω_A	mass fraction-based activity coefficient of the solvent A in the liquid phase with activity $a_A = w_A\Omega_A$
ψ_A	segment fraction of solvent A
ψ_B	segment fraction of polymer B

1.5. References

1968BON   Bondi, A., *Physical Properties of Molecular Crystals, Liquids and Glasses*, J. Wiley & Sons, New York, 1968.

1968KON   Koningsveld, R. and Staverman, A.J., Liquid-liquid phase separation in multicomponent polymer solutions I and II, *J. Polym. Sci.*, Pt. A-2, 6, 305, 325, 1968.

1969WOL   Wolf, B.A., Zur Bestimmung der kritischen Konzentration von Polymerlösungen, *Makromol. Chem.*, 128, 284, 1969.

1971YAM   Yamakawa, H., *Modern Theory of Polymer Solutions*, Harper & Row, New York, 1971.

1972HUG   Huglin, M.B., Ed., *Light Scattering from Polymer Solutions*, Academic Press, New York, 1972.

1972KON   Koningsveld, R., Polymer solutions and fractionation, in *Polymer Science*, Jenkins, E.D., Ed., North-Holland, Amsterdam, 1972, 1047.

1972SCH   Schulz, G.V. and Lechner, M.D., Influence of pressure and temperature, in *Light Scattering from Polymer Solutions*, Huglin, M.B., Ed., Academic Press, New York, 1972, 503.

1974TOM   Tombs, M.P. and Peacock, A.R., *The Osmotic Pressure of Macromolecules*, Oxford University Press, London, 1974.

1975BON   Bonner, D.C., Vapor-liquid equilibria in concentrated polymer solutions, *Macromol. Sci. Rev. Macromol. Chem.*, C13, 263, 1975.

1975CAS   Casassa, E.F. and Berry, G.C., Light scattering from solutions of macromolecules, in *Polymer Molecular Weights*, Marcel Dekker, New York, 1975, Pt. 1, 161.

1980VAN   Van Konynenburg, P.H. and Scott, R.L., Critical lines and phase equilibria in binary van-der-Waals mixtures, *Philos. Trans. Roy. Soc.*, 298, 495, 1980.

1982GLA   Glatter, O. and Kratky, O., Eds., *Small-Angle X-Ray Scattering*, Academic Press, London, 1982.

1986HIG   Higgins, J.S. and Macconachie, A., Neutron scattering from macromolecules in solution, in *Polymer Solutions*, Forsman, W.C., Ed., Plenum Press, New York, 1986, 183.

1986ZOL   Zoller, P., Dilatometry, in *Encyclopedia of Polymer Science and Engineering*, Vol. 5, 2nd ed., Mark, H. et al., Eds., J. Wiley & Sons, New York, 1986, 69.

1987BER   Berry, G.C., Light scattering, in *Encyclopedia of Polymer Science and Engineering*, Vol. 8, 2nd ed., Mark, H. et al., Eds., J. Wiley & Sons, New York, 1986, 721.

1987COO   Cooper, A.R., Molecular weight determination, in *Encyclopedia of Polymer Science and Engineering*, Vol. 10, 2nd ed., Mark, H. et al., Eds., J. Wiley & Sons, New York, 1986, 1.

1987KRA   Kratochvil, P., *Classical Light Scattering from Polymer Solutions*, Elsevier, Amsterdam, 1987.

1989RAE   Rätzsch, M.T. and Kehlen, H., Continuous thermodynamics of polymer systems, *Prog. Polym. Sci.*, 14, 1, 1989.

1987WIG   Wignall, G.D., Neutron scattering, in *Encyclopedia of Polymer Science and Engineering*, Vol. 10, 2nd ed., Mark, H. et al., Eds., J. Wiley & Sons, New York, 1986, 112.

1990BAR   Barton, A.F.M., *CRC Handbook of Polymer-Liquid Interaction Parameters and Solubility Parameters*, CRC Press, Boca Raton, 1990.

1990FUJ   Fujita, H., *Polymer Solutions*, Elsevier, Amsterdam, 1990.

1990KAM   Kamide, K., *Thermodynamics of Polymer Solutions*, Elsevier, Amsterdam, 1990.

1990KRE   [Van] Krevelen, D.W., *Properties of Polymers*, 3rd ed., Elsevier, Amsterdam, 1990.

1990RAE   Rätzsch, M.T. and Wohlfarth, C., Continuous thermodynamics of copolymer systems, *Adv. Polym. Sci.*, 98, 49, 1990.

1991CHU   Chu, B., *Laser Light Scattering*, Academic Press, New York, 1991.

1991BAR   Barth, H.G. and Mays, J.W., Eds., *Modern Methods of Polymer Characterization*, J. Wiley & Sons, New York, 1991, 201.

1992WEN   Wen, H., Elbro, H.S., and Alessi, P., *Polymer Solution Data Collection.* I. Vapor-liquid equilibrium; II. Solvent activity coefficients at infinite dilution; III. Liquid-liquid equlibrium, Chemistry Data Series, Vol. 15, DECHEMA, Frankfurt am Main, 1992.

1993DAN   Danner, R.P. and High, M.S., *Handbook of Polymer Solution Thermodynamics*, American Institute of Chemical Engineers, New York, 1993.

1994MCH   McHugh, M.A. and Krukonis, V.J., *Supercritical Fluid Extraction: Principles and Practice*, 2nd ed., Butterworth Publishing, Stoneham, 1994.

1994WOH   Wohlfarth, C., *Vapour-Liquid Equilibrium Data of Binary Polymer Solutions: Physical Science Data*, 44, Elsevier, Amsterdam, 1994.

1995ZOL   Zoller, P. and Walsh, D.J., *Standard Pressure-Volume-Temperature Data for Polymers*, Technomic Publishing, Lancaster, 1995.

1997KIR   Kiran, E. and Zhuang, W., *Miscibility and Phase Separation of Polymers in Near- and Supercritical Fluids*, ACS Symposium Series 670, 2, 1997.

1999BRA   Brandrup, J., Immergut, E.H., and Grulke, E.A., Eds., *Polymer Handbook*, 4th ed., J. Wiley & Sons, New York, 1999.

1999KIR   Kirby, C.F. and McHugh, M.A., Phase behavior of polymers in supercritical fluid solvents, *Chem. Rev.*, 99, 565, 1999.

1999KLE   Klenin, V.J., *Thermodynamics of Systems Containing Flexible-Chain Polymers*, Elsevier, Amsterdam, 1999.

1999LOP   Lopez, E.R., Coxam, J.-Y., Fernandez, J., and Grolier, J.-P.E., Pressure and temperature dependence of excess enthalpies of methanol + tetraethylene glycol dimethyl ether and methanol + polyethylene glycol dimethyl ether 250, *J. Chem. Eng. Data*, 44, 1409, 1999.

1999PET   Pethrick, R.A. and Dawkins, J.V., Eds., *Modern Techniques for Polymer Characterization*, J. Wiley & Sons, Chichester, 1999.

1999PRA   Prausnitz, J.M., Lichtenthaler, R.N., and de Azevedo, E.G., *Molecular Thermodynamics of Fluid Phase Equilibria*, 3rd ed., Prentice Hall, Upper Saddle River, NJ, 1999.

1999YEL   Yelash, L.V. and Kraska, T., The global phase behaviour of binary mixtures of chain molecules. Theory and application, *Phys. Chem. Chem. Phys.*, 1, 4315, 1999.

2000WOH   Wohlfarth, C., Methods for the measurement of solvent activity of polymer solutions, in *Handbook of Solvents*, Wypych, G., Ed., ChemTec Publishing, Toronto, 2000, 146.

2001KON   Koningsveld, R., Stockmayer, W.H., and Nies, E., *Polymer Phase Diagrams*, Oxford University Press, Oxford, 2001.

2001WOH   Wohlfarth, C., *CRC Handbook of Thermodynamic Data of Copolymer Solutions*, CRC Press, Boca Raton, 2001.

2002CHR   Christov, M. and Dohrn, R., High-pressure fluid phase equilibria. Experimental methods and systems investigated (1994-1999), *Fluid Phase Equil.*, 202, 153, 2002.

2003WOH   Wohlfarth, C., Pressure-volume-temperature relationship for polymer melts, in *CRC Handbook of Chemistry and Physics*, Lide, D.R., Ed., 84th ed., pp. 13-16 to 13-20, 2003.

2004WOH   Wohlfarth, C., *CRC Handbook of Thermodynamic Data of Aqueous Polymer Solutions*, CRC Press, Boca Raton, 2004.

2. VAPOR-LIQUID EQUILIBRIUM (VLE) DATA AND GAS SOLUBILITIES AT ELEVATED PRESSURES

2.1. Binary polymer solutions

Polymer (B):	**butylene succinate/butylene adipate copolymer**						**2000SA1**

Characterization: $M_n/\text{kg.mol}^{-1} = 53$, $M_w/\text{kg.mol}^{-1} = 180$, 20.0 mol% adipate, $T_g/\text{K} = 231$, $T_m/\text{K} = 365$, 25 wt% crystallinity

Solvent (A):	**carbon dioxide**	**CO$_2$**					**124-38-9**

Type of data: gas solubility

$T/\text{K} = 323.15$

P/MPa	1.098	2.097	3.065	4.053	6.016	7.870	9.844
m_A/(g CO$_2$/kg polymer)	12.59	24.18	35.59	47.49	73.64	103.6	150.3

Comments: The isotherm at 323 K is below the melting temperature. Below T_m, only the amorphous parts absorb CO$_2$. Thus, the solubility can be recalculated with respect to the amount of the amorphous parts of PBS.

P/MPa	1.098	2.097	3.065	4.053	6.016	7.870	9.844
m_A/(g CO$_2$/kg amorphous)	16.79	32.24	47.45	63.32	98.19	138.1	200.4

$T/\text{K} = 393.15$

P/MPa	2.082	2.581	4.028	4.610	6.200	6.681	8.053
m_A/(g CO$_2$/kg polymer)	18.39	22.85	35.91	41.22	55.78	60.20	72.80

P/MPa	8.630	11.997	12.460	16.024	16.548	20.036	
m_A/(g CO$_2$/kg polymer)	78.14	108.4	112.7	142.3	146.7	174.1	

$T/\text{K} = 423.15$

P/MPa	2.102	2.676	4.118	4.684	6.098	6.700	8.177
m_A/(g CO$_2$/kg polymer)	14.51	18.51	28.63	32.63	42.64	46.92	57.49

P/MPa	8.699	12.314	16.027	16.572	20.074		
m_A/(g CO$_2$/kg polymer)	61.17	86.49	111.5	115.0	137.4		

$T/\text{K} = 453.15$

P/MPa	2.079	2.600	4.126	4.698	6.070	6.628	8.259
m_A/(g CO$_2$/kg polymer)	11.84	14.83	23.57	26.89	34.85	38.09	47.63

P/MPa	8.973	12.009	12.766	16.139	20.127		
m_A/(g CO$_2$/kg polymer)	51.50	69.20	73.49	92.35	113.6		

Polymer (B):   **butyl methacrylate/N,N-dimethylaminoethyl**
                      methacrylate block copolymer   **2003WAN**
Characterization: M_w/kg.mol^{-1} = 95.7, 59.43 mol% butyl methacrylate
Solvent (A):   **carbon dioxide   CO$_2$**   **124-38-9**

Type of data: gas solubility

T/K = 298.15

P/MPa	0.2	0.3	0.4	0.5	0.6	0.7	0.8	0.9	1.0
w_A	0.0054	0.0090	0.0122	0.0154	0.0185	0.0220	0.0255	0.0286	0.0317

Polymer (B):   **butyl methacrylate/N,N-dimethylaminoethyl**
                      methacrylate block copolymer   **2003WAN**
Characterization: M_w/kg.mol^{-1} = 87.8, 89.54 mol% butyl methacrylate
Solvent (A):   **carbon dioxide   CO$_2$**   **124-38-9**

Type of data: gas solubility

T/K = 298.15

P/MPa	0.2	0.3	0.4	0.5	0.6	0.7	0.8	0.9	1.0
w_A	0.0060	0.0091	0.0121	0.0152	0.0183	0.0216	0.0248	0.0278	0.0311

Polymer (B):   **butyl methacrylate/N,N-dimethylaminoethyl**
                      methacrylate block copolymer   **2003WAN**
Characterization: M_w/kg.mol^{-1} = 90, 93.94 mol% butyl methacrylate
Solvent (A):   **carbon dioxide   CO$_2$**   **124-38-9**

Type of data: gas solubility

T/K = 298.15

P/MPa	0.2	0.3	0.4	0.5	0.6	0.7	0.8	0.9	1.0
w_A	0.0055	0.0082	0.0115	0.0150	0.0183	0.0223	0.0256	0.0291	0.0326

Polymer (B):   **butyl methacrylate/N,N-dimethylaminoethyl**
                      methacrylate block copolymer   **2003WAN**
Characterization: M_w/kg.mol^{-1} = 26.4, 95.13 mol% butyl methacrylate
Solvent (A):   **carbon dioxide   CO$_2$**   **124-38-9**

Type of data: gas solubility

T/K = 298.15

P/MPa	0.2	0.3	0.4	0.5	0.6	0.7	0.8	0.9	1.0
w_A	0.0091	0.0133	0.0170	0.0213	0.0249	0.0291	0.0329	0.0368	0.0406

Polymer (B): **butyl methacrylate/perfluoroalkylethyl acrylate block copolymer** **2003WAN**

Characterization: M_w/kg.mol^{-1} = 18.9, 81.99 mol% butyl methacrylate, perfluoroalkyl = $(CF_2)_{8.6}CF_3$

Solvent (A): **carbon dioxide** CO_2 **124-38-9**

Type of data: gas solubility

T/K = 298.15

P/MPa	0.2	0.3	0.4	0.5	0.6	0.7	0.8	0.9
w_A	0.0054	0.0080	0.0111	0.0151	0.0185	0.0221	0.0249	0.0318

Polymer (B): **butyl methacrylate/perfluoroalkylethyl acrylate block copolymer** **2003WAN**

Characterization: M_w/kg.mol^{-1} = 14.9, 94.50 mol% butyl methacrylate, perfluoroalkyl = $(CF_2)_{8.6}CF_3$

Solvent (A): **carbon dioxide** CO_2 **124-38-9**

Type of data: gas solubility

T/K = 298.15

P/MPa	0.2	0.3	0.4	0.5	0.6	0.7	0.8	0.9
w_A	0.0074	0.0111	0.0149	0.0185	0.0221	0.0258	0.0293	0.0328

Polymer (B): **ethylene/propylene copolymer** **1992CH1**

Characterization: M_n/kg.mol^{-1} = 0.78, M_w/kg.mol^{-1} = 0.79, 50 mol% propene, alternating ethylene/propylene units from complete hydrogenation of polyisoprene

Solvent (A): **1-butene** C_4H_8 **106-98-9**

Type of data: vapor-liquid equilibrium

w_B 0.151 was kept constant

T/K	355.05	373.15	398.15	421.65	424.05	426.35	428.15	432.95	444.45
P/bar	11.5	16.4	27.1	39.8	42.5	44.8	46.5	51.7	63.4

T/K	473.15
P/bar	90.8

Polymer (B): **ethylene/propylene block copolymer** **2000SA2**

Characterization: M_w/kg.mol^{-1} = 240, 4.9 wt% ethene, 90% isotactic, 58.1 wt% crystallinity, impact propylene copolymer, Mitsubishi Chemical, Kurashiki, Japan

Solvent (A): **ethene** C_2H_4 **74-85-1**

continued

continued

Type of data: gas solubility

$T/K = 323.15$

P/MPa	0.3800	0.8987	1.4021	1.9014	2.4131	3.0362
w_A	0.00252	0.00580	0.00898	0.01213	0.01535	0.01927

$T/K = 343.15$

P/MPa	0.3738	0.8009	1.2862	1.7985	2.3093	2.8260
w_A	0.00197	0.00416	0.00664	0.00926	0.01185	0.01447

$T/K = 363.15$

P/MPa	0.3836	0.8533	1.3355	1.8383	2.3135	2.8231
w_A	0.00169	0.00374	0.00583	0.00800	0.01003	0.01219

Polymer (B):   **ethylene/propylene block copolymer**   **2000SA2**
Characterization: M_w/kg.mol^{-1} = 240, 4.9 wt% ethene, 90% isotactic,
                         58.1 wt% crystallinity, impact propylene copolymer,
                         Mitsubishi Chemical, Kurashiki, Japan
Solvent (A):   **propene**   **C$_3$H$_6$**   **115-07-1**

Type of data: gas solubility

$T/K = 323.15$

P/MPa	0.2143	0.6207	1.0232	1.3323	1.6264	1.7866
w_A	0.00664	0.02097	0.03989	0.05663	0.07448	0.08682

$T/K = 343.15$

P/MPa	0.3024	0.7554	1.4385	1.8452	2.2823	2.6448
w_A	0.00736	0.01898	0.03884	0.05417	0.07091	0.08703

$T/K = 363.15$

P/MPa	0.3659	0.9153	1.4844	2.0138	2.5029	3.0951
w_A	0.00628	0.01671	0.02847	0.04058	0.05306	0.07089

Polymer (B):   **ethylene/vinyl acetate copolymer**   **1991FIN**
Characterization: M_n/kg.mol^{-1} = 2.2, M_w/kg.mol^{-1} = 5.7, 29.0 wt% vinyl acetate,
                         Leuna-Werke, Germany
Solvent (A):   **ethene**   **C$_2$H$_4$**   **74-85-1**

Type of data: gas solubility

$T/K = 443.15$

P/MPa	1.991	3.041	5.882	8.972	12.864	14.415	16.768	20.788	24.318
w_A	0.0116	0.0184	0.0657	0.0538	0.0777	0.0884	0.1005	0.1207	0.1430

P/MPa	28.143
w_A	0.1619

Polymer (B):   **ethylene/vinyl acetate copolymer**   **1991FIN**
Characterization: M_n/kg.mol^{-1} = 13.1, M_w/kg.mol^{-1} = 41.3, 34.2 wt% vinyl acetate,
     Leuna-Werke, Germany
Solvent (A):   **ethene** C_2H_4 **74-85-1**

Type of data: gas solubility

T/K = 473.15

P/MPa	2.695	4.683	6.464	6.551	6.779	8.847	11.356	14.006	16.135
w_A	0.0135	0.0264	0.0336	0.0337	0.0373	0.0462	0.0577	0.0652	0.0752

P/MPa	22.361	29.520
w_A	0.1091	0.1323

Polymer (B):   **ethylene/vinyl acetate copolymer**   **1991FIN**
Characterization: M_n/kg.mol^{-1} = 2.2, M_w/kg.mol^{-1} = 5.7, 29.0 wt% vinyl acetate,
     Leuna-Werke, Germany
Solvent (A):   **vinyl acetate** $C_4H_6O_2$ **108-05-4**

Type of data: gas solubility

T/K = 413.15

P/MPa	0.152	0.169	0.186	0.273	0.288	0.371	0.405	0.435	0.444
w_A	0.042	0.077	0.079	0.114	0.115	0.205	0.245	0.237	0.318

P/MPa	0.464
w_A	0.299

T/K = 443.15

P/MPa	0.125	0.144	0.165	0.166	0.185	0.265	0.494	0.557	0.567
w_A	0.028	0.030	0.047	0.050	0.065	0.071	0.116	0.127	0.129

P/MPa	0.886	1.02	1.14
w_A	0.322	0.456	0.684

T/K = 473.15

P/MPa	0.179	0.211	0.234	0.425	0.495	0.543	0.788	0.935
w_A	0.024	0.037	0.039	0.055	0.056	0.086	0.122	0.155

Polymer (B):   **nylon 6**   **2003XUQ**
Characterization: ρ (298 K) = 1.140 g/cm^3, 1020C, Mitsubishi, Japan
Solvent (A):   **carbon dioxide** CO_2 **124-38-9**

Type of data: gas solubility

T/K = 313.15

P/MPa	8	10	12	14	16
w_A	0.0177	0.0172	0.0083	0.0105	0.0132

Polymer (B):	nylon 6				**2003XUQ**

Characterization: ρ (298 K) = 1.140 g/cm^3, 1020C, Mitsubishi, Japan

Solvent (A):	styrene	C_8H_8			**100-42-5**

Type of data: vapor solubility

$T/K = 313.15$

P/MPa	8	10	12	14	16
w_A	0.0104	0.0059	0.0024	0.0024	0.0035

Polymer (B):	polybutadiene			**2002JOU**

Characterization: M_w/kg.mol^{-1} = 420, Aldrich Chem. Co., Inc., Milwaukee, WI

Solvent (A):	cyclohexane	C_6H_{12}		**110-82-7**

Type of data: vapor pressure

w_B	0.050	was kept constant

T/K	363.15	373.15	383.15	393.15	403.15	413.15	423.15	433.15	443.15
P_A/MPa	0.5	0.6	0.6	0.8	0.9	1.0	1.2	1.4	1.6

T/K	453.15	463.15	473.15
P_A/MPa	1.9	2.1	2.5

Polymer (B):	poly(butylene succinate)					**2000SA1**

Characterization: M_n/kg.mol^{-1} = 29, M_w/kg.mol^{-1} = 140,
                        branched, T_g/K = 243, T_m/K = 388, 35 wt% crystallinity

Solvent (A):	carbon dioxide	CO_2				**124-38-9**

Type of data: gas solubility

$T/K = 323.15$

P/MPa	1.025	2.209	3.068	4.245	6.272	8.008	10.129
m_A/(g CO$_2$/kg polymer)	9.84	20.63	28.26	38.79	56.81	73.00	107.6

$T/K = 353.15$

P/MPa	1.148	2.103	3.158	4.218	6.169	8.050	10.451
m_A/(g CO$_2$/kg polymer)	8.76	15.23	22.43	29.61	43.32	56.01	72.50

Comments: The isotherms at 323 and 353 are below the melting temperature where only the amorphous parts absorb CO_2. The original paper provides a column where the solubility is also calculated with respect to the amount of the amorphous parts of PBS.

$T/K = 393.15$

P/MPa	2.351	2.947	4.141	4.725	6.060	6.599	8.076
m_A/(g CO$_2$/kg polymer)	21.74	27.08	37.76	43.08	55.21	60.14	73.57

P/MPa	8.928	12.076	12.606	16.035	16.544	19.987
m_A/(g CO$_2$/kg polymer)	81.30	109.8	114.6	144.0	148.1	176.1

continued

continued

$T/K = 423.15$

P/MPa	2.221	2.935	4.065	4.705	6.035	6.572	8.242
m_A/(g CO_2/kg polymer)	15.01	19.93	27.77	32.16	41.45	45.29	57.10

P/MPa	8.696	12.036	12.572	16.038	16.541	20.095
m_A/(g CO_2/kg polymer)	60.41	83.09	87.33	110.7	113.9	135.5

$T/K = 453.15$

P/MPa	2.133	2.798	4.132	4.627	6.039	6.626	8.037
m_A/(g CO_2/kg polymer)	11.92	15.66	23.15	26.04	34.11	37.36	45.34

P/MPa	8.570	12.007	12.592	16.071	16.574	20.144
m_A/(g CO_2/kg polymer)	48.40	68.30	71.53	90.82	93.67	112.6

Polymer (B):   **poly(butyl methacrylate)**   **2003WAN**
Characterization: M_w/kg.mol^{-1} = 13.6
Solvent (A):   **carbon dioxide**   **CO_2**   **124-38-9**

Type of data: gas solubility

$T/K = 298.15$

P/MPa	0.2	0.3	0.4	0.5	0.6	0.7	0.8	0.9
w_A	0.0078	0.0125	0.0171	0.0212	0.0263	0.0314	0.0365	0.0415

Polymer (B):   **poly(butyl methacrylate)**   **1990WAN, 1994WAN**
Characterization: M_η/kg.mol^{-1} = 126, Aldrich Chem. Co., Inc., Milwaukee, WI
Solvent (A):   **carbon dioxide**   **CO_2**   **124-38-9**

Type of data: gas solubility

$T/K = 313.2$

P/MPa	0.549	0.971	1.01	1.49	1.65	2.06	2.32	2.82	2.96
w_A	0.0147	0.0284	0.0269	0.0400	0.0469	0.0551	0.0627	0.0756	0.0835

P/MPa	3.65	3.78	4.35	4.74	5.01	5.62	5.70	6.41	6.46
w_A	0.104	0.105	0.125	0.138	0.145	0.172	0.169	0.197	0.208

P/MPa	7.23	7.77
w_A	0.232	0.252

$T/K = 333.2$

P/MPa	0.691	0.971	1.26	1.67	1.94	2.34	2.67	3.01	3.43
w_A	0.0148	0.0220	0.0270	0.0365	0.0411	0.0506	0.0564	0.0649	0.0744

P/MPa	3.73	4.23	4.36	4.92	5.05	5.68	5.74	6.54	7.47
w_A	0.0802	0.0919	0.0941	0.108	0.110	0.128	0.127	0.148	0.172

P/MPa	8.12	8.67
w_A	0.195	0.204

continued

continued

$T/K = 353.2$

P/MPa	0.701	1.09	1.30	1.76	1.94	2.48	2.63	3.10	3.26
w_A	0.0133	0.0214	0.0236	0.0328	0.0354	0.0461	0.0477	0.0572	0.0591

P/MPa	3.78	3.95	4.55	4.63	5.26	5.34	5.60	5.98	6.82
w_A	0.0693	0.0711	0.0833	0.0822	0.0966	0.0941	0.0987	0.111	0.125

P/MPa	7.47	8.24	8.98	9.68	10.2
w_A	0.142	0.155	0.173	0.187	0.199

Polymer (B):	**polycarbonate bisphenol-A**		**1999KEL**
Characterization:	–		
Solvent (A):	**carbon dioxide**	CO_2	**124-38-9**

Type of data: gas solubility

$T/K = 293.15$

P/MPa	0.992	1.994	3.458	3.976	4.495	5.032	5.836
w_A	0.0321	0.0588	0.0897	0.0984	0.1075	0.1151	0.1282

Polymer (B):	**polycarbonate bisphenol-A**		**2004TAN**
Characterization:	M_w/kg.mol^{-1} = 64, Acros		
Solvent (A):	**carbon dioxide**	CO_2	**124-38-9**

Type of data: gas solubility

T/K	313.15	313.15	313.15	323.15	323.15	323.15	333.15	333.15	333.15
P/MPa	20	30	40	20	30	40	20	30	40
w_A	0.1241	0.1344	0.1470	0.1171	0.1313	0.1459	0.1087	0.1298	0.1432

Polymer (B):	**poly(chlorotrifluoroethylene)**		**1999WEB**
Characterization:	Modern Plastics, Springfield, MA		
Solvent (A):	**carbon dioxide**	CO_2	**124-38-9**

Type of data: gas solubility

$T/K = 313.15$ P/MPa   10.5 w_A 0.0416

Polymer (B):	**poly(2,6-dimethyl-1,4-phenylene ether)**	**2002SAT**
Characterization:	M_n/kg.mol^{-1} = 22, M_w/kg.mol^{-1} = 47,	
	T_g/K = 477.1, Asahi Chemicals Inc., Kawasaki, Japan	
Solvent (A):	**carbon dioxide** CO_2	**124-38-9**

Type of data: gas solubility

$T/K = 373.15$

continued

continued

P/MPa	2.125	4.118	6.098	8.051	12.043	16.023	19.964
w_A	0.02894	0.04675	0.06102	0.07354	0.09978	0.12540	0.15421

$T/K = 423.15$

P/MPa	2.150	4.150	6.040	8.133	12.121	16.077
w_A	0.01438	0.02546	0.03419	0.04373	0.06326	0.08317

$T/K = 473.15$

P/MPa	2.097	4.155	6.120	8.048	12.122	16.072
w_A	0.00869	0.01678	0.02414	0.03136	0.04805	0.06013

Polymer (B):	**poly(2,6-dimethyl-1,4-phenylene ether)**	**2002SAT**
Characterization:	$M_n/\text{kg.mol}^{-1} = 24$, $M_w/\text{kg.mol}^{-1} = 63$,	
	$T_g/K = 479.0$, Asahi Chemicals Inc., Kawasaki, Japan	
Solvent (A):	**carbon dioxide** $\quad$ **CO$_2$**	**124-38-9**

Type of data: gas solubility

$T/K = 373.15$

P/MPa	2.058	4.003	8.100	10.111	11.965	13.935	15.993	17.933	20.006
w_A	0.03459	0.05113	0.07517	0.08534	0.09585	0.10727	0.12086	0.13498	0.14852

$T/K = 423.15$

P/MPa	1.983	4.105	8.056	10.061	12.108	14.078	16.098	18.076	20.076
w_A	0.01486	0.02653	0.04350	0.05299	0.06378	0.07373	0.08340	0.09299	0.10261

$T/K = 473.15$

P/MPa	2.179	4.142	8.085	10.082	12.078	14.063	16.032	18.106	20.208
w_A	0.00699	0.01560	0.03310	0.04134	0.05000	0.00182	0.06606	0.07502	0.08289

Polymer (B):	**poly(2,6-dimethyl-1,4-phenylene ether)**	**1994HAN**
Characterization:	–	
Solvent (A):	**carbon dioxide** $\quad$ **CO$_2$**	**124-38-9**

Type of data: T_g at saturation gas solubility

P/atm	0.0	13.6	27.2	34.0	40.8	54.4	61.2
T_g/K	489.15	484.55	476.75	475.25	468.95	466.15	457.55

Polymer (B):	**poly(dimethylsiloxane)**	**2002FLI**
Characterization:	$M_n/\text{kg.mol}^{-1} = 161$, $M_w/\text{kg.mol}^{-1} = 188$,	
	Polymer Source, Inc., Quebec	
Solvent (A):	**carbon dioxide** $\quad$ **CO$_2$**	**124-38-9**

Type of data: gas solubility and swelling

continued

continued

$T/K = 296.15$

P/MPa	2.14	4.24	6.24	8.27	10.45
w_A	0.1205	0.2107	0.3632	0.3867	0.4089
$(\Delta V/V)/(\%)$	9.31	24.03	83.05	94.29	104.77

$T/K = 323.15$

P/MPa	1.38	2.83	4.17	5.62	7.00	8.31	9.83	11.17
w_A	0.0438	0.0842	0.1161	0.1488	0.1811	0.2163	0.2553	0.2842
$(\Delta V/V)/(\%)$	3.31	7.25	11.77	17.81	25.53	35.19	48.70	59.58

Polymer (B):	**poly(dimethylsiloxane)**	**1999WEB**
Characterization:	cross-linked in the laboratory by using a sample from GE-Plastics, Electrical Insulation Suppliers, Atlanta, GA	
Solvent (A):	**carbon dioxide** CO_2	**124-38-9**

Type of data: gas solubility

$T/K = 313.15$ P/MPa   10.5 w_A 0.1019

Polymer (B):	**poly(dimethylsiloxane)**	**1998VIN**
Characterization:	United Chemical Technologies, Inc., Petrarch Silanes & Silicones, cross-linked using benzoyl peroxide	
Solvent (A):	**carbon dioxide** CO_2	**124-38-9**

Type of data: swelling at equilbrium pressure

$T/K = 314.65$

P/bar	1.0	6.6	12.0	33.0	33.4	54.3	55.3	57.7	65.7
V/V_0	1.012	1.025	1.040	1.112	1.129	1.217	1.194	1.240	1.309

P/bar	68.9	77.3	80.9	83.3	89.0	96.9	103.9	108.1	112.4
V/V_0	1.344	1.449	1.473	1.511	1.557	1.634	1.671	1.682	1.718

P/bar	115.2	120.8	122.0	128.1	131.3	140.9	148.2	150.4	154.2
V/V_0	1.713	1.758	1.771	1.792	1.817	1.861	1.899	1.889	1.931

P/bar	158.6	163.5	171.8	172.5
V/V_0	1.950	1.983	2.027	2.038

Polymer (B):	**poly(dimethylsiloxane)**	**2002PFO**
Characterization:	M_n/kg.mol^{-1} = 20, M_w/kg.mol^{-1} = 60, Baysilone M10000, GE Bayer Silicones, Germany	
Solvent (A):	**chlorodifluoromethane** $CHClF_2$	**75-45-6**

Type of data: gas solubility

$T/K = 298.15$

continued

P/kPa	2.8	8.7	16.2	32.7	61.3	115.0
w_A	0.00068	0.00223	0.00423	0.00899	0.01640	0.03126
a_A	0.0033	0.0100	0.0186	0.0376	0.0701	0.1303

$T/\text{K} = 343.15$

P/kPa	30.5	66.5	109.7	151.0	181.1	213.5
w_A	0.00315	0.00742	0.01224	0.01689	0.02030	0.02405
a_A	0.0153	0.0331	0.0544	0.0744	0.0889	0.1044

Polymer (B):	**polyester (hyperbranched, aliphatic)**	**2003SEI**
Characterization:	$M_n/\text{kg.mol}^{-1} = 1.6$, $M_w/\text{kg.mol}^{-1} = 2.1$,	
	hydroxyl functional hyperbranched polyesters produced from poly-alcohol cores and hydroxy acids, 16 OH groups per macromolecule, hydroxyl no. = 490-520 mg KOH/g, acid no. = 5-9 mg KOH/g, Boltorn H20, Perstorp Specialty Chemicals AB, Perstorp, Sweden	
Solvent (A):	**ethanol** $\qquad$ **C_2H_6O**	**64-17-5**

Type of data: vapor pressure

$w_B = 0.10 = \text{constant}$

T/K	316.5	323.2	329.0	336.6	343.0	351.1	359.6	371.9	386.2
P_A/MPa	0.036	0.047	0.058	0.077	0.096	0.127	0.174	0.237	0.377

T/K	400.4	415.0	429.9	439.1	451.0	464.0	472.1
P_A/MPa	0.515	0.740	1.083	1.340	1.756	2.309	2.709

Polymer (B):	**polyester (hyperbranched, aliphatic)**	**2003SEI**
Characterization:	$M_n/\text{kg.mol}^{-1} = 1.6$, $M_w/\text{kg.mol}^{-1} = 2.1$,	
	Boltorn H20, for details see the above data set	
Solvent (A):	**water** $\qquad$ **H_2O**	**7732-18-5**

Type of data: vapor pressure

$w_B = 0.10 = \text{constant}$

T/K	348.2	353.7	361.2	366.5	370.1	375.5	381.0	386.6	393.4
P_A/MPa	0.045	0.052	0.069	0.083	0.095	0.113	0.135	0.163	0.203

T/K	398.3	405.4	411.4	415.4
P_A/MPa	0.235	0.291	0.346	0.387

Polymer (B):	**polyester (hyperbranched, aliphatic)**	**2003SEI**
Characterization:	$M_n/\text{kg.mol}^{-1} = 2.8$, $M_w/\text{kg.mol}^{-1} = 5.1$,	
	hydroxyl functional hyperbranched polyesters produced from poly-alcohol cores and hydroxy acids, 64 OH groups per macromolecule, hydroxyl no. = 470-500 mg KOH/g, acid no. = 7-11 mg KOH/g, Boltorn H40, Perstorp Specialty Chemicals AB, Perstorp, Sweden	
Solvent (A):	**water** $\qquad$ **H_2O**	**7732-18-5**

continued

continued

Type of data: vapor pressure

$w_B = 0.10$ = constant

T/K	350.6	358.4	363.1	368.2	374.0	378.2	383.1	390.9	403.3
P_A/MPa	0.046	0.062	0.073	0.087	0.107	0.123	0.145	0.185	0.271

T/K	408.8	413.2	418.2	423.2	428.7	433.4
P_A/MPa	0.320	0.363	0.418	0.477	0.553	0.622

Polymer (B):   **polyethylene**   **1992WOH**
Characterization: M_n/kg.mol^{-1} = 1.94, M_w/kg.mol^{-1} = 5.37,
                         LDPE-wax, ρ = 0.9238 g/cm^3, Leuna-Werke, Germany
Solvent (A):   **1-butene**   **C$_4$H$_8$**   **106-98-9**

Type of data: gas solubility

T/K = 493.15

P/MPa	0.526	1.096	1.606	2.067	2.861	4.585	4.869	5.433	7.614
w_A	0.0142	0.0357	0.0532	0.0688	0.0943	0.1543	0.1603	0.1860	0.2706

Comments: High-pressure fluid phase equilibrium data for this system are given in Chapter 4.

Polymer (B):   **polyethylene**   **2001MOO**
Characterization: M_n/kg.mol^{-1} = 22, M_w/kg.mol^{-1} = 104, ρ = 0.923 g/cm^3,
                         50.4% crystallinity, LDPE
Solvent (A):   **1-butene**   **C$_4$H$_8$**   **106-98-9**

Type of data: gas solubility

T/K = 302.95

P/kPa	119.968	228.904	286.130
m_A/(g/g polymer)	0.01764	0.04359	0.06633
m_A/(g/g amorphous polymer)	0.03556	0.08788	0.13372

T/K = 322.45

P/kPa	197.188	359.214	528.823
m_A/(g/g polymer)	0.02339	0.05141	0.11382
m_A/(g/g amorphous polymer)	0.04715	0.10365	0.22948

T/K = 341.95

P/kPa	292.335	570.192	819.090
m_A/(g/g polymer)	0.02566	0.06770	0.15770
m_A/(g/g amorphous polymer)	0.05173	0.13649	0.31794

T/K = 361.05

P/kPa	526.066	872.869	1243.804
m_A/(g/g polymer)	0.04693	0.10323	0.26743
m_A/(g/g amorphous polymer)	0.09462	0.20812	0.53917

Polymer (B): **polyethylene** **2001MOO**
Characterization: M_n/kg.mol^{-1} = 11.5, M_w/kg.mol^{-1} = 110.5, ρ = 0.954 g/cm^3,
70.2% crystallinity, HDPE
Solvent (A): **1-butene** **C$_4$H$_8$** **106-98-9**

Type of data: gas solubility

T/K = 303.15

P/kPa	122.726	233.041	290.956
m_A/(g/g polymer)	0.01036	0.02288	0.03284
m_A/(g/g amorphous polymer)	0.03477	0.07677	0.11019

T/K = 322.45

P/kPa	197.188	389.551	564.676
m_A/(g/g polymer)	0.01087	0.02535	0.04961
m_A/(g/g amorphous polymer)	0.03649	0.08051	0.16648

T/K = 322.75

P/kPa	212.357	350.940	540.544
m_A/(g/g polymer)	0.01271	0.02220	0.04525
m_A/(g/g amorphous polymer)	0.04266	0.07448	0.15183

T/K = 342.05

P/kPa	281.304	550.197	836.327
m_A/(g/g polymer)	0.01294	0.02617	0.05072
m_A/(g/g amorphous polymer)	0.04344	0.08781	0.17022

T/K = 361.05

P/kPa	580.534	913.548	1225.878
m_A/(g/g polymer)	0.01805	0.03515	0.06289
m_A/(g/g amorphous polymer)	0.06057	0.11794	0.21103

Polymer (B): **polyethylene** **2001MOO**
Characterization: M_n/kg.mol^{-1} = 25.9, M_w/kg.mol^{-1} = 85.2, ρ = 0.917 g/cm^3,
47.0% crystallinity, LLDPE with 1-butene
Solvent (A): **1-butene** **C$_4$H$_8$** **106-98-9**

Type of data: gas solubility

T/K = 303.15

P/kPa	117.899	228.904	286.820
m_A/(g/g polymer)	0.02075	0.05094	0.07919
m_A/(g/g amorphous polymer)	0.03915	0.09612	0.14941

T/K = 322.65

P/kPa	203.394	365.419	532.960
m_A/(g/g polymer)	0.02574	0.05797	0.13351
m_A/(g/g amorphous polymer)	0.04858	0.10938	0.25191

continued

continued

$T/K = 342.05$

P/kPa	290.956	560.539	819.780
m_A/(g/g polymer)	0.03121	0.07302	0.17059
m_A/(g/g amorphous polymer)	0.05889	0.13777	0.32187

$T/K = 361.15$

P/kPa	559.850	881.143	1241.046
m_A/(g/g polymer)	0.05130	0.10614	0.24769
m_A/(g/g amorphous polymer)	0.09680	0.20026	0.46734

Polymer (B):	**polyethylene**		**2001MOO**
Characterization:	M_n/kg.mol^{-1} = 40.4, M_w/kg.mol^{-1} = 86.5, ρ = 0.885 g/cm^3, 18.5% crystallinity, LLDPE with 1-hexene		
Solvent (A):	**1-butene** **C$_4$H$_8$**		**106-98-9**

Type of data: gas solubility

$T/K = 303.15$

P/kPa	118.589	229.594	283.372
m_A/(g/g polymer)	0.04011	0.11308	0.19134
m_A/(g/g amorphous polymer)	0.04922	0.13874	0.23477

$T/K = 322.65$

P/kPa	209.599	393.687	535.718
m_A/(g/g polymer)	0.05225	0.14751	0.39621
m_A/(g/g amorphous polymer)	0.06411	0.18099	0.48615

$T/K = 342.05$

P/kPa	287.509	573.639	666.028
m_A/(g/g polymer)	0.05853	0.16911	0.23187
m_A/(g/g amorphous polymer)	0.07182	0.20750	0.28450

$T/K = 342.35$

P/kPa	285.441	589.497	814.264
m_A/(g/g polymer)	0.05840	0.17304	0.41314
m_A/(g/g amorphous polymer)	0.07165	0.21232	0.50692

Polymer (B):	**polyethylene**	**2004SOL**
Characterization:	HDPE, ρ = 0.955 g/cm^3	
Solvent (A):	**carbon dioxide** **CO$_2$**	**124-38-9**

Type of data: gas solubility

$T/K = 293.15$

P/bar	22.1	22.1	24.0	24.0	31.3	31.3	39.1	39.1
w_A	0.00884	0.00894	0.00964	0.00975	0.01215	0.01225	0.01468	0.01468

continued

continued

T/K = 305.15

P/bar	18.8	18.8	29.9	31.4	31.6	31.6	37.4	42.5	42.5
w_A	0.00649	0.00655	0.00946	0.00970	0.00990	0.00979	0.01088	0.01205	0.01215

T/K = 313.15

P/bar	20.4	20.4	29.8	29.8	40.7	40.7
w_A	0.00685	0.00706	0.00956	0.00959	0.01215	0.01235

T/K = 323.15

P/bar	22.5	22.5	29.3	29.3	40.0	40.0
w_A	0.00695	0.00700	0.00831	0.00849	0.01049	0.01049

Polymer (B): **polyethylene** 2004DAV
Characterization: M_w/kg.mol^{-1} = 250, LDPE, Dow Chemical
Solvent (A): **carbon dioxide** **CO$_2$** 124-38-9

Type of data: gas solubility

T/K = 423.15

P/psia	95	227	326	435	488
w_A	0.005	0.011	0.015	0.020	0.023

Polymer (B): **polyethylene** 1999SAT
Characterization: M_w/kg.mol^{-1} = 8.2, M_w/kg.mol^{-1} = 111, HDPE,
Nippon Petrochemicals, Kawasaki, Japan
Solvent (A): **carbon dioxide** **CO$_2$** 124-38-9

Type of data: gas solubility

T/K = 433.15

P/MPa	6.936	11.066	13.644	15.286	16.347	17.453
w_A	0.0391	0.0695	0.0888	0.1024	0.1095	0.1166

T/K = 453.15

P/MPa	7.055	11.326	14.013	15.762	16.896	18.123
w_A	0.0338	0.0629	0.0817	0.0913	0.0986	0.1070

T/K = 473.15

P/MPa	6.608	10.731	13.344	15.034	17.019
w_A	0.0309	0.0522	0.0694	0.0803	0.0936

Polymer (B): **polyethylene** 1997SUR
Characterization: M_n/kg.mol^{-1} = 76, LDPE, MI = 65, ρ = 0.919 g/cm^3,
DSM, Geleen, The Netherlands
Solvent (A): **cyclopentane** **C$_5$H$_{10}$** 287-92-3

continued

continued

Type of data: vapor pressure

$T/K = 425.65$

w_B	0.9513	0.8873	0.7981	0.7013	0.6038	0.5069
P_A/bar	1.76	3.56	5.79	7.63	9.02	10.10

$T/K = 474.15$

w_B	0.9587	0.9085	0.8108	0.6992	0.5921	0.5032
P_A/bar	2.68	5.74	10.72	15.53	19.44	22.41

Polymer (B):   **polyethylene**   **1977CHE**
Characterization: M_n/kg.mol^{-1} = 31.7, M_w/kg.mol^{-1} = 248.7, M_z/kg.mol^{-1} = 2378,
                        LDPE, ρ = 0.9188 g/cm^3, Union Carbide Corp., NY
Solvent (A):   **ethene**   **C$_2$H$_4$**   **74-85-1**

Type of data: gas solubility

$T/K = 399.15$

P/atm	4.40	7.80	11.2	14.6	18.0	21.4	24.8	28.2	31.6
w_A	0.0018	0.0037	0.0055	0.0075	0.0107	0.0136	0.0158	0.0175	0.0198

P/atm	35.0	38.4	41.8	45.2	48.6	52.0	55.4	58.8	62.2
w_A	0.0221	0.0242	0.0255	0.0285	0.0305	0.0330	0.0359	0.0381	0.0417

P/atm	65.6	69.0
w_A	0.0438	0.0469

$T/K = 413.15$

P/atm	4.40	7.80	11.2	14.6	18.0	21.4	24.8	28.2	31.6
w_A	0.0015	0.0034	0.0048	0.0068	0.0087	0.0112	0.0131	0.0151	0.0166

P/atm	35.0	38.4	41.8	45.2	48.6	52.0	55.4	58.8	62.2
w_A	0.0189	0.0208	0.0235	0.0250	0.0277	0.0296	0.0328	0.0341	0.0370

P/atm	65.6	69.0
w_A	0.0382	0.0415

$T/K = 428.15$

P/atm	4.40	7.80	11.2	14.6	18.0	21.4	24.8	28.2	31.6
w_A	0.0013	0.0029	0.0039	0.0055	0.0074	0.0090	0.0105	0.0120	0.0146

P/atm	35.0	38.4	41.8	45.2	48.6	52.0	55.4	58.8	62.2
w_A	0.0164	0.0178	0.0207	0.0222	0.0242	0.0265	0.0286	0.0304	0.0329

P/atm	65.6	69.0
w_A	0.0352	0.0372

Polymer (B):   **polyethylene**   **1985ROU**
Characterization: M_n/kg.mol^{-1} = 2.2, M_w/kg.mol^{-1} = 2.4,
                        HDPE, Barex 2000, CDF Chimie, France
Solvent (A):   **ethene**   **C$_2$H$_4$**   **74-85-1**

continued

continued

Type of data: gas solubility/bubble pressures/densities

T/K = 403.1

w_A	0.0275	0.0637	0.0773	0.1182	0.1189	0.1284	0.1482
P/MPa	3.39	8.06	10.01	15.71	15.49	17.23	22.14
ρ/(g/cm^3)	0.768	0.747	0.740	0.719	0.716	0.711	0.703

T/K = 433.1

w_A	0.0275	0.0637	0.0773	0.1182	0.1189	0.1284	0.1482
P/MPa	3.89	9.17	11.49	17.72	17.46	19.45	24.33
ρ/(g/cm^3)	0.747	0.723	0.723	0.703	0.701	0.697	0.687

T/K = 463.1

w_A	0.0275	0.0637	0.0772	0.0773	0.1182	0.1189	0.1284	0.1482
P/MPa	4.35	10.24	12.64	12.76	19.35	19.09	21.17	26.13
ρ/(g/cm^3)	0.733	0.712	0.706	0.706	0.688	0.685	0.685	0.673

T/K = 493.1

w_A	0.0275	0.0637	0.1189	0.1284
P/MPa	4.73	11.02	20.32	22.49

Polymer (B):	**polyethylene**	**1985ROU**
Characterization:	M_n/kg.mol^{-1} = 12.5, M_w/kg.mol^{-1} = 41, HDPE, ZMP16, CDF Chimie, France	
Solvent (A):	**ethene** **C$_2$H$_4$**	**74-85-1**

Type of data: gas solubility/bubble pressures/densities

T/K	403.1	463.1	523.1
w_A	0.0127	0.0127	0.0127
P/MPa	1.52	1.91	2.28
ρ/(g/cm^3)	0.7724	0.7247	

Polymer (B):	**polyethylene**	**1987KOB**
Characterization:	M_n/kg.mol^{-1} = 3.65, M_w/kg.mol^{-1} = 7.6, LDPE-wax	
Solvent (A):	**ethene** **C$_2$H$_4$**	**74-85-1**

Type of data: gas solubility

T/K = 393.2

w_B	0.958	0.920	0.870	0.835	0.795	0.755	0.715	0.665	0.605
P/MPa	4.90	9.80	19.61	29.42	39.22	49.03	58.84	68.54	78.45

w_B	0.535	0.455	0.365	0.270
P/MPa	88.25	98.06	107.82	117.67

continued

continued

$T/K = 433.2$

w_B	0.958	0.920	0.865	0.820	0.780	0.735	0.690	0.630	0.560
P/MPa	4.90	9.80	19.61	29.42	39.22	49.03	58.84	68.54	78.45

w_B	0.485	0.395	0.300
P/MPa	88.25	98.06	107.82

$T/K = 473.2$

w_B	0.958	0.920	0.855	0.805	0.765	0.710	0.655	0.585	0.510
P/MPa	4.90	9.80	19.61	29.42	39.22	49.03	58.84	68.54	78.45

w_B	0.430	0.345
P/MPa	88.25	98.06

Polymer (B):	**polyethylene**		**1987KOB**
Characterization:	M_n/kg.mol^{-1} = 4.95, M_w/kg.mol^{-1} = 11.0, LDPE		
Solvent (A):	**ethene**	**C$_2$H$_4$**	**74-85-1**

Type of data: gas solubility

$T/K = 393.2$

w_B	0.958	0.920	0.870	0.835	0.800	0.765	0.725	0.685	0.640
P/MPa	4.90	9.80	19.61	29.42	39.22	49.03	58.84	68.54	78.45

w_B	0.585	0.525	0.380
P/MPa	88.25	98.06	117.67

$T/K = 433.2$

w_B	0.958	0.920	0.865	0.825	0.785	0.745	0.700	0.655	0.600
P/MPa	4.90	9.80	19.61	29.42	39.22	49.03	58.84	68.54	78.45

w_B	0.540	0.475	0.305
P/MPa	88.25	98.06	117.67

$T/K = 473.2$

w_B	0.958	0.920	0.860	0.815	0.765	0.715	0.660	0.605	0.485
P/MPa	4.90	9.80	19.61	29.42	39.22	49.03	58.84	68.54	78.45

w_B	0.454	0.415
P/MPa	88.25	98.06

Polymer (B):	**polyethylene**		**2001MOO**
Characterization:	M_n/kg.mol^{-1} = 22, M_w/kg.mol^{-1} = 104, ρ = 0.923 g/cm^3,		
	50.4% crystallinity, LDPE		
Solvent (A):	**ethene**	**C$_2$H$_4$**	**74-85-1**

Type of data: gas solubility

$T/K = 300.75$

continued

continued

P/kPa	783.238	2154.594	3547.323
m_A/(g/g polymer)	0.00430	0.01163	0.02022
m_A/(g/g amorphous polymer)	0.00867	0.02346	0.04078

T/K = 320.85

P/kPa	805.990	2154.594	3530.086
m_A/(g/g polymer)	0.00389	0.01076	0.01792
m_A/(g/g amorphous polymer)	0.00785	0.02169	0.03613

T/K = 340.85

P/kPa	785.996	2138.736	3512.850
m_A/(g/g polymer)	0.00356	0.00971	0.01607
m_A/(g/g amorphous polymer)	0.00717	0.01958	0.03240

T/K = 360.95

P/kPa	795.648	2140.804	3526.639
m_A/(g/g polymer)	0.00320	0.00921	0.01606
m_A/(g/g amorphous polymer)	0.00645	0.01856	0.03237

Polymer (B):	**polyethylene**		**2001MOO**
Characterization:	M_n/kg.mol^{-1} = 11.5, M_w/kg.mol^{-1} = 110.5, ρ = 0.954 g/cm^3, 70.2% crystallinity, HDPE		
Solvent (A):	**ethene**	**C$_2$H$_4$**	**74-85-1**

Type of data: gas solubility

T/K = 300.85

P/kPa	859.080	2186.309	3561.113
m_A/(g/g polymer)	0.00276	0.00697	0.01074
m_A/(g/g amorphous polymer)	0.00928	0.02339	0.03606

T/K = 320.75

P/kPa	864.595	2147.699	3568.697
m_A/(g/g polymer)	0.00226	0.00553	0.00929
m_A/(g/g amorphous polymer)	0.00759	0.01855	0.03117

T/K = 340.75

P/kPa	852.185	2148.389	3534.223
m_A/(g/g polymer)	0.00308	0.00450	0.00731
m_A/(g/g amorphous polymer)	0.01033	0.01510	0.02453

T/K = 361.05

P/kPa	852.185	2175.278	3540.428
m_A/(g/g polymer)	0.00136	0.00351	0.00554
m_A/(g/g amorphous polymer)	0.00458	0.01178	0.01860

Polymer (B):	**polyethylene**		**2001MOO**
Characterization:	M_n/kg.mol^{-1} = 25.9, M_w/kg.mol^{-1} = 85.2, ρ = 0.917 g/cm^3, 47.0% crystallinity, LLDPE with 1-butene		
Solvent (A):	**ethene**	**C$_2$H$_4$**	**74-85-1**

Type of data: gas solubility

T/K = 303.05

P/kPa	819.780	2173.209	3531.465
m_A/(g/g polymer)	0.00395	0.01090	0.01793
m_A/(g/g amorphous polymer)	0.00746	0.02056	0.03383

T/K = 322.75

P/kPa	807.369	2210.441	3539.739
m_A/(g/g polymer)	0.00464	0.01052	0.01672
m_A/(g/g amorphous polymer)	0.00876	0.01985	0.03154

T/K = 342.25

P/kPa	808.748	2210.441	3545.944
m_A/(g/g polymer)	0.00331	0.00873	0.01429
m_A/(g/g amorphous polymer)	0.00625	0.01647	0.02697

T/K = 361.05

P/kPa	810.817	2244.914	3605.928
m_A/(g/g polymer)	0.00306	0.00872	0.01493
m_A/(g/g amorphous polymer)	0.00578	0.01645	0.02816

Polymer (B):	**polyethylene**		**2001MOO**
Characterization:	M_n/kg.mol^{-1} = 40.4, M_w/kg.mol^{-1} = 86.5, ρ = 0.885 g/cm^3, 18.5% crystallinity, LLDPE with 1-hexene		
Solvent (A):	**ethene**	**C$_2$H$_4$**	**74-85-1**

Type of data: gas solubility

T/K = 300.85

P/kPa	814.954	2169.762	3547.323
m_A/(g/g polymer)	0.00810	0.02359	0.04077
m_A/(g/g amorphous polymer)	0.00994	0.02895	0.05003

T/K = 320.85

P/kPa	838.396	2169.073	3545.255
m_A/(g/g polymer)	0.00798	0.02121	0.03613
m_A/(g/g amorphous polymer)	0.00979	0.02603	0.04433

T/K = 340.85

P/kPa	814.954	2180.104	3526.639
m_A/(g/g polymer)	0.00724	0.01999	0.03317
m_A/(g/g amorphous polymer)	0.00889	0.02453	0.04070

Polymer (B):	polyethylene	2004DAV
Characterization:	M_w/kg.mol^{-1} = 250, LDPE, Dow Chemical	
Solvent (A):	ethene C_2H_4	74-85-1

Type of data: gas solubility

T/K = 423.15

P/psia	266	473	553	633	763
w_A	0.011	0.020	0.022	0.025	0.031

Polymer (B):	polyethylene	2002JOU
Characterization:	M_w/kg.mol^{-1} = 125, Aldrich Chem. Co., Inc., Milwaukee, WI	
Solvent (A):	n-heptane C_7H_{16}	142-82-5

Type of data: vapor pressure

w_B	0.024	was kept constant

T/K	393.15	408.15	423.15	438.15	443.15	453.15
P_A/MPa	0.6	0.7	0.9	1.1	1.2	1.2

Comments: Liquid-liquid equilibrium data for this system are given in Chapter 3.

Polymer (B):	polyethylene	1990KEN
Characterization:	M_n/kg.mol^{-1} = 8, M_w/kg.mol^{-1} = 177, M_z/kg.mol^{-1} = 1000, HDPE, DSM, Geleen, The Netherlands	
Solvent (A):	n-hexane C_6H_{14}	110-54-3

Type of data: vapor-liquid equilibrium

w_B	0.0053	0.0053	0.0053	0.0053	0.0095	0.0095	0.0095	0.0095	0.0954
T/K	392.62	394.62	396.61	402.54	396.90	397.94	398.93	399.90	402.62
P/bar	4.0	4.2	4.4	5.0	4.5	4.6	4.7	4.8	5.8

w_B	0.0954	0.0954	0.0954	0.0954	0.0954	0.1313	0.1313	0.1313	0.1313
T/K	404.58	406.57	414.54	418.51	420.51	414.94	417.93	419.93	421.87
P/bar	6.0	6.2	7.1	7.7	8.0	6.6	7.1	7.4	7.6

w_B	0.1313	0.1313
T/K	423.69	425.66
P/bar	7.9	8.2

Comments: Liquid-liquid equilibrium and liquid-liquid-vapor equilibrium data for this system are given in Chapter 3.

Polymer (B):	polyethylene	2001TOR
Characterization:	M_n/kg.mol^{-1} = 43, M_w/kg.mol^{-1} = 105, M_z/kg.mol^{-1} = 190, HDPE, DSM, Geleen, The Netherlands	
Solvent (A):	n-hexane C_6H_{14}	110-54-3

continued

continued

Type of data: vapor-liquid equilibrium

w_B	0.2128	0.1382	0.0813	0.0138	0.0138
T/K	446.2	436.5	431.3	423.4	433.6
P/MPa	1.05	0.92	0.77	0.68	0.78

Comments: LLE and LLVE data for this system are given in Chapter 3.

Polymer (B):　　　**polyethylene**　　　　　　　　　　　　**2004CHE**
Characterization:　M_n/kg.mol^{-1} = 14.4, M_w/kg.mol^{-1} = 15.5, completely hydrogenated
　　　　　　　　polybutadiene, Scientific Polymer Products, Inc., Ontario, NY
Solvent (A):　　　**n-hexane**　　　　　**C$_6$H$_{14}$**　　　　　**110-54-3**

Type of data: vapor-liquid equilibrium

w_B	0.0906	0.0906	0.0906	0.0906	0.0906
T/K	373.21	393.21	413.12	433.14	443.14
P/MPa	0.4	0.6	0.9	1.1	1.3

Comments: LLE and LLVE data for this system are given in Chapter 3.

Polymer (B):　　　**polyethylene**　　　　　　　　　　　　**2004CHE**
Characterization:　M_n/kg.mol^{-1} = 82, M_w/kg.mol^{-1} = 108, completely hydrogenated
　　　　　　　　polybutadiene, Scientific Polymer Products, Inc., Ontario, NY
Solvent (A):　　　**n-hexane**　　　　　**C$_6$H$_{14}$**　　　　　**110-54-3**

Type of data: vapor-liquid equilibrium

w_B	0.0829	0.0829	0.0829
T/K	373.23	393.18	413.15
P/MPa	0.4	0.7	0.8

Comments: LLE and LLVE data for this system are given in Chapter 3.

Polymer (B):　　　**polyethylene**　　　　　　　　　　　　**2004CHE**
Characterization:　M_n/kg.mol^{-1} = 23.3, M_w/kg.mol^{-1} = 60.4, M_z/kg.mol^{-1} = 100.7,
　　　　　　　　metallocene LLDPE, unspecified comonomer, industrial source
Solvent (A):　　　**n-hexane**　　　　　**C$_6$H$_{14}$**　　　　　**110-54-3**

Type of data: vapor-liquid equilibrium

w_B	0.0822	0.0822
T/K	393.09	413.15
P/MPa	0.6	0.9

Comments: LLE and LLVE data for this system are given in Chapter 3.

Polymer (B):   **polyethylene**   **2001MOO**
Characterization: $M_n/\text{kg.mol}^{-1} = 11.5$, $M_w/\text{kg.mol}^{-1} = 110.5$, $\rho = 0.954$ g/cm^3,
                          70.2% crystallinity, HDPE
Solvent (A):   **1-hexene**   **C$_6$H$_{12}$**   **592-41-6**

Type of data: vapor solubility

$T/\text{K} = 342.25$

P/kPa	51.021	67.568	106.178
m_A/(g/g polymer)	0.02077	0.03031	0.07508
m_A/(g/g amorphous polymer)	0.06971	0.10173	0.25196

$T/\text{K} = 361.05$

P/kPa	71.705	128.931	190.294
m_A/(g/g polymer)	0.01842	0.04169	0.10961
m_A/(g/g amorphous polymer)	0.06181	0.13991	0.36783

Polymer (B):   **polyethylene**   **2001MOO**
Characterization: $M_n/\text{kg.mol}^{-1} = 22$, $M_w/\text{kg.mol}^{-1} = 104$, $\rho = 0.923$ g/cm^3,
                          50.4% crystallinity, LDPE
Solvent (A):   **1-hexene**   **C$_6$H$_{12}$**   **592-41-6**

Type of data: vapor solubility

$T/\text{K} = 342.25$

P/kPa	55.158	68.947	110.315
m_A/(g/g polymer)	0.05371	0.08019	0.27483
m_A/(g/g amorphous polymer)	0.10829	0.16168	0.55409

$T/\text{K} = 361.15$

P/kPa	75.842	132.378	191.673
m_A/(g/g polymer)	0.07117	0.17575	0.50437
m_A/(g/g amorphous polymer)	0.14348	0.35434	1.01687

Polymer (B):   **polyethylene**   **2001MOO**
Characterization: $M_n/\text{kg.mol}^{-1} = 25.9$, $M_w/\text{kg.mol}^{-1} = 85.2$, $\rho = 0.917$ g/cm^3,
                          47.0% crystallinity, LLDPE with 1-butene
Solvent (A):   **1-hexene**   **C$_6$H$_{12}$**   **592-41-6**

Type of data: vapor solubility

$T/\text{K} = 342.05$

P/kPa	46.884	76.531	113.763
m_A/(g/g polymer)	0.04459	0.09503	0.30252
m_A/(g/g amorphous polymer)	0.08413	0.17930	0.57080

$T/\text{K} = 361.25$

P/kPa	78.600	135.826	178.573
m_A/(g/g polymer)	0.05581	0.14184	0.32805
m_A/(g/g amorphous polymer)	0.10530	0.26762	0.61896

Polymer (B):	**polyethylene**		**2001MOO**
Characterization:	M_n/kg.mol^{-1} = 40.4, M_w/kg.mol^{-1} = 86.5, ρ = 0.885 g/cm^3,		
	18.5% crystallinity, LLDPE with 1-hexene		
Solvent (A):	**1-hexene**	**C$_6$H$_{12}$**	**592-41-6**

Type of data: vapor solubility

T/K = 342.25

P/kPa	51.021	75.152	91.010
m_A/(g/g polymer)	0.10675	0.22058	0.46488
m_A/(g/g amorphous polymer)	0.13098	0.27065	0.53490

Polymer (B):	**polyethylene**		**2004SOL**
Characterization:	HDPE, ρ = 0.955 g/cm^3		
Solvent (A):	**methane**	**CH$_4$**	**74-82-8**

Type of data: gas solubility

T/K = 293.15

P/bar	50.1	50.1	106.0	106.0	142.2	142.2
w_A	0.00296	0.00267	0.00515	0.00512	0.00601	0.00634

T/K = 305.15

P/bar	48.3	55.5	103.8	105.0	150.0	151.0	151.0	157.1	157.1
w_A	0.00196	0.00215	0.00311	0.00310	0.00357	0.00409	0.00419	0.00413	0.00388

T/K = 313.15

P/bar	49.6	49.6	105.0	105.0	135.2	135.2
w_A	0.00244	0.00245	0.00411	0.00427	0.00484	0.00490

T/K = 323.15

P/bar	50.2	50.2	50.9	50.9	105.4	105.4	106.3	106.3	128.6
w_A	0.00217	0.00223	0.00220	0.00247	0.00353	0.00377	0.00380	0.00387	0.00396

P/bar	128.6	142.2	142.2
w_A	0.00424	0.00422	0.00440

Polymer (B):	**polyethylene**		**1981PAR**
Characterization:	M_n/kg.mol^{-1} = 12.4, M_w/kg.mol^{-1} = 89.6, ρ = 0.964 g/cm^3,		
	77.5% crystallinity, HDPE, Marlex EMN 6030		
Solvent (A):	**2-methylpropane**	**C$_4$H$_{10}$**	**75-28-5**

Type of data: gas solubility

T/K = 338.75

P/bar	0.97	1.03	2.74	4.55	6.20	7.91	9.13
w_A	0.0022	0.0023	0.0070	0.0122	0.0179	0.0246	0.0313

continued

continued

$T/K = 347.05$

P/bar	1.03	2.76	4.51	6.24	7.94	9.67			
w_A	0.0024	0.0065	0.0106	0.0155	0.0212	0.0277			

$T/K = 355.35$

P/bar	1.00	2.77	4.53	6.25	7.97	9.72	11.41		
w_A	0.0020	0.0056	0.0092	0.0129	0.0171	0.0223	0.0277		

$T/K = 366.45$

P/bar	1.01	2.82	4.57	6.29	8.01	10.75	11.47		
w_A	0.0017	0.0047	0.0079	0.0112	0.0146	0.0183	0.0224		

Polymer (B):   **polyethylene**   **1983MEY**
Characterization: M_n/kg.mol^{-1} = 14, M_w/kg.mol^{-1} = 94, ρ = 0.951 g/cm^3, HDPE
Solvent (A):   **2-methylpropane**   **C$_4$H$_{10}$**   **75-28-5**

Type of data: gas solubility

$T/K = 422.0$

P/Torr	99.9	129	197	335	646	867	1095	1351	1496
w_A	0.00056	0.00101	0.00118	0.00202	0.00376	0.00513	0.00650	0.00818	0.00911

$T/K = 477.6$

P/Torr	58.1	101	185	202	438	654	926	1369	1649
w_A	0.00021	0.00038	0.00066	0.00079	0.00165	0.00246	0.00344	0.00528	0.00591

$T/K = 533.2$

P/Torr	45.0	106	200	421	727	1015	1262	1454	1722
w_A	0.00012	0.00025	0.00053	0.00112	0.00196	0.00276	0.00353	0.00395	0.00434

Polymer (B):   **polyethylene**   **1983MEY**
Characterization: M_n/kg.mol^{-1} = 18, M_w/kg.mol^{-1} = 109, ρ = 0.919 g/cm^3, LDPE
Solvent (A):   **2-methylpropane**   **C$_4$H$_{10}$**   **75-28-5**

Type of data: gas solubility

$T/K = 422.0$

P/Torr	67.6	139	300	682	1120	1380
w_A	0.00037	0.00072	0.00168	0.00388	0.00579	0.00820

$T/K = 477.6$

P/Torr	72.5	162	350	774	1310	1566
w_A	0.00024	0.00054	0.00125	0.00280	0.00491	0.00595

$T/K = 533.2$

P/Torr	83.2	179	387	895	1453	1745
w_A	0.00022	0.00056	0.00131	0.00203	0.00365	0.00420

Polymer (B):	polyethylene				1995LEE
Characterization:	M_w/kg.mol^{-1} = 800, LDPE				
Solvent (A):	nitrogen		N_2		7727-37-9

Type of data: gas solubility

T/K = 394.3

P/atm	10	20	50	100	125
w_A	0.00209	0.00392	0.00621	0.01002	0.01118
φ_A	0.00192	0.00360	0.00572	0.00924	0.01031

T/K = 408.2

P/atm	10	20	50	100	125
w_A	0.00229	0.00424	0.00771	0.01248	0.01374
φ_A	0.00209	0.00386	0.00701	0.01135	0.01251

T/K = 422.1

P/atm	10	20	50	100	125
w_A	0.00233	0.00454	0.00878	0.01420	0.01615
φ_A	0.00210	0.00409	0.00790	0.01277	0.01453

T/K = 436.0

P/atm	10	20	50	100	125
w_A	0.00252	0.00482	0.00974	0.01535	0.01764
φ_A	0.00225	0.00429	0.00866	0.01365	0.01570

T/K = 449.9

P/atm	10	20	50	100	125
w_A	0.00272	0.00518	0.01059	0.01729	0.01981
φ_A	0.00240	0.00457	0.00932	0.01522	0.01744

Polymer (B):	polyethylene				1999SAT
Characterization:	M_n/kg.mol^{-1} = 8.2, M_w/kg.mol^{-1} = 111, HDPE, Nippon Petrochemicals, Kawasaki, Japan				
Solvent (A):	nitrogen		N_2		7727-37-9

Type of data: gas solubility

T/K = 433.15

P/MPa	2.541	4.278	5.463	7.935	10.089	11.555	14.629	
w_A	0.00257	0.00430	0.00549	0.00773	0.00945	0.01144	0.01444	

T/K = 453.15

P/MPa	3.743	6.330	8.192	9.401	11.809			
w_A	0.00371	0.00652	0.00842	0.00957	0.01245			

T/K = 473.15

P/MPa	2.818	4.704	5.958	6.792	8.386	10.695	12.682	15.214
w_A	0.00287	0.00516	0.00666	0.00754	0.00975	0.01170	0.01472	0.01695

Polymer (B):	**polyethylene**				**2004DAV**
Characterization:	$M_w/\text{kg.mol}^{-1}$ = 250, LDPE, Dow Chemical				
Solvent (A):	**nitrogen**	**N$_2$**			**7727-37-9**

Type of data: gas solubility

$T/\text{K} = 423.15$

P/psia	922	953
w_A	0.004	0.005

Polymer (B):	**polyethylene**			**2001TOR**
Characterization:	$M_n/\text{kg.mol}^{-1}$ = 43, $M_w/\text{kg.mol}^{-1}$ = 105, $M_z/\text{kg.mol}^{-1}$ = 190, HDPE, DSM, Geleen, The Netherlands			
Solvent (A):	**1-octene**	**C$_8$H$_{16}$**		**111-66-0**

Type of data: vapor-liquid equilibrium

$T/\text{K} = 503$

w_B	0.1791	0.0897	0.0439	0.0198
P/MPa	0.97	0.94	0.95	0.95

$T/\text{K} = 513$

w_B	0.1791	0.0897	0.0439
P/MPa	1.14	1.13	1.11

Comments: LLE and LLVE data for this system are given in Chapter 3.

Polymer (B):	**polyethylene**					**1997SUR**
Characterization:	$M_n/\text{kg.mol}^{-1}$ = 76, LDPE, MI = 65, ρ = 0.919 g/cm^3, DSM, Geleen, The Netherlands					
Solvent (A):	**n-pentane**	**C$_5$H$_{12}$**				**109-66-0**

Type of data: vapor pressure

$T/\text{K} = 423.65$

w_D	0.9477	0.9054	0.8088	0.7049	0.6274	0.5191
P_A/bar	2.47	5.36	9.48	12.26	13.63	15.01

$T/\text{K} = 474.15$

w_B	0.9705	0.9202	0.8410	0.7417	0.6404	0.5571
P_A/bar	2.64	7.16	13.61	20.86	26.80	30.58

Polymer (B):	**polyethylene**			**1997SUR**
Characterization:	$M_n/\text{kg.mol}^{-1}$ = 76, LDPE, MI = 65, ρ = 0.919 g/cm^3, DSM, Geleen, The Netherlands			
Solvent (A):	**3-pentanol**	**C$_5$H$_{12}$O**		**584-02-1**

continued

continued

Type of data: vapor pressure

$T/K = 423.15$

w_B	0.9536	0.8929	0.7894	0.7036	0.6128	0.5134
P_A/bar	0.83	1.53	2.12	2.48	2.69	2.81

$T/K = 473.15$

w_B	0.9600	0.9095	0.8136	0.6415	0.5027
P_A/bar	1.40	3.08	5.21	7.55	8.32

Polymer (B):	**polyethylene**	**1997SUR**
Characterization:	M_n/kg.mol^{-1} = 76, LDPE, MI = 65, ρ = 0.919 g/cm^3, DSM, Geleen, The Netherlands	
Solvent (A):	**3-pentanone** **C$_5$H$_{10}$O**	**96-22-0**

Type of data: vapor pressure

$T/K = 425.15$

w_B	0.9538	0.8961	0.7950	0.7049	0.6002	0.5054
P_A/bar	0.78	1.61	2.51	2.90	3.21	3.38

$T/K = 477.15$

w_B	0.9597	0.9071	0.8171	0.7113	0.6054	0.5078
P_A/bar	1.55	3.21	5.73	7.66	8.77	9.37

Polymer (B):	**polyethylene**	**1997SUR**
Characterization:	M_n/kg.mol^{-1} = 76, LDPE, MI = 65, ρ = 0.919 g/cm^3, DSM, Geleen, The Netherlands	
Solvent (A):	**1-pentene** **C$_5$H$_{10}$**	**109-67-1**

Type of data: vapor pressure

$T/K = 423.65$

w_B	0.9741	0.9096	0.8092	0.6704	0.5286
P_A/bar	1.58	5.01	9.40	14.05	16.50

$T/K = 474.15$

w_B	0.9602	0.9088	0.8082	0.6972	0.5840
P_A/bar	4.20	9.36	17.57	25.63	32.74

Polymer (B):	**polyethylene**	**1983MEY**
Characterization:	M_n/kg.mol^{-1} = 14, M_w/kg.mol^{-1} = 94, ρ = 0.951 g/cm^3, HDPE	
Solvent (A):	**propane** **C$_3$H$_8$**	**74-98-6**

Type of data: gas solubility

continued

continued

$T/K = 422.0$

$P/$Torr	124	250	479	793	1182	1337
w_A	0.00036	0.00073	0.00140	0.00242	0.00338	0.00370

$T/K = 477.6$

$P/$Torr	119	250	589	815	1093	1443
w_A	0.00023	0.00049	0.00120	0.00169	0.00226	0.00291

$T/K = 533.2$

$P/$Torr	84.5	280	487	905	1233	1447
w_A	0.00013	0.00048	0.00082	0.00155	0.00215	0.00261

Polymer (B):	**polyethylene**	**1997SUR**
Characterization:	$M_n/$kg.mol^{-1} = 76, LDPE, MI = 65, ρ = 0.919 g/cm^3, DSM, Geleen, The Netherlands	
Solvent (A):	**propyl acetate** $C_5H_{10}O_2$	**109-60-4**

Type of data: vapor pressure

$T/K = 426.15$

w_B	0.9521	0.8891	0.7831	0.6457	0.5005
$P_A/$bar	0.86	1.78	2.77	3.39	3.77

$T/K = 474.15$

w_B	0.9621	0.8963	0.8030	0.6666	0.5117
$P_A/$bar	1.59	3.77	6.31	8.77	10.39

Polymer (B):	**polyethylene**	**1997SUR**
Characterization:	$M_n/$kg.mol^{-1} = 76, LDPE, MI = 65, ρ = 0.919 g/cm^3, DSM, Geleen, The Netherlands	
Solvent (A):	**2-propylamine** C_3H_9N	**75-31-0**

Type of data: vapor pressure

$T/K = 427.15$

w_B	0.95478	0.8959	0.8140	0.6776	0.5245
$P_A/$bar	4.32	8.90	13.53	17.76	19.95

$T/K = 475.15$

w_B	0.96168	0.90917	0.8370	0.7163	0.5692
$P_A/$bar	4.56	10.76	18.85	29.77	38.37

Polymer (B):	**poly(ethylene glycol)**	**2004WEI**
Characterization:	$M_w/$kg.mol^{-1} = 0.2, Hoechst AG, Frankfurt/M., Germany	
Solvent (A):	**carbon dioxide** CO_2	**124-38-9**

continued

continued

Type of data: gas solubility

$T/K = 303.15$

P/bar	22	44	69	78	98	157	190	221	242
w_A	0.077	0.114	0.198	0.208	0.214	0.224	0.233	0.250	0.253

$T/K = 323.15$

P/bar	23	37	60	81	116	148	194	241	278
w_A	0.057	0.081	0.128	0.160	0.188	0.203	0.215	0.230	0.236

$T/K = 353.15$

P/bar	32	40	61	80	102	127	174	190	218
w_A	0.052	0.061	0.079	0.107	0.134	0.152	0.165	0.182	0.192

P/bar	258	285
w_A	0.210	0.217

$T/K = 373.15$

P/bar	25	37	51	66	78	90	107	123	143
w_A	0.036	0.048	0.058	0.077	0.085	0.100	0.114	0.130	0.142

P/bar	158	198	237	252	285
w_A	0.152	0.170	0.190	0.197	0.203

Polymer (B):	**poly(ethylene glycol)**		**2000WIE, 2004WEI**
Characterization:	M_w/kg.mol^{-1} = 1.5, Hoechst AG, Frankfurt/M., Germany		
Solvent (A):	**carbon dioxide**	**CO$_2$**	**124-38-9**

Type of data: gas solubility

$T/K = 316.15$

P/bar	20	41	61	82	116	137	181	232	284
w_A	0.041	0.090	0.148	0.198	0.232	0.238	0.262	0.281	0.295

$T/K = 323.15$

P/bar	21	41	54	78	118	157	203	240	271
w_A	0.034	0.090	0.117	0.169	0.200	0.215	0.263	0.298	0.300

$T/K = 333.15$

P/bar	21	45	74	106	138	211	230	286
w_A	0.034	0.080	0.143	0.187	0.215	0.249	0.253	0.260

$T/K = 353.15$

P/bar	20	41	61	79	98	130	162	202	270
w_A	0.021	0.050	0.081	0.108	0.137	0.156	0.179	0.206	0.245

Comments: Other gas solubilities are given together with the cloud points in Chapter 4.

Polymer (B):	poly(ethylene glycol)						**2000WIE, 2004WEI**	
Characterization:	M_w/kg.mol^{-1} = 4.0, Hoechst AG, Frankfurt/M., Germany							
Solvent (A):	carbon dioxide		CO_2				**124-38-9**	

Type of data: gas solubility

T/K = 328.15

P/bar	43	60	81	138	182	221	260	282
w_A	0.098	0.119	0.158	0.205	0.238	0.260	0.271	0.286

T/K = 333.15

P/bar	5	9	20	34	40	61	80	115	139
w_A	0.013	0.022	0.044	0.062	0.073	0.109	0.147	0.162	0.198

P/bar	220	286
w_A	0.249	0.266

T/K = 353.15

P/bar	6	16	24	31	44	54	66	79	117
w_A	0.011	0.022	0.037	0.042	0.054	0.073	0.089	0.106	0.133

P/bar	140	181	219	265
w_A	0.157	0.193	0.215	0.234

T/K = 373.15

P/bar	11	20	32	40	44	53	58	81	101
w_A	0.011	0.015	0.032	0.042	0.047	0.054	0.062	0.087	0.102

P/bar	149	178	220	257
w_A	0.157	0.159	0.188	0.197

Comments: Other gas solubilities are given together with the cloud points in Chapter 4.

Polymer (B):	poly(ethylene glycol)		**2000WIE**
Characterization:	M_w/kg.mol^{-1} = 8.0, Hoechst AG, Frankfurt/M., Germany		
Solvent (A):	carbon dioxide	CO_2	**124-38-9**

Comments: The gas solubilities are given together with the cloud points in Chapter 4.

Polymer (B):	poly(ethylene glycol)		**2000WIE**
Characterization:	M_w/kg.mol^{-1} = 1.5, Hoechst AG, Frankfurt/M., Germany		
Solvent (A):	nitrogen	N_2	**7727-37-9**

Type of data: gas solubility

T/K = 323.15

P/bar	49	62	87	98	110	137	188
w_B	0.9994	0.9990	0.9986	0.9982	0.9978	0.9971	0.9961

continued

continued

T/K = 333.15

P/bar	88	112	140	172	187
w_B	0.9981	0.9976	0.9968	0.9959	0.9954

T/K = 353.15

P/bar	65	76	89	115	140	183	192
w_B	0.9983	0.9981	0.9978	0.9972	0.9965	0.9951	0.9948

Polymer (B):   **poly(ethylene glycol)**   **2000WIE, 2004WEI**
Characterization: M_w/kg.mol^{-1} = 4.0, Hoechst AG, Frankfurt/M., Germany
Solvent (A):   **nitrogen**   **N$_2$**   **7727-37-9**

Type of data: gas solubility

T/K = 338.15

P/bar	69	81	120	138	139	175	200
w_B	0.9990	0.9984	0.9971	0.9965	0.9965	0.9958	0.9953

T/K = 353.15

P/bar	80	100	123	145	180	200
w_B	0.9979	0.9973	0.9964	0.9959	0.9947	0.9944

T/K = 373.15

P/bar	62	84	104	118	137	181	199
w_B	0.9982	0.9975	0.9966	0.9962	0.9955	0.9939	0.9932

Polymer (B):   **poly(ethylene glycol)**   **2000WIE**
Characterization: M_w/kg.mol^{-1} = 8.0, Hoechst AG, Frankfurt/M., Germany
Solvent (A):   **nitrogen**   **N$_2$**   **7727-37-9**

Type of data: gas solubility

T/K = 353.15

P/bar	82	95	116	143	178	199
w_B	0.9976	0.9972	0.9968	0.9962	0.9951	0.9932

T/K = 373.15

P/bar	58	78	99	120	138	176	199
w_B	0.9986	0.9972	0.9968	0.9961	0.9955	0.9941	0.9932

Polymer (B):   **poly(ethylene glycol)**   **2004WEI**
Characterization: M_w/kg.mol^{-1} = 0.2, Hoechst AG, Frankfurt/M., Germany
Solvent (A):   **propane**   **C$_3$H$_8$**   **74-98-6**

Type of data: gas solubility

continued

continued

T/K = 393.15

P/bar	7	20	31	39	51	62	83	120	162
w_A	0.0023	0.0113	0.0209	0.0262	0.0330	0.0364	0.0394	0.0424	0.0468

Polymer (B): **poly(ethylene glycol)** **2004WEI**
Characterization: M_w/kg.mol^{-1} = 0.6, Hoechst AG, Frankfurt/M., Germany
Solvent (A): **propane** **C$_3$H$_8$** **74-98-6**

Type of data: gas solubility

T/K = 393.15

P/bar	38	49	78	114	156	198
w_A	0.033	0.048	0.057	0.059	0.063	0.066

Polymer (B): **poly(ethylene glycol)** **2000WIE**
Characterization: M_w/kg.mol^{-1} = 1.5, Hoechst AG, Frankfurt/M., Germany
Solvent (A): **propane** **C$_3$H$_8$** **74-98-6**

Type of data: gas solubility

T/K = 393.15

P/bar	10	21	30	41	50	66
w_A	0.012	0.023	0.032	0.045	0.052	0.061

Comments: Other solubilities are given together with the cloud points in Chapter 4.

Polymer (B): **poly(ethylene glycol)** **2000WIE, 2004WEI**
Characterization: M_w/kg.mol^{-1} = 4.0, Hoechst AG, Frankfurt/M., Germany
Solvent (A): **propane** **C$_3$H$_8$** **74-98-6**

Type of data: gas solubility

T/K = 333.15

P/bar	20	23	31	50	120	160	220		
w_A	0.042	0.050	0.052	0.053	0.056	0.057	0.058		

T/K = 353.15

P/bar	11	21	52	64	80	110	120	160	250
w_A	0.020	0.038	0.059	0.061	0.058	0.064	0.061	0.062	0.064

T/K = 373.15

P/bar	9	20	30	40	62	90	150	200	250
w_A	0.014	0.026	0.040	0.053	0.062	0.070	0.073	0.076	0.077

continued

continued

$T/K = 393.15$

P/bar	11	20	30	40	50	60	70	100	110
w_A	0.010	0.021	0.034	0.044	0.054	0.060	0.065	0.072	0.076

P/bar	140	190	250
w_A	0.077	0.088	0.094

Comments: Other gas solubilities are given together with the cloud points in Chapter 4.

Polymer (B):   **poly(ethylene glycol)**   **2000WIE, 2004WEI**
Characterization: M_w/kg.mol^{-1} = 8.0, Hoechst AG, Frankfurt/M., Germany
Solvent (A):   **propane**   **C$_3$H$_8$**   **74-98-6**

Type of data: gas solubility

$T/K = 393.15$

P/bar	11	20	30	46	59	68	110	150
w_A	0.013	0.024	0.037	0.054	0.062	0.067	0.075	0.077

Comments: Other solubilities are given together with the cloud points in Chapter 4.

Polymer (B):   **poly(ethylene glycol) dimethyl ether**   **2004WEI**
Characterization: M_w/kg.mol^{-1} = 0.2, Hoechst AG, Frankfurt/M., Germany
Solvent (A):   **carbon dioxide**   **CO$_2$**   **124-38-9**

Type of data: liquid-gas equilibrium (first line – liquid, second line – gas phase)

$T/K = 298.15$

P/bar	10	20	30	41	49	57
w_A	0.0517	0.1255	0.2344	0.3064	0.4073	0.5604

P/bar	10	20	30	41	49	57
w_A	0.9731	0.9787	0.9632	0.9812	0.9843	0.9863

$T/K = 311.15$

P/bar	10	20	40	61	75
w_A	0.0185	0.0689	0.1626	0.3220	0.5175

P/bar	10	20	40	61	75
w_A	0.9864		0.9754	0.9870	0.9855

$T/K = 351.15$

P/bar	20	41	67	105	127	142	151	165
w_A	0.0507	0.1127	0.2164	0.3152	0.3991	0.4593	0.4645	0.5260

P/bar				105	127	142	151	165
w_A				0.9757	0.9798	0.9735	0.9567	0.9397

continued

continued

$T/K = 373.15$

P/bar	14	42	68	106	126	150	167	178	196
w_A	0.0288	0.0810	0.1451	0.2300	0.2988	0.3573	0.4008	0.4490	0.4977

P/bar	13	41	68	106	126	150	167	178	196
w_A	0.9955	0.9614	0.9628	0.9775	0.9789	0.9667	0.9638	0.9490	0.9178

$T/K = 393.15$

P/bar	20	45	70	106	131	165	181	198
w_A	0.0364	0.0733	0.1304	0.2014	0.2556	0.3240	0.3545	0.3909

P/bar				106	131	165	181	198
w_A				0.9768	0.9696	0.9675	0.9533	0.9557

Polymer (B):	**poly(ethylene glycol) dimethyl ether**	**2004WEI**
Characterization:	M_w/kg.mol^{-1} = 2.0, Hoechst AG, Frankfurt/M., Germany	
Solvent (A):	**carbon dioxide** $\quad$ **CO$_2$**	**124-38-9**

Type of data: gas solubility

$T/K = 316.15$

P/bar	20	41	61	82	116	137	181	211	232
w_A	0.041	0.090	0.148	0.198	0.232	0.238	0.262	0.262	0.281

$T/K = 333.15$

P/bar	21	45	74	106	138	211	230	286
w_A	0.034	0.080	0.143	0.187	0.215	0.249	0.253	0.260

$T/K = 353.15$

P/bar	20	41	61	79	98	130	162	202	270
w_A	0.021	0.050	0.081	0.108	0.137	0.156	0.179	0.206	0.245

Polymer (B):	**poly(ethylene glycol) dimethyl ether**	**1995GAI**
Characterization:	PEGDM mixture used for the Selexol process	

diethylene glycol dimethylether	2.07 wt%
diethylene glycol monomethylether	0.16 wt%
triethylene glycol dimethylether	21.01 wt%
triethylene glycol monomethylether	0.72 wt%
tetraethylene glycol dimethylether	33.68 wt%
tetraethylene glycol monomethylether	0.67 wt%
pentaethylene glycol dimethylether	26.39 wt%
pentaethylene glycol monomethylether	0.38 wt%
hexaethylene glycol dimethylether	12.78 wt%
hexaethylene glycol monomethylether	0.10 wt%
heptaethylene glycol dimethylether	1.87 wt%

Solvent (A):	**carbon dioxide** $\quad$ **CO$_2$**	**124-38-9**

Type of data: gas solubility

continued

continued

T/K = 298.15

P/MPa	0.486	0.800	1.064	1.530	2.037	2.553	3.141	3.557	4.023
x_A	0.1110	0.1745	0.2208	0.2947	0.3725	0.4364	0.5060	0.5523	0.6162

T/K = 313.15

P/MPa	0.456	0.790	1.034	1.550	2.067	2.543	3.080	3.617	4.418
x_A	0.0813	0.1384	0.1753	0.2426	0.2955	0.3554	0.4116	0.4621	0.5438

T/K = 333.15

P/MPa	0.527	0.760	1.023	1.550	2.067	2.726	3.019	3.597	4.104
x_A	0.0641	0.0998	0.1263	0.1821	0.2347	0.2955	0.3270	0.3820	0.4292

Polymer (B): **poly(ethylene glycol) dimethyl ether** **1995GAI**
Characterization: PEGDM mixture used for the Selexol process
see data set above
Solvent (A): **hydrogen** **H$_2$** **1333-74-0**

Type of data: gas solubility

T/K = 298.15

P/MPa	1.023	1.419	1.692	1.925	2.260	2.543	3.060	3.577	4.104
x_A	0.0048	0.0067	0.0078	0.0089	0.0105	0.0119	0.0141	0.0165	0.0189

T/K = 313.15

P/MPa	1.044	1.398	1.571	1.945	2.401	2.624	3.101	3.567	4.063
x_A	0.0052	0.0069	0.0077	0.0095	0.0118	0.0129	0.0151	0.0174	0.0197

T/K = 333.15

P/MPa	1.003	1.358	1.601	1.834	2.270	2.503	2.969	3.445	4.043
x_A	0.0053	0.0072	0.0084	0.0096	0.0120	0.0131	0.0156	0.0180	0.0210

Polymer (B): **poly(ethylene glycol) dimethyl ether** **1998HER**
Characterization: M_n/kg.mol^{-1} = 0.283, Aldrich Chem. Co., Inc., Milwaukee, WI

triethylene glycol dimethylether	14.19 wt%
tetraethylene glycol dimethylether	16.97 wt%
pentaethylene glycol dimethylether	22.76 wt%
hexaethylene glycol dimethylether	21.93 wt%
heptaethylene glycol dimethylether	12.99 wt%
octaethylene glycol dimethylether	8.51 wt%
nonaethylene glycol dimethylether	2.64 wt%

Solvent (A): **methanol** **CH$_4$O** **67-56-1**

Type of data: vapor pressure

continued

continued

T/K = 343.15

x_B	0.6010	0.5201	0.4443	0.3572	0.2833	0.1928	0.1173	0.0392	0.0297
P_A/kPa	43.11	53.89	62.71	74.10	84.45	96.83	109.82	118.84	121.51

T/K = 353.15

x_B	0.6010	0.5201	0.4443	0.3572	0.2833	0.1928	0.1173	0.0392	0.0297
P_A/kPa	60.64	76.14	89.34	105.49	120.70	139.16	157.80	172.01	175.50

T/K = 363.15

x_B	0.6008	0.5201	0.4443	0.3572	0.2833	0.1928	0.1173	0.0392	0.0297
P_A/kPa	83.42	104.85	123.26	146.98	168.25	195.19	222.17	243.60	246.59

T/K = 373.15

x_B	0.6009	0.5201	0.4443	0.3572	0.2833	0.1928	0.1173	0.0392	0.0297
P_A/kPa	112.61	141.17	167.20	200.29	231.07	268.50	305.54	337.30	340.69

T/K = 383.15

x_B	0.9159	0.8413	0.7629	0.6803	0.6009	0.5201	0.4443	0.3572	0.2833
P_A/kPa	28.46	54.23	83.00	115.11	148.96	186.39	222.04	267.55	307.61

x_B	0.1928	0.1173	0.0392	0.0297
P_A/kPa	361.77	412.33	457.40	462.35

T/K = 393.15

x_B	0.9159	0.8413	0.7629	0.6803	0.6009	0.5201	0.4443	0.3572	0.2833
P_A/kPa	36.70	69.87	106.95	149.12	193.10	241.40	289.66	350.85	408.64

x_B	0.1928	0.1173	0.0392	0.0297
P_A/kPa	477.72	546.34	609.59	617.00

T/K = 403.15

x_B	0.9159	0.8413	0.7629	0.6803	0.6009	0.5201	0.4443	0.3572	0.2833
P_A/kPa	46.90	88.45	135.60	189.55	246.06	307.07	371.52	452.18	529.20

x_B	0.1928	0.1173	0.0392	0.0297
P_A/kPa	621.44	712.08	799.30	811.56

T/K = 413.15

x_B	0.9159	0.8413	0.7628	0.6803	0.6009	0.5200	0.4443	0.3571	0.2833
P_A/kPa	58.86	110.32	169.50	237.39	306.80	382.60	467.78	573.95	673.97

x_B	0.1928	0.1173	0.0392
P_A/kPa	795.58	914.30	914.30

T/K = 423.15

x_B	0.8413	0.7628	0.6803	0.6009	0.5200	0.4442	0.3571	0.2833	0.1928
P_A/kPa	135.70	208.70	292.61	377.15	467.80	581.00	717.44	846.21	1004.6

Comments: Additional data below 343 K can be found in the original source.

Polymer (B):	**poly(ethylene glycol) dimethyl ether**		**1995GAI**
Characterization:	PEGDM mixture used for the Selexol process		

diethylene glycol dimethylether	2.07 wt%
diethylene glycol monomethylether	0.16 wt%
triethylene glycol dimethylether	21.01 wt%
triethylene glycol monomethylether	0.72 wt%
tetraethylene glycol dimethylether	33.68 wt%
tetraethylene glycol monomethylether	0.67 wt%
pentaethylene glycol dimethylether	26.39 wt%
pentaethylene glycol monomethylether	0.38 wt%
hexaethylene glycol dimethylether	12.78 wt%
hexaethylene glycol monomethylether	0.10 wt%
heptaethylene glycol dimethylether	1.87 wt%

Solvent (A):	**nitrogen**	**N_2**	**7727-37-9**

Type of data: gas solubility

$T/K = 298.15$

P/MPa	1.044	1.368	1.581	1.834	2.047	2.331	2.574	3.010	3.547
x_A	0.0058	0.0076	0.0089	0.0103	0.0114	0.0130	0.0143	0.0167	0.0197

$T/K = 313.15$

P/MPa	1.084	1.409	1.571	1.783	2.037	2.432	2.918	3.466	3.770
x_A	0.0066	0.0085	0.0094	0.0108	0.0124	0.0146	0.0175	0.0206	0.0224

$T/K = 333.15$

P/MPa	1.074	1.297	1.530	1.844	2.057	2.543	2.837	3.182	3.597
x_A	0.0071	0.0087	0.0101	0.0122	0.0136	0.0168	0.0186	0.0210	0.0235

Polymer (B):	**poly(ethylene glycol) monododecyl ether-**		
	b-poly(propylene glycol) diblock copolymer		**2002LIU**
Characterization:	M_n/kg.mol^{-1} = 0.67, 3 EO und 6 PPO units, Ls-36 Surfactant,		
	Henkel AG, Germany		
Solvent (A):	**carbon dioxide**	**CO_2**	**124-38-9**

Type of data: gas solubility

$T/K = 308.15$

P/MPa	11.10	14.50	16.73	18.71	19.74
c_B/mol.l^{-1}	0.01	0.02	0.03	0.04	0.05

$T/K = 318.15$

P/MPa	15.31	17.56	19.78	21.33
c_B/mol.l^{-1}	0.01	0.02	0.03	0.04

Polymer (B):	**poly(ethylene glycol) monododecyl ether-**	
	b-poly(propylene glycol) diblock copolymer	**2002LIU**
Characterization:	M_n/kg.mol^{-1} = 0.65, 4 EO und 5 PPO units, Ls-45 Surfactant,	
	Henkel AG, Germany	
Solvent (A):	**carbon dioxide** **CO$_2$**	**124-38-9**

Type of data: gas solubility

T/K = 308.15

P/MPa	12.97	16.73	19.32	21.05
c_B/mol.l^{-1}	0.01	0.02	0.03	0.04

T/K = 318.15

P/MPa	16.46	20.11	22.37
c_B/mol.l^{-1}	0.01	0.02	0.03

Polymer (B):	**polyisobutylene**	**2002JOU**
Characterization:	M_n/kg.mol^{-1} = 600, M_w/kg.mol^{-1} = 1000,	
	Aldrich Chem. Co., Inc., Milwaukee, WI	
Solvent (A):	**n-heptane** **C$_7$H$_{16}$**	**142-82-5**

Type of data: vapor pressure

w_B 0.025   was kept constant

T/K	363.15	373.15	383.95	398.25	403.15	413.15	423.15	433.15
P_A/MPa	0.4	0.5	0.5	0.6	0.7	0.9	1.0	1.1

Comments: Liquid-liquid equilibrium data for this system are given in Chapter 3.

Polymer (B):	**polyisoprene**	**1997ZHA**
Characterization:	M_n/kg.mol^{-1} = 100, synthesized in the laboratory	
Solvent (A):	**carbon dioxide** **CO$_2$**	**124-38-9**

Type of data: gas solubility and swelling

T/K = 308.15

P/MPa	1.379	2.758	4.137	5.516	6.895	8.274	10.342
w_A	0.0256	0.0481	0.0720	0.0975	0.1197	0.1582	0.2019
$(\Delta V/V)$/(%)	2.25	3.75	6.07	8.69	10.6	13.9	17.0

Polymer (B):	**poly(methyl methacrylate)**	**1997ZHA**
Characterization:	M_n/kg.mol^{-1} = 69, synthesized in the laboratory	
Solvent (A):	**carbon dioxide** **CO$_2$**	**124-38-9**

Type of data: gas solubility and swelling

continued

continued

$T/K = 308.15$

P/MPa	1.379	2.758	4.137	5.516	6.895	8.274	10.342
w_A	0.0381	0.0650	0.0897	0.1119	0.1379	0.1590	0.1817
$(\Delta V/V)/(\%)$	3.15	5.63	8.52	11.1	13.6	15.7	17.1

Polymer (B):   **poly(methyl methacrylate)**   **1991WI1, 1991WI2**
Characterization: $M_w/\text{kg.mol}^{-1} = 90$, $T_g/K = 378$,
     Scientific Polymer Products, Inc., Ontario, NY
Solvent (A):   **carbon dioxide**   **CO_2**   **124-38-9**

Type of data: gas solubility and swelling at T_g

T_g/K	331.95	315.15	305.85
P/atm	39	40	39
$c_A/(\text{cm}^3(\text{STP})/\text{g polymer})$	37	50	61
$(\Delta V/V)/(\%)$	5.8	8.3	10.4

Polymer (B):   **poly(methyl methacrylate)**   **1996HAN**
Characterization: $M_n/\text{kg.mol}^{-1} = 100$, $M_w/\text{kg.mol}^{-1} = 200$,
     Röhm and Haas
Solvent (A):   **carbon dioxide**   **CO_2**   **124-38-9**

Type of data: T_g at saturation gas solubility

P/atm	0.0	8.8	18.2	22.3	29.5	37.2
T_g/K	363.85	353.65	342.35	336.85	332.85	323.95

Polymer (B):   **poly(methyl methacrylate)**   **1998EDW**
Characterization: –
Solvent (A):   **carbon dioxide**   **CO_2**   **124-38-9**

Type of data: gas solubility

$T/K = 263.15$

P/atm	15
$c_A/(\text{cm}^3(\text{STP})/\text{g polymer})$	46.1

$T/K = 273.15$

P/atm	15	30
$c_A/(\text{cm}^3(\text{STP})/\text{g polymer})$	40.2	78.4

$T/K = 283.15$

P/atm	15	30
$c_A/(\text{cm}^3(\text{STP})/\text{g polymer})$	29.4	67.7

continued

continued

$T/\text{K} = 293.15$

P/atm	15	30	45			
$c_A/(\text{cm}^3(\text{STP})/\text{g polymer})$	25.0	55.8	74.9			

$T/\text{K} = 303.15$

P/atm	15	30	45	60	75	
$c_A/(\text{cm}^3(\text{STP})/\text{g polymer})$	22.0	42.1	60.7	88.6	115.3	

$T/\text{K} = 313.15$

P/atm	15	30	45	60	75	90
$c_A/(\text{cm}^3(\text{STP})/\text{g polymer})$	19.2	36.6	48.9	72.4	90.4	129.7

$T/\text{K} = 323.15$

P/atm	15	30	45	60	75	90
$c_A/(\text{cm}^3(\text{STP})/\text{g polymer})$	15.6	28.2	40.6	59.3	76.7	99.1

$T/\text{K} = 333.15$

P/atm	15	30	45	60	75	90
$c_A/(\text{cm}^3(\text{STP})/\text{g polymer})$	12.6	23.3	34.4	50.4	60.3	74.7

$T/\text{K} = 343.15$

P/atm	30	45	60	90	
$c_A/(\text{cm}^3(\text{STP})/\text{g polymer})$	19.9	29.9	43.5	59.3	

$T/\text{K} = 353.15$

P/atm	15	30	45	60	75	90
$c_A/(\text{cm}^3(\text{STP})/\text{g polymer})$	8.6	16.9	25.3	36.6	47.0	55.7

$T/\text{K} = 363.15$

P/atm	30	60	90	
$c_A/(\text{cm}^3(\text{STP})/\text{g polymer})$	14.9	32.5	50.0	

$T/\text{K} = 373.15$

P/atm	15	30	45	60	75	90
$c_A/(\text{cm}^3(\text{STP})/\text{g polymer})$	6.7	13.4	20.2	28.3	33.8	46.0

$T/\text{K} = 383.15$

P/atm	30	60	90	
$c_A/(\text{cm}^3(\text{STP})/\text{g polymer})$	11.9	24.9	36.4	

$T/\text{K} = 393.15$

P/atm	15	30	45	60	75	90
$c_A/(\text{cm}^3(\text{STP})/\text{g polymer})$	5.4	10.9	17.0	22.9	28.1	37.2

$T/\text{K} = 413.15$

P/atm	15	30	45	60	75	90
$c_A/(\text{cm}^3(\text{STP})/\text{g polymer})$	4.5	9.4	14.1	19.4	22.7	30.5

continued

continued

$T/K = 433.15$

P/atm	15	30	45	60	75	90
c_A/(cm^3(STP)/g polymer)	4.1	8.1	12.2	16.5	19.5	21.3

$T/K = 453.15$

P/atm	15	30	45	60	75	90
c_A/(cm^3(STP)/g polymer)	3.4	7.2	10.8	13.8	17.3	17.4

Polymer (B): **poly(methyl methacrylate)** **2003WAN**
Characterization: –
Solvent (A): **carbon dioxide** **CO$_2$** **124-38-9**

Type of data: gas solubility

$T/K = 298.15$

P/MPa	0.2	0.3	0.4
w_A	0.0152	0.0218	0.0269

Polymer (B): **poly(methyl methacrylate)** **1999WEB**
Characterization: Laird Plastics, Atlanta, GA
Solvent (A): **carbon dioxide** **CO$_2$** **124-38-9**

Type of data: gas solubility

$T/K = 313.15$ P/MPa   10.5 w_A 0.1817

Polymer (B): **poly(methyl methacrylate)** **1996HAN**
Characterization: M_n/kg.mol^{-1} = 100, M_w/kg.mol^{-1} = 200, Röhm and Haas
Solvent (A): **ethene** **C$_2$H$_4$** **74-85-1**

Type of data: T_g at saturation gas solubility

P/atm	0.0	11.0	21.8	21.8	36.3	54.5
T_g/K	363.85	353.35	347.85	346.85	339.95	330.45

Polymer (B): **poly(methyl methacrylate)** **1996HAN**
Characterization: M_n/kg.mol^{-1} = 100, M_w/kg.mol^{-1} = 200, Röhm and Haas
Solvent (A): **methane** **CH$_4$** **74-82-8**

Type of data: T_g at saturation gas solubility

P/atm	0.0	11.4	24.5	48.4	97.0	152	218
T_g/K	363.85	361.15	360.35	357.45	353.95	348.65	346.65

Polymer (B): **polypropylene** 1999SAT
Characterization: $M_w/\text{kg.mol}^{-1} = 52$, $M_w/\text{kg.mol}^{-1} = 220$,
Idemitsu Petrochemical, Ichihara, Japan
Solvent (A): **carbon dioxide** CO_2 124-38-9

Type of data: gas solubility

$T/\text{K} = 453.15$

P/MPa	7.083	11.443	14.212	16.037	17.242
w_A	0.0400	0.0739	0.0999	0.1169	0.1249

Polymer (B): **polypropylene** 1999SAT
Characterization: $M_w/\text{kg.mol}^{-1} = 64$, $M_w/\text{kg.mol}^{-1} = 451$,
Nippon Petrochemicals, Kawasaki, Japan
Solvent (A): **carbon dioxide** CO_2 124-38-9

Type of data: gas solubility

$T/\text{K} = 433.15$

P/MPa	7.400	11.803	14.558	16.355	17.529
w_A	0.0479	0.0866	0.1113	0.1267	0.1370

$T/\text{K} = 453.15$

P/MPa	5.419	7.160	8.797	10.953	11.558	12.352	14.889	16.170	17.376
w_A	0.0313	0.0422	0.0561	0.0730	0.0780	0.0840	0.1038	0.1155	0.1242

$T/\text{K} = 473.15$

P/MPa	6.204	10.113	12.637	14.296	15.397
w_A	0.0293	0.0570	0.0766	0.0883	0.0980

Polymer (B): **polypropylene** 1999SAT
Characterization: $M_w/\text{kg.mol}^{-1} = 64$, $M_w/\text{kg.mol}^{-1} = 451$,
Nippon Petrochemicals, Kawasaki, Japan
Solvent (A): **nitrogen** N_2 7727-37-9

Type of data: gas solubility

$T/\text{K} = 453.15$

P/MPa	4.233	7.115	9.031	10.898	12.699	14.938	17.999
w_A	0.00437	0.00752	0.00965	0.01173	0.01351	0.01675	0.01892

$T/\text{K} = 473.15$

P/MPa	4.013	6.726	8.545	9.878	11.998	14.819	17.838
w_A	0.00450	0.00848	0.01048	0.01206	0.01491	0.01960	0.02200

Polymer (B):   **polypropylene**   **2000SA2**
Characterization: $M_w/\text{kg.mol}^{-1} = 190$, 98.5% isotactic, 66.9 wt% crystallinity,
                               Mitsubishi Chemical, Kurashiki, Japan
Solvent (A):   **propene**   **C_3H_6**   **115-07-1**

Type of data: gas solubility

$T/\text{K} = 323.15$

P/MPa	0.2761	0.3256	0.6050	0.6405	1.1073	1.1720	1.4685	1.5334	1.6131
w_A	0.00862	0.01021	0.01869	0.01969	0.03474	0.03760	0.04754	0.05008	0.05341

P/MPa	1.7000	1.8215	1.9223
w_A	0.05673	0.06299	0.07023

$T/\text{K} = 343.15$

P/MPa	0.3226	0.6268	0.9238	1.0580	1.5331	2.1616	2.4071	2.8347
w_A	0.00661	0.01302	0.01911	0.02182	0.03285	0.04929	0.05592	0.07129

$T/\text{K} = 363.15$

P/MPa	0.3621	0.6992	1.2676	1.7470	2.3472	2.7945	3.6261
w_A	0.00550	0.01057	0.01959	0.02781	0.03889	0.04818	0.06491

Polymer (B):   **polystyrene**   **2003WAN**
Characterization: $M_w/\text{kg.mol}^{-1} = 50$, $T_g/\text{K} = 373.2$
Solvent (A):   **carbon dioxide**   **CO_2**   **124-38-9**

Type of data: gas solubility

$T/\text{K} = 298.15$

P/MPa	0.4	0.5	0.6	0.7	0.8	0.9
w_A	0.0064	0.0081	0.0092	0.0103	0.0110	0.0117

Polymer (B):   **polystyrene**   **1997ZHA**
Characterization: $M_\eta/\text{kg.mol}^{-1} = 71$, synthesized in the laboratory
Solvent (A):   **carbon dioxide**   **CO_2**   **124-38-9**

Type of data: gas solubility and swelling

$T/\text{K} = 308.15$

P/MPa	1.379	2.758	4.137	5.516	6.895	8.274	10.342
w_A	0.0181	0.0318	0.0475	0.0547	0.0559	0.0426	0.0527
$(\Delta V/V)/(\%)$	1.58	2.77	4.14	5.48	7.11	7.93	8.61

Polymer (B):   **polystyrene**   **1996SAT**
Characterization: $M_n/\text{kg.mol}^{-1} = 70.3$, $M_w/\text{kg.mol}^{-1} = 187$,
                               Mitsui Toatsu Chemicals, Inc., Japan
Solvent (A):   **carbon dioxide**   **CO_2**   **124-38-9**

continued

continued

Type of data: gas solubility

$T/K = 373.15$

P/MPa	3.617	5.860	7.921	11.928	14.348	15.908	17.476	18.554
w_A	0.0221	0.0347	0.0462	0.0724	0.0850	0.0909	0.0994	0.1037

$T/K = 413.15$

P/MPa	7.290	10.721	11.594	11.994	14.316	14.738	16.110	17.087	18.097
w_A	0.0335	0.0500	0.0548	0.0568	0.0678	0.0666	0.0769	0.0789	0.0852

P/MPa	19.739	20.036
w_A	0.0919	0.0949

$T/K = 453.15$

P/MPa	2.472	6.276	10.188	12.704	14.379	15.444	17.372
w_A	0.0093	0.0214	0.0374	0.0488	0.0539	0.0562	0.0643

Polymer (B):	**polystyrene**		**2001HIL**
Characterization:	M_n/kg.mol^{-1} = 70.4, M_w/kg.mol^{-1} = 190,		
	Lacqrene 1450N, ATOFINA, France		
Solvent (A):	**carbon dioxide**	**CO$_2$**	**124-38-9**

Type of data: gas solubility and swelling

$T/K = 338.22$

P/MPa	3.71	7.54	10.25	11.96	16.54	18.33	24.52
w_A	0.03581	0.06942	0.08684	0.09596	0.12205	0.12786	0.13510
$(\Delta V/V)/(\%)$	3.860	7.300	9.440				

$T/K = 362.50$

P/MPa	4.30	8.09	11.47	14.15	16.50	18.92	21.60	24.65
w_A	0.03627	0.05866	0.08056	0.09271	0.10826	0.11540	0.12188	0.12690
$(\Delta V/V)/(\%)$	4.16	7.00	9.19	10.94	12.03	12.69	13.13	13.34

$T/K = 383.22$

P/MPa	6.55	11.04	14.51	17.66	20.42	23.55	27.03	30.61
w_A	0.03472	0.05488	0.06907	0.08483	0.09514	0.10423	0.10962	0.12137
$(\Delta V/V)/(\%)$	5.24	7.14	8.73	9.74	10.46	11.03	11.27	12.04

P/MPa	34.38	38.63	42.81
w_A	0.12603	0.13073	0.13835
$(\Delta V/V)/(\%)$	12.32	12.86	13.05

$T/K = 402.51$

P/MPa	5.14	9.39	21.48	33.05	44.41
w_A	0.01441	0.04225	0.08759	0.11887	0.12914

Polymer (B):	polystyrene							2001SA2
Characterization:	M_n/kg.mol^{-1} = 107, M_w/kg.mol^{-1} = 330,							
	Asahi Chemicals Inc., Kawasaki, Japan							
Solvent (A):	carbon dioxide CO$_2$							124-38-9

Type of data: gas solubility

T/K = 373.15

P/MPa	2.068	2.627	4.093	4.629	6.038	6.502	8.054	8.585	12.105
w_A	0.0126	0.0159	0.0247	0.0279	0.0362	0.0388	0.0475	0.0506	0.0690

P/MPa	12.631	16.059	20.067
w_A	0.0717	0.0898	0.1086

T/K = 423.15

P/MPa	2.159	2.689	4.073	4.732	6.049	6.530	8.071	8.566	12.046
w_A	0.00896	0.0112	0.0168	0.0194	0.0247	0.0267	0.0327	0.0346	0.0473

P/MPa	12.593	16.143
w_A	0.0492	0.0606

T/K = 473.15

P/MPa	2.167	2.884	4.232	4.795	6.108	6.685	8.027	8.813	12.165
w_A	0.0071	0.0094	0.0137	0.0155	0.0197	0.0215	0.0257	0.0282	0.0381

P/MPa	12.659	16.040	20.151
w_A	0.0397	0.0489	0.0587

Polymer (B):	polystyrene			1991WI1, 1991WI2
Characterization:	M_w/kg.mol^{-1} = 250, T_g/K = 373,			
	Scientific Polymer Products, Inc., Ontario, NY			
Solvent (A):	carbon dioxide CO$_2$			124-38-9

Type of data: gas solubility and swelling at T_g

T_g/K	338.15	323.15	308.15
P/atm	36	48	60
c_A/(cm^3(STP)/g polymer)	19	30	46
$(\Delta V/V)$/(%)	4.1	5.5	7.4

Polymer (B):	polystyrene			2000SA3
Characterization:	M_n/kg.mol^{-1} = 70.3, M_w/kg.mol^{-1} = 187,			
	Mitsui Toatsu Chemicals, Inc., Japan			
Solvent (A):	1-chloro-1,1-difluoroethane C$_2$H$_3$ClF$_2$			75-68-3

Type of data: gas solubility

T/K = 347.92

P/MPa	0.091	0.156	0.341	0.506	0.525	0.644	0.678
w_A	0.0098	0.0135	0.0401	0.0638	0.0661	0.0858	0.0912

continued

continued

$T/K = 373.62$

P/MPa	0.124	0.185	0.474	0.728	0.743	0.940	0.989
w_A	0.0089	0.0131	0.0365	0.0590	0.0603	0.0794	0.0837

$T/K = 422.15$

P/MPa	0.194	0.291	0.758	1.207	1.225	1.605	1.674
w_A	0.0074	0.0113	0.0302	0.0503	0.0502	0.0679	0.0696

$T/K = 471.42$

P/MPa	0.263	0.402	1.030	1.666	1.726	2.289	2.353
w_A	0.0063	0.0099	0.0258	0.0430	0.0438	0.0588	0.0594

Polymer (B):   **polystyrene**   **1986GOR**
Characterization: Dow Styron 430U
Solvent (A):   **chlorodifluoromethane**   **CHClF₂**   **75-45-6**

Type of data: vapor solubility

P_A/P_{0A}	0.3557	0.2668	0.1353
T/K	323.15	358.15	443.15
w_A	0.0699	0.0580	0.0415
φ_A	0.0661	0.0607	0.0554

Polymer (B):   **polystyrene**   **1986GOR**
Characterization: Dow Styron 430U
Solvent (A):   **dichlorodifluoromethane**   **CCl₂F₂**   **75-71-8**

Type of data: vapor solubility

P_A/P_{0A}	0.6486	0.3360	0.1477
T/K	323.15	358.15	443.15
w_A	0.0514	0.0523	0.0392
φ_A	0.04392	0.04838	0.04302

Polymer (B):   **polystyrene**   **1986GOR**
Characterization: Dow Styron 430U
Solvent (A):   **1,2-dichloro-1,1,2,2-tetrafluoroethane**   **C₂Cl₂F₄**   **76-14-2**

Type of data: vapor solubility

P_A/P_{0A}	0.6966	0.3453	0.0885
T/K	323.15	358.15	443.15
w_A	0.00802	0.00430	0.00587
φ_A	0.006192	0.003540	0.005581

Polymer (B):   **polystyrene**   2000SA3
Characterization: M_n/kg.mol^{-1} = 70.3, M_w/kg.mol^{-1} = 187,
                      Mitsui Toatsu Chemicals, Inc., Japan
Solvent (A):   **1,1-difluoroethane   C$_2$H$_4$F$_2$**   75-37-6

Type of data: gas solubility

T/K = 348.15

P/MPa	0.283	0.485	0.678	0.689	0.829	0.991	1.200
w_A	0.0122	0.0212	0.0308	0.0310	0.0381	0.0479	0.0594

T/K = 373.13

P/MPa	0.367	0.639	0.889	0.904	1.090	1.322	1.603
w_A	0.0107	0.0188	0.0272	0.0274	0.0337	0.0426	0.0521

T/K = 423.14

P/MPa	0.529	0.938	1.289	1.314	1.592	1.964	2.375
w_A	0.0087	0.0157	0.0222	0.0224	0.0274	0.0352	0.0422

T/K = 473.09

P/MPa	0.677	1.218	1.659	1.707	2.059	2.568	3.116
w_A	0.0074	0.0136	0.0192	0.0192	0.0238	0.0310	0.0365

Polymer (B):   **polystyrene**   1948NEW
Characterization: –
Solvent (A):   **ethene   C$_2$H$_4$**   74-85-1

Type of data: gas solubility

T/K = 443.2

P/atm	49.0	77.5	92.5	92.5
c_A/(cm^3/g polymer)	5.27	8.05	9.20	10.30

Polymer (B):   **polystyrene**   1948NEW
Characterization: –
Solvent (A):   **hydrogen   H$_2$**   1333-74-0

Type of data: gas solubility

T/K = 443.2

P/atm	80.0	142.0	224.5	233.0	243.0	306.5
c_A/(cm^3/g polymer)	2.89	5.35	8.48	9.42	9.66	11.95

Polymer (B):   **polystyrene**   1963LUN
Characterization: –
Solvent (A):   **methane   CH$_4$**   74-82-8

continued

continued

Type of data: gas solubility

$T/K = 373.35$

P/atm	67.1	239.0	319.8
c_A/(cm^3/g polymer)	8.15	22.15	32.12

$T/K = 398.57$

P/atm	49.7	214.1	299.2	343.6
c_A/(cm^3/g polymer)	4.98	18.82	25.54	27.94

$T/K = 428.57$

P/atm	47.6	161.7	241.3
c_A/(cm^3/g polymer)	5.44	15.59	23.45

$T/K = 461.55$

P/atm	122.2	198.2	249.8	313.6
c_A/(cm^3/g polymer)	14.19	21.49	26.07	31.45

Polymer (B):	**polystyrene**	**1997END**
Characterization:	M_n/kg.mol^{-1} = 143, M_w/kg.mol^{-1} = 405,	
	BASF AG, Ludwigshafen, Germany	
Solvent (A):	**methylcyclohexane** **C$_7$H$_{14}$**	**108-87-2**

Type of data: vapor-liquid equilibrium

w_B	0.060	0.060	0.060	0.060	0.060	0.060	0.060	0.060	0.060
T/K	492.32	494.74	497.21	499.68	502.15	504.32	506.80	509.18	511.46
P/bar	10.6	11.0	11.5	12.0	12.5	13.0	13.5	14.1	14.6

w_B	0.065	0.065	0.065	0.065	0.065	0.065	0.065	0.065	0.065
T/K	489.72	492.21	494.69	497.20	499.45	502.12	504.64	507.12	509.61
P/bar	10.0	10.5	10.9	11.4	11.9	12.4	12.9	13.5	14.0

w_B	0.065	0.065	0.065	0.065	0.098	0.098	0.098	0.098	0.098
T/K	512.10	514.59	517.04	519.50	488.73	491.25	493.81	496.06	498.56
P/bar	14.6	15.1	15.7	16.6	9.5	9.9	10.3	10.9	11.3

w_B	0.098	0.098	0.098	0.098	0.098	0.098	0.098	0.098	0.098
T/K	498.69	501.30	503.80	506.47	508.41	508.55	511.16	513.88	516.07
P/bar	11.4	11.9	12.3	12.9	13.4	13.3	14.0	14.6	15.0

w_B	0.098	0.152	0.152	0.152	0.152	0.152	0.152	0.152	0.152
T/K	518.60	489.74	492.32	494.53	496.07	498.70	501.30	503.81	506.46
P/bar	15.8	9.0	9.5	9.9	10.2	10.9	11.3	11.9	12.5

w_B	0.152	0.152	0.152	0.152	0.152	0.187	0.187	0.187	0.187
T/K	508.42	511.15	513.88	516.05	518.84	489.74	492.31	494.54	496.06
P/bar	12.8	13.4	14.0	14.5	15.0	8.7	9.2	9.6	9.9

w_B	0.187	0.187	0.187	0.187	0.187	0.187	0.187	0.187	0.187
T/K	498.69	501.30	503.80	506.47	508.41	511.16	513.88	516.08	518.85
P/bar	10.6	11.0	11.7	12.2	12.5	13.0	13.6	14.1	14.6

Comments: Cloud-points and spinodal points for this system are given in Chapter 3.

Polymer (B):   **polystyrene**   **1992WAN**
Characterization: $M_w/\text{kg.mol}^{-1} = 69.3$, $M_w/\text{kg.mol}^{-1} = 187$
Solvent (A):   **nitrogen**   **N$_2$**   **7727-37-9**

Type of data: gas solubility

$T/\text{K} = 323.2$

P/MPa	2.0	5.0	8.0	11.0	14.0	17.0	20.0	23.0	25.0
w_A	0.0020	0.0037	0.0050	0.0062	0.0073	0.0084	0.0095	0.0105	0.0112

$T/\text{K} = 353.2$

P/MPa	2.0	5.0	8.0	11.0	14.0	17.0	20.0	23.0	25.0
w_A	0.0016	0.0034	0.0050	0.0066	0.0080	0.0095	0.0109	0.0123	0.0132

$T/\text{K} = 393.2$

P/MPa	2.0	5.0	8.0	11.0	14.0	17.0	20.0	23.0	25.0
w_A	0.0017	0.0037	0.0056	0.0074	0.0092	0.0109	0.0126	0.0143	0.0155

Polymer (B):   **polystyrene**   **1999SAT**
Characterization: $M_w/\text{kg.mol}^{-1} = 70.3$, $M_w/\text{kg.mol}^{-1} = 187$,
                                Mitsui Toatsu Chemicals, Inc., Japan
Solvent (A):   · **nitrogen**   **N$_2$**   **7727-37-9**

Type of data: gas solubility

$T/\text{K} = 313.15$

P/MPa	4.934	8.378	10.771	12.450	16.136
w_A	0.00220	0.00438	0.00591	0.00627	0.00818

$T/\text{K} = 333.15$

P/MPa	5.037	6.149	7.882	9.868	11.034	11.674	13.457	16.542
w_A	0.00182	0.00215	0.00305	0.00355	0.00426	0.00428	0.00484	0.00630

$T/\text{K} = 353.15$

P/MPa	2.989	3.645	5.053	5.295	6.476	7.945	9.042	9.535	11.660
w_A	0.00106	0.00131	0.00168	0.00180	0.00232	0.00295	0.00308	0.00339	0.00395

P/MPa	11.770	14.071	17.521
w_A	0.00417	0.00497	0.00627

Polymer (B):   **polystyrene**   **2001HIL**
Characterization: $M_n/\text{kg.mol}^{-1} = 70.4$, $M_w/\text{kg.mol}^{-1} = 190$,
                                Lacqrene 1450N, ATOFINA, France
Solvent (A):   **nitrogen**   **N$_2$**   **7727-37-9**

Type of data: gas solubility and swelling

$T/\text{K} = 313.11$

continued

continued

P/MPa	3.11	5.32	6.91	9.15	10.88	14.26	18.17	22.11
w_A	0.00217	0.00341	0.00415	0.00484	0.00593	0.00753	0.00916	0.01033
$(\Delta V/V)/(\%)$	0.102	0.219	0.300	0.408	0.487	0.627	0.772	0.937

P/MPa	26.65	31.73	46.77	69.47
w_A	0.01236	0.01466	0.02099	0.02535
$(\Delta V/V)/(\%)$	1.08	1.23	1.60	2.22

$T/K = 333.23$

P/MPa	3.05	6.43	9.82	13.47	16.43	20.29	24.34	28.14
w_A	0.00109	0.00235	0.00425	0.00580	0.00689	0.00834	0.01063	0.01247
$(\Delta V/V)/(\%)$	0.149	0.270	0.380	0.480	0.557	0.646	0.726	0.796

P/MPa	32.54	37.31	41.54	45.84	50.11
w_A	0.01310	0.01482	0.01660	0.01815	0.02035
$(\Delta V/V)/(\%)$	0.872	0.948	1.012	1.078	1.135

$T/K = 353.15$

P/MPa	3.29	7.00	10.42	13.95	17.26	21.38	25.35	30.26
w_A	0.00157	0.00284	0.00414	0.00494	0.00615	0.00709	0.00849	0.00911
$(\Delta V/V)/(\%)$	0.211	0.280	0.325	0.365	0.398	0.430	0.458	0.485

P/MPa	35.10	40.12	44.76	49.27	53.94	58.40	62.50
w_A	0.01011	0.01151	0.01261	0.01436	0.01578	0.01674	0.01781
$(\Delta V/V)/(\%)$	0.515	0.556	0.567	0.590	0.612	0.630	0.661

Polymer (B):	**polystyrene**	**1996SAT**
Characterization:	$M_w/\text{kg.mol}^{-1} = 70.3$, $M_w/\text{kg.mol}^{-1} = 187$,	
	Mitsui Toatsu Chemicals, Inc., Japan	
Solvent (A):	**nitrogen** **N₂**	**7727-37-9**

Type of data: gas solubility

$T/K = 373.15$

P/MPa	9.467	12.165	14.008	16.113	17.815
w_A	0.00379	0.00485	0.00556	0.00645	0.00710

$T/K = 413.15$

P/MPa	6.452	8.247	9.574	11.902	13.991	18.011
w_A	0.00333	0.00393	0.00446	0.00606	0.00669	0.00853

$T/K = 453.15$

P/MPa	7.092	9.020	10.057	10.296	11.136	15.031	16.170
w_A	0.00392	0.00523	0.00577	0.00588	0.00661	0.00890	0.00973

Polymer (B):	**polystyrene**				**1981ALB**
Characterization:	M_n/kg.mol^{-1} = 2.33, M_w/kg.mol^{-1} = 3.46,				
	BASF Telomerstyrol				
Solvent (A):	**sulfur dioxide**		**SO$_2$**		**7446-09-5**

Type of data: vapor pressure

w_B 0.648   was kept constant

T/K	292.95	283.65	275.55	264.45	254.05	242.85
P_A/MPa	0.2828	0.2118	0.1578	0.1013	0.0668	0.0376
P_A/P_{0A}	0.8829	0.9145	0.9255	0.9414	0.9970	0.9691

Comments: More phase equilibrium data for this system are given in Chapter 4.

Polymer (B):	**polystyrene**				**2000SA3**
Characterization:	M_n/kg.mol^{-1} = 70.3, M_w/kg.mol^{-1} = 187,				
	Mitsui Toatsu Chemicals, Inc., Japan				
Solvent (A):	**1,1,1,2-tetrafluoroethane**		**C$_2$H$_2$F$_4$**		**811-97-2**

Type of data: gas solubility

T/K = 348.16

P/MPa	0.333	0.496	0.745	0.863	1.150	1.308
w_A	0.0108	0.0155	0.0232	0.0264	0.0363	0.0424

T/K = 373.16

P/MPa	0.414	0.609	0.911	1.057	1.420	1.621
w_A	0.0092	0.0133	0.0202	0.0228	0.0311	0.0364

T/K = 423.15

P/MPa	0.552	0.811	1.214	1.421	1.919	2.199
w_A	0.0075	0.0110	0.0169	0.0187	0.0255	0.0303

T/K = 473.16

P/MPa	0.682	1.000	1.501	1.757	2.384	2.748
w_A	0.0067	0.0097	0.0151	0.0167	0.0229	0.0273

Polymer (B):	**polystyrene**		**1986GOR**
Characterization:	Dow Styron 430U		
Solvent (A):	**trichlorofluoromethane**	**CCl$_3$F**	**75-69-4**

Type of data: vapor solubility

P_A/P_{0A}	0.5789	0.4930	0.4132
T/K	323.15	358.15	403.15
w_A	0.1586	0.1521	0.1414
φ_A	0.1236	0.1244	0.1207

Polymer (B):	polystyrene-b-polybutadiene-b-polystyrene triblock copolymer	2003WAN
Characterization:	M_w/kg.mol^{-1} = 100, 70 mol% butadiene	
Solvent (A):	carbon dioxide CO$_2$	124-38-9

Type of data: gas solubility

T/K = 298.15

P/MPa	0.2	0.3	0.4	0.5	0.6	0.7	0.8	0.9	1.0
w_A	0.0043	0.0064	0.0081	0.0105	0.0120	0.0140	0.0159	0.0180	0.0202

Polymer (B):	polystyrene-b-polyisoprene diblock copolymer	1997ZHA
Characterization:	M_η/kg.mol^{-1} = 100, 25 mol% isoprene, synthesized in the laboratory	
Solvent (A):	carbon dioxide CO$_2$	124-38-9

Type of data: gas solubility and swelling

T/K = 308.15

P/MPa	1.379	2.758	4.137	5.516	6.895	8.274	10.342
w_A	0.0202	0.0349	0.0534	0.0646	0.0763	0.0858	0.0934
$(\Delta V/V)$/%	1.75	2.87	4.39	6.18	7.72	9.11	11.0

Polymer (B):	polystyrene-b-polyisoprene diblock copolymer	1997ZHA
Characterization:	M_η/kg.mol^{-1} = 100, 50 mol% isoprene, synthesized in the laboratory	
Solvent (A):	carbon dioxide CO$_2$	124-38-9

Type of data: gas solubility and swelling

T/K = 308.15

P/MPa	1.379	2.758	4.137	5.516	6.895	8.274	10.342
w_A	0.0210	0.0397	0.0592	0.0777	0.0848	0.0958	0.1119
$(\Delta V/V)$/%	1.84	3.23	5.21	7.01	8.54	11.1	12.0

Polymer (B):	polystyrene-b-polyisoprene diblock copolymer	1997ZHA
Characterization:	M_η/kg.mol^{-1} = 100, 75 mol% isoprene, synthesized in the laboratory	
Solvent (A):	carbon dioxide CO$_2$	124-38-9

Type of data: gas solubility and swelling

T/K = 308.15

P/MPa	1.379	2.758	4.137	5.516	6.895	8.274	10.342
w_A	0.0208	0.0384	0.0652	0.0839	0.1007	0.1197	0.1416
$(\Delta V/V)$/%	2.39	3.44	5.45	7.84	9.84	12.4	15.1

Polymer (B):   **polystyrene-b-poly(methyl methacrylate)**
                         diblock copolymer   **1997ZHA**
Characterization: M_η/kg.mol^{-1} = 75, 18 mol% methyl methacrylate,
                         synthesized in the laboratory
Solvent (A):   **carbon dioxide**   **CO$_2$**   **124-38-9**

Type of data: gas solubility and swelling

continued

T/K = 308.15

P/MPa	1.379	2.758	4.137	5.516	6.895	8.274	10.342
w_A	0.0221	0.0388	0.0561	0.0667	0.0733	0.0767	0.0784
$(\Delta V/V)$/%	1.89	3.34	5.01	6.61	8.40	9.48	10.3

Polymer (B):   **polystyrene-b-poly(methyl methacrylate)**
                         diblock copolymer   **1997ZHA**
Characterization: M_η/kg.mol^{-1} = 75, 46 mol% methyl methacrylate,
                         synthesized in the laboratory
Solvent (A):   **carbon dioxide**   **CO$_2$**   **124-38-9**

Type of data: gas solubility and swelling

T/K = 308.15

P/MPa	1.379	2.758	4.137	5.516	6.895	8.274	10.342
w_A	0.0277	0.0481	0.0666	0.0774	0.0908	0.0950	0.1031
$(\Delta V/V)$/%	2.78	4.44	5.85	7.68	9.68	12.1	13.2

Polymer (B):   **polystyrene-b-poly(methyl methacrylate)**
                         diblock copolymer   **1997ZHA**
Characterization: M_η/kg.mol^{-1} = 81, 56 mol% methyl methacrylate,
                         synthesized in the laboratory
Solvent (A):   **carbon dioxide**   **CO$_2$**   **124-38-9**

Type of data: gas solubility and swelling

T/K = 308.15

P/MPa	1.379	2.758	4.137	5.516	6.895	8.274	10.342
w_A	0.0300	0.0522	0.0638	0.0850	0.1047	0.1150	0.1319
$(\Delta V/V)$/%	2.52	4.49	6.76	8.86	11.0	12.6	13.7

Polymer (B): **polystyrene-b-poly(methyl methacrylate)**
diblock copolymer **2003WAN**
Characterization: M_w/kg.mol^{-1} = 12.1, 58.55 mol% methyl methacrylate
Solvent (A): **carbon dioxide** CO_2 **124-38-9**

Type of data: gas solubility

T/K = 298.15

P/MPa	0.2	0.3	0.4
w_A	0.0224	0.0329	0.0424

Polymer (B): **polystyrene-b-poly(vinyl pyridine)**
diblock copolymer **1997ZHA**
Characterization: M_η/kg.mol^{-1} = 70, 44 mol% vinyl pyridine,
synthesized in the laboratory
Solvent (A): **carbon dioxide** CO_2 **124-38-9**

Type of data: gas solubility and swelling

T/K = 308.15

P/MPa	1.379	2.758	4.137	5.516	6.895	8.274	10.342
w_A	0.0207	0.0350	0.0513	0.0658	0.0777	0.0830	0.0950
$(\Delta V/V)$/%	2.62	4.49	5.81	6.91	8.43	9.05	10.5

Polymer (B): **polystyrene-b-poly(vinyl pyridine)**
diblock copolymer **1997ZHA**
Characterization: M_η/kg.mol^{-1} = 90, 77 mol% vinyl pyridine,
synthesized in the laboratory
Solvent (A): **carbon dioxide** CO_2 **124-38-9**

Type of data: gas solubility and swelling

T/K = 308.15

P/MPa	1.379	2.758	4.137	5.516	6.895	8.274	10.342
w_A	0.0224	0.0368	0.0515	0.0691	0.0832	0.1023	0.1150
$(\Delta V/V)$/%	3.09	5.21	6.68	7.73	8.94	9.78	13.4

Polymer (B): **poly(vinyl acetate)** **1990WAN, 1994WAN**
Characterization: M_η/kg.mol^{-1} = 158, Aldrich Chem. Co., Inc., Milwaukee, WI
Solvent (A): **carbon dioxide** CO_2 **124-38-9**

Type of data: gas solubility

T/K = 313.2

continued

continued

P/MPa	0.294	0.667	1.16	1.86	2.41	2.52	3.16	3.29	4.07
w_A	0.0113	0.0249	0.0428	0.0686	0.0953	0.0940	0.118	0.129	0.159

P/MPa	4.19	4.74	5.23	5.32	5.88	6.03	6.44	7.18
w_A	0.158	0.182	0.209	0.210	0.251	0.256	0.282	0.327

T/K = 333.2

P/MPa	0.814	1.41	1.99	2.69	3.15	3.52	4.02	4.27	4.96
w_A	0.0235	0.0383	0.0535	0.0720	0.0891	0.0948	0.112	0.115	0.136

P/MPa	5.12	5.71	6.12	6.99	7.93	8.88
w_A	0.144	0.158	0.175	0.205	0.241	0.285

T/K = 353.2

P/MPa	0.843	1.50	2.18	2.18	2.90	3.03	3.51	3.96	4.30
w_A	0.0193	0.0335	0.0476	0.0516	0.0625	0.0689	0.0752	0.0888	0.0914

P/MPa	4.64	5.12	5.46	5.70	6.32	7.27	8.05	9.22	10.1
w_A	0.103	0.110	0.121	0.124	0.143	0.168	0.189	0.224	0.244

Polymer (B):	**poly(vinyl acetate)**		**1990TAK**
Characterization:	M_η/kg.mol^{-1} = 158, Aldrich Chem. Co., Inc., Milwaukee, WI		
Solvent (A):	**carbon dioxide**	**CO$_2$**	**124-38-9**

Type of data: gas solubility

T/K = 313.2

P/MPa	1.258	1.449	2.650	3.467	3.846	4.583	5.659	6.885	8.755
w_A	0.0416	0.0532	0.0936	0.1264	0.1429	0.1723	0.2123	0.2789	0.3909

T/K = 323.2

P/MPa	0.898	1.819	1.906	3.309	4.698	5.655	6.519	7.861
w_A	0.0254	0.0485	0.0533	0.0978	0.1397	0.1696	0.2073	0.2668

Type of data: swelling (volume dilation)

T/K = 313.2

P/MPa		2.374	3.522	4.210	6.364	6.955	8.020	8.893
$(\Delta V/V)$/%		10.09	19.36	20.69	37.22	41.88	52.74	55.44

T/K = 323.2

P/MPa		1.725	3.793	4.804	5.444	6.187	6.850	8.403
$(\Delta V/V)$/%		5.99	15.21	19.12	23.97	25.85	31.18	39.01

Polymer (B): **poly(vinyl acetate)** **2001SA2**
Characterization: M_n/kg.mol^{-1} = 167, Aldrich Chem. Co., Inc., Milwaukee, WI
Solvent (A): **carbon dioxide** **CO_2** **124-38-9**

Type of data: gas solubility

T/K = 313.15

P/MPa	0.199	0.459	0.858	1.271	1.949	2.375	2.818	3.260	3.589
w_A	0.0055	0.0144	0.0274	0.0410	0.0639	0.0785	0.0941	0.1100	0.1220

P/MPa	3.979	4.449	4.978	5.407	5.916	6.375	6.876
w_A	0.1366	0.1545	0.1753	0.1927	0.2141	0.2341	0.2576

T/K = 333.15

P/MPa	2.263	4.104	6.010	7.772	9.694
w_A	0.0531	0.0979	0.1463	0.1931	0.2495

T/K = 353.15

P/MPa	2.211	4.199	6.110	8.015	9.876
w_A	0.0384	0.0737	0.1086	0.1433	0.1765

T/K = 373.15

P/MPa	2.257	6.389	10.194	13.688	17.449
w_A	0.0296	0.0892	0.1453	0.1920	0.2303

Polymer (B): **poly(vinyl pyridine)** **1997ZHA**
Characterization: M_η/kg.mol^{-1} = 65, synthesized in the laboratory
Solvent (A): **carbon dioxide** **CO_2** **124-38-9**

Type of data: gas solubility and swelling

T/K = 308.15

P/MPa	1.379	2.758	4.137	5.516	6.895	8.274	10.342
w_A	0.0252	0.0388	0.0538	0.0734	0.0967	0.1446	0.1518
$(\Delta V/V)$/%	3.52	6.19	7.48	8.42	9.52	10.5	15.0

2.2. Table of binary systems where data were published only in graphical form as phase diagrams or related figures

Polymer (B)	Solvent (A)	Ref.
Carboxymethylcellulose		
	carbon dioxide	1999KIK
Cellulose acetate		
	carbon dioxide	1978STE
	carbon dioxide	1989BER
Cellulose acetate butyrate		
	carbon dioxide	1996SH2
Chloropolyethylene		
	carbon dioxide	1998AUB
Ethylcellulose		
	carbon dioxide	1999KIK
Ethylene/propylene copolymer		
	ethane	2001TSU
	ethene	1994YOO
	ethene	2001TSU
	propane	2001TSU
	propene	1994YOO
	propene	2001TSU
Ethylene/vinyl acetate copolymer		
	carbon dioxide	1994KAM
	carbon dioxide	2002SHI
	methane	1994KAM
	nitrogen	1994KAM
Natural rubber		
	carbon dioxide	2003WAN

Polymer (B)	Solvent (A)	Ref.
Nylon-66		
	carbon dioxide	1996SH1
	carbon dioxide	2003WAN
Poly(acrylic acid)		
	carbon dioxide	2003KIK
Polyamide-11		
	carbon dioxide	2001MAR
Polyarylene		
	carbon dioxide	1998AUB
Poly(benzyl methacrylate)		
	carbon dioxide	1996WAN
Polybutadiene		
	argon	1989KA2
	carbon dioxide	1989KA2
	carbon dioxide	1997KRY
	carbon dioxide	2003KOG
	carbon dioxide	2003RAM
	ethane	1989KA2
	ethene	1989KA2
	helium	1989KA2
	hydrogen	1989KA2
	methane	1989KA2
	nitrogen	1989KA2
	oxygen	1989KA2
Poly(butylene terephthalate)		
	carbon dioxide	1988THU
	carbon dioxide	1991SCH
	carbon monoxide	1991SCH
	helium	1991SCH
	methane	1991SCH
	neon	1991SCH
	nitrogen	1991SCH
	water	1991SCH
Poly(butyl methacrylate)		
	carbon dioxide	2002NIK

Polymer (B)	Solvent (A)	Ref.
Poly(ε-caprolactone)		
	carbon dioxide	2003COT
Polycarbonate bisphenol-A		
	carbon dioxide	1976KOR
	carbon dioxide	1986FLE
	carbon dioxide	1986KA2
	carbon dioxide	1987SAD
	carbon dioxide	1987WIS
	carbon dioxide	1989BER
	carbon dioxide	1993CON
	carbon dioxide	1996SH2
	carbon dioxide	1998AUB
	carbon dioxide	1998ZHA
	carbon dioxide	2003ALE
	chlorodifluoromethane	1986GOR
	chlorotrifluoromethane	1986GOR
	dichlorodifluoromethane	1986GOR
	1,2-dichloro-1,1,2,2-tetrafluoroethane	1986GOR
	methane	1987SAD
Polycarbonate hexafluorobisphenol-A		
	carbon dioxide	1993CON
Polycarbonate tetramethylene bisphenol-A		
	carbon dioxide	1993CON
Poly(chlorotrifluoroethylene)		
	carbon dioxide	1999WEB
Poly(2-chloroxylylene)		
	carbon dioxide	1998AUB
Poly(2,3-dichloroxylylene)		
	carbon dioxide	1998AUB
Poly(2,6-dimethyl-1,4-phenylene oxide)		
	carbon dioxide	1982MOR
	carbon dioxide	1996CON
	carbon dioxide	1996SH2
	carbon dioxide	2003KIK
	chlorodifluoromethane	1986GOR
	chlorotrifluoromethane	1986GOR
	dichlorodifluoromethane	1986GOR
	1,2-dichloro-1,1,2,2-tetrafluoroethane	1986GOR

Polymer (B)	Solvent (A)	Ref.
Poly(dimethylsilmethylene)		
	carbon dioxide	1993SHA
	methane	1993SHA
	propane	1993SHA
Poly(dimethylsiloxane)		
	carbon dioxide	1986FLE
	carbon dioxide	1986SHA
	carbon dioxide	1989SHI
	carbon dioxide	1991BRI
	carbon dioxide	1991POP
	carbon dioxide	1994GAR
	carbon dioxide	1997KRY
	carbon dioxide	1998WES
	carbon dioxide	1999ANG
	carbon dioxide	1999ROY
	carbon dioxide	1999WEB
	carbon dioxide	2001SIR
	carbon dioxide	2002FLI
	ethane	1991POP
	ethane	1999ANG
	methane	1986SHA
	methane	1991POP
	methane	1999ANG
	nitrogen	1991POP
	nitrogen	1999ANG
	oxygen	1999ANG
	propane	1986SHA
	propane	1999ANG
Polyetherimide		
	carbon dioxide	1996SH2
Poly(ether sulfone)		
	carbon dioxide	1998BOE
	dinitrogen oxide	1998BOE
	methane	1998BOE
Polyethylene		
	argon	1987KOL
	n-butane	1974DOS
	n-butane	1975HOR
	n-butane	1987CAS
	1-butene	1974DOS
	1-butene	1992WOH
	carbon dioxide	1986HIR

Polymer (B)	Solvent (A)	Ref.
Polyethylene (*continued*)		
	carbon dioxide	1986KA2
	carbon dioxide	1996SH1
	carbon dioxide	2002AR1
	carbon dioxide	2002AR2
	chlorodifluoromethane	1986GOR
	chlorotrifluoromethane	1986GOR
	dichlorodifluoromethane	1986GOR
	1,2-dichloro-1,1,2,2-tetrafluoroethane	1986GOR
	ethane	1987HEU
	ethene	1978CHE
	ethene	1984FON
	ethene	1984KOB
	ethene	1987HEU
	ethene	1992WOH
	ethene	1995LOO
	ethene	2000FEN
	fluorotrichloromethane	1975HOR
	n-hexane	1973BRO
	n-hexane	1987CAS
	n-hexane	1990KEN
	4-methyl-1-pentene	1992WOH
	2-methylpropane	1974DOS
	2-methylpropene	1974DOS
	methane	1963LUN
	methane	1977BON
	nitrogen	1962LUN
	nitrogen	1963LUN
	nitrogen	1975BON
	nitrogen	1977BON
	nitrogen	1978CHE
	nitrogen	1986KA2
	nitrogen	1986KOL
	nitrogen	1987KOL
	n-pentane	1987CAS
	2,2,4-trimethylpentane	1973BRO
Poly(ethylene glycol)		
	carbon dioxide	1997WEI
	carbon dioxide	2001HAT
Poly(ethylene oxide)		
	carbon dioxide	2001HAT

Polymer (B)	Solvent (A)	Ref.
Poly(ethylene terephthalate)		
	argon	1964VIE
	carbon dioxide	1964VIE
	carbon dioxide	1991LAM
	carbon dioxide	1996SH1
	carbon dioxide	1996SH2
	carbon dioxide	2000BRA
	methane	1964VIE
	nitrogen	1964VIE
	oxygen	1964VIE
Poly(ethyl methacrylate)		
	argon	1989CHI
	carbon dioxide	1981KOR
	carbon dioxide	1989CHI
	carbon dioxide	1989KA1
	carbon dioxide	1994CON
	methane	1989CHI
	nitrogen	1989CHI
	propane	1987GOR
Polyimide		
	carbon dioxide	1988SAD
	carbon dioxide	1998AUB
Polyisobutylene		
	methane	1969LUN
	methane	1977BON
	n-pentane	2002JOU
Polyisoprene		
	carbon dioxide	2003RAM
Poly(methyl methacrylate)		
	carbon dioxide	1981KOR
	carbon dioxide	1987WIS
	carbon dioxide	1989BER
	carbon dioxide	1992CO1
	carbon dioxide	1992CO2
	carbon dioxide	1994CON
	carbon dioxide	1994GOE
	carbon dioxide	1996SH2
	carbon dioxide	1997VIN
	carbon dioxide	1998AUB

Polymer (B)	Solvent (A)	Ref.
Poly(methyl methacrylate) (*continued*)		
	carbon dioxide	1998KAM
	carbon dioxide	1999KIK
	carbon dioxide	1999WEB
	carbon dioxide	2002NIK
	carbon dioxide	2002SIR
	carbon dioxide	2003ALE
	carbon dioxide	2003SHI
Poly(methyloctylsiloxane)		
	carbon dioxide	1986SHA
	methane	1986SHA
	propane	1986SHA
Poly(4-methyl-1-pentene)		
	carbon dioxide	1998AUB
Poly(methyltrifluoropropylsiloxane)		
	carbon dioxide	1986SHA
	methane	1986SHA
	propane	1986SHA
Poly(oxymethylene)		
	carbon dioxide	1996SH1
Poly(phenylmethylsiloxane)		
	carbon dioxide	1986SHA
	methane	1986SHA
	propane	1986SHA
Polypropylene		
	n-butane	1994HOR
	n-butane	2001NAG
	carbon dioxide	1996SH1
	chlorodifluoromethane	1986GOR
	chlorotrifluoromethane	1986GOR
	dichlorodifluoromethane	1986GOR
	1,2-dichloro-1,1,2,2-tetrafluoroethane	1986GOR
	ethene	1994HOR
	n-heptane	1994HOR
	n-pentane	1994HOR
	propene	1994HOR
	propene	2001SA1
	propene	2001TSU
	propene	2003PAL

Polymer (B)	Solvent (A)	Ref.
Polystyrene		
	carbon dioxide	1982MOR
	carbon dioxide	1982TOI
	carbon dioxide	1986CAR
	carbon dioxide	1987SAD
	carbon dioxide	1989BER
	carbon dioxide	1989HAC
	carbon dioxide	1992CO2
	carbon dioxide	1996CON
	carbon dioxide	1997KRY
	carbon dioxide	1998AUB
	carbon dioxide	1998WON
	carbon dioxide	1998ZHA
	carbon dioxide	2000FEN
	carbon dioxide	2003ALE
	carbon dioxide	2003KOG
	carbon dioxide	2003NIK
	carbon dioxide	2003SHI
	carbon dioxide	2004CAO
	1,1,1-chlorodifluoroethane	1999HON
	chlorodifluoromethane	1986GOR
	chlorodifluoromethane	1999HON
	chlorotrifluoromethane	1986GOR
	dichlorodifluoromethane	1986GOR
	1,2-dichloro-1,1,2,2-tetrafluoroethane	1986GOR
	1,1-difluoroethane	1994GAR
	1,1-difluoroethane	1998HON
	1,1-difluoroethane	1999HON
	methane	1962LUN
	methane	1963LUN
	methane	1987SAD
	nitrogen	2000FEN
	nitrogen	2000HIL
	nitrogen	1990KUM
	1,1,1,2-tetrafluoroethane	1998WON
	1,1,1,2-tetrafluoroethane	1998ZHA
	1,1,1,2-tetrafluoroethane	1999HON
	1,1,1-trifluoroethane	1999HON
Polysulfone		
	carbon dioxide	1986KA2
	carbon dioxide	1996SH2
	carbon dioxide	1998BOE
	dinitrogen oxide	1998BOE
	ethane	1986KA2
	methane	1986KA2
	methane	1998BOE
	nitrogen	1986KA2

Polymer (B)	Solvent (A)	Ref.
Polytetrafluoroethylene		
	carbon dioxide	1991BRI
	carbon dioxide	1996SH1
Poly(tetramethylsilhexylenesiloxane)		
	carbon dioxide	1993SHA
	methane	1993SHA
	propane	1993SHA
Poly(1-trimethyl-1-propyne)		
	carbon dioxide	1994POP
	methane	1994POP
Polyurethane and precursors		
	carbon dioxide	1996SH2
Poly(vinyl acetate)		
	carbon dioxide	1988MAS
	carbon dioxide	2003PAL
	carbon dioxide	2003WAN
Poly(vinyl benzoate)		
	argon	1986KA1
	carbon dioxide	1986KA1
	carbon dioxide	1989BER
	nitrogen	1986KA1
Poly(vinyl chloride)		
	carbon dioxide	1996SH2
	carbon dioxide	1998WON
	carbon dioxide	1998ZHA
	carbon dioxide	2001MUT
	chlorodifluoromethane	1986GOR
	chlorotrifluoromethane	1986GOR
	dichlorodifluoromethane	1986GOR
	1,2-dichloro-1,1,2,2-tetrafluoroethane	1986GOR
Poly(vinylidene fluoride)		
	carbon dioxide	1996SH1
	carbon dioxide	2003SHE
Poly(vinyl methyl ether)		
	carbon dioxide	1989HAC

Polymer (B)	Solvent (A)	Ref.
Poly(1-vinyl-2-pyrrolidinone)	carbon dioxide	1999KIK
Polyxylylene	carbon dioxide	1998AUB
Styrene/butadiene copolymer	carbon dioxide	2003KOG
Styrene/methyl methacrylate copolymer	carbon dioxide	1990RA1
Tetrafluoroethylene/2,2-bistrifluoromethyl-4,5-difluoro-1,3-dioxole copolymer		
	n-butane	1999BON
	n-butane	2002DEA
	carbon dioxide	1999BON
	carbon dioxide	2002DEA
	ethane	1999BON
	ethane	2002DEA
	helium	1999BON
	methane	1999BON
	methane	2002DEA
	nitrogen	1999BON
	nitrogen	2002DEA
	oxygen	1999BON
	oxygen	2002DEA
	propane	1999BON
	propane	2002DEA
	tetrafluoroethane	1999BON
	tetrafluoroethane	2002DEA
	tetrafluoromethane	1999BON
	tetrafluoromethane	2002DEA
Vinyl acetate/vinylidene cyanide copolymer	carbon dioxide	1988HAC
Vinyl acetate/1-vinyl-2-pyrrolidinone copolymer	carbon dioxide	2003KIK

2.3. Ternary and quaternary polymer solutions

Polymer (B):	**ethylene/norbornene copolymer**		**2003LEE**

Characterization: M_n/kg.mol^{-1} = 34.8, M_w/kg.mol^{-1} = 83.3, 50 mol% ethylene, T_g/K = 408.15, Union Chemical Laboratories, Taiwan

Solvent (A):	**ethene**	**C$_2$H$_4$**	**74-85-1**
Solvent (C):	**toluene**	**C$_7$H$_8$**	**108-88-3**
Solvent (D):	**bicyclo[2,2,1]-2-heptene (norbornene)**	**C$_7$H$_{10}$**	**498-66-8**

Type of data: gas solubility

Comments: The mass fractions of B/C/D belong to the feed liquid copolymer solution.

T/K = 323.15

$w_B/w_C/w_D$ = 0.04/0.40/0.56

P/bar	5.1	9.9	14.6	19.2	23.8
m_A/(g/g feed liquid copolymer solution)	0.0160	0.0321	0.0502	0.0698	0.0915

$w_B/w_C/w_D$ = 0.16/0.35/0.49

P/bar	6.4	10.8	14.6	19.1	25.2
m_A/(g/g feed liquid copolymer solution)	0.0157	0.0299	0.0436	0.0590	0.0836

$w_B/w_C/w_D$ = 0.28/0.30/0.42

P/bar	5.7	9.4	16.0	20.6	23.3
m_A/(g/g feed liquid copolymer solution)	0.0112	0.0230	0.0376	0.0518	0.0661

T/K = 373.15

$w_B/w_C/w_D$ = 0.04/0.40/0.56

P/bar	6.1	10.7	15.4	20.5	23.7
m_A/(g/g feed liquid copolymer solution)	0.0111	0.0224	0.0333	0.0465	0.0563

$w_B/w_C/w_D$ = 0.16/0.35/0.49

P/bar	5.6	9.9	13.7	19.4	25.6
m_A/(g/g feed liquid copolymer solution)	0.0088	0.0169	0.0251	0.0380	0.0530

T/K = 423.15

$w_B/w_C/w_D$ = 0.04/0.40/0.56

P/bar	7.0	10.5	14.8	19.9	25.3
m_A/(g/g feed liquid copolymer solution)	0.0055	0.0118	0.0179	0.0288	0.0385

$w_B/w_C/w_D$ = 0.16/0.35/0.49

P/bar	6.8	10.7	15.4	20.0	23.9
m_A/(g/g feed liquid copolymer solution)	0.0047	0.0100	0.0171	0.0250	0.0313

Polymer (B):	**ethylene/norbornene copolymer**		**2003LEE**

Characterization: $M_n/\text{kg.mol}^{-1} = 34.8$, $M_w/\text{kg.mol}^{-1} = 83.3$, 50 mol% ethylene, $T_g/\text{K} = 408.15$, Union Chemical Laboratories, Taiwan

Solvent (A):	**ethene**	**C_2H_4**	**74-85-1**
Solvent (C):	**toluene**	**C_7H_8**	**108-88-3**
Solvent (D):	**bicyclo[2,2,1]-2-heptene (norbornene)**	**C_7H_{10}**	**498-66-8**

Type of data: vapor-liquid equilibrium

$T/\text{K} = 323.15$

	liquid phase				vapor phase		
P/bar	x_A	x_B	x_C	x_D	y_A	y_C	y_D
5.1	0.0526	0.0001	0.3995	0.5478	0.9625	0.0088	0.0287
9.9	0.1001	0.0001	0.3794	0.5204	0.9767	0.0057	0.0176
14.6	0.1481	0.0001	0.3592	0.4926	0.9860	0.0037	0.0103
19.2	0.1948	0.0001	0.3395	0.4656	0.9876	0.0030	0.0094
23.8	0.2406	0.0001	0.3202	0.4391	0.9893	0.0029	0.0078
6.4	0.0583	0.0005	0.3968	0.5444	0.9660	0.0093	0.0247
10.8	0.1057	0.0004	0.3768	0.5171	0.9852	0.0034	0.0114
14.6	0.1471	0.0004	0.3594	0.4931	0.9867	0.0033	0.0100
19.1	0.1894	0.0004	0.3416	0.4686	0.9900	0.0021	0.0079
25.2	0.2486	0.0004	0.3166	0.4344	0.9908	0.0024	0.0068
5.7	0.0494	0.0010	0.4003	0.5493	0.9622	0.0122	0.0256
9.4	0.0958	0.0010	0.3807	0.5225	0.9757	0.0072	0.0171
16.0	0.1479	0.0009	0.3588	0.4924	0.9855	0.0043	0.0102
20.6	0.1928	0.0008	0.3399	0.4665	0.9886	0.0032	0.0082
23.3	0.2339	0.0008	0.3227	0.4426	0.9902	0.0032	0.0066

$T/\text{K} = 373.15$

	liquid phase				vapor phase		
P/bar	x_A	x_B	x_C	x_D	y_A	y_C	y_D
6.1	0.0371	0.0001	0.4147	0.5481	0.8305	0.0681	0.1014
10.7	0.0719	0.0001	0.3997	0.5283	0.9012	0.0276	0.0712
15.4	0.1032	0.0001	0.3863	0.5104	0.9257	0.0208	0.0535
20.5	0.1385	0.0001	0.3710	0.4904	0.9411	0.0167	0.0422
23.7	0.1630	0.0001	0.3605	0.4764	0.9476	0.0153	0.0371
5.6	0.0337	0.0005	0.4090	0.5568	0.8305	0.0762	0.0933
9.9	0.0627	0.0005	0.3968	0.5400	0.8919	0.0614	0.0467
13.7	0.0904	0.0005	0.3850	0.5241	0.9210	0.0371	0.0419
19.4	0.1306	0.0004	0.3680	0.5010	0.9402	0.0258	0.0340
25.6	0.1734	0.0004	0.3499	0.4763	0.9500	0.0238	0.0262

continued

continued

$T/K = 423.15$

	liquid phase				vapor phase		
P/bar	x_A	x_B	x_C	x_D	y_A	y_C	y_D
7.0	0.0185	0.0001	0.4149	0.5665	0.4610	0.1940	0.3450
10.5	0.0393	0.0001	0.4061	0.5545	0.6322	0.1223	0.2455
14.8	0.0583	0.0001	0.3980	0.5436	0.7084	0.0942	0.1974
19.9	0.0906	0.0001	0.3844	0.5249	0.7833	0.0692	0.1475
25.3	0.1177	0.0001	0.3729	0.5093	0.8162	0.0469	0.1369
6.8	0.0181	0.0005	0.4140	0.5674	0.4725	0.1784	0.3491
10.7	0.0382	0.0005	0.4055	0.5558	0.6479	0.1113	0.2408
15.4	0.0632	0.0005	0.3950	0.5413	0.7411	0.0820	0.1769
20.0	0.0866	0.0005	0.3851	0.5278	0.7878	0.0655	0.1467
23.9	0.1102	0.0005	0.3752	0.5141	0.8170	0.0559	0.1271

Polymer (B):   **ethylene/propylene block copolymer**   **2000SA2**
Characterization: M_w/kg.mol^{-1} = 240, 4.9 wt% ethene, 90% isotactic,
58.1 wt% crystallinity, impact propylene copolymer,
Mitsubishi Chemical, Kurashiki, Japan
Solvent (A):   **ethene**   **C_2H_4**   **74-85-1**
Solvent (C):   **n-hexane**   **C_6H_{14}**   **110-54-3**

Type of data: weight fraction-based partition coefficient at infinite dilution, $K^\infty = w_C^{gas}/w_C^{liq}$

$T/K = 323.15$

P/MPa	0.6	1.2	1.8
K^∞	30.6	15.2	12.0

$T/K = 343.15$

P/MPa	0.6	1.2	1.8	2.4
K^∞	43.9	20.7	15.9	13.1

$T/K = 363.15$
P/MPa	0.6	1.2	1.8	2.4
K^∞	59.7	31.9	23.2	17.4

Comments: The weight fraction-based partition coefficients of n-hexane at infinite dilution are given
for the equilibrated system of C_2H_4 + polymer at binary system equilibrium temperature,
pressure, and concentration (please see Chapter 2.1. for the binary data).

Polymer (B):   **ethylene/propylene block copolymer**   **2000SA2**
Characterization: M_w/kg.mol^{-1} = 240, 4.9 wt% ethene, 90% isotactic,
58.1 wt% crystallinity, impact propylene copolymer,
Mitsubishi Chemical, Kurashiki, Japan

continued

continued

Solvent (A):	**propene**	C_3H_6	**115-07-1**
Solvent (C):	**n-hexane**	C_6H_{14}	**110-54-3**

Type of data: weight fraction-based partition coefficient at infinite dilution, $K^\infty = w_C^{gas}/w_C^{liq}$

$T/K = 323.15$

P/MPa	1.2	1.6	1.8
K^∞	12.4	9.4	8.6

$T/K = 343.15$

P/MPa	1.2	1.6	1.8	2.4
K^∞	21.2	15.7	14.1	11.2

$T/K = 363.15$

P/MPa	1.2	1.8	2.4
K^∞	38.2	20.4	14.7

Comments: The weight fraction-based partition coefficients of n-hexane at infinite dilution are given for the equilibrated system of C_3H_6 + polymer at binary system equilibrium temperature, pressure, and concentration (please see Chapter 2.1. for the binary data).

Polymer (B):	**poly(butyl methacrylate)**		**2000BY1**
Characterization:	Aldrich Chem. Co., Inc., Milwaukee, WI		
Solvent (A):	**carbon dioxide**	CO_2	**124-38-9**
Solvent (C):	**butyl methacrylate**	$C_8H_{14}O_2$	**97-88-1**

Type of data: vapor-liquid equilibrium

w_A	0.400		w_B	0.05		w_C	0.550	were kept constant

T/K	339.05	323.95	312.95
P/bar	95	78	65

Comments: High-pressure fluid phase equilibrium data for this system are given in Chapter 4.

Polymer (B):	**polybutadiene**		**2002JOU**
Characterization:	M_w/kg.mol^{-1} = 420, Aldrich Chem. Co., Inc., Milwaukee, WI		
Solvent (A):	**carbon dioxide**	CO_2	**124-38-9**
Solvent (C):	**cyclohexane**	C_6H_{12}	**110-82-7**

Type of data: gas/vapor solubility

w_A	0.081	0.081	0.081	0.081	0.081	0.081	0.081	0.081	0.081
w_B	0.050	0.050	0.050	0.050	0.050	0.050	0.050	0.050	0.050
w_C	0.869	0.869	0.869	0.869	0.869	0.869	0.869	0.869	0.869
T/K	353.15	363.15	373.15	383.15	393.15	403.15	413.15	423.15	433.15
P/MPa	2.6	2.9	3.4	3.6	3.7	4.4	4.5	4.9	5.2

continued

continued

w_A	0.081	0.081	0.081	0.081	0.172	0.172	0.172	0.172	0.172
w_B	0.050	0.050	0.050	0.050	0.050	0.050	0.050	0.050	0.050
w_C	0.869	0.869	0.869	0.869	0.778	0.778	0.778	0.778	0.778
T/K	443.15	453.15	463.15	473.15	353.15	363.15	373.15	383.15	393.15
P/MPa	5.7	6.1	6.5	6.6	5.3	5.5	6.6	7.2	7.7

w_A	0.172	0.172	0.240	0.240	0.240	0.240	0.240	0.300
w_B	0.050	0.050	0.050	0.050	0.050	0.050	0.050	0.050
w_C	0.778	0.778	0.710	0.710	0.710	0.710	0.710	0.650
T/K	403.15	413.15	353.15	363.15	373.15	383.15	393.15	353.15
P/MPa	8.1	8.5	6.5	6.9	7.7	8.5	9.1	7.6

Comments: High-pressure fluid phase equilibrium data for this system are given in Chapter 4.

Polymer (B):	**polybutadiene**		**1999INO**
Characterization:	$M_n/kg.mol^{-1} = 159$, $M_w/kg.mol^{-1} = 165$,		
	low *cis*-butadiene rubber		
Solvent (A):	**carbon dioxide**	**CO_2**	**124-38-9**
Solvent (C):	**n-hexane**	**C_6H_{14}**	**110-54-3**

Type of data: weight fraction-based partition coefficient at infinite dilution, $K^\infty = w_C^{gas}/w_C^{liq}$

$T/K = 318.2$

$\rho(CO_2)/(g/cm^3)$	0.142	0.161	0.183	0.221	0.241	0.282	0.353	
K^∞	0.301	0.304	0.322	0.390	0.456	0.548	0.967	

$T/K = 333.2$

$\rho(CO_2)/(g/cm^3)$	0.125	0.140	0.162	0.184	0.204	0.303	0.351	0.396
K^∞	0.400	0.403	0.450	0.472	0.519	0.880	1.040	1.315

$T/K = 358.2$

$\rho(CO_2)/(g/cm^3)$	0.107	0.144	0.166	0.196	0.261
K^∞	0.714	0.719	0.823	0.877	1.013

Comments: The values of K^∞ are given for the equilibrated system of CO_2 + polymer at binary
system equilibrium temperature, density, pressure, and concentration.

Polymer (B):	**polybutadiene**		**2002JOU**
Characterization:	$M_w/kg.mol^{-1} = 420$, Aldrich Chem. Co., Inc., Milwaukee, WI		
Solvent (A):	**carbon dioxide**	**CO_2**	**124-38-9**
Solvent (C):	**toluene**	**C_7H_8**	**108-88-3**

Type of data: gas/vapor solubility

w_A	0.108	0.108	0.108	0.108	0.108	0.108	0.108	0.108	0.108
w_B	0.059	0.059	0.059	0.059	0.059	0.059	0.059	0.059	0.059
w_C	0.833	0.833	0.833	0.833	0.833	0.833	0.833	0.833	0.833
T/K	323.15	343.15	363.15	383.15	403.15	423.15	443.15	463.15	473.15
P/MPa	1.9	2.9	4.0	4.8	5.4	6.1	6.9	7.6	8.0

continued

continued

w_A	0.186	0.186	0.186	0.186	0.186	0.186	0.186	0.186	0.278
w_B	0.059	0.059	0.059	0.059	0.059	0.059	0.059	0.059	0.059
w_C	0.755	0.755	0.755	0.755	0.755	0.755	0.755	0.755	0.663
T/K	323.15	343.15	363.15	383.15	403.15	423.15	443.15	453.15	323.15
P/MPa	2.8	3.7	4.4	6.3	7.9	9.2	10.6	11.0	4.9

w_A	0.278	0.278	0.278	0.278	0.337	0.337	0.337	0.337
w_B	0.059	0.059	0.059	0.059	0.059	0.059	0.059	0.059
w_C	0.663	0.663	0.663	0.663	0.604	0.604	0.604	0.604
T/K	343.15	363.15	383.15	403.15	323.15	343.15	363.15	383.15
P/MPa	6.3	8.3	10.3	11.9	6.6	7.9	9.9	11.4

Comments: High-pressure fluid phase equilibrium data for this system are given in Chapter 4.

Polymer (B):   **poly(2,6-dimethyl-1,4-phenylene ether)**   **2002SAT**
Characterization: M_n/kg.mol^{-1} = 22, M_w/kg.mol^{-1} = 47,
                          T_g/K = 477.1, Asahi Chemicals Inc., Kawasaki, Japan
Solvent (A):   **carbon dioxide   CO$_2$**   **124-38-9**
Polymer (C):   **polystyrene**
Characterization: M_n/kg.mol^{-1} = 106, M_w/kg.mol^{-1} = 331,
                          T_g/K = 381.4, Asahi Chemicals Inc., Kawasaki, Japan

Type of data: gas solubility

$w_B/w_C = 25/75$ was kept constant

T/K = 373.15

P/MPa	2.127	4.107	6.051	8.095	11.887	15.965	20.110
w_A	0.01356	0.02551	0.03706	0.04849	0.06670	0.08206	0.09478

T/K = 423.15

P/MPa	2.113	4.125	6.023	8.047	12.023	15.995	20.013
w_A	0.00987	0.01913	0.02767	0.03641	0.05229	0.06532	0.07693

T/K = 473.15

P/MPa	2.219	4.113	6.111	8.088	12.092	16.065	20.060
w_A	0.00680	0.01393	0.02133	0.02849	0.04239	0.05451	0.06582

$w_B/w_C = 50/50$ was kept constant

T/K = 373.15

P/MPa	2.147	4.087	6.085	8.108	12.059	16.012	19.789
w_A	0.01687	0.02896	0.03977	0.05122	0.07187	0.09238	0.11116

T/K = 423.15

P/MPa	2.167	4.126	6.091	8.115	12.110	16.059	20.104
w_A	0.01020	0.01950	0.02868	0.03794	0.05452	0.06850	0.08005

T/K = 473.15

P/MPa	2.148	4.148	6.127	8.110	12.117	16.144	20.046
w_A	0.00830	0.01587	0.02313	0.03050	0.04453	0.05648	0.06640

Polymer (B):   **poly(dimethylsiloxane)**   **1998VIN**
Characterization: United Chemical Technologies, Inc., Petrarch Silanes & Silicones,
                         cross-linked using benzoyl peroxide
Solvent (A):   **carbon dioxide   CO$_2$**   **124-38-9**
Solvent (C):   **1,1,1,3,3,3-hexadeutero-2-propanone   C$_3$D$_6$O**   **666-52-4**

Type of data: partition coefficient at equilibrium swelling, $K_c = c_C$ (in swollen PDMS)/c_C (in CO$_2$)

T/K = 314.65

The concentration of C$_3$D$_6$O in CO$_2$ is 0.040 M.

P/bar	10.2	15.6	22.6	29.5	41.1	51.1	58.8	65.5	72.8
K_c	81.593	77.151	106.08	77.390	44.694	33.904	26.644	21.346	12.686

P/bar	79.7	85.2	91.5	96.5	105.0	111.2	173.8
K_c	8.975	4.778	3.652	2.990	2.099	1.963	0.207

The concentration of C$_3$D$_6$O in CO$_2$ is 0.012 M.

P/bar	82.8	84.4	86.6	90.2	91.7	96.5	98.5	105.0	111.0
K_c	5.376	4.437	3.428	2.046	1.934	1.625	1.433	1.287	1.026

The concentration of C$_3$D$_6$O in CO$_2$ is 0.200 M.

P/bar	80.1	83.0	84.1	86.1	89.4	91.7	96.0	103.4	116.0
K_c	6.903	3.764	2.996	2.647	1.793	1.603	1.542	1.533	0.789

P/bar	125.5	138.5	152.5	173.5
K_c	0.705	0.329	0.146	0.225

Polymer (B):   **poly(dimethylsiloxane)**   **1998VIN**
Characterization: United Chemical Technologies, Inc., Petrarch Silanes & Silicones,
                         cross-linked using benzoyl peroxide
Solvent (A):   **carbon dioxide   CO$_2$**   **124-38-9**
Solvent (C):   **methanol   CH$_4$O**   **67-56-1**

Type of data: swelling at equilbrium pressure

T/K = 314.65

The concentration of CH$_4$O in CO$_2$ is 0.0892 M.

P/bar	82.7	89.4	96.8	102.6	110.0	116.2	123.6	131.1	138.1
V/V_0	1.638	1.605	1.636	1.651	1.707	1.739	1.760	1.822	1.857

P/bar	146.2	152.8	159.9	168.3	173.3
V/V_0	1.912	1.941	1.952	1.992	2.021

The concentration of CH$_4$O in CO$_2$ is 0.131 M.

P/bar	84.2	85.3	86.8	88.2	89.5	91.6	93.5	99.0	105.3
V/V_0	1.717	1.687	1.652	1.627	1.610	1.611	1.620	1.643	1.674

P/bar	112.0	118.0	125.5	132.2	138.3	146.1	153.3	162.0	171.3
V/V_0	1.713	1.749	1.792	1.815	1.853	1.883	1.935	1.970	1.998

continued

continued

The concentration of CH_4O in CO_2 is 0.670 M.

P/bar	83.9	85.0	86.6	92.1	96.0	99.4	102.1	108.5	112.6
V/V_0	1.879	1.817	1.800	1.767	1.760	1.776	1.784	1.809	1.848

P/bar	118.6	125.1	131.7	141.5	149.2	160.8	172.5
V/V_0	1.877	1.935	1.976	2.042	2.089	2.176	2.236

The concentration of CH_4O in CO_2 is its concentration at saturated conditions.

P/bar	0.0	1.0	14.1	20.5	34.4	41.6	41.9	52.3	56.3
V/V_0	1.028	1.057	1.104	1.111	1.172	1.231	1.248	1.368	1.408

P/bar	63.2	64.4	68.8	68.8	70.4	72.8	75.9	76.5	77.7
V/V_0	1.554	1.596	1.717	1.664	1.671	1.901	1.978	2.128	2.204

P/bar	78.5	79.2	80.3	80.4	80.9	81.1	81.6	82.1	83.0
V/V_0	2.214	2.256	2.211	2.268	2.275	2.193	2.228	2.165	1.988

Comments: The equilibrium swelling data in pure CO_2 are given above in Chapter 2.1.

Polymer (B):	**poly(dimethylsiloxane)**		**1998VIN**
Characterization:	United Chemical Technologies, Inc., Petrarch Silanes & Silicones, cross-linked using benzoyl peroxide		
Solvent (A):	**carbon dioxide**	**CO₂**	**124-38-9**
Solvent (C):	**methanol-d4**	**CD₄O**	**811-98-3**

Type of data: partition coefficient at equilibrium swelling, $K_c = c_C$ (in swollen PDMS)$/c_C$ (in CO_2)

$T/K = 314.65$

The concentration of CD_4O in CO_2 is 0.0480 M.

P/bar	52.1	58.8	63.5	67.5	69.9	71.8	74.4	78.5	81.0
K_c	13.186	15.680	9.806	8.356	8.104	6.526	5.763	4.891	4.292

P/bar	84.0	91.1	96.4	103.9	112.4	118.7	126.8	132.9	139.4
K_c	2.717	2.759	2.750	2.284	1.700	1.613	2.059	2.108	1.515

P/bar	147.4	154.3	175.4
K_c	1.361	1.823	1.416

The concentration of CD_4O in CO_2 is 0.0890 M.

P/bar	71.9	80.6	84.1	88.3	91.4	93.7	97.2	104.0	110.1
K_c	19.337	7.311	5.092	3.679	2.508	2.712	2.776	2.555	2.259

P/bar	120.0	126.8	132.9	138.6	146.7	151.5	159.2	166.4	174.4
K_c	1.883	1.983	2.015	1.525	1.473	1.477	1.605	1.448	1.421

The concentration of CD_4O in CO_2 is 0.130 M.

P/bar	78.1	79.3	81.8	83.7	85.9	87.8	91.0	96.5	104.1
K_c	13.889	11.517	8.101	6.102	4.814	3.982	2.904	2.649	2.372

P/bar	111.0	120.8	125.1	132.6	141.5	147.3	153.8	165.6	176.1
K_c	2.404	1.815	1.737	1.979	1.932	1.532	1.698	1.340	1.253

Polymer (B):   **poly(dimethylsiloxane)**   **1998VIN**
Characterization: United Chemical Technologies, Inc., Petrarch Silanes & Silicones,
                        cross-linked using benzoyl peroxide
Solvent (A):   **carbon dioxide**   **CO_2**   **124-38-9**
Solvent (C):   **2-propanol**   **C_3H_8O**   **67-63-0**

Type of data: swelling at equilbrium pressure

T/K = 314.65

The concentration of C_3H_8O in CO_2 is 0.160 M.

P/bar	81.2	82.7	87.0	96.3	110.2	123.7	143.0	157.2	173.1
V/*V*$_0$	1.790	1.721	1.610	1.649	1.738	1.810	1.914	1.985	2.109

The concentration of C_3H_8O in CO_2 is 0.743 M.

P/bar	81.1	87.1	95.0	100.9	107.2	117.8	132.3	153.9	171.7
V/*V*$_0$	2.147	2.092	2.065	2.058	2.086	2.178	2.261	2.519	2.643

The concentration of C_3H_8O in CO_2 is its concentration at saturated conditions.

P/bar	1.0	1.0	4.4	5.9	12.5	14.1	17.9	22.5	30.0
V/*V*$_0$	1.910	1.967	2.240	2.187	2.832	2.567	3.268	3.439	3.887

P/bar	35.0	41.4	49.6	55.7	56.6	60.8	64.2	69.1	69.3
V/*V*$_0$	4.184	5.440	6.362	6.749	7.201	7.515	8.417	8.561	7.354

P/bar	73.4	76.7	76.9	78.5	80.2
V/*V*$_0$	5.149	3.108	2.917	2.296	2.234

Comments: The equilibrium swelling data in pure CO_2 are given above in Chapter 2.1.

Polymer (B):   **poly(dimethylsiloxane)**   **1998VIN**
Characterization: United Chemical Technologies, Inc., Petrarch Silanes & Silicones,
                        cross-linked using benzoyl peroxide
Solvent (A):   **carbon dioxide**   **CO_2**   **124-38-9**
Solvent (C):   **2-propanol-d8**   **C_3D_8O**   **22739-76-0**

Type of data: partition coefficient at equilibrium swelling, $K_c = c_C$ (in swollen PDMS)/c_C (in CO_2)

T/K = 314.65

The concentration of C_3D_8O in CO_2 is 0.160 M.

P/bar	80.6	82.4	84.0	86.7	91.4	97.7	104.6	111.7	124.5
K_c	17.349	9.142	6.062	4.155	2.578	2.085	1.998	2.001	1.715

P/bar	139.1	156.1	174.2
K_c	1.305	1.281	1.321

The concentration of C_3D_8O in CO_2 is 0.01 M.

P/bar	8.5	15.5	21.9	29.0	38.4	44.5	51.1	58.6	64.4
K_c	35.033	57.339	78.837	87.014	105.269	105.004	90.562	63.572	40.862

continued

continued

P/bar	71.9	78.9	83.9	86.3	86.7	91.1	94.2	97.9	105.0
K_c	21.143	9.474	5.246	4.432	2.282	2.635	2.305	1.978	1.861

P/bar	112.7	115.8	119.0	127.8	128.0	142.5	143.9	160.2	174.2
K_c	1.606	1.981	1.387	1.870	2.058	2.059	1.346	1.454	1.355

Polymer (B):	**poly(dimethylsiloxane)**	**1998VIN**

Characterization: United Chemical Technologies, Inc., Petrarch Silanes & Silicones, cross-linked using benzoyl peroxide

Solvent (A):	**carbon dioxide**	**CO_2**	**124-38-9**
Solvent (C):	**2-propanone**	**C_3H_6O**	**67-64-1**

Type of data: swelling at equilbrium pressure

$T/K = 314.65$

The concentration of C_3H_6O in CO_2 is 0.0402 M.

P/bar	1.0	8.2	16.8	22.7	31.9	39.2	49.9	61.4	78.1
V/V_0	1.507	1.583	1.538	1.445	1.343	1.346	1.357	1.383	1.521

P/bar	84.8	90.6	96.6	106.9	116.3	130.2	142.7	156.3	171.6
V/V_0	1.580	1.586	1.624	1.686	1.749	1.832	1.904	1.969	2.048

The concentration of C_3H_6O in CO_2 is 0.203 M.

P/bar	77.4	79.8	82.1	83.9	85.2	87.1	91.2	90.9	99.0
V/V_0	1.969	1.830	1.661	1.623	1.597	1.584	1.588	1.584	1.606

P/bar	101.4	110.1	120.7	143.3	155.4	172.1
V/V_0	1.630	1.671	1.721	1.858	1.932	2.038

The concentration of C_3H_6O in CO_2 is 0.915 M.

P/bar	79.3	82.4	84.9	87.9	92.0	99.7	170.5
V/V_0	2.199	2.210	2.228	2.216	2.206	2.263	2.512

The concentration of C_3H_6O in CO_2 is its concentration at saturated conditions.

P/bar	1.0	1.0	6.5	13.2	20.7	26.9	29.4	34.7	36.6
V/V_0	1.897	1.735	1.986	2.413	2.873	3.567	4.039	4.483	4.913

P/bar	41.4	44.1	50.2	51.0	55.2	60.2	66.4	70.0	72.3
V/V_0	5.315	5.510	6.011	5.809	5.778	4.982	3.487	2.702	2.442

P/bar	72.8	73.9	74.8
V/V_0	2.294	2.298	2.115

Comments: The equilibrium swelling data in pure CO_2 are given above in Chapter 2.1.

Polymer (B):	**poly(dimethylsiloxane)**	**1989SHI**

Characterization: silicone rubber, cross-linked

Solvent (A):	**carbon dioxide**	**CO_2**	**124-38-9**
Solvent (C):	**toluene**	**C_7H_8**	**108-88-3**

continued

continued

Type of data: vapor-liquid equilibrium

Comments: The sorption of toluene is given in the equilibrated system of CO_2 + PDMS at binary
system equilibrium temperature, pressure, and concentration (these binary data were
published in the original source in diagrams only).

$T/K = 308.15$

P/bar	40.5	60.8	60.8	60.8	70.9	70.9	81.0	81.0
m_C/(g/g polymer)	0.06427	0.04390	0.04694	0.04245	0.03012	0.03159	0.00910	0.00871

P/bar	81.0	101.3	101.3	101.3	121.6	121.6	141.8	141.8
m_C/(g/g polymer)	0.00762	0.00662	0.00583	0.00683	0.00680	0.00718	0.00715	0.00634

P/bar	141.8	202.6	253.3
m_C/(g/g polymer)	0.00654	0.00490	0.00537

$T/K = 343.15$

P/bar	40.5	60.8	70.9	81.0	101.3	121.6	141.8	202.6
m_C/(g/g polymer)	0.01535	0.01580	0.01534	0.01381	0.00706	0.00735	0.00601	0.00421

Polymer (B):	**polyethylene**		**1990KEN**
Characterization:	M_n/kg.mol^{-1} = 8, M_w/kg.mol^{-1} = 177, M_z/kg.mol^{-1} = 1000,		
	HDPE, DSM, Geleen, The Netherlands		
Solvent (A):	**n-hexane**	**C$_6$H$_{14}$**	**110-54-3**
Solvent (C):	**nitrogen**	**N$_2$**	**7727-37-9**

Type of data: vapor-liquid equilibrium

w_A	0.99218	0.99218	0.99218	0.99218	0.97801	0.97801	0.97801	0.93668	0.93668
w_B	0.00557	0.00557	0.00557	0.00557	0.01970	0.01970	0.01970	0.06110	0.06110
w_C	0.00225	0.00225	0.00225	0.00225	0.00229	0.00229	0.00229	0.00222	0.00222
T/K	396.71	398.67	400.68	402.67	396.64	398.65	400.63	401.78	403.80
P/bar	8.6	8.8	9.0	9.1	8.8	8.9	9.1	9.9	10.0

w_A	0.93668	0.93668	0.91660	0.91660	0.91660	0.91660	0.87440	0.87440	0.87440
w_B	0.06110	0.06110	0.08180	0.08180	0.08180	0.08180	0.12350	0.12350	0.12350
w_C	0.00222	0.00222	0.00160	0.00160	0.00160	0.00160	0.00210	0.00210	0.00210
T/K	405.79	407.75	407.59	410.06	412.55	417.54	417.51	419.98	422.51
P/bar	10.2	10.4	9.9	10.2	10.5	11.1	11.4	11.7	12.1

w_A	0.86616	0.86616	0.86616	0.86616	0.98317	0.98317	0.98317	0.97797	0.97797
w_B	0.13160	0.13160	0.13160	0.13160	0.00443	0.00443	0.00443	0.00973	0.00973
w_C	0.00224	0.00224	0.00224	0.00224	0.01240	0.01240	0.01240	0.01230	0.01230
T/K	417.54	419.98	422.51	424.98	384.67	385.69	386.65	384.28	386.24
P/bar	11.4	11.7	12.1	12.4	25.6	25.7	25.7	25.8	25.9

w_A	0.97797	0.97637	0.97637	0.97637	0.96690	0.96690	0.96690	0.96600	0.96600
w_B	0.00973	0.01093	0.01093	0.01093	0.02060	0.02060	0.02060	0.02130	0.02130
w_C	0.01230	0.01270	0.01270	0.01270	0.01250	0.01250	0.01250	0.01270	0.01270
T/K	388.21	384.67	386.68	388.15	384.10	386.13	388.11	384.75	386.70
P/bar	26.0	26.4	26.4	26.5	26.4	26.4	26.5	26.7	26.8

continued

continued

w_A	0.96600	0.92590	0.92590	0.92590	0.90640	0.90640	0.90640	0.90640	0.88390
w_B	0.02130	0.06140	0.06140	0.06140	0.08120	0.08120	0.08120	0.08120	0.10390
w_C	0.01270	0.01270	0.01270	0.01270	0.01240	0.01240	0.01240	0.01240	0.01220
T/K	388.63	394.62	396.19	397.32	394.62	396.19	400.03	401.97	400.03
P/bar	26.9	27.6	27.7	27.8	28.4	28.5	28.7	28.8	28.9

w_A	0.88390	0.88390	0.92975	0.92975	0.92975	0.93350	0.93350	0.93350	0.98115
w_B	0.10390	0.10390	0.05980	0.05980	0.05980	0.06040	0.06040	0.06040	0.00986
w_C	0.01220	0.01220	0.01045	0.01045	0.01045	0.00610	0.00610	0.00610	0.00899
T/K	402.02	404.01	392.04	394.62	396.61	401.94	403.95	405.91	387.28
P/bar	29.0	29.1	24.2	24.3	24.4	16.6	16.7	16.8	20.0

w_A	0.98115	0.98115	0.89158	0.89158	0.89158	0.88122	0.88122	0.88122
w_B	0.00986	0.00986	0.10340	0.10340	0.10340	0.11110	0.11110	0.11110
w_C	0.00899	0.00899	0.00502	0.00502	0.00502	0.00768	0.00768	0.00768
T/K	390.69	392.69	407.80	409.83	411.83	406.59	408.64	412.57
P/bar	20.2	20.4	16.0	16.1	16.2	20.8	20.9	21.2

Comments: High-pressure fluid phase demixing data for this system are given in Chapter 4.

Polymer (B):   **polyethylene**   **2001TOR**
Characterization: $M_n/$kg.mol^{-1} = 43, $M_w/$kg.mol^{-1} = 105, $M_z/$kg.mol^{-1} = 190,
                        HDPE, DSM, Geleen, The Netherlands
Solvent (A):   **n-hexane**   **C$_6$H$_{14}$**   **110-54-3**
Solvent (C):   **1-octene**   **C$_8$H$_{16}$**   **111-66-0**

Type of data: vapor-liquid equilibrium

w_A	0.7385	0.7385	0.5932	0.5932	0.3210	0.3210	0.3210	0.2156	0.2156
w_B	0.1794	0.1794	0.1441	0.1441	0.1971	0.1971	0.1971	0.1324	0.1324
w_C	0.0821	0.0821	0.2627	0.2627	0.4819	0.4819	0.4819	0.6520	0.6520
T/K	441.7	451.9	452.6	462.4	472.5	482.7	493.1	482.6	493.1
P/MPa	0.92	1.10	0.93	1.10	1.00	1.22	1.46	1.02	1.25

Comments: Liquid-liquid equilibrium and liquid-liquid-vapor equilibrium data for this system
                    are given in Chapter 3.

Polymer (B):   **poly(ethylene glycol) dimethyl ether**   **1995GAI**
Characterization: PEGDM mixture used for the Selexol process

diethylene glycol dimethylether	2.07 wt%
diethylene glycol monomethylether	0.16 wt%
triethylene glycol dimethylether	21.01 wt%
triethylene glycol monomethylether	0.72 wt%
tetraethylene glycol dimethylether	33.68 wt%
tetraethylene glycol monomethylether	0.67 wt%
pentaethylene glycol dimethylether	26.39 wt%
pentaethylene glycol monomethylether	0.38 wt%
hexaethylene glycol dimethylether	12.78 wt%
hexaethylene glycol monomethylether	0.10 wt%
heptaethylene glycol dimethylether	1.87 wt%

continued

continued

Solvent (A):	carbon dioxide	CO_2	124-38-9
Solvent (C):	diisopropanolamine	$C_6H_{15}NO_2$	110-97-4
Solvent (D):	water	H_2O	7732-18-5

Type of data: gas solubility

Comments: The feed liquid mixture is given by $w_B = 0.85 + w_C = 0.10 + w_D = 0.05$.

$T/K = 298.15$

P/MPa	0.385	0.730	1.327	1.753	2.077	2.533	3.030	3.587	4.053
x_A	0.0798	0.1386	0.2100	0.2584	0.2856	0.3224	0.3650	0.4139	0.4600

$T/K = 313.15$

P/MPa	0.517	0.963	1.317	1.814	2.148	2.614	2.908	3.567	4.063
x_A	0.0753	0.1374	0.1748	0.2211	0.2474	0.2772	0.2976	0.3366	0.3716

$T/K = 333.15$

P/MPa	0.547	0.922	1.327	1.854	2.290	2.553	2.928	3.526	4.083
x_A	0.0558	0.0879	0.1215	0.1557	0.1814	0.1958	0.2188	0.2562	0.2996

Polymer (B):	**poly(ethylene glycol) dimethyl ether**		**1995GAI**
Characterization:	PEGDM mixture used for the Selexol process		
	see data set above		
Solvent (A):	hydrogen	H_2	1333-74-0
Solvent (C):	diisopropanolamine	$C_6H_{15}NO_2$	110-97-4
Solvent (D):	water	H_2O	7732-18-5

Type of data: gas solubility

Comments: The feed liquid mixture is given by $w_B = 0.85 + w_C = 0.10 + w_D = 0.05$.

$T/K = 298.15$

P/MPa	1.023	1.510	1.672	2.057	2.361	2.634	3.009	3.405	3.769
x_A	0.0022	0.0031	0.0035	0.0042	0.0049	0.0055	0.0062	0.0070	0.0078

$T/K = 313.15$

P/MPa	1.013	1.398	1.601	2.097	2.341	2.564	2.999	3.263	3.587
x_A	0.0023	0.0032	0.0036	0.0048	0.0054	0.0058	0.0068	0.0074	0.0081

$T/K = 333.15$

P/MPa	1.023	1.419	1.702	1.935	2.280	2.564	2.979	3.212	3.840
x_A	0.0027	0.0038	0.0045	0.0052	0.0060	0.0068	0.0079	0.0086	0.0102

Polymer (B):	**poly(ethylene glycol) dimethyl ether**		**1995GAI**
Characterization:	PEGDM mixture used for the Selexol process		
	see data set above		
Solvent (A):	nitrogen	N_2	7727-37-9

continued

continued

| **Solvent (C):** | **diisopropanolamine** | $C_6H_{15}NO_2$ | **110-97-4** |
| **Solvent (D):** | **water** | H_2O | **7732-18-5** |

Type of data: gas solubility

Comments: The feed liquid mixture is given by $w_B = 0.85 + w_C = 0.10 + w_D = 0.05$.

$T/K = 298.15$

P/MPa	1.034	1.439	1.925	2.239	2.564	2.817	3.081	3.405	3.699
x_A	0.0031	0.0043	0.0058	0.0067	0.0077	0.0085	0.0092	0.0103	0.0111

$T/K = 313.15$

P/MPa	1.023	1.419	1.672	1.935	2.219	2.604	2.717	3.091	3.526
x_A	0.0035	0.0049	0.0057	0.0067	0.0076	0.0089	0.0096	0.0106	0.0121

$T/K = 333.15$

P/MPa	1.013	1.409	1.804	1.996	2.169	2.462	2.888	3.101	3.506
x_A	0.0037	0.0053	0.0067	0.0074	0.0080	0.0091	0.0106	0.0115	0.0129

Polymer (B):	**poly(ethyl methacrylate)**		**2000BY2**
Characterization:	Aldrich Chem. Co., Inc., Milwaukee, WI		
Solvent (A):	**carbon dioxide**	CO_2	**124-38-9**
Solvent (C):	**ethyl methacrylate**	$C_6H_{10}O_2$	**97-63-2**

Type of data: vapor-liquid equilibrium

w_A	0.493	0.493	0.493	0.493	0.493
w_B	0.05	0.05	0.05	0.05	0.05
w_C	0.457	0.457	0.457	0.457	0.457
T/K	298.75	306.85	308.35	309.75	316.85
P/bar	44.3	52.7	54.5	56.2	62.4

Comments: High-pressure fluid phase equilibrium data for this system are given in Chapter 4.

Polymer (B):	**polyisobutylene**		**2002JOU**
Characterization:	M_n/kg.mol^{-1} = 600, M_w/kg.mol^{-1} = 1000,		
	Aldrich Chem. Co., Inc., Milwaukee, WI		
Solvent (A):	**carbon dioxide**	CO_2	**124-38-9**
Solvent (C):	**n-heptane**	C_7H_{16}	**142-82-5**

Type of data: vapor-liquid equilibrium

w_A	0.088		w_B	0.025		w_C	0.887	were kept constant

T/K	323.15	338.15	353.15	373.15	393.15
P/MPa	1.9	2.0	2.3	3.0	3.0

Comments: High-pressure fluid phase equilibrium data for this system are given in Chapter 4.

Polymer (B): **polypropylene** **2000OLI**
Characterization: M_n/kg.mol^{-1} = 48.9, M_w/kg.mol^{-1} = 245, 92% isotactic,
Polibrasil Resinas S.A., Brazil
Solvent (A): **n-butane** **C$_4$H$_{10}$** **106-97-8**
Solvent (C): **toluene** **C$_7$H$_8$** **108-88-3**

Type of data: vapor-liquid equilibrium

w_A	0.8905	0.8905	0.8905	0.8905	0.8508	0.8508	0.8508	0.8508	0.8012
w_B	0.0996	0.0996	0.0996	0.0996	0.0983	0.0983	0.0983	0.0983	0.0992
w_C	0.0099	0.0099	0.0099	0.0099	0.0509	0.0509	0.0509	0.0509	0.0996
T/K	373.15	368.15	363.15	358.15	373.15	368.15	363.15	358.15	383.15
P/bar	18.1	15.9	14.5	12.9	17.5	15.8	14.3	12.6	19.9

w_A	0.8012	0.8012	0.8012	0.8012	0.8012
w_B	0.0992	0.0992	0.0992	0.0992	0.0992
w_C	0.0996	0.0996	0.0996	0.0996	0.0996
T/K	373.15	368.15	363.15	358.15	353.15
P/bar	16.0	15.0	13.9	12.0	10.7

Comments: Liquid-liquid equilibrium data are given in Chapter 3.

Polymer (B): **polypropylene** **2000OLI**
Characterization: M_n/kg.mol^{-1} = 48.9, M_w/kg.mol^{-1} = 245, 92% isotactic,
Polibrasil Resinas S.A., Brazil
Solvent (A): **1-butene** **C$_4$H$_8$** **106-98-9**
Solvent (C): **toluene** **C$_7$H$_8$** **108-88-3**

Type of data: vapor-liquid equilibrium

w_A	0.8910	0.8512	0.8512	0.8512	0.8022	0.8022	0.8022	0.8022
w_B	0.0986	0.0974	0.0974	0.0974	0.0993	0.0993	0.0993	0.0993
w_C	0.0104	0.0514	0.0514	0.0514	0.0985	0.0985	0.0985	0.0985
T/K	363.15	368.15	363.15	358.15	373.15	368.15	363.15	358.15
P/bar	17.0	17.5	15.3	12.9	18.3	16.2	14.0	11.8

Comments: Liquid-liquid equilibrium data are given in Chapter 3.

Polymer (B): **poly(vinyl acetate)** **1991WAN, 1993WAN**
Characterization: M_η/kg.mol^{-1} = 158, Aldrich Chem. Co., Inc., Milwaukee, WI
Solvent (A): **carbon dioxide** **CO$_2$** **124-38-9**
Solvent (C): **benzene** **C$_6$H$_6$** **71-43-2**

Type of data: weight fraction-based partition coefficient at infinite dilution, $K^\infty = w_C{}^{gas}/w_C{}^{liq}$

T/K = 313.2

P/MPa	2.04	3.01	4.33	4.95	5.93	6.87	7.85
K^∞	0.14	0.12	0.11	0.12	0.14	0.16	0.25

continued

continued

$T/K = 333.2$

P/MPa	1.88	2.72	3.81	4.81	5.01	6.16	7.18	7.95
K^∞	0.38	0.29	0.23	0.21	0.22	0.23	0.25	0.29

Comments: The weight fraction-based partition coefficients of benzene at infinite dilution are given for the equilibrated system of CO_2 + polymer at binary system equilibrium temperature, pressure, and concentration (please see Chapter 2.1. for the binary data).

Polymer (B):	**poly(vinyl acetate)**		**1991WAN, 1993WAN**
Characterization:	M_η/kg.mol^{-1} = 158, Aldrich Chem. Co., Inc., Milwaukee, WI		
Solvent (A):	**carbon dioxide**	**CO_2**	**124-38-9**
Solvent (C):	**toluene**	**C_7H_8**	**108-88-3**

Type of data: weight fraction-based partition coefficient at infinite dilution, $K^\infty = w_C^{gas}/w_C^{liq}$

$T/K = 313.2$

P/MPa	1.97	2.86	2.92	2.95	3.91	4.68	5.79	6.85	7.73
K^∞	0.099	0.072	0.078	0.075	0.066	0.071	0.079	0.11	0.18

$T/K = 333.2$

P/MPa	2.03	3.04	3.98	4.02	5.00	5.98	6.96	6.97	7.97
K^∞	0.19	0.15	0.13	0.14	0.14	0.15	0.17	0.17	0.20

Comments: The weight fraction-based partition coefficients of toluene at infinite dilution are given for the equilibrated system of CO_2 + polymer at binary system equilibrium temperature, pressure, and concentration (please see Chapter 2.1. for the binary data).

Polymer (B):	**starch**		**1996SIN**
Characterization:	degermed yellow cornmeal, Lauhoff Grain Co., Danville, IL		
Solvent (A):	**carbon dioxide**	**CO_2**	**124-38-9**
Solvent (C):	**water**	**H_2O**	**7732-18-5**

Type of data: gas solubility

$T/K = 343.15$

w_A	0.0037	0.0086	0.0121	0.0162	0.0037	0.0080	0.0131	0.0171	0.0035
w_B	0.6057	0.6014	0.6006	0.5982	0.6057	0.6031	0.6000	0.5975	0.6378
w_C	0.3906	0.3900	0.3873	0.3856	0.3906	0.3889	0.3869	0.3854	0.3587
P/bar	26.39	58.41	92.11	118.31	25.41	58.76	92.82	117.58	22.92

w_A	0.0074	0.0117	0.0152	0.0037	0.0080	0.0127	0.0171	0.0035	0.0075
w_B	0.6353	0.6325	0.6303	0.6376	0.6349	0.6319	0.6291	0.6527	0.6501
w_C	0.3573	0.3558	0.3545	0.3587	0.3571	0.3554	0.3538	0.3438	0.3424
P/bar	55.78	90.56	114.24	22.00	56.62	92.67	117.99	24.09	57.61

w_A	0.0118	0.0157	0.0035	0.0076	0.0121	0.0167
w_B	0.6473	0.6447	0.6527	0.6500	0.6471	0.6441
w_C	0.3409	0.3396	0.3438	0.3424	0.3408	0.3392
P/bar	91.14	116.85	23.07	58.27	93.88	116.82

2.4. Table of ternary or quaternary systems where data were published only in graphical form as phase diagrams or related figures

Polymer (B)	Second/third/fourth component	Ref.
Ethylene/propylene copolymer		
	ethene and propene	1994YOO
Polycarbonate bisphenol-A		
	carbon dioxide and polystyrene	1997KAT
Polycarbonate bisphenol chloral		
	carbon dioxide and poly(methyl methacrylate)	1990RA2
	methane and poly(methyl methacrylate)	1990RA2
Poly(cyanopropylmethylsiloxane)		
	carbon dioxide and methanol	1999BRA
	carbon dioxide and 2-propanol	1999BRA
Poly(2,6-dimethyl-1,4-phenylene oxide)		
	carbon dioxide and polystyrene	1982MOR
	carbon dioxide and polystyrene	1996CON
Poly(dimethylsiloxane)		
	carbon dioxide and methanol	1998VIN
	carbon dioxide and methanol	1998WES
	carbon dioxide and methanol	1999BRA
	carbon dioxide and 2-propanol	1998VIN
	carbon dioxide and 2-propanol	1999BRA
	carbon dioxide and 2-propanone	1998VIN
	carbon dioxide and 2-propanone	1998WES
	carbon dioxide and toluene	1998WES
Polyethylene		
	ethene and 4-methyl-1-pentene	1992WOH
	n-hexane and nitrogen	1990KEN

Polymer (B)	Second/third/fourth component	Ref.
Poly(ethylene oxide)-b-poly(propylene oxide)-b-poly(ethylene oxide) triblock copolymer		
	carbon dioxide and p-xylene	2002ZHA
Poly(methyl methacrylate)		
	carbon dioxide and bisphenol chloral polycarbonate	1990RA2
	methane and bisphenol chloral polycarbonate	1990RA2
	carbon dioxide and water	1997VIN
	carbon dioxide and water	1998KAZ
Polypropylene		
	ethene and propene	1994HOR
Polystyrene		
	benzene and carbon dioxide	1989SAS
	benzene and carbon dioxide	1990SAS
	carbon dioxide and polycarbonate bisphenol-A	1997KAT
	carbon dioxide and poly(2,6-dimethyl-1,4-phenylene oxide)	1982MOR
	carbon dioxide and poly(2,6-dimethyl-1,4-phenylene oxide)	1996CON
	carbon dioxide and poly(vinyl methyl ether)	1989HAC
	carbon dioxide and poly(vinyl methyl ether)	1998MOK
	carbon dioxide and pyrene	2004CAO
Poly(vinyl acetate)		
	benzene and carbon dioxide	1988MAS
	benzene and carbon dioxide	1989SAS
	benzene and carbon dioxide	1990SAS
	benzene and carbon dioxide	1992HAT
	carbon dioxide and toluene	1992HAT
Poly(vinyl methyl ether)		
	carbon dioxide and polystyrene	1989HAC
	carbon dioxide and polystyrene	1998MOK

2.5. References

1948NEW   Newitt, D.M. and Weale, K.E., Solution and diffusion of gases in polystyrene at high pressures, *J. Chem. Soc. London*, 1541, 1948.

1962LUN   Lundberg, J.L., Wilk, M.B., and Huyett, M.J., Estimation of diffusivities and solubilities from sorption studies, *J. Polym. Sci.*, 57, 275, 1962.

1963LUN   Lundberg, J.L., Wilk, M.B., and Huyett, M.J., Sorption studies using automation and computation, *Ind. Eng. Chem. Fundam.*, 2, 37, 1963.

1964VIE   Vieth, W.R., Alcalay, H.H., and Frabetti, A.J., Solution of gases in oriented poly(ethylene terephthalate), *J. Appl. Polym. Sci.*, 8, 2125, 1964.

1969LUN   Lundberg, J.L., Mooney, E.J., and Rogers, C.E., Diffusion and solubility of methane in polyisobutylene, *J. Polym. Sci.: Part A-2*, 7, 947, 1969.

1973BRO   Brockmeier, N.F., Carlson, R.E., and McCoy, R.W., Gas chromatographic determination of thermodynamic properties of polymer solutions at high pressure, *AIChE-J.*, 19, 1133, 1973.

1974DOS   DosSantos, M.L., Correa, N.F., and Leitao, D.M., Interaction of polyethylene with isobutane, isobutylene, 1-butene and normal butane, *J. Coll. Interface Sci.*, 47, 621, 1974.

1975BON   Bonner, D.C. and Cheng, Y., A new method for determination of equilibrium sorption of gases by polymers at elevated temperatures and pressures, *J. Polym. Sci.: Polym. Lett. Ed.*, 13, 259, 1975.

1975HOR   Horacek, H., Gleichgewichtsdrücke, Löslichkeit und Mischbarkeit des Systems Polyethylen niedriger Dichte und Kohlenwasserstoffen bzw. halogenierten Kohlenwasserstoffen, *Makromol. Chem., Suppl.*, 1, 415, 1975.

1976KOR   Koros, W.J., Paul, D.R., and Rocha, A.A., Carbon dioxide sorption and transport in polycarbonate, *J. Polym. Sci.: Polym. Phys. Ed.*, 14, 687, 1976.

1977BON   Bonner, D.G., Solubility of supercritical gases in polymers, *Polym. Eng. Sci.*, 17, 65, 1977.

1977CHE   Cheng, Y.L. and Bonner, D.C., Solubility of ethylene in liquid, low-density polyethylene to 69 atmospheres, *J. Polym. Sci.: Polym. Phys. Ed.*, 15, 593, 1977.

1978CHE   Cheng, Y.L. and Bonner, D.C., Solubility of nitrogen and ethylene in molten, low-density polyethylene to 69 atmospheres, *J. Polym. Sci.: Polym. Phys. Ed.*, 16, 319, 1978.

1978STE   Stern, S.A. and DeMeringo, A.H., Solubility of carbon dioxide in cellulose acetate at elevated pressure, *J. Polym. Sci.: Polym. Phys. Ed.*, 16, 735, 1978.

1981ALB   Albihn, P. and Kubat, J., Phase equilibria in the two-phase system polystyrene-liquid sulphur dioxide, *Brit. Polym. J.*, 13, 137, 1981.

1981KOR   Koros, W.J., Smith, G.N., and Stannett, V., High-pressure sorption of carbon dioxide in solvent-cast poly(methyl methacrylate) and poly(ethyl methacrylate) films, *J. Appl. Polym. Sci.*, 26, 159, 1981.

1981PAR   Parrish, WM.R., Solubility of isobutane in two high-density polyethylene polymer fluffs, *J. Appl. Polym. Sci.*, 26, 2279, 1981.

1982MOR   Morel, G. and Paul, D.R., CO_2 sorption and transport in miscible poly(phenylene oxide)/polystyrene blends, *J. Membrane Sci.*, 10, 273, 1982.

1982TOI   Toit, K. and Paul, D.R., Effect of polystyrene molecular weight on the carbon dioxide sorption isotherm, *Macromolecules*, 15, 1104, 1982.

1983MEY   Meyer, J.A. and Blanks, R.F., Solubility of isobutane and propane in polyethylene at high temperatures and low pressures, *J. Appl. Polym. Sci.*, 28, 725, 1983.

1984FON   Fonin, M.F., Saltanova, V.B., Anishchuk, V.V., Evdokimova, T.N., Shvyd'ko, T.I., and Gordeev, V.K., Solubility of ethylene and its diffusion in freshly prepared high-pressure polyethylene (Russ.), *Plast.Massy*, 7, 16, 1984.

1984KOB   Kobyakov, V.M. and Zernov, V.S., Solubility of ethylene in polyethylene at high pressures and temperatures (Russ.), in *Sintez, Svoistva, Pererab. Poliolefinov*, 1984, 60.

1985ROU   Rousseaux, P., Richon, D., and Renon, H., Ethylene-polyethylene mixtures, saturated liquid densities and bubble pressures up to 26.1 MPa and 493.1 K, *J. Polym. Sci.: Polym. Chem. Ed.*, 23, 1771 1985.

1986CAR   Carfagna, C., Nicodemo, L., Nicolais, L., and Campanile, G., CO_2 sorption in uniaxially drawn atactic polystyrene, *J.Polym.Sci.: Part B: Polym.Phys.*, 24, 1805, 1986.

1986FLE   Fleming, G.K. and Koros, W.J., Dilation of polymers by sorption of carbon dioxide at elevated pressures. 1. Silicone rubber and unconditioned polycarbonate, *Macromolecules*, 19, 2285, 1986.

1986GOR   Gorski, R.A., Ramsey, R.B., and Dishart, K.T., Physical properties of blowing agent polymer systems - I. Solubility of fluorocarbon blowing agents in thermoplastic resins, *J. Cell. Plast.*, 22, 21, 1986.

1986HIR   Hirose, T., Mizoguchi, K., and Kamiya, Y., Dilation of polyethylene by sorption of carbon dioxide, *J. Polym. Sci.: Part B: Polym. Phys.*, 24, 2107, 1986.

1986KA1   Kamiya, Y., Mizoguchi, K., Naito, Y., and Hirose, T., Gas sorption in poly(vinyl benzoate), *J. Polym. Sci.: Part B: Polym. Phys.*, 24, 535, 1986.

1986KA2   Kamiya, Y., Hirose, T., Mizoguchi, K., and Naito, Y., Gravimetric study of high-pressure sorption of gases in polymers, *J. Polym. Sci.: Part B: Polym. Phys.*, 24, 1525, 1986.

1986KOL   Kolmacka, J. and Smilek, P., Methods and problems in the measurement of gas solubility in polymer melts II. (Czech.), *Plasty Kaucuk*, 23, 13, 1986.

1986SHA   Shah, V.M., Hardy, B.J., and Stern, S.A., Solubility of carbon dioxide, methane, and propane in silicone polymers. Effect of polymer side chains, *J. Polym. Sci.: Part B: Polym. Phys.*, 24, 2033, 1986.

1987CAS   Castro, E.F., Gonzo, E.E., and Gottifredi, J.C., Thermodynamics of the absorption of hydrocarbon vapors in polyethylene films, *J. Membrane Sci.*, 31, 235, 1987.

1987GOR   Goradia, U.B. and Spencer, H.G., Sorption of propane in poly(ethyl methacrylate) near T_g, *J. Appl. Polym. Sci.*, 33, 1525, 1987.

1987HEU   Heuer, T., Untersuchungen zur Löslichkeit von Gasen in Polymerschmelzen, *Dissertation*, TH Leuna-Merseburg, 1987.

1987KOB   Kobyakov, V.M., Kogan, V.B., and Zernov, V.S., Metod otsenki dannykh o parozhidkostnom ravnovesii pri vysokikh davleniyakh, *Zh. Prikl. Khim.*, 60, 81, 1987.

1987KOL   Kolmacka, J. and Smilek, P., Solubility of gases in a polyethylene melt and its importance for the manufacture of plastic foams by a physical method, *Plasty Kaucuk*, 24, 65, 1987.

1987SAD   Sada, E., Kumazawa, H., Yakushij, H., Bamba, Y., Sakata, K., and Wang, S.-T., Sorption and diffusion of gases in glassy polymers, *Ind. Eng. Chem. Res.*, 26, 433, 1987.

1987WIS   Wissinger, R.G. and Paulaitis, M.E., Swelling and sorption in polymer-CO_2 mixtures at elevated pressures, *J. Polym. Sci.: Part B: Polym. Phys.*, 25, 2497, 1987.

1988HAC   Hachisuka, H., Tsujita, Y., Takizawa, A., and Kinoshita, T., CO_2 sorption properties and enthalpy relaxation in alternating copoly(vinylidene cyanide-vinyl acetate)s, *Polymer*, 29, 2050, 1988.

1988MAS   Masuoka, H., Takishima, S., and Wang, N.-H., Supercritical fluid extraction of high-boiling materials from polymers, *Rep. Asahi Glass Found. Ind. Technol.*, 52, 275, 1988.

1988SAD   Sada, E., Kumazawa, H., and Xu, P., Sorption and diffusion of carbon dioxide in polyimide films, *J. Appl. Polym. Sci.*, 35, 1497, 1988.

1988THU   Thuy, L.P. and Springer, J., Carbon dioxide sorption in poly(butylene terephthalate), *Colloid Polym. Sci.*, 266, 614, 1988.

1989BER   Berens, A.R. and Huvard, G.S., Interaction of polymers with near-critical carbon dioxide, *ACS Symp. Ser.*, 406, 207, 1989.

1989CHI   Chiou, J.S. and Paul, D.R., Gas sorption and permeation in poly(ethyl methacrylate), *J. Membrane Sci.*, 45, 167, 1989.

1989HAC   Hachisuka, H., Sato, T., Tsujita, Y., Takizawa, A., and Kinoshita, T., CO_2 gas sorption properties in single-phase and phase-separated polystyrene/poly(vinyl methyl ether) blends, *Polym. J.*, 21, 417, 1989.

1989KA1   Kamiya, Y., Mizoguchi, K., Hirose, T., and Naito, Y., Sorption and dilation in poly(ethyl methacrylate)-carbon dioxide system, *J. Polym. Sci.: Part B: Polym. Phys.*, 27, 879, 1989.

1989KA2   Kamiya, Y., Naito, Y., and Mizoguchi, K., Sorption and partial molar volume of gases in polybutadiene, *J. Polym. Sci.: Part B: Polym. Phys.*, 27, 2243, 1989.

1989SAS   Sasaki, M., Takishima, S., and Masuoka, H., Supercritical carbon dioxide extraction of benzene in poly(vinyl acetate) and polystyrene, *Sekiyu Gakkaishi*, 32, 67, 1989.

1989SHI   Shim, J.-J. and Johnston, K.P., Adjustable solute distribution between polymers and super-critical fluids, *AIChE-J.*, 35, 1097, 1989.

1990HAT   Hattori, K., Wang, N., Takishima, S., Masioka, H., Chromatographic measurement of vapor-liquid equilibrium ratios of organic solvents in supercritical CO_2 + poly(vinyl acetate) system, in *Solvent Extraction 1990*, Ed. T. Sekine, Elsevier Sci. Publ., 1990, 1671.

1990KEN   Kennis, H.A.J., Loos, Th.W. de, DeSwaan Arons, J., Van der Haegen, R., and Kleintjens, L.A., The influence of nitrogen on the liquid-liquid phase behaviour of the system n-hexane-polyethylene: experimental results and predictions with the mean-field lattice-gas model, (experimental data by Th.W. de Loos), *Chem. Eng. Sci.*, 45, 1875, 1990.

1990KUM   Kumar, V. and Suh, N.P., A process for making microcellular thermoplastic parts, *Polym. Eng. Sci.*, 30, 1323, 1990.

1990RA1   Raymond, P.C. and Paul, D.R., Sorption and transport of pure gases in random styrene/methyl methacrylate copolymers, *J. Polym. Sci.: Part B: Polym. Phys.*, 28, 2079, 1990.

1990RA2   Raymond, P.C. and Paul, D.R., Sorption and transport of pure CO_2 and CH_4 in miscible bisphenol chloral polycarbonate/poly(methyl methacrylate) blends, *J. Polym. Sci.: Part B: Polym. Phys.*, 28, 2103, 1990.

1990SAS   Sasaki, M., Takishima, S., and Masuoka, H., Supercritical carbon dioxide extraction of benzene in poly(vinyl acetate) and polystyrene (Part 2), *Sekiyu Gakkaishi*, 33, 304, 1990.

1990TAK   Takashima, S., Nakamura, K., Sasaki, M., and Masuoka, H., Dilatation and solubility in carbon dioxide + poly(vinyl acetate) system at high pressures, *Sekiyu Gakkaishi*, 33, 332, 1990.

1990WAN   Wang, N.-H., Takishima, S., and Masuoka, H., Solubility measurements of gas in polymer by piezoelectric-quartz sorption method at high pressures and its correlation, *Kogaku Ronbunshu*, 16, 931, 1990.

1991BRI   Briscoe, B.J. and Zakaria, S., Interaction of CO_2 gas with silicone elastomer at high ambient pressures, *J. Polym. Sci.: Part B: Polym. Phys.*, 29, 989, 1991.

1991FIN   Finck, U., Phasengleichgewichte in Monomer + Copolymer - Systemen, *Diploma Paper*, TH Leuna-Merseburg, 1991.

1991LAM   Lambert, S.M. and Paulaitis, M.E., Crystallization of poly(ethylene terephthalate) induced by carbon dioxide sorption at elevated pressures, *J. Supercrit. Fluids*, 4, 15, 1991.

1991POP   Pope, D.S., Sanchez, I.C., Koros, W.J., and Fleming, G.K., Statistical thermodynamic interpretation of sorption/dilation behavior of gases in silicone rubber, *Macromolecules*, 24, 1779, 1991.

1991SCH   Schultze, J.D., Zhou, Z., and Springer, J., Gas sorption in poly(butylene terephthalate). A critical test of the influence of molecular gas data, *Angew. Makromol. Chem.*, 185-186, 265, 1991.

1991WAN   Wang, N.H., Hattori, K., Takashima, S., and Masuoka, H., Measurement and prediction of vapor-liquid equilibrium solutes at infinite dilution in carbon dioxide + poly(vinyl acetate) system at high pressures, *Kagaku Kogaku Ronbunshu*, 17, 1138, 1991.

1991WI1   Wissinger, R.G. and Paulaitis, M.E., Glass transitions in polymer/CO_2 mixtures at elevated pressures, *J. Polym. Sci.: Part B: Polym. Phys.*, 29, 631, 1991.

1991WI2   Wissinger, R.G. and Paulaitis, M.E., Molecular thermodynamic model for sorption and swelling in glassy polymer-CO_2 systems at elevated pressures, *Ind. Eng. Chem. Res.*, 29, 842, 1991.

1992CO1   Condo, P.D. and Johnston, K.P., Retrograde vitrification of polymers with compressed fluid diluents: experimental confirmation, *Macromolecules*, 25, 6730, 1992.

1992CO2   Condo, P.D., Sanchez, I.C., Panayiotou, C.G., and Johnston, K.P., Glass transition behavior including retrograde vitrification of polymers with compressed fluid diluents, *Macromolecules*, 25, 6119, 1991.

1992HAT   Hattori, K., Wang, N., Takishima, S., and Masuoka, H., Chromatographic measurement of vapor-liquid equilibrium ratios of organic solvents in supercritical CO_2 + poly(vinyl acetate) system, in *Solvent Extraction 1990*, Ed. T. Sekine, Elsevier Sci. Publ., 1992, 1671.

1992WAN   Wang, N.-H., Ishida, S., Takishima, S., and Masuoka, H., Sorption measurement of nitrogen in polystyrene at high pressure and its correlation by a modified dual-sorption model, *Kagaku Kogaku Ronbunshu*, 18, 226, 1992.

1992WOH   Wohlfarth, C., Finck, U., Schultz, R., and Heuer, T., Investigation of phase equilibria in mixtures composed of ethene, 1-butene, 4-methyl-1-pentene and a polyethylene wax, *Angew. Makromol. Chem.*, 198, 91, 1992.

1993CON   Conforti, R.M. and Barbari, T.A., A thermodynamic analysis of gas sorption-desorption hystereses in glassy polymers, *Macromolecules*, 26, 5209, 1993.

1993SHA   Shah, V.M., Hardy, B.J., and Stern, S.A., Solubility of carbon dioxide, methane, and propane in silicone polymers. Effect of polymer backbone chains, *J. Polym. Sci.: Part B: Polym. Phys.*, 31, 313, 1993.

1993WAN   Wang, N. and Masuoka, H., Measurement and prediction of vapor-liquid equilibrium ratios for solutes at infinite dilution in CO_2 + polyvinylacetate system at high pressures, *Dalian Ligong Daxue Xuebao*, 33, Suppl. 2, 175, 1993.

1994CON   Condo, P.D. and Johnston, K.P., *In situ* measurement of the glass transition temperature of polymers with compressed fluids, *J. Polym. Sci.: Part B: Polym. Phys.*, 32, 523, 1994.

1994GAR   Garg, A., Gulari, E., and Manke, C.W., Thermodynamics of polymer melts swollen with supercritical gases, *Macromolecules*, 27, 5643, 1994.

1994GOE   Goel, S.K. and Beckman, E.J., Generation of microcellular polymeric foams using supercritical carbon dioxide. I. Effect of pressure and temperature on nucleation, *Polym. Eng. Sci.*, 34, 1137, 1994.

1994HAN   Handa, Y.P., Lampron, S., and O'Neill, M.L., On the plasticization of poly(2,6-dimethyl phenylene oxide) by CO_2, *J. Polym. Sci.: Part B: Polym. Phys.*, 32, 2549, 1994.

1994HOR   Horacek, H., Extrapolation of solubility data of monomeric, oligomeric and polymeric hydrocarbons, *Macromol. Chem. Phys.*, 195, 3381, 1994.

1994KAM   Kamiya, Y., Naito, Y., and Borubon, D., Sorption and partial molar volumes of gases in poly(ethylene-*co*-vinyl acetate), *J. Polym. Sci.: Part B: Polym. Phys.*, 32, 281, 1994.

1994POP   Pope, D.S., Koros, W.J., and Hopfenberg, H.B., Sorption and dilation of poly(1-trimethyl-1-propyne) by carbon dioxide and methane, *Macromolecules*, 27, 5839, 1994.

1994WAN   Wang, N.-H., Takishima, S., and Masuoka, H., Measurement and correlation of solubility of a high pressure gas in a polymer by piezoelectric quartz sorption: CO_2 + PVAc and CO_2 + PBMA systems, *Int. Chem. Eng.*, 34, 255, 1994.

1994YOO   Yoon, J.S., Chung, C.Y., and Lee, I.H., Solubility and diffusion coefficient of gaseous ethylene and α-olefin in ethylene/α-olefin random copolymers, *Eur. Polym. J.*, 30, 1209, 1994.

1995GAI   Gainar, I. and Anitescu, G., The solubility of CO_2, N_2, and H_2 in a mixture of dimethyl ether polyethylene glycols at high pressures, *Fluid Phase Equil.*, 109, 281, 1995.

1995LEE   Lee, J.G. and Flumerfelt, W., Nitrogen solubilities in low-density polyethylene at high temperatures and high pressures, *J. Appl. Polym. Sci.*, 58, 2213, 1995.

1995LOO   Loos, Th.W. de, Poot, W., and Lichtenthaler, R.N., The influence of branching on high-pressure vapor-liquid equilibria in systems of ethylene and polyethylene (experimental data by Th.W. de Loos), *J. Supercrit. Fluids*, 8, 282, 1995.

1996CON   Conforti, R.M., Barbari, T.A., and Pozo de Fernandes, M.E., Enthalpy of mixing for a glassy polymer blend from CO_2 sorption and dilation measurements, *Macromolecules*, 29, 6629, 1996.

1996HAN   Handa, Y.P., Kruus, P., and O'Neill, M., High-pressure calorimetric study of plastification of poly(methyl methacrylate) by methane, ethylene, and carbon dioxide, *J. Polym. Sci.: Part B: Polym. Phys.*, 34, 2635, 1996.

1996SAT   Sato, Y., Yurugi, M., Fujiwara, K., Takishima, S., and Masuoka, H., Solubilities of carbon dioxide and nitrogen in polystyrene under high temperature and pressure, *Fluid Phase Equil.*, 125, 129, 1996.

1996SH1   Shieh, Y.-T., Su, J.-H., Manivannan, G., Lee, P.H.C., Sawan, S.P., and Spall, W.D., Interaction of supercritical carbon dioxide with polymers. I. Crystalline polymers, *J. Appl. Polym. Sci.*, 59, 695, 1996.

1996SH2   Shieh, Y.-T., Su, J.-H., Manivannan, G., Lee, P.H.C., Sawan, S.P., and Spall, W.D. Interaction of supercritical carbon dioxide with polymers. II. Amorphous polymers, *J. Appl. Polym. Sci.*, 59, 707, 1996.

1996SIN   Singh, B., Rizvi, S.S.H., and Harriott, P., Measurement of diffusivity and solubility of carbon dioxide in gelatinized starch at elevated pressures, *Ind. Eng. Chem. Res.*, 35, 4457, 1996.

1996WAN   Wang, J.-S., Naito, Y., and Kamiya, Y., Effect of the penetrant-induced isothermal glass transition on sorption, dilation, and diffusion behavior of polybenzylmethacrylate/CO$_2$, *J. Polym. Sci.: Part B: Polym. Phys.*, 34, 2027, 1996.

1997END   Enders, S. and Loos, Th.W. de, Pressure dependence of the phase behaviour of polystyrene in methylcyclohexane (experimental data by S. Enders), *Fluid Phase Equil.*, 139, 335, 1997.

1997KAT   Kato, S., Tsujita, Y., Yoshimizu, H., Kinoshita, T., and Higgins, J.S., Characterization and CO$_2$ sorption behaviour of polystyrene/polycarbonate blend system, *Polymer*, 38, 2807, 1997.

1997KRY   Krykin, M.A., Bondar, V.I., Kukharsky, Yu.M., and Tarasov, A.V., Gas sorption and diffusion processes in polymer matrices at high pressures, *J. Polym. Sci.: Part B: Polym. Phys.*, 35, 1339, 1997.

1997SUR   Surana, R.K., Danner, R.P., DeHaan, A.B., and Beckers, N., New technique to measure high-pressure and high-temperature polymer-solvent vapor-liquid equilibrium, *Fluid Phase Equil.*, 139, 361, 1997.

1997VIN   Vincent, M.F., Kazarian, S.G., and Eckert, C.A., Tunable diffusion of D$_2$O in CO$_2$-swollen poly(methyl methacrylate) films, *AIChE-J.*, 43, 1838, 1997.

1997WEI   Weidner, E., Wiesmet, V., Knez, Z., and Skerget, M., Phase equilibrium (solid-liquid-gas) in polyethyleneglycol-carbon dioxide systems, *J. Supercrit. Fluids*, 10, 139, 1997.

1997ZHA   Zhang, Y., Gangwani, K.K., and Lemert, R.M., Sorption and swelling of block copolymers in the presence of supercritical carbon dioxide, *J. Supercrit. Fluids*, 11, 115, 1997.

1998AUB   Aubert, J.H., Solubility of carbon dioxide in polymers by the quartz crystal microbalance technique, *J. Supercrit. Fluids*, 11, 163, 1998.

1998BOE   Böhning, M. and Springer, J., Sorptive dilation and relaxational processes in glassy polymer/gas systems - I. Poly(sulfone) and poly(ether sulfone), *Polymer*, 39, 5183, 1998.

1998EDW   Edwards, R.R., Tao, Y., Xu, S., Wells, P.S., Yun, K.S., and Parcher, J.F., Chromatographic investigation of CO$_2$-polymer interactions at near-critical conditions, *J. Phys. Chem. B*, 102, 1287, 1998.

1998HER   Herraiz, J., Shen, S., and Coronas, A., Vapor-liquid equilibria for methanol + poly(ethylene glycol)-250 dimethyl ether, *J. Chem. Eng. Data*, 43, 191, 1998.

1998HON   Hong, S.-U. and Pretel, E.J., Predicting the solubility of 1,1-difluoroethane in polystyrene using the perturbed soft chain theory, *Polym. Int.*, 45, 55, 1998.

1998KAM   Kamiya, Y., Miziguchi, K., Terada, K., Fujiwara, Y., and Wang, J.-S., CO$_2$ sorption and dilation of poly(methyl methacrylate), *Macromolecules*, 31, 472, 1998.

1998KAZ   Kazarian, S.G., Vincent, M.F., and Eckert, C.A., Partitioning of solutes and cosolvents between supercritical CO$_2$ and polymer phases, *J. Supercrit. Fluids*, 13, 107, 1998.

1998MOK   Mokdad, A., Dubault, A., and Monnerie, L., Sorption and diffusion of carbon dioxide in single-phase polystyrene/poly(vinyl methyl ether) blends, *J. Polym. Sci.: Part B: Polym. Phys.*, 34, 2723, 1998.

1998VIN   Vincent, M.F., Kazarian, S.G., West, B.L., Berkner, J.A., Bright, F.V., Liotta, C.L., and Eckert, C.A., Cosolvent effects of modified supercritical carbon dioxide on cross-linked poly(dimethylsiloxane), *J. Phys. Chem. B*, 102, 2176, 1998.

1998WES   West, B.L., Bush, D., Brantley, N.H., Vincent, M.F., Kazarian, S.G., and Eckert, C.A., Modeling the effects of cosolvent-modified supercritical fluids on polymers with a lattice fluid equation of state, *Ind. Eng. Chem. Res.*, 37, 3305, 1998.

1998WON   Wong, B., Zhang, Z., and Handa, Y.P., High-precision gravimetric technique for determining the solubility and diffusivity of gases in polymers, *J. Polym. Sci.: Part B: Polym. Phys.*, 36, 2025, 1998.

1998ZHA   Zhang, Z. and Handa, Y.P., An *in situ* study of plasticization of polymers by high-pressure gases, *J. Polym. Sci.: Part B: Polym. Phys.*, 36, 977, 1998.

1999ANG   Angelis, M.G.de, Merkel, T.C., Bondar, V.I., Freeman, B.D., Doghieri, F., and Sartim G.C., Hydrocarbon and fluorocarbon solubility and dilation in poly(dimethylsiloxane): comparison of experimental data with predictions of the Sanchez-Lacombe equation of state, *J. Polym. Sci.: Part B: Polym. Phys.*, 37, 3011, 1999.

1999BON   Bondar, V.I., Freeman, B.D., and Yampolskii, Yu.P., Sorption of gases and vapors in an amorphous glassy perfluorodioxole copolymer, *Macromolecules*, 32, 6163, 1999.

1999BRA   Brantley, N.H., Bush, D., Kazarian, S.G., and Eckert, C.A., Spectroscopic measurement of solute and cosolvent partitioning between supercritical CO_2 and polymers, *J. Phys. Chem. B*, 103, 10007, 1999.

1999INO   Inomata, H., Honma, Y., Imahori, M., and Arai, K., Fundamental study of de-solventing polymer solutions with supercritical CO_2, *Fluid Phase Equil.*, 158-160, 857, 1999.

1999KIK   Kikic, I., Lora, M., Cortesi, A., and Sist, P., Sorption of CO_2 in biocompatible polymers: experimental data and qualitative interpretation, *Fluid Phase Equil.*, 158-160, 913, 1999.

1999HON   Hong, S.U., Albouy, A., and Duda, J.L., Measurement and prediction of blowing agent solubility in polystyrene at supercritical conditions, *Cell. Polym.*, 18, 301, 1999.

1999KEL   Keller, J.U., Rave, H., and Staudt, R., Measurement of gas absorption in a swelling polymeric material by a combined gravimetric-dynamic method, *Macromol. Chem. Phys.*, 200, 2269, 1999.

1999ROY   Royer, J.R., DeSimone, J.M., and Khan, S.A., Carbon dioxide-induced swelling of poly(dimethylsiloxane), *Macromolecules*, 32, 8965, 1999.

1999SAT   Sato, Y., Fujiwara, K., Takikawa, T., Takishima, S., and Masuoka, H., Solubilities and diffusion coefficients of carbon dioxide and nitrogen in polypropylene, high-density polyethylene, and polystyrene under high pressures and temperatures, *Fluid Phase Equil.*, 162, 261, 1999.

1999WEB   Webb, K.F. and Teja, A.S., Solubility and diffusion of carbon dioxide in polymers, *Fluid Phase Equil.*, 158-160, 1029, 1999.

2000BRA   Brantley, N.H., Kazarian, S.G., and Eckert, C.A., *In situ* FTIR measurement of carbon dioxide sorption into poly(ethylene terephthalate) at elevated pressures, *J. Appl. Polym. Sci.*, 77, 764, 2000.

2000BY1   Byun, H.-S. and McHugh, M.A., Impact of "free" monomer concentration on the phase behavior of supercritical carbon dioxide-polymer mixtures, *Ind. Eng. Chem. Res.*, 39, 4658, 2000.

2000BY2   Byun, H.-S. and Choi, T.-H., Effect of monomer concentration on the phase behavior of supercritical carbon dioxide-poly(ethyl methacrylate) mixture, *J. Korean Ind. Eng. Chem.*, 11, 396, 2000.

2000FEN   Feng, W., Wen, H., Xu, Z., and Wang, W., Comparison of perturbed hard-sphere-chain theory with statistical associating fluid theory for square-well fluids, *Ind. Eng. Chem. Res.*, 39, 2559, 2000.

2000HIL   Hilic, S., Padua, A.A.H., and Grolier, J.-P.E., Simultaneous measurement of the solubility of gases in polymers and of the associated volume change, *Rev. Sci. Instrum.*, 71, 4236, 2000.

2000OLI   Oliveira, J.V., Dariva, C., and Pinto, J.C., High-pressure phase equilibria for polypropylene-hydrocarbon systems, *Ind. Eng. Chem. Res.*, 39, 4627, 2000.

2000SA1   Sato, Y., Takikawa, T., Sorakubo, A., Takishima, S., Masuoka, H., and Imaizumi, M., Solubility and diffusion coefficient of carbon dioxide in biodegradable polymers, *Ind. Eng. Chem. Res.*, 39, 4813, 2000.

2000SA2   Sato, Y., Tsuboi, A., Sorakubo, A., Takishima, S., Masuoka, H., and Ishikawa, T., Vapor-liquid equilibrium ratios for hexane at infinite dilution in ethylene + impact polypropylene copolymer and propylene + impact polypropylene copolymer, *Fluid Phase Equil.*, 170, 49, 2000.

2000SA3   Sato, Y., Iketani, T., Takishima, S., and Masuoka, H., Solubility of hydrofluorocarbon (HFC-134a, HFC-152a) and hydrochlorofluorocarbon (HCFC-142b) blowing agents in polystyrene, *Polym. Eng. Sci.*, 40, 1369, 2000.

2000WIE   Wiesmet, V., Weidner, E., Behme, S., Sadowski, G., and Arlt, W., Measurement and modelling of high-pressure phase equilibria in the systems polyethyleneglycol (PEG)-propane, PEG-nitrogen and PEG-carbon dioxide, *J. Supercrit. Fluids*, 17, 1, 2000.

2001HAT   Hata, K., The sorption properties of supercritical carbon dioxide in poly(ethylene glycol) and poly(ethylene oxide) with high-pressure thermogravimetry-differential thermal analysis (Jap.), *Kobunshi Ronbunshu*, 58, 714, 2001.

2001HIL   Hilic, S., Boyer, S.A.E., Padua, A.A.H., and Grolier, J.-P.E., Simultaneous measurement of the solubility of nitrogen and carbon dioxide in polystyrene and of the associated polymer swelling, *J. Polym. Sci.: Part B: Polym. Phys.*, 39, 2063, 2001.

2001MAR   Martinache, J.D., Royer, J.R., Siripurapu, S., Henon, F.E., Genzer, J., Khan, S.A., and Carbonell, R.G., Processing of polyamide-11 with supercritical carbon dioxide, *Ind. Eng. Chem. Res.*, 40, 5570, 2001.

2001MCC   McCabe, C., Galindo, A., Garcia-Lisbona, M.N., and Jackson, G., Examining the absorption (vapor-liquid equilibria) of short-chain hydrocarbons in low-density polyethylene with SAFT-VR approach, *Ind. Eng. Chem. Res.*, 40, 3835, 2001.

2001MOO   Moore, S.J. and Wanke, S.E., Solubility of ethylene, 1-butene and 1-hexene in polyethylenes (experimental data by S.E. Wanke), *Chem. Eng. Sci.*, 56, 4121, 2001.

2001MUT   Muth, O., Hirth, Th., and Vogel, O., Investigation of sorption and diffusion of supercritical carbon dioxide into poly(vinyl chloride), *J. Supercrit. Fluids*, 19, 299, 2001.

2001NAG   Naguib, H. E., Richards, E. V., Pop-Iliev, R., Xu, X., and Park, C. B., Measurement of the swelling of polypropylene melts with dissolved butane, *Annual Technical Conference - Society of Plastics Engineers*, 59th ,Vol. 2, 1537, 2001.

2001SA1   Sato, Y., Yurugi, M., Yamabiki, T., Takishima, S., and Masuoka, H., Solubility of propylene in semicrystalline polypropylene, *J. Appl. Polym. Sci.*, 79, 1134, 2001.

2001SA2   Sato, Y., Takikawa, T., Takishima, S., and Masuoka, H., Solubilities and diffusion coefficients of carbon dioxide in poly(vinyl acetate) and polystyrene, *J. Supercrit. Fluids*, 19, 187, 2001.

2001TOR   Tork, T., Measurement and calculation of phase equilibria in polyolefin/solvent systems (Ger.), *Dissertation*, TU Berlin, 2001.

2001SIR   Sirard, S.M., Green, P.F., and Johnston, K.P., Spectroscopic ellipsometry investigation of the swelling of poly(dimethylsiloxane) thin films with high pressure carbon dioxide, *J. Phys. Chem. B*, 105, 766, 2001.

2001TSU   Tsuboi, A., Kolar, P., Ishikawa, T., Kamiya, Y., and Masuoka, H., Sorption and partial molar volumes of C_2 and C_3 hydrocarbons in polypropylene copolymers, *J. Polym. Sci.: Part B: Polym. Phys.*, 39, 1255, 2001.

2002AR1   Areerat, S., Hayata, Y., Katsumoto, R., Kegaswaa, T., Engami, H., and Ohshima, M., Solubility of carbon dioxide in polyethylene/titanium dioxide composite under high pressure and temperature, *J. Appl. Polym. Sci.*, 86, 282, 2002.

2002AR2   Areerat, S., Nagata, T., and Ohshima, M., Measurement and prediction of LDPE/CO_2 solution viscosity, *Polym. Eng. Sci.*, 42, 2234, 2002.

2002DEA   DeAngelis, M.G., Merkel, T.C., Bondar, V.I., Freeman, B.D., Doghieri, F., and Sarti, G.C., Gas sorption and dilation in poly(2,2-bistrifluoromethyl-4,5-difluoro-1,3-dioxole-*co*-tetra-fluoroethylene): comparison of experimental data with predictions of the nonequilibrium lattice fluid model, *Macromolecules*, 35, 1276, 2002.

2002FLI   Flichy, N.M.B., Kazarian, S.G., Lawrence, C.J., and Briscoe, B.J., An ATR-IR study of poly(dimethylsiloxane) under high-pressure carbon dioxide: simultaneous measurement of sorption and swelling, *J. Phys. Chem. B*, 106, 754, 2002.

2002JOU   Joung, S.N., Park, J.-U., Kim, S.Y., and Yoo, K.-P., High-pressure phase behavior of polymer-solvent systems with addition of supercritical CO_2 at temperatures from 323.15 K to 503.15 K, *J. Chem. Eng. Data*, 47, 270, 2002.

2002LIU   Liu, J., Han, B., Wang, Z., Zhang, J., Li, G., and Yang, G., Solubility of Ls-36 and Ls-45 surfactants in supercritical CO_2 and loading water in the CO_2/water/surfactant system, *Langmuir*, 18, 3086, 2002.

2002NIK   Nikitin, L.N., Said-Galiyev, E.E., Vinokur, R.A., Khokhlov, A.R., Gallayamov, M.O., and Schaumburg, K., Poly(methyl methacrylate) and poly(butyl methacrylate) swelling in supercritical carbon dioxide, *Macromolecules*, 35, 934, 2002.

2002PFO   Pfohl, O., Riebesell, C., and Dohrn, R., Measurement and calculation of phase equilibria in the system n-pentane + poly(dimethylsiloxane) at 308.15-423.15 K, *Fluid Phase Equil.*, 202, 289, 2002.

2002SAT   Sato, Y., Takikawa, T., Yamane, M., Takishima, S., and Masuoka, H., Solubility of carbon dioxide in PPO and PPO/PS blends, *Fluid Phase Equil.*, 194-197, 847, 2002.

2002SHI   Shieh, Y.-T. and Lin, Y.-G., Equilibrium solubility of CO_2 in rubbery EVA over a wide pressure range: effects of carbonyl group content and crystallinity, *Polymer*, 43, 1849, 2002.

2002SIR   Sirard, S.M., Ziegler, K.J., Sanchez, I.C., Green, P.F., and Johnston, K.P., Anomalous properties of poly(methyl methacrylate) thin films in supercritical carbon dioxide, *Macromolecules*, 35, 1928, 2002.

2002ZHA   Zhang, R., Liu, J., He, J., Han, B., Zhang, X, Liu, Z., Jiang, T., and Hu, G., Compressed CO_2-assisted formation of reverse micelles of PEO-PPO-PEO copolymer, *Macromolecules*, 35, 7869, 2002.

2003ALE   Alessi, P., Cortesi, A., Kikic, I., and Vecchione, F., Plasticization of polymers with supercritical carbon dioxide: experimental determination of glass-transition temperatures, *J. Appl. Polym. Sci.*, 88, 2189, 2003.

2003COT   Cotugno, S., DiMaio, E., Ciardiello, C., Iannace, S., Mensitieri, G., and Nicolais, L., Sorption thermodynamics and mutual diffusivity of carbon dioxide in molten polycaprolactone, *Ind. Eng. Chem. Res.*, 42, 4398, 2003.

2003DIM   Dimos, V., Anastasiadis, A., Chasiotis, A., and Kiparissides, C., An experimental and theoretical investigation of solubility and diffusion of ethylene in semi-crystalline PE at elevated pressures and temperatures, *J. Appl. Polym. Sci.*, 87, 953, 2003.

2003KIK   Kikic, I., Vecchione, F., Alessi, P., Cortesi, A., Eva, F., and Elvassore, N., Polymer plasticization using supercritical carbon dioxide: experiment and modeling, *Ind. Eng. Chem. Res.*, 42, 3022, 2003.

2003KOG   Koga, T., Seo, Y.-S., Shin, K., Zhang, Y., Rafailovich, M.H., Sokolov, J.C., Chu, B., and Satija, S.K., The role of elasticity in the anomalous swelling of polymer thin films in density fluctuating supercritical fluids, *Macromolecules*, 36, 5236, 2003.

2003LEE   Lee, L.-S., Shih, R.-F., Ou, H.-J., and Lee, T.-S., Solubility of ethylene in mixtures of toluene, norbornene, and cyclic olefin copolymer at various temperatures and pressures, *Ind. Eng. Chem. Res.*, 42, 6977, 2003.

2003NIK   Nikitin, L.N., Gallayamov, M.O., Vinokur, R.A., Nikolaec, A.Yu., Said-Galiyev, E.E., and Khokhlov, A.R., Swelling and impregnation of polystyrene using supercritical carbon dioxide, *J. Supercrit. Fluids*, 26, 263, 2003.

2003PAL   Palamara, J.E., Davis, P.K., Suriyapraphadilok, U., Danner, R.P., Duda, J.L., Kitzhoffer, R.J., and Zielinski, J.M., A static sorption technique for vapor solubility measurements, *Ind. Eng. Chem. Res.*, 42, 1557, 2003.

2003RAM   Ramachandrarao, V.S., Vogt, B.D., Gupta, R.R., and Watkins, J.J., Effect of carbon dioxide sorption on the phase behavior of weakly interacting polymer mixtures: solvent-induced segregation in deuterated polybutadiene/polyisoprene blends, *J. Polym. Sci.: Part B: Polym. Phys.*, 41, 3114, 2003.

2003SEI   Seiler, M., Rolker, J., and Arlt, W., Phase behavior and thermodynamic phenomena of hyperbranched polymer solutions, *Macromolecules*, 36, 2085, 2003.

2003SHE   Shenoy, S.L., Fujiwara, T., and Wynne, K.J., Quantifying plasticization and melting behavior of poly(vinylidene fluoride) in supercritical CO_2 utilizing a linear variable differential transformer, *Macromolecules*, 36, 3380, 2003.

2003SHI   Shieh, Y.-T. and Liu, K.-H., The effect of carbonyl group on sorption of CO_2 in glassy polymers, *J. Supercrit. Fluids*, 25, 261, 2003.

2003WAN   Wang, S., Peng, C., Li, K., Wang, J., Shi, J., and Liu, H., Measurement of solubilities of CO_2 in polymers by quartz crystal micobalance (Chin.), *J. Chem. Ind. Eng. (China)*, 54, 141, 2003.

2003WI1   Wind, J.D., Sirard, S.M., Paul, D.R., Green, P.F., Johnston, K.P., and Koros, W.J.: Carbon dioxide-induced plasticization of polyimide membranes: pseudo-equilibrium relationships of diffusion, sorption, and swelling, *Macromolecules*, 36, 6433, 2003.

2003WI2   Wind, J.D., Sirard, S.M., Paul, D.R., Green, P.F., Johnston, K.P., and Koros, W.J.: Relaxation dynamics of CO_2 diffusion, sorption, and polymer swelling for plasticized polyimide membranes, *Macromolecules*, 36, 6442, 2003.

2003XUQ   Xu, Q., Chang, Y., He, J., Han, B., and Liu, Y., Supercritical CO_2-assisted synthesis of poly(acrylic acid)/nylon-6 and polystyrene/nylon-6 blends, *Polymer*, 44, 5449, 2003.

2004CAO   Cao, T., Johnston, K.P., and Webber, S.E., CO_2-Enhanced transport of small molecules in thin films: a fluorescence study, *Macromolecules*, 37, 1897, 2004.

2004CHE   Chen, X., Yasuda, K., Sato, Y., Takishima, S., and Masuoka, H., Measurement and correlation of phase equilibria of ethylene + n-hexane + metallocene polyethylene at temperatures between 373 and 473 K and at pressures up to 20 MPa, *Fluid Phase Equil.*, 215, 105, 2004.

2004DAV   Davis, P.K., Lundy, G.D., Palamara, J.E., Duda, J.L., and Danner, R.P., New pressure-decay techniques to study gas sorption and diffusion in polymers at elevated pressures, *Ind. Eng. Chem. Res.*, 43, 1537, 2004.

2004SOL   Solms, N. von, Nielsen, J.K., Hassager, O., Rubin, A., Dandekar, A.Y., Andersen, S.I., and Stenby, E.H., Direct measurement of gas solubilities in polymers with a high-pressure microbalance, *J. Appl. Polym. Sci.*, 91, 1476, 2004.

2004TAN   Tang, M., Du, T.-B., and Chen, Y.-P., Sorption and diffusion of supercritical carbon dioxide in polycarbonate, *J. Supercrit. Fluids*, 28, 207, 2004.

2004WEI   Weidner, E. and Wiesmet, V., Phase equilibrium (solid-liquid-gas) in binary systems of poly(ethylene glycol)s, poly(ethylene glycol) dimethyl ether with carbon dioxide, propane, and nitrogen, in *Thermodynamic Properties of Complex Fluid Mixtures*, Wiley-VCH, Deutsche Forschungsgemeinschaft, Ed. G. Maurer, 2004, 511.

3. LIQUID-LIQUID EQUILIBRIUM (LLE) DATA OF POLYMER SOLUTIONS AT ELEVATED PRESSURES

3.1. Cloud-point and/or coexistence curves of quasibinary solutions

Polymer (B):	ethylene/vinyl acetate copolymer				1999BEY

Characterization: M_n/kg.mol^{-1} = 74, M_w/kg.mol^{-1} = 285, 70.0 wt% vinyl acetate, Scientific Polymer Products, Inc., Ontario, NY

Solvent (A):	cyclopentane	C$_5$H$_{10}$	287-92-3

Type of data: cloud points (LCST behavior)

w_B	0.06	0.06	0.06	0.06	0.06	0.06
T/K	462.55	480.05	494.45	507.05	518.35	530.35
P/bar	34.6	60.7	80.9	97.4	113.3	125.3

Polymer (B):	ethylene/vinyl acetate copolymer				1999BEY

Characterization: M_n/kg.mol^{-1} = 74, M_w/kg.mol^{-1} = 285, 70.0 wt% vinyl acetate, Scientific Polymer Products, Inc., Ontario, NY

Solvent (A):	cyclopentene	C$_5$H$_8$	142-29-0

Type of data: cloud points (LCST behavior)

w_B	0.06	0.06	0.06	0.06	0.06	0.06	0.06	0.06
T/K	475.75	475.95	485.25	486.35	496.75	497.95	505.65	507.15
P/bar	37.3	37.6	51.9	53.4	68.4	70.1	80.7	83.2

Polymer (B):	N-isopropylacrylamide/1-deoxy-1-methacryl-amido-D-glucitol copolymer				2002RE1

Characterization: M_n/kg.mol^{-1} = 28.6, M_w/kg.mol^{-1} = 56, 13.3 mol% glucitol, synthesized in the laboratory by radical polymerization

Solvent (A):	water	H$_2$O	7732-18-5

Type of data: cloud points (LCST behavior)

w_B	0.0102	0.0174	0.0233	0.0263	0.0397	0.0550	0.0102	0.0102	0.0102
T/K	413.2	384.8	368.5	358.1	353.0	366.0	428.6	450.3	465.0
P/bar	1.00	1.00	1.00	1.00	1.00	1.00	2.90	3.87	4.48

w_B	0.0174	0.0174	0.0233	0.0233	0.0233	0.0233	0.0233	0.0397	0.0397
T/K	401.6	432.5	391.3	395.0	419.0	433.5	453.0	377.8	400.0
P/bar	2.27	3.87	2.10	2.27	3.58	5.71	8.57	2.57	3.65

w_B	0.0550	0.0550	0.0550	0.0550	0.0550	0.0550	0.0550	0.0550
T/K	373.0	388.5	392.2	403.8	411.5	421.6	440.2	451.6
P/bar	1.62	2.26	2.45	2.90	3.58	4.70	7.43	8.84

Polymer (B):   *N*-isopropylacrylamide/1-deoxy-1-methacryl-
amido-D-glucitol copolymer   **2002RE1**

Characterization: M_n/kg.mol^{-1} = 51.6, M_w/kg.mol^{-1} = 110, 13.7 mol% glucitol,
synthesized in the laboratory by radical polymerization

Solvent (A):   **water**   **H_2O**   **7732-18-5**

Type of data: cloud points (LCST behavior)

w_B	0.01001	0.03914	0.0400	0.08810	0.0070	0.0070	0.0070	0.0070	0.0070
T/K	321.1	315.6	313.7	315.1	320.9	321.8	322.7	323.4	323.8
P/bar	1.0	1.0	1.0	1.0	20.0	100.0	200.0	300.0	500.0

w_B	0.0070	0.0230	0.0230	0.0230	0.0230	0.0230	0.0230
T/K	324.0	317.6	319.0	320.3	320.7	321.9	322.4
P/bar	600.0	20.0	200.0	300.0	400.0	600.0	700.0

Polymer (B):   *N*-isopropylacrylamide/1-deoxy-1-methacryl-
amido-D-glucitol copolymer   **2002RE1**

Characterization: M_n/kg.mol^{-1} = 145, M_w/kg.mol^{-1} = 432, 14.0 mol% glucitol,
synthesized in the laboratory by radical polymerization

Solvent (A):   **water**   **H_2O**   **7732-18-5**

Type of data: cloud points (LCST behavior)

w_B	0.0050	0.0100	0.0305	0.0432	0.0530	0.0983	0.1103	0.1294	0.0038
T/K	313.1	311.3	310.1	309.8	309.3	308.6	308.0	307.3	311.7
P/bar	1.0	1.0	1.0	1.0	1.0	1.0	1.0	1.0	3.5

w_B	0.0038	0.0038	0.0038	0.0038	0.0038	0.0038	0.0038	0.0038	0.0432
T/K	312.0	312.4	313.3	313.6	313.7	314.3	314.3	314.0	310.0
P/bar	62.0	110.0	200.0	300.0	400.0	500.0	600.0	700.0	10.0

w_B	0.0432	0.0432	0.0432	0.0432	0.0432	0.0432	0.0432	0.0432	0.0432
T/K	310.1	310.3	310.5	310.8	311.1	311.4	311.8	311.8	312.0
P/bar	40.0	70.0	100.0	150.0	200.0	250.0	300.0	350.0	400.0

w_B	0.0432	0.0432	0.0432	0.0432
T/K	312.1	312.3	312.3	312.0
P/bar	450.0	550.0	600.0	700.0

Polymer (B):   **penta(ethylene glycol) monoheptyl ether**   **1992SAS**
Characterization: M_n/kg.mol^{-1} = 0.34, surfactant C_7E_5, Bachem, purity > 98 wt%
Solvent (A):   **n-dodecane**   **$C_{12}H_{26}$**   **112-40-3**

Type of data: cloud points

w_B	0.0499	0.0499	0.0499	0.0499	0.0499	0.0499	0.0499	0.0857	0.0857
T/K	273.71	275.64	277.35	279.18	281.00	282.71	284.51	277.92	279.31
P/MPa	9.80	24.00	36.80	51.00	65.20	79.60	95.00	13.20	23.40

continued

continued

w_B	0.0857	0.0857	0.0857	0.0857	0.0857	0.0999	0.0999	0.0999	0.0999
T/K	280.42	281.97	283.65	285.78	287.59	280.05	282.03	284.17	286.20
P/MPa	31.20	43.40	55.45	69.05	85.25	14.65	28.35	43.95	58.75

w_B	0.0999	0.0999	0.1680	0.1680	0.1680	0.1680	0.1680	0.1680	0.1680
T/K	288.20	290.08	279.57	281.63	283.57	285.66	287.58	289.57	291.52
P/MPa	73.45	88.05	7.95	22.95	37.35	52.95	67.55	82.85	98.35

w_B	0.2037	0.2037	0.2037	0.2037	0.2037	0.2037	0.2037	0.3010	0.3010
T/K	279.68	281.66	283.54	285.57	287.55	289.51	291.48	279.45	281.56
P/MPa	8.22	22.45	36.35	51.45	66.35	81.55	97.05	11.95	27.45

w_B	0.3010	0.3010	0.3010	0.3010	0.4013	0.4013	0.4013	0.4013	0.4013
T/K	283.48	285.51	287.51	289.47	277.39	279.19	281.06	283.75	285.38
P/MPa	41.75	57.05	72.55	88.75	7.55	21.55	35.75	55.55	68.25

w_B	0.4013	0.4013	0.4681	0.4681	0.4681	0.4681	0.4681	0.4681	0.6025
T/K	287.14	289.05	276.54	278.43	280.29	282.47	285.76	288.09	273.38
P/MPa	82.25	97.45	8.45	22.35	36.55	53.10	78.45	97.05	6.15

w_B	0.6025	0.6025	0.6025	0.6025	0.6025	0.6025
T/K	274.50	276.36	278.23	280.03	281.50	283.66
P/MPa	14.35	29.15	43.45	58.25	70.65	88.95

Polymer (B):	**penta(ethylene glycol) monoheptyl ether**	**1992SAS**
Characterization:	M_n/kg.mol^{-1} = 0.34, surfactant C$_7$E$_5$, Bachem, purity > 98 wt%	
Solvent (A):	**water** $\quad$ **H$_2$O**	**7732-18-5**

Type of data: cloud points

w_B	0.0100	0.0100	0.0100	0.0100	0.0100	0.0302	0.0302	0.0302	0.0302
T/K	341.74	342.82	343.92	345.93	348.12	341.34	342.28	343.02	345.04
P/MPa	6.35	15.34	25.65	46.20	71.60	8.35	15.45	21.45	39.35

w_B	0.0302	0.0302	0.0302	0.0799	0.0799	0.0799	0.0799	0.0799	0.0799
T/K	346.98	347.88	349.98	339.94	340.96	341.91	343.15	344.14	345.06
P/MPa	59.25	70.15	96.15	1.88	9.85	17.55	28.65	38.15	47.45

w_B	0.0799	0.0799	0.0987	0.0987	0.0987	0.0987	0.0987	0.0987	0.0987
T/K	347.03	348.93	340.88	341.93	342.97	345.08	346.89	347.77	348.75
P/MPa	70.25	95.95	7.95	16.55	25.75	46.15	66.75	77.65	91.15

w_B	0.1595	0.1595	0.1595	0.1595	0.1595	0.1595	0.1595	0.1862	0.1862
T/K	340.81	341.84	342.85	344.87	346.94	348.05	349.40	340.56	342.65
P/MPa	6.65	15.05	23.85	43.15	66.15	80.05	99.15	2.83	19.65

w_B	0.1862	0.1862	0.1862	0.2488	0.2488	0.2488	0.2488	0.2488	0.2488
T/K	344.78	346.67	348.65	340.79	341.91	342.89	344.06	345.84	347.81
P/MPa	39.45	59.75	84.75	1.03	9.75	17.95	28.55	46.25	68.65

w_B	0.2488	0.2488	0.3489	0.3489	0.3489	0.3489	0.3489	0.3489	0.3489
T/K	348.80	349.79	343.87	345.05	345.92	347.86	349.80	350.75	351.88
P/MPa	81.45	95.35	10.55	20.65	28.55	47.95	70.55	83.05	99.35

continued

continued

w_B	0.4006	0.4006	0.4006	0.4006	0.4006	0.4006	0.4006	0.5029	0.5029
T/K	345.04	346.19	347.36	347.88	350.71	351.73	353.05	349.77	350.70
P/MPa	5.95	15.45	25.35	30.05	58.45	70.35	87.45	7.15	14.45

w_B	0.5029	0.5029	0.5029	0.5029	0.5029	0.5029
T/K	351.74	352.61	354.53	356.43	357.44	358.37
P/MPa	23.15	30.85	49.35	69.55	81.55	93.15

Polymer (B):   **poly(butyl methacrylate)**   **1984SAN, 1986SAN**
Characterization: $M_n/kg.mol^{-1} = 427$, $M_w/kg.mol^{-1} = 470$,
     Röhm GmbH, Darmstadt, Germany
Solvent (A):   **ethanol**   **C_2H_6O**   **64-17-5**

Type of data: cloud points

w_B 0.075   was kept constant

T/K	315.35	314.15	313.15	312.15	311.15	310.15
P/bar	1	105	198	400	625	1054

Polymer (B):   **poly(dimethylsiloxane)**   **1972ZE1**
Characterization: $M_\eta/kg.mol^{-1} = 626$, Dow-Corning
Solvent (A):   **n-butane**   **C_4H_{10}**   **106-97-8**

Type of data: cloud points

$T/K = 392.95$ $w_B = 0.0369$ $(dT/dP)_{P=1}/K.bar^{-1} = +0.92$

Polymer (B):   **poly(dimethylsiloxane)**   **1972ZE1**
Characterization: $M_\eta/kg.mol^{-1} = 1.2$, 10 cSt viscosity, Dow-Corning
Solvent (A):   **ethane**   **C_2H_6**   **74-84-0**

Type of data: cloud points

$T/K = 280.65$ $w_B = 0.366$ $(dT/dP)_{P=1}/K.bar^{-1} = +0.92$

Polymer (B):   **poly(dimethylsiloxane)**   **1972ZE1**
Characterization: $M_\eta/kg.mol^{-1} = 3.2$, 50 cSt viscosity, Dow-Corning
Solvent (A):   **ethane**   **C_2H_6**   **74-84-0**

Type of data: cloud points

$T/K = 273.15$ $w_B = 0.262$ $(dT/dP)_{P=1}/K.bar^{-1} = +0.90$

Polymer (B):   **poly(dimethylsiloxane)**   **1972ZE1**
Characterization: M_η/kg.mol^{-1} = 14.2, 292 cSt viscosity, Dow-Corning
Solvent (A):   **ethane**   **C$_2$H$_6$**   **74-84-0**

Type of data: cloud points

T/K = 272.15 w_B = 0.0364   (dT/dP)$_{P=1}$/K.bar^{-1} = +0.89

Polymer (B):   **poly(dimethylsiloxane)**   **1972ZE1**
Characterization: M_η/kg.mol^{-1} = 626, Dow-Corning
Solvent (A):   **ethane**   **C$_2$H$_6$**   **74-84-0**

Type of data: cloud points

T/K = 259.65 w_B = 0.0651   (dT/dP)$_{P=1}$/K.bar^{-1} = +0.85

Polymer (B):   **poly(dimethylsiloxane)**   **1972ZE1**
Characterization: M_η/kg.mol^{-1} = 203, Dow-Corning
Solvent (A):   **propane**   **C$_3$H$_8$**   **74-98-6**

Type of data: cloud points

T/K = 340.15 w_D = 0.0400   (dT/dP)$_{P=1}$/K.bar^{-1} – +0.92

Polymer (B):   **poly(dimethylsiloxane)**   **1972ZE1**
Characterization: M_η/kg.mol^{-1} = 626, Dow-Corning
Solvent (A):   **propane**   **C$_3$H$_8$**   **74-98-6**

Type of data: cloud points

T/K = 337.75 w_B = 0.0402   (dT/dP)$_{P=1}$/K.bar^{-1} = +0.91

Polymer (B):   **polyethylene**   **2000BEH**
Characterization: M_n/kg.mol^{-1} = 13, M_w/kg.mol^{-1} = 89, M_z/kg.mol^{-1} = 600,
                 LDPE, Stamylan, DSM, Geleen, The Netherlands
Solvent (A):   **cyclohexane**   **C$_6$H$_{12}$**   **110-82-7**

Type of data: cloud points

w_B	0.0766	0.0766	0.0766	0.0766	0.0766	0.1083	0.1083	0.1083
T/K	514.49	517.75	521.00	522.25	526.45	515.38	520.75	526.19
P/bar	26.4	32.9	39.7	41.8	48.6	25.60	30.43	38.43

Polymer (B):   **polyethylene**   **2002HOR**
Characterization: M_n/kg.mol^{-1} = 13, M_w/kg.mol^{-1} = 89, M_z/kg.mol^{-1} = 600,
                      LDPE, Stamylan, DSM, Geleen, The Netherlands
Solvent (A):   **cyclohexane**   **C$_6$H$_{12}$**   **110-82-7**

Type of data: cloud points

w_B 0.103   was kept constant

T/K	554.06	560.69	564.14	570.03	574.22	578.33	584.85	589.48	592.73
P/bar	62	68	73	78	82	86	91	95	98

T/K	598.74	603.65	610.30	614.57	619.14	628.86
P/bar	103	107	112	115	118	124

Polymer (B):   **polyethylene**   **1999BEY**
Characterization: M_n/kg.mol^{-1} = 20.1, M_w/kg.mol^{-1} = 106
Solvent (A):   **cyclopentane**   **C$_5$H$_{10}$**   **287-92-3**

Type of data: cloud points (LCST behavior)

w_B	0.06	0.06	0.06	0.06	0.06	0.06	0.06	0.06	0.06
T/K	482.95	483.35	492.45	492.95	501.65	502.55	513.65	526.05	528.75
P/bar	47.4	48.5	61.7	62.3	74.5	75.4	90.4	106.4	110.4

Polymer (B):   **polyethylene**   **1999BEY**
Characterization: M_n/kg.mol^{-1} = 20.1, M_w/kg.mol^{-1} = 106
Solvent (A):   **cyclopentene**   **C$_5$H$_8$**   **142-29-0**

Type of data: cloud points (LCST behavior)

w_B	0.06	0.06	0.06	0.06	0.06	0.06	0.06
T/K	482.85	483.95	488.55	489.15	498.05	498.75	510.35
P/bar	39.4	41.0	47.8	48.6	62.3	63.1	79.6

Polymer (B):   **polyethylene**   **2002JOU**
Characterization: M_w/kg.mol^{-1} = 125, Aldrich Chem. Co., Inc., Milwaukee, WI
Solvent (A):   **n-heptane**   **C$_7$H$_{16}$**   **142-82-5**

Type of data: cloud points

w_B 0.024   was kept constant

T/K	463.15	473.15	483.15	493.15	503.15
P/MPa	2.0	3.2	4.6	5.7	7.1

Comments: Vapor-liquid equilibrium data for this system are given in Chapter 2.

Polymer (B):	**polyethylene**							**2004SCH**

Characterization: M_n/kg.mol^{-1} = 6.28, M_w/kg.mol^{-1} = 6.5, linear, completely hydrogenated polybutadiene, Polymer Source, Inc., Dorval, Quebec

Solvent (A): **n-hexane** C_6H_{14} 110-54-3

Type of data: cloud points

w_B	0.018	0.018	0.018	0.018	0.018	0.018	0.0355	0.0355	0.0355
T/K	470.15	479.15	488.15	498.15	505.15	515.15	464.15	473.65	480.15
P/bar	30	41	48	54	60	69	19	31	38

w_B	0.0355	0.0355	0.0355	0.0355	0.051	0.051	0.051	0.051	0.0564
T/K	488.15	497.15	507.15	519.15	472.15	489.15	499.15	508.15	482.15
P/bar	49	58	67	76	28	45	55	66	68

w_B	0.0564	0.0564	0.0564	0.0564	0.065	0.065	0.083	0.083	0.083
T/K	492.15	502.15	514.15	522.15	487.15	496.15	479.15	488.15	497.15
P/bar	75	81	87	97	35	44	28	37	48

w_B	0.083	0.083	0.085	0.085	0.085	0.085	0.085	0.085	0.090
T/K	507.65	516.15	466.15	474.15	483.15	492.15	500.15	510.15	494.15
P/bar	56	63	34	43	52	57	63	72	67

w_B	0.090	0.090	0.090	0.103	0.103	0.103	0.103	0.155	0.155
T/K	504.15	514.15	524.15	489.15	496.15	506.15	515.15	502.15	509.15
P/bar	78	88	100	45	58	66	75	43	45

w_B	0.155	0.279	0.279	0.279
T/K	521.15	518.15	527.65	536.65
P/bar	57	38	50	54

Type of data: critical points

$\varphi_{B, crit}$	0.138	0.138	0.138
T/K	451	463	476
P/bar	20	40	60

Polymer (B):	**polyethylene**							**1990KEN**

Characterization: M_n/kg.mol^{-1} = 8, M_w/kg.mol^{-1} = 177, M_z/kg.mol^{-1} = 1000, HDPE, DSM, Geleen, The Netherlands

Solvent (A): **n-hexane** C_6H_{14} 110-54-3

Type of data: cloud points

w_B	0.0053	0.0053	0.0053	0.0053	0.0053	0.0055	0.0055	0.0055	0.0055
T/K	412.50	419.93	422.43	427.40	434.77	412.60	415.07	417.59	420.03
P/bar	16.3	29.0	32.2	40.2	51.0	23.1	27.1	31.1	34.6

w_B	0.0055	0.0055	0.0055	0.0095	0.0095	0.0095	0.0095	0.0095	0.0095
T/K	422.50	427.49	432.47	411.76	414.28	416.83	419.29	421.80	421.78
P/bar	38.1	45.6	52.6	18.2	21.2	26.1	30.8	33.1	33.7

continued

continued

w_B	0.0095	0.0095	0.0095	0.0199	0.0199	0.0199	0.0199	0.0199	0.0199
T/K	426.80	429.78	436.70	412.52	417.54	422.51	424.93	427.46	432.36
P/bar	40.8	45.1	55.2	15.2	23.4	30.9	35.0	38.3	46.0

w_B	0.0199	0.0199	0.0298	0.0298	0.0298	0.0298	0.0298	0.0298	0.0298
T/K	437.33	442.29	412.49	417.46	419.93	422.45	424.93	427.41	429.88
P/bar	52.7	59.8	11.3	19.3	23.4	27.2	31.2	35.2	38.6

w_B	0.0298	0.0298	0.0298	0.0423	0.0423	0.0423	0.0423	0.0423	0.0423
T/K	432.39	437.39	442.36	412.52	417.51	419.98	422.48	424.95	427.46
P/bar	42.6	49.8	57.2	7.6	16.1	19.5	23.7	27.8	32.1

w_B	0.0423	0.0423	0.0423	0.0423	0.0498	0.0498	0.0498	0.0498	0.0498
T/K	429.91	432.44	434.90	437.39	417.54	419.95	422.51	425.01	427.48
P/bar	35.2	38.9	42.7	46.3	13.8	17.3	21.8	25.6	29.2

w_B	0.0498	0.0612	0.0612	0.0612	0.0612	0.0612	0.0612	0.0612	0.0612
T/K	432.44	417.45	419.93	422.43	424.90	427.40	427.38	429.86	432.36
P/bar	36.8	10.4	14.5	18.4	22.2	26.2	26.4	30.1	34.0

w_B	0.0612	0.0612	0.0666	0.0666	0.0666	0.0666	0.0666	0.0666	0.0666
T/K	437.33	439.82	421.63	423.11	424.64	426.59	428.09	429.54	431.55
P/bar	41.0	44.6	14.6	17.0	19.2	22.1	24.4	26.5	29.8

w_B	0.0666	0.0666	0.0666	0.0820	0.0820	0.0820	0.0820	0.0820	0.0954
T/K	436.51	439.47	441.43	421.75	423.72	425.69	427.67	429.62	422.52
P/bar	37.1	41.4	44.3	11.3	14.2	17.2	20.4	23.2	9.8

w_B	0.0954	0.0954	0.0954	0.0954	0.0954	0.0954	0.0954	0.0954	0.0954
T/K	424.82	427.30	429.78	432.29	433.79	434.77	437.25	439.82	442.31
P/bar	13.3	17.3	21.1	24.9	27.3	28.7	32.4	36.3	39.9

w_B	0.0954	0.0954	0.0954	0.1313	0.1313	0.1313	0.1313	0.1313	0.1313
T/K	444.80	447.29	452.26	429.15	432.45	434.11	437.41	439.03	446.55
P/bar	43.5	47.0	54.3	10.3	15.5	18.2	23.1	25.4	36.6

w_B	0.1313	0.1313
T/K	451.46	457.15
P/bar	43.7	52.0

Type of data: coexistence data (liquid-liquid-vapor three phase equilibrium)

w_B	0.0055	0.0053	0.0095	0.0199	0.0298	0.0423	0.0498	0.0612	0.0666
T/K	401.45	405.75	404.35	407.05	409.55	411.85	413.75	415.55	417.35
P/bar	4.8	5.5	5.3	5.8	6.3	6.8	7.1	7.4	7.7

w_B	0.0820	0.0954	0.1313
T/K	419.85	421.55	428.15
P/bar	8.0	8.2	8.5

Comments: Vapor-liquid equilibrium data for this system are given in Chapter 2.

Polymer (B): **polyethylene** **2004CHE**
Characterization: M_n/kg.mol^{-1} = 14.4, M_w/kg.mol^{-1} = 15.5, completely hydrogenated
polybutadiene, Scientific Polymer Products, Inc., Ontario, NY
Solvent (A): **n-hexane** **C$_6$H$_{14}$** **110-54-3**

Type of data: cloud points

w_B	0.0322	0.0322	0.0322	0.0824	0.0824	0.0824	0.0906	0.0906	0.0906
T/K	453.17	463.11	473.15	453.07	463.11	473.12	453.11	463.17	473.09
P/MPa	1.7	3.0	4.3	2.1	3.4	4.8	2.0	3.4	4.7

w_B	0.0984	0.0984	0.0984	0.1187	0.1187	0.1187	0.1379	0.1379	0.1379
T/K	453.18	463.15	473.22	453.21	463.10	473.18	453.17	463.15	473.20
P/MPa	2.0	3.3	4.7	1.9	3.2	4.5	1.8	3.2	4.5

w_B	0.0906	0.0906	0.0906
T/K	453.13	463.14	473.12
P/MPa	1.5	1.8	2.0
	VLLE	VLLE	VLLE

Comments: Vapor-liquid equilibrium data for this system are given in Chapter 2.

Polymer (B): **polyethylene** **2000BEH**
Characterization: M_n/kg.mol^{-1} = 22.5, M_w/kg.mol^{-1} = 58,
HDPE, metallocene product, BASF AG, Germany
Solvent (A): **n-hexane** **C$_6$H$_{14}$** **110-54-3**

Type of data: cloud points

w_B	0.0497	0.0497	0.0497	0.0497	0.0497	0.0497	0.0497	0.0497	0.0497
T/K	430.52	438.64	446.86	455.16	463.47	471.86	479.83	488.05	496.17
P/bar	15.5	28.4	40.5	52.3	63.9	74.1	83.8	93.9	102.4

w_B	0.0497	0.1037	0.1037	0.1037	0.1037	0.1037	0.1037	0.1037	0.1037
T/K	503.21	430.94	437.29	445.42	453.63	462.15	470.24	478.48	486.85
P/bar	110.0	10.1	20.0	31.0	42.1	53.9	64.9	75.3	84.9

w_B	0.1037	0.1037	0.1508	0.1508	0.1508	0.1508	0.1508	0.1508	0.1508
T/K	495.09	503.07	434.00	439.84	446.61	446.86	454.32	461.36	469.38
P/bar	94.6	103.1	9.5	16.9	26.6	27.0	37.3	46.7	55.5

w_B	0.1508	0.1508	0.1508	0.1508	0.1508
T/K	475.84	482.86	490.05	497.37	503.34
P/bar	62.3	73.6	81.8	90.1	96.9

Polymer (B): **polyethylene** **2000BEH**
Characterization: M_n/kg.mol^{-1} = 20, M_w/kg.mol^{-1} = 210,
HDPE, BASF AG, Germany
Solvent (A): **n-hexane** **C$_6$H$_{14}$** **110-54-3**

Type of data: cloud points

continued

continued

w_B	0.0488	0.0488	0.0488	0.0488	0.0488	0.0488	0.0488	0.0488	0.0488
T/K	436.15	440.77	446.20	452.04	456.57	461.96	467.33	472.64	478.75
P/bar	42.3	47.1	54.2	61.2	66.8	74.5	81.4	88.4	96.4

w_B	0.0488	0.0488	0.0488	0.0488	0.1097	0.1097	0.1097	0.1097	0.1097
T/K	484.21	487.39	495.09	504.53	423.75	430.13	435.31	441.82	448.99
P/bar	103.0	107.1	114.3	126.2	21.6	29.9	35.1	43.8	51.7

w_B	0.1097	0.1097	0.1097	0.1097	0.1097	0.1097	0.1097	0.1097	0.1097
T/K	455.25	462.02	462.15	469.19	476.63	483.67	490.59	496.95	503.85
P/bar	59.3	67.2	67.4	75.5	83.7	94.4	94.4	105.7	113.5

Polymer (B):	**polyethylene**		**2000BEH**
Characterization:	$M_n/kg.mol^{-1} = 20$, $M_w/kg.mol^{-1} = 585$,		
	HDPE, BASF AG, Germany		
Solvent (A):	**n-hexane**	**C_6H_{14}**	**110-54-3**

Type of data: cloud points

w_B	0.0477	0.0477	0.0477	0.0477	0.0477	0.0477	0.0477	0.0477	0.0477
T/K	423.48	429.84	436.07	443.68	451.38	459.90	467.99	475.57	483.40
P/bar	25.6	34.1	43.3	52.4	62.5	74.2	84.9	94.8	104.2

w_B	0.0477	0.0477	0.0477	0.0994	0.0994	0.0994	0.0994	0.0994	0.0994
T/K	491.11	498.15	504.39	422.96	431.18	438.64	447.27	454.83	462.95
P/bar	111.7	120.3	127.0	8.8	19.8	31.0	40.6	51.6	62.5

w_B	0.0994	0.0994	0.0994	0.0994	0.0994	0.1482	0.1482	0.1482	0.1482
T/K	470.39	478.90	487.54	495.36	502.94	438.88	439.03	446.08	452.85
P/bar	72.6	83.7	94.5	103.3	111.5	12.7	13.0	24.9	36.7

w_B	0.1482	0.1482	0.1482	0.1482	0.1482	0.1482	0.1482		
T/K	460.29	467.60	474.64	481.93	488.98	496.56	503.61		
P/bar	47.6	58.9	68.2	77.7	86.0	94.5	102.4		

Polymer (B):	**polyethylene**		**2004CHE**
Characterization:	$M_n/kg.mol^{-1} = 23.3$, $M_w/kg.mol^{-1} = 60.4$, $M_z/kg.mol^{-1} = 100.7$,		
	metallocene LLDPE, unspecified comonomer, industrial source		
Solvent (A):	**n-hexane**	**C_6H_{14}**	**110-54-3**

Type of data: cloud points

w_B	0.0049	0.0049	0.0049	0.0049	0.0148	0.0148	0.0148	0.0148	0.0364
T/K	443.13	453.13	463.18	473.22	443.25	453.15	463.14	473.05	443.49
P/MPa	2.5	3.9	5.2	6.6	2.6	4.1	5.5	6.6	2.6

w_B	0.0364	0.0364	0.0364	0.0822	0.0822	0.0822	0.0888	0.0888	0.0888
T/K	453.54	463.58	473.55	433.15	453.10	473.14	443.34	453.36	463.37
P/MPa	4.0	5.4	6.6	1.6	4.5	7.2	2.1	3.4	4.9

continued

continued

w_B	0.0888	0.1083	0.1083	0.1083	0.1083	0.1535	0.1535	0.1535	0.1535
T/K	473.41	443.42	453.38	463.42	473.45	443.03	453.04	463.05	473.01
P/MPa	6.3	1.9	3.2	4.7	6.1	1.4	2.8	4.1	5.4

w_B	0.1952	0.1952	0.1952
T/K	452.98	463.09	473.21
P/MPa	2.4	3.8	5.1

w_B	0.0822	0.0822	0.0822	
T/K	433.18	453.13	473.13	
P/MPa	1.1	1.5	1.9	(three VLLE data points)

Comments: Vapor-liquid equilibrium data for this system are given in Chapter 2.

Polymer (B):	**polyethylene**		**2001TOR**
Characterization:	$M_n/kg.mol^{-1} = 43$, $M_w/kg.mol^{-1} = 105$, $M_z/kg.mol^{-1} = 190$,		
	HDPE, DSM, Geleen, The Netherlands		
Solvent (A):	**n-hexane**	C_6H_{14}	**110-54-3**

Type of data: cloud points

w_B	0.2128	0.2128	0.2128	0.2128	0.2128	0.2128	0.2128	0.2128	0.2128
T/K	446.2	457.4	467.6	478.0	457.4	467.6	478.0	488.2	498.4
P/MPa	1.05	1.33	1.60	1.88	2.29	3.78	5.16	6.45	7.65
	VLE	VLLE	VLLE	VLLE	LLE	LLE	LLE	LLE	LLE

w_B	0.2128	0.2128	0.1382	0.1382	0.1382	0.1382	0.1382	0.1382	0.1382
T/K	508.5	519.0	436.5	446.9	456.2	467.5	446.9	456.2	467.5
P/MPa	8.79	9.87	0.92	1.06	1.26	1.52	1.91	3.25	4.81
	LLE	LLE	VLE	VLLE	VLLE	VLLE	LLE	LLE	LLE

w_B	0.1382	0.1382	0.1382	0.1382	0.1382	0.0813	0.0813	0.0813	0.0813
T/K	477.7	488.1	498.4	508.5	519.0	431.3	441.2	451.4	461.9
P/MPa	6.13	7.44	8.67	9.80	10.89	0.77	0.93	1.14	1.38
	LLE	LLE	LLE	LLE	LLE	VLE	VLLE	VLLE	VLLE

w_B	0.0813	0.0813	0.0813	0.0813	0.0813	0.0813	0.0813	0.0813	0.0582
T/K	441.2	451.4	461.9	472.0	482.5	492.9	502.9	513.0	441.2
P/MPa	1.48	3.00	4.50	5.86	7.16	8.45	9.59	10.60	1.63
	LLE	LLE	LLE	LLE	LLE	LLE	LLE	LLE	LLE

w_B	0.0582	0.0582	0.0582	0.0582	0.0221	0.0221	0.0221	0.0221	0.0221
T/K	451.5	461.9	472.2	482.5	464.0	474.4	482.7	492.6	502.9
P/MPa	3.15	4.63	6.01	7.29	5.11	6.49	7.53	8.73	9.90
	LLE	LLE	LLE	LLE	LLE	LLE	LLE	LLE	LLE

w_B	0.0221	0.0138	0.0138	0.0138	0.0138	0.0138	0.0138	0.0138	0.0138
T/K	513.3	423.4	433.6	443.4	453.6	462.9	473.4	483.5	493.8
P/MPa	10.98	0.68	0.78	2.15	3.64	4.92	6.30	7.57	8.76
	LLE	VLE	VLE	LLE	LLE	LLE	LLE	LLE	LLE

w_B	0.0138	0.0138
T/K	503.	513.4
P/MPa	9.82	10.86
	LLE	LLE

Polymer (B):   **polyethylene**   **2004SCH**
Characterization: M_n/kg.mol^{-1} = 61.0, M_w/kg.mol^{-1} = 67.1, linear, completely
                          hydrogenated polybutadiene, Polymer Source, Inc., Dorval, Quebec
Solvent (A):   **n-hexane**   **C$_6$H$_{14}$**   **110-54-3**

Type of data: cloud points

w_B	0.0207	0.0207	0.0207	0.0207	0.0207	0.0207	0.0374	0.0374	0.0374
T/K	436.15	446.15	453.15	463.15	481.15	490.15	423.15	436.65	446.15
P/bar	15	27	37	52	72	94	22	30	46

w_B	0.0374	0.0374	0.0374	0.0374	0.050	0.050	0.069	0.069	0.069
T/K	454.15	463.15	471.15	478.15	428.15	437.15	436.15	454.65	464.15
P/bar	59	71	81	91	22	35	22	49	58

w_B	0.069	0.106	0.106	0.106	0.106	0.106	0.106	0.118	0.118
T/K	473.15	448.65	458.65	466.15	474.15	481.65	493.15	443.55	451.35
P/bar	76	42	54	60	70	80	91	23	34

w_B	0.118	0.118	0.118	0.118	0.285	0.285	0.285	0.285	0.285
T/K	469.65	478.15	487.15	496.15	479.15	488.15	498.15	507.65	517.15
P/bar	63	71	77	92	54	69	83	93	101

Type of data: critical points

$\varphi_{B,\,crit}$	0.026	0.025	0.024
T/K	429.0	444.0	461.0
P/bar	20	40	60

Polymer (B):   **polyethylene**   **2004CHE**
Characterization: M_n/kg.mol^{-1} = 82, M_w/kg.mol^{-1} = 108, completely hydrogenated
                          polybutadiene, Scientific Polymer Products, Inc., Ontario, NY
Solvent (A):   **n-hexane**   **C$_6$H$_{14}$**   **110-54-3**

Type of data: cloud points

w_B	0.0076	0.0076	0.0076	0.0076	0.0186	0.0186	0.0186	0.0186	0.0310
T/K	443.67	453.14	463.18	473.12	443.15	453.16	463.15	473.15	443.15
P/MPa	3.8	5.3	6.6	7.9	3.9	5.4	6.7	8.0	4.2

w_B	0.0310	0.0310	0.0310	0.0536	0.0536	0.0536	0.0536	0.0829	0.0829
T/K	453.16	463.16	473.16	443.17	453.15	463.17	473.15	433.13	453.13
P/MPa	5.6	6.9	8.2	4.0	5.4	6.7	8.0	2.4	5.3

w_B	0.0829	0.0886	0.0886	0.0886	0.0886	0.1026	0.1026	0.1026	0.1026
T/K	473.13	443.24	453.16	463.16	473.17	443.08	453.15	463.16	473.19
P/MPa	8.0	3.8	5.3	6.6	7.9	3.8	5.1	6.5	7.8

w_B	0.1310	0.1310	0.1310	0.1310
T/K	443.14	453.14	463.18	473.15
P/MPa	3.6	5.0	6.4	7.7

w_B	0.0829	0.0829	0.0829
T/K	433.13	453.13	473.12
P/MPa	1.1	1.5	2.0	(three VLLE data points)

Comments: Vapor-liquid equilibrium data for this system are given in Chapter 2.

Polymer (B):	**polyethylene**							**2004SCH**
Characterization:	$M_n/\text{kg.mol}^{-1} = 322$, $M_w/\text{kg.mol}^{-1} = 383$, linear, completely hydrogenated polybutadiene, Polymer Source, Inc., Dorval, Quebec							
Solvent (A):	**n-hexane**			$\mathbf{C_6H_{14}}$				**110-54-3**

Type of data: cloud points

w_B	0.00422	0.00422	0.00422	0.00422	0.00422	0.00422	0.00422	0.00422	0.006
T/K	437.65	447.35	456.15	465.65	475.05	484.75	494.55	503.75	434.15
P/bar	50	62	75	87	99	110	119.5	129.5	41

w_B	0.006	0.006	0.006	0.006	0.006	0.006	0.0175	0.0175	0.0175
T/K	442.15	452.15	461.15	469.15	478.15	487.15	453.15	461.15	469.15
P/bar	54	66	79	90	101	111	69	81	91

w_B	0.0175	0.0175	0.0175	0.0428	0.0428	0.0428	0.0428	0.0428	0.0428
T/K	478.15	488.15	496.65	430.15	437.15	445.15	454.15	462.15	473.15
P/bar	104	115	125	29	43	57	69	83	94

w_B	0.0428	0.0428	0.060	0.060	0.060	0.060	0.060	0.060	0.060
T/K	482.15	492.15	429.85	437.55	448.15	456.15	464.85	474.65	483.65
P/bar	107	117	34	48	63	74	85	96	107

w_B	0.060	0.0838	0.0838	0.0838	0.0838	0.0838	0.0838	0.0838	0.127
T/K	491.95	444.15	452.65	460.15	468.15	477.15	486.15	495.15	442.15
P/bar	115	46	58	69	80	93	101	114	32

w_B	0.127	0.127	0.127	0.127	0.127	0.127	0.237	0.237	0.237
T/K	453.15	461.15	472.15	479.15	489.15	497.15	467.65	482.85	494.45
P/bar	43	56	74	86	98	112	47	66	83

Type of data: critical points

$\varphi_{B,\,crit}$	0.012	0.012	0.013
T/K	416.2	431.4	446.3
P/bar	20	40	60

Polymer (B):	**polyethylene**							**2001TOR**
Characterization:	$M_n/\text{kg.mol}^{-1} = 43$, $M_w/\text{kg.mol}^{-1} = 105$, $M_z/\text{kg.mol}^{-1} = 190$, HDPE, DSM, Geleen, The Netherlands							
Solvent (A):	**1-octene**			$\mathbf{C_8H_{16}}$				**111-66-0**

Type of data: cloud points

w_B	0.1791	0.1791	0.1791	0.1791	0.1791	0.1791	0.1791	0.0897	0.0897
T/K	503.3	513.4	524.3	533.9	524.3	533.9	544.8	503.2	512.8
P/MPa	0.97	1.14	1.35	1.57	1.38	2.30	3.30	0.94	1.13
	VLE	VLE	VLLE	VLLE	LLE	LLE	LLE	VLE	VLE

w_B	0.0897	0.0897	0.0897	0.0897	0.0897	0.0439	0.0439	0.0439	0.0439
T/K	523.8	535.0	523.8	534.9	544.3	503.3	513.4	524.5	533.4
P/MPa	1.35	1.60	1.93	3.07	3.95	0.95	1.11	1.34	1.54
	VLLE	VLLE	LLE	LLE	LLE	VLE	VLE	VLLE	VLLE

continued

continued

w_B	0.0439	0.0439	0.0439	0.0198	0.0198	0.0198	0.0198	0.0198	0.0198
T/K	524.4	533.4	544.1	503.3	513.3	523.9	533.5	523.9	533.5
P/MPa	2.19	3.09	4.10	0.95	1.14	1.35	1.58	2.22	3.16
	LLE	LLE	LLE	VLE	VLLE	VLLE	VLLE	LLE	LLE

w_B	0.0198
T/K	544.5
P/MPa	4.16
	LLE

Polymer (B):	**polyethylene**	**2002YEO**

Characterization: $M_n/\text{kg.mol}^{-1} = 11.3$, $M_w/\text{kg.mol}^{-1} = 41.7$,
Pressure Chemical Company, Pittsburgh, PA

Solvent (A):	**n-pentane**	**C_5H_{12}**	**109-66-0**

Type of data: cloud points

$T/K = 403.15$

w_B	0.003	0.010	0.015	0.020	0.030	0.035	0.037	0.039	0.040
P/bar	97.2	113.1	112.4	109.6	108.9	111.7	109.6	111.7	111.7

w_B	0.043	0.045	0.050	0.060	0.070
P/bar	108.2	109.6	108.9	102.0	101.3

critical concentration: $w_{B, crit} = 0.0395$

$T/K = 423.15$

w_B	0.003	0.010	0.015	0.020	0.030	0.035	0.037	0.039	0.040
P/bar	124.8	137.9	137.9	133.7	135.1	137.2	135.1	136.5	136.5

w_B	0.043	0.045	0.050	0.060	0.070
P/bar	131.7	133.7	131.0	127.5	127.5

critical concentration: $w_{B, crit} = 0.0395$

$T/K = 443.15$

w_B	0.003	0.010	0.015	0.020	0.030	0.035	0.037	0.039	0.040
P/bar	153.7	160.0	160.6	157.9	159.3	158.6	157.2	156.5	158.6

w_B	0.043	0.045	0.050	0.060	0.070
P/bar	155.1	156.5	153.7	150.3	149.6

critical concentration: $w_{B, crit} = 0.0395$

Polymer (B):	**polyethylene**	**2002YEO**

Characterization: $M_n/\text{kg.mol}^{-1} = 13.9$, $M_w/\text{kg.mol}^{-1} = 27.7$,
Pressure Chemical Company, Pittsburgh, PA

Solvent (A):	**n-pentane**	**C_5H_{12}**	**109-66-0**

Type of data: cloud points

continued

continued

$T/K = 403.15$

w_B	0.003	0.020	0.040	0.060	0.062	0.065	0.068	0.070	0.075
P/bar	46.2	78.6	70.3	66.2	66.2	66.9	64.1	63.4	62.7

w_B	0.080	0.100	0.110
P/bar	62.0	47.5	42.7

critical concentration: $w_{B, crit} = 0.069$

$T/K = 423.15$

w_B	0.003	0.020	0.040	0.060	0.062	0.065	0.068	0.070	0.075
P/bar	79.3	108.9	100.6	93.1	92.4	92.4	90.3	89.6	88.9

w_B	0.080	0.100	0.110
P/bar	88.2	75.8	70.3

critical concentration: $w_{B, crit} = 0.069$

$T/K = 443.15$

w_B	0.003	0.020	0.040	0.060	0.062	0.065	0.068	0.070	0.075
P/bar	115.8	130.3	126.2	120.0	119.3	117.9	117.2	115.8	115.1

w_B	0.080	0.100	0.110
P/bar	113.1	102.7	95.8

critical concentration: $w_{B, crit} = 0.069$

Polymer (B):	**poly(ethylene oxide)**	**1992COO**
Characterization:	M_n/kg.mol^{-1} = 14.7, M_w/kg.mol^{-1} = 19.7, Pressure Chemical Company, Pittsburgh, PA	
Solvent (A):	**water** H_2O	**7732-18-5**

Type of data: cloud points

c_B/g.dl^{-1}	1.0	20.0
T/K	295.65	295.65
P/MPa	557	539

Comments: The complete phase behavior is given in the original source in graphs.

Polymer (B):	**poly(ethylene oxide)**	**1992COO**
Characterization:	M_n/kg.mol^{-1} = 248, M_w/kg.mol^{-1} = 270, American Polymer Standards	
Solvent (A):	**water** H_2O	**7732-18-5**

Type of data: cloud points

c_B/g.dl^{-1}	0.3	1.0
T/K	295.65	295.65
P/MPa	430	435

Comments: The complete phase behavior is given in the original source in graphs.

Polymer (B):	**polyisobutylene**		**1972ZE1**

Characterization: $M_\eta/\text{kg.mol}^{-1} = 6.0$, laboratory fractions, $M_w/M_n = 1.14$ by GPC

Solvent (A):	**n-butane**	**C$_4$H$_{10}$**	**106-97-8**

Type of data: cloud points

$T/K = 344.35$	$w_B = 0.0079$	$(\mathrm{d}T/\mathrm{d}P)_{P=1}/\text{K.bar}^{-1} = +0.47$
$T/K = 337.85$	$w_B = 0.0272$	$(\mathrm{d}T/\mathrm{d}P)_{P=1}/\text{K.bar}^{-1} = +0.50$
$T/K = 321.85$	$w_B = 0.0466$	$(\mathrm{d}T/\mathrm{d}P)_{P=1}/\text{K.bar}^{-1} = +0.58$

Polymer (B):	**polyisobutylene**		**1972ZE1**

Characterization: $M_\eta/\text{kg.mol}^{-1} = 703$, laboratory fractions, $M_w/M_n = 1.14$ by GPC

Solvent (A):	**n-butane**	**C$_4$H$_{10}$**	**106-97-8**

Type of data: cloud points

$T/K = 264.75$	$w_B = 0.0114$	$(\mathrm{d}T/\mathrm{d}P)_{P=1}/\text{K.bar}^{-1} = +0.37$

Polymer (B):	**polyisobutylene**		**1972ZE1**

Characterization: $M_\eta/\text{kg.mol}^{-1} = 1660$, laboratory fractions, $M_w/M_n = 1.14$ by GPC

Solvent (A):	**n-butane**	**C$_4$H$_{10}$**	**106-97-8**

Type of data: cloud points

$T/K = 253.85$	$w_B = 0.0061$	$(\mathrm{d}T/\mathrm{d}P)_{P=1}/\text{K.bar}^{-1} = +0.37$
$T/K = 253.85$	$w_B = 0.0112$	$(\mathrm{d}T/\mathrm{d}P)_{P=1}/\text{K.bar}^{-1} = +0.37$
$T/K = 253.85$	$w_B = 0.0466$	$(\mathrm{d}T/\mathrm{d}P)_{P=1}/\text{K.bar}^{-1} = +0.37$

Polymer (B):	**polyisobutylene**		**2002JOU**

Characterization: $M_n/\text{kg.mol}^{-1} = 600$, $M_w/\text{kg.mol}^{-1} = 1000$,
                       Aldrich Chem. Co., Inc., Milwaukee, WI

Solvent (A):	**n-heptane**	**C$_7$H$_{16}$**	**142-82-5**

Type of data: cloud points

w_B	0.025	was kept constant				
T/K	443.15	453.15	463.95	473.25	483.15	493.15
P/MPa	1.3	2.5	3.9	5.4	6.7	8.3

Comments: Vapor-liquid equilibrium data for this system are given in Chapter 2.

Polymer (B):	**polyisobutylene**		**1972ZE1**

Characterization: $M_\eta/\text{kg.mol}^{-1} = 1660$, laboratory fractions, $M_w/M_n = 1.14$ by GPC

Solvent (A):	**n-hexane**	**C$_6$H$_{14}$**	**110-54-3**

Type of data: cloud points

$T/K = 409.65$	$w_B = 0.0204$	$(\mathrm{d}T/\mathrm{d}P)_{P=1}/\text{K.bar}^{-1} = +0.61$

Polymer (B): polyisobutylene   **1972ZE1**
Characterization: M_η/kg.mol^{-1} = 6.0, laboratory fractions, M_w/M_n = 1.14 by GPC
Solvent (A): **2-methylbutane**   **C$_6$H$_{12}$**   **78-78-4**

Type of data: cloud points

T/K = 387.85 w_B = 0.0235   (dT/dP)$_{P=1}$/K.bar^{-1} = +0.55

Polymer (B): polyisobutylene   **1972ZE1**
Characterization: M_η/kg.mol^{-1} = 703, laboratory fractions, M_w/M_n = 1.14 by GPC
Solvent (A): **2-methylbutane**   **C$_6$H$_{12}$**   **78-78-4**

Type of data: cloud points

T/K = 330.55 w_B = 0.0220   (dT/dP)$_{P=1}$/K.bar^{-1} = +0.45

Polymer (B): polyisobutylene   **1972ZE1**
Characterization: M_η/kg.mol^{-1} = 1660, laboratory fractions, M_w/M_n = 1.14 by GPC
Solvent (A): **2-methylbutane**   **C$_6$H$_{12}$**   **78-78-4**

Type of data: cloud points

T/K = 323.35 w_B = 0.0220   (dT/dP)$_{P=1}$/K.bar^{-1} = +0.44

Polymer (B): polyisobutylene   **1972ZE1**
Characterization: M_η/kg.mol^{-1} = 6.0, laboratory fractions, M_w/M_n = 1.14 by GPC
Solvent (A): **n-pentane**   **C$_5$H$_{12}$**   **109-66-0**

Type of data: cloud points

T/K = 403.55 w_B = 0.0226   (dT/dP)$_{P=1}$/K.bar^{-1} = +0.57

Polymer (B): polyisobutylene   **1972ZE1**
Characterization: M_η/kg.mol^{-1} = 703, laboratory fractions, M_w/M_n = 1.14 by GPC
Solvent (A): **n-pentane**   **C$_5$H$_{12}$**   **109-66-0**

Type of data: cloud points

T/K = 353.55 w_B = 0.0247   (dT/dP)$_{P=1}$/K.bar^{-1} = +0.42

Polymer (B): polyisobutylene   **1972ZE1**
Characterization: M_η/kg.mol^{-1} = 1660, laboratory fractions, M_w/M_n = 1.14 by GPC
Solvent (A): **n-pentane**   **C$_5$H$_{12}$**   **109-66-0**

Type of data: cloud points

T/K = 347.35 w_B = 0.0243   (dT/dP)$_{P=1}$/K.bar^{-1} = +0.45

Polymer (B):   **polyisobutylene**   **1972ZE1**
Characterization: M_η/kg.mol^{-1} = 2.2, laboratory fractions, M_w/M_n = 1.14 by GPC
Solvent (A):   **propane**   **C$_3$H$_8$**   **74-98-6**

Type of data: cloud points

T/K = 291.55 w_B = 0.0135 $(dT/dP)_{P=1}$/K.bar^{-1} = +0.33
T/K = 279.85 w_B = 0.0374 $(dT/dP)_{P=1}$/K.bar^{-1} = +0.48

Polymer (B):   **poly(*N*-isopropylacrylamide)**   **2004SHI**
Characterization: synthesized in the laboratory
Solvent (A):   **deuterium oxide**   **D$_2$O**   **7789-20-0**

Type of data: cloud points

c_B/mol.l^{-1} = 0.690 T/K = 308.45 − 5.99 10^{-4} $(P/\text{MPa} - 48.2)^2$

Polymer (B):   **poly(*N*-isopropylacrylamide)**   **2002RE1**
Characterization: M_n/kg.mol^{-1} = 70, M_w/kg.mol^{-1} = 144,
                               synthesized in the laboratory by radical polymerization
Solvent (A):   **water**   **H$_2$O**   **7732-18-5**

Type of data: cloud points (LCST behavior)

w_B	0.0035	0.0035	0.0035	0.0035	0.0035	0.0035	0.0035	0.0035	0.0307
T/K	313.0	313.7	314.2	314.9	315.1	315.6	315.8	315.4	310.2
P/bar	100.0	202.0	354.5	507.5	558.5	609.5	660.5	711.5	20.5

w_B	0.0307	0.0307	0.0307	0.0307	0.0307	0.0307	0.0307	0.0307	0.0307
T/K	311.2	311.7	312.5	313.1	313.2	313.4	313.4	313.3	313.0
P/bar	148.5	262.5	365.0	426.0	477.0	507.5	538.5	558.5	609.5

Polymer (B):   **poly(*N*-isopropylacrylamide)**   **2002RE1**
Characterization: M_n/kg.mol^{-1} = 120, M_w/kg.mol^{-1} = 258,
                               synthesized in the laboratory by radical polymerization
Solvent (A):   **water**   **H$_2$O**   **7732-18-5**

Type of data: cloud points (LCST behavior)

w_B	0.0039	0.0039	0.0039	0.0039	0.0039	0.0039	0.0039	0.0039	0.0295
T/K	307.2	307.5	308.1	308.4	308.4	308.4	308.2	308.0	306.6
P/bar	20.55	50.00	148.50	252.50	354.50	426.00	507.50	609.50	20.55

w_B	0.0295	0.0295	0.0295	0.0295	0.0295	0.0295	0.0295
T/K	307.0	307.7	308.1	308.1	308.0	307.9	307.9
P/bar	50.00	148.50	283.50	354.50	426.00	507.50	609.50

Polymer (B):   **poly(*N*-isopropylacrylamide)**   **2001GOM**
Characterization: M_n/kg.mol^{-1} = 301.5, M_w/kg.mol^{-1} = 615,
     synthesized in the laboratory by radical polymerization
Solvent (A):   **water**   **H$_2$O**   **7732-18-5**

Type of data: cloud points (LCST behavior)

w_B	0.01030	0.02011	0.02502	0.03574	0.03701	0.04167	0.04167	0.04167	0.04167
T/K	307.20	306.74	306.65	306.46	305.59	305.50	305.60	305.70	305.73
P/MPa	0.1	0.1	0.1	0.1	0.1	0.1	1.0	2.5	3.0

w_B	0.04167	0.04167	0.04167	0.04167	0.04167	0.05646	0.06402	0.06774	0.07670
T/K	305.86	305.90	308.12	308.27	308.32	306.10	306.25	306.40	306.35
P/MPa	4.0	5.0	20.0	30.0	40.0	0.1	0.1	0.1	0.1

w_B	0.08839	0.1057	0.1146	0.1474	0.1757
T/K	306.46	305.89	306.20	305.84	305.72
P/MPa	0.1	0.1	0.1	0.1	0.1

Type of data: spinodal points (LCST behavior)

w_B	0.02011	0.03574	0.04167	0.04167	0.04167	0.04167	0.04167	0.04167	0.04529
T/K	321.6	312.1	309.2	308.5	309.2	308.2	307.6	307.4	310.7
P/MPa	0.10	0.10	0.10	1.0	2.5	3.0	4.0	5.0	0.10

w_B	0.05646	0.06402	0.06774	0.07670	0.08839	0.1146	0.1474	0.1757
T/K	308.9	307.8	308.7	309.1	308.7	309.1	308.7	310.9
P/MPa	0.10	0.10	0.10	0.10	0.10	0.10	0.10	0.10

Polymer (B):   **poly(*N*-isopropylacrylamide)**   **2002RE1**
Characterization: M_n/kg.mol^{-1} = 301.5, M_w/kg.mol^{-1} = 615,
     synthesized in the laboratory by radical polymerization
Solvent (A):   **water**   **H$_2$O**   **7732-18-5**

Type of data: cloud points (LCST behavior)

w_B	0.0103	0.0201	0.0250	0.0357	0.0370	0.0417	0.0417	0.0417	0.0417
T/K	307.2	306.7	306.6	306.5	305.6	305.5	305.6	305.7	305.7
P/bar	1.0	1.0	1.0	1.0	1.0	1.0	10.0	25.0	30.0

w_B	0.0417	0.0417	0.0417	0.0417	0.0417	0.0417	0.0453	0.0565	0.0640
T/K	305.9	305.9	306.7	308.0	308.0	308.0	306.3	306.1	306.2
P/bar	40.0	50.0	100.0	200.0	300.0	400.0	1.0	1.0	1.0

w_B	0.0677	0.0767	0.0884	0.1057	0.1146	0.1474	0.1757
T/K	306.4	306.3	306.5	305.9	306.2	305.8	305.7
P/bar	1.0	1.0	1.0	1.0	1.0	1.0	1.0

Polymer (B):   **poly(*N*-isopropylacrylamide)**   **2004SHI**
Characterization: synthesized in the laboratory
Solvent (A):   **water**   **H$_2$O**   **7732-18-5**

Type of data: cloud points

c_B/mol.l^{-1} = 0.690 $T/K = 306.75 - 5.59\ 10^{-4}\ (P/\text{MPa} - 51.7)^2$

Polymer (B):	**polystyrene**							**1997DES**

Characterization: $M_n/\text{kg.mol}^{-1} = 2.3$, $M_w/\text{kg.mol}^{-1} = 2.5$,
Pressure Chemical Company, Pittsburgh, PA

Solvent (A):	**acetaldehyde**		**C₂H₄O**					**75-07-0**

Solvent (A):   **acetaldehyde** C_2H_4O **75-07-0**

Type of data: cloud points

w_B	0.1090	0.1090	0.1090	0.1090	0.1090	0.1090	0.1090	0.1090	0.1090
T/K	276.17	276.18	276.17	276.67	276.67	277.17	277.18	277.18	277.67
P/MPa	2.91	2.92	2.92	2.50	2.51	2.02	2.02	2.05	1.55

w_B	0.1090	0.1090	0.1090	0.1090	0.1090	0.1090	0.1090	0.1090	0.1374
T/K	277.67	277.67	278.47	278.48	278.67	278.66	278.97	278.97	281.66
P/MPa	1.57	1.58	0.79	0.74	0.61	0.62	0.34	0.32	2.69

w_B	0.1374	0.1374	0.1374	0.1374	0.1374	0.1374	0.1374	0.1374	0.1374
T/K	281.66	281.66	282.16	282.16	282.16	282.16	282.66	282.66	282.66
P/MPa	2.65	2.67	2.27	2.28	2.26	2.27	1.83	1.81	1.81

w_B	0.1374	0.1374	0.1374	0.1374	0.1374	0.1374	0.1374	0.1374	0.1709
T/K	283.16	283.16	283.16	283.97	283.97	284.16	284.16	284.32	284.16
P/MPa	1.36	1.38	1.37	0.60	0.61	0.46	0.46	0.31	2.96

w_B	0.1709	0.1709	0.1709	0.1709	0.1709	0.1709	0.1709	0.1709	0.1709
T/K	284.16	284.16	284.67	284.67	284.67	285.17	285.16	285.17	285.67
P/MPa	2.96	2.97	2.42	2.41	2.42	1.89	1.89	1.91	1.39

w_B	0.1709	0.1709	0.1709	0.1709	0.1709	0.1709	0.1709	0.1709	0.1709
T/K	285.67	285.87	285.86	286.26	286.26	286.48	286.47	286.66	286.66
P/MPa	1.38	1.14	1.16	0.73	0.74	0.54	0.54	0.34	0.35

w_B	0.1766	0.1766	0.1766	0.1766	0.1766	0.1766	0.1766	0.1766	0.1766
T/K	290.07	290.07	290.07	291.17	291.17	291.38	291.38	289.67	289.67
P/MPa	1.51	1.54	1.50	0.74	0.73	0.65	0.65	1.75	1.72

w_B	0.1766	0.1766	0.1766	0.1766	0.1766	0.1986	0.1986	0.1986	0.1986
T/K	289.46	289.46	288.76	288.76	288.67	286.67	286.66	287.17	287.17
P/MPa	1.93	1.90	2.41	2.43	2.68	3.10	3.10	2.65	2.67

w_B	0.1986	0.1986	0.1986	0.1986	0.1986	0.1986	0.1986	0.1986	0.1986
T/K	287.66	287.66	287.66	288.17	288.17	288.67	288.67	288.97	288.97
P/MPa	2.25	2.25	2.25	1.82	1.83	1.39	1.41	1.19	1.16

w_B	0.1986	0.1986	0.1986	0.1986	0.2076	0.2076	0.2076	0.2076	0.2076
T/K	289.97	289.97	290.27	290.27	281.66	282.16	282.67	283.16	283.16
P/MPa	0.46	0.46	0.22	0.23	2.93	2.55	2.18	1.80	1.75

w_B	0.2076	0.2225	0.2225	0.2225	0.2225	0.2225	0.2225	0.2225	0.2225
T/K	284.68	290.67	290.68	291.17	291.18	291.66	291.66	292.17	292.17
P/MPa	0.43	2.96	3.02	2.64	2.64	2.24	2.23	1.90	1.88

w_B	0.2225	0.2225	0.2225	0.2225	0.2225	0.2225	0.2225	0.2225	0.2329
T/K	292.67	292.67	293.16	293.16	294.16	294.16	294.67	294.68	286.17
P/MPa	1.54	1.56	1.23	1.24	0.58	0.59	0.29	0.29	2.82

continued

continued

w_B	0.2329	0.2329	0.2329	0.2329	0.2329	0.2329	0.2329	0.2329	0.2329
T/K	286.17	286.66	287.17	287.66	287.66	287.67	287.66	288.17	288.17
P/MPa	2.83	2.38	1.99	1.65	1.63	1.63	1.62	1.29	1.28

w_B	0.2329	0.2329	0.2329	0.2329	0.2329	0.2329	0.2631	0.2631	0.2631
T/K	288.17	289.16	289.16	289.57	289.57	289.77	285.68	286.17	286.17
P/MPa	1.26	0.62	0.62	0.30	0.30	0.17	3.08	2.70	2.71

w_B	0.2631	0.2631	0.2631	0.2631	0.2631	0.2631	0.2631	0.2631	0.2631
T/K	286.66	286.66	286.66	287.17	287.17	287.67	287.67	288.17	288.17
P/MPa	2.33	2.31	2.31	1.95	1.94	1.53	1.52	1.15	1.13

w_B	0.2631	0.2631	0.2631	0.2631	0.3363	0.3363	0.3363	0.3363	0.3363
T/K	288.67	288.67	289.16	289.16	284.17	284.17	284.68	284.68	285.17
P/MPa	0.77	0.78	0.43	0.43	2.97	2.96	2.52	2.52	2.11

w_B	0.3363	0.3363	0.3363	0.3363	0.3363	0.3363	0.3363	0.3363	0.3363
T/K	285.17	285.68	285.68	286.17	286.17	286.66	286.66	287.17	287.17
P/MPa	2.10	1.61	1.61	1.27	1.28	0.87	0.88	0.49	0.50

Type of data: spinodal points

w_B	0.1090	0.1090	0.1090	0.1090	0.1090	0.1090	0.1090	0.1090	0.1090
T/K	276.17	276.18	276.17	276.67	276.67	277.17	277.18	277.18	277.67
P/MPa	2.77	2.81	2.78	2.40	2.41	1.89	1.90	1.92	1.44

w_B	0.1090	0.1090	0.1090	0.1090	0.1090	0.1090	0.1090	0.1090	0.1374
T/K	277.67	277.67	278.47	278.48	278.67	278.66	278.97	278.97	281.66
P/MPa	1.46	1.45	0.70	0.65	0.53	0.51	0.22	0.19	2.51

w_B	0.1374	0.1374	0.1374	0.1374	0.1374	0.1374	0.1374	0.1374	0.1374
T/K	281.66	281.66	282.16	282.16	282.16	282.16	282.66	282.66	282.66
P/MPa	2.51	2.51	2.12	2.09	2.10	2.12	1.68	1.66	1.67

w_B	0.1374	0.1374	0.1374	0.1374	0.1374	0.1374	0.1374	0.1374	0.1709
T/K	283.16	283.16	283.16	283.97	283.97	284.16	284.16	284.32	284.16
P/MPa	1.21	1.21	1.23	0.47	0.50	0.34	0.35	0.18	2.83

w_B	0.1709	0.1709	0.1709	0.1709	0.1709	0.1709	0.1709	0.1709	0.1709
T/K	284.16	284.16	284.67	284.67	284.67	285.17	285.16	285.17	285.67
P/MPa	2.84	2.82	2.28	2.28	2.29	1.74	1.76	1.77	1.24

w_B	0.1709	0.1709	0.1709	0.1709	0.1709	0.1709	0.1709	0.1709	0.1709
T/K	285.67	285.87	285.86	286.26	286.26	286.48	286.47	286.66	286.66
P/MPa	1.21	1.00	1.01	0.62	0.63	0.42	0.41	0.21	0.22

w_B	0.1766	0.1766	0.1766	0.1766	0.1766	0.1766	0.1766	0.1766	0.1766
T/K	290.07	290.07	290.07	291.17	291.17	291.38	291.38	289.67	289.67
P/MPa	1.33	1.40	1.32	0.57	0.60	0.53	0.50	1.61	1.61

w_B	0.1766	0.1766	0.1766	0.1766	0.1766	0.1986	0.1986	0.1986	0.1986
T/K	289.46	289.46	288.76	288.76	288.67	286.67	286.66	287.17	287.17
P/MPa	1.80	1.76	2.25	2.24	2.49	2.96	2.97	2.48	2.51

w_B	0.1986	0.1986	0.1986	0.1986	0.1986	0.1986	0.1986	0.1986	0.1986
T/K	287.66	287.66	287.66	288.17	288.17	288.67	288.67	288.97	288.97
P/MPa	2.10	2.12	2.13	1.71	1.69	1.28	1.28	1.05	1.02

continued

continued

w_B	0.1986	0.1986	0.1986	0.1986	0.2076	0.2076	0.2076	0.2076	0.2076
T/K	289.97	289.97	290.27	290.27	281.66	282.16	282.67	283.16	283.16
P/MPa	0.33	0.34	0.09	0.10	2.85	2.43	2.06	1.65	1.60

w_B	0.2076	0.2225	0.2225	0.2225	0.2225	0.2225	0.2225	0.2225	0.2225
T/K	284.68	290.67	290.68	291.17	291.18	291.66	291.66	292.17	292.17
P/MPa	0.32	2.91	2.93	2.55	2.55	2.15	2.15	1.80	1.81

w_B	0.2225	0.2225	0.2225	0.2225	0.2225	0.2225	0.2225	0.2225	0.2329
T/K	292.67	292.67	293.16	293.16	294.16	294.16	294.67	294.68	286.17
P/MPa	1.50	1.52	1.16	1.19	0.52	0.53	0.22	0.22	2.71

w_B	0.2329	0.2329	0.2329	0.2329	0.2329	0.2329	0.2329	0.2329	0.2329
T/K	286.17	286.66	287.17	287.66	287.66	287.67	287.66	288.17	288.17
P/MPa	2.72	2.27	1.91	1.58	1.57	1.56	1.53	1.22	1.19

w_B	0.2329	0.2329	0.2329	0.2329	0.2329	0.2329	0.2631	0.2631	0.2631
T/K	288.17	289.16	289.16	289.57	289.57	289.77	285.68	286.17	286.17
P/MPa	1.18	0.57	0.54	0.25	0.25	0.10	2.98	2.59	2.60

w_B	0.2631	0.2631	0.2631	0.2631	0.2631	0.2631	0.2631	0.2631	0.2631
T/K	286.66	286.66	286.66	287.17	287.17	287.67	287.67	288.17	288.17
P/MPa	2.26	2.22	2.20	1.84	1.84	1.42	1.42	1.01	1.03

w_B	0.2631	0.2631	0.2631	0.2631	0.3363	0.3363	0.3363	0.3363	0.3363
T/K	288.67	288.67	289.16	289.16	284.17	284.17	284.68	284.68	285.17
P/MPa	0.73	0.70	0.33	0.37	2.88	2.90	2.43	2.45	2.02

w_B	0.3363	0.3363	0.3363	0.3363	0.3363	0.3363	0.3363	0.3363	0.3363
T/K	285.17	285.68	285.68	286.17	286.17	286.66	286.66	287.17	287.17
P/MPa	2.02	1.54	1.53	1.18	1.20	0.79	0.79	0.41	0.42

Polymer (B):	**polystyrene**							**1997DES**
Characterization:	M_n/kg.mol^{-1} = 3.1, M_w/kg.mol^{-1} = 3.25,							
	Pressure Chemical Company, Pittsburgh, PA							
Solvent (A):	**acetaldehyde**		**C$_2$H$_4$O**					**75-07-0**

Type of data: cloud points

w_B	0.0736	0.0736	0.0736	0.0736	0.0736	0.0736	0.0736	0.0736	0.0736
T/K	286.16	286.65	287.16	287.65	288.66	289.66	290.15	290.67	290.86
P/MPa	4.32	3.91	3.51	3.10	2.33	1.64	1.30	0.96	0.84

w_B	0.0736	0.0736	0.0736	0.0892	0.0892	0.0892	0.0892	0.0892	0.0892
T/K	291.16	291.46	291.76	294.68	295.17	295.67	296.67	296.67	297.16
P/MPa	0.67	0.46	0.26	3.25	3.01	2.72	2.19	2.20	1.87

w_B	0.0892	0.0892	0.0892	0.0892	0.0892	0.1058	0.1058	0.1058	0.1058
T/K	297.65	298.16	299.15	299.56	299.76	297.25	297.05	296.96	296.85
P/MPa	1.60	1.35	0.83	0.62	0.52	0.15	0.26	0.31	0.36

w_B	0.1058	0.1058	0.1058	0.1058	0.1058	0.1058	0.1058	0.1058	0.1058
T/K	296.76	296.66	296.56	296.45	296.35	295.17	294.16	293.16	292.17
P/MPa	0.41	0.47	0.53	0.59	0.63	1.27	1.86	2.47	3.05

continued

continued

w_B	0.1058	0.1058	0.1184	0.1184	0.1184	0.1184	0.1184	0.1184	0.1184
T/K	291.66	291.15	295.66	296.17	296.66	297.66	298.16	299.15	299.66
P/MPa	3.40	3.78	3.34	3.02	2.74	2.19	1.92	1.40	1.17

w_B	0.1184	0.1184	0.1184	0.1184	0.1184	0.1184	0.1415	0.1415	0.1415
T/K	300.65	301.15	301.34	301.56	301.75	301.95	309.14	309.14	309.65
P/MPa	0.65	0.48	0.40	0.29	0.24	0.16	4.18	4.20	3.95

w_B	0.1415	0.1415	0.1415	0.1415	0.1415	0.1415	0.1415	0.1415	0.1415
T/K	310.14	312.15	314.14	317.14	318.65	320.63	321.63	322.63	323.63
P/MPa	3.71	3.00	2.38	1.60	1.21	0.81	0.60	0.44	0.29

w_B	0.1415	0.1561	0.1561	0.1561	0.1561	0.1561	0.1561	0.1561	0.1561
T/K	324.14	315.64	316.64	317.64	318.65	320.63	323.12	326.14	327.63
P/MPa	0.23	4.03	3.72	3.38	3.10	2.57	2.00	1.43	1.21

w_B	0.1561	0.1561	0.1561	0.1561	0.1561	0.1654	0.1654	0.1654	0.1654
T/K	330.62	331.62	333.11	334.12	335.12	314.64	315.64	316.64	317.65
P/MPa	0.80	0.70	0.57	0.46	0.41	3.80	3.45	3.20	2.93

w_B	0.1654	0.1654	0.1654	0.1654	0.1654	0.1654	0.1654	0.1654	0.1654
T/K	318.65	320.63	322.63	325.14	328.63	329.62	330.63	331.63	333.12
P/MPa	2.62	2.13	1.72	1.28	0.80	0.70	0.57	0.48	0.35

w_B	0.1654	0.1654	0.1676	0.1676	0.1676	0.1676	0.1676	0.1676	0.1676
T/K	334.22	334.71	289.85	290.67	291.16	292.17	293.16	294.16	294.37
P/MPa	0.25	0.23	4.08	3.52	3.19	2.55	1.96	1.40	1.26

w_B	0.1676	0.1676	0.1676	0.1676	0.1676	0.1676	0.1676	0.1702	0.1702
T/K	295.46	295.56	295.75	295.96	296.17	296.35	296.56	275.17	275.66
P/MPa	0.69	0.64	0.53	0.45	0.32	0.23	0.13	4.11	3.43

w_B	0.1702	0.1702	0.1702	0.1702	0.1702	0.1702	0.1702	0.1702	0.1702
T/K	276.17	277.17	277.66	277.76	278.26	278.47	278.66	278.76	278.87
P/MPa	2.90	1.89	1.39	1.30	0.84	0.65	0.50	0.40	0.32

w_B	0.1702	0.1786	0.1786	0.1786	0.1786	0.1786	0.1786	0.1786	0.1786
T/K	278.96	270.17	270.68	271.17	271.67	272.67	273.06	273.67	273.76
P/MPa	0.25	4.56	3.95	3.33	2.74	1.66	1.24	0.65	0.55

w_B	0.1786	0.1786	0.1786	0.1932	0.1932	0.1932	0.1932	0.1932	0.1932
T/K	273.87	273.97	274.06	282.15	283.15	284.15	284.67	285.16	285.67
P/MPa	0.42	0.32	0.21	4.11	3.27	2.45	2.06	1.68	1.32

w_B	0.1932	0.1932	0.1932	0.1932	0.1932	0.1932	0.2069	0.2069	0.2069
T/K	286.46	286.66	287.07	287.16	287.35	287.46	312.64	313.13	314.13
P/MPa	0.72	0.60	0.33	0.27	0.14	0.08	4.35	4.18	3.84

w_B	0.2069	0.2069	0.2069	0.2069	0.2069	0.2069	0.2069	0.2069	0.2069
T/K	315.64	317.13	319.14	321.13	323.13	325.64	326.64	327.32	327.93
P/MPa	3.26	2.81	2.21	1.71	1.26	0.78	0.60	0.49	0.40

w_B	0.2069	0.2253	0.2253	0.2253	0.2253	0.2253	0.2253	0.2253	0.2253
T/K	328.43	316.91	317.64	318.65	320.63	323.13	325.64	328.13	329.63
P/MPa	0.33	4.12	3.89	3.58	3.02	2.41	1.88	1.46	1.20

continued

continued

w_B	0.2253	0.2253	0.2253	0.2253	0.2253	0.2253	0.2253	0.2461	0.2461
T/K	333.12	334.11	335.61	337.62	339.10	340.10	341.10	316.14	317.14
P/MPa	0.78	0.67	0.55	0.39	0.30	0.22	0.17	3.89	3.58

w_B	0.2461	0.2461	0.2461	0.2461	0.2461	0.2461	0.2461	0.2461	0.2461
T/K	318.13	320.14	322.13	325.14	328.13	331.62	333.12	334.12	335.12
P/MPa	3.30	2.76	2.29	1.70	1.23	0.79	0.64	0.55	0.45

w_B	0.2461	0.3196	0.3196	0.3196	0.3196	0.3196	0.3196	0.3196	0.3196
T/K	337.12	316.64	317.14	318.13	320.13	322.13	324.14	327.13	330.12
P/MPa	0.34	4.55	4.35	4.02	3.43	2.90	2.44	1.85	1.37

w_B	0.3196	0.3196	0.3196	0.3196	0.3196	0.3196
T/K	333.62	336.12	338.12	340.10	341.10	342.10
P/MPa	0.92	0.69	0.54	0.39	0.32	0.27

Type of data: spinodal points

w_B	0.0736	0.0736	0.0736	0.0736	0.0736	0.0736	0.0736	0.0736	0.0736
T/K	286.16	286.65	287.16	287.65	288.66	289.66	290.15	290.86	291.16
P/MPa	4.18	3.79	3.37	3.02	2.27	1.56	1.23	0.79	0.56

w_B	0.0736	0.0736	0.1058	0.1058	0.1058	0.1058	0.1058	0.1058	0.1058
T/K	291.46	291.76	297.25	297.05	296.96	296.85	296.76	296.66	296.45
P/MPa	0.38	0.19	0.06	0.19	0.21	0.26	0.30	0.35	0.52

w_B	0.1058	0.1058	0.1058	0.1058	0.1058	0.1058	0.1415	0.1415	0.1415
T/K	295.17	294.16	293.16	292.18	291.66	291.15	309.14	309.65	310.14
P/MPa	1.12	1.75	2.39	2.90	3.25	3.63	3.94	3.74	3.63

w_B	0.1415	0.1415	0.1415	0.1415	0.1415	0.1415	0.1415	0.1415	0.1561
T/K	312.15	314.14	317.14	318.64	321.63	322.63	323.63	324.14	315.64
P/MPa	2.88	2.30	1.50	1.14	0.50	0.34	0.19	0.15	3.90

w_B	0.1561	0.1561	0.1561	0.1561	0.1561	0.1561	0.1561	0.1561	0.1561
T/K	316.64	317.64	318.65	320.63	323.12	326.14	327.63	330.62	333.11
P/MPa	3.57	3.28	2.96	2.49	1.94	1.32	1.14	0.76	0.50

w_B	0.1561	0.1561	0.1654	0.1654	0.1654	0.1654	0.1654	0.1654	0.1654
T/K	334.12	335.12	314.64	315.65	316.64	317.65	318.66	320.63	322.63
P/MPa	0.36	0.28	3.69	3.40	3.12	2.81	2.59	2.04	1.66

w_B	0.1654	0.1654	0.1654	0.1654	0.1654	0.1654	0.1654	0.1676	0.1676
T/K	325.14	329.62	330.63	331.63	333.12	334.22	334.71	289.85	290.67
P/MPa	1.23	0.68	0.53	0.43	0.27	0.19	0.17	3.95	3.33

w_B	0.1676	0.1676	0.1676	0.1676	0.1676	0.1676	0.1676	0.1676	0.1676
T/K	291.16	292.17	293.16	294.16	294.37	295.46	295.56	295.75	295.96
P/MPa	3.13	2.49	1.89	1.29	1.18	0.63	0.58	0.48	0.35

w_B	0.1676	0.1676	0.1702	0.1702	0.1702	0.1702	0.1702	0.1702	0.1702
T/K	296.17	296.35	275.17	275.66	276.17	277.17	277.66	277.76	278.26
P/MPa	0.24	0.11	3.95	3.32	2.76	1.76	1.30	1.22	0.75

w_B	0.1702	0.1702	0.1702	0.1702	0.1702	0.1786	0.1786	0.1786	0.1786
T/K	278.47	278.66	278.76	278.87	278.96	270.17	270.68	271.17	271.67
P/MPa	0.53	0.38	0.29	0.22	0.18	4.47	3.86	3.20	2.63

continued

continued

w_B	0.1786	0.1786	0.1786	0.1786	0.1786	0.1786	0.1786	0.1932	0.1932
T/K	272.67	273.06	273.67	273.76	273.87	273.97	274.06	282.15	283.15
P/MPa	1.52	1.16	0.54	0.41	0.35	0.23	0.15	3.95	3.05

w_B	0.1932	0.1932	0.1932	0.1932	0.1932	0.1932	0.1932	0.1932	0.2069
T/K	284.16	284.67	285.16	285.67	286.66	287.07	287.16	287.35	312.64
P/MPa	2.31	1.95	1.51	1.18	0.48	0.20	0.14	0.03	4.27

w_B	0.2069	0.2069	0.2069	0.2069	0.2069	0.2069	0.2069	0.2069	0.2069
T/K	313.13	314.13	315.64	317.14	319.14	321.13	323.13	325.64	326.64
P/MPa	4.08	3.73	3.19	2.71	2.16	1.63	1.16	0.71	0.56

w_B	0.2069	0.2069	0.2069	0.2253	0.2253	0.2253	0.2253	0.2253	0.2253
T/K	327.32	327.93	328.43	316.90	317.64	318.65	320.63	323.13	325.64
P/MPa	0.45	0.32	0.25	4.05	3.81	3.51	2.95	2.34	1.82

w_B	0.2253	0.2253	0.2253	0.2253	0.2253	0.2253	0.2253	0.2253	0.2253
T/K	328.13	329.63	333.12	334.11	335.61	337.62	339.10	340.10	341.10
P/MPa	1.38	1.14	0.73	0.62	0.46	0.32	0.23	0.18	0.11

w_B	0.2461	0.2461	0.2461	0.2461	0.2461	0.2461	0.2461	0.2461	0.2461
T/K	316.14	317.14	318.13	320.14	322.13	325.14	328.13	331.62	333.12
P/MPa	3.69	3.43	3.19	2.66	2.12	1.55	1.13	0.64	0.51

w_B	0.2461	0.2461	0.2461	0.3196	0.3196	0.3196	0.3196	0.3196	0.3196
T/K	334.12	335.12	337.12	316.64	317.14	318.13	320.13	322.13	324.14
P/MPa	0.37	0.28	0.19	4.45	4.25	3.93	3.35	2.82	2.33

w_B	0.3196	0.3196	0.3196	0.3196	0.3196	0.3196	0.3196	0.3196
T/K	327.13	330.12	333.62	336.12	338.12	340.10	341.10	342.10
P/MPa	1.77	1.29	0.90	0.63	0.47	0.33	0.25	0.17

Polymer (B):	**polystyrene**		**1997DES, 1997REB**
Characterization:	M_n/kg.mol^{-1} = 4.0, M_w/kg.mol^{-1} = 4.14,		
	Pressure Chemical Company, Pittsburgh, PA		
Solvent (A):	**acetaldehyde**	**C_2H_4O**	**75-07-0**

Type of data: cloud points

w_B	0.0563	0.0563	0.0563	0.0563	0.0563	0.0563	0.0563	0.0563	0.0563
T/K	268.07	268.65	269.16	270.19	270.95	271.55	272.07	272.57	428.05
P/MPa	4.94	4.34	3.68	2.33	1.60	0.93	0.53	0.10	5.29

w_B	0.0563	0.0563	0.0703	0.0703	0.0703	0.0703	0.0703	0.0703	0.0703
T/K	427.05	426.23	293.61	295.21	298.19	300.63	302.21	302.79	304.31
P/MPa	4.37	3.41	5.58	4.80	3.19	2.11	1.59	1.26	0.80

w_B	0.0703	0.0703	0.0703	0.0703	0.0985	0.0985	0.0985	0.0985	0.0985
T/K	304.59	305.19	305.47	306.02	323.61	326.12	328.65	332.09	337.08
P/MPa	0.71	0.55	0.40	0.28	4.22	3.65	3.18	2.70	2.14

w_B	0.0985	0.0985	0.0985	0.0985	0.0985	0.0985	0.0985	0.0985	0.0985
T/K	342.07	348.10	353.13	356.60	361.59	363.61	368.08	371.12	375.11
P/MPa	1.89	1.78	1.81	1.92	2.01	2.08	2.46	2.64	3.07

continued

continued

w_B	0.0985	0.0985	0.1115	0.1115	0.1115	0.1115	0.1115	0.1115	0.1115
T/K	380.13	385.12	281.66	282.62	283.73	284.70	285.47	286.57	286.72
P/MPa	3.58	4.04	4.39	3.63	2.81	2.04	1.50	0.68	0.56

w_B	0.1115	0.1115	0.1115	0.1115	0.1115	0.1115	0.1115	0.1253	0.1253
T/K	287.02	287.20	413.23	412.69	412.22	411.75	411.11	278.68	279.57
P/MPa	0.41	0.27	4.27	3.64	3.11	2.54	2.09	5.64	4.77

w_B	0.1253	0.1253	0.1253	0.1253	0.1253	0.1253	0.1253	0.1253	0.1253
T/K	280.74	281.75	282.75	283.68	284.14	284.45	284.74	285.01	285.19
P/MPa	3.75	2.91	2.08	1.35	0.99	0.74	0.53	0.29	0.24

w_B	0.1253	0.1253	0.1253	0.1253	0.1253	0.1419	0.1419	0.1419	0.1419
T/K	414.25	413.66	413.54	413.22	412.84	280.17	281.19	282.17	283.17
P/MPa	4.05	3.44	3.00	2.53	2.18	5.80	4.83	3.87	3.12

w_B	0.1419	0.1419	0.1419	0.1419	0.1419	0.1419	0.1419	0.1419	0.1419
T/K	284.18	285.07	285.66	286.27	286.61	286.76	287.19	415.13	414.13
P/MPa	2.35	1.59	1.24	0.89	0.63	0.48	0.22	4.39	3.79

w_B	0.1419	0.1419	0.1419	0.1630	0.1630	0.1630	0.1630	0.1630	0.1630
T/K	413.15	412.56	412.19	306.08	307.09	309.15	311.05	313.14	315.11
P/MPa	3.14	2.61	2.21	4.78	4.37	3.65	2.98	2.40	1.87

w_B	0.1630	0.1630	0.1630	0.1630	0.1630	0.1630	0.1630	0.1630	0.1630
T/K	317.14	319.62	320.63	322.11	323.53	324.20	389.62	389.12	387.11
P/MPa	1.37	0.87	0.70	0.51	0.26	0.17	4.88	4.58	4.01

w_B	0.1630	0.1630	0.1630	0.1630	0.1630	0.1700	0.1700	0.1700	0.1700
T/K	385.11	383.10	380.16	377.14	376.08	305.02	306.15	308.10	310.09
P/MPa	3.51	2.97	2.26	1.53	1.19	4.70	4.29	3.49	2.80

w_B	0.1700	0.1700	0.1700	0.1700	0.1700	0.1700	0.1700	0.1700	0.1700
T/K	312.03	313.95	315.22	317.02	317.99	319.10	319.93	320.64	391.13
P/MPa	2.13	1.56	1.34	0.89	0.69	0.50	0.34	0.20	4.85

w_B	0.1700	0.1700	0.1700	0.1700	0.1700	0.1700	0.1700	0.1700	0.1731
T/K	390.20	388.18	386.15	384.19	382.15	380.18	378.10	376.09	304.18
P/MPa	4.48	3.98	3.49	3.05	2.61	2.11	1.66	1.23	4.74

w_B	0.1731	0.1731	0.1731	0.1731	0.1731	0.1731	0.1731	0.1731	0.1731
T/K	305.13	307.02	308.99	311.12	313.02	314.13	316.57	317.06	318.07
P/MPa	4.27	3.50	2.83	2.21	1.61	1.38	0.79	0.67	0.45

w_B	0.1731	0.1731	0.1731	0.1731	0.1731	0.1731	0.1731	0.1731	0.1731
T/K	305.13	307.02	308.99	311.12	313.02	314.13	316.57	317.06	318.07
P/MPa	4.19	3.44	2.74	2.10	1.53	1.33	0.72	0.60	0.37

w_B	0.1731	0.1731	0.1870	0.1870	0.1870	0.1870	0.1870	0.1870	0.1870
T/K	318.62	319.03	306.14	308.19	310.02	312.12	314.16	316.11	318.15
P/MPa	0.33	0.25	4.91	4.09	3.41	2.77	2.20	1.71	1.28

w_B	0.1870	0.1870	0.1870	0.1870	0.1870	0.1870	0.1870	0.1870	0.1870
T/K	320.59	322.07	323.99	393.13	391.16	389.13	386.15	383.14	380.09
P/MPa	0.79	0.52	0.28	5.35	4.76	4.21	3.46	2.77	2.10

continued

continued

w_B	0.1870	0.2201	0.2201	0.2201	0.2201	0.2201	0.2201	0.2201	0.2201
T/K	377.14	303.08	305.14	307.11	309.12	311.16	313.18	315.18	316.18
P/MPa	1.43	4.91	4.08	3.30	2.59	1.94	1.36	0.88	0.65

w_B	0.2201	0.2201	0.2201	0.2201	0.2201	0.2201	0.2201	0.2201	0.2201
T/K	317.31	318.23	393.15	391.12	389.19	386.15	383.13	380.11	377.13
P/MPa	0.45	0.23	4.98	4.41	3.91	3.22	2.56	1.93	1.31

w_B	0.2201	0.2201	0.2201	0.2201	0.2201	0.2201	0.2201	0.2201	0.2201
T/K	317.31	318.23	393.15	391.12	389.19	386.15	383.13	380.11	377.13
P/MPa	0.36	0.16	4.79	4.29	3.75	3.13	2.44	1.76	1.15

Type of data: spinodal points

w_B	0.0563	0.0563	0.0563	0.0563	0.0563	0.0563	0.0563	0.0563	
T/K	268.07	268.65	269.16	270.19	270.95	271.55	272.07	272.57	
P/MPa	4.85	4.24	3.59	2.25	1.54	0.89	0.47	0.03	

w_B	0.0563	0.0563	0.0703	0.0703	0.0703	0.0703	0.0703	0.0703	0.0703
T/K	427.05	426.23	293.61	295.21	298.19	300.63	302.21	302.79	304.31
P/MPa	4.30	3.36	5.53	4.74	3.10	2.05	1.51	1.18	0.77

w_B	0.0703	0.0703	0.0703	0.0703	0.0985	0.0985	0.0985	0.0985	0.0985
T/K	304.59	305.19	305.47	306.02	323.61	326.12	328.65	332.09	337.08
P/MPa	0.65	0.49	0.34	0.21	4.13	3.56	3.09	2.64	2.10

w_B	0.0985	0.0985	0.0985	0.0985	0.0985	0.0985	0.0985	0.0985	0.0985
T/K	342.07	348.10	353.13	356.60	361.59	363.61	368.08	371.12	375.11
P/MPa	1.85	1.71	1.76	1.90	1.96	2.05	2.38	2.61	3.02

w_B	0.0985	0.0985	0.1115	0.1115	0.1115	0.1115	0.1115	0.1115	0.1115
T/K	380.13	385.12	281.66	282.62	283.73	284.70	285.47	286.57	286.72
P/MPa	3.56	3.99	4.28	3.54	2.71	1.93	1.41	0.60	0.47

w_B	0.1115	0.1115	0.1115	0.1115	0.1115	0.1115	0.1115	0.1253	0.1253
T/K	287.02	287.20	413.23	412.69	412.22	411.75	411.11	278.68	279.57
P/MPa	0.39	0.21	4.25	3.61	3.04	2.50	2.06	5.48	4.67

w_B	0.1253	0.1253	0.1253	0.1253	0.1253	0.1253	0.1253	0.1253	0.1253
T/K	280.74	281.75	282.75	283.68	284.14	284.45	284.74	285.01	285.19
P/MPa	3.61	2.79	1.95	1.25	0.92	0.66	0.46	0.21	0.17

w_B	0.1253	0.1253	0.1253	0.1253	0.1253	0.1419	0.1419	0.1419	0.1419
T/K	414.25	413.66	413.54	413.22	412.84	280.17	281.19	282.17	283.17
P/MPa	3.94	3.37	2.91	2.41	2.07	5.68	4.76	3.80	3.03

w_B	0.1419	0.1419	0.1419	0.1419	0.1419	0.1419	0.1419	0.1419	0.1419
T/K	284.18	285.07	285.66	286.27	286.61	286.76	287.19	415.13	414.13
P/MPa	2.28	1.52	1.13	0.78	0.58	0.42	0.16	4.34	3.74

w_B	0.1419	0.1419	0.1419	0.1630	0.1630	0.1630	0.1630	0.1630	0.1630
T/K	413.15	412.56	412.19	306.08	307.09	309.15	311.05	313.14	315.11
P/MPa	3.08	2.53	2.17	4.69	4.29	3.57	2.91	2.33	1.77

w_B	0.1630	0.1630	0.1630	0.1630	0.1630	0.1630	0.1630	0.1630	0.1630
T/K	317.14	319.62	320.63	322.11	323.53	324.20	389.62	389.12	387.11
P/MPa	1.32	0.82	0.64	0.44	0.19	0.11	4.79	4.49	3.97

continued

continued

w_B	0.1630	0.1630	0.1630	0.1630	0.1630	0.1700	0.1700	0.1700	0.1700
T/K	385.11	383.10	380.16	377.14	376.08	305.02	306.15	308.10	310.09
P/MPa	3.47	2.91	2.23	1.48	1.15	4.62	4.19	3.39	2.70

w_B	0.1700	0.1700	0.1700	0.1700	0.1700	0.1700	0.1700	0.1700	0.1700
T/K	312.03	313.95	315.22	317.02	317.99	319.10	319.93	320.64	391.13
P/MPa	2.07	1.50	1.28	0.85	0.64	0.45	0.27	0.14	4.78

w_B	0.1700	0.1700	0.1700	0.1700	0.1700	0.1700	0.1700	0.1700	0.1731
T/K	390.20	388.18	386.15	384.19	382.15	380.18	378.10	376.09	304.18
P/MPa	4.45	3.93	3.42	3.00	2.55	2.08	1.62	1.19	4.62

w_B	0.1731	0.1731	0.1731	0.1731	0.1731	0.1731	0.1731	0.1731	0.1731
T/K	305.13	307.02	308.99	311.12	313.02	314.13	316.57	317.06	318.07
P/MPa	4.19	3.44	2.74	2.10	1.53	1.33	0.72	0.60	0.37

w_B	0.1731	0.1731	0.1870	0.1870	0.1870	0.1870	0.1870	0.1870	0.1870
T/K	318.62	319.03	306.14	308.19	310.02	312.12	314.16	316.11	318.15
P/MPa	0.27	0.18	4.84	4.01	3.33	2.68	2.14	1.67	1.22

w_B	0.1870	0.1870	0.1870	0.1870	0.1870	0.1870	0.1870	0.1870	0.1870
T/K	320.59	322.07	323.99	393.13	391.16	389.13	386.15	383.14	380.09
P/MPa	0.75	0.47	0.23	5.30	4.72	4.14	3.43	2.70	2.05

w_B	0.1870	0.2201	0.2201	0.2201	0.2201	0.2201	0.2201	0.2201	0.2201
T/K	377.14	303.08	305.14	307.11	309.12	311.16	313.18	315.18	316.18
P/MPa	1.38	4.77	3.95	3.22	2.50	1.89	1.29	0.81	0.58

Polymer (B):	**polystyrene**		**1981WOL**
Characterization:	M_n/kg.mol^{-1} = 545.5, M_w/kg.mol^{-1} = 600,		
	Pressure Chemical Company, Pittsburgh, PA		
Solvent (A):	**cyclohexane**	**C$_6$H$_{12}$**	**110-82-7**

Type of data: cloud points (UCST behavior)

w_B 0.060   was kept constant at the critical concentration

T/K	301.34	301.11	300.95	300.96	301.12	301.43	301.68
P/bar	1	37	88	152	187	260	300

Polymer (B):	**polystyrene**		**1999BUN**
Characterization:	M_n/kg.mol^{-1} = 93, M_w/kg.mol^{-1} = 101.4, M_z/kg.mol^{-1} = 111.9,		
	BASF AG, Germany		
Solvent (A):	**cyclohexane**	**C$_6$H$_{12}$**	**110-82-7**

Type of data: cloud points

w_B 0.114   was kept constant

T/K	496.35	498.35	501.45	505.45	511.05	515.15	522.45	530.95
P/bar	20.6	23.5	27.6	33.9	43.6	51.0	61.0	74.0

continued

continued

Type of data: coexistence data

$T/K = 530.95$

Total feed concentration of the homogeneous system before demixing: $w_B = 0.1138$.

$P/$ bar	w_B gel phase	w_B sol phase	
74.0		0.1138	(cloud point)
63.8	0.2553	0.0184	
55.5	0.2749	0.0073	
41.9	0.3500		
32.0	0.3600		

Polymer (B): **polystyrene** **2002HOR**
Characterization: $M_n/\text{kg.mol}^{-1} = 93$, $M_w/\text{kg.mol}^{-1} = 101.4$, $M_z/\text{kg.mol}^{-1} = 111.9$,
BASF AG, Germany
Solvent (A): **cyclohexane** C_6H_{12} **110-82-7**

Type of data: cloud points

w_B 0.107 was kept constant

T/K	496.09	498.61	501.22	503.56	506.14	508.65	511.10	513.56	516.10
P/bar	24	26	32	34	38	41	44	48	51

Polymer (B): **polystyrene** **2003JI1, 2004JIA**
Characterization: $M_n/\text{kg.mol}^{-1} = 135$, $M_w/\text{kg.mol}^{-1} = 270$,
Pressure Chemical Company, Pittsburgh, PA
Solvent (A): ***trans*-decalin** $C_{10}H_{18}$ **493-02-7**

Type of data: cloud points (UCST behavior)

φ_B	0.042	0.084	0.128	0.172	0.216	0.042	0.084	0.128	0.172
T/K	279.0	282.0	284.9	284.0	273.0	279.7	283.0	285.5	284.3
P/bar	1	1	1	1	1	100	100	100	100

φ_B	0.216	0.042	0.084	0.128	0.172	0.216	0.042	0.084	0.128
T/K	274.8	280.7	283.7	286.1	284.9	275.9	281.5	284.4	286.6
P/bar	100	200	200	200	200	200	300	300	300

φ_B	0.172	0.216	0.042	0.084	0.128	0.172	0.216	0.042	0.084
T/K	285.5	277.0	282.3	285.1	287.2	286.1	278.1	283.2	285.7
P/bar	300	300	400	400	400	400	400	500	500

continued

continued

φ_B	0.128	0.172	0.216	0.042	0.084	0.128	0.172	0.216	0.042
T/K	287.8	286.7	279.2	284.0	286.5	288.4	287.3	280.3	284.8
P/bar	500	500	500	600	600	600	600	600	700

φ_B	0.084	0.128	0.172	0.216	0.042	0.084	0.128	0.172	0.216
T/K	287.1	289.0	287.9	281.5	285.7	287.8	289.6	288.5	282.6
P/bar	700	700	700	700	800	800	800	800	800

Comments: The volume fractions were determined at 1 bar.

Polymer (B):   **polystyrene**   1986SAE
Characterization: M_n/kg.mol^{-1} = 15.8, M_w/kg.mol^{-1} = 16.7,
     Pressure Chemical Company, Pittsburgh, PA
Solvent (A):   **diethyl oxalate**   **$C_6H_{10}O_4$**   **95-92-1**

Type of data: cloud points (UCST behavior)

T/K = 261.96 φ_B = 0.1578 $(dT/dP)_{P=1}$/K.atm^{-1} = +0.00126

Polymer (B):   **polystyrene**   1986SAE
Characterization: M_n/kg.mol^{-1} = 47.2, M_w/kg.mol^{-1} = 50.0,
     Pressure Chemical Company, Pittsburgh, PA
Solvent (A):   **diethyl oxalate**   **$C_6H_{10}O_4$**   **95-92-1**

Type of data: cloud points (UCST behavior)

T/K = 280.05 φ_B = 0.1302 $(dT/dP)_{P=1}$/K.atm^{-1} = −0.0045

Polymer (B):   **polystyrene**   1986SAE
Characterization: M_n/kg.mol^{-1} = 100, M_w/kg.mol^{-1} = 110,
     Pressure Chemical Company, Pittsburgh, PA
Solvent (A):   **diethyl oxalate**   **$C_6H_{10}O_4$**   **95-92-1**

Type of data: cloud points (UCST behavior)

T/K = 292.59 φ_B = 0.0927 $(dT/dP)_{P=1}$/K.atm^{-1} = −0.0105

Polymer (B):   **polystyrene**   1986SAE
Characterization: M_n/kg.mol^{-1} = 545.5, M_w/kg.mol^{-1} = 600,
     Pressure Chemical Company, Pittsburgh, PA
Solvent (A):   **diethyl oxalate**   **$C_6H_{10}O_4$**   **95-92-1**

Type of data: cloud points (UCST behavior)

T/K = 309.96 φ_B = 0.0427 $(dT/dP)_{P=1}$/K.atm^{-1} = −0.019

Polymer (B):	polystyrene	1992SZY
Characterization:	M_n/kg.mol^{-1} = 6.6, M_w/kg.mol^{-1} = 7.8,	
	Scientific Polymer Products, Inc., Ontario, NY	
Solvent (A):	1,1,1,3,3,3-hexadeutero-2-propanone $\quad$ C$_3$D$_6$O	666-52-4

Type of data: cloud points

w_B	0.072	0.090	0.110	0.130	0.149	0.162	0.293
φ_B =	0.060	0.075	0.092	0.109	0.125	0.137	0.253
T/K	277.63	281.52	284.58	286.07	286.75	286.85	284.72
$(dT/dP)_{P=0}$/K.MPa^{-1}	−1.70	−1.40	−1.30	−1.27	−1.25	−1.26	−1.31

Type of data: spinodal points

w_B	0.072	0.090	0.110	0.162	0.293
φ_B =	0.060	0.075	0.092	0.137	0.253
T/K	276.60	279.73	283.30	286.50	284.01
$(dT/dP)_{P=0}$/K.MPa^{-1}	−1.63	−1.35	−1.23	−1.27	−1.26

Polymer (B):	polystyrene	1992SZY
Characterization:	M_n/kg.mol^{-1} = 10.75, M_w/kg.mol^{-1} = 11.5,	
	Scientific Polymer Products, Inc., Ontario, NY	
Solvent (A):	1,1,1,3,3,3-hexadeutero-2-propanone $\quad$ C$_3$D$_6$O	666-52-4

Type of data: cloud points

w_B	0.090	0.140	0.220	0.240
φ_B =	0.075	0.115	0.185	0.203
T/K	280.95	284.14	283.23	283.02
$(dT/dP)_{P=0}$/K.MPa^{-1}	−1.25	−1.29	−1.29	−1.32

Type of data: spinodal points

w_B	0.090	0.140	0.220	0.240
φ_B =	0.075	0.115	0.185	0.203
T/K	280.48	284.00	283.23	282.93
$(dT/dP)_{P=0}$/K.MPa^{-1}	−1.28	−1.43	−1.29	−1.37

Polymer (B):	polystyrene	1995LUS
Characterization:	M_n/kg.mol^{-1} = 12.75, M_w/kg.mol^{-1} = 13.5,	
	Pressure Chemical Company, Pittsburgh, PA	
Solvent (A):	1,1,1,3,3,3-hexadeutero-2-propanone $\quad$ C$_3$D$_6$O	666-52-4

Type of data: cloud points

w_B	0.101	0.101	0.101	0.101	0.101	0.101	0.101	0.101	0.114
T/K	328.09	332.09	334.55	336.41	337.92	339.34	380.82	379.15	331.96
P/MPa	2.33	1.65	1.05	0.81	0.56	0.45	1.34	0.92	2.25

continued

continued

w_B	0.114	0.114	0.114	0.114	0.114	0.114	0.114	0.114	0.123
T/K	334.66	337.41	340.13	342.71	378.77	380.57	371.78	374.31	332.14
P/MPa	1.72	1.43	1.04	0.92	1.22	1.39	0.93	1.01	2.67

w_B	0.123	0.123	0.123	0.123	0.123	0.123	0.123	0.123	0.123
T/K	333.44	337.59	341.36	344.53	346.66	350.05	352.68	358.35	364.62
P/MPa	2.35	1.64	1.27	1.11	0.86	0.85	0.90	1.31	1.78

w_B	0.123	0.131	0.131	0.131	0.131	0.131	0.131	0.131	0.131
T/K	368.55	334.51	339.31	343.41	345.83	351.16	356.50	358.36	363.62
P/MPa	2.46	2.20	1.59	1.18	1.13	0.84	0.82	0.91	1.06

w_B	0.131	0.131	0.131	0.147	0.147	0.147	0.147	0.147	0.147
T/K	366.71	369.29	372.79	332.28	336.35	340.25	344.21	345.44	348.79
P/MPa	1.08	1.28	1.68	2.48	1.81	1.25	1.01	0.84	0.72

w_B	0.147	0.147	0.147	0.147	0.147	0.147	0.147	0.163	0.163
T/K	351.74	352.11	354.97	358.01	363.66	367.13	371.58	336.06	338.37
P/MPa	0.51	0.49	0.45	0.45	0.53	0.71	0.92	1.77	1.45

w_B	0.163	0.163	0.163	0.163	0.163	0.163	0.163	0.163	0.194
T/K	341.10	345.26	347.69	352.92	358.34	362.11	366.28	368.08	337.28
P/MPa	1.09	0.74	0.59	0.39	0.37	0.41	0.51	0.59	1.54

w_B	0.194	0.194	0.194	0.194	0.194	0.194	0.194	0.204	0.204
T/K	341.73	340.81	344.61	344.60	342.87	339.09	333.17	330.56	333.13
P/MPa	1.05	1.12	0.75	0.76	0.88	1.31	2.24	2.77	2.24

w_B	0.204	0.204	0.204	0.204	0.204	0.204	0.204	0.204	0.204
T/K	333.13	338.53	341.31	346.48	349.67	355.46	358.33	362.83	365.49
P/MPa	2.23	1.38	1.11	0.66	0.49	0.32	0.33	0.40	0.49

w_B	0.204	0.204	0.204	0.229	0.229	0.229	0.229	0.229	0.229
T/K	369.33	372.62	379.32	329.85	332.52	335.22	340.82	343.52	346.05
P/MPa	0.68	0.88	1.39	2.49	2.01	1.54	0.83	0.58	0.39

w_B	0.229	0.229	0.229	0.229	0.229	0.229			
T/K	348.33	370.51	372.95	376.66	379.86	384.97			
P/MPa	0.26	0.55	0.69	0.97	1.23	1.72			

Type of data: spinodal points

w_B	0.101	0.101	0.101	0.101	0.101	0.101	0.101	0.101	0.114
T/K	328.09	332.09	334.55	336.41	337.92	339.34	380.82	379.15	331.96
P/MPa	1.25	0.73	−0.25	−0.45	−0.52	−0.81	0.34	0.01	1.55

w_B	0.114	0.114	0.114	0.114	0.114	0.114	0.114	0.114	0.123
T/K	334.66	337.41	340.13	342.71	378.77	380.57	371.78	374.31	332.14
P/MPa	0.85	0.64	−0.06	−0.13	0.58	0.70	0.24	0.29	0.79

w_B	0.123	0.123	0.123	0.123	0.123	0.123	0.123	0.123	0.123
T/K	333.44	337.59	341.36	344.53	346.66	350.05	352.68	358.35	364.62
P/MPa	0.54	0.29	0.04	−0.005	−0.18	−0.11	−0.04	0.35	0.74

w_B	0.123	0.131	0.131	0.131	0.131	0.131	0.131	0.131	0.131
T/K	368.55	334.51	339.31	343.41	345.83	351.16	356.50	358.36	363.62
P/MPa	1.06	1.65	1.11	0.68	0.48	−0.24	0.006	0.03	−0.04

continued

continued

w_B	0.131	0.131	0.131	0.147	0.147	0.147	0.147	0.147	0.147
T/K	366.71	369.29	372.79	332.28	336.35	340.25	344.21	345.44	348.79
P/MPa	0.17	0.47	0.36	2.08	1.07	0.64	−0.16	0.51	0.47

w_B	0.147	0.147	0.147	0.147	0.147	0.147	0.147	0.163	0.163
T/K	351.74	352.11	354.97	358.01	363.66	367.13	371.58	336.06	338.37
P/MPa	0.31	0.13	0.17	0.29	0.37	0.55	0.13	1.44	1.18

w_B	0.163	0.163	0.163	0.163	0.163	0.163	0.163	0.163	0.194
T/K	341.10	345.26	347.69	352.92	358.34	362.11	366.28	368.08	337.28
P/MPa	0.74	0.47	0.41	0.19	0.15	0.21	0.40	0.46	1.44

w_B	0.194	0.194	0.194	0.194	0.194	0.194	0.194	0.204	0.204
T/K	341.73	340.81	344.61	344.60	342.87	339.09	333.17	330.56	333.13
P/MPa	0.95	1.04	0.67	0.68	0.80	1.23	2.15	2.63	2.13

w_B	0.204	0.204	0.204	0.204	0.204	0.204	0.204	0.204	0.204
T/K	333.13	338.53	341.31	346.48	349.67	355.46	358.33	362.83	365.49
P/MPa	2.14	1.32	0.95	0.53	0.36	0.20	0.19	0.28	0.39

w_B	0.204	0.204	0.204	0.229	0.229	0.229	0.229	0.229	0.229
T/K	369.33	372.62	379.32	329.85	332.52	335.22	340.82	348.33	370.51
P/MPa	0.56	0.75	1.28	1.51	1.12	0.81	0.04	−0.32	−0.05

w_B	0.229	0.229	0.229	0.229
T/K	372.95	376.66	379.86	384.97
P/MPa	0.02	0.26	0.42	0.69

critical concentration: $w_{B, crit} = 0.200$

Polymer (B):	**polystyrene-d8**		**1995LUS**
Characterization:	M_n/kg.mol^{-1} = 10.3, M_w/kg.mol^{-1} = 10.5,		
	completely deuterated, Polymer Laboratories, Amherst, MA		
Solvent (A):	**1,1,1,3,3,3-hexadeutero-2-propanone**	**C$_3$D$_6$O**	**666-52-4**

Type of data: cloud points

w_B	0.219	0.219	0.219	0.219
T/K	266.69	267.22	268.05	268.78
P/MPa	2.61	2.03	1.22	0.47

Polymer (B):	**polystyrene-d8**		**1995LUS**
Characterization:	M_n/kg.mol^{-1} = 25.4, M_w/kg.mol^{-1} = 26.9,		
	completely deuterated, Polymer Laboratories, Amherst, MA		
Solvent (A):	**1,1,1,3,3,3-hexadeutero-2-propanone**	**C$_3$D$_6$O**	**666-52-4**

Type of data: cloud points

w_B	0.120	0.120	0.120	0.120	0.120	0.120	0.120	0.140	0.140
T/K	330.21	336.97	343.05	350.07	357.01	363.63	370.11	330.34	337.04
P/MPa	4.04	3.22	2.79	2.57	2.61	2.62	3.18	4.45	3.59

continued

continued

w_B	0.140	0.140	0.140	0.140	0.140	0.150	0.150	0.150	0.150
T/K	343.20	350.05	356.97	363.50	369.96	330.76	337.14	343.16	350.01
P/MPa	3.12	2.87	2.66	3.07	3.40	4.52	3.70	3.22	2.96

w_B	0.150	0.150	0.150	0.160	0.160	0.160	0.160	0.160	0.160
T/K	357.04	363.44	370.06	336.24	336.23	342.24	348.99	355.63	363.30
P/MPa	2.96	3.13	3.46	3.86	3.83	3.33	3.04	3.00	3.18

w_B	0.160	0.170	0.170	0.170	0.170	0.170	0.170	0.185	0.185
T/K	369.95	337.45	343.67	350.09	356.66	363.67	370.10	329.96	336.98
P/MPa	3.51	3.67	3.30	3.05	3.05	3.23	3.57	4.04	3.23

w_B	0.185	0.185	0.185	0.185	0.185	0.193	0.193	0.193	0.193
T/K	343.41	350.03	356.99	363.52	369.66	330.21	337.13	343.25	349.95
P/MPa	2.84	2.67	2.75	3.02	3.36	3.95	3.16	2.80	2.62

w_B	0.193	0.193	0.193	0.205	0.205	0.205	0.205	0.205	0.205
T/K	357.07	363.73	370.00	357.26	363.67	370.07	329.97	336.99	343.12
P/MPa	2.71	2.93	3.32	2.65	2.89	3.27	3.86	3.11	2.73

w_B	0.205	0.212	0.212	0.212	0.212	0.212	0.212	0.212	0.217
T/K	349.98	330.01	336.89	342.35	349.87	357.01	363.25	369.96	330.04
P/MPa	2.57	3.85	3.08	2.70	2.50	2.60	2.82	3.22	3.33

w_B	0.217	0.217	0.217	0.217	0.217	0.217	0.225	0.225	0.225
T/K	337.00	343.13	350.01	357.16	363.54	370.02	329.98	336.89	342.98
P/MPa	2.83	2.57	2.47	2.55	2.79	3.18	3.59	2.85	2.51

w_B	0.225	0.225	0.225	0.225	0.239	0.239	0.239	0.239	0.239
T/K	350.08	357.37	363.48	370.03	331.21	341.82	345.75	350.19	357.17
P/MPa	2.38	2.47	2.73	3.13	3.12	2.33	2.22	2.21	2.31

w_B	0.239	0.239	0.259	0.259	0.259	0.259	0.259	0.259	0.259
T/K	363.37	369.78	337.02	340.82	347.67	352.05	358.22	363.07	368.57
P/MPa	2.56	2.97	1.86	1.69	1.54	1.58	1.79	2.05	2.41

Type of data: spinodal points

w_B	0.140	0.140	0.140	0.140	0.140	0.150	0.150	0.150	0.150
T/K	330.34	337.04	343.20	363.50	369.96	330.76	337.14	343.16	350.01
P/MPa	4.17	3.26	2.69	2.79	3.11	4.29	3.50	3.01	2.66

w_B	0.150	0.150	0.150	0.160	0.160	0.160	0.160	0.160	0.160
T/K	357.04	363.44	370.06	336.24	336.23	342.24	348.99	355.63	363.30
P/MPa	2.82	2.99	3.29	3.55	3.64	3.22	2.90	2.67	3.05

w_B	0.160	0.170	0.170	0.170	0.170	0.170	0.170	0.185	0.185
T/K	369.95	337.45	343.67	350.09	356.66	363.67	370.10	329.96	336.98
P/MPa	3.39	3.43	3.04	2.64	2.63	3.07	3.37	3.83	3.04

w_B	0.185	0.185	0.185	0.185	0.185	0.193	0.193	0.193	0.193
T/K	343.41	350.03	356.99	363.52	369.66	330.21	337.13	343.25	349.95
P/MPa	2.65	2.52	2.56	2.76	3.18	3.81	3.06	2.66	2.52

w_B	0.193	0.193	0.193	0.205	0.205	0.205	0.205	0.205	0.205
T/K	357.07	363.73	370.00	357.26	363.67	370.07	329.97	336.99	343.12
P/MPa	2.59	2.85	3.21	2.51	2.78	3.17	3.75	2.98	2.61

continued

continued

w_B	0.205	0.212	0.212	0.212	0.212	0.212	0.212	0.212	0.217
T/K	349.98	330.01	336.89	342.35	349.87	357.01	363.25	369.96	330.04
P/MPa	2.43	3.66	2.91	2.56	2.40	2.43	2.70	3.06	3.15

w_B	0.217	0.217	0.217	0.217	0.217	0.225	0.225	0.225	0.225
T/K	337.00	350.01	357.16	363.54	370.02	329.98	342.98	350.08	357.37
P/MPa	2.60	2.23	2.39	2.59	3.07	3.23	2.17	2.01	2.18

w_B	0.225	0.225	0.239	0.239	0.239	0.239	0.239
T/K	363.48	370.03	331.21	341.82	357.17	363.37	369.78
P/MPa	2.55	2.88	2.68	1.89	1.88	2.12	2.60

critical concentration: $w_{B, crit} = 0.193$

Polymer (B):	**polystyrene**	**1972ZE2**
Characterization:	$M_w/kg.mol^{-1} = 51$, $M_w/M_n < 1.1$,	
	Pressure Chemical Company, Pittsburgh, PA	
Solvent (A):	**methyl acetate** $C_3H_6O_2$	**79-20-9**

Type of data: cloud points

$T/K = 423.35$ $w_B = 0.023 - 0.024$ $(dT/dP)_{P=1}/K.bar^{-1} = +0.47$ (LCST behavior)

Polymer (B):	**polystyrene**	**1972ZE2**
Characterization:	$M_w/kg.mol^{-1} = 97.2$, $M_w/M_n < 1.1$,	
	Pressure Chemical Company, Pittsburgh, PA	
Solvent (A):	**methyl acetate** $C_3H_6O_2$	**79-20-9**

Type of data: cloud points

$T/K = 275.15$ $w_B = 0.023 - 0.024$ $(dT/dP)_{P=1}/K.bar^{-1} = -0.018$ (UCST behavior)
$T/K = 415.15$ $w_B = 0.023 - 0.024$ $(dT/dP)_{P=1}/K.bar^{-1} = +0.45$ (LCST behavior)

Polymer (B):	**polystyrenc**	**1972ZE2**
Characterization:	$M_w/kg.mol^{-1} = 160$, $M_w/M_n < 1.1$,	
	Pressure Chemical Company, Pittsburgh, PA	
Solvent (A):	**methyl acetate** $C_3H_6O_2$	**79-20-9**

Type of data: cloud points

$T/K = 281.45$ $w_B = 0.023 - 0.024$ $(dT/dP)_{P=1}/K.bar^{-1} = -0.041$ (UCST behavior)
$T/K = 409.95$ $w_B = 0.023 - 0.024$ $(dT/dP)_{P=1}/K.bar^{-1} = +0.44$ (LCST behavior)

Polymer (B):	**polystyrene**	**1972ZE2**
Characterization:	$M_w/kg.mol^{-1} = 498$, $M_w/M_n < 1.1$,	
	Pressure Chemical Company, Pittsburgh, PA	
Solvent (A):	**methyl acetate** $C_3H_6O_2$	**79-20-9**

Type of data: cloud points

$T/K = 294.15$ $w_B = 0.023 - 0.024$ $(dT/dP)_{P=1}/K.bar^{-1} = -0.056$ (UCST behavior)
$T/K = 400.55$ $w_B = 0.023 - 0.024$ $(dT/dP)_{P=1}/K.bar^{-1} = +0.46$ (LCST behavior)

Polymer (B):	**polystyrene**	**1972ZE2**
Characterization:	$M_w/\text{kg.mol}^{-1} = 670$, $M_w/M_n < 1.1$,	
	Pressure Chemical Company, Pittsburgh, PA	
Solvent (A):	**methyl acetate** **$C_3H_6O_2$**	**79-20-9**

Type of data: cloud points

$T/\text{K} = 296.45$	$w_B = 0.023 - 0.024$	$(dT/dP)_{P=1}/\text{K.bar}^{-1} = -0.060$	(UCST behavior)
$T/\text{K} = 397.85$	$w_B = 0.023 - 0.024$	$(dT/dP)_{P=1}/\text{K.bar}^{-1} = +0.47$	(LCST behavior)

Polymer (B):	**polystyrene**	**1972ZE2**
Characterization:	$M_w/\text{kg.mol}^{-1} = 860$, $M_w/M_n < 1.1$,	
	Pressure Chemical Company, Pittsburgh, PA	
Solvent (A):	**methyl acetate** **$C_3H_6O_2$**	**79-20-9**

Type of data: cloud points

$T/\text{K} = 299.25$	$w_B = 0.023 - 0.024$	$(dT/dP)_{P=1}/\text{K.bar}^{-1} = -0.073$	(UCST behavior)
$T/\text{K} = 396.15$	$w_B = 0.023 - 0.024$	$(dT/dP)_{P=1}/\text{K.bar}^{-1} = +0.47$	(LCST behavior)

Polymer (B):	**polystyrene**	**1972ZE2**
Characterization:	$M_w/\text{kg.mol}^{-1} = 1800$, $M_w/M_n < 1.1$,	
	Pressure Chemical Company, Pittsburgh, PA	
Solvent (A):	**methyl acetate** **$C_3H_6O_2$**	**79-20-9**

Type of data: cloud points

$T/\text{K} = 303.15$	$w_B = 0.023 - 0.024$	$(dT/dP)_{P=1}/\text{K.bar}^{-1} = -0.082$	(UCST behavior)
$T/\text{K} = 391.75$	$w_B = 0.023 - 0.024$	$(dT/dP)_{P=1}/\text{K.bar}^{-1} = +0.48$	(LCST behavior)

Polymer (B):	**polystyrene**	**1991SZY**
Characterization:	$M_n/\text{kg.mol}^{-1} = 10.75$, $M_w/\text{kg.mol}^{-1} = 11.5$,	
	Scientific Polymer Products, Inc., Ontario, NY	
Solvent (A):	**methylcyclohexane** **C_7H_{14}**	**108-87-2**

Type of data: cloud points (UCST behavior)

w_B	0.092	0.140
T/K	293	294
$(dT/dP)_{P=0}/\text{K.MPa}^{-1}$	−0.22	−0.26

Polymer (B):	**polystyrene**	**1993HOS**
Characterization:	$M_n/\text{kg.mol}^{-1} = 9.63$, $M_w/\text{kg.mol}^{-1} = 10.2$,	
	Pressure Chemical Company, Pittsburgh, PA	
Solvent (A):	**methylcyclohexane** **C_7H_{14}**	**108-87-2**

continued

continued

Type of data: cloud points (UCST behavior)

w_B 0.1972   was kept constant

T/K	285.72	285.58	285.29	285.00	284.58	284.38	284.06	283.92	283.80
P/MPa	0.10	1.18	3.43	6.28	11.28	14.51	19.91	24.61	28.83

T/K	283.73	283.70	283.73	283.83	283.93	284.03	284.16	284.30	284.48
P/MPa	34.32	40.31	45.11	50.50	55.31	59.53	63.84	67.86	72.37

T/K	284.65	284.83	285.03
P/MPa	76.59	80.41	84.93

critical double point: $T_{dcrit}/K = 283.71$ $P_{dcrit}/MPa = 38.3$

Polymer (B):	**polystyrene**	**1993HOS**

Characterization: $M_n/kg.mol^{-1} = 15.2$, $M_w/kg.mol^{-1} = 16.1$,
                        Pressure Chemical Company, Pittsburgh, PA

Solvent (A):	**methylcyclohexane** C_7H_{14}	**108-87-2**

Type of data: cloud points (UCST behavior)

w_B	0.0786	0.0786	0.0786	0.0786	0.0786	0.0786	0.0786	0.0786	0.0786
T/K	295.69	295.37	295.05	294.75	294.47	294.17	293.91	293.63	293.24
P/MPa	0.10	1.86	3.82	5.79	7.55	9.81	11.96	14.32	18.93

w_B	0.0786	0.0786	0.0786	0.0786	0.0786	0.0786	0.0786	0.0786	0.0786
T/K	293.03	292.82	292.65	292.51	292.38	292.32	292.30	292.33	292.39
P/MPa	21.97	25.20	28.93	33.05	38.15	43.44	49.13	53.64	58.15

w_B	0.0786	0.0786	0.0786	0.0786	0.0998	0.0998	0.0998	0.0998	0.0998
T/K	292.45	292.55	292.73	292.88	296.05	294.96	294.55	294.28	294.08
P/MPa	63.06	67.67	72.86	77.47	0.10	6.67	9.32	11.57	13.53

w_B	0.0998	0.0998	0.0998	0.0998	0.0998	0.0998	0.0998	0.0998	0.0998
T/K	293.82	293.54	293.30	293.11	292.81	292.67	292.60	292.53	292.55
P/MPa	15.98	18.93	22.56	27.36	32.75	37.27	42.66	48.05	52.86

w_B	0.0998	0.0998	0.0998	0.0998	0.0998	0.0998	0.0998	0.1287	0.1287
T/K	292.63	292.68	292.81	292.95	293.13	293.35	293.51	296.32	295.17
P/MPa	57.86	62.27	67.27	72.77	77.57	83.55	87.28	0.10	6.86

w_B	0.1287	0.1287	0.1287	0.1287	0.1287	0.1287	0.1287	0.1287	0.1287
T/K	294.90	294.58	294.47	294.10	293.89	293.59	293.42	293.20	293.03
P/MPa	8.83	10.98	12.75	15.30	17.85	21.57	24.61	28.24	32.36

w_B	0.1287	0.1287	0.1287	0.1287	0.1287	0.1287	0.1287	0.1287	0.1287
T/K	292.92	292.83	292.78	292.78	292.81	292.89	292.96	293.10	293.24
P/MPa	36.48	41.48	46.78	51.68	56.29	60.90	65.51	70.41	74.73

w_B	0.1287	0.1721	0.1721	0.1721	0.1721	0.1721	0.1721	0.1721	0.1721
T/K	293.34	296.21	295.91	295.44	295.12	294.85	294.47	294.12	293.88
P/MPa	78.26	0.10	1.86	4.22	6.47	8.53	11.57	14.12	16.87

continued

continued

w_B	0.1721	0.1721	0.1721	0.1721	0.1721	0.1721	0.1721	0.1721	0.1721
T/K	293.67	293.43	293.23	293.05	292.89	292.78	292.70	292.68	292.68
P/MPa	19.42	22.36	25.99	29.62	33.44	37.76	42.46	47.27	51.78

w_B	0.1721	0.1721	0.1721	0.1721	0.1721	0.1721	0.2044	0.2044	0.2044
T/K	292.72	292.78	292.85	292.95	293.07	293.21	296.27	295.92	295.48
P/MPa	56.19	60.61	64.82	68.84	72.96	77.28	0.10	1.77	4.31

w_B	0.2044	0.2044	0.2044	0.2044	0.2044	0.2044	0.2044	0.2044	0.2044
T/K	295.25	294.96	294.70	294.40	294.13	293.90	293.68	293.44	293.16
P/MPa	5.88	7.85	9.81	12.26	14.51	16.77	19.61	22.85	27.16

w_B	0.2044	0.2044	0.2044	0.2044	0.2044	0.2044	0.2044	0.2044	0.2044
T/K	293.02	292.85	292.77	292.70	292.70	292.75	292.81	292.92	293.05
P/MPa	31.19	35.79	40.70	46.29	51.78	56.88	61.78	66.69	71.78

w_B	0.2044	0.2044	0.2044	0.2461	0.2461	0.2461	0.2461	0.2461	0.2461
T/K	293.20	293.35	293.51	295.79	294.57	294.23	294.00	293.73	293.47
P/MPa	76.49	81.20	86.00	0.10	7.06	9.12	11.38	13.73	15.98

w_B	0.2461	0.2461	0.2461	0.2461	0.2461	0.2461	0.2461	0.2461	0.2461
T/K	293.22	292.95	292.68	292.52	292.35	292.24	292.16	292.20	292.20
P/MPa	18.83	22.85	27.07	31.97	36.19	41.78	48.44	49.03	53.84

w_B	0.2461	0.2461	0.2461	0.2461	0.2461	0.2461	0.2461	0.2952	0.2952
T/K	292.25	292.35	292.45	292.57	292.74	292.87	293.03	294.79	293.70
P/MPa	58.55	63.74	68.65	73.55	78.16	82.08	85.51	0.10	6.57

w_B	0.2952	0.2952	0.2952	0.2952	0.2952	0.2952	0.2952	0.2952	0.2952
T/K	293.42	293.20	292.93	292.70	292.44	292.23	291.95	291.73	291.55
P/MPa	8.53	10.59	12.75	15.20	18.24	21.28	25.60	30.60	35.50

w_B	0.2952	0.2952	0.2952	0.2952	0.2952	0.2952	0.2952	0.2952	0.2952
T/K	291.45	291.40	291.40	291.45	291.53	291.65	291.78	291.92	292.13
P/MPa	40.40	45.50	50.41	54.92	61.78	66.78	71.69	75.90	81.49

critical double point: $T_{dcrit}/K = 291.43$ $P_{dcrit}/MPa = 48.6$

Polymer (B):	**polystyrene**	**1993HOS**
Characterization:	$M_n/kg.mol^{-1} = 33$, $M_w/kg.mol^{-1} = 34.9$,	
	Pressure Chemical Company, Pittsburgh, PA	
Solvent (A):	**methylcyclohexane C_7H_{14}**	**108-87-2**

Type of data: cloud points (UCST behavior)

w_B	0.1292	was kept constant

T/K	308.87	308.48	308.10	307.80	307.44	307.08	306.73	306.33	305.86
P/MPa	0.10	1.27	2.65	3.82	5.20	6.67	8.24	9.90	12.36

T/K	305.51	305.23	304.77	304.53	304.22	303.90	303.58	303.25	302.97
P/MPa	14.42	16.38	19.61	21.97	24.42	27.65	31.77	36.68	43.44

T/K	302.76	302.71	302.71	302.81	302.91	303.05	303.25
P/MPa	51.39	58.94	65.51	72.37	78.36	84.53	90.61

critical double point: $T_{dcrit}/K = 302.86$ $P_{dcrit}/MPa = 61.6$

Polymer (B):	**polystyrene**							**1993WEL**

Characterization: M_n/kg.mol^{-1} = 21.4, M_w/kg.mol^{-1} = 22.0

Solvent (A): **methylcyclohexane** **C$_7$H$_{14}$** 108-87-2

Type of data: spinodal points (UCST behavior)

w_B	0.055	0.055	0.055	0.055	0.055	0.055	0.055	0.055	0.055
T/K	298.20	297.99	297.39	297.11	296.64	296.24	295.75	295.41	294.93
P/bar	5.42	9.64	43.97	62.05	90.96	120.48	160.84	195.78	253.61

w_B	0.055	0.055	0.055	0.055	0.055	0.094	0.094	0.094	0.094
T/K	294.61	294.34	294.29	294.32	294.51	300.25	299.75	299.45	299.05
P/bar	315.66	402.41	481.32	551.81	682.53	18.07	39.76	56.63	80.12

w_B	0.094	0.094	0.094	0.094	0.094	0.094	0.094	0.094	0.094
T/K	298.68	298.35	298.02	297.53	297.08	296.86	296.68	296.13	296.17
P/bar	100.60	124.70	146.38	152.41	239.15	283.13	304.22	509.04	653.61

w_B	0.200	0.200	0.200	0.200	0.200	0.200	0.200	0.200	0.200
T/K	301.09	300.76	300.47	300.09	300.13	299.42	298.93	298.41	298.39
P/bar	24.10	28.31	42.17	62.05	63.25	100.00	134.94	184.34	189.16

w_B	0.200	0.200	0.200	0.200	0.200
T/K	297.74	297.34	297.09	296.89	297.04
P/bar	256.63	333.73	424.70	533.73	680.12

critical point: $w_{B, crit}$ = 0.200 and T_{crit}/K = 300.85 at P/bar = 1.

Polymer (B):	**polystyrene**							**1994VA1**

Characterization: M_n/kg.mol^{-1} = 16.5, M_w/kg.mol^{-1} = 17.5,
Pressure Chemical Company, Pittsburgh, PA

Solvent (A): **methylcyclohexane** **C$_7$H$_{14}$** 108-87-2

Type of data: cloud points

w_B	0.15	0.15	0.15	0.15	0.15	0.15	0.15	0.15	0.15
T/K	296.751	296.044	295.337	294.739	294.433	294.200	293.892	293.953	294.203
P/bar	48.54	90.14	144.85	210.13	249.33	309.56	385.50	638.82	743.95

w_B	0.15	0.27	0.27	0.27	0.27	0.27	0.27	0.27	0.27
T/K	294.490	296.418	295.687	295.080	294.603	294.257	293.944	293.603	293.653
P/bar	828.10	53.06	98.49	151.24	201.30	249.66	306.13	428.80	645.79

w_B	0.27	0.27	0.30	0.30	0.30	0.30	0.30	0.30	0.30
T/K	293.944	294.323	296.448	295.705	294.949	294.449	293.935	293.690	293.492
P/bar	758.94	857.25	37.87	76.82	134.13	185.03	252.98	311.90	377.40

w_B	0.30	0.30	0.30	0.09374	0.12389	0.15492	0.18487	0.21412	0.23366
T/K	293.444	293.694	294.041	295.053	295.673	295.885	295.925	295.954	295.915
P/bar	620.16	726.60	822.60	100	100	100	100	100	100

w_B	0.24524	0.27350	0.30267	0.09374	0.12389	0.15492	0.18487	0.21412	0.23366
T/K	295.822	295.623	295.406	294.014	294.675	294.853	294.871	294.882	294.857
P/bar	100	100	100	200	200	200	200	200	200

continued

continued

w_B	0.24524	0.15492	0.18487	0.21412	0.23366	0.24524	0.15492	0.18487	0.21412
T/K	294.727	294.853	294.871	294.882	294.857	294.727	294.853	294.871	294.882
P/bar	200	200	200	200	200	200	200	200	200

w_B	0.23366	0.24524	0.18487	0.21412	0.23366	0.24524	0.27350	0.30267	0.09111
T/K	294.857	294.727	293.915	293.928	293.912	293.804	293.631	293.425	293.105
P/bar	200	200	400	400	400	400	400	400	500

w_B	0.12100	0.14958	0.17954	0.20815	0.23883	0.26815	0.29882	0.09374	0.12389
T/K	293.615	293.800	293.844	293.846	293.717	293.556	293.417	293.409	293.890
P/bar	500	500	500	500	500	500	500	700	700

w_B	0.15492	0.18487	0.21412	0.23366	0.24524	0.27350	0.30267	0.09374	0.12389
T/K	294.039	294.045	294.059	294.031	293.991	293.792	293.578	293.726	294.200
P/bar	700	700	700	700	700	700	700	800	800

w_B	0.15492	0.18487	0.21412	0.23366	0.24524	0.27350	0.30267	0.09208	0.12272
T/K	294.353	294.344	294.362	294.330	294.223	294.117	293.944	293.744	294.234
P/bar	800	800	800	800	800	800	800	800	800

w_B	0.18131	0.23075	0.24135	0.27272	0.30123	0.14889	0.20957		
T/K	294.393	294.366	294.249	294.137	293.969	294.389	294.402		
P/bar	800	800	800	800	800	800	800		

Type of data: spinodal points

w_B	0.120	0.120	0.120	0.120	0.121	0.121	0.121	0.121	0.121
T/K	293.950	292.858	293.510	293.950	295.762	293.850	293.750	293.850	293.950
P/bar	173.70	500.00	800.00	896.52	1.00	190.05	854.92	876.42	895.35

w_B	0.148	0.150	0.150	0.150	0.150	0.150	0.150	0.150	0.150
T/K	294.025	297.026	296.753	296.046	294.742	294.192	293.950	293.910	293.750
P/bar	800.00	1.00	15.73	50.77	155.01	229.50	278.00	283.15	317.31

w_B	0.150	0.150	0.150	0.150	0.150	0.150	0.150	0.151	0.151
T/K	293.465	293.750	293.935	293.950	294.198	294.487	295.352	293.850	293.850
P/bar	500.00	710.59	767.42	779.15	855.49	896.34	989.29	305.54	737.38

w_B	0.151	0.178	0.180	0.180	0.180	0.181	0.181	0.181	0.181
T/K	293.950	297.536	293.950	293.758	293.950	293.950	293.850	293.850	293.950
P/bar	777.24	1.00	361.10	500.00	656.00	365.40	423.39	594.39	657.92

w_B	0.181	0.198	0.210	0.210	0.210	0.210	0.210	0.210	0.230
T/K	294.307	293.750	297.656	293.950	293.850	293.856	293.850	293.950	293.950
P/bar	800.00	198.38	1.00	382.00	496.32	500.00	545.18	645.00	378.36

w_B	0.230	0.230	0.230	0.240	0.240	0.240	0.240	0.240	0.240
T/K	293.850	293.850	293.950	293.950	293.950	293.850	293.750	293.717	293.750
P/bar	438.57	603.26	656.78	339.77	343.30	384.87	448.77	500.00	595.03

w_B	0.240	0.240	0.240	0.268	0.268	0.268	0.269	0.269	0.269
T/K	293.850	293.950	293.950	293.750	293.362	293.750	293.950	293.950	293.850
P/bar	664.64	702.90	705.41	294.68	500.00	763.80	258.52	260.73	789.57

w_B	0.269	0.270	0.270	0.270	0.270	0.270	0.270	0.270	0.270
T/K	293.950	296.398	295.693	295.070	294.605	294.276	293.953	293.850	293.658
P/bar	823.87	35.94	81.43	130.13	177.20	222.59	255.95	286.81	733.15

continued

continued

w_B	0.270	0.270	0.270	0.270	0.271	0.300	0.300	0.300	0.300
T/K	293.874	293.950	293.923	294.325	296.981	296.200	295.707	294.939	294.403
P/bar	800.00	820.69	830.18	904.40	1.00	1.00	32.20	86.88	129.89

w_B	0.300	0.300	0.300	0.300	0.300	0.300	0.300	0.300	0.300
T/K	293.952	293.950	293.850	293.707	293.750	293.498	292.920	293.442	293.459
P/bar	169.00	181.00	194.26	202.98	203.43	264.54	500.00	800.00	807.83

w_B	0.300	0.300	0.300	0.300
T/K	293.750	293.746	293.850	293.950
P/bar	880.57	893.29	903.48	925.00

Type of data: cloud points (hour glass shaped)

T/K = 293.75

w_B	0.09346	0.12511	0.12575	0.09717	0.30657	0.27626	0.24389	0.27265	0.30324
P/bar	242.30	390.04	628.77	807.35	750.55	691.49	458.09	366.24	307.93

Type of data: cloud points (hour glass shaped)

T/K = 293.75

w_B	0.09047	0.12062	0.12013	0.08932	0.29871	0.26886	0.23907	0.26842	0.29852
P/bar	797.75	617.66	379.31	237.63	298.77	358.60	448.76	678.21	739.72

Type of data: cloud points (hour glass shaped)

T/K = 293.85

w_B	0.09098	0.12146	0.15103	0.18096	0.18206	0.15130	0.12167	0.09115	0.29925
P/bar	829.50	672.32	608.93	530.56	501.39	433.99	346.08	225.87	280.15

w_B	0.26965	0.23964	0.23028	0.21022	0.20997	0.23038	0.23969	0.26923	0.29875
P/bar	333.30	384.09	437.78	497.12	544.41	604.05	664.62	723.37	774.25

Type of data: cloud points (LCP and UCP branch)

T/K = 293.95

w_B	0.09764	0.12742	0.15562	0.18612	0.21626	0.23594	0.24565	0.27551	0.30541
P/bar	867.74	722.59	667.50	646.92	650.97	664.81	712.02	761.05	803.69

w_B	0.09764	0.12742	0.15562	0.18612	0.21626	0.23594	0.24565	0.27551	0.30541
P/bar	215.73	322.72	373.59	391.89	395.86	386.19	346.92	310.12	265.02

Type of data: cloud points (LCP and UCP branch)

T/K = 293.95

w_B	0.09048	0.12000	0.15048	0.18044	0.20992	0.22902	0.23962	0.26936	0.29906
P/bar	851.13	717.32	658.03	639.52	635.14	655.21	703.69	754.97	821.17

w_B	0.09048	0.12000	0.15048	0.18044	0.20992	0.22902	0.23962	0.26936	0.29906
P/bar	207.21	314.20	365.48	383.05	383.37	376.80	339.77	304.79	258.84

continued

continued

Type of data: cloud points (LCP and UCP branch)

$T/K = 293.95$

w_B	0.09071	0.12052	0.15092	0.18080	0.21015	0.24062	0.27109	0.29993
P/bar	852.80	719.97	658.02	642.46	637.21	704.12	758.15	822.49

w_B	0.09071	0.12052	0.15092	0.18080	0.21015	0.24062	0.27109	
P/bar	208.48	316.63	366.78	386.02	384.63	342.01	304.55	

Type of data: cloud points (LCP and UCP branch)

$T/K = 294.15$

w_B	0.09764	0.12742	0.15562	0.18612	0.21626	0.23594	0.24565	0.27551	0.30541
P/bar	907.31	784.78	745.01	734.67	733.53	739.27	775.31	812.97	852.68

w_B	0.09764	0.12742	0.15562	0.18612	0.21626	0.23594	0.24565	0.27551	0.30541
P/bar	189.20	279.93	314.50	324.57	326.57	317.03	293.78	270.23	232.38

Type of data: cloud points (LCP and UCP branch)

$T/K = 294.35$

w_B	0.09764	0.12742	0.15562	0.18612	0.21626	0.23594	0.24565	0.27551	0.30541
P/bar	955.28	842.03	805.16	802.03	797.84	804.64	831.39	859.88	891.35

w_B	0.09764	0.12742	0.15562	0.18612	0.21626	0.23594	0.24565	0.27551	0.30541
P/bar	161.74	246.23	274.68	280.64	282.68	276.23	257.06	235.71	204.83

Type of data: cloud points (UCP behavior)

$T/K = 295.15$

w_B	0.09764	0.12742	0.15562	0.18612	0.21626	0.23594	0.24565	0.27551	0.30541
P/bar	92.17	151.23	168.52	172.56	176.52	168.00	160.15	141.73	122.22

Type of data: cloud points (UCP behavior)

$T/K = 296.55$

w_B	0.09764	0.12742	0.15562	0.18612	0.21626	0.23594	0.24565	0.27551	0.30541
P/bar	11.69	44.21	59.46	62.41	66.45	62.94	57.16	40.75	32.56

Polymer (B):	**polystyrene**	**1997END**
Characterization:	M_n/kg.mol^{-1} = 143, M_w/kg.mol^{-1} = 405,	
	BASF AG, Ludwigshafen, Germany	
Solvent (A):	**methylcyclohexane C_7H_{14}**	**108-87-2**

Type of data: cloud points (UCST behavior, measured by PPICS)

w_B	0.0067	0.0067	0.0067	0.0067	0.0067	0.0067	0.0067	0.0067	0.0067
T/K	331.18	330.54	330.06	329.47	329.02	328.57	328.04	327.23	326.43
P/bar	17.71	19.64	20.60	23.92	30.00	37.97	47.02	60.22	74.35

continued

continued

w_B	0.0067	0.0067	0.0067	0.0067	0.0067	0.0067	0.0067	0.0067	0.0067
T/K	325.95	325.50	325.04	324.51	323.82	323.12	322.63	322.29	321.79
P/bar	84.72	92.67	108.28	122.42	143.15	160.12	174.25	187.45	211.95
w_B	0.0067	0.0067	0.0067	0.0067	0.0067	0.0067	0.0067	0.0067	0.0067
T/K	321.47	321.04	320.61	320.33	319.54	319.01	318.41	318.05	317.24
P/bar	225.15	243.05	270.39	297.72	331.65	362.75	393.85	417.41	462.65
w_B	0.018	0.018	0.018	0.018	0.018	0.018	0.018	0.018	0.018
T/K	331.16	330.65	330.05	329.58	329.25	328.78	328.10	327.56	326.81
P/bar	12.55	13.26	19.64	29.31	36.41	44.64	55.16	66.85	82.16
w_B	0.018	0.018	0.018	0.018	0.018	0.018	0.018	0.018	0.018
T/K	326.02	324.96	324.09	323.36	322.98	322.23	321.65	321.15	320.48
P/bar	101.19	127.38	157.74	181.35	196.63	217.61	247.92	269.92	294.99
w_B	0.018	0.018	0.018	0.018	0.018	0.025	0.025	0.025	0.025
T/K	320.00	319.12	318.54	317.64	317.11	331.38	330.80	330.13	329.50
P/bar	325.99	356.25	386.56	426.54	456.84	11.57	18.18	29.14	40.89
w_B	0.025	0.025	0.025	0.025	0.025	0.025	0.025	0.025	0.025
T/K	329.02	328.49	327.67	326.99	326.37	325.88	325.40	324.94	324.17
P/bar	45.25	52.50	73.34	80.95	99.21	115.74	123.63	145.09	169.36
w_B	0.025	0.025	0.025	0.025	0.025	0.025	0.025	0.025	0.025
T/K	323.41	322.93	322.39	321.68	321.01	320.40	320.13	319.70	319.48
P/bar	182.49	196.35	223.73	253.42	279.26	300.20	312.85	337.30	346.84
w_B	0.025	0.025	0.025	0.025	0.025	0.025	0.0458	0.0458	0.0458
T/K	319.13	318.62	318.16	317.74	317.09	316.18	331.22	330.73	330.33
P/bar	364.87	393.05	414.95	442.06	490.77	592.86	27.39	34.03	38.51
w_B	0.0458	0.0458	0.0458	0.0458	0.0458	0.0458	0.0458	0.0458	0.0458
T/K	329.71	329.35	328.60	328.51	327.92	327.41	326.85	326.17	325.63
P/bar	51.45	61.25	71.41	73.00	87.75	102.49	117.62	137.17	152.61
w_B	0.0458	0.0458	0.0482	0.0482	0.0482	0.0482	0.0482	0.0482	0.0482
T/K	325.23	324.84	332.61	332.11	331.63	331.26	330.98	330.23	329.49
P/bar	160.12	170.49	9.84	14.09	19.33	27.87	30.81	44.51	57.70
w_B	0.0482	0.0482	0.0482	0.0482	0.0482	0.0482	0.0482	0.0482	0.0482
T/K	328.66	328.30	327.77	327.44	326.98	326.50	325.87	324.89	324.44
P/bar	73.68	84.18	91.10	112.05	126.19	139.38	153.52	174.26	194.05
w_B	0.0482	0.0482	0.0482	0.0482	0.0482	0.0482	0.0482	0.0482	0.0482
T/K	324.26	323.40	322.92	322.45	322.18	321.59	320.96	320.52	320.22
P/bar	201.59	218.55	235.52	256.06	273.62	299.65	329.25	354.85	376.00
w_B	0.0482	0.0482	0.0482	0.0482	0.0482	0.0482	0.0482	0.0482	0.0482
T/K	320.15	319.94	319.41	318.92	318.51	317.96	317.53	317.11	316.94
P/bar	373.11	388.53	410.81	435.32	459.50	500.35	558.78	613.44	658.68
w_B	0.0482	0.0482	0.0482	0.0682	0.0682	0.0682	0.0682	0.0682	0.0682
T/K	316.84	316.79	316.68	331.79	331.34	330.90	329.89	329.11	328.22
P/bar	692.61	719.94	754.81	19.69	19.69	29.12	49.85	57.39	78.12

continued

continued

w_B	0.0682	0.0682	0.0682	0.0682	0.0682	0.0682	0.0682	0.0682	0.0682
T/K	327.73	327.41	326.45	325.90	325.30	324.79	324.12	323.57	323.25
P/bar	84.72	91.32	126.19	143.15	163.89	177.08	197.82	209.13	218.55

w_B	0.0682	0.0682	0.0682	0.0682	0.0682	0.0682	0.0682	0.0682	0.0682
T/K	322.50	321.92	321.43	321.06	320.52	319.85	319.35	318.93	318.40
P/bar	243.05	273.21	293.95	318.45	352.38	376.88	409.40	439.80	478.50

w_B	0.0682	0.0682
T/K	318.12	317.82
P/bar	517.31	551.88

Type of data: spinodal points (UCST behavior, measured by PPICS)

w_B	0.0067	0.0067	0.0067	0.0067	0.0067	0.0067	0.0067	0.0067	0.0067
T/K	331.18	330.54	330.06	329.47	329.02	328.57	328.04	327.23	326.43
P/bar	2.93	4.02	2.60	5.92	12.00	19.97	29.02	42.22	56.35

w_B	0.0067	0.0067	0.0067	0.0067	0.0067	0.0067	0.0067	0.0067	0.0067
T/K	325.95	325.50	325.04	324.51	323.82	323.12	322.63	322.29	321.79
P/bar	66.72	74.67	90.28	104.42	125.15	142.12	156.26	169.45	193.95

w_B	0.0067	0.0067	0.0067	0.0067	0.0067	0.0067	0.0067	0.0067	0.0067
T/K	321.47	321.04	320.61	320.33	319.54	319.01	318.41	318.05	317.24
P/bar	209.18	225.06	252.39	277.93	313.65	342.31	375.85	399.41	444.65

w_B	0.018	0.018	0.018	0.018	0.018	0.018	0.018	0.018	0.018
T/K	331.16	330.65	330.05	329.58	329.25	328.78	328.10	327.56	326.81
P/bar	2.38	5.13	11.66	18.20	26.93	32.94	44.94	58.03	70.04

w_B	0.018	0.018	0.018	0.018	0.018	0.018	0.018	0.018	0.018
T/K	326.02	324.96	324.09	323.36	322.98	322.23	321.65	321.15	320.48
P/bar	91.32	115.33	150.25	169.89	184.08	204.81	237.31	252.28	276.29

w_B	0.018	0.018	0.018	0.018	0.018	0.025	0.025	0.025	0.025
T/K	320.00	319.12	318.54	317.64	317.11	331.38	330.80	330.13	329.50
P/bar	306.84	327.38	361.96	408.50	441.62	4.86	10.42	20.83	32.64

w_B	0.025	0.025	0.025	0.025	0.025	0.025	0.025	0.025	0.025
T/K	329.02	328.49	327.67	326.99	326.37	325.88	325.40	324.94	324.17
P/bar	38.19	42.36	65.28	72.92	93.05	107.64	113.89	138.19	158.33

w_B	0.025	0.025	0.025	0.025	0.025	0.025	0.025	0.025	0.025
T/K	323.41	322.93	322.39	321.68	321.01	320.40	320.13	319.70	319.48
P/bar	165.11	188.69	214.13	229.58	271.75	288.89	302.00	325.00	333.33

w_B	0.025	0.025	0.025	0.025	0.025	0.025	0.0458	0.0458	0.0458
T/K	319.13	318.62	318.16	317.74	317.09	316.18	331.22	330.73	330.33
P/bar	352.08	382.64	403.47	431.81	477.78	578.47	21.99	26.95	32.92

w_B	0.0458	0.0458	0.0458	0.0458	0.0458	0.0458	0.0458	0.0458	0.0458
T/K	329.71	329.35	328.60	328.51	327.92	327.41	326.85	326.17	325.63
P/bar	46.25	55.72	66.36	68.09	82.72	97.20	112.07	130.72	146.59

w_B	0.0458	0.0458	0.0482	0.0482	0.0482	0.0482	0.0482	0.0482	0.0482
T/K	325.23	324.84	332.61	332.11	331.63	331.26	330.98	330.23	329.49
P/bar	153.54	164.65	8.61	9.13	13.09	23.02	25.00	43.19	56.00

continued

continued

w_B	0.0482	0.0482	0.0482	0.0482	0.0482	0.0482	0.0482	0.0482	0.0482
T/K	328.66	328.30	327.77	327.44	326.98	326.50	325.87	324.89	324.44
P/bar	72.14	83.06	89.37	107.34	120.24	134.13	149.01	167.86	184.72
w_B	0.0482	0.0482	0.0482	0.0482	0.0482	0.0482	0.0482	0.0482	0.0482
T/K	324.26	323.40	322.92	322.45	322.18	321.59	320.96	320.52	320.22
P/bar	193.65	212.5	228.37	252.82	271.57	297.84	325.89	350.68	371.23
w_B	0.0482	0.0482	0.0482	0.0482	0.0482	0.0482	0.0482	0.0482	0.0482
T/K	320.15	319.94	319.41	318.92	318.51	317.96	317.53	317.11	316.94
P/bar	365.28	385.73	400.99	427.78	457.89	493.25	551.79	605.36	646.63
w_B	0.0482	0.0482	0.0482	0.0682	0.0682	0.0682	0.0682	0.0682	0.0682
T/K	316.84	316.79	316.68	331.79	331.34	330.90	329.89	329.11	328.22
P/bar	682.74	711.51	746.23	15.77	17.06	24.80	46.73	54.46	74.45
w_B	0.0682	0.0682	0.0682	0.0682	0.0682	0.0682	0.0682	0.0682	0.0682
T/K	327.73	327.41	326.45	325.90	325.30	324.79	324.12	323.57	323.25
P/bar	82.19	86.06	123.46	138.29	159.35	174.40	193.75	204.07	213.74
w_B	0.0682	0.0682	0.0682	0.0682	0.0682	0.0682	0.0682	0.0682	0.0682
T/K	322.50	321.92	321.43	321.06	320.52	319.85	319.35	318.93	318.40
P/bar	239.53	268.55	291.12	313.69	349.16	374.30	405.26	436.85	472.97
w_B	0.0682	0.0682							
T/K	318.12	317.82							
P/bar	514.24	549.06							

Type of data: cloud points (UCST behavior)

w_B	0.011	0.011	0.011	0.011	0.011	0.011	0.011	0.011	0.011
T/K	330.17	329.93	329.88	329.57	329.38	329.30	329.00	328.84	328.77
P/bar	7.0	12.0	13.2	19.2	23.0	24.2	30.6	33.6	35.3
w_B	0.011	0.011	0.011	0.011	0.011	0.011	0.011	0.011	0.011
T/K	328.49	328.37	328.26	327.96	327.80	327.67	327.39	327.32	327.03
P/bar	40.8	43.5	45.8	51.6	55.3	57.7	64.2	65.6	71.3
w_B	0.011	0.011	0.011	0.011	0.011	0.011	0.011	0.011	0.011
T/K	326.71	326.67	326.62	326.47	326.38	326.17	326.15	326.09	325.97
P/bar	78.8	79.8	81.0	84.0	86.0	90.8	91.3	92.5	95.0
w_B	0.011	0.060	0.060	0.060	0.060	0.060	0.060	0.060	0.060
T/K	325.63	331.66	331.47	331.28	331.09	330.91	330.71	330.50	330.30
P/bar	103.3	23.0	26.0	28.4	31.8	34.4	37.6	41.1	44.6
w_B	0.060	0.060	0.060	0.060	0.060	0.060	0.060	0.060	0.060
T/K	330.09	329.96	329.72	329.51	329.30	329.13	328.94	328.78	328.52
P/bar	48.3	50.5	55.0	58.7	62.7	66.2	70.1	73.3	78.5
w_B	0.060	0.060	0.060	0.060	0.060	0.060	0.060	0.120	0.120
T/K	328.29	328.11	327.90	327.74	327.52	327.31	327.12	329.96	329.76
P/bar	83.7	87.5	92.4	96.0	101.0	106.5	111.0	11.6	14.8
w_B	0.120	0.120	0.120	0.120	0.120	0.120	0.120	0.120	0.120
T/K	329.55	329.35	329.14	328.94	328.75	328.56	328.35	328.16	327.97
P/bar	17.7	21.0	24.3	28.0	31.0	34.3	38.3	42.3	45.8

continued

continued

w_B	0.120	0.120	0.120	0.120	0.120	0.120	0.120	0.120	0.120
T/K	327.75	327.57	327.38	327.19	326.98	326.79	326.57	326.39	326.16
P/bar	51.0	54.7	58.8	63.2	68.0	72.5	78.0	82.3	88.5

w_B	0.120	0.120	0.187	0.187	0.187	0.187	0.187	0.187	0.187
T/K	325.98	325.78	327.54	327.33	327.12	326.92	326.73	326.55	326.35
P/bar	93.6	99.0	10.1	13.2	17.0	19.7	23.3	27.0	31.0

w_B	0.187	0.187	0.187	0.187	0.187	0.187	0.187	0.187	0.187
T/K	326.15	325.97	325.75	325.55	325.34	325.14	324.94	324.78	324.58
P/bar	35.1	39.5	44.5	49.6	54.8	60.2	66.0	71.0	77.0

w_B	0.187	0.187	0.187
T/K	324.35	324.13	323.91
P/bar	84.0	91.0	99.0

Type of data: cloud points (LCST behavior)

w_B	0.065	0.065	0.065	0.065	0.065	0.065	0.065	0.065	0.065
T/K	499.45	502.12	504.64	507.12	509.61	512.10	514.59	517.04	519.50
P/bar	18.0	21.2	24.5	27.5	30.6	33.6	37.05	39.8	42.9

w_B	0.187	0.187	0.187	0.187	0.187	0.187	0.187
T/K	503.80	506.47	508.41	511.16	513.88	516.08	518.85
P/bar	13.2	16.1	18.4	21.7	24.7	27.6	30.8

Type of data: coexistence data (liquid-liquid-vapor three phase equilibrium)

w_B	0.0113	0.0516	0.0600	0.0655	0.0989	0.1207	0.1520	0.18699
T/K	500.92	495.35	493.67	493.90	492.92	496.35	498.78	502.30
P/bar	12.78	11.47	10.81	10.77	9.82	10.68	10.84	11.29

Polymer (B):	**polystyrene**	**1998KOA**
Characterization:	M_n/kg.mol^{-1} = 29.1, M_w/kg.mol^{-1} = 31.6,	
	Aldrich Chem. Co., Inc., Milwaukee, WI	
Solvent (A):	**methylcyclohexane** C_7H_{14}	**108-87-2**

Type of data: cloud points

w_B	0.0323	0.0323	0.0323	0.0323	0.0323	0.0323	0.0323	0.0494	0.0494
T/K	299.38	298.99	298.58	298.25	297.92	297.60	297.14	301.30	301.01
P/MPa	1.05	3.05	5.05	7.05	9.05	11.05	13.05	1.05	2.05

w_B	0.0494	0.0494	0.0494	0.0494	0.0494	0.0494	0.0494	0.0494	0.0494
T/K	300.78	300.60	300.41	300.22	300.03	299.85	299.68	299.50	299.19
P/MPa	3.05	4.05	5.05	6.05	7.05	8.05	9.05	10.05	12.05

w_B	0.0494	0.0726	0.0726	0.0726	0.0726	0.0726	0.0726	0.0726	0.0726
T/K	298.95	302.53	302.29	302.04	301.82	301.61	301.41	301.25	301.05
P/MPa	14.05	1.05	2.05	3.05	4.05	5.05	6.05	7.05	8.05

w_B	0.0726	0.0726	0.0726	0.0726	0.0882	0.0882	0.0882	0.0882	0.0882
T/K	300.88	300.73	300.42	300.13	303.46	303.39	302.93	302.72	302.47
P/MPa	9.05	10.05	12.05	14.05	1.05	2.05	3.05	4.05	5.05

continued

continued

w_B	0.0882	0.0882	0.0882	0.0882	0.0882	0.0882	0.0882	0.1097	0.1097
T/K	302.37	302.10	301.90	301.79	301.52	301.24	300.90	304.05	303.48
P/MPa	6.05	7.05	8.05	9.05	10.05	12.05	14.05	1.05	3.05

w_B	0.1097	0.1097	0.1097	0.1097	0.1097	0.1281	0.1281	0.1281	0.1281
T/K	303.06	302.69	302.25	301.88	301.52	304.44	303.88	303.45	302.97
P/MPa	5.05	7.05	9.05	11.05	13.05	1.05	3.05	5.05	7.05

w_B	0.1281	0.1281	0.1281	0.1281	0.1519	0.1519	0.1519	0.1519	0.1519
T/K	302.56	302.24	301.93	301.68	305.03	304.44	304.00	303.53	303.11
P/MPa	9.05	11.05	13.05	14.05	1.05	3.05	5.05	7.05	9.05

w_B	0.1519	0.1519	0.1519	0.1821	0.1821	0.1821	0.1821	0.2060	0.2060
T/K	302.62	302.35	302.17	305.45	304.41	304.00	303.53	305.42	304.95
P/MPa	11.05	13.05	14.05	1.05	5.05	7.05	9.05	1.05	3.05

w_B	0.2060	0.2060	0.2060	0.2060	0.2060	0.2251	0.2251	0.2251	0.2251
T/K	304.53	303.89	303.52	303.15	302.80	305.43	304.96	304.35	303.91
P/MPa	5.05	7.05	9.05	11.05	13.05	1.05	3.05	5.05	7.05

w_B	0.2251	0.2251	0.2251	0.2634	0.2634	0.2634	0.2634	0.2634	0.2634
T/K	303.45	303.08	302.75	305.03	304.44	303.90	303.51	303.02	302.70
P/MPa	9.05	11.05	13.05	1.05	3.05	5.05	7.05	9.05	11.05

w_B	0.2634
T/K	302.32
P/MPa	13.05

Comments: Cloud points for mixtures of this system + carbon dioxide are given in Chapter 4.

Polymer (B):	**polystyrene**	**1998KOA**

Characterization: $M_n/kg.mol^{-1} = 64$, $M_w/kg.mol^{-1} = 250$,
                        Novacor Technology and Research Corporation

Solvent (A):	**methylcyclohexane** C_7H_{14}	**108-87-2**

Type of data: cloud points

w_B	0.0211	0.0211	0.0211	0.0211	0.0211	0.0211	0.0330	0.0330	0.0330
T/K	329.81	329.35	328.50	327.87	327.24	326.63	330.76	330.37	329.87
P/MPa	3.95	4.90	6.55	7.95	9.25	10.85	2.40	3.15	4.00

w_B	0.0330	0.0330	0.0330	0.0330	0.0511	0.0511	0.0511	0.0511	0.0511
T/K	329.01	328.09	327.43	326.35	330.91	330.35	329.46	328.62	327.71
P/MPa	5.65	7.60	9.05	11.80	1.45	2.40	3.95	5.60	7.60

w_B	0.0511	0.0511	0.0511	0.0648	0.0648	0.0648	0.0648	0.0648	0.0648
T/K	326.91	326.23	325.75	330.11	330.09	329.04	328.98	327.83	326.85
P/MPa	9.50	11.30	12.60	2.00	2.00	3.90	4.00	6.30	8.60

w_B	0.0648	0.0648	0.0648	0.0971	0.0971	0.0971	0.0971	0.0971	0.0971
T/K	326.37	325.83	325.26	329.34	328.59	327.81	327.05	326.30	325.47
P/MPa	9.75	11.15	12.85	1.60	3.05	4.55	6.10	7.70	9.80

w_B	0.0971	0.1360	0.1360	0.1360	0.1360	0.1360	0.1360	0.1360	0.1360
T/K	324.86	328.36	328.29	327.68	327.33	326.63	325.86	325.15	324.47
P/MPa	11.55	1.50	1.60	3.00	3.50	4.85	6.55	8.25	10.10

Polymer (B):   **polystyrene**   **1995LUS**
Characterization: $M_n/\text{kg.mol}^{-1} = 3.91$, $M_w/\text{kg.mol}^{-1} = 4.14$,
                         Pressure Chemical Company, Pittsburgh, PA
Solvent (A):   **methylcyclopentane C_6H_{12}**   **96-37-7**

Type of data: cloud points

w_B	0.181	0.181
T/K	259.49	259.62
P/MPa	3.70	1.30

Polymer (B):   **polystyrene**   **1995LUS**
Characterization: $M_n/\text{kg.mol}^{-1} = 12.75$, $M_w/\text{kg.mol}^{-1} = 13.5$,
                         Pressure Chemical Company, Pittsburgh, PA
Solvent (A):   **methylcyclopentane C_6H_{12}**   **96-37-7**

Type of data: cloud points

w_B	0.165	0.165	0.165	0.165	0.165
T/K	291.30	291.43	291.53	291.68	291.78
P/MPa	2.00	1.49	1.27	0.79	0.45

Polymer (B):   **polystyrene**   **1995LUS**
Characterization: $M_n/\text{kg.mol}^{-1} = 21.45$, $M_w/\text{kg.mol}^{-1} = 22.1$,
                         Polymer Laboratories, Amherst, MA
Solvent (A):   **methylcyclopentane C_6H_{12}**   **96-37-7**

Type of data: cloud points

w_B	0.153	0.153	0.153	0.153	0.153
T/K	299.99	300.12	300.33	300.57	300.75
P/MPa	2.59	2.24	1.64	1.02	0.51

Polymer (B):   **polystyrene**   **1995LUS**
Characterization: $M_n/\text{kg.mol}^{-1} = 23.6$, $M_w/\text{kg.mol}^{-1} = 25.0$,
                         Pressure Chemical Company, Pittsburgh, PA
Solvent (A):   **methylcyclopentane C_6H_{12}**   **96-37-7**

Type of data: cloud points

w_B	0.142	0.142	0.142	0.142	0.148	0.148	0.148	0.148	0.157
T/K	302.58	302.93	303.20	303.38	302.60	302.88	303.18	303.46	302.60
P/MPa	2.20	1.36	0.67	0.23	2.31	1.59	0.84	0.19	2.32
w_B	0.157	0.157	0.157	0.164	0.164	0.164	0.164	0.164	0.165
T/K	302.87	303.24	303.44	302.53	302.76	302.98	303.19	303.43	302.64
P/MPa	1.67	0.77	0.32	2.90	2.25	1.67	1.15	0.55	2.33

continued

continued

w_B	0.165	0.165	0.165	0.177	0.177	0.177	0.177	0.188	0.188
T/K	302.93	303.25	303.47	302.68	302.98	303.28	303.51	302.60	302.90
P/MPa	1.60	0.90	0.38	2.27	1.51	0.83	0.28	2.13	1.45

w_B	0.188	0.188
T/K	303.16	303.41
P/MPa	0.84	0.26

critical concentration: $w_{B, crit}$ = 0.170

Polymer (B):	**polystyrene**	**1995LUS**
Characterization:	M_n/kg.mol^{-1} = 100.3, M_w/kg.mol^{-1} = 106.3,	
	Pressure Chemical Company, Pittsburgh, PA	
Solvent (A):	**methylcyclopentane** C_6H_{12}	**96-37-7**

Type of data: cloud points

w_B	0.155	0.155	0.155	0.155	0.155
T/K	325.19	325.42	325.96	326.38	327.01
P/MPa	2.45	2.15	1.63	1.16	0.55

Polymer (B):	**polystyrene-d8**	**1995LUS**
Characterization:	M_n/kg.mol^{-1} = 25.4, M_w/kg.mol^{-1} = 26.9,	
	completely deuterated, Polymer Laboratories, Amherst, MA	
Solvent (A):	**methylcyclopentane** C_6H_{12}	**96-37-7**

Type of data: cloud points

w_B	0.179	0.179	0.179	0.179	0.179
T/K	302.48	302.90	303.11	303.20	303.39
P/MPa	2.52	1.51	0.95	0.70	0.23

Polymer (B):	**polystyrene**	**2000DE1**
Characterization:	M_n/kg.mol^{-1} = 12.74, M_w/kg.mol^{-1} = 13.5,	
	Pressure Chemical Company, Pittsburgh, PA	
Solvent (A):	**nitroethane** $C_2H_5NO_2$	**79-24-3**

Type of data: cloud points

w_B	0.3145	0.3145	0.3145	0.3145	0.3145	0.2867	0.2867	0.2867	0.2867
T/K	277.16	277.27	277.37	277.45	277.55	277.47	277.58	277.67	277.77
P/MPa	4.96	3.71	2.46	1.54	0.36	6.04	4.72	3.71	2.41

w_B	0.2867	0.2867	0.2634	0.2634	0.2634	0.2634	0.2634	0.2634	0.2348
T/K	277.85	277.93	278.08	278.18	278.17	278.29	278.38	278.53	278.59
P/MPa	1.47	0.51	5.56	4.36	4.45	3.16	2.04	0.33	5.33

w_B	0.2348	0.2348	0.2348	0.2348	0.2242	0.2242	0.2242	0.2242	0.2242
T/K	278.69	278.79	278.88	279.01	278.96	279.05	279.16	279.25	279.37
P/MPa	4.14	3.00	1.76	0.27	5.11	4.05	2.66	1.61	0.26

continued

continued

w_B	0.2048	0.2048	0.2048	0.2048	0.2048	0.1799	0.1799	0.1799	0.1799
T/K	279.05	279.16	279.26	279.32	279.42	279.16	279.23	279.36	279.43
P/MPa	4.79	3.49	2.26	1.56	0.40	5.22	4.35	2.83	2.00

w_B	0.1562	0.1562	0.1562	0.1562	0.1340	0.1340	0.1340	0.1340	0.1340
T/K	279.32	279.41	279.49	279.63	279.30	279.40	279.49	279.61	279.75
P/MPa	5.45	4.28	3.31	1.59	5.42	4.28	3.22	1.89	0.15

w_B	0.1179	0.1179	0.1179	0.1801	0.1801	0.1801	0.1801	0.1801	0.1611
T/K	279.19	279.30	279.40	279.19	279.30	279.40	279.45	279.54	279.19
P/MPa	5.72	4.40	3.20	4.59	3.21	2.09	1.52	0.37	5.24

w_B	0.1611	0.1611	0.1611	0.1611	0.1385	0.1385	0.1385	0.1385	0.1385
T/K	279.30	279.39	279.49	279.60	279.49	279.60	279.70	279.80	279.96
P/MPa	3.89	2.72	1.56	0.30	5.76	4.65	3.52	2.26	0.38

w_B	0.1215	0.1215	0.1215	0.0999	0.0999	0.0999	0.0999	0.0999	0.0833
T/K	279.49	279.39	279.50	279.39	279.49	279.60	279.70	279.80	279.30
P/MPa	5.72	5.56	4.39	5.37	4.21	2.94	1.65	0.37	4.99

w_B	0.0833	0.0833	0.0833	0.0833	0.0714	0.0714	0.0714	0.0714	0.0625
T/K	279.40	279.49	279.61	279.66	277.27	277.37	277.46	277.57	276.28
P/MPa	3.84	2.75	1.38	0.67	5.30	4.17	2.96	1.56	4.75

w_B	0.0625	0.0625	0.0625
T/K	276.38	276.47	276.58
P/MPa	3.53	2.54	1.07

Type of data: spinodal points

w_B	0.3145	0.3145	0.3145	0.3145	0.3145	0.2867	0.2867	0.2867	0.2867
T/K	277.16	277.27	277.37	277.45	277.55	277.47	277.58	277.67	277.77
P/MPa	4.85	3.59	2.36	1.45	0.29	5.88	4.59	3.56	2.29

w_B	0.2867	0.2867	0.2634	0.2634	0.2634	0.2634	0.2634	0.2634	0.2348
T/K	277.85	277.93	278.08	278.18	278.17	278.29	278.38	278.53	278.59
P/MPa	1.35	0.36	5.42	4.25	4.32	3.05	1.93	0.23	5.14

w_B	0.2348	0.2348	0.2348	0.2348	0.2242	0.2242	0.2242	0.2242	0.2242
T/K	278.69	278.79	278.88	279.01	278.96	279.05	279.16	279.25	279.37
P/MPa	3.98	2.81	1.58	0.14	4.78	3.68	2.33	1.28	−0.01

w_B	0.2048	0.2048	0.2048	0.2048	0.2048	0.1799	0.1799	0.1799	0.1799
T/K	279.05	279.16	279.26	279.32	279.42	279.16	279.23	279.36	279.43
P/MPa	4.44	3.12	1.95	1.25	0.10	5.04	4.12	2.61	1.78

w_B	0.1562	0.1562	0.1562	0.1562	0.1340	0.1340	0.1340	0.1340	0.1340
T/K	279.32	279.41	279.49	279.63	279.30	279.40	279.49	279.61	279.75
P/MPa	5.18	4.05	3.14	1.39	5.19	4.05	3.00	1.62	−0.08

w_B	0.1179	0.1179	0.1179	0.1801	0.1801	0.1801	0.1801	0.1801	0.1611
T/K	279.19	279.30	279.40	279.19	279.30	279.40	279.45	279.54	279.19
P/MPa	5.52	4.11	2.97	4.32	2.87	1.81	1.26	0.15	5.06

w_B	0.1611	0.1611	0.1611	0.1611	0.1385	0.1385	0.1385	0.1385	0.1385
T/K	279.30	279.39	279.49	279.60	279.49	279.60	279.70	279.80	279.96
P/MPa	3.70	2.58	1.38	0.11	5.54	4.45	3.28	2.09	0.24

continued

continued

w_B	0.1215	0.1215	0.1215	0.0999	0.0999	0.0999	0.0999	0.0999	0.0833
T/K	279.49	279.39	279.50	279.39	279.49	279.60	279.70	279.80	279.30
P/MPa	5.63	5.28	4.16	4.95	3.82	2.51	1.34	−0.02	4.47

w_B	0.0833	0.0833	0.0833	0.0833	0.0714	0.0714	0.0714	0.0714	0.0625
T/K	279.40	279.49	279.61	279.66	277.27	277.37	277.46	277.57	276.28
P/MPa	3.25	2.27	0.99	0.19	4.92	3.81	2.66	1.27	4.51

w_B	0.0625	0.0625	0.0625
T/K	276.38	276.47	276.58
P/MPa	3.22		0.80

Polymer (B):	**polystyrene**		**2000DE1**
Characterization:	M_n/kg.mol^{-1} = 29.0, M_w/kg.mol^{-1} = 30.74,		
	Pressure Chemical Company, Pittsburgh, PA		
Solvent (A):	**nitroethane** $C_2H_5NO_2$		**79-24-3**

Type of data: cloud points

w_B	0.3649	0.3649	0.3649	0.3649	0.3649	0.1977	0.1977	0.1977	0.1977
T/K	289.05	289.10	289.15	289.26	289.42	296.25	296.44	296.55	296.65
P/MPa	4.49	4.02	3.67	2.84	1.66	5.29	3.70	3.17	2.57

w_B	0.1977	0.1977	0.1977	0.1977	0.1654	0.1654	0.1654	0.1654	0.1654
T/K	296.79	296.88	296.95	296.98	296.65	296.74	296.84	297.04	297.24
P/MPa	1.57	1.03	0.66	0.48	4.90	4.35	3.64	2.39	1.22

w_B	0.1654	0.1503	0.1503	0.1503	0.1503	0.1503	0.1503	0.1503	0.1503
T/K	297.35	296.65	296.74	296.84	297.05	297.24	297.35	297.45	297.49
P/MPa	0.49	5.64	5.05	4.44	3.08	1.90	1.26	0.63	0.34

w_B	0.1503	0.1503	0.1393	0.1393	0.1393	0.1393	0.1393	0.1393	0.1393
T/K	297.54	297.54	296.61	296.66	296.77	296.98	297.05	297.15	297.25
P/MPa	0.13	0.05	4.83	4.49	3.80	2.41	1.94	1.36	0.76

w_B	0.1393	0.1393	0.1388	0.1388	0.1388	0.1388	0.1388	0.1388	0.1388
T/K	297.30	297.35	296.65	296.75	296.95	297.05	297.15	297.24	297.35
P/MPa	0.46	0.12	4.93	4.32	3.09	2.45	1.82	1.22	0.56

w_B	0.1388	0.1388	0.1126	0.1126	0.1126	0.1126	0.1126	0.1126	0.1126
T/K	297.40	297.43	296.48	296.58	296.65	296.86	296.96	297.05	297.10
P/MPa	0.30	0.10	5.25	4.62	4.17	2.72	2.18	1.59	1.28

w_B	0.1126	0.1126	0.1126	0.0977	0.0977	0.0977	0.0977	0.0977	0.0977
T/K	297.17	297.25	297.27	296.33	296.44	296.55	296.74	296.89	297.00
P/MPa	0.79	0.39	0.23	4.79	4.16	3.46	2.21	1.26	0.57

w_B	0.0977	0.0944	0.0944	0.0944	0.0944	0.0944	0.0944	0.0944	0.0944
T/K	297.06	296.24	296.36	296.46	296.67	296.73	296.86	296.92	296.99
P/MPa	0.14	5.17	4.39	3.75	2.43	1.98	1.22	0.84	0.43

w_B	0.0944	0.0861	0.0861	0.0861	0.0861	0.0861	0.0861	0.0861	0.0861
T/K	297.02	296.06	296.17	296.25	296.45	296.64	296.79	296.81	296.87
P/MPa	0.23	5.33	4.62	3.98	2.74	1.56	0.53	0.44	0.04

continued

continued

w_B	0.0727	0.0727	0.0727	0.0727	0.0727	0.0727	0.0622	0.0622	0.0622
T/K	295.57	295.65	295.75	295.95	296.16	296.25	294.89	294.95	295.08
P/MPa	4.85	4.28	3.67	2.33	0.95	0.36	4.89	4.47	3.60

w_B	0.0622	0.0622	0.0622	0.0622
T/K	295.26	295.43	295.49	295.60
P/MPa	2.34	1.18	0.73	0.03

Type of data: spinodal points

w_B	0.3649	0.3649	0.3649	0.3649	0.3649	0.1977	0.1977	0.1977	0.1977
T/K	289.05	289.10	289.15	289.26	289.42	296.25	296.44	296.55	296.65
P/MPa	4.27	3.72	3.48	2.63	1.46	5.02	3.49	3.03	2.32

w_B	0.1977	0.1977	0.1977	0.1977	0.1654	0.1654	0.1654	0.1654	0.1654
T/K	296.79	296.88	296.95	296.98	296.65	296.74	296.84	297.04	297.24
P/MPa	1.41	0.86	0.50	0.29	4.65	3.98	3.34	2.05	0.87

w_B	0.1654	0.1503	0.1503	0.1503	0.1503	0.1503	0.1503	0.1503	0.1503
T/K	297.35	296.65	296.74	296.84	297.05	297.24	297.35	297.45	297.49
P/MPa	0.20	5.35	4.75	4.16	2.75	1.57	1.04	0.33	0.10

w_B	0.1503	0.1503	0.1393	0.1393	0.1393	0.1393	0.1393	0.1393	0.1393
T/K	297.54	297.54	296.61	296.66	296.77	296.98	297.05	297.15	297.25
P/MPa	−0.18	−0.19	4.49	4.24	3.54	2.15	1.70	1.14	0.57

w_B	0.1393	0.1393	0.1388	0.1388	0.1388	0.1388	0.1388	0.1388	0.1388
T/K	297.30	297.35	296.65	296.75	296.95	297.05	297.15	297.24	297.35
P/MPa	0.24	−0.08	4.72	4.12	2.74	2.11	1.54	1.01	0.20

w_B	0.1388	0.1388	0.1126	0.1126	0.1126	0.1126	0.1126	0.1126	0.1126
T/K	297.40	297.43	296.48	296.58	296.65	296.86	296.96	297.05	297.10
P/MPa	0.07	−0.20	4.98	4.27	3.92	2.52	1.93	1.28	1.06

w_B	0.1126	0.1126	0.1126	0.0977	0.0977	0.0977	0.0977	0.0977	0.0977
T/K	297.17	297.25	297.27	296.33	296.44	296.55	296.74	296.89	297.00
P/MPa	0.52	0.05	−0.03		3.99	3.28	2.06	1.07	0.40

w_B	0.0977	0.0944	0.0944	0.0944	0.0944	0.0944	0.0944	0.0944	0.0944
T/K	297.06	296.24	296.36	296.46	296.67	296.73	296.86	296.92	296.99
P/MPa		0.06	4.25	3.52	2.34	1.89	1.06	0.60	

w_B	0.0944	0.0861	0.0861	0.0861	0.0861	0.0861	0.0861	0.0861	0.0861
T/K	297.02	296.06	296.17	296.25	296.45	296.64	296.79	296.81	296.87
P/MPa	−0.07	5.29	4.52	3.86	2.64	1.49	0.40	0.35	−0.11

w_B	0.0727	0.0727	0.0727	0.0727	0.0727	0.0727	0.0622	0.0622	0.0622
T/K	295.57	295.65	295.75	295.95	296.16	296.25	294.89	294.95	295.08
P/MPa	4.69	4.19	3.53	2.17	0.83	0.23	4.79	4.30	3.45

w_B	0.0622	0.0622	0.0622	0.0622
T/K	295.26	295.43	295.49	295.60
P/MPa	2.19	1.03		−0.07

Polymer (B):	polystyrene						2000DE1

Characterization: M_n/kg.mol^{-1} = 86.6, M_w/kg.mol^{-1} = 90.1,
Pressure Chemical Company, Pittsburgh, PA

Solvent (A):	nitroethane	$C_2H_5NO_2$					79-24-3

Type of data: cloud points

w_B	0.2265	0.2265	0.2265	0.2265	0.2265	0.2265	0.2265	0.1911	0.1911
T/K	309.43	309.43	309.74	309.73	310.14	310.45	310.68	312.43	312.83
P/MPa	4.96	4.96	3.89	3.90	2.42	1.35	0.44	5.27	3.80

w_B	0.1911	0.1911	0.1911	0.1651	0.1651	0.1651	0.1651	0.1651	0.1454
T/K	313.14	313.44	313.84	314.13	314.45	314.75	315.16	315.39	315.44
P/MPa	2.74	1.63	0.22	5.18	4.04	2.90	1.42	0.56	3.68

w_B	0.1454	0.1454	0.1454	0.1390	0.1390	0.1390	0.1390	0.1390	0.1245
T/K	315.75	316.05	316.39	314.84	315.15	315.44	315.65	316.00	315.15
P/MPa	2.64	1.53	0.31	4.68	3.54	2.47	1.69	0.44	4.66

w_B	0.1245	0.1245	0.1245	0.1245	0.1059	0.1059	0.1059	0.1059	0.1059
T/K	315.44	315.74	316.05	316.30	315.44	315.34	315.74	316.05	316.64
P/MPa	3.67	2.54	1.39	0.54	4.63	5.03	3.57	2.42	0.28

w_B	0.0922	0.0922	0.0922	0.0922	0.0922	0.0633	0.0633	0.0633	0.0633
T/K	315.44	315.75	316.05	316.34	316.74	315.34	315.74	316.05	316.34
P/MPa	4.98	3.86	2.75	1.62	0.14	5.22	3.68	2.56	1.52

w_B	0.0633	0.0519	0.0519	0.0519	0.0519	0.0519	0.0519	0.0440	0.0440
T/K	316.65	315.04	315.03	315.44	315.74	316.05	316.45	314.84	315.15
P/MPa	0.47	5.21	5.23	3.78	2.69	1.55	0.13	4.33	3.08

w_B	0.0440	0.0440	0.0440	0.0337	0.0337	0.0337	0.0337	0.0337	
T/K	315.44	315.55	315.84	313.63	314.04	314.34	314.54	314.94	
P/MPa	1.95	1.47	0.40	4.98	3.51	2.41	1.71	0.19	

Type of data: spinodal points

w_B	0.2265	0.2265	0.2265	0.2265	0.2265	0.2265	0.2265	0.1911	0.1911
T/K	309.43	309.43	309.74	309.73	310.14	310.45	310.68	312.43	312.83
P/MPa	4.80	4.81	3.75	3.74	2.27	1.21	0.31	5.09	3.63

w_B	0.1911	0.1911	0.1911	0.1651	0.1651	0.1651	0.1651	0.1651	0.1454
T/K	313.14	313.44	313.84	314.13	314.45	314.75	315.16	315.39	315.44
P/MPa	2.56	1.47	0.07	5.06	3.94	2.79	1.31	0.48	3.51

w_B	0.1454	0.1454	0.1454	0.1390	0.1390	0.1390	0.1390	0.1390	0.1245
T/K	315.75	316.05	316.39	314.84	315.15	315.44	315.65	316.00	315.15
P/MPa	2.48	1.36	0.14	4.54	3.39	2.32	1.56	0.35	4.50

w_B	0.1245	0.1245	0.1245	0.1245	0.1059	0.1059	0.1059	0.1059	0.1059
T/K	315.44	315.74	316.05	316.30	315.44	315.34	315.74	316.05	316.64
P/MPa	3.51	2.34	1.22	0.43	4.43	4.78	3.32	2.22	0.11

w_B	0.0922	0.0922	0.0922	0.0922	0.0922	0.0633	0.0633	0.0633	0.0633
T/K	315.44	315.75	316.05	316.34	316.74	315.34	315.74	316.05	316.34
P/MPa	4.68	3.59	2.47	1.36	−0.11	5.06	3.50	2.39	1.35

continued

continued

w_B	0.0633	0.0519	0.0519	0.0519	0.0519	0.0519	0.0519	0.0440	0.0440
T/K	316.65	315.04	315.03	315.44	315.74	316.05	316.45	314.84	315.15
P/MPa	0.32	5.05	5.04	3.63	2.52	1.40	−0.00	4.14	2.80

w_B	0.0440	0.0440	0.0440	0.0337	0.0337	0.0337	0.0337	0.0337
T/K	315.44	315.55	315.84	313.63	314.04	314.34	314.54	314.94
P/MPa	1.72	1.22	0.12	4.71	3.22	2.18	1.49	−0.04

Polymer (B):	**polystyrene**		**2000DE1**
Characterization:	M_n/kg.mol^{-1} = 125.4, M_w/kg.mol^{-1} = 129.2,		
	Pressure Chemical Company, Pittsburgh, PA		
Solvent (A):	**nitroethane**	**$C_2H_5NO_2$**	**79-24-3**

Type of data: cloud points

w_B	0.1322	0.1322	0.1322	0.1322	0.1322	0.1212	0.1212	0.1212	0.1212
T/K	321.54	321.94	322.34	322.74	323.03	321.84	322.23	322.63	322.94
P/MPa	5.44	4.03	2.67	1.34	0.27	4.70	3.45	2.12	1.15

w_B	0.1212	0.1119	0.1119	0.1119	0.1119	0.1119	0.0929	0.0929	0.0929
T/K	323.11	321.94	322.34	322.74	322.84	323.24	321.84	321.93	322.03
P/MPa	0.53	4.61	3.26	1.91	1.59	0.29	4.27	3.92	3.62

w_B	0.0929	0.0929	0.0929	0.0929	0.0929	0.0929	0.0812	0.0812	0.0812
T/K	322.22	322.43	322.63	322.84	323.03	323.20	321.63	321.84	322.03
P/MPa	3.06	2.39	1.76	1.17	0.58	0.11	5.52	4.90	4.30

w_B	0.0812	0.0812	0.0812	0.0812	0.0645	0.0645	0.0645	0.0645	0.0645
T/K	322.44	322.84	323.03	323.44	321.72	322.14	322.53	322.84	323.13
P/MPa	3.10	1.92	1.38	0.29	5.61	4.25	3.02	2.10	1.23

w_B	0.0645	0.0535	0.0535	0.0535	0.0535	0.0535	0.0535	0.0457	0.0457
T/K	323.44	321.43	321.94	322.34	322.54	322.85	323.14	321.04	321.44
P/MPa	0.37	5.73	4.06	2.82	2.21	1.30	0.41	6.13	4.77

w_B	0.0457	0.0457	0.0457	0.0457	0.0457	0.0355	0.0355	0.0355	0.0355
T/K	321.63	322.03	322.43	322.74	322.93	320.74	321.13	321.54	321.94
P/MPa	4.21	2.86	1.65	0.76	0.16	5.84	4.58	3.27	1.98

w_B	0.0355	0.0355	0.0307	0.0307	0.0307	0.0307	0.0307	0.0307	0.0250
T/K	322.13	322.39	319.64	320.04	320.44	320.73	320.94	321.23	318.35
P/MPa	1.35	0.60	5.37	4.14	2.89	1.99	1.24	0.36	4.55

w_B	0.0250	0.0250	0.0250	0.0250
T/K	318.74	319.15	319.44	319.75
P/MPa	3.43	2.23	1.26	0.37

Type of data: spinodal points

w_B	0.1322	0.1322	0.1322	0.1322	0.1322	0.1212	0.1212	0.1212	0.1212
T/K	321.54	321.94	322.34	322.74	323.03	321.84	322.23	322.63	322.94
P/MPa	5.27	3.89	2.50	1.20	0.13	4.52	3.24	1.90	0.93

continued

continued

w_B	0.1212	0.1119	0.1119	0.1119	0.1119	0.1119	0.0929	0.0929	0.0929
T/K	323.11	321.94	322.34	322.74	322.84	323.24	321.84	321.93	322.03
P/MPa	0.40	4.34	3.02	1.67	1.37	0.13	3.92	3.63	3.33

w_B	0.0929	0.0929	0.0929	0.0929	0.0929	0.0929	0.0812	0.0812	0.0812
T/K	322.22	322.43	322.63	322.84	323.03	323.20	321.63	321.84	322.03
P/MPa	2.75	2.07	1.48	0.88	0.33	−0.15	5.37	4.70	4.12

w_B	0.0812	0.0812	0.0812	0.0812	0.0645	0.0645	0.0645	0.0645	0.0645
T/K	322.44	322.84	323.03	323.44	321.72	322.14	322.53	322.84	323.13
P/MPa	2.89	1.69	1.19	0.11	5.46	3.93	2.83	1.84	0.94

w_B	0.0645	0.0535	0.0535	0.0535	0.0535	0.0535	0.0535	0.0457	0.0457
T/K	323.44	321.43	321.94	322.34	322.54	322.85	323.14	321.04	321.44
P/MPa	0.11	5.68	3.98	2.73	2.13	1.22	0.34	6.09	4.69

w_B	0.0457	0.0457	0.0457	0.0457	0.0457	0.0355	0.0355	0.0355	0.0355
T/K	321.63	322.03	322.43	322.74	322.93	320.74	321.13	321.54	321.94
P/MPa	4.10	2.76	1.57	0.68	0.10	5.68	4.44	3.12	1.84

w_B	0.0355	0.0355	0.0307	0.0307	0.0307	0.0307	0.0307	0.0307	0.0250
T/K	322.13	322.39	319.64	320.04	320.44	320.73	320.94	321.23	318.35
P/MPa	1.23	0.49	5.20	3.96	2.66	1.80	1.09	0.22	4.39

w_B	0.0250	0.0250	0.0250	0.0250
T/K	318.74	319.15	319.44	319.75
P/MPa	3.28	2.04	1.11	0.20

Polymer (B):	**polystyrene**			**2000DE1**

Characterization:　$M_n/kg.mol^{-1} = 86.6$, $M_w/kg.mol^{-1} = 90.1$,
Pressure Chemical Company, Pittsburgh, PA

Solvent (A):	**nitroethane-d5**	**$C_2D_5NO_2$**	**57817-88-6**

Type of data:　cloud points

w_B	0.2291	0.2291	0.2291	0.2291	0.2291	0.2291	0.2291	0.2291	0.1844
T/K	336.41	336.41	336.52	336.61	336.82	337.11	337.43	337.74	338.53
P/MPa	4.32	4.25	3.90	3.60	3.02	2.21	1.37	0.58	4.17

w_B	0.1844	0.1844	0.1844	0.1844	0.1543	0.1543	0.1543	0.1543	0.1543
T/K	338.75	338.94	339.24	339.84	339.02	339.25	339.55	339.85	340.47
P/MPa	3.56	2.96	2.12	0.53	4.17	3.54	2.77	2.03	0.48

w_B	0.1326	0.1326	0.1326	0.1326	0.1326	0.1205	0.1205	0.1205	0.1205
T/K	339.31	339.62	339.95	340.25	340.67	339.35	339.35	339.66	339.50
P/MPa	4.17	3.33	2.45	1.63	0.57	4.57	4.54	3.70	2.91

w_B	0.1205	0.1205	0.0995	0.0995	0.0995	0.0995	0.0995	0.0848	0.0848
T/K	340.34	340.93	339.47	339.86	340.26	340.65	341.11	339.16	339.66
P/MPa	1.92	0.49	4.52	3.50	2.56	1.58	0.49	4.87	3.53

w_B	0.0848	0.0848	0.0848	0.0738	0.0738	0.0738	0.0738	0.0738	0.0669
T/K	340.07	340.44	340.92	339.05	339.54	340.06	340.25	340.85	338.96
P/MPa	2.57	1.62	0.44	4.81	3.51	2.26	1.80	0.37	4.60

continued

continued

w_B	0.0669	0.0669	0.0669	0.0669
T/K	339.48	339.87	340.24	340.63
P/MPa	3.33	2.33	1.34	0.47

Type of data: spinodal points

w_B	0.2291	0.2291	0.2291	0.2291	0.2291	0.2291	0.2291	0.2291	0.1844
T/K	336.41	336.41	336.52	336.61	336.82	337.11	337.43	337.74	338.53
P/MPa	4.25	4.14	3.79	3.53	2.92	2.14	1.25	0.52	4.06

w_B	0.1844	0.1844	0.1844	0.1844	0.1543	0.1543	0.1543	0.1543	0.1543
T/K	338.75	338.94	339.24	339.84	339.02	339.25	339.55	339.85	340.47
P/MPa	3.40	2.86	2.03	0.44	3.99	3.29	2.60	1.88	0.27

w_B	0.1326	0.1326	0.1326	0.1326	0.1326	0.1205	0.1205	0.1205	0.1205
T/K	339.31	339.62	339.95	340.25	340.67	339.35	339.35	339.66	339.50
P/MPa	3.90	3.08	2.24	1.44	0.44	4.34	4.35	3.47	2.71

w_B	0.1205	0.1205	0.0995	0.0995	0.0995	0.0995	0.0995	0.0848	0.0848
T/K	340.34	340.93	339.47	339.86	340.26	340.65	341.11	339.16	339.66
P/MPa	1.73	0.29	4.24	3.26	2.30	1.38	0.23	4.66	3.29

w_B	0.0848	0.0848	0.0848	0.0738	0.0738	0.0738	0.0738	0.0738	0.0669
T/K	340.07	340.44	340.92	339.05	339.54	340.06	340.25	340.85	338.96
P/MPa	2.33	1.44	0.21	4.66	3.28	1.99	1.63	0.11	4.45

w_B	0.0669	0.0669	0.0669	0.0669
T/K	339.48	339.87	340.24	340.63
P/MPa	3.05	2.08	1.11	0.25

Polymer (B):	**polystyrene-d8**	**2000DE1**

Characterization: M_n/kg.mol^{-1} = 26.4, M_w/kg.mol^{-1} = 27.2,
completely deuterated, Polymer Laboratories, Amherst, MA

Solvent (A):	**nitroethane**	**C$_2$H$_5$NO$_2$**	**79-24-3**

Type of data: cloud points

w_B	0.2362	0.2362	0.2362	0.2362	0.2362	0.2362	0.2121	0.2121	0.2121
T/K	268.58	268.58	268.68	268.77	268.88	269.01	268.68	268.68	268.78
P/MPa	5.26	5.19	4.07	3.02	1.78	0.32	5.11	5.11	3.99

w_B	0.2121	0.2121	0.2121	0.1866	0.1866	0.1866	0.1866	0.1866	0.1866
T/K	268.89	268.98	269.11	268.89	268.88	268.98	269.07	269.19	269.28
P/MPa	2.75	1.66	0.34	4.95	4.94	3.84	2.75	1.49	0.48

w_B	0.1644	0.1644	0.1644	0.1644	0.1644	0.1644	0.1453	0.1453	0.1453
T/K	269.07	269.07	269.19	269.28	269.35	269.42	269.07	269.07	269.18
P/MPa	4.44	4.43	3.25	2.14	1.40	0.48	4.52	4.52	3.21

w_B	0.1453	0.1453	0.1453	0.1395	0.1395	0.1395	0.1395	0.1395	0.1395
T/K	269.28	269.34	269.42	269.28	269.28	269.38	269.49	269.54	269.62
P/MPa	2.09	1.41	0.54	4.59	4.55	3.34	1.98	1.33	0.39

continued

continued

w_B	0.1202	0.1202	0.1202	0.1202	0.1202	0.1202	0.1021	0.1021	0.1021
T/K	269.18	269.18	269.29	269.38	269.49	269.57	268.98	268.98	269.07
P/MPa	4.73	4.71	3.52	2.45	1.28	0.39	4.46	4.48	3.32

w_B	0.1021	0.1021	0.1021	0.0874	0.0874	0.0874	0.0874	0.0874	0.0874
T/K	269.17	269.27	269.34	268.54	268.54	268.64	268.74	268.84	268.93
P/MPa	2.33	1.23	0.35	5.06	5.07	3.98	2.77	1.65	0.60

w_B	0.0764	0.0764	0.0764	0.0764	0.0764	0.0764
T/K	268.13	268.13	268.25	268.34	268.43	268.51
P/MPa	4.88	4.90	3.58	2.42	1.41	0.49

Type of data: spinodal points

w_B	0.2362	0.2362	0.2362	0.2362	0.2362	0.2362	0.2121	0.2121	0.2121
T/K	268.58	268.58	268.68	268.77	268.88	269.01	268.68	268.68	268.78
P/MPa	5.13	5.06	3.91	2.90	1.66	0.26	4.93	4.99	3.81

w_B	0.2121	0.2121	0.2121	0.1866	0.1866	0.1866	0.1866	0.1866	0.1866
T/K	268.89	268.98	269.11	268.89	268.88	268.98	269.07	269.19	269.28
P/MPa	2.57	1.49	0.15	4.62	4.68	3.52	2.51	1.23	0.21

w_B	0.1644	0.1644	0.1644	0.1644	0.1644	0.1644	0.1453	0.1453	0.1453
T/K	269.07	269.07	269.19	269.28	269.35	269.42	269.07	269.07	269.18
P/MPa	4.05	4.03	2.88	1.86	1.10	0.11	4.23	4.23	2.93

w_B	0.1453	0.1453	0.1453	0.1395	0.1395	0.1395	0.1395	0.1395	0.1395
T/K	269.28	269.34	269.42	269.28	269.28	269.38	269.49	269.54	269.62
P/MPa	1.88	1.22	0.33	4.28	4.22	3.03	1.66	1.01	0.17

w_B	0.1202	0.1202	0.1202	0.1202	0.1202	0.1202	0.1021	0.1021	0.1021
T/K	269.18	269.18	269.29	269.38	269.49	269.57	268.98	268.98	269.07
P/MPa	4.60	4.58	3.41	2.36	1.19	0.29	4.35	4.39	3.22

w_B	0.1021	0.1021	0.1021	0.0874	0.0874	0.0874	0.0874	0.0874	0.0874
T/K	269.17	269.27	269.34	268.54	268.54	268.64	268.74	268.84	268.93
P/MPa	2.22	1.09	0.26	4.94	4.92	3.84	2.64	1.48	0.49

w_B	0.0764	0.0764	0.0764	0.0764	0.0764	0.0764
T/K	268.13	268.13	268.25	268.34	268.43	268.51
P/MPa	4.76	4.78	3.48	2.29	1.31	0.39

Polymer (B):	**polystyrene-d8**	**2000DE1**

Characterization: $M_n/kg.mol^{-1} = 83.5$, $M_w/kg.mol^{-1} = 85.2$,
completely deuterated, Polymer Laboratories, Amherst, MA

Solvent (A):	**nitroethane**	**$C_2H_5NO_2$**	**79-24-3**

Type of data: cloud points

w_B	0.2273	0.2273	0.2273	0.2273	0.2273	0.2273	0.2273	0.2273	0.1992
T/K	285.98	285.98	286.19	286.29	286.38	286.59	286.65	286.85	286.49
P/MPa	5.25	5.27	4.08	3.53	3.02	1.81	1.50	0.46	4.65

continued

continued

w_B	0.1992	0.1992	0.1992	0.1992	0.1992	0.1773	0.1773	0.1773	0.1773
T/K	286.49	286.69	286.90	287.09	287.29	286.99	286.99	287.20	287.40
P/MPa	4.65	3.64	2.49	1.40	0.46	4.74	4.74	3.53	2.43
w_B	0.1773	0.1773	0.1522	0.1522	0.1522	0.1522	0.1522	0.1522	0.1315
T/K	287.61	287.77	287.30	287.30	287.49	287.71	287.92	288.06	287.49
P/MPa	1.35	0.47	4.80	4.82	3.72	2.52	1.33	0.61	4.77
w_B	0.1315	0.1315	0.1315	0.1315	0.1315	0.1158	0.1158	0.1158	0.1158
T/K	287.49	287.70	287.91	288.08	288.27	287.49	287.50	287.70	287.92
P/MPa	4.80	3.56	2.37	1.45	0.41	4.87	4.86	3.61	2.38
w_B	0.1158	0.1158	0.1124	0.1124	0.1124	0.1124	0.1124	0.0995	0.0995
T/K	288.08	288.25	288.06	288.07	288.09	288.12	288.14	287.40	287.40
P/MPa	1.51	0.53	0.67	0.61	0.51	0.36	0.25	4.61	4.62
w_B	0.0995	0.0995	0.0995	0.0995	0.0840	0.0840	0.0840	0.0840	0.0840
T/K	287.61	287.80	287.96	288.11	287.30	287.30	287.49	287.71	287.85
P/MPa	3.42	2.30	1.37	0.56	4.65	4.66	3.49	2.28	1.46
w_B	0.0840	0.0703	0.0703	0.0703	0.0703	0.0703	0.0703	0.0531	0.0531
T/K	288.06	286.99	287.00	287.21	287.40	287.56	287.75	286.28	286.28
P/MPa	0.27	4.82	4.82	3.53	2.42	1.45	0.38	4.92	4.92
w_B	0.0531	0.0531	0.0531	0.0531					
T/K	286.50	286.70	286.85	287.05					
P/MPa	3.64	2.43	1.54	0.34					

Type of data: spinodal points

w_B	0.2273	0.2273	0.2273	0.2273	0.2273	0.2273	0.2273	0.2273	0.1992
T/K	285.98	285.98	286.19	286.29	286.38	286.59	286.65	286.85	286.49
P/MPa	5.08	5.09	3.91	3.23	2.80	1.67	1.31	0.28	4.53
w_B	0.1992	0.1992	0.1992	0.1992	0.1992	0.1773	0.1773	0.1773	0.1773
T/K	286.49	286.69	286.90	287.09	287.29	286.99	286.99	287.20	287.40
P/MPa	4.54	3.49	2.36	1.30	0.36	4.65	4.65	3.40	2.34
w_B	0.1773	0.1773	0.1522	0.1522	0.1522	0.1522	0.1522	0.1522	0.1315
T/K	287.61	287.77	287.30	287.30	287.49	287.71	287.92	288.06	287.49
P/MPa	1.25	0.40	4.76	4.76	3.64	2.41	1.25	0.54	4.62
w_B	0.1315	0.1315	0.1315	0.1315	0.1315	0.1158	0.1158	0.1158	0.1158
T/K	287.49	287.70	287.91	288.08	288.27	287.49	287.50	287.70	287.92
P/MPa	4.65	3.37	2.17	1.30	0.24	4.61	4.61	3.34	2.14
w_B	0.1158	0.1158	0.1124	0.1124	0.1124	0.1124	0.1124	0.0995	0.0995
T/K	288.08	288.25	288.06	288.07	288.09	288.12	288.14	287.40	287.40
P/MPa	1.29	0.32	0.43	0.33	0.23	0.08	0.02	4.40	4.41
w_B	0.0995	0.0995	0.0995	0.0995	0.0840	0.0840	0.0840	0.0840	0.0840
T/K	287.61	287.80	287.96	288.11	287.30	287.30	287.49	287.71	287.85
P/MPa	3.24	2.08	1.16	0.39	4.52	4.58	3.38	2.14	1.35
w_B	0.0840	0.0703	0.0703	0.0703	0.0703	0.0703	0.0703	0.0531	0.0531
T/K	288.06	286.99	287.00	287.21	287.40	287.56	287.75	286.28	286.28
P/MPa	0.15	4.72	4.72	3.42	2.31	1.37	0.26	4.80	4.81

continued

continued

w_B	0.0531	0.0531	0.0531	0.0531
T/K	286.50	286.70	286.85	287.05
P/MPa	3.52	2.28	1.41	0.22

Polymer (B):	**polystyrene-d8**								**2000DE1**

Characterization: $M_n/\text{kg.mol}^{-1} = 83.5$, $M_w/\text{kg.mol}^{-1} = 85.2$,
completely deuterated, Polymer Laboratories, Amherst, MA

Solvent (A):	**nitroethane-d5**		**$C_2D_5NO_2$**						**57817-88-6**

Type of data: cloud points

w_B	0.1964	0.1964	0.1964	0.1964	0.1964	0.1964	0.1964	0.1964	0.1964
T/K	308.08	308.09	308.18	308.28	308.48	308.76	309.06	309.25	309.65
P/MPa	6.28	6.28	5.85	5.38	4.69	3.58	2.45	1.72	0.24

w_B	0.1741	0.1741	0.1741	0.1741	0.1741	0.1519	0.1519	0.1519	0.1519
T/K	309.65	309.95	310.26	310.51	310.83	310.46	310.78	311.03	311.23
P/MPa	4.85	3.61	2.60	1.67	0.47	4.59	3.33	2.28	1.57

w_B	0.1519	0.1330	0.1330	0.1330	0.1330	0.1330	0.1237	0.1237	0.1237
T/K	311.51	310.68	310.86	311.18	311.37	311.65	310.87	310.87	311.18
P/MPa	0.47	4.53	3.78	2.57	1.70	0.63	4.37	4.36	3.19

w_B	0.1237	0.1237	0.1237	0.1183	0.1183	0.1183	0.1183	0.1183	0.1069
T/K	311.37	311.59	311.81	310.77	310.99	311.27	311.49	311.78	310.90
P/MPa	2.30	1.43	0.61	4.48	3.56	2.43	1.61	0.50	4.72

w_B	0.1069	0.1069	0.1069	0.1069	0.0927	0.0927	0.0927	0.0927	0.0927
T/K	311.18	311.38	311.59	311.92	310.77	311.09	311.27	311.49	311.83
P/MPa	3.53	2.63	1.81	0.55	4.60	3.37	2.59	1.83	0.47

w_B	0.0807	0.0807	0.0807	0.0807	0.0807	0.0715	0.0715	0.0715	0.0715
T/K	310.57	310.77	311.08	311.38	311.73	310.60	310.78	311.10	311.30
P/MPa	5.12	4.34	3.05	1.89	0.50	4.45	3.60	2.35	1.56

Type of data: spinodal points

w_B	0.1964	0.1964	0.1964	0.1964	0.1964	0.1964	0.1964	0.1964	0.1964
T/K	308.08	308.09	308.18	308.28	308.48	308.76	309.06	309.25	309.65
P/MPa	6.10	6.22	5.63	5.23	4.45	3.45	2.24	1.58	0.05

w_B	0.1741	0.1741	0.1741	0.1741	0.1741	0.1519	0.1519	0.1519	0.1519
T/K	309.65	309.95	310.26	310.51	310.83	310.46	310.78	311.03	311.23
P/MPa	4.65	3.48	2.44	1.50	0.27	4.46	3.20	2.13	1.44

w_B	0.1519	0.1330	0.1330	0.1330	0.1330	0.1330	0.1237	0.1237	0.1237
T/K	311.51	310.68	310.86	311.18	311.37	311.65	310.87	310.87	311.18
P/MPa	0.37	4.36	3.59	2.43	1.49	0.42	4.10	4.12	2.90

w_B	0.1237	0.1237	0.1237	0.1183	0.1183	0.1183	0.1183	0.1183	0.1069
T/K	311.37	311.59	311.81	310.77	310.99	311.27	311.49	311.78	310.90
P/MPa	2.01	1.12	0.35	4.13	3.26	2.16	1.36	0.25	4.50

continued

continued

w_B	0.1069	0.1069	0.1069	0.1069	0.0927	0.0927	0.0927	0.0927	0.0927
T/K	311.18	311.38	311.59	311.92	310.77	311.09	311.27	311.49	311.83
P/MPa	3.24	2.37	1.54	0.27	4.37	3.10	2.38	1.51	0.22

w_B	0.0807	0.0807	0.0807	0.0807	0.0807	0.0715	0.0715	0.0715	0.0715
T/K	310.57	310.77	311.08	311.38	311.73	310.60	310.78	311.10	311.30
P/MPa	4.93	4.08	2.82	1.66	0.26	4.31	3.47	2.12	1.33

Polymer (B):   **polystyrene**   1991SZY
Characterization: $M_n/kg.mol^{-1} = 6.57$, $M_w/kg.mol^{-1} = 7.82$,
     Scientific Polymer Products, Inc., Ontario, NY
Solvent (A):   **2-propanone**   **C₃H₆O**   67-64-1

Type of data: cloud points (UCST behavior)

$w_B = 0.20$ $T/K = 270$ $(dT/dP)_{P=0}/K.MPa^{-1} = -1.10$

Polymer (B):   **polystyrene**   1991SZY
Characterization: $M_n/kg.mol^{-1} = 12.75$, $M_w/kg.mol^{-1} = 13.5$,
     Pressure Chemical Company, Pittsburgh, PA
Solvent (A):   **2-propanone**   **C₃H₆O**   67-64-1

Type of data: cloud points (UCST behavior)

w_B	0.19	0.20	0.23
T/K	287	286	285
$(dT/dP)_{P=0}/K.MPa^{-1}$	-1.70	-1.75	-1.76

Polymer (B):   **polystyrene**   1991SZY
Characterization: $M_n/kg.mol^{-1} = 12.75$, $M_w/kg.mol^{-1} = 13.5$,
     Pressure Chemical Company, Pittsburgh, PA
Solvent (A):   **2-propanone**   **C₃H₆O**   67-64-1

Type of data: cloud points (LCST behavior)

$w_B = 0.20$ $T/K = 400$ $(dT/dP)_{P=0}/K.MPa^{-1} = +6.53$

Polymer (B):   **polystyrene**   1992SZY
Characterization: $M_n/kg.mol^{-1} = 6.6$, $M_w/kg.mol^{-1} = 7.8$,
     Scientific Polymer Products, Inc., Ontario, NY
Solvent (A):   **2-propanone**   **C₃H₆O**   67-64-1

Type of data: cloud points

w_B	0.077	0.090	0.112	0.145	0.170	0.200	0.325
$\varphi_B =$	0.059	0.069	0.087	0.113	0.133	0.158	0.265
T/K	248.40	249.75	252.60	254.20	254.53	254.46	251.65
$(dT/dP)_{P=1}/K.MPa^{-1}$	-1.10	-0.95	-0.76	-0.70	-0.73	-0.78	-0.74

continued

3. Liquid-Liquid Equilibrium (LLE) Data

continued

Type of data: spinodal points

w_B		0.077	0.090	0.112	0.145	0.170	0.200	0.325
$\varphi_B =$		0.059	0.069	0.087	0.113	0.133	0.158	0.265
T/K		247.40	249.50	251.75	253.70	254.29	254.00	251.43
$(dT/dP)_{P=1}/K.MPa^{-1}$		−0.90	−0.76			−0.68		−0.70

Polymer (B):	**polystyrene**	**1992SZY**
Characterization:	$M_n/kg.mol^{-1} = 10.75$, $M_w/kg.mol^{-1} = 11.5$,	
	Scientific Polymer Products, Inc., Ontario, NY	
Solvent (A):	**2-propanone** C_3H_6O	**67-64-1**

Type of data: cloud points

w_B		0.090	0.146	0.164	0.218	0.292
$\varphi_B =$		0.068	0.112	0.126	0.171	0.234
T/K		245.15	247.92	248.18	247.52	246.09
$(dT/dP)_{P=0}/K.MPa^{-1}$		−0.74	−0.74	−0.75	−0.75	−0.72

Type of data: spinodal points

w_B		0.090	0.146	0.164	0.218	0.292
$\varphi_B =$		0.068	0.112	0.126	0.171	0.234
T/K		244.87	247.77	247.86	247.52	246.00
$(dT/dP)_{P=0}/K.MPa^{-1}$		−0.78	−0.79	−0.73	−0.75	−0.80

Polymer (B):	**polystyrene**	**1995LUS**
Characterization:	$M_n/kg.mol^{-1} = 7.34$, $M_w/kg.mol^{-1} = 8.0$,	
	Pressure Chemical Company, Pittsburgh, PA	
Solvent (A):	**2-propanone** C_3H_6O	**67-64-1**

Type of data: cloud points

w_B	0.222	0.222	0.222	0.222	0.222
T/K	253.77	253.45	253.19	252.98	252.83
P/MPa	0.07	0.34	0.72	1.02	1.25

critical concentration: $w_{B, crit} = 0.220$

Polymer (B):	**polystyrene**	**1995LUS**
Characterization:	$M_n/kg.mol^{-1} = 11.35$, $M_w/kg.mol^{-1} = 11.7$,	
	Polymer Laboratories, Amherst, MA	
Solvent (A):	**2-propanone** C_3H_6O	**67-64-1**

Type of data: cloud points

w_B	0.210	0.210	0.210	0.210
T/K	273.75	274.16	274.68	274.91
P/MPa	1.42	1.07	0.61	0.41

critical concentration: $w_{B, crit} = 0.220$

Polymer (B):	**polystyrene**							**1995LUS**
Characterization:	M_n/kg.mol^{-1} = 12.75, M_w/kg.mol^{-1} = 13.5,							
	Pressure Chemical Company, Pittsburgh, PA							
Solvent (A):	**2-propanone**		**C$_3$H$_6$O**					**67-64-1**

Type of data: cloud points

w_B	0.102	0.102	0.102	0.102	0.102	0.102	0.107	0.107	0.107
T/K	281.41	281.42	281.83	282.19	282.42	282.78	283.19	282.37	281.91
P/MPa	1.08	1.11	0.92	0.64	0.65	0.31	0.62	1.13	1.24

w_B	0.107	0.107	0.163	0.163	0.163	0.163	0.163	0.163	0.163
T/K	281.60	281.30	287.13	287.11	286.65	286.44	285.96	285.49	285.27
P/MPa	1.27	1.47	0.08	0.09	0.30	0.42	0.70	0.97	1.12

w_B	0.163	0.197	0.197	0.197	0.197	0.212	0.212	0.212	0.212
T/K	284.71	285.60	285.26	284.89	284.60	286.12	285.42	285.26	284.50
P/MPa	1.45	0.40	0.59	0.84	0.98	0.05	0.44	0.56	0.99

w_B	0.212	0.212	0.229	0.229	0.229	0.229	0.229	0.229	0.229
T/K	284.11	283.63	285.77	285.35	285.17	284.79	284.52	284.14	283.80
P/MPa	1.21	1.55	0.07	0.31	0.41	0.64	0.82	1.04	1.28

w_B	0.264	0.264	0.264	0.264	0.264	0.264	0.264	0.264	0.264
T/K	285.10	285.04	284.82	284.60	284.39	284.07	283.64	283.32	283.23
P/MPa	0.07	0.09	0.20	0.33	0.46	0.66	0.90	1.10	1.16

Type of data: spinodal points

w_B	0.102	0.102	0.102	0.102	0.102	0.102	0.107	0.107	0.107
T/K	281.41	281.42	281.83	282.19	282.42	282.78	283.19	282.37	281.91
P/MPa	0.84	0.94	0.73	0.47	0.39	0.05	0.27	0.80	0.97

w_B	0.107	0.107	0.163	0.163	0.163	0.163	0.163	0.163	0.197
T/K	281.60	281.30	286.65	286.44	285.96	285.49	285.27	284.71	285.60
P/MPa	1.06	1.24	−0.09	0.07	0.25	0.62	0.69	1.06	0.30

w_B	0.197	0.197	0.197	0.212	0.212	0.212	0.212	0.212	0.229
T/K	285.26	284.89	284.60	285.42	285.26	284.50	284.11	283.63	285.35
P/MPa	0.51	0.72	0.88	0.34	0.43	0.88	1.12	1.41	0.21

w_B	0.229	0.229	0.229	0.229	0.229	0.264	0.264	0.264	0.264
T/K	285.17	284.79	284.52	284.14	283.80	284.82	284.60	284.39	284.07
P/MPa	0.32	0.53	0.70	0.92	1.12	−0.11	0.04	0.17	0.35

w_B	0.264	0.264	0.264
T/K	283.64	283.32	283.23
P/MPa	0.65	0.81	0.85

critical concentration: $w_{B, crit}$ = 0.220

Polymer (B):   **polystyrene**   **1995LUS**

Characterization: $M_n/\text{kg.mol}^{-1} = 21.45$, $M_w/\text{kg.mol}^{-1} = 22.1$,

Polymer Laboratories, Amherst, MA

Solvent (A):   **2-propanone**   **C₃H₆O**   **67-64-1**

Type of data: cloud points

w_B	0.111	0.111	0.111	0.111	0.111	0.111	0.111	0.111	0.111
T/K	321.22	322.81	322.82	326.73	329.25	330.88	365.52	367.63	371.30
P/MPa	1.56	1.18	1.19	0.74	0.47	0.31	0.55	0.75	0.99
w_B	0.111	0.111	0.146	0.146	0.146	0.146	0.146	0.146	0.146
T/K	375.80	378.48	325.87	327.52	330.55	333.20	337.06	340.93	342.45
P/MPa	1.49	1.78	1.55	1.33	0.98	0.75	0.47	0.33	0.32
w_B	0.146	0.146	0.146	0.146	0.146	0.146	0.146	0.146	0.146
T/K	344.05	351.54	353.15	355.25	358.39	362.31	366.80	371.68	374.10
P/MPa	0.26	0.29	0.30	0.45	0.60	0.86	1.18	1.59	1.82
w_B	0.158	0.158	0.158	0.158	0.158	0.158	0.158	0.158	0.158
T/K	325.17	329.19	332.14	340.37	344.55	348.88	354.68	359.23	366.57
P/MPa	1.81	1.29	0.98	0.48	0.39	0.38	0.51	0.71	1.19
w_B	0.158	0.163	0.163	0.163	0.163	0.163	0.163	0.163	0.163
T/K	372.72	327.82	330.90	334.09	337.43	340.81	344.46	346.55	349.25
P/MPa	1.73	1.56	1.20	0.92	0.68	0.55	0.48	0.47	0.47
w_B	0.163	0.163	0.163	0.163	0.163	0.169	0.169	0.169	0.169
T/K	352.96	358.40	362.91	367.21	372.39	327.38	327.25	330.92	334.48
P/MPa	0.54	0.75	1.00	1.30	1.73	1.45	1.43	1.01	0.71
w_B	0.169	0.169	0.169	0.169	0.169	0.169	0.169	0.174	0.174
T/K	337.35	341.57	346.73	351.19	357.06	367.75	375.89	328.21	331.83
P/MPa	0.54	0.39	0.35	0.41	0.59	1.29	2.00	1.38	0.97
w_B	0.174	0.174	0.174	0.174	0.174	0.174	0.174	0.174	0.174
T/K	335.82	339.91	342.81	347.19	351.33	356.23	361.75	367.39	373.22
P/MPa	0.68	0.45	0.37	0.34	0.40	0.58	0.85	1.25	1.75
w_B	0.185	0.185	0.185	0.185	0.185	0.185	0.185	0.185	0.185
T/K	329.45	332.27	335.50	338.67	342.72	345.75	350.24	353.05	358.41
P/MPa	1.23	0.96	0.72	0.55	0.40	0.37	0.38	0.44	0.67
w_B	0.185	0.185	0.185	0.213	0.213	0.213	0.213	0.213	0.213
T/K	363.93	368.03	373.76	326.70	329.17	332.66	335.29	342.10	344.41
P/MPa	0.99	1.29	1.79	1.61	1.31	0.95	0.77	0.46	0.44
w_B	0.213	0.213	0.213	0.213	0.213	0.213	0.249	0.249	0.249
T/K	346.66	352.30	357.91	364.35	369.74	372.95	324.78	328.07	331.15
P/MPa	0.45	0.53	0.73	1.11	1.52	1.80	1.59	1.14	0.79
w_B	0.249	0.249	0.249	0.249	0.249	0.249	0.249	0.273	0.273
T/K	334.45	339.12	354.98	358.42	364.97	371.79	376.92	321.19	324.14
P/MPa	0.51	0.27	0.30	0.46	0.86	1.40	1.89	1.71	1.30
w_B	0.273	0.273	0.273	0.273	0.273	0.273			
T/K	327.28	329.31	362.71	365.18	373.24	378.35			
P/MPa	0.87	0.63	0.47	0.64	1.29	1.83			

Polymer (B):	polystyrene		1995LUS
Characterization:	$M_n/\text{kg.mol}^{-1} = 21.45$, $M_w/\text{kg.mol}^{-1} = 22.1$,		
	Polymer Laboratories, Amherst, MA		
Solvent (A):	2-propanone	C_3H_6O	67-64-1

Type of data: spinodal points

w_B	0.111	0.111	0.111	0.111	0.111	0.111	0.111	0.111	0.111
T/K	321.22	322.81	322.82	326.73	329.25	330.88	365.52	367.63	371.30
P/MPa	0.95	0.85	0.88	0.14	0.14	0.37	0.32	0.32	0.71

w_B	0.111	0.111	0.146	0.146	0.146	0.146	0.146	0.146	0.146
T/K	375.80	378.48	325.87	330.55	333.20	340.93	342.45	344.05	351.54
P/MPa	1.17	1.32	0.78	0.27	0.09	0.28	0.22	0.29	0.22

w_B	0.146	0.146	0.146	0.146	0.146	0.158	0.158	0.158	0.158
T/K	355.25	358.39	362.31	366.80	374.10	325.17	329.19	332.14	340.37
P/MPa	0.09	0.24	0.20	0.59	1.39	1.39	0.88	0.47	0.11

w_B	0.158	0.158	0.158	0.158	0.163	0.163	0.163	0.163	0.163
T/K	348.88	354.68	359.23	366.57	327.82	330.90	334.09	337.43	340.81
P/MPa	0.07	0.31	0.50	1.04	1.34	0.89	0.64	0.42	0.18

w_B	0.163	0.163	0.163	0.163	0.163	0.163	0.169	0.169	0.169
T/K	344.46	349.25	358.40	362.91	367.21	372.39	327.38	327.25	330.92
P/MPa	0.18	0.23	0.32	0.12	0.25	0.18	0.87	1.56	1.08

w_B	0.169	0.169	0.169	0.169	0.169	0.169	0.169	0.169	0.174
T/K	334.48	337.35	341.57	346.73	351.19	357.06	367.75	375.89	328.21
P/MPa	0.14	0.17	0.49	0.74	1.10	1.57	0.97	1.00	0.63

w_B	0.174	0.174	0.174	0.174	0.174	0.174	0.174	0.174	0.174
T/K	331.83	335.82	339.91	342.81	347.19	351.33	356.23	361.75	367.39
P/MPa	0.67	0.33	0.20	0.17	0.13	0.17	0.30	0.64	1.06

w_B	0.174	0.185	0.185	0.185	0.185	0.185	0.185	0.185	0.185
T/K	373.22	329.45	332.27	335.50	338.67	342.72	345.75	350.24	353.05
P/MPa	1.57	1.09	0.77	0.49	0.34	0.19	0.18	0.24	0.31

w_B	0.185	0.185	0.185	0.185	0.213	0.213	0.213	0.213	0.213
T/K	358.41	363.93	368.03	373.76	326.70	329.17	332.66	335.29	342.10
P/MPa	0.51	0.83	1.15	1.62	1.47	1.12	0.60	0.59	0.25

w_B	0.213	0.213	0.213	0.213	0.213	0.213	0.213	0.249	0.249
T/K	344.41	346.66	352.30	357.91	364.35	369.74	372.95	324.78	328.07
P/MPa	0.27	0.17	0.34	0.61	0.92	1.35	1.65	0.80	0.28

w_B	0.249	0.249	0.249	0.249	0.249	0.249	0.249	0.249	0.273
T/K	331.15	334.45	339.12	354.98	358.42	364.97	371.79	376.92	321.19
P/MPa	0.25	0.19	0.28	0.34	0.14	0.27	0.95	1.49	1.46

w_B	0.273	0.273	0.273	0.273	0.273	0.273	0.273		
T/K	324.14	327.28	329.31	362.71	365.18	373.24	378.35		
P/MPa	1.17	0.73	0.47	0.19	0.38	1.15	1.63		

critical concentration: $w_{B, \text{crit}} = 0.22$

| **Polymer (B):** | **polystyrene-d8** | **1995LUS** |

Characterization: $M_n/\text{kg.mol}^{-1} = 10.3$, $M_w/\text{kg.mol}^{-1} = 10.5$,
completely deuterated, Polymer Laboratories, Amherst, MA

Solvent (A): **2-propanone** **C$_3$H$_6$O** **67-64-1**

Type of data: cloud points

w_B	0.209	0.209	0.209	0.209
T/K	234.95	234.60	235.98	235.25
P/MPa	1.55	2.02	0.15	1.11

| **Polymer (B):** | **polystyrene-d8** | **1995LUS** |

Characterization: $M_n/\text{kg.mol}^{-1} = 25.4$, $M_w/\text{kg.mol}^{-1} = 26.9$,
completely deuterated, Polymer Laboratories, Amherst, MA

Solvent (A): **2-propanone** **C$_3$H$_6$O** **67-64-1**

Type of data: cloud points

w_B	0.132	0.132	0.132	0.132	0.157	0.157	0.157	0.181	0.181
T/K	274.49	275.63	276.62	277.47	275.91	276.65	277.79	276.13	276.79
P/MPa	1.89	1.31	0.73	0.29	1.69	1.30	0.71	1.89	1.53

w_B	0.181	0.214	0.214	0.214	0.214	0.214	0.233	0.233	0.233
T/K	279.62	275.36	276.70	277.34	277.93	279.30	275.02	277.09	278.16
P/MPa	1.14	2.57	1.83	1.49	1.14	0.47	2.48	1.41	0.88

w_B	0.233	0.253	0.253	0.253	0.253	0.270	0.270	0.270	0.270
T/K	279.03	274.58	275.75	277.78	278.61	273.48	274.46	275.94	277.36
P/MPa	0.45	2.23	1.73	0.71	0.35	2.25	1.76	1.03	0.31

w_B	0.300	0.300	0.300	0.300
T/K	270.82	271.26	271.87	272.42
P/MPa	1.11	0.84	0.57	0.28

| **Polymer (B):** | **polystyrene** | **2002RE2** |

Characterization: $M_n/\text{kg.mol}^{-1} = 15.8$, $M_w/\text{kg.mol}^{-1} = 16.6$,
Scientific Polymer Products, Inc., Ontario, NY

Solvent (A): **2-propanone** **C$_3$H$_6$O** **67-64-1**

Type of data: cloud points (negative pressures were measured using a Berthelot tube technique)

$w_B = 0.180$ was kept constant

T/K	257.4	258.1	258.2	258.9	262.9	265.0	273.0	280.0	283.0
P/bar	767.0	634.3	600.0	525.0	425.0	370.0	255.0	158.6	125.5

T/K	289.0	293.0	298.0	303.0	307.5	308.0	313.0	323.0	336.5
P/bar	82.5	62.5	32.5	0.4	−15.2	−20.2	−26.1	−32.4	−32.8

T/K	351.5	368.0	380.0	403.5	417.0	430.5
P/bar	−26.1	−10.2	4.4	43.0	61.5	97.0

Polymer (B):	**polystyrene**							**2002RE2**

Characterization: $M_n/\text{kg.mol}^{-1} = 23.8$, $M_w/\text{kg.mol}^{-1} = 24.7$,
Scientific Polymer Products, Inc., Ontario, NY

Solvent (A):	**2-propanone**		**C₃H₆O**					**67-64-1**

Type of data: cloud points (negative pressures were measured using a Berthelot tube technique)

$w_B = 0.018$ was kept constant

T/K	251.0	251.5	253.0	253.0	256.3	258.0	263.0	265.0	273.0
P/bar	700.00	590.00	536.50	630.00	455.00	403.00	296.80	260.00	156.00

T/K	277.3	281.8	286.0	293.4	295.0	298.5	299.8	301.0	301.6
P/bar	113.00	78.80	49.30	10.00	0.28	−21.60	−27.70	−26.40	−29.40

T/K	301.9	303.2	304.0	308.0	368.0	375.0	381.0	383.0	390.3
P/bar	−30.30	−35.10	−41.00	−42.00	−26.70	−15.60	−12.40	−6.20	5.70

T/K	401.0	413.0	422.5	433.0
P/bar	20.00	40.00	50.00	73.00

Polymer (B):	**polystyrene**							**1994IMR**

Characterization: $M_n/\text{kg.mol}^{-1} = 70.3$, $M_w/\text{kg.mol}^{-1} = 187$,
Mitsui Toatsu Chemicals, Inc., Japan

Solvent (A):	**propionitrile**		**C₃H₅N**					**107-12-0**

Type of data: cloud points

w_B	0.044	0.044	0.044	0.063	0.063	0.063	0.063	0.063	0.104
T/K	316.75	323.15	330.05	335.05	337.35	339.95	341.55	345.55	346.65
P/MPa	3.23	0.02	−1.95	3.77	1.81	0.03	−0.95	−1.15	2.60

w_B	0.104	0.104	0.104	0.104	0.201	0.201	0.201	0.201	0.201
T/K	351.65	359.35	371.15	372.00	347.15	350.65	353.10	356.15	358.55
P/MPa	1.37	0.95	−0.40	−0.50	2.01	1.80	1.39	0.70	0.32

w_B	0.201	0.201	0.201	0.201	0.201	0.201	0.201	0.201	0.201
T/K	362.05	369.80	372.35	377.55	391.55	392.75	396.75	401.10	420.80
P/MPa	0.21	−0.36	−0.42	−0.51	−0.48	−0.31	−0.27	0.01	1.45

Comments: The negative pressures were measured using a Berthelot tube technique.

Polymer (B):	**polystyrene**							**1996LUS**

Characterization: $M_n/\text{kg.mol}^{-1} = 7.34$, $M_w/\text{kg.mol}^{-1} = 8.0$,
Pressure Chemical Company, Pittsburgh, PA

Solvent (A):	**propionitrile**		**C₃H₅N**					**107-12-0**

Type of data: cloud points

w_B	0.1300	0.1300	0.1300	0.1300	0.1300	0.1300	0.1600	0.1600	0.1600
T/K	267.151	267.623	267.948	268.421	268.936	268.397	268.856	269.421	269.981
P/MPa	3.518	2.869	2.366	1.511	0.917	0.508	3.682	2.903	2.045

continued

continued

w_B	0.1600	0.1600	0.1600	0.1800	0.1800	0.1800	0.1800	0.1800	0.1800
T/K	270.454	270.934	271.222	267.979	268.447	268.999	268.503	269.986	270.522
P/MPa	1.430	0.721	0.369	4.058	3.390	2.688	1.866	1.255	0.427

w_B	0.1900	0.1900	0.1900	0.1900	0.1900	0.1900	0.2100	0.2100	0.2100
T/K	268.939	269.449	269.938	270.553	271.086	271.395	268.077	268.470	268.923
P/MPa	4.139	3.412	2.658	1.979	1.267	0.906	3.025	2.602	4.196

w_B	0.2100	0.2100	0.2100	0.2100	0.2100	0.2200	0.2200	0.2200	0.2200
T/K	269.513	269.977	270.491	270.960	271.482	268.994	268.436	270.010	270.486
P/MPa	3.780	2.820	2.048	1.532	0.840	4.086	3.622	2.461	2.274

w_B	0.2200	0.2200	0.2400	0.2400	0.2400	0.2400	0.2400	0.2400	0.2400
T/K	270.988	271.483	268.877	269.461	270.028	270.518	270.930	271.494	271.950
P/MPa	1.346	0.749	4.896	4.305	3.623	2.952	2.319	1.613	1.180

w_B	0.2400	0.2500	0.2500	0.2500	0.2500	0.2500	0.2600	0.2600	0.2600
T/K	272.245	269.072	269.576	270.128	271.130	271.525	269.531	269.954	270.493
P/MPa	0.688	3.971	3.267	2.763	1.249	0.845	4.404	3.874	3.162

w_B	0.2600	0.2600	0.2600	0.2600	0.2800	0.2800	0.2800	0.2800	0.2800
T/K	271.082	271.468	271.926	272.497	269.840	270.538	270.957	271.445	271.979
P/MPa	2.399	1.901	1.328	0.558	4.802	3.565	2.914	2.111	1.649

w_B	0.2800	0.2800	0.2800	0.3000	0.3000	0.3000	0.3000	0.3000	0.3000
T/K	272.465	272.465	272.784	268.445	269.027	269.466	270.143	270.641	271.018
P/MPa	0.922	1.137	0.742	5.566	4.813	4.417	3.424	2.562	2.075

w_B	0.3000	0.3000	0.3300	0.3300	0.3300	0.3300	0.3300	0.3300	0.3300
T/K	271.438	271.933	267.859	268.475	268.972	269.397	269.948	270.515	271.032
P/MPa	1.653	1.022	4.575	3.748	3.230	2.672	1.964	1.409	0.718

w_B	0.3591	0.3591	0.3591	0.3591	0.3591	0.3591	0.3591
T/K	268.225	268.594	268.977	269.516	270.019	270.425	270.995
P/MPa	4.441	3.952	3.393	2.711	1.981	1.421	0.654

Type of data: spinodal points

w_B	0.1300	0.1300	0.1300	0.1300	0.1300	0.1300	0.1600	0.1600	0.1600
T/K	267.151	267.623	267.948	268.421	268.936	268.397	268.856	269.421	269.981
P/MPa	3.256	4.479	2.004	1.337	0.595	−0.197	3.076	2.304	1.666

w_B	0.1600	0.1600	0.1600	0.1800	0.1800	0.1800	0.1800	0.1800	0.1800
T/K	270.454	270.934	271.222	267.979	268.447	268.999	268.503	269.986	270.522
P/MPa	0.843	0.225	0.185	3.442	2.705	1.800	1.183	0.483	0.036

w_B	0.1900	0.1900	0.1900	0.1900	0.1900	0.1900	0.2100	0.2100	0.2100
T/K	268.939	269.449	269.938	270.553	271.086	271.395	268.077	268.470	268.923
P/MPa	3.880	3.007	2.365	1.585	0.820	0.498	2.204	2.070	3.689

w_B	0.2100	0.2100	0.2100	0.2100	0.2100	0.2200	0.2200	0.2200	0.2200
T/K	269.513	269.977	270.491	270.960	271.482	268.994	268.436	270.010	270.486
P/MPa		2.325	1.547	0.752	0.194	3.860	3.077	2.359	1.755

w_B	0.2200	0.2200	0.2400	0.2400	0.2400	0.2400	0.2400	0.2400	0.2400
T/K	270.988	271.483	268.877	269.461	270.028	270.518	270.930	271.494	271.950
P/MPa	1.105	0.462	4.691	3.420	3.118	2.020	1.596	0.800	0.258

continued

continued

w_B	0.2400	0.2500	0.2500	0.2500	0.2500	0.2500	0.2600	0.2600	0.2600
T/K	272.245	269.072	269.576	270.128	271.130	271.525	269.531	269.954	270.493
P/MPa	0.128	3.785	3.075	2.181	0.783	0.381	4.214	3.680	2.923

w_B	0.2600	0.2600	0.2600	0.2600	0.2800	0.2800	0.2800	0.2800	0.2800
T/K	271.082	271.468	271.926	272.497	269.840	270.538	270.957	271.445	271.979
P/MPa	2.187	1.662	0.988	0.279	4.369	3.224	2.689	1.882	1.404

w_B	0.2800	0.2800	0.2800	0.3000	0.3000	0.3000	0.3000	0.3000	0.3000
T/K	272.465	272.465	272.784	268.445	269.027	269.466	270.143	270.641	271.018
P/MPa			0.277	4.339	0.701	3.811	1.633	1.652	1.349

w_B	0.3000	0.3000	0.3300	0.3300	0.3300	0.3300	0.3300	0.3300	0.3300
T/K	271.438	271.933	267.859	268.475	268.972	269.397	269.948	270.515	271.032
P/MPa	0.964	0.223	3.622	2.053	2.356	1.947	1.144	−0.153	−0.756

w_B	0.3591	0.3591	0.3591	0.3591	0.3591	0.3591	0.3591
T/K	268.225	268.594	268.977	269.516	270.019	270.425	270.995
P/MPa	3.870	1.465	2.669	−0.458	−0.750	−0.247	−0.959

Polymer (B):	**polystyrene**	**1996LUS**

Characterization: $M_n/kg.mol^{-1} = 12.75$, $M_w/kg.mol^{-1} = 13.5$,
Pressure Chemical Company, Pittsburgh, PA

Solvent (A):	**propionitrile**	**C_3H_5N**	**107-12-0**

Type of data: cloud points

w_B	0.1600	0.1600	0.1600	0.1600	0.1600	0.1906	0.1906	0.1906	0.1906
T/K	307.292	307.952	308.506	308.947	309.475	308.495	308.969	309.496	309.974
P/MPa	2.554	2.133	1.900	1.487	1.256	1.778	1.459	1.194	0.921

w_B	0.1906	0.2256	0.2256	0.2256	0.2256	0.2256
T/K	310.369	305.873	306.207	307.297	308.274	309.210
P/MPa	0.744	3.351	3.035	2.300	1.674	1.098

Type of data: spinodal points

w_B	0.1600	0.1600	0.1600	0.1600	0.1600	0.1906	0.1906	0.1906	0.1906
T/K	307.292	307.952	308.506	308.947	309.475	308.495	308.969	309.496	309.974
P/MPa	2.252	1.841	1.600	1.304	1.066	1.140	0.796	0.447	0.080

w_B	0.1906	0.2256	0.2256	0.2256	0.2256	0.2256
T/K	310.369	305.873	306.207	307.297	308.274	309.210
P/MPa	0.181	2.909	2.712	2.034	1.484	1.018

Polymer (B):	**polystyrene**	**1996LUS**

Characterization: $M_n/kg.mol^{-1} = 15.9$, $M_w/kg.mol^{-1} = 16.7$,
Scientific Polymer Products, Inc., Ontario, NY

Solvent (A):	**propionitrile (75% deuterated at CH_2)**		
		$C_3H_5N/C_3H_3D_2N$	**107-12-0/24300-23-0**

Type of data: cloud points

continued

continued

w_B	0.1000	0.1000	0.1000	0.1000	0.1000	0.1000	0.1000	0.1000	0.1000
T/K	323.698	323.695	330.240	330.239	337.046	337.039	337.037	445.521	445.513
P/MPa	5.540	5.539	2.922	2.964	0.549	0.445	0.514	1.015	0.882

w_B	0.1000	0.1000	0.1000	0.1000	0.1499	0.1499	0.1499	0.1499	0.1499
T/K	445.513	454.443	463.595	463.592	330.263	330.274	338.051	338.047	343.830
P/MPa	0.820	1.858	2.894	2.834	5.203	5.211	2.282	2.249	0.575

w_B	0.1499	0.1499	0.1499	0.1499	0.1499	0.1499	0.1499	0.2000	0.2000
T/K	343.827	436.837	436.823	445.574	445.555	454.465	454.469	330.240	330.247
P/MPa	0.582	0.716	0.765	1.744	1.735	2.778	2.812	5.238	5.253

w_B	0.2000	0.2000	0.2000	0.2000	0.2000	0.2000	0.2000	0.2000	0.2000
T/K	336.994	336.992	343.826	343.832	436.736	436.730	445.503	445.498	454.431
P/MPa	2.686	2.664	0.659	0.665	0.791	0.803	1.903	1.897	3.008

w_B	0.2402	0.2402	0.2402	0.2402	0.2402	0.2402	0.2402	0.2402	0.2402
T/K	330.365	330.351	330.355	330.343	335.772	335.766	341.227	341.219	342.888
P/MPa	3.865	4.352	4.666	4.714	2.837	2.628	1.061	0.967	0.580

w_B	0.2533	0.2533	0.2533	0.2533	0.2533	0.2533	0.2533	0.2533	0.2533
T/K	330.270	337.948	340.070	340.049	436.726	436.737	445.534	445.527	454.467
P/MPa	4.519	1.847	1.170	1.186	0.712	0.706	1.819	1.820	3.012

w_B	0.2533	0.2610	0.2610	0.2610	0.2610	0.2610	0.2610	0.2610	0.2610
T/K	454.453	462.646	462.497	454.442	454.446	447.535	447.536	441.496	436.768
P/MPa	3.014	3.972	3.975	2.975	2.965	2.122	2.156	1.366	0.843

w_B	0.2610	0.2610	0.2610	0.2610	0.2610	0.2610	0.2610	0.2610	0.2610
T/K	436.755	325.361	325.333	330.352	330.346	335.723	335.723	341.393	341.386
P/MPa	0.815	5.693	5.476	3.577	3.607	2.207	2.140	0.443	0.421

Type of data: spinodal points

w_B	0.1000	0.1000	0.1000	0.1000	0.1000	0.1000	0.1000	0.1000	0.1000
T/K	323.698	323.695	330.240	330.239	337.046	337.039	337.037	445.521	445.513
P/MPa	3.503	4.817	1.695	0.751	−4.110	−2.901	−2.867	0.445	−4.122

w_B	0.1000	0.1000	0.1000	0.1000	0.1499	0.1499	0.1499	0.1499	0.1499
T/K	445.513	454.443	463.595	463.592	330.263	330.274	338.051	338.047	343.830
P/MPa	−0.104	1.328	2.332	2.469	3.994	4.045	−0.607	0.744	−2.058

w_B	0.1499	0.1499	0.1499	0.1499	0.1499	0.1499	0.1499	0.2000	0.2000
T/K	343.827	436.837	436.823	445.574	445.555	454.465	454.469	330.240	330.247
P/MPa	−1.279	−0.367	−3.810	0.551	−1.090	2.052	0.082	4.914	5.007

w_B	0.2000	0.2000	0.2000	0.2000	0.2000	0.2000	0.2000	0.2000	0.2000
T/K	336.994	336.992	343.826	343.832	436.736	436.730	445.503	445.498	454.431
P/MPa	1.975	2.148	0.113	0.286	0.471	0.554	1.492	1.518	2.591

w_B	0.2402	0.2402	0.2402	0.2402	0.2402	0.2402	0.2402	0.2402	0.2402
T/K	330.365	330.351	330.355	330.343	335.772	335.766	341.227	341.219	342.888
P/MPa	−0.504	3.703	4.459	4.543	1.976	2.489	0.623	0.828	0.359

continued

continued

w_B	0.2533	0.2533	0.2533	0.2533	0.2533	0.2533	0.2533	0.2533	0.2533
T/K	330.270	337.948	340.070	340.049	436.726	436.737	445.534	445.527	454.467
P/MPa		1.074	0.545	0.637	0.281	0.463	1.240	1.291	2.010

w_B	0.2533	0.2533	0.2610	0.2610	0.2610	0.2610	0.2610	0.2610	0.2610
T/K	454.453	462.646	462.497	454.442	454.446	447.535	447.536	441.496	436.768
P/MPa	2.301	3.406	3.497	2.319	2.326	1.077	1.264	0.510	0.603

w_B	0.2610	0.2610	0.2610	0.2610	0.2610	0.2610	0.2610	0.2610	0.2610
T/K	436.755	325.361	325.333	330.352	330.346	335.723	335.723	341.393	341.386
P/MPa	0.344	5.134	5.411	1.571	2.304	1.828	1.436	−0.371	−0.157

Polymer (B):	**polystyrene**		**1996LUS**
Characterization:	M_n/kg.mol^{-1} = 21.45, M_w/.g.mol^{-1} = 22.1,		
	Polymer Laboratories, Amherst, MA		
Solvent (A):	**propionitrile**	**C$_3$H$_5$N**	**107-12-0**

Type of data: cloud points

w_B	0.0600	0.0600	0.0600	0.0600	0.0600	0.0600	0.0600	0.0600	0.0600
T/K	331.770	331.771	337.301	337.300	433.404	433.385	445.537	454.400	454.407
P/MPa	2.413	2.049	0.893	0.534	1.044	0.987	2.223	3.527	3.433

w_B	0.0900	0.0900	0.0900	0.0900	0.0900	0.0900	0.0900	0.0900	0.0900
T/K	336.255	336.243	344.053	344.053	351.039	351.037	421.594	421.573	421.571
P/MPa	4.261	4.383	2.451	2.075	0.676	0.617	0.920	0.955	1.070

w_B	0.0900	0.0900	0.0900	0.0900	0.0900	0.0900	0.0900	0.0900	0.1076
T/K	428.369	428.371	436.952	436.953	445.765	445.767	454.720	454.712	337.250
P/MPa	1.454	1.713	2.403	2.184	3.425	3.430	4.388	4.403	4.400

w_B	0.1076	0.1076	0.1076	0.1076	0.1076	0.1076	0.1076	0.1076	0.1076
T/K	337.250	344.098	344.100	351.095	351.082	351.083	355.350	355.351	420.023
P/MPa	4.353	3.033	2.973	1.946	1.736	1.377	1.009	0.861	1.086

w_B	0.1076	0.1076	0.1076	0.1076	0.1076	0.1076	0.1076	0.1076	0.1076
T/K	428.443	428.439	437.035	437.038	437.036	445.843	445.842	454.781	454.778
P/MPa	1.956	1.930	2.804	2.667	2.922	3.708	3.838	4.900	4.926

w_B	0.1300	0.1300	0.1300	0.1300	0.1300	0.1300	0.1300	0.1600	0.1600
T/K	337.180	337.179	337.173	344.101	344.102	351.080	351.072	337.157	337.142
P/MPa	4.889	4.804	4.633	2.668	2.905	1.375	1.362	5.387	5.140

w_B	0.1600	0.1600	0.1600	0.1600	0.1600	0.1600	0.1600	0.1600	0.1600
T/K	337.138	343.960	343.961	343.965	350.936	350.939	357.997	357.991	360.818
P/MPa	5.109	3.149	3.193	3.084	1.545	1.750	0.568	0.602	0.241

w_B	0.1600	0.1600	0.1600	0.1600	0.1600	0.1600	0.1600	0.1600	0.1600
T/K	360.815	410.364	410.302	419.885	419.884	428.336	428.344	436.922	436.920
P/MPa	0.182	0.616	0.488	1.144	1.179	1.940	1.909	2.878	2.854

w_B	0.1600	0.1600	0.1600	0.1600	0.1899	0.1899	0.1899	0.1899	0.1899
T/K	445.755	445.746	454.689	454.691	336.998	336.994	343.823	343.823	350.807
P/MPa	3.850	3.817	4.729	4.848	4.956	4.915	3.093	3.030	1.606

continued

continued

w_B	0.1899	0.1899	0.1899	0.1899	0.1899	0.1899	0.1899	0.1899	0.1899
T/K	350.806	355.097	355.095	357.836	357.831	360.710	410.135	410.099	419.836
P/MPa	1.601	0.951	0.929	0.576	0.569	0.241	0.352	0.357	1.134

w_B	0.1899	0.1899	0.1899	0.1899	0.1899	0.1899	0.1899	0.1899	0.1899
T/K	419.829	428.285	428.284	436.891	436.895	445.680	445.682	454.629	454.635
P/MPa	1.153	1.968	1.947	2.834	2.818	3.777	3.823	4.881	4.834

w_B	0.2199	0.2199	0.2199	0.2199	0.2199	0.2199	0.2199	0.2199	0.2199
T/K	336.988	336.992	343.807	343.804	350.815	350.801	355.202	355.197	359.272
P/MPa	4.741	5.059	3.274	3.130	1.745	1.659	1.055	1.128	0.476

w_B	0.2199	0.2199	0.2199	0.2199	0.2199	0.2199	0.2199	0.2199	0.2199
T/K	359.268	411.469	411.480	419.692	419.662	428.099	428.101	436.717	436.694
P/MPa	0.502	0.542	0.599	1.424	1.241	1.991	2.026	2.821	2.883

w_B	0.2199	0.2199	0.2199	0.2199	0.2499	0.2499	0.2499	0.2499	0.2499
T/K	445.482	445.484	454.423	454.416	336.980	336.977	343.788	343.813	350.806
P/MPa	3.859	3.922	4.847	4.729	4.546	4.539	2.699	2.680	1.281

w_B	0.2499	0.2499	0.2499	0.2499	0.2499	0.2499	0.2499	0.2499	0.2499
T/K	350.798	355.232	355.226	357.840	357.841	414.134	414.392	419.714	419.704
P/MPa	1.308	0.620	0.584	0.269	0.275	0.498	0.577	1.001	0.995

w_B	0.2499	0.2499	0.2499	0.2499	0.2499	0.2499	0.2499	0.2499	0.2750
T/K	428.141	428.149	436.737	436.713	445.540	445.539	454.457	454.465	337.076
P/MPa	1.857	1.822	2.688	2.699	3.650	3.660	4.732	4.704	4.304

w_B	0.2750	0.2750	0.2750	0.2750	0.2750	0.2750	0.2750	0.2750	0.2750
T/K	337.069	343.873	343.877	345.600	345.581	350.878	350.880	353.764	353.758
P/MPa	4.294	2.477	2.460	2.033	2.042	0.978	0.985	0.489	0.470

w_B	0.2750	0.2750	0.2750	0.2750	0.2750	0.2750	0.2750	0.2750	0.2750
T/K	416.490	416.452	419.623	419.624	428.112	428.111	436.712	436.673	445.469
P/MPa	0.546	0.537	0.935	0.951	1.765	1.758	2.644	2.629	3.582

w_B	0.2750	0.2750	0.2750
T/K	445.463	454.396	454.380
P/MPa	3.608	4.670	4.671

Type of data: spinodal points

w_B	0.0600	0.0600	0.0600	0.0600	0.0600	0.0600	0.0600	0.0600	0.0600
T/K	331.770	331.771	337.301	337.300	433.404	433.385	445.537	454.400	454.407
P/MPa	1.324	1.593	−0.239	0.017	0.188	−0.248	0.760	2.990	2.663

w_B	0.0900	0.0900	0.0900	0.0900	0.0900	0.0900	0.0900	0.0900	0.0900
T/K	336.255	336.243	344.053	344.053	351.039	351.037	421.594	421.573	421.571
P/MPa	3.706	3.699	1.450	1.632	0.355	0.286	0.460	0.499	0.524

w_B	0.0900	0.0900	0.0900	0.0900	0.0900	0.0900	0.0900	0.0900	0.1076
T/K	428.369	428.371	436.952	436.953	445.765	445.767	454.720	454.712	337.250
P/MPa	1.173	1.100	2.068	2.140	3.093	3.098	4.210	4.175	3.679

w_B	0.1076	0.1076	0.1076	0.1076	0.1076	0.1076	0.1076	0.1076	0.1076
T/K	337.250	344.098	344.100	351.095	351.082	351.083	355.350	355.351	420.023
P/MPa	3.863	1.572	1.640	−0.457	1.113	0.682	−0.190	−0.326	0.539

continued

continued

w_B	0.1076	0.1076	0.1076	0.1076	0.1076	0.1076	0.1076	0.1076	0.1076
T/K	428.443	428.439	437.035	437.038	437.036	445.843	445.842	454.781	454.778
P/MPa	1.318	1.482	2.272	1.928	2.502	3.327	3.274	4.218	4.418

w_B	0.1300	0.1300	0.1300	0.1300	0.1300	0.1300	0.1300	0.1600	0.1600
T/K	337.180	337.179	337.173	344.101	344.102	351.080	351.072	337.157	337.142
P/MPa	4.137	3.640	3.614	2.078	1.532	0.308	1.042	4.759	4.415

w_B	0.1600	0.1600	0.1600	0.1600	0.1600	0.1600	0.1600	0.1600	0.1600
T/K	337.138	343.960	343.961	343.965	350.936	350.939	357.997	357.991	360.818
P/MPa	4.642	2.519	2.577	2.399	1.091	0.824	−0.122	0.025	−0.239

w_B	0.1600	0.1600	0.1600	0.1600	0.1600	0.1600	0.1600	0.1600	0.1600
T/K	360.815	410.364	410.302	419.885	419.884	428.336	428.344	436.922	436.920
P/MPa	−0.170	−0.070	−0.011	0.705	0.612	1.774	1.799	2.561	2.595

w_B	0.1600	0.1600	0.1600	0.1600	0.1899	0.1899	0.1899	0.1899	0.1899
T/K	445.755	445.746	454.689	454.691	336.998	336.994	343.823	343.823	350.807
P/MPa	3.387	3.505	3.782	4.573	4.505	4.668	2.689	2.831	1.340

w_B	0.1899	0.1899	0.1899	0.1899	0.1899	0.1899	0.1899	0.1899	0.1899
T/K	350.806	355.097	355.095	357.836	357.831	360.710	410.135	410.099	419.836
P/MPa	1.426	0.626	0.764	0.392	0.391	0.054	0.183	0.163	0.910

w_B	0.1899	0.1899	0.1899	0.1899	0.1899	0.1899	0.1899	0.1899	0.1899
T/K	419.829	428.285	428.284	436.891	436.895	445.680	445.682	454.629	454.635
P/MPa	0.908	1.547	1.682	2.431	2.618	3.591	3.473	4.641	4.622

w_B	0.2199	0.2199	0.2199	0.2199	0.2199	0.2199	0.2199	0.2199	0.2199
T/K	336.988	336.992	343.807	343.804	350.815	350.801	355.202	355.197	359.272
P/MPa	4.736	4.752	2.745	2.865	1.382	1.490	0.720	0.739	0.292

w_B	0.2199	0.2199	0.2199	0.2199	0.2199	0.2199	0.2199	0.2199	0.2199
T/K	359.268	411.469	411.480	419.692	419.662	428.099	428.101	436.717	436.694
P/MPa	0.244	0.263	0.325	1.001	0.950	1.757	1.702	2.693	2.642

w_B	0.2199	0.2199	0.2199	0.2199	0.2499	0.2499	0.2499	0.2499	0.2499
T/K	445.482	445.484	454.423	454.416	336.980	336.977	343.788	343.813	350.806
P/MPa	3.630	3.691	4.658	4.708	4.413	4.348	2.568	2.492	1.122

w_B	0.2499	0.2499	0.2499	0.2499	0.2499	0.2499	0.2499	0.2499	0.2499
T/K	350.798	355.232	355.226	357.840	357.841	414.134	414.392	419.714	419.704
P/MPa	1.126	0.437	0.504	0.165	0.157	0.428	0.421	0.871	0.895

w_B	0.2499	0.2499	0.2499	0.2499	0.2499	0.2499	0.2499	0.2499	0.2750
T/K	428.141	428.149	436.737	436.713	445.540	445.539	454.457	454.465	337.076
P/MPa	1.665	1.628	2.553	2.581	3.579	3.574	4.589	4.615	3.066

w_B	0.2750	0.2750	0.2750	0.2750	0.2750	0.2750	0.2750	0.2750	0.2750
T/K	337.069	343.873	343.877	345.600	345.581	350.878	350.880	353.764	353.758
P/MPa	3.203	0.915	1.384	0.998	0.835	−0.072	0.000	−0.694	−0.525

w_B	0.2750	0.2750	0.2750	0.2750	0.2750	0.2750	0.2750	0.2750	0.2750
T/K	416.490	416.452	419.623	419.624	428.112	428.111	436.712	436.673	445.469
P/MPa	−0.060	−0.007	0.530	−0.150	0.869	0.586	1.964	2.289	3.316

w_B	0.2750	0.2750	0.2750
T/K	445.463	454.396	454.380
P/MPa	3.305	4.278	4.007

Polymer (B):	**polystyrene**							**1996LUS**

Characterization: $M_n/\text{kg.mol}^{-1} = 21.45$, $M_w/.\text{g.mol}^{-1} = 22.1$,
Polymer Laboratories, Amherst, MA

Solvent (A): **propionitrile (48% deuterated at CH$_2$)**
$$C_3H_5N/C_3H_3D_2N \qquad 107\text{-}12\text{-}0/24300\text{-}23\text{-}0$$

Type of data: cloud points

w_B	0.0800	0.0800	0.0800	0.0800	0.0800	0.0800	0.0800	0.0800	0.0800
T/K	357.740	357.741	364.997	364.993	372.356	372.360	379.893	379.888	387.404
P/MPa	3.687	3.572	2.253	2.670	1.582	1.547	1.306	1.058	0.913

w_B	0.0800	0.0800	0.0800	0.0800	0.0800	0.0800	0.0800	0.0800	0.0800
T/K	387.413	395.342	395.341	403.261	403.256	411.385	411.387	419.576	419.550
P/MPa	0.956	0.836	0.911	0.999	0.952	1.301	1.613	1.865	1.820

w_B	0.0800	0.0800	0.0800	0.0800	0.0800	0.0800	0.1101	0.1101	0.1101
T/K	427.986	427.979	436.500	436.496	445.337	445.331	357.758	357.760	365.093
P/MPa	2.442	2.653	3.267	3.221	4.257	4.172	5.301	5.141	3.882

w_B	0.1101	0.1101	0.1101	0.1101	0.1101	0.1101	0.1101	0.1101	0.1101
T/K	365.090	372.385	372.383	380.021	380.015	387.561	387.560	395.380	395.383
P/MPa	3.782	3.423	3.031	2.269	2.364	2.172	2.038	1.815	1.939

w_B	0.1101	0.1101	0.1101	0.1101	0.1101	0.1101	0.1101	0.1101	0.1101
T/K	403.309	403.290	411.400	411.397	411.393	419.617	419.610	428.072	428.067
P/MPa	1.885	1.872	2.510	2.338	1.591	2.763	2.728	3.441	3.520

w_B	0.1101	0.1101	0.1101	0.1101	0.1200	0.1200	0.1200	0.1200	0.1200
T/K	436.644	436.648	445.454	445.456	357.802	357.800	365.036	365.032	365.033
P/MPa	4.157	4.086	5.234	5.289	5.495	5.274	3.549	4.109	4.973

w_B	0.1200	0.1447	0.1447	0.1447	0.1447	0.1447	0.1447	0.1447	0.1447
T/K	365.042	365.034	365.039	372.386	372.391	379.919	379.920	387.571	387.572
P/MPa	3.935	4.785	4.768	3.573	3.491	2.842	2.670	2.583	2.472

w_B	0.1447	0.1447	0.1447	0.1447	0.1447	0.1447	0.1447	0.1447	0.1447
T/K	395.378	395.380	403.301	403.303	411.419	411.421	419.613	419.608	428.062
P/MPa	2.513	2.350	2.543	2.624	2.724	2.161	3.116	3.078	3.534

w_B	0.1447	0.1447	0.1447	0.1447	0.1447	0.1500	0.1500	0.1500	0.1500
T/K	428.064	436.687	436.684	445.457	445.458	357.796	357.794	365.057	365.038
P/MPa	3.658	4.475	4.358	5.160	5.159	5.693	5.807	4.340	4.329

w_B	0.1500	0.1500	0.1500	0.1500	0.1500	0.1500	0.1500	0.1500	0.1500
T/K	372.399	372.390	379.932	379.930	387.551	387.550	395.365	395.370	403.281
P/MPa	3.342	3.359	2.656	2.674	2.305	2.300	2.155	2.142	2.245

w_B	0.1500	0.1500	0.1500	0.1500	0.1500	0.1500	0.1500	0.1500	0.1500
T/K	403.280	411.410	411.409	411.404	419.630	419.629	419.628	428.095	428.092
P/MPa	2.244	2.580	2.620	2.579	2.986	3.015	2.948	3.544	3.595

w_B	0.1500	0.1500	0.1500	0.1500	0.1500	0.1799	0.1799	0.1799	0.1799
T/K	436.673	436.675	445.478	445.474	445.471	357.783	357.786	365.044	365.038
P/MPa	4.254	4.296	5.091	5.016	5.116	5.770	5.510	4.360	4.312

continued

continued

w_B	0.1799	0.1799	0.1799	0.1799	0.1799	0.1799	0.1799	0.1799	0.1799
T/K	372.394	372.391	379.930	379.930	387.575	387.575	395.401	395.393	403.293
P/MPa	3.393	3.047	2.754	2.796	2.232	2.282	2.201	2.085	2.141

w_B	0.1799	0.1799	0.1799	0.1799	0.1799	0.1799	0.1799	0.1799	0.1799
T/K	403.286	411.442	411.435	419.663	419.656	428.114	428.114	436.551	436.557
P/MPa	2.221	2.485	2.518	2.957	2.944	3.608	3.586	4.288	4.332

w_B	0.1799	0.1799	0.1799	0.2101	0.2101	0.2101	0.2101	0.2101	0.2101
T/K	445.492	445.484	445.491	357.790	357.791	365.030	365.043	372.407	372.406
P/MPa	5.110	5.100	4.893	5.199	5.179	3.898	3.859	2.902	2.888

w_B	0.2101	0.2101	0.2101	0.2101	0.2101	0.2101	0.2101	0.2101	0.2101
T/K	379.933	379.930	387.540	387.541	395.392	395.390	403.305	403.303	411.427
P/MPa	2.254	2.255	1.911	1.888	1.778	1.785	1.901	1.927	2.198

w_B	0.2101	0.2101	0.2101	0.2101	0.2101	0.2101	0.2101	0.2101	0.2101
T/K	411.427	419.658	419.655	428.105	428.104	436.673	436.670	445.498	445.493
P/MPa	2.181	2.655	2.661	3.255	3.302	3.993	3.986	4.817	4.810

w_B	0.2400	0.2400	0.2400	0.2400	0.2400	0.2400	0.2400	0.2400	0.2400
T/K	357.803	357.802	365.054	365.051	372.419	372.416	379.949	379.942	387.470
P/MPa	4.785	4.774	3.474	3.465	2.527	2.534	1.931	1.927	1.596

w_B	0.2400	0.2400	0.2400	0.2400	0.2400	0.2400	0.2400	0.2400	0.2400
T/K	387.467	395.430	395.425	403.326	403.322	411.443	411.441	419.684	419.684
P/MPa	1.594	1.518	1.512	1.648	1.614	1.964	1.965	2.443	2.452

w_B	0.2400	0.2400	0.2400	0.2400	0.2400	0.2400	0.2726	0.2726	0.2726
T/K	428.131	428.133	436.696	436.693	445.529	445.535	350.782	350.783	357.901
P/MPa	3.037	3.072	3.802	3.794	4.751	4.792	4.841	4.879	3.343

w_B	0.2726	0.2726	0.2726	0.2726	0.2726	0.2726	0.2726	0.2726	0.2726
T/K	357.895	365.668	365.646	372.512	372.510	380.038	380.031	387.713	387.690
P/MPa	3.325	2.139	2.089	1.362	1.334	0.790	0.782	0.578	0.573

w_B	0.2726	0.2726	0.2726	0.2726	0.2726	0.2726	0.2726	0.2726	0.2726
T/K	395.511	395.508	403.446	403.441	411.723	411.703	419.644	419.644	428.302
P/MPa	0.628	0.633	0.885	0.886	1.308	1.348	1.848	1.845	2.526

w_B	0.2726	0.2726	0.2726	0.2726	0.2726	0.3073	0.3073		
T/K	428.297	436.690	436.684	445.492	445.485	343.805	343.803		
P/MPa	2.521	3.343	3.292	4.234	4.123	4.981	4.984		

Type of data: spinodal points

w_B	0.0800	0.0800	0.0800	0.0800	0.0800	0.0800	0.0800	0.0800	0.0800
T/K	357.740	357.741	364.997	364.993	372.356	372.360	379.893	379.888	387.404
P/MPa	3.059	3.354	2.133	2.043	1.293	1.309	0.578	0.753	0.455

w_B	0.0800	0.0800	0.0800	0.0800	0.0800	0.0800	0.0800	0.0800	0.0800
T/K	387.413	395.342	395.341	403.261	403.256	411.385	411.387	419.576	419.550
P/MPa	0.189	0.356	0.186	0.577	0.531	0.955	0.738	1.317	1.545

w_B	0.0800	0.0800	0.0800	0.0800	0.0800	0.0800	0.1101	0.1101	0.1101
T/K	427.986	427.979	436.500	436.496	445.337	445.331	357.758	357.760	365.093
P/MPa	2.101	2.077	2.737	2.560	3.746	3.747	5.139	5.020	3.691

continued

continued

w_B	0.1101	0.1101	0.1101	0.1101	0.1101	0.1101	0.1101	0.1101	0.1101
T/K	365.090	372.385	372.383	380.021	380.015	387.561	387.560	395.380	395.383
P/MPa	3.655	1.673	2.769	2.026	2.069	1.726	1.731	1.588	1.239

w_B	0.1101	0.1101	0.1101	0.1101	0.1101	0.1101	0.1101	0.1101	0.1101
T/K	403.309	403.290	411.400	411.397	411.393	419.617	419.610	428.072	428.067
P/MPa	1.712	1.635	1.798	1.234	1.884	2.330	2.416	2.889	3.106

w_B	0.1101	0.1101	0.1101	0.1101	0.1200	0.1200	0.1200	0.1200	0.1200
T/K	436.644	436.648	445.454	445.456	357.802	357.800	365.036	365.032	365.033
P/MPa	3.797	3.845	4.696	4.198	4.609	3.342	2.014	3.902	3.828

w_B	0.1200	0.1447	0.1447	0.1447	0.1447	0.1447	0.1447	0.1447	0.1447
T/K	365.042	365.034	365.039	372.386	372.391	379.919	379.920	387.571	387.572
P/MPa	3.821	4.308	4.311	3.100	3.137	2.475	2.527	1.972	1.993

w_B	0.1447	0.1447	0.1447	0.1447	0.1447	0.1447	0.1447	0.1447	0.1447
T/K	395.378	395.380	403.301	403.303	411.419	411.421	419.613	419.608	428.062
P/MPa	1.672	2.079	1.973	2.048	2.111	2.373	2.871	2.741	3.362

w_B	0.1447	0.1447	0.1447	0.1447	0.1447	0.1500	0.1500	0.1500	0.1500
T/K	428.064	436.687	436.684	445.457	445.458	357.796	357.794	365.057	365.038
P/MPa	3.306	4.181	4.019	4.835	4.897	5.602	5.475	4.142	4.169

w_B	0.1500	0.1500	0.1500	0.1500	0.1500	0.1500	0.1500	0.1500	0.1500
T/K	372.399	372.390	379.932	379.930	387.551	387.550	395.365	395.370	403.281
P/MPa	3.032	3.047	2.429	2.346	2.074	2.046	1.878	1.903	2.037

w_B	0.1500	0.1500	0.1500	0.1500	0.1500	0.1500	0.1500	0.1500	0.1500
T/K	403.280	411.410	411.409	411.404	419.630	419.629	419.628	428.095	428.092
P/MPa	2.029	2.382	2.204	2.222	2.631	2.776	2.731	3.294	3.276

w_B	0.1500	0.1500	0.1500	0.1500	0.1500	0.1799	0.1799	0.1799	0.1799
T/K	436.673	436.675	445.478	445.474	445.471	357.783	357.786	365.044	365.038
P/MPa	3.941	4.052	4.841	4.778	4.805	5.475	5.422	3.998	4.008

w_B	0.1799	0.1799	0.1799	0.1799	0.1799	0.1799	0.1799	0.1799	0.1799
T/K	372.394	372.391	379.930	379.930	387.575	387.575	395.401	395.393	403.293
P/MPa	2.751	3.046	2.062	2.424	1.949	1.965	1.835	1.846	1.958

w_B	0.1799	0.1799	0.1799	0.1799	0.1799	0.1799	0.1799	0.1799	0.1799
T/K	403.286	411.442	411.435	419.663	419.656	428.114	428.114	436.551	436.557
P/MPa	1.954	2.203	2.192	2.633	2.672	3.354	3.313	3.990	3.961

w_B	0.1799	0.1799	0.1799	0.2101	0.2101	0.2101	0.2101	0.2101	0.2101
T/K	445.492	445.484	445.491	357.790	357.791	365.030	365.043	372.407	372.406
P/MPa	4.820	4.843	4.500	4.938	4.997	3.597	3.652	2.6250	2.761

w_B	0.2101	0.2101	0.2101	0.2101	0.2101	0.2101	0.2101	0.2101	0.2101
T/K	379.933	379.930	387.540	387.541	395.392	395.390	403.305	403.303	411.427
P/MPa	2.109	2.070	1.697	1.757	1.657	1.679	1.744	1.723	2.124

w_B	0.2101	0.2101	0.2101	0.2101	0.2101	0.2101	0.2101	0.2101	0.2101
T/K	411.427	419.658	419.655	428.105	428.104	436.673	436.670	445.498	445.493
P/MPa		2.527	2.508	3.119	3.081	3.837	3.835	4.709	4.713

continued

continued

w_B	0.2400	0.2400	0.2400	0.2400	0.2400	0.2400	0.2400	0.2400	0.2400
T/K	357.803	357.802	365.054	365.051	372.419	372.416	379.949	379.942	387.470
P/MPa	3.863	3.711	2.925	2.885	1.850	1.669	0.808	0.968	0.323

w_B	0.2400	0.2400	0.2400	0.2400	0.2400	0.2400	0.2400	0.2400	0.2400
T/K	387.467	395.430	395.425	403.326	403.322	411.443	411.441	419.684	419.684
P/MPa	0.097	−0.144	0.373	0.579	−0.119	1.225	0.985	0.497	1.566

w_B	0.2400	0.2400	0.2400	0.2400	0.2400	0.2400	0.2726	0.2726	0.2726
T/K	428.131	428.133	436.696	436.693	445.529	445.535	350.782	350.783	357.901
P/MPa	2.303	1.513	2.048	2.533	4.146	2.331	3.881	3.195	1.081

w_B	0.2726	0.2726	0.2726	0.2726	0.2726	0.2726	0.2726	0.2726	0.2726
T/K	357.895	365.668	365.646	372.512	372.510	380.038	380.031	387.713	387.690
P/MPa	0.787	0.551	1.166	−0.653	−0.137	−0.217	−0.354	−0.448	0.045

w_B	0.2726	0.2726	0.2726	0.2726	0.2726	0.2726	0.2726	0.2726	0.2726
T/K	395.511	395.508	403.446	403.441	411.723	411.703	419.644	419.644	428.302
P/MPa	0.195	0.269	0.290	0.174	−0.087	−0.232	1.166	1.211	1.924

w_B	0.2726	0.2726	0.2726	0.2726	0.2726	0.3073	0.3073		
T/K	428.297	436.690	436.684	445.492	445.485	343.805	343.803		
P/MPa	1.284	1.494	2.093	3.297	3.256	4.370	4.278		

Polymer (B):	**polystyrene**	**1996LUS**
Characterization:	M_n/kg.mol^{-1} = 21.45, M_w/.g.mol^{-1} = 22.1,	
	Polymer Laboratories, Amherst, MA	
Solvent (A):	**propionitrile (63.55% deuterated at CH$_2$)**	
	$C_3H_5N/C_3H_3D_2N$	107-12-0/24300-23-0

Type of data: cloud points

w_B	0.1800	0.1800	0.1800	0.1800	0.1800	0.1800	0.1800	0.1800	0.1800
T/K	372.424	372.414	379.970	379.948	387.582	387.588	395.407	395.408	403.104
P/MPa	5.410	5.410	4.542	4.553	3.998	3.977	3.686	3.706	3.612

w_B	0.1800	0.1800	0.1800	0.1800	0.2098	0.2098	0.2098	0.2098	0.2098
T/K	403.112	411.471	411.469	419.685	372.404	372.415	372.415	372.415	380.378
P/MPa	3.620	3.759	3.766	4.106	4.879	5.401	5.385	4.957	3.992

w_B	0.2098	0.2098	0.2098	0.2098	0.2098	0.2098	0.2098	0.2098	0.2098
T/K	380.344	387.507	387.501	395.424	395.416	403.327	403.325	403.325	411.445
P/MPa	4.004	3.510	3.566	3.340	3.324	3.379	3.625	3.322	3.554

w_B	0.2098	0.2098	0.2098	0.2098	0.2098	0.2098	0.2098	0.2098	0.2098
T/K	411.443	419.645	419.644	428.072	428.080	436.643	436.646	445.447	445.446
P/MPa	3.458	4.211	4.290	4.736	4.737	5.381	5.424	6.063	6.116

w_B	0.2098	0.2401	0.2401	0.2401	0.2401	0.2401	0.2401	0.2401	0.2401
T/K	419.650	365.080	365.078	372.387	372.388	380.020	380.015	387.610	387.607
P/MPa	4.423	5.712	5.711	4.510	4.509	3.644	3.659	3.121	3.115

w_B	0.2401	0.2401	0.2401	0.2401	0.2401	0.2401	0.2401	0.2401	0.2401
T/K	395.885	395.866	403.342	403.343	411.695	411.666	419.709	419.700	428.139
P/MPa	2.872	2.866	2.847	2.856	3.052	3.019	3.505	3.520	4.026

continued

continued

w_B	0.2401	0.2401	0.2401	0.2401	0.2401	0.2401	0.2700	0.2700	0.2700
T/K	428.143	436.720	436.724	445.520	445.520	445.522	365.092	365.087	372.436
P/MPa	4.054	4.717	4.709	5.515	5.536	5.485	5.552	5.597	4.408

w_B	0.2700	0.2700	0.2700	0.2700	0.2700	0.2700	0.2700	0.2700	0.2700
T/K	372.430	379.984	379.988	387.570	387.570	395.407	395.407	403.304	403.303
P/MPa	4.348	3.512	3.514	2.952	2.958	2.683	2.716	2.674	2.684

w_B	0.2700	0.2700	0.2700	0.2700	0.2700	0.2700	0.2700	0.2700	0.2700
T/K	411.453	411.449	419.630	419.630	428.111	428.110	436.743	436.723	445.577
P/MPa	2.944	2.927	3.330	3.310	3.859	3.835	4.525	4.521	5.350

w_B	0.2700	0.2964	0.2964	0.2964	0.2964	0.2964	0.2964	0.2964	0.2964
T/K	445.528	360.685	360.684	365.226	365.183	372.427	372.428	372.431	379.999
P/MPa	5.325	6.047	6.067	5.150	5.140	3.691	3.775	3.817	3.008

w_B	0.2964	0.2964	0.2964	0.2964	0.2964	0.2964	0.2964	0.2964	0.2964
T/K	379.998	387.592	387.596	395.420	395.425	403.292	403.283	411.438	411.424
P/MPa	3.024	2.549	2.550	2.339	2.342	2.398	2.362	2.595	2.602

w_B	0.2964	0.2964	0.2964	0.2964	0.2964	0.2964	0.2964	0.2964	0.2964
T/K	419.485	419.472	428.097	428.098	436.487	436.485	445.468	445.437	454.411
P/MPa	2.981	2.989	3.532	3.559	4.396	4.356	5.259	5.169	6.119

Type of data: spinodal points

w_B	0.1800	0.1800	0.1800	0.1800	0.1800	0.1800	0.1800	0.1800	0.1800
T/K	372.424	372.414	379.970	379.948	387.582	387.588	395.407	395.408	403.104
P/MPa	5.105	5.268	4.203	4.395	3.741	3.800	3.423	3.403	3.432

w_B	0.1800	0.1800	0.1800	0.1800	0.2098	0.2098	0.2098	0.2098	0.2098
T/K	403.112	411.471	411.469	419.685	372.404	372.415	372.415	372.415	380.378
P/MPa	3.372	3.563	3.583	3.902	4.693	4.715	4.710	4.714	3.808

w_B	0.2098	0.2098	0.2098	0.2098	0.2098	0.2098	0.2098	0.2098	0.2098
T/K	380.344	387.507	387.501	395.424	395.416	403.327	403.325	403.325	411.445
P/MPa	3.853	3.352	3.389	3.084	3.098	3.051	3.118	3.123	3.294

w_B	0.2098	0.2098	0.2098	0.2098	0.2098	0.2098	0.2098	0.2098	0.2098
T/K	411.443	419.645	419.644	428.072	428.080	436.643	436.646	445.447	445.446
P/MPa	3.288	4.051	3.994	4.505	4.528	5.202	5.174	5.951	5.929

w_B	0.2098	0.2401	0.2401	0.2401	0.2401	0.2401	0.2401	0.2401	0.2401
T/K	419.650	365.080	365.078	372.387	372.388	380.020	380.015	387.610	387.607
P/MPa	4.154	5.586	5.549	4.291	4.420	3.412	3.489	2.865	2.948

w_B	0.2401	0.2401	0.2401	0.2401	0.2401	0.2401	0.2401	0.2401	0.2401
T/K	395.885	395.866	403.342	403.343	411.695	411.666	419.709	419.700	428.139
P/MPa	2.647	2.706	2.657	2.661	2.829	2.911	3.321	3.339	3.906

w_B	0.2401	0.2401	0.2401	0.2401	0.2401	0.2401	0.2700	0.2700	0.2700
T/K	428.143	436.720	436.724	445.520	445.520	445.522	365.092	365.087	372.436
P/MPa		4.575	4.400	5.371	5.315	5.0290	4.561	4.941	3.007

w_B	0.2700	0.2700	0.2700	0.2700	0.2700	0.2700	0.2700	0.2700	0.2700
T/K	372.430	379.984	379.988	387.570	387.570	395.407	395.407	403.304	403.303
P/MPa	3.786	2.944	2.421	2.422	2.423	1.903	2.539	2.157	2.252

continued

continued

w_B	0.2700	0.2700	0.2700	0.2700	0.2700	0.2700	0.2700	0.2700	0.2700
T/K	411.453	411.449	419.630	419.630	428.111	428.110	436.743	436.723	445.577
P/MPa	2.297	2.506	2.724	2.891	3.340	3.333	3.844	3.803	4.487

w_B	0.2700	0.2964	0.2964	0.2964	0.2964	0.2964	0.2964	0.2964	0.2964
T/K	445.528	360.685	360.684	365.226	365.183	372.427	372.428	372.431	379.999
P/MPa	4.661	5.559	5.001	3.842	2.937	2.299	1.875	2.998	1.921

w_B	0.2964	0.2964	0.2964	0.2964	0.2964	0.2964	0.2964	0.2964	0.2964
T/K	379.998	387.592	387.596	395.420	395.425	403.292	403.283	411.438	411.424
P/MPa	2.240	1.297	1.125	1.557	1.221	1.005	1.628	1.920	1.609

w_B	0.2964	0.2964	0.2964	0.2964	0.2964	0.2964	0.2964	0.2964	0.2964
T/K	419.485	419.472	428.097	428.098	436.487	436.485	445.468	445.437	454.411
P/MPa	2.187	1.591	2.971	3.021	3.491	3.662	3.286	3.225	5.652

Polymer (B):	**polystyrene**	**1996LUS**
Characterization:	$M_n/kg.mol^{-1} = 21.45$, $M_w/.g.mol^{-1} = 22.1$,	
	Polymer Laboratories, Amherst, MA	
Solvent (A):	**propionitrile (75% deuterated at CH_2)**	
	$C_3H_5N/C_3H_3D_2N$	107-12-0/24300-23-0

Type of data: cloud points

w_B	0.1399	0.1399	0.1399	0.1399	0.1399	0.1399	0.1399	0.1399	0.1399
T/K	373.109	373.104	373.105	380.614	380.623	388.293	388.293	388.293	388.293
P/MPa	5.777	5.735	5.555	4.700	5.179	3.590	4.146	3.389	4.268

w_B	0.1399	0.1399	0.1399	0.1399	0.1399	0.1399	0.1399	0.1399	0.1399
T/K	396.073	396.076	403.969	403.971	412.105	412.145	420.378	420.372	428.821
P/MPa	3.794	3.648	3.725	3.651	3.845	3.741	4.223	3.960	4.385

w_B	0.1399	0.1399	0.1399	0.1399	0.1700	0.1700	0.1700	0.1700	0.1700
T/K	428.847	437.480	437.450	446.484	373.092	373.092	373.095	380.624	380.623
P/MPa	4.616	5.349	5.181	5.891	5.711	5.715	5.698	4.744	4.733

w_B	0.1700	0.1700	0.1700	0.1700	0.1700	0.1700	0.1700	0.1700	0.1700
T/K	388.244	388.245	396.093	396.085	404.061	404.031	413.878	420.348	420.350
P/MPa	4.092	4.100	3.743	3.746	3.629	3.633	3.803	4.084	4.094

w_B	0.1700	0.1700	0.1700	0.1700	0.1700	0.1700	0.2202	0.2202	0.2202
T/K	428.803	428.802	437.412	437.415	446.239	446.209	372.913	372.917	372.915
P/MPa	4.490	4.497	5.099	5.102	5.860	5.855	5.851	5.919	5.835

w_B	0.2202	0.2202	0.2202	0.2202	0.2202	0.2202	0.2202	0.2202	0.2202
T/K	380.578	380.575	388.239	388.241	396.057	396.063	403.989	403.989	403.989
P/MPa	4.838	4.871	4.105	4.367	3.885	3.905	3.762	3.820	3.775

w_B	0.2202	0.2202	0.2202	0.2202	0.2202	0.2202	0.2202	0.2202	0.2202
T/K	403.989	412.142	412.127	420.351	420.337	428.830	428.830	437.450	428.830
P/MPa	3.720	3.826	3.670	4.119	4.095	4.655	4.655	5.143	4.573

w_B	0.2202	0.2299	0.2299	0.2299	0.2299	0.2299	0.2299	0.2299	0.2299
T/K	437.431	372.488	372.492	379.987	387.629	387.629	395.570	395.567	403.590
P/MPa	5.101	6.065	6.082	5.157	4.407	4.262	3.907	3.919	3.799

continued

continued

w_B	0.2299	0.2299	0.2299	0.2299	0.2299	0.2299	0.2299	0.2299	0.2299
T/K	403.606	411.920	411.938	420.160	420.145	428.615	428.604	437.210	437.189
P/MPa	3.729	3.894	3.895	4.171	4.124	4.632	4.699	5.213	5.132

w_B	0.2588	0.2588	0.2588	0.2588	0.2588	0.2588	0.2588	0.2588	0.2588
T/K	380.009	380.009	387.600	387.593	395.371	395.384	403.281	403.320	411.465
P/MPa	4.883	4.868	4.189	4.218	3.789	3.826	3,694	3.742	3.823

w_B	0.2588	0.2588	0.2588	0.2588	0.2588	0.2588	0.2588
T/K	411.469	419.692	419.679	428.165	428.154	436.707	436.691
P/MPa	3.814	4.079	4.082	4.542	4.561	5.134	5.139

Type of data: spinodal points

w_B	0.1399	0.1399	0.1399	0.1399	0.1399	0.1399	0.1399	0.1399	0.1399
T/K	373.109	373.104	373.105	380.614	380.623	388.293	388.293	388.293	388.293
P/MPa	5.344	5.154	5.306	4.279	4.337	3.590	3.611	3.389	3.325

w_B	0.1399	0.1399	0.1399	0.1399	0.1399	0.1399	0.1399	0.1399	0.1399
T/K	396.073	396.076	403.969	403.971	412.105	412.145	420.378	420.372	428.821
P/MPa	3.272	3.400	3.187	2.980	3.476	3.424	3.715	3.812	4.241

w_B	0.1399	0.1399	0.1399	0.1399	0.1700	0.1700	0.1700	0.1700	0.1700
T/K	428.847	437.480	437.450	446.484	373.092	373.092	373.095	380.624	380.623
P/MPa	4.378	4.730	4.764	5.561	5.560	5.553	5.592	4.4960	4.509

w_B	0.1700	0.1700	0.1700	0.1700	0.1700	0.1700	0.1700	0.1700	0.1700
T/K	388.244	388.245	396.093	396.085	404.061	404.031	413.878	420.348	420.350
P/MPa	3.875	3.916	3.481	3.528	3.400	3.369	3.564	3.896	3.858

w_B	0.1700	0.1700	0.1700	0.1700	0.1700	0.1700	0.2202	0.2202	0.2202
T/K	428.803	428.802	437.412	437.415	446.239	446.209	372.913	372.917	372.915
P/MPa	4.264	4.319	4.903	4.924	5.653	5.683	5.515	5.652	5.729

w_B	0.2202	0.2202	0.2202	0.2202	0.2202	0.2202	0.2202	0.2202	0.2202
T/K	380.578	380.575	388.239	388.241	396.057	396.063	403.989	403.989	403.989
P/MPa	4.725	4.729	4.108	4.013	3.629	3.717	3.606	3.585	3.628

w_B	0.2202	0.2202	0.2202	0.2202	0.2202	0.2202	0.2202	0.2202	0.2202
T/K	403.989	412.142	412.127	420.351	420.337	428.830	428.830	437.450	428.830
P/MPa	3.631	3.738	3.743	3.991	4.015	4.492	4.492	5.059	4.456

w_B	0.2202	0.2299	0.2299	0.2299	0.2299	0.2299	0.2299	0.2299	0.2299
T/K	437.431	372.488	372.492	379.987	387.629	387.629	395.570	395.567	403.590
P/MPa	5.072	5.976	5.999	4.778	4.247	4.192	3.806	3.818	3.678

w_B	0.2299	0.2299	0.2299	0.2299	0.2299	0.2299	0.2299	0.2299	0.2299
T/K	403.606	411.920	411.938	420.160	420.145	428.615	428.604	437.210	437.189
P/MPa	3.689	3.790	3.782	4.066	4.052	4.531	4.545	5.129	5.096

w_B	0.2588	0.2588	0.2588	0.2588	0.2588	0.2588	0.2588	0.2588	0.2588
T/K	380.009	380.009	387.600	387.593	395.371	395.384	403.281	403.320	411.465
P/MPa	4.625	4.677	4.025	3.884	3.715	3.673	3.529	3.546	3.721

w_B	0.2588	0.2588	0.2588	0.2588	0.2588	0.2588	0.2588
T/K	411.469	419.692	419.679	428.165	428.154	436.707	436.691
P/MPa	3.737	4.019	4.052	4.441	4.436	5.045	5.035

Polymer (B):	**polystyrene**							**1996LUS**

Characterization: $M_n/\text{kg.mol}^{-1} = 23.6$, $M_w/\text{kg.mol}^{-1} = 25.0$,
Pressure Chemical Company, Pittsburgh, PA

Solvent (A):	**propionitrile**			**C₃H₅N**				**107-12-0**

Type of data: cloud points

w_B	0.1687	0.1687	0.1687	0.1687	0.1687	0.1687	0.1687	0.1687	0.1687
T/K	436.982	436.984	428.529	420.105	420.101	413.475	403.751	403.747	397.076
P/MPa	4.593	4.560	3.946	3.229	3.170	2.763	2.275	2.269	2.077
w_B	0.1687	0.1687	0.1687	0.1687	0.1687	0.1687	0.1687	0.1687	0.1687
T/K	397.085	387.998	387.989	380.095	380.110	380.107	372.753	372.723	366.403
P/MPa	2.044	1.997	2.001	2.174	1.401	2.176	2.601	2.607	3.198
w_B	0.1687	0.1687	0.1687	0.1799	0.1799	0.1799	0.1799	0.1799	0.1799
T/K	366.009	358.121	358.116	365.993	365.996	373.340	373.345	380.978	380.974
P/MPa	3.237	4.333	4.405	4.905	4.924	4.151	3.995	3.521	3.434
w_B	0.1799	0.1799	0.1799	0.1799	0.1799	0.1799	0.1799	0.1799	0.1799
T/K	388.677	388.669	396.524	396.529	396.529	396.529	404.434	404.432	412.646
P/MPa	3.140	2.902	3.080	3.046	2.955	3.013	3.299	3.212	3.456
w_B	0.1799	0.1799	0.1799	0.1799	0.1799	0.1799	0.1799	0.1999	0.1999
T/K	412.646	420.921	420.904	429.373	429.330	437.974	437.977	437.501	437.500
P/MPa	3.388	3.956	3.871	4.423	4.230	5.206	5.240	5.261	5.333
w_B	0.1999	0.1999	0.1999	0.1999	0.1999	0.1999	0.1999	0.1999	0.1999
T/K	437.510	428.865	428.857	420.705	420.674	412.373	412.350	404.353	404.339
P/MPa	5.477	4.726	4.522	4.183	3.909	3.583	3.589	3.244	3.286
w_B	0.1999	0.1999	0.1999	0.1999	0.1999	0.1999	0.1999	0.1999	0.1999
T/K	396.419	396.415	388.628	381.025	381.030	381.030	373.334	373.328	366.011
P/MPa	3.130	3.049	3.073	3.366	4.238	3.367	3.971	3.999	4.599
w_B	0.1999	0.1999	0.2200	0.2200	0.2200	0.2200	0.2200	0.2200	0.2200
T/K	366.011	366.016	365.570	365.567	373.014	373.018	380.604	380.601	388.270
P/MPa	4.717	4.790	4.655	4.630	3.779	3.736	3.156	3.165	2.861
w_B	0.2200	0.2200	0.2200	0.2200	0.2200	0.2200	0.2200	0.2200	0.2200
T/K	388.259	396.102	396.100	403.934	403.933	412.096	412.100	420.560	420.572
P/MPa	2.864	2.800	2.813	2.913	2.932	3.235	3.221	3.710	3.699
w_B	0.2200	0.2200	0.2200	0.2200	0.2308	0.2459	0.2459	0.2459	0.2459
T/K	428.894	428.892	437.472	437.482	437.064	437.175	437.181	428.700	428.636
P/MPa	4.297	4.293	5.005	5.023	4.683	4.651	4.676	3.932	4.050
w_B	0.2459	0.2459	0.2459	0.2459	0.2459	0.2459	0.2459	0.2459	0.2459
T/K	420.286	420.277	420.214	412.202	412.045	404.140	404.135	396.033	396.035
P/MPa	3.452	3.474	3.473	3.035	3.002	2.523	2.446	2.314	2.341
w_B	0.2459	0.2459	0.2459	0.2459	0.2459	0.2459	0.2459	0.2459	0.2459
T/K	388.125	388.129	380.387	380.385	372.911	372.914	365.658	365.666	358.452
P/MPa	2.433	2.422	2.748	2.762	3.258	3.228	4.060	4.232	5.406

continued

continued

w_B	0.2459	0.2459	0.2798	0.2798	0.2798	0.2798	0.2798	0.2798	0.2798
T/K	358.449	358.453	437.980	429.468	429.462	429.464	420.935	420.934	412.739
P/MPa	5.451	5.388	3.211	2.832	2.759	2.532	1.932	1.881	1.004

w_B	0.2798	0.2798	0.2798	0.2971	0.2971	0.2971	0.2971	0.2971	0.2971
T/K	412.710	344.822	344.826	438.267	438.256	429.743	429.731	429.728	421.267
P/MPa	0.969	0.644	0.714	3.767	3.849	3.112	2.980	2.976	2.213

w_B	0.2971	0.2971	0.2971	0.2971	0.2971	0.2971	0.2971	0.2971	0.2971
T/K	421.249	413.028	413.027	413.024	404.747	404.741	396.954	358.883	358.883
P/MPa	2.192	1.618	1.600	1.670	0.963	0.979	0.637	0.757	0.778

w_B	0.2971	0.2971	0.2971	0.2971	0.2971	0.2971	0.2971	0.2971	0.2971
T/K	351.969	351.969	344.974	344.967	406.035	412.749	420.973	429.415	437.999
P/MPa	1.693	1.698	2.988	3.066	0.707	1.360	2.665	2.767	3.618

Type of data: spinodal points

w_B	0.1687	0.1687	0.1687	0.1687	0.1687	0.1687	0.1687	0.1687	0.1687
T/K	436.982	436.984	428.529	420.105	420.101	413.475	403.751	403.747	397.076
P/MPa	4.075	3.939	2.479	2.155	2.935	2.118	1.765	1.771	1.411

w_B	0.1687	0.1687	0.1687	0.1687	0.1687	0.1687	0.1687	0.1687	0.1687
T/K	397.085	387.998	387.989	380.095	380.110	380.107	372.753	372.723	366.403
P/MPa	1.748	1.569	1.615	1.886	−0.373	1.739	2.217	2.135	2.904

w_B	0.1687	0.1687	0.1687	0.1799	0.1799	0.1799	0.1799	0.1799	0.1799
T/K	366.009	358.121	358.116	365.993	365.996	373.340	373.345	380.978	380.974
P/MPa	2.979	4.216	3.897	4.729	4.760	3.769	3.905	3.245	3.280

w_B	0.1799	0.1799	0.1799	0.1799	0.1799	0.1799	0.1799	0.1799	0.1799
T/K	388.677	388.669	396.524	396.529	396.529	396.529	404.434	404.432	412.646
P/MPa	2.964	3.015	2.815	2.894	2.909	2.895	2.912	2.979	3.200

w_B	0.1799	0.1799	0.1799	0.1799	0.1799	0.1799	0.1799	0.1999	0.1999
T/K	412.646	420.921	420.904	429.373	429.330	437.974	437.977	437.501	437.500
P/MPa	3.252	3.794	3.811	4.423	4.376	5.097	5.093	5.180	5.091

w_B	0.1999	0.1999	0.1999	0.1999	0.1999	0.1999	0.1999	0.1999	0.1999
T/K	437.510	428.865	428.857	420.705	420.674	412.373	412.350	404.353	404.339
P/MPa	5.194	4.351	4.393	3.708	3.861	3.311	3.328	3.006	3.058

w_R	0.1999	0.1999	0.1999	0.1999	0.1999	0.1999	0.1999	0.1999	0.1999
T/K	396.419	396.415	388.628	381.025	381.030	381.030	373.334	373.328	366.011
P/MPa	2.841	2.943	2.976	3.294	3.341	3.312	3.866	3.846	4.753

w_B	0.1999	0.1999	0.2200	0.2200	0.2200	0.2200	0.2200	0.2200	0.2200
T/K	366.011	366.016	365.570	365.567	373.014	373.018	380.604	380.601	388.270
P/MPa	4.740	4.719	3.941	3.904	2.780	3.340	2.656	2.842	2.443

w_B	0.2200	0.2200	0.2200	0.2200	0.2200	0.2200	0.2200	0.2200	0.2200
T/K	388.259	396.102	396.100	403.934	403.933	412.096	412.100	420.560	420.572
P/MPa	2.393	2.186	2.313	2.514	2.572	2.862	2.852	3.355	3.361

w_B	0.2200	0.2200	0.2200	0.2200	0.2308	0.2459	0.2459	0.2459	0.2459
T/K	428.894	428.892	437.472	437.482	437.064	437.175	437.181	428.700	428.636
P/MPa	3.848	3.773	4.513	4.497	4.644	3.676	3.596	1.903	3.113

continued

continued

w_B	0.2459	0.2459	0.2459	0.2459	0.2459	0.2459	0.2459	0.2459	0.2459
T/K	420.286	420.277	420.214	412.202	412.045	404.140	404.135	396.033	396.035
P/MPa	1.939	2.197	1.978	0.418	0.075	−1.632	−2.350	−3.799	−2.979
w_B	0.2459	0.2459	0.2459	0.2459	0.2459	0.2459	0.2459	0.2459	0.2459
T/K	388.125	388.129	380.387	380.385	372.911	372.914	365.658	365.666	358.452
P/MPa	−2.024	−1.773	−2.851	−2.208	−1.389	−1.877	−2.593	−1.551	1.046
w_B	0.2459	0.2459	0.2798	0.2798	0.2798	0.2798	0.2798	0.2798	0.2798
T/K	358.449	358.453	437.980	429.468	429.462	429.464	420.935	420.934	412.739
P/MPa	−1.537	−1.562	3.092	2.003	1.453	2.316	0.451	1.073	−4.083
w_B	0.2798	0.2798	0.2798	0.2971	0.2971	0.2971	0.2971	0.2971	0.2971
T/K	412.710	344.822	344.826	438.267	438.256	429.743	429.731	429.728	421.267
P/MPa	−1.355	−0.010	−0.119	3.627	3.711	2.884	2.865	2.864	2.152
w_B	0.2971	0.2971	0.2971	0.2971	0.2971	0.2971	0.2971	0.2971	0.2971
T/K	421.249	413.028	413.027	413.024	404.747	404.741	396.954	358.883	358.883
P/MPa	2.117	−2.544	0.959	0.725	−0.169	−0.153	−1.035	0.111	−0.306
w_B	0.2971	0.2971	0.2971	0.2971	0.2971	0.2971	0.2971	0.2971	0.2971
T/K	351.969	351.969	344.974	344.967	406.035	412.749	420.973	429.415	437.999
P/MPa	1.036	1.225	2.440	1.469	−0.756	0.352	1.426	2.522	3.348

Polymer (B):   **polystyrene-d8**   **1996LUS**
Characterization: M_n/kg.mol^{-1} = 25.4, M_w/kg.mol^{-1} = 26.7,
                          completely deuterated, Polymer Laboratories, Amherst, MA
Solvent (A):   **propionitrile**   **C_3H_5N**   **107-12-0**

Type of data: cloud points

w_B	0.1100	0.1100	0.1100	0.1100	0.1100	0.1100	0.1100	0.1100	0.1100
T/K	298.411	298.417	300.909	300.912	304.551	304.550	304.548	433.682	433.647
P/MPa	3.855	3.881	2.595	2.622	0.970	0.884	0.945	0.722	0.716
w_B	0.1100	0.1100	0.1100	0.1100	0.1100	0.1100	0.1100	0.1100	0.1100
T/K	436.659	436.648	445.435	445.434	445.430	454.325	454.326	463.495	463.490
P/MPa	1.141	1.135	2.199	2.282	2.249	3.555	3.552	4.830	4.842
w_B	0.1100	0.1100	0.1401	0.1401	0.1401	0.1401	0.1401	0.1401	0.1401
T/K	463.484	463.487	298.413	298.407	301.744	301.746	305.074	307.000	306.987
P/MPa	4.757	4.673	4.434	4.431	2.763	2.740	1.272	0.495	0.501
w_B	0.1401	0.1401	0.1401	0.1401	0.1401	0.1401	0.1401	0.1401	0.1401
T/K	431.523	431.515	436.657	436.664	445.455	445.436	454.385	454.377	463.517
P/MPa	0.613	0.607	1.309	1.286	2.487	2.478	3.733	3.706	4.969
w_B	0.1401	0.1699	0.1699	0.1699	0.1699	0.1699	0.1699	0.1699	0.1699
T/K	463.506	299.564	299.418	301.930	301.903	305.638	305.560	308.075	308.082
P/MPa	4.999	4.609	4.914	3.627	3.589	2.294	1.825	0.860	1.037
w_B	0.1699	0.1699	0.1699	0.1699	0.1699	0.1699	0.1699	0.1699	0.1699
T/K	431.307	431.300	436.635	436.638	445.431	445.424	454.336	455.293	455.293
P/MPa	0.720	0.825	1.418	1.550	2.791	2.760	3.906	3.987	3.981

continued

continued

w_B	0.1699	0.1999	0.1999	0.1999	0.1999	0.1999	0.1999	0.1999	0.1999
T/K	463.488	298.698	298.691	301.242	301.215	304.599	304.609	306.836	306.866
P/MPa	5.198	4.400	4.386	3.097	3.104	1.548	1.572	0.674	0.689

w_B	0.1999	0.1999	0.1999	0.1999	0.1999	0.1999	0.1999	0.1999	0.1999
T/K	433.304	433.303	436.649	436.646	445.412	445.426	454.364	454.359	463.517
P/MPa	0.943	0.991	1.460	1.469	2.663	2.653	3.876	3.881	5.222

w_B	0.1999	0.2301	0.2301	0.2301	0.2301	0.2301	0.2301	0.2301	0.2301
T/K	463.520	294.662	294.663	294.674	298.524	298.527	298.522	301.023	301.014
P/MPa	5.207	4.966	5.030	5.011	2.894	3.026	3.087	1.854	1.889

w_B	0.2301	0.2301	0.2301	0.2301	0.2301	0.2301	0.2301	0.2301	0.2301
T/K	303.396	303.411	303.408	304.645	304.647	431.698	431.701	436.840	436.841
P/MPa	1.239	1.075	1.465	1.047	1.147	0.657	0.672	1.360	1.377

w_B	0.2301	0.2301	0.2301	0.2301	0.2301	0.2301	0.2581	0.2581	0.2581
T/K	445.825	445.819	454.766	454.759	463.931	463.929	294.668	294.670	298.405
P/MPa	2.626	2.600	3.842	3.846	5.346	5.407	5.183	5.193	3.291

w_B	0.2581	0.2581	0.2581	0.2581	0.2581	0.2581	0.2581	0.2581	0.2581
T/K	298.404	300.847	300.853	304.522	304.533	431.546	431.558	435.041	435.039
P/MPa	3.258	2.104	2.169	0.578	0.574	0.559	0.575	1.056	1.060

w_B	0.2581	0.2581	0.2581	0.2581	0.2581	0.2581	0.2581	0.2581	
T/K	439.829	439.790	445.466	445.456	454.364	454.367	463.608	463.610	
P/MPa	1.704	1.708	2.503	2.497	3.694	3.753	5.042	4.984	

Type of data: spinodal points

w_B	0.1100	0.1100	0.1100	0.1100	0.1100	0.1100	0.1100	0.1100	0.1100
T/K	298.411	298.417	300.909	300.912	304.551	304.550	304.548	433.682	433.647
P/MPa	2.884	3.316	1.645	1.592	−0.103	0.147	0.198	−0.098	−0.246

w_B	0.1100	0.1100	0.1100	0.1100	0.1100	0.1100	0.1100	0.1100	0.1100
T/K	436.659	436.648	445.435	445.434	445.430	454.325	454.326	463.495	463.490
P/MPa	0.314	0.246	1.254	1.282	1.141	3.019	2.795	4.043	4.017

w_B	0.1100	0.1100	0.1401	0.1401	0.1401	0.1401	0.1401	0.1401	0.1401
T/K	463.484	463.487	298.413	298.407	301.744	301.746	305.074	307.000	306.987
P/MPa	3.137	2.334	4.206	4.195	2.407	2.546	0.998	0.261	0.272

w_B	0.1401	0.1401	0.1401	0.1401	0.1401	0.1401	0.1401	0.1401	0.1401
T/K	431.523	431.515	436.657	436.664	445.455	445.436	454.385	454.377	463.517
P/MPa	0.279	0.378	0.958	1.061	2.323	2.260	3.533	3.514	4.666

w_B	0.1401	0.1699	0.1699	0.1699	0.1699	0.1699	0.1699	0.1699	0.1699
T/K	463.506	299.564	299.418	301.930	301.903	305.638	305.560	308.075	308.082
P/MPa	4.721	4.579	4.687	3.327	3.347	1.566	1.664	0.659	0.715

w_B	0.1699	0.1699	0.1699	0.1699	0.1699	0.1699	0.1699	0.1699	0.1699
T/K	431.307	431.300	436.635	436.638	445.431	445.424	454.336	455.293	455.293
P/MPa	0.613	0.554	1.343	1.215	2.469	2.448	3.678	3.623	3.631

w_B	0.1699	0.1999	0.1999	0.1999	0.1999	0.1999	0.1999	0.1999	0.1999
T/K	463.488	298.698	298.691	301.242	301.215	304.599	304.609	306.836	306.866
P/MPa	5.014	3.980	4.111	2.791	2.801	1.200	1.382	0.458	0.432

continued

continued

w_B	0.1999	0.1999	0.1999	0.1999	0.1999	0.1999	0.1999	0.1999	0.1999
T/K	433.304	433.303	436.649	436.646	445.412	445.426	454.364	454.359	463.517
P/MPa	0.848	0.839	1.303	1.335	2.513	2.503	3.794	3.799	5.014

w_B	0.1999	0.2301	0.2301	0.2301	0.2301	0.2301	0.2301	0.2301	0.2301
T/K	463.520	294.662	294.663	294.674	298.524	298.527	298.522	301.023	301.014
P/MPa	5.036	1.051	3.672	−2.596	−1.244	1.298	1.044	−9.850	−3.048

w_B	0.2301	0.2301	0.2301	0.2301	0.2301	0.2301	0.2301	0.2301	0.2301
T/K	303.396	303.411	303.408	304.645	304.647	431.698	431.701	436.840	436.841
P/MPa	−2.994	−6.011	0.470	−2.238	−1.003	0.473	0.496	1.216	1.199

w_B	0.2301	0.2301	0.2301	0.2301	0.2301	0.2301	0.2581	0.2581	0.2581
T/K	445.825	445.819	454.766	454.759	463.931	463.929	294.668	294.670	298.405
P/MPa	2.389	2.389	3.595	3.623	4.663	4.910	−1.464	1.175	−3.335

w_B	0.2581	0.2581	0.2581	0.2581	0.2581	0.2581	0.2581	0.2581	0.2581
T/K	298.404	300.847	300.853	304.522	304.533	431.546	431.558	435.041	435.039
P/MPa	−1.763	−1.739	−1.038	−3.486	−2.691	0.404	0.320	0.755	0.811

w_B	0.2581	0.2581	0.2581	0.2581	0.2581	0.2581	0.2581	0.2581
T/K	439.829	439.790	445.466	445.456	454.364	454.367	463.608	463.610
P/MPa	1.500	1.411	2.260	2.286	3.225	3.308	4.716	4.714

Polymer (B):	**poly(vinyl acetate)**	**1999BEY**
Characterization:	M_n/kg.mol^{-1} = 52.7, M_w/kg.mol^{-1} = 124.8, Aldrich Chem. Co., Inc., Milwaukee, WI	
Solvent (A):	**cyclopentane** C_5H_{10}	287-92-3

Type of data: cloud points (LCST behavior)

w_B	0.06	0.06	0.06	0.06	0.06	0.06
T/K	516.55	519.35	524.95	531.45	536.25	541.15
P/bar	316.0	302.0	278.0	254.0	246.0	243.0

Polymer (B):	**poly(vinyl acetate)**	**1999BEY**
Characterization:	M_n/kg.mol^{-1} = 52.7, M_w/kg.mol^{-1} = 124.8, Aldrich Chem. Co., Inc., Milwaukee, WI	
Solvent (A):	**cyclopentene** C_5H_8	142-29-0

Type of data: cloud points (LCST behavior)

w_B	0.06	0.06	0.06	0.06	0.06	0.06
T/K	445.65	451.75	460.45	469.55	478.45	490.85
P/bar	270.6	168.2	115.9	97.2	84.2	98.3

Polymer (B):   **poly(vinyl chloride)**   **1985GE1**
Characterization: M_n/kg.mol^{-1} = 64.7, M_w/kg.mol^{-1} = 75,
                              fractionated in the laboratory
Solvent (A):   **1,2-dimethylbenzene**   **C$_8$H$_{10}$**   **95-47-6**

Type of data: cloud points

w_B 0.08   was kept constant

T/K	323.45	320.55	316.45	312.45	309.15
P/bar	1	250	500	750	1000

Polymer (B):   **poly(vinyl chloride)**   **1985GE1**
Characterization: M_n/kg.mol^{-1} = 31.4, M_w/kg.mol^{-1} = 37,
                              fractionated in the laboratory
Solvent (A):   **phenetole**   **C$_8$H$_{10}$O**   **103-73-1**

Type of data: cloud points

w_B 0.10   was kept constant

T/K	295.65	292.85	290.15	288.45	287.05
P/bar	1	250	500	750	1000

Polymer (B):   **poly(vinyl chloride)**   **1985GE1**
Characterization: M_n/kg.mol^{-1} = 64.7, M_w/kg.mol^{-1} = 75,
                              fractionated in the laboratory
Solvent (A):   **phenetole**   **C$_8$H$_{10}$O**   **103-73-1**

Type of data: cloud points

w_B 0.08   was kept constant

T/K	317.65	313.35	311.05	309.85	309.05
P/bar	1	250	500	750	1000

3.2. Table of binary systems where data were published only in graphical form as phase diagrams or related figures

Polymer (B)	Solvent (A)	Ref.
Ethylene/1-butene copolymer		
	n-heptane	1996LOO
Ethylene/1-hexene copolymer		
	n-heptane	1996LOO
Ethylene/methyl acrylate copolymer		
	n-hexane	1994LOS
Ethylene/4-methyl-1-pentene copolymer		
	n-heptane	1996LOO
Ethylene/1-octene copolymer		
	cyclohexane	1996LOO
	n-heptane	1996LOO
	n-hexane	1996LOO
	2-methylpentane	1996LOO
Ethylene/propylene copolymer		
	n-heptane	1996LOO
Ethylene/vinyl acetate copolymer		
	cyclopentane	1999BEY
	cyclopentane	2000BEY
	cyclopentene	1999BEY
	cyclopentene	2000BEY
Hydroxypropylcellulose		
	water	2001KUN
N-Isopropylacrylamide/acrylic acid copolymer		
	water	1993OTA
	water	2000YAM

Polymer (B)	Solvent (A)	Ref.
N-Isopropylacrylamide/1-deoxy-1-methacryl-amide-D-glucitol copolymer	water	2001GOM
N-Isopropylacrylamide/4-pentenoic acid copolymer	water	1999KUN
Poly(butyl methacrylate)	ethanol	1986SAN
Poly(decyl methacrylate)	cyclopentane	1983MA2
	n-heptane	1983MA2
	n-hexane	1983MA2
	n-pentane	1983MA2
	toluene	1983MA2
	2,2,4-trimethylpentane	1983MA1
	2,2,4-trimethylpentane	1983MA2
	2,2,4-trimethylpentane	1997WOL
Poly(dimethylsiloxane)	n-butane	1972ZE1
	ethane	1972ZE1
	propane	1972ZE1
Polyethylene	n-butane	1963EHR
	n-butane	1975HOR
	n-butane	1994KI2
	cyclopentane	1999BEY
	cyclopentane	2000BEY
	cyclopentene	1999BEY
	cyclopentene	2000BEY
	dichlorodifluoroethene	1975HOR
	dichlorodifluoromethane	1975HOR
	fluorotrichloromethane	1975HOR
	n-heptane	1973HAM
	n-heptane	1975HOR
	n-heptane	1980KLE
	n-heptane	1996LOO
	n-hexane	1973HAM
	n-hexane	1975HOR
	n-hexane	1980KLE
	n-hexane	1989HAE
	n-hexane	1990KEN
	n-hexane	1996LOO

Polymer (B)	Solvent (A)	Ref.
Polyethylene (*continued*)		
	2-methylpropane	1975HOR
	n-octane	1973HAM
	n-octane	1975HOR
	n-octane	1980KLE
	n-octane	1996LOO
	n-pentane	1963EHR
	n-pentane	1973HAM
	n-pentanc	1975HOR
	n-pentane	1992KIR
	n-pentane	1993KI2
	n-pentane	1994KI2
	n-pentane	1998ZHU
	n-pentane	2003ZH2
	perfluoro-2-butene	1975HOR
	1,1,2-trichlorotrifluoroethane	1975HOR
Poly(ethylene glycol)		
	water	1976SA1
Poly(ethylene oxide)		
	water	1992COO
Polyisobutylene		
	n-butane	1972ZE1
	n-hexane	1972ZE1
	2-methylbutane	1965ALL
	2-methylbutane	1972ZE1
	n-pentane	2002JOU
	n-pentane	1972ZE1
	n-pentane	1997WOL
	propane	1972ZE1
Poly(*N*-isopropylacrylamide)		
	deuterium oxide	2004SHI
	water	1993OTA
	water	1997KUN
	water	1999KUN
	water	2004SHI
Poly(methacrylic acid)		
	water	1974TAN
Polypeptide		
	water	2003YAM

Polymer (B)	Solvent (A)	Ref.
Polypropylene		
	n-pentane	1998KI2
Poly(propylene glycol)		
	water	1974TAN
Poly(propylene oxide)		
	propane	1972ZE2
Polystyrene		
	acetaldehyde	1997REB
	tert-butyl acetate	1976SA2
	cyclohexane	1962HAM
	cyclohexane	1975SAE
	cyclohexane	1980HOS
	cyclohexane	1981WOL
	cyclohexane	1997WOL
	cyclohexane	1999HOO
	cyclohexane	2000KOI
	cyclopentane	1978ISH
	cyclopentane	1981WOL
	cyclopentane	1997WOL
	trans-decalin	1977WO1
	trans-decalin	1977WO2
	trans-decalin	2003JI1
	trans-decalin	2004JIA
	n-decane	2001IMR
	n-decane	2002IMR
	diethyl ether	1976SA2
	diethyl ether	1997WOL
	diethyl malonate	1981WOL
	diethyl oxalate	1986SAE
	1,4-dimethylcyclohexane	1999IM1
	1,4-dimethylcyclohexane	2002IMR
	n-dodecane	2001IMR
	n-dodecane	2002IMR
	dodecyl acetate	2000IMR
	ethyl formate	1997IMR
	1,1,1,3,3,3-hexadeutero-2-propanone	1991SZY
	n-hexane	1997XIO
	methyl acetate	1965MYR
	methyl acetate	1972ZE2
	methyl acetate	1997IMR
	methylcyclohexane	1988KI1
	methylcyclohexane	1991KIE
	methylcyclohexane	1993HOS

Polymer (B)	Solvent (A)	Ref.
Polystyrene (*continued*)		
	methylcyclohexane	1993WEL
	methylcyclohexane	1996DOB
	methylcyclohexane	1996IMR
	methylcyclohexane	1997END
	methylcyclohexane	1998SZY
	methylcyclohexane	1999HOO
	methylcyclohexane	2000XIO
	methylcyclopentane	1991SZY
	nitroethane	2000DE2
	n-octane	2001IMR
	n-octane	2002IMR
	n-pentane	1988KI1
	n-pentane	1988KI2
	n-pentane	1988KI3
	n-pentane	1988KI4
	n-pentane	1991KIE
	n-pentane	1994KI1
	n-pentane	2001LI4
	1-phenyldecane	1981WOL
	1-phenyldecane	1997WOL
	2-propanone	1972ZE2
	2-propanone	1991SZY
	2-propanone	1993REB
	2-propanone	1999IM1
	propionitrile	1998IMR
	n-tetradecane	2001IMR
	n-tetradecane	2002IMR
Polytetrafluoroethylene		
	perfluorocarbon ether	1995TUM
	perfluorodecalin	1995TUM
	n-tetradecafluorohexane	1995TUM
	1,1,2-trichloro-1,2,2-trifluoroethane	1995TUM
Poly(vinyl acetate)		
	cyclopentane	1999BEY
	cyclopentane	2000BEY
	cyclopentene	1999BEY
	cyclopentene	2000BEY
Poly(*N*-vinylisobutyramide)		
	water	1997KUN
	water	1999KUN

Polymer (B)	Solvent (A)	Ref.
N-Vinylacetylamide/vinyl acetate copolymer	water	2003SET
N-Vinylformamide/vinyl acetate copolymer	water	2003SET
N-Vinylisobutyramide/N-vinylamine copolymer	water	2000KUN

3.3. Cloud-point and/or coexistence curves of quasiternary and/or quasiquaternary solutions

Polymer (B): polyethylene **2001TOR**
Characterization: M_n/kg.mol^{-1} = 43, M_w/kg.mol^{-1} = 105, M_z/kg.mol^{-1} = 190,
HDPE, DSM, Geleen, The Netherlands
Solvent (A): **n-hexane** **C$_6$H$_{14}$** **110-54-3**
Solvent (C): **1-octene** **C$_8$H$_{16}$** **111-66-0**

Type of data: cloud points

w_A	0.7385	0.7385	0.7385	0.7385	0.7385	0.7385	0.7385	0.7385	0.7385
w_B	0.1794	0.1794	0.1794	0.1794	0.1794	0.1794	0.1794	0.1794	0.1794
w_C	0.0821	0.0821	0.0821	0.0821	0.0821	0.0821	0.0821	0.0821	0.0821
T/K	441.7	451.9	462.3	472.6	462.4	472.6	483.0	493.1	503.3
P/MPa	0.92	1.10	1.33	1.55	2.47	3.44	5.22	6.45	7.61
	VLE	VLE	VLLE	VLLE	LLE	LLE	LLE	LLE	LLE

w_A	0.7385	0.7385	0.5932	0.5932	0.5932	0.5932	0.5932	0.5932	0.5932
w_B	0.1794	0.1794	0.1441	0.1441	0.1441	0.1441	0.1441	0.1441	0.1441
w_C	0.0821	0.0821	0.2627	0.2627	0.2627	0.2627	0.2627	0.2627	0.2627
T/K	513.5	523.2	452.6	462.4	472.6	483.0	472.6	483.	493.4
P/MPa	8.71	9.71	0.93	1.10	1.32	1.58	2.07	3.44	4.72
	LLE	LLE	VLE	VLE	VLLE	VLLE	LLE	LLE	LLE

w_A	0.5932	0.5932	0.5932	0.3210	0.3210	0.3210	0.3210	0.3210	0.3210
w_B	0.1441	0.1441	0.1441	0.1971	0.1971	0.1971	0.1971	0.1971	0.1971
w_C	0.2627	0.2627	0.2627	0.4819	0.4819	0.4819	0.4819	0.4819	0.4819
T/K	503.5	513.3	523.9	472.5	482.7	493.1	503.1	503.1	513.4
P/MPa	5.99	7.09	8.22	1.00	1.22	1.46	1.69	2.52	3.67
	LLE	LLE	LLE	VLE	VLE	VLE	VLLE	LLE	LLE

w_A	0.3210	0.3210	0.2156	0.2156	0.2156	0.2156	0.2156	0.2156	0.2156
w_B	0.1971	0.1971	0.1324	0.1324	0.1324	0.1324	0.1324	0.1324	0.1324
w_C	0.4819	0.4819	0.6520	0.6520	0.6520	0.6520	0.6520	0.6520	0.6520
T/K	523.3	534.0	482.6	493.1	503.2	513.4	503.2	513.2	523.1
P/MPa	4.72	5.83	1.02	1.25	1.40	1.66	1.73	2.85	3.84
	LLE	LLE	VLE	VLE	VLLE	VLLE	LLE	LLE	LLE

w_A	0.2156	0.2156
w_B	0.1324	0.1324
w_C	0.6520	0.6520
T/K	534.4	545.0
P/MPa	4.99	5.98
	LLE	LLE

Polymer (B):	poly(ethylene oxide)-b-poly(dimethylsiloxane)							**2003JI2**

	diblock copolymer							

Characterization: M_w/kg.mol^{-1} = 1.8, 77 mol% ethylene oxide, EO27-b-DMS8

Solvent (A):	**toluene**	C_7H_8						**108-88-3**

Polymer (C):   **poly(ethylene oxide)**

Characterization: M_w/kg.mol^{-1} = 35, Fluka AG, Buchs, Switzerland

Type of data: spinodal points

w_A	0.695	0.695	0.695	0.695	0.695	0.438	0.438	0.438	0.438
w_B	0.169	0.169	0.169	0.169	0.169	0.386	0.386	0.386	0.386
w_C	0.172	0.172	0.172	0.172	0.172	0.176	0.176	0.176	0.176
T/K	309.2	308.2	307.2	306.2	305.2	318.2	317.2	316.2	315.2
P/bar	588	441	342	112	23	617	483	364	159

w_A	0.438	0.255	0.255	0.255	0.255	0.255	0.209	0.209	0.209
w_B	0.386	0.558	0.558	0.558	0.558	0.558	0.626	0.626	0.626
w_C	0.176	0.187	0.187	0.187	0.187	0.187	0.165	0.165	0.165
T/K	314.2	328.2	327.2	326.2	325.2	324.2	333.2	332.2	331.2
P/bar	21	160	223	265	321	383	85	114	163

w_A	0.209	0.209	0.209	0.209	0.209
w_B	0.626	0.626	0.626	0.626	0.626
w_C	0.165	0.165	0.165	0.165	0.165
T/K	330.2	329.2	328.2	327.2	326.2
P/bar	212	258	322	369	408

Polymer (B):	polypropylene							**2000OLI**

Characterization: M_n/kg.mol^{-1} = 48.9, M_w/kg.mol^{-1} = 245, 92% isotactic,

                    Polibrasil Resinas S.A., Brazil

Solvent (A):	**n-butane**	C_4H_{10}						**106-97-8**
Solvent (C):	**toluene**	C_7H_8						**108-88-3**

Type of data: cloud points

w_A	0.8905	0.8905	0.8905	0.8905	0.8905	0.8508	0.8508	0.8508	0.8508
w_B	0.0996	0.0996	0.0996	0.0996	0.0996	0.0983	0.0983	0.0983	0.0983
w_C	0.0099	0.0099	0.0099	0.0099	0.0099	0.0509	0.0509	0.0509	0.0509
T/K	423.15	413.15	403.15	393.15	383.15	423.15	413.15	403.15	393.15
P/bar	108.4	92.3	75.3	56.4	36.9	99.0	84.1	66.9	49.1

w_A	0.8508	0.8012	0.8012	0.8012	0.8012
w_B	0.0983	0.0992	0.0992	0.0992	0.0992
w_C	0.0509	0.0996	0.0996	0.0996	0.0996
T/K	383.15	423.15	413.15	403.15	393.15
P/bar	30.8	87.2	68.8	51.5	33.2

Comments: Vapor-liquid equilibrium data are given in Chapter 2.

Polymer (B): **polypropylene** **2000OLI**
Characterization: M_n/kg.mol^{-1} = 48.9, M_w/kg.mol^{-1} = 245, 92% isotactic,
Polibrasil Resinas S.A., Brazil

Solvent (A): **1-butene** **C$_4$H$_8$** **106-98-9**
Solvent (C): **toluene** **C$_7$H$_8$** **108-88-3**

Type of data: cloud points

w_A	0.8910	0.8910	0.8910	0.8910	0.8910	0.8910	0.8910	0.8512	0.8512
w_B	0.0986	0.0986	0.0986	0.0986	0.0986	0.0986	0.0986	0.0974	0.0974
w_C	0.0104	0.0104	0.0104	0.0104	0.0104	0.0104	0.0104	0.0514	0.0514
T/K	423.15	413.15	403.15	393.15	383.15	373.15	368.15	423.15	413.15
P/bar	127.5	111.0	95.4	75.2	56.6	35.6	26.3	117.7	97.2

w_A	0.8512	0.8512	0.8512	0.8512	0.8022	0.8022	0.8022	0.8022	0.8022
w_B	0.0974	0.0974	0.0974	0.0974	0.0993	0.0993	0.0993	0.0993	0.0993
w_C	0.0514	0.0514	0.0514	0.0514	0.0985	0.0985	0.0985	0.0985	0.0985
T/K	403.15	393.15	383.15	373.15	423.15	413.15	403.15	393.15	383.15
P/bar	78.8	60.6	41.6	21.0	102.8	82.6	66.4	46.2	26.5

Comments: Vapor-liquid equilibrium data are given in Chapter 2.

Polymer (B): **polystyrene** **2000BEH**
Characterization: M_n/kg.mol^{-1} = 93, M_w/kg.mol^{-1} = 101.4, M_z/kg.mol^{-1} = 111.9,
BASF AG, Germany

Solvent (A): **cyclohexane** **C$_6$H$_{12}$** **110-82-7**
Polymer (C): **polyethylene**
Characterization: M_n/kg.mol^{-1} = 13, M_w/kg.mol^{-1} = 89, M_z/kg.mol^{-1} = 600,
LDPE, Stamylan, DSM, Geleen, The Netherlands

Type of data: cloud points (UCST behavior)

w_A	0.9131	0.9131	0.9131	0.9131	0.9131	0.9156	0.9156	0.9156	0.9156
w_B	0.0797	0.0797	0.0797	0.0797	0.0797	0.0718	0.0718	0.0718	0.0718
w_C	0.0072	0.0072	0.0072	0.0072	0.0072	0.0126	0.0126	0.0126	0.0126
T/K	401.29	403.75	403.96	405.45	404.75	435.05	433.85	433.31	434.92
P/bar	154.1	98.3	64.3	40.5	15.4	241.7	155.5	100.1	52.2

w_A	0.9156
w_B	0.0718
w_C	0.0126
T/K	436.37
P/bar	21.4

Type of data: cloud points (LCST behavior)

w_A	0.9131	0.9131	0.9131	0.9131	0.9131	0.9131	0.9131	0.9156	0.9156
w_B	0.0797	0.0797	0.0797	0.0797	0.0797	0.0797	0.0797	0.0718	0.0718
w_C	0.0072	0.0072	0.0072	0.0072	0.0072	0.0072	0.0072	0.0126	0.0126
T/K	495.51	501.21	506.00	510.65	516.02	520.97	527.38	465.74	470.30
P/bar	29.8	37.3	41.5	47.2	53.7	59.4	67.1	11.1	13.7

continued

continued

w_A	0.9156	0.9156	0.9156	0.9156	0.9156	
w_B	0.0718	0.0718	0.0718	0.0718	0.0718	
w_C	0.0126	0.0126	0.0126	0.0126	0.0126	
T/K	475.97	480.61	486.19	491.25	496.44	506.12
P/bar	15.5	19.3	24.2	30.1	35.5	46.0

Type of data: cloud points (1:1 polymer blend)

w_A	0.9302		w_B	0.0349		w_C	0.0349	were kept constant	
T/K	445.15	449.85	454.98	460.09	465.17	469.07	470.30	471.86	473.65
P/bar	228.1	136.6	91.1	71.2	61.0	60.6	54.1	56.8	56.3

T/K	476.70	485.80	496.17	506.27	516.76	527.38
P/bar	55.6	56.8	61.9	69.5	78.7	88.8

Polymer (B):	**polystyrene**	**2002FRI**
Characterization:	$M_n/kg.mol^{-1} = 1.8$, $M_w/kg.mol^{-1} = 1.93$,	
	synthesized in the laboratory	
Solvent (A):	**1,2-dichlorobenzene $C_6H_4Cl_2$**	**95-50-1**
Polymer (C):	**polybutadiene (deuterated)**	
Characterization:	$M_n/kg.mol^{-1} = 1.9$, $M_w/kg.mol^{-1} = 2.1$, 46% 1,4- (unspecified)	
	and 54% 1,2-content, synthesized in the laboratory	

Type of data: binodal points

Comments: The concentration of the polymer mixture is kept constant close to the critical
concentration of the polymer blend, i.e., $\varphi_C/\varphi_B = 0.477/0.523$.

φ_A	0.00	0.00	0.00	0.00	0.00	0.05	0.05	0.05	0.05
P/MPa	0.1	50	100	150	200	0.1	50	100	150
T/K	358.65	361.97	365.05	368.21	372.12	344.94	348.59	351.75	354.90

φ_A	0.05	0.20	0.20	0.20	0.20	0.20	0.20
P/MPa	200	0.1	50	100	150	180	200
T/K	358.05	292.47	294.49	298.10	299.45	302.84	301.60

Type of data: spinodal points

φ_A	0.00	0.00	0.00	0.00	0.00	0.05	0.05	0.05	0.05
P/MPa	0.1	50	100	150	200	0.1	50	100	150
T/K	355.750	359.483	362.259	365.682	369.246	343.625	346.447	349.770	353.040

φ_A	0.05	0.20	0.20	0.20	0.20	0.20	0.20
P/MPa	200	0.1	50	100	150	180	200
T/K	356.253	291.126	293.477	295.811	298.476	301.210	301.226

Polymer (B):	polystyrene			**2002FRI**

Characterization: M_n/kg.mol^{-1} = 1.7, M_w/kg.mol^{-1} = 1.82,
synthesized in the laboratory

Solvent (A):	1,2-dichlorobenzene $C_6H_4Cl_2$	**95-50-1**

Polymer (C):	polybutadiene (deuterated)

Characterization: M_n/kg.mol^{-1} = 2.1, M_w/kg.mol^{-1} = 2.3, 40% 1,4-*cis*-, 53% 1,4-*trans*-,
and 7% 1,2-content, synthesized in the laboratory

Type of data: binodal points

Comments: The concentration of the polymer mixture is kept constant close to the critical
concentration of the polymer blend, i.e., φ_C/φ_B = 0.43/0.57.

φ_A	0.00	0.00	0.00	0.00
P/MPa	0.1	50	100	150
T/K	338.31	342.15	343.10	349.72

Type of data: spinodal points

φ_A	0.00	0.00	0.00	0.00	0.05	0.05	0.05	0.05	0.05
P/MPa	0.1	50	100	150	0.1	50	100	150	200
T/K	336.534	340.520	344.352	349.116	331.779	335.483	339.267	342.280	345.696

φ_A	0.20	0.20	0.20	0.20	0.20
P/MPa	0.1	50	100	150	200
T/K	295.388	298.722	301.750	304.939	308.190

Polymer (B):	polystyrene			**1995LUS**

Characterization: M_n/kg.mol^{-1} = 23.6, M_w/kg.mol^{-1} = 25,
Pressure Chemical Company, Pittsburgh, PA

Solvent (A):	methylcyclopentane C_6H_{12}		**96-37-7**
Solvent (C):	dodecadeuteromethylcyclopentane	C_6D_{12}	**144120-51-4**

Type of data: cloud points

Comments: The mole fraction ratio of C_6D_{12}/C_6H_{12} = 0.943 is kept constant.

w_B	0.140	0.140	0.140	0.140	0.140
T/K	308.58	308.83	309.21	309.51	309.76
P/MPa	2.88	2.38	1.62	1.05	0.59

Polymer (B):	polystyrene			**1995LUS**

Characterization: M_n/kg.mol^{-1} = 100.3, M_w/kg.mol^{-1} = 106.3,
Pressure Chemical Company, Pittsburgh, PA

Solvent (A):	methylcyclopentane C_6H_{12}		**96-37-7**
Solvent (C):	dodecadeuteromethylcyclopentane	C_6D_{12}	**144120-51-4**

Type of data: cloud points

Comments: The mole fraction ratio of C_6D_{12}/C_6H_{12} = 0.943 is kept constant.

w_B	0.134	0.134	0.134	0.134
T/K	336.23	336.75	337.35	338.45
P/MPa	2.13	1.74	1.36	0.64

Polymer (B):	**polystyrene-d8**	**1995LUS**
Characterization:	M_n/kg.mol^{-1} = 25.4, M_w/kg.mol^{-1} = 26.9,	
	completely deuterated, Polymer Laboratories, Amherst, MA	
Solvent (A):	**methylcyclopentane** C_6H_{12}	**96-37-7**
Solvent (C):	**dodecadeuteromethylcyclopentane** C_6D_{12}	**144120-51-4**

Type of data: cloud points

Comments: The mole fraction ratio of C_6D_{12}/C_6H_{12} = 0.943 is kept constant.

w_B	0.137	0.137	0.137	0.137
T/K	307.74	308.04	308.32	308.67
P/MPa	2.25	1.64	1.09	0.42

Polymer (B):	**polystyrene**	**1995LUS**
Characterization:	M_n/kg.mol^{-1} = 7.3, M_w/kg.mol^{-1} = 8.0,	
	Pressure Chemical Company, Pittsburgh, PA	
Solvent (A):	**2-propanone** C_3H_6O	**67-64-1**
Solvent (C):	**1,1,1,3,3,3-hexadeutero-2-propanone** C_3D_6O	**666-52-4**

Type of data: cloud points

Comments: The mole fraction ratio of C_3D_6O/C_3H_6O = 0.546 is kept constant.

w_B	0.219	0.219	0.219	0.219	0.219	0.219
T/K	271.41	271.42	271.27	271.26	270.95	270.67
P/MPa	0.24	0.25	0.42	0.42	0.75	1.07

critical concentration: $w_{B, crit}$ = 0.210

Comments: The mole fraction ratio of C_3D_6O/C_3H_6O = 0.670 is kept constant.

w_B	0.211	0.211	0.211	0.211	0.211	0.211	0.211	0.211	0.211
T/K	275.75	275.78	275.72	275.55	275.35	275.10	274.90	274.76	274.54
P/MPa	0.06	0.06	0.07	0.18	0.38	0.63	0.84	1.02	1.27

critical concentration: $w_{B, crit}$ = 0.208

Polymer (B):	**polystyrene**	**1995LUS**
Characterization:	M_n/kg.mol^{-1} = 12.75, M_w/kg.mol^{-1} = 13.5,	
	Pressure Chemical Company, Pittsburgh, PA	
Solvent (A):	**2-propanone** C_3H_6O	**67-64-1**
Solvent (C):	**1,1,1,3,3,3-hexadeutero-2-propanone** C_3D_6O	**666-52-4**

Comments: The mole fraction ratio of C_3D_6O/C_3H_6O = 0.483 is kept constant.

Type of data: cloud points

w_B	0.205	0.205	0.205	0.205	0.205
T/K	302.77	303.49	304.93	306.86	307.34
P/MPa	1.89	1.63	1.27	0.63	0.37

continued

continued

Type of data: spinodal points

w_B	0.205	0.205	0.205	0.205	0.205
T/K	302.77	303.49	304.93	306.86	307.34
P/MPa	1.85	1.58	1.23	0.59	0.32

critical concentration: $w_{B, crit} = 0.200$

Comments: The mole fraction ratio of $C_3D_6O/C_3H_6O = 0.719$ is kept constant.

Type of data: cloud points

w_B	0.207	0.207	0.207	0.207	0.207	0.207
T/K	313.76	318.51	319.15	316.83	315.61	314.73
P/MPa	1.99	0.58	0.42	1.00	1.39	1.66

Type of data: spinodal points

w_B	0.207	0.207	0.207	0.207	0.207	0.207
T/K	313.76	318.51	319.15	316.83	315.61	314.73
P/MPa	1.95	0.55	0.38	0.97	1.34	1.62

critical concentration: $w_{B, crit} = 0.200$

Comments: The mole fraction ratio of $C_3D_6O/C_3H_6O = 0.952$ is kept constant.

Type of data: cloud points

w_B	0.206	0.206	0.206	0.206	0.206	0.206	0.206	0.206	0.206
T/K	330.82	330.82	331.84	332.91	333.91	335.96	338.37	340.36	343.28
P/MPa	2.11	2.09	1.86	1.67	1.53	1.15	0.90	0.61	0.34

w_B	0.206	0.206	0.206	0.206
T/K	373.09	375.93	378.69	384.87
P/MPa	0.59	0.77	0.99	1.57

Type of data: spinodal points

w_B	0.206	0.206	0.206	0.206	0.206	0.206	0.206	0.206	0.206
T/K	330.82	330.82	331.84	332.91	333.91	335.96	338.37	340.36	343.28
P/MPa	1.91	1.91	1.73	1.55	1.35	1.05	0.71	0.51	0.23

w_B	0.206	0.206	0.206	0.206
T/K	373.09	375.93	378.69	384.87
P/MPa	0.45	0.69	0.92	1.49

critical concentration: $w_{B, crit} = 0.200$

Polymer (B):	**polystyrene**		**1995LUS**
Characterization:	$M_n/kg.mol^{-1} = 12.75$, $M_w/kg.mol^{-1} = 13.5$,		
	Pressure Chemical Company, Pittsburgh, PA		
Solvent (A):	**2-propanone**	**C_3H_6O**	**67-64-1**
Solvent (C):	**1,1,1,3,3,3-hexadeutero-2-propanone**	**C_3D_6O**	**666-52-4**

continued

continued

Polymer (D):   **polystyrene**
Characterization: $M_n/\text{kg.mol}^{-1} = 21.45$, $M_w/\text{kg.mol}^{-1} = 22.1$,
                          Polymer Laboratories, Amherst, MA

Comments: The mole fraction ratio of $C_3D_6O/C_3H_6O = 0.501$ is kept constant.
                  The mole fraction ratio of polymer(D)/polymer(B) = 0.500 is kept constant.

Type of data: cloud points

w_{B+D}	0.210	0.210	0.210	0.210	0.210	0.210	0.210	0.210	0.210
T/K	335.83	340.15	343.69	347.37	351.31	353.49	358.40	362.30	368.51
P/MPa	2.55	2.10	1.81	1.62	1.51	1.47	1.52	1.63	1.90

w_{B+D}	0.210
T/K	378.34
P/MPa	2.58

Type of data: spinodal points

w_{B+D}	0.210	0.210	0.210	0.210	0.210	0.210	0.210	0.210	0.210
T/K	335.83	340.15	343.69	347.37	351.31	353.49	358.40	362.30	368.51
P/MPa	2.46	2.01	1.76	1.56	1.45	1.39	1.49	1.56	1.85

w_{B+D}	0.210
T/K	378.34
P/MPa	2.53

critical concentration: $w_{B+D, \text{crit}} = 0.210$

Comments: The mole fraction ratio of $C_3D_6O/C_3H_6O = 0.501$ is kept constant.
                  The mole fraction ratio of polymer(D)/polymer(B) = 0.501 is kept constant.

Type of data: cloud points

w_{B+D}	0.103	0.103	0.103	0.103	0.103	0.103	0.103	0.103	0.103
T/K	325.99	329.97	335.66	341.31	347.11	356.39	364.63	373.45	381.86
P/MPa	2.35	1.76	1.05	0.73	0.54	0.59	0.79	1.35	2.20

w_{B+D}	0.265	0.265	0.265	0.265	0.265	0.265	0.265	0.265
T/K	324.00	328.57	334.53	338.01	366.05	373.78	379.64	384.82
P/MPa	2.39	1.53	0.71	0.38	0.43	1.00	1.52	2.03

Type of data: spinodal points

w_{B+D}	0.103	0.103	0.103	0.103	0.103	0.103	0.103	0.103	0.103
T/K	325.99	329.97	335.66	341.31	347.11	356.39	364.63	373.45	381.86
P/MPa	1.33	1.00	0.46	0.01	0.35	0.27	0.50		1.63

w_{B+D}	0.265	0.265	0.265	0.265	0.265	0.265	0.265	0.265
T/K	324.00	328.57	334.53	338.01	366.05	373.78	379.64	384.82
P/MPa	2.22	1.33	0.62	0.13	0.22	0.84	1.42	1.77

critical concentration: $w_{B+D, \text{crit}} = 0.210$

continued

continued

Comments: The mole fraction ratio of $C_3D_6O/C_3H_6O = 0.501$ is kept constant.
                        The mole fraction ratio of polymer(D)/polymer(B) $= 0.502$ is kept constant.

Type of data: cloud points

w_{B+D}	0.126	0.126	0.126	0.126	0.126	0.126	0.126	0.126	0.126
T/K	331.06	335.10	339.51	343.35	350.12	354.21	356.36	364.60	372.36
P/MPa	2.11	1.65	1.29	1.05	0.67	0.81	0.92	1.15	1.63

w_{B+D}	0.126	0.145	0.145	0.145	0.145	0.145	0.145	0.145	0.145
T/K	380.49	331.95	335.12	337.71	341.63	345.55	351.27	356.14	365.16
P/MPa	2.36	2.51	2.07	1.75	1.43	1.22	1.07	1.13	1.37

w_{B+D}	0.145	0.145	0.167	0.167	0.167	0.167	0.167	0.167	0.167
T/K	372.07	379.54	331.56	335.91	341.02	345.32	350.63	353.45	356.39
P/MPa	1.80	2.40	2.58	1.96	1.48	1.22	1.08	1.07	1.13

w_{B+D}	0.167	0.167	0.167
T/K	365.41	372.96	360.01
P/MPa	1.39	1.66	2.43

Type of data: spinodal points

w_{B+D}	0.126	0.126	0.126	0.126	0.126	0.126	0.126	0.126	0.167
T/K	331.06	335.10	339.51	343.35	350.12	364.60	372.36	380.49	331.56
P/MPa	1.81	1.34	0.67	0.45	0.36	0.52	1.26	1.66	2.21

w_{B+D}	0.167	0.167	0.167	0.167	0.167	0.167	0.167	0.167	0.167
T/K	335.91	341.02	345.32	350.63	353.45	356.39	365.41	372.96	360.01
P/MPa	1.72	1.36	1.13	0.96	0.96	1.00	1.32	1.75	2.32

critical concentration: $w_{B+D, crit} = 0.210$

Comments: The mole fraction ratio of $C_3D_6O/C_3H_6O = 0.500$ is kept constant.
                        The mole fraction ratio of polymer(D)/polymer(B) $= 0.503$ is kept constant.

Type of data: cloud points

w_{B+D}	0.245	0.245	0.245	0.245	0.245	0.245	0.245	0.245
T/K	332.25	336.80	343.55	349.28	358.29	363.91	372.41	379.29
P/MPa	2.53	1.93	1.34	1.09	1.06	1.22	1.69	2.25

Type of data: spinodal points

w_{B+D}	0.245	0.245	0.245	0.245	0.245	0.245	0.245	0.245
T/K	332.25	336.80	343.55	349.28	358.29	363.91	372.41	379.29
P/MPa	2.45	1.82	1.27	0.99	0.97	1.07	1.62	2.00

critical concentration: $w_{B+D, crit} = 0.210$

Comments: The mole fraction ratio of $C_3D_6O/C_3H_6O = 0.501$ is kept constant.
                        The mole fraction ratio of polymer(D)/polymer(B) $= 0.505$ is kept constant.

Type of data: cloud points

w_{B+D}	0.186	0.186	0.186	0.186	0.186	0.186	0.186	0.186	0.186
T/K	334.35	337.99	341.51	346.27	351.43	354.56	358.37	365.30	372.69
P/MPa	2.27	1.85	1.55	1.27	1.14	1.12	1.16	1.41	1.83

continued

continued

w_{B+D}	0.186	0.236	0.236	0.236	0.236	0.236	0.236	0.236	0.236
T/K	379.72	331.15	335.67	338.49	342.30	347.37	351.79	354.86	358.45
P/MPa	2.40	3.17	2.48	2.16	1.82	1.53	1.40	1.36	1.39

w_{B+D}	0.236	0.236	0.236
T/K	365.87	371.73	376.61
P/MPa	1.63	1.97	2.32

Type of data: spinodal points

w_{B+D}	0.186	0.186	0.186	0.186	0.186	0.186	0.186	0.186	0.186
T/K	334.35	337.99	341.51	346.27	351.43	354.56	358.37	365.30	372.69
P/MPa	2.09	1.72	1.45	1.19	1.05	1.05	1.11	1.35	1.79

w_{B+D}	0.186	0.236	0.236	0.236	0.236	0.236	0.236	0.236	0.236
T/K	379.72	331.15	335.67	338.49	342.30	347.37	351.79	354.86	358.45
P/MPa	2.33	3.02	2.34	2.03	1.71	1.43	1.29	1.27	1.30

w_{B+D}	0.236	0.236	0.236
T/K	365.87	371.73	376.61
P/MPa	1.57	1.89	2.27

critical concentration: $w_{B+D, crit} = 0.210$

Polymer (B):	**polystyrene**	**1996LUS**
Characterization:	M_n/kg.mol^{-1} = 23.6, M_w/kg.mol^{-1} = 25.0,	
	Pressure Chemical Company, Pittsburgh, PA	
Solvent (A):	**propionitrile** C_3H_5N	**107-12-0**
Polymer (C):	**polystyrene-d8**	
Characterization:	M_n/kg.mol^{-1} = 25.4, M_w/kg.mol^{-1} = 26.7,	
	completely deuterated, Polymer Laboratories, Amherst, MA	

Comments: The mole fraction ratio of polymer(C)/polymer(B) = 0.50 is kept constant.

Type of data: cloud points

w_{B+C}	0.1300	0.1300	0.1300	0.1300	0.1300	0.1300	0.1300	0.1300	0.1300
T/K	330.309	330.309	330.305	330.305	337.064	337.052	343.871	343.869	351.509
P/MPa	5.017	5.009	4.991	4.996	3.032	3.036	1.604	1.583	0.396

w_{B+C}	0.1300	0.1300	0.1300	0.1300	0.1300	0.1300	0.1300	0.1300	0.1300
T/K	351.391	411.479	411.466	411.477	419.828	419.828	419.828	419.828	419.814
P/MPa	0.402	0.563	0.620	0.699	1.439	1.452	1.476	1.445	1.432

w_{B+C}	0.1300	0.1300	0.1300	0.1300	0.1300	0.1300	0.1300	0.1300	0.1300
T/K	419.814	428.203	428.197	428.197	436.813	436.813	436.817	436.817	436.817
P/MPa	1.436	2.269	2.160	2.165	3.198	3.201	3.195	3.224	3.202

w_{B+C}	0.1300	0.1300	0.1300	0.1300	0.1300	0.1300	0.1300	0.1300	0.1300
T/K	436.817	436.817	445.655	445.650	445.650	445.650	454.681	454.681	454.681
P/MPa	3.191	3.198	4.182	4.202	4.204	4.202	5.246	5.252	5.261

continued

continued

w_{B+C}	0.1300	0.1300	0.1300	0.1599	0.1599	0.1599	0.1599	0.1599	0.1599
T/K	454.658	454.658	454.658	330.291	330.290	337.001	336.998	343.823	343.821
P/MPa	5.273	5.271	5.270	4.938	4.934	3.105	3.018	1.638	1.547

w_{B+C}	0.1599	0.1599	0.1599	0.1599	0.1599	0.1599	0.1599	0.1599	0.1599
T/K	350.797	350.796	411.521	411.522	419.731	419.733	428.235	428.234	436.879
P/MPa	0.509	0.483	0.684	0.689	1.479	1.520	2.356	2.305	3.254

w_{B+C}	0.1599	0.1599	0.1599	0.1599	0.1599	0.1599	0.1599	0.1599	0.1900
T/K	436.875	436.875	436.875	445.699	445.699	445.698	454.664	454.674	330.206
P/MPa	3.138	3.158	3.274	4.353	4.267	4.253	5.276	5.376	5.257

w_{B+C}	0.1900	0.1900	0.1900	0.1900	0.1900	0.1900	0.1900	0.1900	0.1900
T/K	330.206	337.173	337.133	343.716	343.716	351.032	351.030	411.488	411.485
P/MPa	5.453	3.364	3.257	1.992	1.923	0.864	0.764	0.904	1.073

w_{B+C}	0.1900	0.1900	0.1900	0.1900	0.1900	0.1900	0.1900	0.1900	0.1900
T/K	420.694	420.679	428.290	428.292	436.890	436.888	445.686	445.684	454.553
P/MPa	1.878	1.809	2.509	2.585	3.459	3.464	4.439	4.503	5.515

w_{B+C}	0.1900	0.2198	0.2198	0.2198	0.2198	0.2198	0.2198	0.2198	0.2198
T/K	454.575	333.385	333.385	333.415	333.397	340.168	340.181	347.027	345.281
P/MPa	5.559	4.889	4.836	4.657	4.684	3.069	2.770	1.419	1.366

w_{B+C}	0.2198	0.2198	0.2198	0.2198	0.2198	0.2198	0.2198	0.2198	0.2198
T/K	350.852	350.851	411.434	411.433	411.432	419.661	419.655	428.097	428.086
P/MPa	0.375	0.302	0.691	0.879	0.659	1.359	1.458	2.227	2.069

w_{B+C}	0.2198	0.2198	0.2198	0.2198	0.2198	0.2198	0.2499	0.2499	0.2499
T/K	436.660	436.655	445.476	445.468	454.374	454.380	334.038	333.672	333.672
P/MPa	3.172	3.277	4.378	4.166	5.327	5.342	4.333	4.341	4.337

w_{B+C}	0.2499	0.2499	0.2499	0.2499	0.2499	0.2499	0.2499	0.2499	0.2499
T/K	340.277	340.222	346.716	346.689	352.653	352.728	416.714	423.091	423.027
P/MPa	2.553	2.529	1.185	1.177	0.197	0.177	0.685	1.251	1.251

w_{B+C}	0.2499	0.2499	0.2499	0.2499	0.2499	0.2499	0.2499	0.2499	
T/K	431.439	431.438	440.118	440.116	449.061	449.009	457.927	458.073	
P/MPa	2.096	2.108	2.996	3.006	4.015	4.080	5.175	5.118	

Type of data: spinodal points

w_{B+C}	0.1300	0.1300	0.1300	0.1300	0.1300	0.1300	0.1300	0.1300	0.1300
T/K	330.309	330.309	330.305	330.305	337.064	337.052	343.871	343.869	351.509
P/MPa	4.421	4.447	4.381	4.311	2.206	2.349	0.810	0.937	−0.456

w_{B+C}	0.1300	0.1300	0.1300	0.1300	0.1300	0.1300	0.1300	0.1300	0.1300
T/K	351.391	411.479	411.466	411.477	419.828	419.828	419.828	419.828	419.814
P/MPa	−0.319	−0.041	0.062	0.178	0.783	0.831	0.398	0.942	0.974

w_{B+C}	0.1300	0.1300	0.1300	0.1300	0.1300	0.1300	0.1300	0.1300	0.1300
T/K	419.814	428.203	428.197	428.197	436.813	436.813	436.817	436.817	436.817
P/MPa	0.918	1.773	1.676	1.565	2.829	2.776	2.804	2.698	2.698

w_{B+C}	0.1300	0.1300	0.1300	0.1300	0.1300	0.1300	0.1300	0.1300	0.1300
T/K	436.817	436.817	445.655	445.650	445.650	445.650	454.681	454.681	454.681
P/MPa	2.879	2.755	3.876	3.854	3.821	3.862	4.946	4.7900	4.641

continued

continued

w_{B+C}	0.1300	0.1300	0.1300	0.1599	0.1599	0.1599	0.1599	0.1599	0.1599
T/K	454.658	454.658	454.658	330.291	330.290	337.001	336.998	343.823	343.821
P/MPa	4.655	4.709	4.714	4.761	4.805	2.576	2.859	1.232	1.427

w_{B+C}	0.1599	0.1599	0.1599	0.1599	0.1599	0.1599	0.1599	0.1599	0.1599
T/K	350.797	350.796	411.521	411.522	419.731	419.733	428.235	428.234	436.879
P/MPa	0.233	0.287	0.580	0.549	1.327	1.221	1.899	1.947	3.036

w_{B+C}	0.1599	0.1599	0.1599	0.1599	0.1599	0.1599	0.1599	0.1599	0.1900
T/K	436.875	436.875	436.875	445.699	445.699	445.698	454.664	454.674	330.206
P/MPa	3.138	3.138	2.898	3.778	3.9990	4.010	5.199	5.071	4.926

w_{B+C}	0.1900	0.1900	0.1900	0.1900	0.1900	0.1900	0.1900	0.1900	0.1900
T/K	330.206	337.173	337.133	343.716	343.716	351.032	351.030	411.488	411.485
P/MPa	4.951	2.582	2.930	1.236	1.452	0.178	0.358	0.4800	0.329

w_{B+C}	0.1900	0.1900	0.1900	0.1900	0.1900	0.1900	0.1900	0.1900	0.1900
T/K	420.694	420.679	428.290	428.292	436.890	436.888	445.686	445.684	454.553
P/MPa	1.438	1.275	2.104	2.1230	3.081	3.001	4.0440	4.0080	4.964

w_{B+C}	0.1900	0.2198	0.2198	0.2198	0.2198	0.2198	0.2198	0.2198	0.2198
T/K	454.575	333.385	333.385	333.415	333.397	340.168	340.181	347.027	345.281
P/MPa	4.971	4.608	4.606	4.545	4.474	2.092	2.664	1.095	1.222

w_{B+C}	0.2198	0.2198	0.2198	0.2198	0.2198	0.2198	0.2198	0.2198	0.2198
T/K	350.852	350.851	411.434	411.433	411.432	419.661	419.655	428.097	428.086
P/MPa	0.084	0.096	0.412	0.126	0.356	1.266	1.169	2.041	2.106

w_{B+C}	0.2198	0.2198	0.2198	0.2198	0.2198	0.2198	0.2499	0.2499	0.2499
T/K	436.660	436.655	445.476	445.468	454.374	454.380	334.038	333.672	333.672
P/MPa	2.933	2.913	3.887	3.986	5.073	5.005	3.740	3.883	3.950

w_{B+C}	0.2499	0.2499	0.2499	0.2499	0.2499	0.2499	0.2499	0.2499	0.2499
T/K	340.277	340.222	346.716	346.689	352.653	352.728	416.714	423.091	423.027
P/MPa	−0.180	1.919	0.279	0.462	−0.407	−0.254	0.262	0.938	1.007

w_{B+C}	0.2499	0.2499	0.2499	0.2499	0.2499	0.2499	0.2499	0.2499	
T/K	431.439	431.438	440.118	440.116	449.061	449.009	457.927	458.073	
P/MPa	1.719	1.550	2.780	2.629	3.910	3.334	4.484	4.760	

Polymer (B):	**poly(vinyl chloride)**		**1985GE2**
Characterization:	M_n/kg.mol^{-1} = 16.7, M_w/kg.mol^{-1} = 20, fractionated in the laboratory		
Solvent (A):	**tetrahydrofuran**	**C$_4$H$_8$O**	**109-99-9**
Solvent (C):	**water**	**H$_2$O**	**7732-18-5**

Type of data: cloud points (LCST behavior)

w_A	0.8145	0.8145	0.8145	0.8094	0.8094	0.8094	0.8065	0.8065	0.8065
w_B	0.0795	0.0795	0.0795	0.0790	0.0790	0.0790	0.0787	0.0787	0.0787
w_C	0.1060	0.1060	0.1060	0.1116	0.1116	0.1116	0.1148	0.1148	0.1148
T/K	283.85	286.85	290.25	295.85	299.05	303.45	303.55	306.75	311.55
P/bar	1	500	1000	1	500	1000	1	500	1000

Polymer (B):   **poly(vinyl chloride)**   **1985GE2**
Characterization: $M_n/\text{kg.mol}^{-1} = 31.4$, $M_w/\text{kg.mol}^{-1} = 37$, fractionated in the laboratory
Solvent (A):   **tetrahydrofuran**   **C$_4$H$_8$O**   **109-99-9**
Solvent (C):   **water**   **H$_2$O**   **7732-18-5**

Type of data: cloud points (LCST behavior)

w_A	0.8194	0.8194	0.8194	0.8151	0.8151	0.8151	0.8097	0.8097	0.8097
w_B	0.0800	0.0800	0.0800	0.0796	0.0796	0.0796	0.0789	0.0789	0.0789
w_C	0.1006	0.1006	0.1006	0.1053	0.1053	0.1053	0.1114	0.1114	0.1114
T/K	282.85	286.95	291.35	293.25	298.25	302.55	307.15	310.25	315.65
P/bar	1	500	1000	1	500	1000	1	500	1000

Polymer (B):   **poly(vinyl chloride)**   **1985GE2**
Characterization: $M_n/\text{kg.mol}^{-1} = 58.8$, $M_w/\text{kg.mol}^{-1} = 70$, fractionated in the laboratory
Solvent (A):   **tetrahydrofuran**   **C$_4$H$_8$O**   **109-99-9**
Solvent (C):   **water**   **H$_2$O**   **7732-18-5**

Type of data: cloud points (LCST behavior)

w_A	0.8546	0.8546	0.8546	0.8498	0.8498	0.8498	0.8547	0.8547	0.8547
w_B	0.0200	0.0200	0.0200	0.0200	0.0200	0.0200	0.0199	0.0199	0.0199
w_C	0.1254	0.1254	0.1254	0.1302	0.1302	0.1302	0.1254	0.1254	0.1254
T/K	279.65	283.95	292.65	293.65	302.05	309.65	309.65	319.15	327.15
P/bar	1	500	1000	1	500	1000	1	500	1000

w_A	0.8439	0.8439	0.8439	0.8389	0.8389	0.8389	0.8334	0.8334	0.8334
w_B	0.0407	0.0407	0.0407	0.0405	0.0405	0.0405	0.0403	0.0403	0.0403
w_C	0.1154	0.1154	0.1154	0.1206	0.1206	0.1206	0.1263	0.1263	0.1263
T/K	284.05	289.25	296.05	299.25	305.25	311.55	310.95	319.75	326.15
P/bar	1	500	1000	1	500	1000	1	500	1000

w_A	0.8338	0.8338	0.8338	0.8292	0.8292	0.8292	0.8245	0.8245	0.8245
w_B	0.0613	0.0613	0.0613	0.0609	0.0609	0.0609	0.0607	0.0607	0.0607
w_C	0.1049	0.1049	0.1049	0.1099	0.1099	0.1099	0.1148	0.1148	0.1148
T/K	285.25	287.85	293.15	296.75	301.65	306.05	308.75	314.35	320.45
P/bar	1	500	1000	1	500	1000	1	500	1000

w_A	0.8244	0.8244	0.8244	0.8200	0.8200	0.8200	0.8150	0.8150	0.8150
w_B	0.0952	0.0952	0.0952	0.0999	0.0999	0.0999	0.1056	0.1056	0.1056
w_C	0.0952	0.0952	0.0952	0.0999	0.0999	0.0999	0.1056	0.1056	0.1056
T/K	275.45	277.55	280.35	287.95	290.15	294.45	299.85	304.65	308.45
P/bar	1	500	1000	1	500	1000	1	500	1000

3.4. Table of ternary or quaternary systems where data were published only in graphical form as phase diagrams or related figures

Polymer (B)	Second/third/fourth component	Ref.
Hydroxypropylcellulose		
	water and potassium chloride	2001KUN
	water and potassium iodide	2001KUN
	water and potassium isocyanate	2001KUN
	water and potassium sulfate	2001KUN
Penta(ethylene glycol) monoheptyl ether		
	n-dodecane and water	1992SAS
Polybutadiene		
	1,2-dichlorobenzene and polystyrene	2002FRI
Poly(dimethylsiloxane)		
	n-pentane and polyethylene	2002KIR
Polyethylene		
	n-hexane and nitrogen	1990KEN
	n-hexane and n-octane	1989HAE
	n-pentane and poly(dimethylsiloxane)	2002KIR
Poly(N-isopropylacrylamide)		
	water and magnesium chloride	1998SUW
	water and potassium chloride	1998SUW
	water and potassium iodide	1998SUW
	water and potassium iodide	1999KUN
	water and potassium sulfate	1998SUW
	water and sodium isocyanate	1998SUW
	water and sodium isocyanate	1999KUN
Polypeptide		
	potassium chloride and water	2003YAM
	potassium iodide and water	2003YAM
	potassium sulfate and water	2003YAM
	sodium chloride and water	2003YAM

Polymer (B)	Second/third/fourth component	Ref.
Polystyrene		
	acetone and diethyl ether	1976WOL
	cyclohexane and water	1991DEE
	1,2-dichlorobenzene and polybutadiene	2002FRI
	n-heptane and methylcyclohexane	1999IM2
Poly(*N*-vinylisobutyramide)		
	water and magnesium chloride	1998SUW
	water and potassium chloride	1998SUW
	water and potassium chloride	1999KUN
	water and potassium iodide	1998SUW
	water and potassium iodide	1999KUN
	water and potassium isocyanate	1998SUW
	water and potassium sulfate	1998SUW
	water and sodium iodide	1998SUW
	water and sodium isocyanate	1998SUW
	water and sodium isocyanate	1999KUN
	water and sodium sulfate	1999KUN
N-Vinylformamide/vinyl acetate copolymer		
	water and potassium chloride	2003SET
	water and potassium iodide	2003SET
	water and potassium sulfate	2003SET

3.5. Lower critical (LCST) and/or upper critical (UCST) solution temperatures at elevated pressures

Note: LCST values are measured in many cases at the vapor pressure of the solvent, i.e., a high pressure that is often near the critical pressure of the pure solvent. Nevertheless, this table presents only LCST data for polymer solutions where a pressure dependence was explicitly investigated.

Polymer	M_n kg/mol	M_w kg/mol	Solvent	$P/$ MPa	UCST/ K	LCST/ K	Ref.
Polystyrene							
	583	670	*tert*-butyl acetate	0.1	270.8	393.7	1976SA2
	583	670	*tert*-butyl acetate	1	269.7	400.2	1976SA2
	583	670	*tert*-butyl acetate	2	268.6	407.1	1976SA2
	583	670	*tert*-butyl acetate	3	267.5	413.7	1976SA2
	583	670	*tert*-butyl acetate	4	266.4	420.2	1976SA2
	583	670	*tert*-butyl acetate	5	265.5		1976SA2
	1320	1450	*tert*-butyl acetate	0.1	276.7	387.1	1976SA2
	1320	1450	*tert*-butyl acetate	1	275.3	393.8	1976SA2
	1320	1450	*tert*-butyl acetate	2	273.8	400.6	1976SA2
	1320	1450	*tert*-butyl acetate	3	272.4	407.3	1976SA2
	1320	1450	*tert*-butyl acetate	4	271.1	414.0	1976SA2
	1320	1450	*tert*-butyl acetate	5	270.0		1976SA2
	2455	2700	*tert*-butyl acetate	0.1	280.7	382.5	1976SA2
	2455	2700	*tert*-butyl acetate	1	278.8	388.8	1976SA2
	2455	2700	*tert*-butyl acetate	2	277.2		1976SA2
	2455	2700	*tert*-butyl acetate	3	275.8		1976SA2
	2455	2700	*tert*-butyl acetate	4	274.4		1976SA2
	2455	2700	*tert*-butyl acetate	5	273.3		1976SA2
	3140	3450	*tert*-butyl acetate	0.1	281.8	381.2	1976SA2
	3140	3450	*tert*-butyl acetate	1	280.0	388.1	1976SA2
	3140	3450	*tert*-butyl acetate	2	278.3	394.9	1976SA2
	3140	3450	*tert*-butyl acetate	3	276.8	401.7	1976SA2
	3140	3450	*tert*-butyl acetate	4	275.4	408.4	1976SA2
	3140	3450	*tert*-butyl acetate	5	274.2		1976SA2
	infinite		*tert*-butyl acetate	0.1	296.1	359.3	1976SA2
	infinite		*tert*-butyl acetate	1	293.1	366.3	1976SA2
	infinite		*tert*-butyl acetate	2	290.7	372.3	1976SA2
	infinite		*tert*-butyl acetate	3	288.5	380.4	1976SA2
	infinite		*tert*-butyl acetate	4	286.8	387.0	1976SA2
	infinite		*tert*-butyl acetate	5	285.5		1976SA2
	34.9	37.0	cyclohexane	0.1	285.41		1975SAE
	34.9	37.0	cyclohexane	1	285.43		1975SAE
	34.9	37.0	cyclohexane	2	285.46		1975SAE
	34.9	37.0	cyclohexane	3	285.49		1975SAE

Polymer	M_n kg/mol	M_w kg/mol	Solvent	P/ MPa	UCST/ K	LCST/ K	Ref.
Polystyrene (*continued*)							
	34.9	37.0	cyclohexane	4	285.53		1975SAE
	34.9	37.0	cyclohexane	5	285.56		1975SAE
	104	110	cyclohexane	0.1	294.38		1975SAE
	104	110	cyclohexane	1	295.36		1975SAE
	104	110	cyclohexane	2	294.35		1975SAE
	104	110	cyclohexane	3	294.36		1975SAE
	104	110	cyclohexane	4	294.35		1975SAE
	104	110	cyclohexane	5	294.35		1975SAE
	545.5	600	cyclohexane	0.1	301.34		1981WOL
	545.5	600	cyclohexane	3.7	301.11		1981WOL
	545.5	600	cyclohexane	8.8	300.95		1981WOL
	545.5	600	cyclohexane	15.2	300.96		1981WOL
	545.5	600	cyclohexane	18.7	301.12		1981WOL
	545.5	600	cyclohexane	26.0	301.43		1981WOL
	545.5	600	cyclohexane	30.0	301.68		1981WOL
	583	670	cyclohexane	0.1	300.97		1975SAE
	583	670	cyclohexane	1	300.91		1975SAE
	583	670	cyclohexane	2	300.86		1975SAE
	583	670	cyclohexane	3	300.81		1975SAE
	583	670	cyclohexane	4	300.79		1975SAE
	583	670	cyclohexane	5	300.75		1975SAE
	1320	1450	cyclohexane	0.1	303.27		1975SAE
	1320	1450	cyclohexane	1	303.21		1975SAE
	1320	1450	cyclohexane	2	303.15		1975SAE
	1320	1450	cyclohexane	3	303.10		1975SAE
	1320	1450	cyclohexane	4	303.04		1975SAE
	1320	1450	cyclohexane	5	302.98		1975SAE
	infinite		cyclohexane	0.1	306.51		2003SIP
	infinite		cyclohexane	2	306.36		2003SIP
	infinite		cyclohexane	4	306.22		2003SIP
	infinite		cyclohexane	6	306.12		2003SIP
	infinite		cyclohexane	10	306.07		2003SIP
	infinite		cyclohexane	15	306.01		2003SIP
	infinite		cyclohexane	20	306.08		2003SIP
	infinite		cyclohexane	30	306.63		2003SIP
	infinite		cyclohexane	40	307.65		2003SIP
	infinite		cyclohexane	50	309.39		2003SIP
	583	670	cyclopentane	1	286.0		1978ISH
	583	670	cyclopentane	10	283.0		1978ISH
	583	670	cyclopentane	20	280.6		1978ISH
	583	670	cyclopentane	40	277.8		1978ISH
	583	670	cyclopentanc	60	276.3		1978ISH
	583	670	cyclopentane	70	276.0		1978ISH
	1540	2000	cyclopentane	1	291.0		1978ISH
	1540	2000	cyclopentane	10	287.3		1978ISH
	1540	2000	cyclopentane	20	284.9		1978ISH

Polymer	M_n kg/mol	M_w kg/mol	Solvent	$P/$ MPa	UCST/ K	LCST/ K	Ref.
Polystyrene (*continued*)							
	1540	2000	cyclopentane	40	282.0		1978ISH
	1540	2000	cyclopentane	60	280.3		1978ISH
	1540	2000	cyclopentane	70	280.1		1978ISH
	19.2	20.4	diethyl ether	0.1	228.4	314.5	1976SA2
	19.2	20.4	diethyl ether	1	226.2	320.8	1976SA2
	19.2	20.4	diethyl ether	2	224.0	327.1	1976SA2
	19.2	20.4	diethyl ether	3	222.2	333.2	1976SA2
	19.2	20.4	diethyl ether	4	220.6	339.4	1976SA2
	19.2	20.4	diethyl ether	5	219.2		1976SA2
	11.3	11.6	ethyl formate	1	273.3	443.8	1997IMR
	11.3	11.6	ethyl formate	2	273.0	449.4	1997IMR
	11.3	11.6	ethyl formate	3	272.6	454.7	1997IMR
	11.3	11.6	ethyl formate	4	272.2	460.7	1997IMR
	11.3	11.6	ethyl formate	5	271.9	465.6	1997IMR
	11.3	11.6	ethyl formate	6	271.5	471.0	1997IMR
	11.3	11.6	ethyl formate	7	271.2	476.4	1997IMR
	11.3	11.6	ethyl formate	8	270.8	481.8	1997IMR
	21.4	22.0	ethyl formate	1	295.4	425.3	1997IMR
	21.4	22.0	ethyl formate	2	294.8	430.2	1997IMR
	21.4	22.0	ethyl formate	3	294.2	435.2	1997IMR
	21.4	22.0	ethyl formate	4	293.5	440.1	1997IMR
	21.4	22.0	ethyl formate	5	292.9	445.0	1997IMR
	21.4	22.0	ethyl formate	6	292.3	450.0	1997IMR
	21.4	22.0	ethyl formate	7	291.6	454.9	1997IMR
	21.4	22.0	ethyl formate	8	291.0	459.8	1997IMR
	64.1	66.0	ethyl formate	1	342.4	384.4	1997IMR
	64.1	66.0	ethyl formate	2	338.8	392.4	1997IMR
	64.1	66.0	ethyl formate	3	335.8	399.9	1997IMR
	64.1	66.0	ethyl formate	4	333.4	407.2	1997IMR
	64.1	66.0	ethyl formate	5	331.2	413.8	1997IMR
	64.1	66.0	ethyl formate	6	329.3	419.7	1997IMR
	64.1	66.0	ethyl formate	7	327.6	425.0	1997IMR
	64.1	66.0	ethyl formate	8	326.0	429.6	1997IMR
	377	400	ethyl formate	6	379.6	380.8	1997IMR
	377	400	ethyl formate	7	365.7	395.3	1997IMR
	377	400	ethyl formate	8	361.2	400.1	1997IMR
	8.5	9.0	methylcyclohexane	4.1	329.420		1994VA1
	8.5	9.0	methylcyclohexane	10.04	324.695		1994VA1
	8.5	9.0	methylcyclohexane	16.27	321.734		1994VA1
	8.5	9.0	methylcyclohexane	23.28	319.850		1994VA1
	8.5	9.0	methylcyclohexane	43.32	320.132		1994VA1
	8.5	9.0	methylcyclohexane	49.92	321.043		1994VA1
	8.5	9.0	methylcyclohexane	59.99	324.167		1994VA1
	8.5	9.0	methylcyclohexane	72.22	329.115		1994VA1
	8.5	9.0	methylcyclohexane	90.76	338.472		1994VA1
	16.5	17.5	methylcyclohexane	5.19	296.767		1994VA1

Polymer	M_n kg/mol	M_w kg/mol	Solvent	P/ MPa	UCST/ K	LCST/ K	Ref.
Polystyrene (*continued*)							
	16.5	17.5	methylcyclohexane	5.33	296.781		1994VA1
	16.5	17.5	methylcyclohexane	8.65	296.057		1994VA1
	16.5	17.5	methylcyclohexane	8.71	296.055		1994VA1
	16.5	17.5	methylcyclohexane	13.94	295.406		1994VA1
	16.5	17.5	methylcyclohexane	14.15	295.440		1994VA1
	16.5	17.5	methylcyclohexane	15.96	295.206		1994VA1
	16.5	17.5	methylcyclohexane	16.33	294.272		1994VA1
	16.5	17.5	methylcyclohexane	17.79	294.999		1994VA1
	16.5	17.5	methylcyclohexane	18.31	294.997		1994VA1
	16.5	17.5	methylcyclohexane	21.25	294.689		1994VA1
	16.5	17.5	methylcyclohexane	21.57	294.727		1994VA1
	16.5	17.5	methylcyclohexane	30.67	294.141		1994VA1
	16.5	17.5	methylcyclohexane	30.97	294.160		1994VA1
	16.5	17.5	methylcyclohexane	38.65	293.890		1994VA1
	16.5	17.5	methylcyclohexane	38.97	293.905		1994VA1
	16.5	17.5	methylcyclohexane	62.69	293.904		1994VA1
	16.5	17.5	methylcyclohexane	62.96	293.840		1994VA1
	16.5	17.5	methylcyclohexane	73.56	294.156		1994VA1
	16.5	17.5	methylcyclohexane	73.91	294.226		1994VA1
	16.5	17.5	methylcyclohexane	84.23	294.474		1994VA1
	16.5	17.5	methylcyclohexane	84.66	294.505		1994VA1
	25.9	28.5	methylcyclohexane	5.65	303.870		1994VA1
	25.9	28.5	methylcyclohexane	9.92	302.971		1994VA1
	25.9	28.5	methylcyclohexane	14.28	302.283		1994VA1
	25.9	28.5	methylcyclohexane	18.24	301.699		1994VA1
	25.9	28.5	methylcyclohexane	22.89	301.293		1994VA1
	25.9	28.5	methylcyclohexane	34.46	300.417		1994VA1
	25.9	28.5	methylcyclohexane	48.00	299.863		1994VA1
	25.9	28.5	methylcyclohexane	72.87	300.082		1994VA1
	25.9	28.5	methylcyclohexane	85.40	300.543		1994VA1
	47.2	50.0	methylcyclohexane	7.04	311.383		1994VA1
	47.2	50.0	methylcyclohexane	9.72	310.795		1994VA1
	47.2	50.0	methylcyclohexane	15.47	309.725		1994VA1
	47.2	50.0	methylcyclohexane	23.40	308.698		1994VA1
	47.2	50.0	methylcyclohexane	31.02	307.844		1994VA1
	47.2	50.0	methylcyclohexane	42.10	307.002		1994VA1
	47.2	50.0	methylcyclohexane	51.19	306.645		1994VA1
	47.2	50.0	methylcyclohexane	65.60	306.515		1994VA1
	47.2	50.0	methylcyclohexane	73.89	306.576		1994VA1
	47.2	50.0	methylcyclohexane	84.55	306.715		1994VA1
	47.2	50.0	methylcyclohexane	90.27	306.874		1994VA1
	7.1	7.5	2-propanone	0.5	253.3	442	1996LUS
	7.3	8.0	2-propanone	0.5	260	437	1996LUS
	11.4	11.7	2-propanone	0.5	274.8	421	1996LUS
	12.7	13.5	2-propanone	0.5	285.3	400	1996LUS

3.6. References

1962HAM   Ham, J.S., Bolen, M.C., and Hughes, J.K., The use of high pressure to study polymer-solvent interaction, *J. Polym. Sci.*, 57, 25, 1962.

1963EHR   Ehrlich, P. and Kurpen, J.J., Phase equilibria of polymer-solvent systems at high pressure near their critical loci, polyethylene with n-alkanes, *J. Polym. Sci.: Part A*, 1, 3217, 1963.

1965ALL   Allen, G. and Baker, C.H., Lower critical solution phenomena in polymer-solvent systems, *Polymer*, 6, 181, 1965.

1965MYR   Myrat, C.D. and Rowlinson, J.S., The separation and fractionation of polystyrene at a lower critical solution point, *Polymer*, 6, 645, 1965.

1972ZE1   Zeman, L., Biros, J., Delmas, G., and Patterson, D., Pressure effects in polymer solution phase equilibria. I. The lower critical solution temperature of polyisobutylene and polydimethylsiloxane in lower alkanes, *J. Phys. Chem.*, 76, 1206, 1972.

1972ZE2   Zeman, L. and Patterson, D., Pressure effects in polymer solution phase equilibria. II. Systems showing upper and lower critical solution temperatures, *J. Phys. Chem.*, 76, 1214, 1972.

1973HAM   Hamada, F., Fujisawa, K., and Nakajima, A.: Lower critical solution temperature in linear polyethylene-n-alkane systems, *Polym. J.*, 4, 316, 1973.

1974TAN   Taniguchi, Y., Suzuki, K., and Enomoto, T., The effect of pressure on the cloud point of aqueous polymer solutions, *J. Colloid Interface Sci.*, 46, 511, 1974.

1975HOR   Horacek, H., Gleichgewichtsdrücke, Löslichkeit und Mischbarkeit des Systems Polyethylen niedriger Dichte und Kohlenwasserstoffen bzw. halogenierten Kohlenwasserstoffen, *Makromol. Chem., Suppl.*, 1, 415, 1975.

1975SAE   Saeki, S., Kuwahara, N., Nakata, M., and Kaneko, M., Pressure dependence of upper critical solutions temperatures in the polystyrene-cyclohexane system, *Polymer*, 16, 445, 1975.

1976SA1   Saeki, S., Kuwahara, N., Nakata, M., and Kaneko, M., Upper and lower critical solution temperatures in poly(ethylene glycol) solutions, *Polymer*, 17, 685, 1976.

1976SA2   Saeki, S., Kuwahara, N., and Kaneko, M., Pressure dependence of upper and lower critical solution temperatures in polystyrene solutions, *Macromolecules*, 9, 101, 1976.

1976WOL   Wolf, B.A. and Blaum, G., Pressure influence on true cosolvency, *Makromol. Chem.*, 177, 1073, 1976.

1977WO1   Wolf, B.A. and Jend, R., Pressure and molecular-weight dependence of polymer solubility in the case of *trans*-decahydronaphthalene-polystyrene, *High Temp. High Press.*, 9, 561, 1977.

1977WO2   Wolf, B.A. and Jend, R., Über die Möglichkeiten zur Bestimmung von Mischungsenthalpien und -volumina aus der Molekulargewichtsabhängigkeit der kritischen Entmischungstemperaturen und -drücke am Beispiel des Systems *trans*-Decahydronaphthalin/Polystyrol, *Makromol. Chem.*, 178, 1811, 1977.

1978ISH   Ishizawa, M., Kuwahara, N., Nakata, M., Nagayama, W., and Kaneko, M.: Pressure dependence of upper critical solution temperatures in the system polystyrene-cyclopentane, *Macromolecules*, 11, 871, 1978.

1980HOS   Hosokawa, H., Nakata, M., Dobashi, T., and Kaneko, M., Pressure dependence of the upper critical solution temperature in the system polystyrene + cyclohexane, *Rep. Progr. Polym. Phys. Japan*, 23, 13, 1980.

1980KLE   Kleintjens, L.A. and Koningsveld, R., Liquid-liquid phase separation in multicomponent polymer systems XIX. Mean-field lattice-gas treatment of the system n-alkane/linear polyethylene, *Colloid Polym. Sci.*, 258, 711, 1980.

1981WOL   Wolf, B.A. and Geerissen, H., Pressure dependence of the demixing of polymer solutions determined by viscometry (some experimental data by B. A. Wolf), *Colloid Polym. Sci.*, 259, 1214, 1981.

1983MA1   Maderek, E., Schulz, G.V, and Wolf, B.A, High temperature demixing of poly(decyl methacrylate) solutions in isooctane and its pressure dependence, *Makromol. Chem.*, 184, 1303, 1983.

1983MA2   Maderek, E., Schulz, G.V., and Wolf, B.A., Lower critical solution temperatures of poly(decyl methacrylate) in hydrocarbons, *Eur. Polym. J.*, 19, 963, 1983.

1984SAN   Sander, U., Löslichkeit von Polybutylmethacrylat und Fliessverhalten verdünnter Lösungen, *Diploma Paper*, Johannes Gutenberg University, Mainz, 1984.

1985GE1   Geerissen, H., Roos, J., and Wolf, B.A., Continuous fractionation and solution properties of PVC 3. Pressure dependence of the solubility in single solvents, *Makromol. Chem.*, 186, 769, 1985.

1985GE2   Geerissen, H., Roos, J., and Wolf, B.A., Continuous fractionation and solution properties of PVC 4. Pressure dependence of the solubility in a mixed solvent, *Makromol. Chem.*, 186, 777, 1985.

1986SAE   Saeki, S., Kuwahara, N., Hamano, K., Kenmochi, Y., and Yamaguchi, T., Pressure dependence of upper critical solution temperatures in polymer solutions, *Macromolecules*, 19, 2353, 1986.

1986SAN   Sander, U. and Wolf, B.A., Solubility of poly(n-alkylmethacrylate)s in hydrocarbons and in alcohols, *Angew. Makromol. Chem.*, 139, 149, 1986.

1998IMR   Imre, A. and van Hook, A.W., Liquid-liquid equilibria in polymer solutions at negative pressure, *Chem. Soc. Rev.*, 27, 117, 1998.

1988KI1   Kiepen, F., Streulichtuntersuchungen an den Systemen Oligostyrol/n-Pentan und Polystyrol/ Methylcyclohexan unter hohen Drücken in der Nähe der Mischungslücke, *Dissertation*, Universität Duisburg, 1988.

1988KI2   Kiepen, F. and Borchard, W., Pressure-pulse-induced critical scattering of oligostyrene in n-pentane, *Macromolecules*, 21, 1784, 1988.

1988KI3   Kiepen, F. and Borchard, W., Critical opalescence of polymer solutions at high pressures, *Makromol. Chem.*, 189, 2595, 1988.

1988KI4   Kiepen, F. and Borchard, W., Determination of binodal and spinodal points by means of the pressure pulse-induced critical scattering method, *Integr. Fundam. Polym. Sci. Technol.*, Elsevier, London, 2, 227, 1988.

1989HAE   Haegen, R. van der, Kleintjens, L.A., Opstal, L. van, and Koningsveld, R., Thermodynamics of polymer solutions, *Pure Appl. Chem.*, 61, 159, 1989.

1990KEN   Kennis, H.A.J., Loos, Th.W. de, DeSwaan Arons, J., Van der Haegen, R., and Kleintjens, L.A., The influence of nitrogen on the liquid-liquid phase behaviour of the system n-hexane-polyethylene: experimental results and predictions with the mean-field lattice-gas model (experimental data by Th.W. de Loos from M.Sc. Thesis by H.A.J. Kennis, TU Delft 1987), *Chem. Eng. Sci.*, 45, 1875, 1990.

1991DEE   Dee, G.T., The application of equation-of-state theories to polar-nonpolar liquid mixtures, *J. Supercrit. Fluids*, 4, 152, 1991.

1991KIE   Kiepen, F., Brinkmann, D., Koningsveld, R., and Borchard, W., Phase diagrams in temperature, pressure and concentration-space of polystyrenes in n-pentane and methylcyclohexane near the critical solution point, *Integr. Fundam. Polym. Sci. Technol.*, Elsevier, Amsterdam, 5, 25, 1991.

1991SCH   Schultze, J.D., Zhou, Z., and Springer, J., Gas sorption in poly(butylene terephthalate). A critical test of the influence of gas data, *Angew. Makromol. Chem.*, 185-186, 265, 1991.

1991SZY   Szydlowski, J. and van Hook, W.A., Isotope and pressure effects on liquid-liquid equilibria in polymer solutions. H/D solvent isotope effects in acetone-polystyrene solutions, *Macromolecules*, 24, 4883, 1991.

1992COO   Cook, R.L., King, H.E., and Peiffer, D.G., Pressure-induced crossover from good to poor solvent behavior for polyethylene oxide in water, *Phys. Rev. Lett.*, 69, 3072, 1992.

1992KIR   Kiran, E. and Zhuang, W., Solubility of polyethylene in n-pentane at high pressures, *Polymer*, 33, 5259, 1992.

1992SAS   Sassen, C.L., Gonzales Casielles, A., Loos, Th.W. de, and DeSwaan Arons, J., The influence of pressure and temperature on the phase behaviour of the system H_2O + C_{12} + C_7E_5 and relevant binary subsystems, *Fluid Phase Equil.*, 72, 173, 1992.

1992SZY   Szydlowski, J., Rebelo, L., and van Hook, W.A., A new apparatus for the detection of phase equilibria in polymer solvent systems by light scattering, *Rev. Sci. Instrum.*, 63, 1717, 1992.

1993HOS   Hosokawa, H., Nakata, M., and Dobashi,T., Coexistence curve of polystyrene in methyl-cyclohexane. VII. Coexistence surface and critical double point of binary system in T-p-φ space (experimental data by M. Nakata and T. Dobashi), *J. Chem. Phys.*, 98, 10078, 1993.

1993KI2   Kiran, E., Xiong, Y., and Zhunag, W., Modeling polyethylene solutions in near and super-critical fluids using the Sanchez-Lacombe model, *J. Supercrit. Fluids*, 6, 193, 1993.

1993OTA   Otake, K., Karaki, R., Ebina, T., Yokoyama, C., and Takahashi, S., Pressure effects on the aggregation of poly(*N*-isopropylacrylamide) and poly(*N*-isopropylacrylamide-*co*-acrylic acid) in aqueous solutions, *Macromolecules*, 26, 2194, 1993.

1993REB   Rebelo, L.P. and van Hook, W.A., An unusual phase diagram: the polystyrene-acetone system in its hypercritical region; near tricritical behavior in a pseudo-binary solution, *J. Polym. Sci.: Part B: Polym. Phys.*, 31, 895, 1993.

1993WEL   Wells, P.A., Loos, Th.W. de, and Kleintjens, L.A., Pressure pulsed induced critical scattering: spinodal and binodal curves for the system polystyrene + methylcyclohexane (experimental data by Th.W. de Loos), *Fluid Phase Equil.*, 83, 383, 1993.

1994IMR   Imre, A. and van Hook, W.A., Polymer-solvent demixing under tension. Isotope and pressure effects on liquid-liquid transitions. VII. Propionitrile-polystyrene solutions at negative pressure, *J. Polym. Sci.: Part B: Polym. Phys.*, 32, 2283, 1994.

1994KI1   Kiran, E. and Zhuang, W., A new experimental method to study kinetics of phase separation in high-pressure polymer solutions. Multiple rapid pressure-drop technique – MRPD, *J. Supercrit. Fluids*, 7, 1, 1994.

1994KI2   Kiran, E., Xiong, Y., and Zhuang, W., Effect of polydispersity on the demixing pressures of polyethylene in near- or supercritical alkanes, *J. Supercrit. Fluids*, 7, 283, 1994.

1994LOS   LoStracco, M.A., Lee, S.-H., and McHugh, M.A., Comparison of the effect of density and hydrogen bonding on the cloud point behavior of poly(ethylene-*co*-methyl acrylate)-propane-cosolvent mixtures, *Polymer*, 35, 3272, 1994.

1994VA1   Vanhee, S., Kiepen, F., Brinkmann, D., Borchard, W., Koningsveld, R., and Berghmans, H., The system methylcyclohexane/polystyrene. Experimental critical curves, cloud-point and spinodal isopleths, and their description with a semi-phenomenological treatment, *Makromol. Chem. Phys.*, 195, 759, 1994.

1995LUS   Luszczyk, M., Rebelo, L.P.N., and van Hook, W.A., Isotope and pressure dependence of liquid-liquid equilibria in polymer solutions. 5. and 6. (experimental data by L.P.N. Rebelo), *Macromolecules*, 28, 745, 1995.

1995TUM   Tuminello, W.H., Brill, D.J., Walsh, D.J., and Paulaitis, M.E., Dissolving poly(tetrafluoro-ethylene) in low boiling halocarbons, *J. Appl. Polym. Sci.*, 56, 495, 1995.

1996LOO   Loos, Th.W. de, Graaf, L.J. de, and DeSwaan Arons, J., Liquid-liquid phase separation in linear low density polyethylene-solvent systems, *Fluid Phase Equil.*, 117, 40, 1996.

1996DOB   Dobashi, T., Koshiba, T., and Nakata, M., Coexistence curve of polystyrene in methyl-cyclohexane. IX. Pressure dependence of tricritical point, *J. Chem. Phys.*, 105, 2906, 1996.

1996IMR   Imre, A. and van Hook, W.A., Demixing in polystyrene/methylcyclohexane solutions, *J. Polym. Sci.: Part B: Polym. Sci.*, 34, 751, 1996.

1996LUS   Luszczyk, M. and van Hook, W.A., Isotope and pressure dependence of liquid-liquid equilibria in polymer solutions. 7. Solute and solvent H/D isotope effects in polystyrene-propionitrile solutions, *Macromolecules*, 29, 6612, 1996.

1997DES   DeSousa, H.C., Phase equilibria in polymer + solvent systems: experimental results and modeling (Portug.), *Ph.D. Thesis*, New University of Lisbon, Portugal, 1997.

1997END   Enders, S. and Loos, Th.W. de, Pressure dependence of the phase behaviour of polystyrene in methylcyclohexane (experimental data by S. Enders), *Fluid Phase Equil.*, 139, 335, 1997.

1997IMR   Imre, A. and van Hook, W.A., Continuity of solvent quality in polymer solutions. Poor-solvent to theta-solvent continuity in some polystyrene solutions, *J. Polym. Sci.: Part B: Polym. Phys.*, 35, 1251, 1997.

1997KUN   Kunugi, S., Takano, K., Tanaka, N., Suwa, K., and Akashi, M., Effects of pressure on the behavior of the thermoresponsive polymer poly(N-vinylisobutyramide) (PNVIBA), *Macromolecules*, 30, 4499, 1997.

1997REB   Rebelo, L.P.N., DeSousa, H.C., and van Hook, W.A., Hypercritically enhanced distortion of a phase diagram: The (polystyrene + acetaldehyde) system (experimental data by H.C. De Sousa), *J.Polym.Sci.: Part B: Polym.Phys.*, 35, 631, 1997.

1997WOL   Wolf, B.A., Improvment of polymer solubility: influence of shear and pressure, *Pure Appl. Chem.*, 69, 929, 1997.

1997XIO   Xiong, Y. and Kiran, E., Miscibility, density and viscosity of polystyrene in n-hexane at high pressures, *Polymer*, 38, 5185, 1997.

1998KI2   Kiran, E. and Xiong, Y., Miscibility of isotactic polypropylene in n-pentane and n-pentane + carbon dioxide mixtures at high pressures, *J. Supercrit. Fluids*, 11, 173, 1998.

1998KOA   Koak, N., Loos, Th.W. de, and Heidemann, R.A., Upper-critical-solution-temperature behavior of the system polystyrene + methylcyclohexane. Influence of CO_2 on the liquid-liquid equilibria, *Fluid Phase Equil.*, 145, 311, 1998.

1998SUW   Suwa, K., Yamamoto, K., Akhashi, M., Takano, K., Tanaka, N., and Kunugi, S., Effects of salt on the temperature and pressure responsive properties of poly(N-vinylisobutyramide) aqueous solutions, *Colloid Polym. Sci.*, 276, 529, 1998.

1998SZY   Szydlowski, J. and van Hook, W.A., Liquid-liquid demixing from polystyrene solutions. Studies on temperature and pressure dependeces using dynamic light scattering and neutron scattering, *Fluid Phase Equil.*, 150-151, 687, 1998.

1998ZHU   Zhuang, W. and Kiran, E., Kinetics of pressure-induced phase separation (PIPS) from polymer solutions by time resolved light scattering. Polyethylene + n-pentane, *Polymer*, 39, 2903, 1998.

1999BEY   Beyer, C., Oellrich, L.R., and McHugh, M.A., Effect of copolymer composition and solvent polarity on the phase behavior of mixtures of poly(ethylene-*co*-vinyl acetate) with cyclopentane and cyclopentene (experimental data by C. Beyer), *Chem. Ing. Techn.*, 71, 1306, 1999.

1999BUN   Bungert, B., Komplexe Phasengleichgewichte von Polymerlösungen, *Dissertation*, TU Berlin, Shaker Vlg., Aachen, 1999.

1999HOO   Hook, W.A. van, Wilczura, H., and Rebelo, L.P.N., Dynamic light scattering of polymer/solvent solutions under pressure. Near critical demixing (0.1<P/MPa<200) for polystyrene/cyclohexane and polystyrene/methylcyclohexane, *Macromolecules*, 32, 7299, 1999.

1999IM1   Imre, A.R., Melnichenko, G., and van Hook, W.A., Liquid-liquid equilibria in polystyrene solutions: The general pressure dependence, *Phys. Chem. Chem. Phys.*, 1, 4287, 1999.

1999IM2   Imre, A.R., Melnichenko, G., and van Hook, W.A., A polymer-solvent system with two homogeneous double critical points: Polystyrene (PS)/(n-heptane + methylcyclohexane), *J. Polym. Sci.: Part B: Polym. Phys.*, 37, 2747, 1999.

1999KUN   Kunugi, S., Yamazaki, Y., Takano, K., and Tanaka, N., Effects of ionic additives and ionic comonomers on the temperature and pressure responsive behavior of thermoresponsive polymers in aqueous solutions, *Langmuir*, 15, 4056, 1999.

2000BEH	Behme, S., Thermodynamik von Polymersystemen bei hohen Drucken, *Dissertation*, TU Berlin, 2000.
2000BEY	Beyer, C., Oellrich, L.R., and McHugh, M.A., Effect of copolymer composition and solvent polarity on the phase behavior of mixtures of poly(ethylene-*co*-vinyl acetate) with cyclopentane and cyclopentene, *Chem. Eng. Technol.*, 23, 592, 2000.
2000DE1	DeSousa, H.C. and Rebelo, L.P.N., (Liquid + liquid) equilibria of (polystyrene + nitro-ethane). Molecular weight, pressure, and isotope effects, *J. Chem. Thermodyn.*, 32, 355, 2000.
2000DE2	DeSousa, H.C. and Rebelo, L.P.N., A continuous polydisperse thermodynamic algorithm for a modified Flory-Huggins model: The (polystyrene + nitroethane) example, *J. Polym. Sci.: Part B: Polym. Phys.*, 38, 632, 2000.
2000IMR	Imre, A. and van Hook, W.A., End group effects on liquid-liquid demixing of polystyrene/oligomethylene solutions. Polystyrene/dodecyl acetate solubility, *Macromolecules*, 33, 5308, 2000.
2000KOI	Koizumi, J., Kawashima, Y., Kita, R., Dobashi, T., Hosokawa, H., and Nakata, M., Coexistence curves of polystyrene in cyclohexane near the critical double point in composition-pressure space, *J. Phys. Soc. Japan*, 69, 2543, 2000.
2000KUN	Kunugi, S., Tada, T., Yamazaki, Y., Yamamoto, K., and Akashi, M., Thermodynamic studies on coil-globule transitions of poly(N-vinylisobutyramide-*co*-vinylamine) in aqueous solutions, *Langmuir*, 16, 2042, 2000.
2000OLI	Oliveira, J.V., Dariva, C., and Pinto, J.C., High-pressure phase equilibria for polypropylene-hydrocarbon systems, *Ind. Eng. Chem. Res.*, 39, 4627, 2000.
2000XIO	Xiong, Y. and Kiran, E., Kinetics of pressure-induced phase separation (PIPS) in polystyrene + methylcyclohexane solutions at high pressure, *Polymer*, 41, 3759, 2000.
2000YAM	Yamazaki, Y., Tada, T., and Kunugi, S., Effect of acrylic acid incorporation on the pressure-temperature behavior and the calorimetric properties of poly(*N*-isopropyl-acrylamide) in aqueous solutions, *Colloid Polym. Sci.*, 278, 80, 2000.
2001GOM	Gomes de Azevedo, R., Rebelo, L.P.N., Ramos, A.M., Szydlowski, J., DeSousa, H.C., and Klein, J., Phase behavior of (polyacrylamides + water) solutions: concentration, pressure and isotope effects, *Fluid Phase Equil.*, 185, 189, 2001.
2001IMR	Imre, A.R., Melnichenko, G., van Hook, W.A., and Wolf, B.A., On the effect of pressure on the phase transition of polymer blends and polymer solutions. Oligostyrene/n-alkane systems, *Phys. Chem. Chem. Phys.*, 3, 1063, 2004.
2001KUN	Kunugi, S., Yoshida, D., and Kiminami, H., Effects of pressure on the behavior of (hydroxypropyl)cellulose in aqueous solution, *Colloid Polym. Sci.*, 279, 1139, 2001.
2001LIU	Liu, K. and Kiran, E., Pressure-induced phase separation in polymer solutions: kinetics of phase separation and crossover from nucleation and growth to spinodal decomposition in solutions of polyethylene in n-pentane, *Macromolecules*, 34, 3060, 2001.
2001TOR	Tork, T., Measurement and calculation of phase equilibria in polyolefin/solvent systems, *Dissertation*, TU Berlin, 2001.
2002FRI	Frielinghaus, H., Schwahn, D., Willner, L., and Freed, K.F., Small angle neutron scattering studies of a polybutadiene/polystyrene blend with small additions of ortho-dichlorobenzene for varying temperatures and pressures. II. Phase boundaries and Flory-Huggins parameter (experimental data by H. Frielinghaus), *J. Chem. Phys.*, 116, 2241, 2002.
2002HOR	Horst, M.H. ter, Behme, S., Sadowski, G., and Loos, Th.W. de, The influence of supercritical gases on the phase behavior of polystyrene-cyclohexane and polyethylene-cyclohexane systems: experimental results and modeling with the SAFT equation of state, *J. Supercrit. Fluids*, 23, 181, 2002.
2002IMR	Imre, A.R., van Hook, W., and Wolf, B.A., Liquid-liquid phase equilibria in polymer solutions and polymer mixtures, *Macromol. Symp.*, 181, 363, 2002.
2002KIR	Kiran, E. and Liu, K., The miscibility and phase behavior of polyethylene with poly(dimethylsiloxane) in near critical pentane, *Korean J. Chem. Eng.*, 19, 153, 2002.

2002RE1   Rebelo, L.P.N., Visak, Z.P., DeSousa, H.C., Szydlowski, J., Gomes de Azevedo, R., Ramos, A.M., Najdanovic-Visak, V., Nunes da Ponte, M., and Klein, J., Double critical phenomena in (water + polyacrylamides) solutions, *Macromolecules*, 35, 1887, 2002.

2002RE2   Rebelo, L.P.N., Visak, Z.P., and Szydlowski, J., Metastable critical lines in (acetone + polystyrene) solutions and the continuity of solvent-quality states, *Phys. Chem. Chem. Phys.*, 4, 1046, 2002.

2002YEO   Yeo, S.-D., Kang, I.-S., and Kiran, E., Critical polymer concentrations of polyethylene solutions in pentane, *J. Chem. Eng. Data*, 47, 571, 2002.

2003JI1   Jiang, S., An, L., Jiang, B., and Wolf, B. A., Pressure effects on the thermodynamics of *trans*-decahydronaphthalene/polystyrene polymer solutions: application of the Sanchez-Lacombe lattice fluid theory (experimental data by S. Jiang), *Macromol. Chem. Phys.*, 204, 692, 2003.

2003JI2   Jiang, S., An, L., Jiang, B., and Wolf, B.A., Liquid-liquid phase behavior of toluene/ polyethylene oxide/poly(ethylene oxide-b-dimethylsiloxane) polymer-containing ternary mixtures (experimental data by S. Jiang), *Phys. Chem. Chem. Phys.*, 5, 2066, 2003.

2003SET   Seto, Y., Kameyama, K., Tanaka, N., Kunugi, S., Yamamoto, K., and Akashi, M., High-pressure studies on the coacervation of copoly(N-vinylformamide-vinylacetate) and copoly(N-vinylacetylamide-vinylacetate), *Colloid Polym. Sci.*, 281, 690, 2003.

2003SIP   Siporska, A., Szydlowski, J., and Rebelo, L.P.N., Solvent H/D isotope effects on miscibility and theta-temperature in the polystyrene-cyclohexane system, *Phys. Chem. Chem. Phys.*, 5, 2996, 2003.

2003YAM   Yamaoka, T., Tamura, T., Seto, Y., Tada, T., Kunugi, S., and Tirrell, D.A., Mechanism for the phase transition of a genetically engineered elastin model peptide (VPGIG) 40 in aqueous solution, *Biomacromolecules*, 4, 1680, 2003.

2003ZH1   Zhang, W. and Kiran, E., (p, V, T) Behaviour and miscibility of (polysulfone + THF + carbon dioxide) at high pressures, *J. Chem. Thermodyn.*, 35, 605, 2003.

2003ZH2   Zhang, W., Dindar, C., Bayraktar, Z., and Kiran, E., Phase behavior, density, and crystallization of polyethylene in n-pentane and in n-pentane/CO_2 at high pressures, *J. Appl. Polym. Sci.*, 89, 2201, -2003.

2004CHE   Chen, X., Yasuda, K., Sato, Y., Takishima, S., and Masuoka, H., Measurement and correlation of phase equilibria of ethylene + n-hexane + metallocene polyethylene at temperatures between 373 and 473 K and at pressures up to 20 MPa, *Fluid Phase Equil.*, 215, 105, 2004.

2004JIA   Jiang, S., An, L., Jiang, B., and Wolf, B. A., Temperature and pressure dependence of phase separation of *trans*-decahydronaphthalene/polystyrene solution, *Chem. Phys.*, 298, 37, 2004.

2004SCH   Schnell, M., Stryuk, S., and Wolf, B.A., Liquid/liquid demixing in the system n-hexane/ narrowly distributed linear polyethylene (experimental data by M. Schnell and B.A. Wolf), *Ind. Eng. Chem. Res.*, 43, 2852, 2004.

2004SHI   Shibayama, M., Isono, K., Okabe, S., Karino, T., and Nagao, M., SANS study on pressure-induced phase separation of poly(N-isopropylacrylamide) aqueous solutions and gels, *Macromolecules*, 37, 2909, 2004.

4. HIGH-PRESSURE FLUID PHASE EQUILIBRIUM (HPPE) DATA OF POLYMER SOLUTIONS

4.1. Cloud-point and/or coexistence curves of quasibinary solutions

Polymer (B):	ethylene/acrylic acid copolymer	1992WIN
Characterization:	M_n/kg.mol^{-1} = 37.5, M_w/kg.mol^{-1} = 183, 3.0 wt% acrylic acid	
Solvent (A):	ethene C$_2$H$_4$	74-85-1

Type of data: cloud points

w_B	0.019	0.019	0.019	0.019	0.019	0.082	0.082	0.082	0.082
T/K	413.15	433.15	453.15	473.15	493.15	413.15	433.15	453.15	473.15
P/MPa	196.8	178.4	163.1	152.2	141.7	186.7	166.6	153.1	143.3

w_B	0.082	0.128	0.128	0.128	0.128	0.128	0.186	0.186	0.186
T/K	493.15	413.15	433.15	453.15	473.15	493.15	413.15	433.15	453.15
P/MPa	135.4	181.2	162.8	150.1	140.3	132.2	172.1	154.3	142.5

w_B	0.186	0.186	0.231	0.231	0.231	0.231	0.231	0.295	0.295
T/K	473.15	493.15	413.15	433.15	453.15	473.15	493.15	413.15	433.15
P/MPa	134.0	128.5	165.6	148.9	137.7	129.8	124.2	153.1	139.8

w_B	0.295	0.295	0.295	0.318	0.318	0.318	0.318	0.318	0.338
T/K	453.15	473.15	493.15	413.15	433.15	453.15	473.15	493.15	413.15
P/MPa	130.2	123.1	118.0	150.4	136.6	127.9	121.3	115.8	146.0

w_B	0.338	0.338	0.338	0.338	0.428	0.428	0.428	0.428	0.428
T/K	433.15	453.15	473.15	493.15	413.15	433.15	453.15	473.15	493.15
P/MPa	134.1	125.3	118.9	113.4	127.5	119.4	113.4	108.7	105.1

w_B	0.510	0.510	0.510	0.510	0.510
T/K	413.15	433.15	453.15	473.15	493.15
P/MPa	109.8	105.5	102.0	99.0	96.0

Type of data: coexistence data

T/K = 403.15

Total feed concentration of the copolymer in the homogeneous system: w_B = 0.155.

Demixing pressure	w_A bottom phase	top phase	Fractionation during demixing bottom phase			top phase		
P/ MPa			M_n/ kg/mol	M_w/ kg/mol	M_z/ kg/mol	M_n/ kg/mol	M_w/ kg/mol	M_z/ kg/mol
116.0	0.508	0.982				2.82	8.80	18.92
161.0	0.711	0.951				9.19	31.06	73.72
178.5	0.789	0.918	42.88	176.3	510.7	13.17	52.62	156.1
188.0	0.845	(cloud-point)	37.50	183.0	601.0	(feed sample)		

continued

continued

T/K = 433.15

Total feed concentration of the copolymer in the homogeneous system: w_B = 0.164.

Demixing pressure	w_A bottom phase	top phase	Fractionation during demixing					
			bottom phase			top phase		
P/ MPa			M_n/ kg/mol	M_w/ kg/mol	M_z/ kg/mol	M_n/ kg/mol	M_w/ kg/mol	M_z/ kg/mol
94.0	0.438	0.993				1.5	6.75	13.9
113.0	0.536	0.982				2.5	8.3	17.6
129.0	0.633	0.966				8.2	29.1	69.6
147.7	0.766	0.939	41.5	173.7	472.8	11.6	47.0	146.2
157.5	0.836	(cloud-point)	37.5	183.0	601.0	(feed sample)		

Polymer (B):   **ethylene/acrylic acid copolymer**   **2003BUB**
Characterization: M_n/kg.mol^{-1} = 21.7, M_w/kg.mol^{-1} = 86, 3.5 mol% acrylic acid
Solvent (A):   **ethene**   **C$_2$H$_4$**   **74-85-1**

Type of data: cloud points

w_B 0.03   was kept constant

T/K	433.15	438.15	443.15	448.15	453.15	458.15	463.15	468.15	473.15
P/bar	2815	2645	2535	2425	2270	2135	2040	1950	1830

T/K	483.15	493.15	503.15	513.15	523.15	533.15
P/bar	1725	1610	1520	1435	1370	1299

Polymer (B):   **ethylene/acrylic acid copolymer**   **2003BUB**
Characterization: M_n/kg.mol^{-1} = 19.1, M_w/kg.mol^{-1} = 63.3, 3.8 mol% acrylic acid
Solvent (A):   **ethene**   **C$_2$H$_4$**   **74-85-1**

Type of data: cloud points

w_B 0.03   was kept constant

T/K	438.15	443.15	448.15	453.15	458.15	463.15	468.15	473.15	478.15
P/bar	2880	2730	2615	2465	2355	2220	2085	1995	1895

T/K	483.15	493.15	503.15	513.15	523.15	533.15
P/bar	1810	1685	1570	1470	1395	1325

Polymer (B):	**ethylene/acrylic acid copolymer**							**2003BUB**
Characterization:	M_n/kg.mol^{-1} = 12.6, M_w/kg.mol^{-1} = 42.1, 5.5 mol% acrylic acid							
Solvent (A):	**ethene**	**C$_2$H$_4$**						**74-85-1**

Type of data: cloud points

w_B 0.03   was kept constant

T/K	473.15	478.15	483.15	488.15	493.15	498.15	503.15	513.15	523.15
P/bar	2820	2700	2540	2390	2255	2110	1990	1785	1630

T/K	533.15
P/bar	1490

Polymer (B):	**ethylene/acrylic acid copolymer**							**2003BUB**
Characterization:	M_n/kg.mol^{-1} = 31, M_w/kg.mol^{-1} = 82.3, 5.7 mol% acrylic acid							
Solvent (A):	**ethene**	**C$_2$H$_4$**						**74-85-1**

Type of data: cloud points

w_B 0.03   was kept constant

T/K	483.15	488.15	493.15	498.15	503.15	508.15	513.15	518.15	523.15
P/bar	2585	2435	2295	2150	2040	1930	1820	1730	1660

T/K	533.15
P/bar	1520

Polymer (B):	**ethylene/acrylic acid copolymer**							**1992WIN**
Characterization:	M_n/kg.mol^{-1} = 30, M_w/kg.mol^{-1} = 150, 6.0 wt% acrylic acid							
Solvent (A):	**ethene**	**C$_2$H$_4$**						**74-85-1**

Type of data: cloud points

w_B	0.038	0.038	0.038	0.038	0.057	0.057	0.057	0.057	0.076
T/K	433.15	453.15	473.15	493.15	433.15	453.15	473.15	493.15	433.15
P/MPa	203.4	176.8	157.3	142.0	195.7	170.5	152.3	138.9	191.2

w_B	0.076	0.076	0.076	0.116	0.116	0.116	0.116	0.150	0.150
T/K	453.15	473.15	493.15	433.15	453.15	473.15	493.15	433.15	453.15
P/MPa	167.2	148.9	134.0	189.5	164.2	146.2	132.5	183.9	161.1

w_B	0.150	0.150	0.215	0.215	0.215	0.215	0.260	0.260	0.260
T/K	473.15	493.15	433.15	453.15	473.15	493.15	433.15	453.15	473.15
P/MPa	145.2	130.5	172.4	152.2	135.4	125.4	165.3	146.3	132.6

w_B	0.260	0.295	0.295	0.295	0.295	0.335	0.335	0.335	0.335
T/K	493.15	433.15	453.15	473.15	493.15	433.15	453.15	473.15	493.15
P/MPa	121.8	158.3	142.0	129.8	119.0	147.5	135.9	125.6	115.5

w_B	0.383	0.383	0.383	0.383	0.430	0.430	0.430	0.430	0.505
T/K	433.15	453.15	473.15	493.15	433.15	453.15	473.15	493.15	433.15
P/MPa	137.5	126.9	118.5	112.2	130.1	120.2	112.0	106.3	114.4

continued

continued

w_B	0.505	0.505	0.505
T/K	453.15	473.15	493.15
P/MPa	108.0	103.1	99.2

Type of data: coexistence data

$T/K = 433.15$

Total feed concentration of the copolymer in the homogeneous system: $w_B = 0.1565$.

$P/$ MPa	w_A bottom phase	top phase
124.9	0.5550	0.9870
138.1	0.6130	0.9735
158.6	0.7130	0.9540
172.0	0.7855	0.9300
182.0	0.8435 (cloud-point)	

Polymer (B):   **ethylene/acrylic acid copolymer**   **2002BEY**
Characterization: $M_n/kg.mol^{-1} = 22.7$, $M_w/kg.mol^{-1} = 258$, 6.0 wt% acrylic acid
Exxon Co., Machelen, Belgium
Solvent (A):   **ethene**   **C_2H_4**   **74-85-1**

Type of data: cloud points

w_B	0.05	0.05	0.05	0.05	0.05
T/K	522.65	497.55	472.75	447.05	423.05
P/bar	1302	1394	1539	1779	2153

Polymer (B):   **ethylene/acrylic acid copolymer**   **2003BUB**
Characterization: $M_n/kg.mol^{-1} = 9.3$, $M_w/kg.mol^{-1} = 31.9$, 6.7 mol% acrylic acid
Solvent (A):   **ethene**   **C_2H_4**   **74-85-1**

Type of data: cloud points

w_B	0.03	was kept constant

T/K	488.15	493.15	498.15	503.15	508.15	513.15	518.15	523.15	529.15
P/bar	2960	2760	2555	2370	2200	2050	1910	1800	1695

T/K	533.15
P/bar	1610

Polymer (B):   ethylene/acrylic acid copolymer   **1992WIN**
Characterization: M_n/kg.mol^{-1} = 25, M_w/kg.mol^{-1} = 126, 7.3 wt% acrylic acid
Solvent (A):   ethene   **C₂H₄**   **74-85-1**

Type of data: cloud points

w_B	0.043	0.043	0.043	0.102	0.102	0.102	0.102	0.160	0.160
T/K	453.15	473.15	493.15	443.15	453.15	473.15	493.15	443.15	453.15
P/MPa	192.4	165.6	143.9	196.2	181.4	157.7	140.2	186.0	171.4

w_B	0.160	0.160	0.203	0.203	0.203	0.203	0.272	0.272	0.272
T/K	473.15	493.15	443.15	453.15	473.15	493.15	443.15	453.15	473.15
P/MPa	150.8	135.3	178.2	165.3	146.3	132.5	166.0	156.4	140.6

w_B	0.272	0.330	0.330	0.330	0.330	0.410	0.410	0.410	0.410
T/K	493.15	443.15	453.15	473.15	493.15	443.15	453.15	473.15	493.15
P/MPa	126.8	155.5	147.0	134.0	122.5	141.0	134.3	123.6	114.9

w_B	0.500	0.500	0.500	0.500
T/K	443.15	453.15	473.15	493.15
P/MPa	121.4	117.0	110.8	104.8

Polymer (B):   ethylene/acrylic acid copolymer   **2002BEY**
Characterization: M_n/kg.mol^{-1} = 19.9, M_w/kg.mol^{-1} = 235, 7.5 wt% acrylic acid
                        Exxon Co., Machelen, Belgium
Solvent (A):   ethene   **C₂H₄**   **74-85-1**

Type of data: cloud points

w_B	0.05	0.05	0.05	0.05	0.05
T/K	523.35	497.85	472.35	447.85	434.25
P/bar	1340	1442	1637	1946	2196

Polymer (B):   ethylene/acrylic acid copolymer   **2002BEY**
Characterization: 8.0 wt% acrylic acid, Exxon Co., Machelen, Belgium
Solvent (A):   ethene   **C₂H₄**   **74-85-1**

Type of data: cloud points

w_B	0.05	0.05	0.05	0.05	0.05
T/K	523.05	497.95	472.65	447.65	433.05
P/bar	1349	1469	1692	2037	2367

Polymer (B):   ethylene/acrylic acid copolymer   **2002BEY**
Characterization: M_n/kg.mol^{-1} = 23.4, M_w/kg.mol^{-1} = 227, 9.0 wt% acrylic acid,
                        Exxon Co., Machelen, Belgium
Solvent (A):   ethene   **C₂H₄**   **74-85-1**

Type of data: cloud points

w_B	0.05	0.05	0.05	0.05
T/K	522.65	497.75	471.85	447.55
P/bar	1390	1521	1779	2209

Polymer (B):   **ethylene/acrylic acid copolymer**   **2002BEY**
Characterization: M_n/kg.mol^{-1} = 23.7, M_w/kg.mol^{-1} = 205, 11.0 wt% acrylic acid,
                        Exxon Co., Machelen, Belgium
Solvent (A):   **ethene**   **C$_2$H$_4$**   **74-85-1**

Type of data: cloud points

w_B	0.05	0.05	0.05
T/K	524.05	497.35	472.35
P/bar	1484	1686	2104

Polymer (B):   **ethylene/acrylic acid copolymer**   **2002BEY**
Characterization: 15.0 wt% acrylic acid, Exxon Co., Machelen, Belgium
Solvent (A):   **ethene**   **C$_2$H$_4$**   **74-85-1**

Type of data: cloud points

w_B	0.05	0.05	0.05
T/K	522.05	498.05	475.45
P/bar	1732	2066	2647

Polymer (B):   **ethylene/1-butene copolymer**   **1999CHE**
Characterization: M_n/kg.mol^{-1} = 6.43, M_w/kg.mol^{-1} = 6.69, 8.6 mol% 1-butene,
                        completely hydrogenated polybutadiene, 8.6 mol% 1,2-units,
                        Polymer Source, Inc., Dorval, Quebec
Solvent (A):   **1-butene**   **C$_4$H$_8$**   **106-98-9**

Type of data: cloud points

w_B	0.00075	was kept constant							
T/K	428.65	428.15	420.95	415.95	409.15	401.25	395.05	391.95	383.85
P/bar	90.1	88.8	80.5	72.3	63.5	50.3	40.7	34.6	23.5

T/K	381.55	328.85	327.75	326.85
P/bar	20.8	48.4	89.0	132.3

Comments: The last three data points are temperature-induced phase transitions.

w_B	0.011	was kept constant							
T/K	428.45	428.35	418.85	411.25	402.65	396.75	391.55	381.95	375.35
P/bar	131.6	131.4	119.3	107.3	96.4	85.8	78.8	63.0	52.3

T/K	366.75	362.95	357.65	352.55	350.85	349.35	348.15
P/bar	37.0	31.0	21.5	12.0	58.6	96.0	162.2

Comments: The last three data points are temperature-induced phase transitions.

Polymer (B):	**ethylene/1-butene copolymer**						**1999CHE**	

Characterization: M_n/kg.mol^{-1} = 7.57, M_w/kg.mol^{-1} = 7.87, 29.0 mol% 1-butene, completely hydrogenated polybutadiene, 29 mol% 1,2-units, Polymer Source, Inc., Dorval, Quebec

Solvent (A):	**1-butene**	**C₄H₈**	**106-98-9**

Type of data: cloud points

w_B 0.0011   was kept constant

T/K	429.45	426.75	419.55	415.55	409.95	406.35	399.45	396.95	389.95
P/bar	93.1	87.4	79.8	73.6	66.0	59.9	49.3	45.1	33.7

T/K	387.35	381.45	259.65	258.85	258.35
P/bar	29.0	20.5	31.1	79.6	144.5

Comments: The last three data points are temperature-induced phase transitions.

w_B 0.0078   was kept constant

T/K	428.95	417.15	412.75	408.95	406.95	398.65	396.55	390.95	386.95
P/bar	134.2	122.8	117.1	107.1	106.4	85.3	83.9	70.6	62.1

T/K	378.75	377.05	372.15	367.45	364.95	277.75	276.85
P/bar	44.7	39.4	30.1	23.6	15.1	22.2	192.4

Comments: The last two data points are temperature-induced phase transitions.

Polymer (B):	**ethylene/1-butene copolymer**						**1999CHE**	

Characterization: M_n/kg.mol^{-1} = 115, M_w/kg.mol^{-1} = 120, 87.5 mol% 1-butene, completely hydrogenated polybutadiene, 87.5 mol% 1,2-units, Polymer Source, Inc., Dorval, Quebec

Solvent (A):	**1-butene**	**C₄H₈**	**106-98-9**

Type of data: cloud points

w_B 0.0012   was kept constant

T/K	431.85	430.25	424.25	419.35	416.15	405.75	400.95	394.15	389.45
P/bar	98.4	96.6	86.7	82.3	74.5	58.4	58.4	39.4	29.2

T/K	386.45	384.45	257.35	255.85
P/bar	24.9	20.7	52.3	140.3

Comments: The last two data points are temperature-induced phase transitions.

w_B 0.0091   was kept constant

T/K	429.05	419.75	416.25	415.85	408.05	405.25	396.75	396.25	388.55
P/bar	96.7	81.0	78.2	77.1	64.2	59.2	44.3	43.9	29.2

T/K	386.15	383.15
P/bar	27.2	20.6

Polymer (B):   **ethylene/1-butene copolymer**   **1999CHE**

Characterization: $M_n/\text{kg.mol}^{-1} = 6.43$, $M_w/\text{kg.mol}^{-1} = 6.69$, 8.6 mol% 1-butene, completely hydrogenated polybutadiene, 8.6 mol% 1,2-units, Polymer Source, Inc., Dorval, Quebec

Solvent (A):   **ethene**   **C$_2$H$_4$**   **74-85-1**

Type of data: cloud points

w_B 0.00015   was kept constant

T/K	431.05	430.35	414.65	413.75	398.75	398.55	382.55	367.75	367.55
P/bar	591.1	591.3	596.4	589.1	598.7	596.9	602.2	620.0	621.3

T/K	352.45	352.45	337.65	324.15	317.15	310.35
P/bar	653.1	652.5	683.5	761.4	998.9	1418

Comments: The last two data points are temperature-induced phase transitions.

w_B 0.00096   was kept constant

T/K	425.55	424.55	414.45	411.75	404.15	401.05	391.75	383.75	372.35
P/bar	742.8	745.8	751.8	753.4	763.1	765.2	776.8	790.2	812.5

T/K	361.95	355.05	347.95	346.65
P/bar	855.5	995.2	1280	1471

Comments: The last three data points are temperature-induced phase transitions.

w_B 0.0065   was kept constant

T/K	426.95	426.85	414.85	413.15	400.05	397.05	389.25	387.55	379.65
P/bar	892.8	896.2	923.8	928.8	970.1	977.6	1010	1018	1055

T/K	377.05	370.45	369.75	368.95	368.25
P/bar	1063	1163	1290	1370	1469

Comments: The last three data points are temperature-induced phase transitions.

Polymer (B):   **ethylene/1-butene copolymer**   **1999CHE**

Characterization: $M_n/\text{kg.mol}^{-1} = 7.57$, $M_w/\text{kg.mol}^{-1} = 7.87$, 29.0 mol% 1-butene, completely hydrogenated polybutadiene, 29 mol% 1,2-units, Polymer Source, Inc., Dorval, Quebec

Solvent (A):   **ethene**   **C$_2$H$_4$**   **74-85-1**

Type of data: cloud points

w_B 0.0013   was kept constant

T/K	432.35	419.45	407.15	403.95	388.85	387.85	372.45	372.15	357.55
P/bar	732.5	737.8	739.1	752.0	772.0	770.7	800.0	797.0	833.4

T/K	357.25	345.05	343.35	328.55	327.45	316.65	306.55	295.75	288.35
P/bar	834.2	867.7	870.2	928.9	933.9	985.0	1043	1124	1265

w_B 0.0079   was kept constant

T/K	425.75	412.05	398.55	384.25	383.15	368.65	367.15	353.05	337.75
P/bar	846.8	860.4	884.1	919.3	922.2	965.2	965.4	1024	1092

T/K	323.35	322.65	313.65	307.25
P/bar	1097	1208	1285	1385

Polymer (B):   **ethylene/1-butene copolymer**   **1999CHE**

Characterization: M_n/kg.mol^{-1} = 115, M_w/kg.mol^{-1} = 120, 87.5 mol% 1-butene,
completely hydrogenated polybutadiene, 87.5 mol% 1,2-units,
Polymer Source, Inc., Dorval, Quebec

Solvent (A):   **ethene**   **C₂H₄**   **74-85-1**

Type of data: cloud points

w_B 0.0012   was kept constant

T/K	424.25	409.25	389.35	372.25	353.25	352.45	333.45	318.25	305.05
P/bar	594.1	595.2	596.4	582.4	577.8	578.8	576.2	583.8	586.8

T/K	288.55	281.35	271.75	266.55	261.15	257.65	253.55	249.75	246.05
P/bar	591.8	593.7	597.9	604.6	629.1	645.8	674.2	693.2	711.1

T/K	235.85	230.25	225.15
P/bar	768.1	821.4	862.2

w_B 0.0091   was kept constant

T/K	429.45	407.95	388.05	365.85	350.15	334.95	320.05	300.95	299.35
P/bar	702.9	705.0	709.7	708.3	708.7	712.0	718.5	745.6	754.1

T/K	297.75	292.45	273.75	266.75	260.75	257.45	252.75	245.45	240.15
P/bar	759.6	768.1	818.2	851.6	880.2	901.2	933.7	988.1	1034

T/K	235.25	231.65	229.55
P/bar	1089	1136	1168

Polymer (B):   **ethylene/1-butene copolymer**   **1998HAN**

Characterization: M_n/kg.mol^{-1} = 52.45, M_w/kg.mol^{-1} = 104.9, 5.3 mol% 1-butene,
T_m/K = 385.2, synthesized with metallocene catalyst

Solvent (A):   **propane**   **C₃H₈**   **74-98-6**

Type of data: cloud points

w_B	0.05	0.05	0.05	0.05	0.05
T/K	374.55	398.45	423.55	446.95	471.65
P/bar	592.2	580.8	576.9	576.6	577.6

Polymer (B):   **ethylene/1-butene copolymer**   **1998HAN**

Characterization: M_n/kg.mol^{-1} = 38.35, M_w/kg.mol^{-1} = 76.7, 7.5 mol% 1-butene,
T_m/K = 381.2, synthesized with metallocene catalyst

Solvent (A):   **propane**   **C₃H₈**   **74-98-6**

Type of data: cloud points

w_B	0.01	0.01	0.01	0.01	0.01	0.01	0.05	0.05	0.05
T/K	353.85	374.35	397.95	423.35	447.25	471.65	373.15	398.35	423.05
P/bar	587.1	569.4	560.5	561.7	561.5	563.4	571.9	563.6	561.8

w_B	0.05	0.05
T/K	446.55	473.15
P/bar	562.7	565.6

Polymer (B):   **ethylene/1-butene copolymer**   **1998HAN**

Characterization: $M_n/\text{kg.mol}^{-1} = 50$, $M_w/\text{kg.mol}^{-1} = 94.9$, 8.8 mol% 1-butene,
$T_m/\text{K} = 371.2$, synthesized with metallocene catalyst

Solvent (A):   **propane**   **C$_3$H$_8$**   **74-98-6**

Type of data: cloud points

w_B	0.01	0.01	0.01	0.01	0.01	0.01	0.01	0.05	0.05
T/K	348.35	363.55	373.95	400.35	421.65	446.35	471.05	374.15	397.85
P/bar	580.1	565.8	556.6	556.2	553.4	558.0	561.1	563.5	556.2

w_B	0.05	0.05	0.05
T/K	423.45	447.15	472.05
P/bar	556.4	558.4	561.9

Polymer (B):   **ethylene/1-butene copolymer**   **1998HAN**

Characterization: $M_n/\text{kg.mol}^{-1} = 41.6$, $M_w/\text{kg.mol}^{-1} = 79$, 15.9 mol% 1-butene,
$T_m/\text{K} = 350.2$, synthesized with metallocene catalyst

Solvent (A):   **propane**   **C$_3$H$_8$**   **74-98-6**

Type of data: cloud points

w_B	0.01	0.01	0.01	0.01	0.01	0.01	0.01	0.01	0.05
T/K	313.55	322.95	348.85	373.15	396.55	423.25	446.55	470.35	373.15
P/bar	616.1	523.2	502.6	506.7	500.3	509.1	515.2	523.5	497.4

w_B	0.05	0.05	0.05	0.05
T/K	398.65	422.45	447.65	474.35
P/bar	501.8	508.6	517.0	525.0

Polymer (B):   **ethylene/1-butene copolymer**   **1998HAN**

Characterization: $M_n/\text{kg.mol}^{-1} = 29$, $M_w/\text{kg.mol}^{-1} = 58.1$, 18.0 mol% 1-butene,
$T_m/\text{K} = 335.2$, synthesized with metallocene catalyst

Solvent (A):   **propane**   **C$_3$H$_8$**   **74-98-6**

Type of data: cloud points

w_B	0.01	0.01	0.01	0.01	0.01	0.01	0.01	0.01	0.05
T/K	324.25	334.55	348.15	373.55	398.85	422.15	446.65	471.85	348.25
P/bar	448.3	443.7	442.1	448.4	460.4	471.7	482.7	493.7	448.2

w_B	0.05	0.05	0.05	0.05	0.05
T/K	372.55	395.95	422.95	447.45	469.65
P/bar	453.2	462.7	475.7	487.7	496.5

Polymer (B):   **ethylene/1-butene copolymer**   **1998HAN**

Characterization: $M_n/\text{kg.mol}^{-1} = 79.3$, $M_w/\text{kg.mol}^{-1} = 87.2$, 23.3 mol% 1-butene,
$T_m/\text{K} = 319.2$, completely hydrogenated polybutadiene

Solvent (A):   **propane**   **C$_3$H$_8$**   **74-98-6**

continued

continued

Type of data: cloud points

w_B	0.01	0.01	0.01	0.01	0.01	0.01	0.01	0.05	0.05
T/K	295.45	324.05	348.75	374.65	398.15	423.05	447.65	295.15	324.15
P/bar	376.5	373.1	384.1	401.2	418.3	437.1	450.6	391.0	385.4

w_B	0.05	0.05	0.05	0.05	0.05
T/K	348.85	373.95	397.95	422.95	447.55
P/bar	396.2	411.0	428.2	445.2	460.4

Polymer (B):	**ethylene/1-butene copolymer**		**1998HAN**
Characterization:	M_n/kg.mol^{-1} = 79.8, M_w/kg.mol^{-1} = 87.8, 60.0 mol% 1-butene, amorphous, completely hydrogenated polybutadiene		
Solvent (A):	**propane**	**C$_3$H$_8$**	**74-98-6**

Type of data: cloud points

w_B	0.01	0.01	0.01	0.01	0.01	0.01	0.01	0.01	0.05
T/K	296.55	323.15	349.05	373.95	397.25	423.55	445.25	472.15	333.85
P/bar	82.4	147.3	200.3	244.4	279.7	313.5	338.0	362.2	289.7

w_B	0.05	0.05	0.05	0.05	0.05	0.05
T/K	348.25	372.65	398.75	422.95	447.65	473.05
P/bar	282.6	317.6	355.5	387.8	412.9	431.3

Polymer (B):	**ethylene/butyl acrylate copolymer**		**1996MU1**
Characterization:	M_n/kg.mol^{-1} = 6.9, M_w/kg.mol^{-1} = 34.9, 15.0 wt% butyl acrylate, sample from BASF AG, Germany		
Solvent (A):	**ethene**	**C$_2$H$_4$**	**74-85-1**

Type of data: cloud points

w_B	0.025	0.025	0.025	0.025	0.025	0.050	0.050	0.050	0.050
T/K	522.85	498.05	473.05	448.55	423.85	524.05	498.35	473.85	449.15
P/MPa	117.7	121.5	126.3	132.3	139.6	116.9	120.8	125.5	131.3

w_B	0.050	0.100	0.100	0.100	0.100	0.100	0.150	0.150	0.150
T/K	423.75	522.85	498.05	473.15	448.95	424.15	524.15	498.25	473.55
P/MPa	138.8	121.2	124.1	127.5	132.1	139.4	120.8	122.5	125.1

w_B	0.150	0.150	0.200	0.200	0.200	0.200	0.200
T/K	448.95	424.35	523.35	498.05	473.35	449.15	423.35
P/MPa	128.5	133.6	117.3	118.8	121.5	124.8	128.7

Polymer (B):	**ethylene/butyl acrylate copolymer**		**1996MU1**
Characterization:	M_n/kg.mol^{-1} = 7.7, M_w/kg.mol^{-1} = 38.7, 19.0 wt% butyl acrylate, sample from BASF AG, Germany		
Solvent (A):	**ethene**	**C$_2$H$_4$**	**74-85-1**

Type of data: cloud points

continued

continued

w_B	0.034	0.034	0.034	0.034	0.034	0.050	0.050	0.050	0.050
T/K	523.05	498.55	472.65	448.15	423.15	523.95	497.85	473.05	449.25
P/MPa	116.8	120.6	125.4	131.6	138.4	118.4	121.3	125.3	130.8

w_B	0.050	0.100	0.100	0.100	0.100	0.100	0.150	0.150	0.150
T/K	423.95	523.55	498.85	473.45	449.05	424.25	524.85	498.15	472.65
P/MPa	139.0	121.7	123.6	128.2	131.6	139.4	120.3	122.1	128.4

w_B	0.150	0.150	0.199	0.199	0.199	0.199
T/K	448.95	424.25	523.25	498.55	473.35	448.95
P/MPa	130.6	134.7	120.7	123.0	126.1	129.8

Polymer (B):	**ethylene/butyl acrylate copolymer**	**1999DIE**
Characterization:	$M_n/kg.mol^{-1} = 16.4$, $M_w/kg.mol^{-1} = 294$,	
	1.0 mol% butyl acrylate, synthesized in the laboratory	
Solvent (A):	**ethene** C_2H_4	**74-85-1**

Type of data: cloud points

w_B	0.03	was kept constant

T/K	383.15	393.15	403.15	413.15	423.15
P/bar	1900	1803	1725	1670	1600

Polymer (B):	**ethylene/butyl acrylate copolymer**	**1999DIE**
Characterization:	$M_n/kg.mol^{-1} = 44.3$, $M_w/kg.mol^{-1} = 162$,	
	5.1 mol% butyl acrylate, synthesized in the laboratory	
Solvent (A):	**ethene** C_2H_4	**74-85-1**

Type of data: cloud points

w_B	0.03	was kept constant

T/K	388.15	393.15	393.15	398.15	398.15	403.15	408.15	413.15	418.15
P/bar	1900	1881	1870	1822	1796	1750	1677	1673	1628

T/K	423.15	423.15	433.15	434.15	438.15	443.15	443.15	453.15	463.15
P/bar	1581	1565	1546	1545	1541	1498	1490	1490	1452

T/K	472.15	483.15
P/bar	1419	1392

Polymer (B):	**ethylene/butyl acrylate copolymer**	**1999DIE**
Characterization:	$M_n/kg.mol^{-1} = 41.1$, $M_w/kg.mol^{-1} = 296$,	
	6.7 mol% butyl acrylate, synthesized in the laboratory	
Solvent (A):	**ethene** C_2H_4	**74-85-1**

Type of data: cloud points

w_B	0.03	was kept constant

T/K	363.15	368.15	384.15	403.15	423.15
P/bar	1770	1750	1595	1510	1438

Polymer (B):   **ethylene/butyl acrylate copolymer**   **1999DIE**

Characterization: $M_n/\text{kg.mol}^{-1} = 39.9$, $M_w/\text{kg.mol}^{-1} = 155$,

       8.7 mol% butyl acrylate, synthesized in the laboratory

Solvent (A):   **ethene**   **C$_2$H$_4$**   **74-85-1**

Type of data: cloud points

w_B 0.03   was kept constant

T/K	333.15	343.15	343.15	353.15	353.15	363.15	363.15	373.15	373.15
P/bar	1655	1582	1562	1503	1472	1460	1432	1415	1402

T/K	383.15	384.15	393.15	403.15	403.15	413.15	423.15	423.15	432.15
P/bar	1355	1380	1320	1320	1318	1289	1270	1259	1255

T/K	443.15	443.15	453.15	463.15	473.15	483.15
P/bar	1230	1242	1212	1189	1173	1160

Polymer (B):   **ethylene/butyl acrylate copolymer**   **1999DIE**

Characterization: 17.2 mol% butyl acrylate, synthesized in the laboratory

Solvent (A):   **ethene**   **C$_2$H$_4$**   **74-85-1**

Type of data: cloud points

w_B 0.03   was kept constant

T/K	328.15	331.15	336.15	341.15	347.15	354.15	364.15	383.15	405.15
P/bar	1400	1400	1370	1340	1324	1300	1290	1220	1190

T/K	421.15	443.15
P/bar	1260	1130

Polymer (B):   **ethylene/butyl acrylate copolymer**   **1999DIE**

Characterization: 22.2 mol% butyl acrylate, synthesized in the laboratory

Solvent (A):   **ethene**   **C$_2$H$_4$**   **74-85-1**

Type of data: cloud points

w_B 0.03   was kept constant

T/K	318.15	323.15	333.15	343.15	353.15	363.15	383.15	393.15	403.15
P/bar	1355	1320	1295	1260	1230	1210	1175	1158	1145

T/K	413.15	423.15	433.15	443.15	453.15	463.15	473.15
P/bar	1130	1127	1125	1120	1092	1090	1080

Polymer (B):   **ethylene/butyl acrylate copolymer**   **1999DIE**

Characterization: $M_n/\text{kg.mol}^{-1} = 29.7$, $M_w/\text{kg.mol}^{-1} = 110.5$,

       25.2 mol% butyl acrylate, synthesized in the laboratory

Solvent (A):   **ethene**   **C$_2$H$_4$**   **74-85-1**

Type of data: cloud points

continued

continued

w_B	0.03	was kept constant						
T/K	343.15	363.15	383.15	403.15	403.15	423.15	443.15	463.15	473.15
P/bar	1240	1173	1165	1132	1125	1092	1058	1040	1037

Polymer (B):   **ethylene/butyl acrylate copolymer**   **2004BEC**

Characterization: M_n/kg.mol^{-1} = 49.8, M_w/kg.mol^{-1} = 159.5, 27.0 mol% butyl acrylate, synthesized in the laboratory

Solvent (A):   **ethene**   **C$_2$H$_4$**   **74-85-1**

Type of data: cloud points

w_B	0.05	was kept constant						
T/K	354.15	374.15	393.15	412.15	432.15	452.15	471.15	489.15	510.15
P/MPa	1144	1106	1071	1052	1033	1014	993	977	965

T/K	530.15
P/MPa	953

Polymer (B):   **ethylene/butyl acrylate copolymer**   **1996MU1**

Characterization: M_n/kg.mol^{-1} = 7.9, M_w/kg.mol^{-1} = 39.8, 35.0 wt% butyl acrylate, sample from BASF AG, Germany

Solvent (A):   **ethene**   **C$_2$H$_4$**   **74-85-1**

Type of data: cloud points

w_B	0.025	0.025	0.025	0.025	0.025	0.050	0.050	0.050	0.050
T/K	523.65	498.55	473.05	448.55	423.45	523.75	498.05	472.95	448.05
P/MPa	111.5	115.9	120.4	127.2	134.4	109.4	113.1	117.2	122.3

w_B	0.050	0.100	0.100	0.100	0.100	0.100	0.150	0.150	0.150
T/K	423.25	523.45	498.05	473.15	448.45	423.45	523.85	498.35	472.95
P/MPa	129.7	107.1	109.8	113.1	116.9	121.9	105.9	108.1	111.5

w_B	0.150	0.150	0.199	0.199	0.199	0.199	0.199
T/K	448.05	423.85	523.75	498.05	473.25	449.15	424.05
P/MPa	115.1	119.8	104.9	106.7	109.5	115.4	120.5

Polymer (B):   **ethylene/butyl acrylate copolymer**   **1996MU1**

Characterization: M_n/kg.mol^{-1} = 11.5, M_w/kg.mol^{-1} = 37.2, 67.0 wt% butyl acrylate, synthesized in the laboratory

Solvent (A):   **ethene**   **C$_2$H$_4$**   **74-85-1**

Type of data: cloud points

w_B	0.168	0.168	0.168	0.168	0.168
T/K	522.35	498.15	473.65	448.35	424.65
P/MPa	96.6	98.4	100.1	102.3	104.0

Polymer (B):   **ethylene/butyl acrylate copolymer**   **1996MU1**

Characterization: M_n/kg.mol^{-1} = 13.9, M_w/kg.mol^{-1} = 35.2, 75.0 wt% butyl acrylate, synthesized in the laboratory

Solvent (A):   **ethene**   **C$_2$H$_4$**   **74-85-1**

Type of data: cloud points

w_B	0.069	0.069	0.069	0.069	0.069	0.069
T/K	524.05	499.05	474.45	448.65	423.75	398.75
P/MPa	89.4	90.3	92.9	96.5	99.8	102.1

Polymer (B):   **ethylene/butyl methacrylate copolymer**   **2004BEC**

Characterization: M_n/kg.mol^{-1} = 23.4, M_w/kg.mol^{-1} = 39.8,
12.0 mol% butyl methacrylate, synthesized in the laboratory

Solvent (A):   **ethene**   **C$_2$H$_4$**   **74-85-1**

Type of data: cloud points

w_B 0.05   was kept constant

T/K	373.15	393.15	413.15	433.15	453.15	473.15	488.15	508.15	528.15
P/bar	1250	1200	1151	1114	1089	1062	1041	1021	1000

Polymer (B):   **ethylene/butyl methacrylate copolymer**   **2004BEC**

Characterization: M_n/kg.mol^{-1} = 24.8, M_w/kg.mol^{-1} = 41,
19.0 mol% butyl methacrylate, synthesized in the laboratory

Solvent (A):   **ethene**   **C$_2$H$_4$**   **74-85-1**

Type of data: cloud points

w_B 0.05   was kept constant

T/K	353.15	373.15	393.15	413.15	433.15	453.15	473.15	488.15	508.15
P/bar	1082	1045	1016	993	973	958	946	930	915

T/K	528.15
P/bar	901

Polymer (B):   **ethylene/butyl methacrylate copolymer**   **2004BEC**

Characterization: M_n/kg.mol^{-1} = 24.5, M_w/kg.mol^{-1} = 44.4,
44.0 mol% butyl methacrylate, synthesized in the laboratory

Solvent (A):   **ethene**   **C$_2$H$_4$**   **74-85-1**

Type of data: cloud points

w_B 0.05   was kept constant

T/K	383.15	393.15	413.15	433.15	453.15	473.15	493.15	513.15	533.15
P/bar	885	878	866	857	847	840	834	823	813

Polymer (B):   **ethylene/ethyl acrylate copolymer**   **2004BEC**
Characterization: 4.0 mol% ethyl acrylate, synthesized in the laboratory
Solvent (A):   **ethene**   **C$_2$H$_4$**   **74-85-1**

Type of data: cloud points

w_B 0.05   was kept constant

T/K	384.05	395.15	414.85	435.15	454.15	473.15	492.15	514.15	531.15
P/bar	1631	1568	1471	1394	1336	1288	1247	1207	1175

Polymer (B):   **ethylene/ethyl acrylate copolymer**   **2004BEC**
Characterization: 6.0 mol% ethyl acrylate, synthesized in the laboratory
Solvent (A):   **ethene**   **C$_2$H$_4$**   **74-85-1**

Type of data: cloud points

w_B 0.05   was kept constant

T/K	372.15	393.15	413.15	433.15	452.15	472.55	492.15	511.15	531.35
P/bar	1510	1420	1351	1295	1252	1212	1181	1147	1122

Polymer (B):   **ethylene/ethyl acrylate copolymer**   **2004BEC**
Characterization: M_n/kg.mol^{-1} = 78.4, M_w/kg.mol^{-1} = 156.6,
                      23 mol% ethyl acrylate, synthesized in the laboratory
Solvent (A):   **ethene**   **C$_2$H$_4$**   **74-85-1**

Type of data: cloud points

w_B 0.05   was kept constant

T/K	353.15	364.15	373.75	393.15	413.15	434.15	454.15	473.15	493.65
P/bar	1280	1256	1236	1190	1155	1123	1095	1072	1047

T/K	513.55	533.15
P/bar	1025	1000

Polymer (B):   **ethylene/ethyl acrylate copolymer**   **2004BEC**
Characterization: M_n/kg.mol^{-1} = 64.4, M_w/kg.mol^{-1} = 117,
                      29 mol% ethyl acrylate, synthesized in the laboratory
Solvent (A):   **ethene**   **C$_2$H$_4$**   **74-85-1**

Type of data: cloud points

w_B 0.05   was kept constant

T/K	363.15	373.15	383.15	393.15	413.15	433.15	453.15	473.15	493.15
P/bar	1236	1206	1183	1165	1131	1100	1072	1048	1022

T/K	513.15	533.15
P/bar	1000	982

Polymer (B):   **ethylene/2-ethylhexyl acrylate copolymer**   **1999DIE**
Characterization: M_n/kg.mol^{-1} = 34.9, M_w/kg.mol^{-1} = 107.7,
                  9.3 mol% 2-ethylhexyl acrylate, synthesized in the laboratory
Solvent (A):   **ethene**   **C$_2$H$_4$**   **74-85-1**

Type of data: cloud points

w_B 0.03   was kept constant

T/K	363.15	383.15	403.15	423.15	443.15	462.15	483.15
P/bar	1640	1590	1440	1390	1340	1310	1290

Polymer (B):   **ethylene/2-ethylhexyl acrylate copolymer**   **1999DIE**
Characterization: M_n/kg.mol^{-1} = 21.2, M_w/kg.mol^{-1} = 49.8,
                  17.0 mol% 2-ethylhexyl acrylate, synthesized in the laboratory
Solvent (A):   **ethene**   **C$_2$H$_4$**   **74-85-1**

Type of data: cloud points

w_B 0.03   was kept constant

T/K	363.15	383.15	403.15	423.15	443.15	463.15	483.15
P/bar	1240	1220	1145	1125	1085	1060	1010

Polymer (B):   **ethylene/2-ethylhexyl acrylate copolymer**   **1999DIE**
Characterization: 22.8 mol% 2-ethylhexyl acrylate, synthesized in the laboratory
Solvent (A):   **ethene**   **C$_2$H$_4$**   **74-85-1**

Type of data: cloud points

w_B 0.03   was kept constant

T/K	353.15	363.15	383.15	403.15	423.15	443.15	463.15	483.15
P/bar	1070	1055	1025	1005	995	995	965	945

Polymer (B):   **ethylene/1-hexene copolymer**   **2000CII3**
Characterization: M_n/kg.mol^{-1} = 52.6, M_w/kg.mol^{-1} = 80, 10.6 wt% 1-hexene
Solvent (A):   **ethene**   **C$_2$H$_4$**   **74-85-1**

Type of data: cloud points

w_B	0.148	0.148	0.148	0.148
T/K	393.25	413.15	433.25	453.15
P/MPa	168.0	154.5	146.0	138.9

Polymer (B):   **ethylene/1-hexene copolymer**   **1999KIN, 2001DOE**
Characterization: M_n/kg.mol^{-1} = 60, M_w/kg.mol^{-1} = 129, 16.1 wt% 1-hexene
Solvent (A):   **ethene**   **C$_2$H$_4$**   **74-85-1**

continued

continued

Type of data: cloud points

w_B	0.150	0.150	0.150	0.150	0.150	0.150
T/K	393.15	413.15	433.15	453.15	473.15	493.15
P/MPa	156.0	144.3	137.3	129.1	123.1	118.6

Polymer (B):	**ethylene/1-hexene copolymer**	**2000CH3**

Characterization: M_n/kg.mol^{-1} = 48.1, M_w/kg.mol^{-1} = 103, 35.0 wt% 1-hexene

Solvent (A):   **ethene**   **C_2H_4**   **74-85-1**

Type of data: cloud points

w_B	0.050	0.050	0.050	0.050	0.050	0.050	0.050	0.100	0.100
T/K	393.15	403.15	413.15	423.15	433.15	443.15	453.15	393.15	403.15
P/MPa	136.3	132.6	129.6	127.0	124.9	122.9	121.0	134.8	131.4

w_B	0.100	0.100	0.100	0.100	0.100	0.150	0.150	0.150	0.150
T/K	413.15	423.15	433.15	443.15	453.15	393.15	403.15	413.15	423.15
P/MPa	128.3	125.8	123.7	121.8	120.8	129.4	126.1	123.6	121.5

w_B	0.150	0.150	0.150
T/K	433.15	443.15	453.15
P/MPa	119.8	118.7	119.4

Polymer (B):	**ethylene/1-hexene copolymer**	**1999PAN**

Characterization: M_n/kg.mol^{-1} = 46.5, M_w/kg.mol^{-1} = 102, 8.5 wt% 1-hexene,
MI = 1.2, ρ = 0.912 g/cm^3, T_m/K = 409.95

Solvent (A):   **2-methylpropane**   **C_4H_{10}**   **75-28-5**

Type of data: cloud points

w_B	0.110	0.110	0.110	0.110	0.110	0.110	0.110	0.110	0.110
T/K	362.15	362.25	376.75	377.55	379.35	393.15	393.35	399.45	405.15
P/bar	353.4	353.4	352.8	352.5	353.0	354.2	354.8	356.0	356.5

w_B	0.110	0.110	0.110	0.110
T/K	407.55	410.45	423.05	423.25
P/bar	357.9	358.5	362.8	364.3

Polymer (B):	**ethylene/1-hexene copolymer**	**1999PAN**

Characterization: M_n/kg.mol^{-1} = 54.8, M_w/kg.mol^{-1} = 109.5, 13.6 wt% 1-hexene,
MI = 1.2, ρ = 0.900 g/cm^3, T_m/K = 406.25

Solvent (A):   **2-methylpropane**   **C_4H_{10}**   **75-28-5**

Type of data: cloud points

w_B	0.112	0.112	0.112	0.112	0.112	0.112	0.112	0.112	0.112
T/K	355.65	358.35	368.55	368.55	377.85	378.65	392.05	395.25	405.45
P/bar	317.9	318.7	318.0	318.2	319.8	319.7	324.8	324.6	330.4

w_B	0.112	0.112	0.112	0.112
T/K	407.55	410.25	421.75	424.65
P/bar	331.5	331.3	338.0	338.5

Polymer (B):   ethylene/1-hexene copolymer   **1999PAN**
Characterization: M_n/kg.mol^{-1} = 45.5, M_w/kg.mol^{-1} = 90.8, 13.7 wt% 1-hexene,
MI = 2.2, ρ = 0.900 g/cm^3, T_m/K = 406.25
Solvent (A):   **2-methylpropane**   **C$_4$H$_{10}$**   **75-28-5**

Type of data: cloud points

w_B	0.111	0.111	0.111	0.111	0.111	0.111	0.111	0.111	0.111
T/K	362.65	362.65	373.15	373.25	383.15	383.25	392.55	393.35	407.95
P/bar	312.5	313.1	314.5	313.3	315.8	315.8	321.1	322.9	327.6

w_B	0.111	0.111	0.111
T/K	408.75	423.25	423.45
P/bar	328.7	335.3	336.2

Polymer (B):   ethylene/1-hexene copolymer   **1999PAN**
Characterization: M_n/kg.mol^{-1} = 42.8, M_w/kg.mol^{-1} = 85.5, 14.2 wt% 1-hexene,
MI = 3.5, ρ = 0.900 g/cm^3, T_m/K = 405.95
Solvent (A):   **2-methylpropane**   **C$_4$H$_{10}$**   **75-28-5**

Type of data: cloud points

w_B	0.110	0.110	0.110	0.110	0.110	0.110	0.110	0.110	0.110
T/K	358.25	358.75	368.15	368.95	377.45	380.25	393.25	394.15	406.45
P/bar	305.1	305.7	306.6	307.0	309.4	311.0	316.7	316.4	322.3

w_B	0.110	0.110	0.110
T/K	408.85	423.45	423.85
P/bar	323.6	331.2	331.1

Polymer (B):   ethylene/1-hexene copolymer   **1999PAN**
Characterization: M_n/kg.mol^{-1} = 31.3, M_w/kg.mol^{-1} = 68.8, 14.6wt% 1-hexene,
MI = 7.5, ρ = 0.900 g/cm^3, T_m/K = 405.55
Solvent (A):   **2-methylpropane**   **C$_4$H$_{10}$**   **75-28-5**

Type of data: cloud points

w_B	0.112	0.112	0.112	0.112	0.112	0.112	0.112	0.112	0.112
T/K	357.45	360.35	366.35	368.25	376.65	378.45	393.15	394.05	406.15
P/bar	301.3	301.8	302.5	302.5	306.6	306.5	312.9	313.4	320.0

w_B	0.112	0.112	0.112
T/K	409.15	422.95	423.25
P/bar	320.9	327.5	327.1

Polymer (B):   ethylene/1-hexene copolymer   **1999PAN**
Characterization: M_n/kg.mol^{-1} = 47.5, M_w/kg.mol^{-1} = 94.9, 21.6wt% 1-hexene,
MI = 2.2, ρ = 0.885 g/cm^3, T_m/K = 400.25
Solvent (A):   **2-methylpropane**   **C$_4$H$_{10}$**   **75-28-5**

continued

continued

Type of data: cloud points

w_B	0.109	0.109	0.109	0.109	0.109	0.109	0.109	0.109	0.109
T/K	358.55	359.75	362.75	367.95	368.95	377.55	377.65	391.95	393.85
P/bar	273.0	273.4	273.8	275.9	276.8	280.8	281.4	289.8	290.4

w_B	0.109	0.109	0.109	0.109	0.109
T/K	395.35	407.15	409.15	423.55	423.55
P/bar	292.1	298.9	300.7	309.9	310.2

Polymer (B):	**ethylene/1-hexene copolymer**	**1998HAN**
Characterization:	M_n/kg.mol^{-1} = 42.4, M_w/kg.mol^{-1} = 93.2, 2.6 mol% 1-hexene,	
	T_m/K = 401.2, synthesized with metallocene catalyst	
Solvent (A):	**propane** **C$_3$H$_8$**	**74-98-6**

Type of data: cloud points

w_B	0.05	0.05	0.05	0.05
T/K	400.75	423.55	448.35	471.95
P/bar	591.2	586.0	584.2	584.7

Polymer (B):	**ethylene/1-hexene copolymer**	**1998HAN**
Characterization:	M_n/kg.mol^{-1} = 53, M_w/kg.mol^{-1} = 101, 4.6 mol% 1-hexene,	
	T_m/K = 391.2, synthesized with metallocene catalyst	
Solvent (A):	**propane** **C$_3$H$_8$**	**74-98-6**

Type of data: cloud points

w_B	0.05	0.05	0.05	0.05	0.05
T/K	373.65	398.75	427.55	447.55	471.75
P/bar	575.2	565.8	563.9	565.3	567.4

Polymer (B):	**ethylene/1-hexene copolymer**	**1998HAN**
Characterization:	M_n/kg.mol^{-1} = 24.1, M_w/kg.mol^{-1} = 72.3, 13.0 mol% 1-hexene,	
	T_m/K = 349.2, synthesized with metallocene catalyst	
Solvent (A):	**propane** **C$_3$H$_8$**	**74-98-6**

Type of data: cloud points

w_B	0.05	0.05	0.05	0.05	0.05	0.05
T/K	353.05	378.45	401.75	426.35	448.95	474.95
P/bar	471.5	472.9	479.3	489.3	495.9	505.5

Polymer (B):	**ethylene/1-hexene copolymer**	**1999PAN**
Characterization:	M_n/kg.mol^{-1} = 46.5, M_w/kg.mol^{-1} = 102, 8.5 wt% 1-hexene,	
	MI = 1.2, ρ = 0.912 g/cm^3, T_m/K = 409.95	
Solvent (A):	**propane** **C$_3$H$_8$**	**74-98-6**

continued

continued

Type of data: cloud points

w_B	0.111	0.111	0.111	0.111	0.111	0.111	0.111	0.111	0.111
T/K	367.15	367.95	377.55	378.05	390.75	393.35	407.95	408.65	422.75
P/bar	520.8	521.1	520.0	520.4	520.4	521.2	521.8	522.1	525.3

w_B	0.111
T/K	423.45
P/bar	525.0

Polymer (B):   **ethylene/1-hexene copolymer**   **1999PAN**
Characterization: M_n/kg.mol^{-1} = 54.8, M_w/kg.mol^{-1} = 109.5, 13.6 wt% 1-hexene,
                        MI = 1.2, ρ = 0.900 g/cm^3, T_m/K = 406.25
Solvent (A):   **propane**   **C$_3$H$_8$**   **74-98-6**

Type of data: cloud points

w_B	0.109	0.109	0.109	0.109	0.109	0.109	0.109	0.109	0.109
T/K	357.15	358.75	367.35	367.55	379.95	381.65	390.55	394.05	408.65
P/bar	481.8	483.0	484.3	484.3	486.2	486.2	489.1	490.8	495.3

w_B	0.109	0.109	0.109
T/K	412.15	422.45	422.75
P/bar	496.7	500.1	500.1

Polymer (B):   **ethylene/1-hexene copolymer**   **1999PAN**
Characterization: M_n/kg.mol^{-1} = 45.5, M_w/kg.mol^{-1} = 90.8, 13.7 wt% 1-hexene,
                        MI = 2.2, ρ = 0.900 g/cm^3, T_m/K = 406.25
Solvent (A):   **propane**   **C$_3$H$_8$**   **74-98-6**

Type of data: cloud points

w_B	0.109	0.109	0.109	0.109	0.109	0.109	0.109	0.109	0.109
T/K	356.55	359.35	366.65	368.35	376.25	377.85	393.15	395.55	406.15
P/bar	460.9	461.5	461.8	462.1	463.8	464.3	468.8	468.9	473.9

w_B	0.109	0.109	0.109
T/K	408.15	423.15	424 15
P/bar	474.9	481.0	481.3

Polymer (B):   **ethylene/1-hexene copolymer**   **1999PAN**
Characterization: M_n/kg.mol^{-1} = 42.8, M_w/kg.mol^{-1} = 85.5, 14.2 wt% 1-hexene,
                        MI = 3.5, ρ = 0.900 g/cm^3, T_m/K = 405.95
Solvent (A):   **propane**   **C$_3$H$_8$**   **74-98-6**

Type of data: cloud points

w_B	0.109	0.109	0.109	0.109	0.109	0.109	0.109	0.109	0.109
T/K	353.25	355.65	364.05	365.15	375.15	377.75	390.15	393.55	408.45
P/bar	459.0	459.2	459.9	461.3	463.2	464.1	468.3	469.9	474.4

continued

continued

w_B	0.109	0.109	0.109
T/K	408.55	422.75	423.15
P/bar	474.2	478.2	478.4

Polymer (B):   **ethylene/1-hexene copolymer**   **1999PAN**
Characterization: M_n/kg.mol^{-1} = 47.5, M_w/kg.mol^{-1} = 94.9, 21.6 wt% 1-hexene,
     MI = 2.2, ρ = 0.885 g/cm^3, T_m/K = 400.25
Solvent (A):   **propane**   **C$_3$H$_8$**   **74-98-6**

Type of data: cloud points

w_B	0.110	0.110	0.110	0.110	0.110	0.110	0.110	0.110	0.110
T/K	355.95	358.65	367.45	368.55	376.95	378.35	392.85	395.65	406.25
P/bar	425.1	426.4	429.5	430.5	435.0	435.1	444.4	444.9	451.2

w_B	0.110	0.110	0.110
T/K	409.35	422.85	423.65
P/bar	453.8	459.9	461.5

Polymer (B):   **ethylene/1-hexene copolymer**   **2000CH3**
Characterization: M_n/kg.mol^{-1} = 52.6, M_w/kg.mol^{-1} = 80, 10.6 wt% 1-hexene
Solvent (A):   **propane**   **C$_3$H$_8$**   **74-98-6**

Type of data: cloud points

w_B	0.010	0.010	0.010	0.010	0.051	0.051	0.051	0.051	0.148
T/K	363.15	373.25	413.15	453.15	363.15	373.15	413.15	453.15	373.25
P/MPa	50.3	50.0	49.9	50.7	51.0	50.4	50.4	50.4	50.6

w_B	0.148	0.148
T/K	413.15	453.15
P/MPa	50.4	51.1

Polymer (B):   **ethylene/1-hexene copolymer**   **2000CH2**
Characterization: M_n/kg.mol^{-1} = 51.3, M_w/kg.mol^{-1} = 126, 20.6 wt% 1-hexene
Solvent (A):   **propane**   **C$_3$H$_8$**   **74-98-6**

Type of data: cloud points

T/K = 453.15

w_B	0.018	0.031	0.055	0.066	0.075	0.079	0.089	0.098	0.100
P/bar	446.0	454.0	456.0	459.0	468.0	475.0	478.0	470.0	474.0

w_B	0.108	0.114	0.148
P/bar	471.0	466.0	459.0

Polymer (B):   **ethylene/1-hexene copolymer**   **2000CH3**

Characterization: M_n/kg.mol^{-1} = 48.1, M_w/kg.mol^{-1} = 103, 35.0 wt% 1-hexene

Solvent (A):   **propane**   **C$_3$H$_8$**   **74-98-6**

Type of data: cloud points

w_B	0.150	0.150	0.150	0.150	0.150	0.150
T/K	313.15	343.15	363.05	373.05	413.15	453.15
P/MPa	37.3	38.1	39.1	39.7	42.3	44.6

Polymer (B):   **ethylene/1-hexene copolymer**   **2000CH3**

Characterization: M_n/kg.mol^{-1} = 112, M_w/kg.mol^{-1} = 139, 35.0 wt% 1-hexene

Solvent (A):   **propane**   **C$_3$H$_8$**   **74-98-6**

Type of data: cloud points

w_B	0.009	0.009	0.009	0.009	0.009	0.009	0.152	0.152	0.152
T/K	313.25	353.25	368.25	373.15	413.15	453.15	353.25	373.15	413.15
P/MPa	39.0	36.8	37.6	37.9	40.6	42.9	37.8	39.2	41.5

w_B	0.152
T/K	453.15
P/MPa	43.7

Polymer (B):   **ethylene/methacrylic acid copolymer**   **2003BUB**

Characterization: M_n/kg.mol^{-1} = 25.7, M_w/kg.mol^{-1} = 83.2, 1.0 mol% methacrylic acid

Solvent (A):   **ethene**   **C$_2$H$_4$**   **74-85-1**

Type of data: cloud points

w_B	0.03	was kept constant

T/K	383.15	393.15	403.15	418.15	433.15	453.15	464.15	473.15	483.15
P/bar	2730	2390	2190	1940	1760	1595	1520	1475	1425

T/K	493.15	503.15	513.15	523.15	533.15
P/bar	1375	1345	1320	1285	1260

Polymer (B):   **ethylene/methacrylic acid copolymer**   **2003BUB**

Characterization: 2.2 mol% methacrylic acid, synthesized in the laboratory

Solvent (A):   **ethene**   **C$_2$H$_4$**   **74-85-1**

Type of data: cloud points

w_B	0.03	was kept constant

T/K	420.15	426.15	431.15	438.15	443.15	453.15	463.15	473.15	482.15
P/bar	2450	2295	2190	2060	1975	1840	1730	1645	1570

T/K	493.15	503.15	512.15	522.15	533.15
P/bar	1490	1435	1395	1360	1315

Polymer (B):   **ethylene/methacrylic acid copolymer**   **2003BUB**
Characterization: M_n/kg.mol^{-1} = 17.4, M_w/kg.mol^{-1} = 51.5, 3.4 mol% methacrylic acid
Solvent (A):   **ethene**   **C$_2$H$_4$**   **74-85-1**

Type of data: cloud points

w_B 0.03   was kept constant

T/K	443.15	453.15	463.15	473.15	483.15	493.15	503.15	513.15	523.15
P/bar	2355	2160	1950	1805	1690	1585	1495	1440	1385

T/K	533.15
P/bar	1340

Polymer (B):   **ethylene/methacrylic acid copolymer**   **2003BUB**
Characterization: M_n/kg.mol^{-1} = 21.8, M_w/kg.mol^{-1} = 61.2, 4.8 mol% methacrylic acid
Solvent (A):   **ethene**   **C$_2$H$_4$**   **74-85-1**

Type of data: cloud points

w_B 0.03   was kept constant

T/K	451.15	453.15	468.15	473.15	478.15	483.15	488.15	493.15	503.15
P/bar	2450	2350	2250	2140	2040	1935	1860	1800	1690

T/K	513.15	523.15	533.15
P/bar	1580	1510	1435

Polymer (B):   **ethylene/methacrylic acid copolymer**   **2003BUB**
Characterization: M_n/kg.mol^{-1} = 15.1, M_w/kg.mol^{-1} = 43.9, 5.8 mol% methacrylic acid
Solvent (A):   **ethene**   **C$_2$H$_4$**   **74-85-1**

Type of data: cloud points

w_B 0.03   was kept constant

T/K	473.15	483.15	493.15	503.15	513.15	523.15	533.15
P/bar	2390	2120	1940	1780	1660	1575	1440

Polymer (B):   **ethylene/methacrylic acid copolymer**   **2003BUB**
Characterization: M_n/kg.mol^{-1} = 11.5, M_w/kg.mol^{-1} = 32.2, 7.0 mol% methacrylic acid
Solvent (A):   **ethene**   **C$_2$H$_4$**   **74-85-1**

Type of data: cloud points

w_B 0.03   was kept constant

T/K	494.15	498.15	501.15	512.15	523.15	533.15
P/bar	2320	2190	2055	1805	1625	1480

Polymer (B):   **ethylene/methacrylic acid copolymer**   **2003BUB**
Characterization: 9.0 mol% methacrylic acid, synthesized in the laboratory
Solvent (A):   **ethene**   **C₂H₄**   **74-85-1**

Type of data: cloud points

w_B 0.03   was kept constant

T/K	508.15	513.15	518.15	523.15	528.15	533.15
P/bar	2650	2440	2250	2125	1990	1860

Polymer (B):   **ethylene/methyl acrylate copolymer**   **2003BUB**
Characterization: 5.5 mol% mol% methyl acrylate, synthesized in the laboratory
Solvent (A):   **ethene**   **C₂H₄**   **74-85-1**

Type of data: cloud points

w_B 0.03   was kept constant

T/K	383.15	393.15	403.15	413.15	423.15	433.15	443.15	453.15	463.15
P/bar	1530	1420	1369	1331	1300	1269	1248	1222	1200

T/K	473.15	483.15
P/bar	1182	1165

Polymer (B):   **ethylene/methyl acrylate copolymer**   **2004BEC**
Characterization: M_n/kg.mol⁻¹ = 58.4, M_w/kg.mol⁻¹ = 128.5,
     6.0 mol% methyl acrylate, synthesized in the laboratory
Solvent (A):   **ethene**   **C₂H₄**   **74-85-1**

Type of data: cloud points

w_B 0.05   was kept constant

T/K	373.15	383.15	393.15	403.15	413.15	423.15	433.15	443.15	453.15
P/bar	1642	1583	1533	1492	1448	1412	1381	1354	1324

T/K	463.15	473.15	483.15	493.15
P/bar	1300	1282	1262	1244

Polymer (B):   **ethylene/methyl acrylate copolymer**   **1999DIE**
Characterization: M_n/kg.mol⁻¹ = 52.5, M_w/kg.mol⁻¹ = 165,
     8.0 mol% methyl acrylate, synthesized in the laboratory
Solvent (A):   **ethene**   **C₂H₄**   **74-85-1**

Type of data: cloud points

w_B 0.03   was kept constant

T/K	373.15	383.15	393.15	403.15	413.15	423.15	433.15	443.15	453.15
P/bar	1780	1700	1590	1540	1510	1485	1450	1440	1395

T/K	463.15	483.15
P/bar	1390	1350

Polymer (B):   **ethylene/methyl acrylate copolymer**   **2004BEC**
Characterization: $M_n/\text{kg.mol}^{-1} = 55$, $M_w/\text{kg.mol}^{-1} = 110$,
13 mol% methyl acrylate, synthesized in the laboratory
Solvent (A):   **ethene**   **C_2H_4**   **74-85-1**

Type of data: cloud points

w_B 0.05   was kept constant

T/K	493.15	483.15	473.15	463.15	453.15	443.15	433.15	423.15	413.15
P/bar	1185	1215	1223	1239	1260	1276	1288	1310	1351

T/K	403.15	393.15
P/bar	1388	1408

Polymer (B):   **ethylene/methyl acrylate copolymer**   **1999DIE**
Characterization: $M_n/\text{kg.mol}^{-1} = 55$, $M_w/\text{kg.mol}^{-1} = 150$,
17.9 mol% methyl acrylate, synthesized in the laboratory
Solvent (A):   **ethene**   **C_2H_4**   **74-85-1**

Type of data: cloud points

w_B 0.03   was kept constant

T/K	393.15	403.15	413.15	423.15	433.15	443.15	453.15	463.15	473.15
P/bar	1823	1780	1703	1612	1565	1510	1455	1429	1378

T/K	482.15
P/bar	1360

Polymer (B):   **ethylene/methyl acrylate copolymer**   **1999DIE**
Characterization: $M_n/\text{kg.mol}^{-1} = 59$, $M_w/\text{kg.mol}^{-1} = 158$,
32.9 mol% methyl acrylate, synthesized in the laboratory
Solvent (A):   **ethene**   **C_2H_4**   **74-85-1**

Type of data: cloud points

w_B 0.03   was kept constant

T/K	391.15	395.15	403.15	413.15	423.15	433.15	443.15	453.15	463.15
P/bar	2040	1980	1849	1768	1685	1605	1545	1490	1450

T/K	473.15	483.15
P/bar	1410	1382

Polymer (B):   **ethylene/methyl acrylate copolymer**   **2004BEC**
Characterization: $M_n/\text{kg.mol}^{-1} = 56.3$, $M_w/\text{kg.mol}^{-1} = 112.5$,
44 mol% methyl acrylate, synthesized in the laboratory
Solvent (A):   **ethene**   **C_2H_4**   **74-85-1**

Type of data: cloud points

continued

continued

| w_B | 0.05 | was kept constant |

T/K	393.15	403.15	413.15	423.15	433.15	443.15	453.15	463.15	473.15
P/bar	2115	2010	1923	1854	1790	1728	1679	1631	1585

T/K	483.15	493.15
P/bar	1545	1508

Polymer (B):	**ethylene/methyl acrylate copolymer**	**1996MU1**

Characterization: M_n/kg.mol^{-1} = 17, M_w/kg.mol^{-1} = 75.4,
25 wt% methyl acrylate, synthesized in the laboratory

Solvent (A):	**ethene**	**C$_2$H$_4$**	**74-85-1**

Type of data: cloud points

w_B	0.010	0.010	0.010	0.010	0.010	0.050	0.050	0.050	0.050
T/K	522.65	498.15	473.85	448.55	424.05	523.95	498.85	472.75	449.15
P/MPa	113.7	119.8	123.6	128.8	133.0	115.9	119.4	123.2	127.5

w_B	0.050	0.050	0.100	0.100	0.100	0.100	0.100	0.113	0.113
T/K	423.95	404.15	525.05	498.35	474.35	448.65	424.95	523.15	498.35
P/MPa	134.2	140.0	117.3	117.4	120.2	124.6	129.2	117.2	119.0

w_B	0.113	0.113	0.113	0.149	0.149	0.149	0.149	0.149	0.207
T/K	473.55	448.85	423.45	522.85	498.55	473.25	448.45	423.05	523.95
P/MPa	122.7	125.5	131.2	116.7	118.8	121.2	125.0	130.2	109.4

w_B	0.207	0.207	0.207	0.207
T/K	497.85	473.05	448.85	424.15
P/MPa	112.2	115.5	119.5	124.1

Polymer (B):	**ethylene/methyl acrylate copolymer**	**1996MU1**

Characterization: M_n/kg.mol^{-1} = 33, M_w/kg.mol^{-1} = 109,
58 wt% methyl acrylate, synthesized in the laboratory

Solvent (A):	**ethene**	**C$_2$H$_4$**	**74-85-1**

Type of data: cloud points

w_B	0.010	0.010	0.010	0.010	0.010	0.050	0.050	0.050	0.050
T/K	523.75	498.35	473.75	448.05	423.55	522.95	498.65	473.95	449.05
P/MPa	125.8	133.7	140.6	149.7	161.3	124.9	130.8	138.2	148.5

w_B	0.050	0.050	0.101	0.101	0.101	0.101	0.101	0.150	0.150
T/K	423.95	404.05	524.65	499.15	473.25	448.45	424.05	524.05	498.95
P/MPa	163.1	178.8	123.0	128.4	135.6	145.0	162.0	122.7	127.8

w_B	0.150	0.150	0.150	0.200	0.200	0.200	0.200
T/K	473.45	448.95	423.55	523.75	498.45	473.85	448.45
P/MPa	132.3	140.6	150.6	117.3	122.1	128.2	136.2

Polymer (B): **ethylene/methyl acrylate copolymer** **1996MU1**
Characterization: M_n/kg.mol^{-1} = 42, M_w/kg.mol^{-1} = 110,
68 wt% methyl acrylate, synthesized in the laboratory
Solvent (A): **ethene** **C$_2$H$_4$** **74-85-1**

Type of data: cloud points

w_B	0.011	0.011	0.011	0.011	0.011	0.050	0.050	0.050	0.050
T/K	523.85	498.95	473.55	448.45	423.75	523.25	498.35	473.15	448.85
P/MPa	136.5	145.5	157.8	165.2	206.4	136.1	145.2	157.7	174.4

w_B	0.050	0.050	0.101	0.101	0.101	0.101	0.101	0.150	0.150
T/K	423.85	403.65	523.65	498.25	473.25	448.25	423.15	523.35	498.15
P/MPa	201.3	234.3	133.6	142.0	153.0	167.4	189.3	133.9	141.0

w_B	0.150	0.150	0.150	0.192	0.192	0.192	0.192	0.192
T/K	473.05	448.95	424.35	523.45	497.95	472.65	448.85	423.45
P/MPa	151.8	164.9	184.1	132.0	139.9	150.0	163.2	181.9

Polymer (B): **ethylene/methyl methacrylate copolymer** **2003BUB**
Characterization: 5.8 mol% mol% methyl methacrylate, synthesized in the laboratory
Solvent (A): **ethene** **C$_2$H$_4$** **74-85-1**

Type of data: cloud points

w_B	0.03	was kept constant							

T/K	383.15	393.15	403.15	413.15	423.15	433.15	443.15	453.15	473.15
P/bar	1433	1366	1320	1290	1266	1241	1215	1195	1156

T/K	483.15
P/bar	1140

Polymer (B): **ethylene/methyl methacrylate copolymer** **2004BEC**
Characterization: M_n/kg.mol^{-1} = 41.4, M_w/kg.mol^{-1} = 67.1,
17 mol% methyl methacrylate, synthesized in the laboratory
Solvent (A): **ethene** **C$_2$H$_4$** **74-85-1**

Type of data: cloud points

w_B	0.05	was kept constant				

T/K	383.15	393.15	413.15	433.15	453.15	473.15
P/bar	1465	1435	1354	1292	1237	1208

Polymer (B): **ethylene/methyl methacrylate copolymer** **2004BEC**
Characterization: M_n/kg.mol^{-1} = 10.9, M_w/kg.mol^{-1} = 20,
19 mol% methyl methacrylate, synthesized in the laboratory
Solvent (A): **ethene** **C$_2$H$_4$** **74-85-1**

Type of data: cloud points

continued

continued

| w_B | 0.05 | was kept constant |

T/K	353.15	373.15	393.15	413.15	433.15	453.15	473.15	488.15	508.15
P/bar	1430	1309	1252	1206	1176	1137	1108	1086	1057

T/K	528.15
P/bar	1039

Polymer (B):	**ethylene/methyl methacrylate copolymer**	**2004BEC**
Characterization:	M_n/kg.mol^{-1} = 50, M_w/kg.mol^{-1} = 83.5,	
	35 mol% methyl methacrylate, synthesized in the laboratory	
Solvent (A):	**ethene** $\quad$ C_2H_4	**74-85-1**

Type of data: cloud points

| w_B | 0.05 | was kept constant |

T/K	393.15	413.15	433.15	453.15	473.15	488.15	508.15
P/bar	1500	1419	1341	1286	1238	1202	1152

Polymer (B):	**ethylene/methyl methacrylate copolymer**	**2004BEC**
Characterization:	M_n/kg.mol^{-1} = 28.6, M_w/kg.mol^{-1} = 52.6,	
	42 mol% methyl methacrylate, synthesized in the laboratory	
Solvent (A):	**ethene** $\quad$ C_2H_4	**74-85-1**

Type of data: cloud points

| w_B | 0.05 | was kept constant |

T/K	393.15	403.15	413.15	423.15	443.15	463.15	473.15	483.15	493.15
P/bar	1673	1585	1533	1478	1378	1306	1271	1242	1210

Polymer (B):	**ethylene/methyl methacrylate copolymer**	**2004BEC**
Characterization:	M_n/kg.mol^{-1} = 31.9, M_w/kg.mol^{-1} = 62.3,	
	50 mol% methyl methacrylate, synthesized in the laboratory	
Solvent (A):	**ethene** $\quad$ C_2H_4	**74-85-1**

Type of data: cloud points

| w_B | 0.05 | was kept constant |

T/K	393.15	413.15	453.15	473.15	488.15	508.15
P/bar	1855	1716	1507	1436	1366	1328

Polymer (B):	**ethylene/1-octene copolymer**	**2000CH4**
Characterization:	M_n/kg.mol^{-1} = 51.2, M_w/kg.mol^{-1} = 83,	
	13.9 wt% 1-octene, 4 hexyl branches/100 ethyl units	
Solvent (A):	**ethene** $\quad$ C_2H_4	**74-85-1**

continued

continued

Type of data: cloud points

w_B	0.099	0.099	0.099	0.099	0.099	0.099	0.099	0.099
T/K	453.15	438.15	423.15	413.15	403.15	393.15	388.15	385.15
P/bar	1362	1410	1477	1510	1556	1614	1644	1668

Polymer (B):   **ethylene/1-octene copolymer**   **2000CH4**
Characterization: M_n/kg.mol^{-1} = 54.8, M_w/kg.mol^{-1} = 115,
                                   25.8 wt% 1-octene, 8 hexyl branches/100 ethyl units
Solvent (A):   **ethene**   **C$_2$H$_4$**   **74-85-1**

Type of data: cloud points

w_B	0.105	0.105	0.105	0.105
T/K	453.15	433.15	413.15	393.15
P/bar	1303	1350	1408	1504

Polymer (B):   **ethylene/1-octene copolymer**   **2000CH4**
Characterization: M_n/kg.mol^{-1} = 11.3, M_w/kg.mol^{-1} = 26,
                                   36.5 wt% 1-octene, 12.6 hexyl branches/100 ethyl units
Solvent (A):   **ethene**   **C$_2$H$_4$**   **74-85-1**

Type of data: cloud points

w_B	0.099	0.099	0.099	0.099	0.099	0.099	0.099	0.099	0.099
T/K	453.15	443.15	433.15	423.15	413.15	403.15	393.15	383.15	373.15
P/bar	1097	1116	1135	1156	1182	1210	1246	1286	1332

Polymer (B):   **ethylene/1-octene copolymer**   **2000CH4**
Characterization: M_n/kg.mol^{-1} = 83.3, M_w/kg.mol^{-1} = 120,
                                   38.7 wt% 1-octene, 13.6 hexyl branches/100 ethyl units
Solvent (A):   **ethene**   **C$_2$H$_4$**   **74-85-1**

Type of data: cloud points

w_B	0.047	0.047	0.047	0.047	0.047	0.047	0.047	0.099	0.099
T/K	453.15	443.15	433.15	423.15	413.15	403.15	393.15	453.15	443.15
P/bar	1221	1231	1249	1270	1297	1330	1363	1199	1217

w_B	0.099	0.099	0.099	0.099	0.099	0.124	0.124	0.124	0.124
T/K	433.15	423.15	413.15	403.15	393.15	453.15	443.15	433.15	423.15
P/bar	1264	1254	1281	1311	1345	1168	1182	1194	1215

w_B	0.124	0.124
T/K	413.15	403.15
P/bar	1241	1268

Polymer (B):   **ethylene/1-octene copolymer**   **1998HAN**

Characterization: $M_n/\text{kg.mol}^{-1} = 40.7$, $M_w/\text{kg.mol}^{-1} = 93.6$, 7.6 mol% 1-octene,
     $T_m/\text{K} = 377.2$, synthesized with metallocene catalyst

Solvent (A):   **propane**   **C$_3$H$_8$**   **74-98-6**

Type of data: cloud points

w_B	0.05	0.05	0.05	0.05	0.05
T/K	374.25	399.25	422.65	446.75	471.55
P/bar	531.2	515.5	520.4	524.0	529.5

Polymer (B):   **ethylene/1-octene copolymer**   **2000CH1**

Characterization: $M_n/\text{kg.mol}^{-1} = 51.2$, $M_w/\text{kg.mol}^{-1} = 83$,
     13.9 wt% 1-octene, 4 hexyl branches/100 ethyl units

Solvent (A):   **propane**   **C$_3$H$_8$**   **74-98-6**

Type of data: cloud points

w_B	0.050	0.050	0.050	0.050	0.050	0.050
T/K	453.15	413.15	393.15	373.15	368.15	363.15
P/MPa	53.4	52.8	52.8	53.2	53.5	53.7

Polymer (B):   **ethylene/1-octene copolymer**   **2000CH1**

Characterization: $M_n/\text{kg.mol}^{-1} = 54.8$, $M_w/\text{kg.mol}^{-1} = 115$,
     25.8 wt% 1-octene, 8 hexyl branches/100 ethyl units

Solvent (A):   **propane**   **C$_3$H$_8$**   **74-98-6**

Type of data: cloud points

w_B	0.049	0.049	0.049
T/K	453.15	413.15	373.15
P/MPa	48.5	47.1	46.3

Polymer (B):   **ethylene/1-octene copolymer**   **2000CH1**

Characterization: $M_n/\text{kg.mol}^{-1} = 83.3$, $M_w/\text{kg.mol}^{-1} = 120$,
     38.7 wt% 1-octene, 13.6 hexyl branches/100 ethyl units

Solvent (A):   **propane**   **C$_3$H$_8$**   **74-98-6**

Type of data: cloud points

w_B	0.010	0.010	0.010	0.010	0.010	0.049	0.049	0.049	0.049
T/K	453.15	413.15	373.15	343.15	333.15	453.15	413.15	373.15	343.15
P/MPa	38.6	36.0	33.2	34.8	37.8	40.8	38.4	35.6	33.7

w_B	0.049	0.100	0.100	0.100
T/K	333.15	453.15	413.15	373.15
P/MPa	33.6	40.4	37.9	35.5

Polymer (B):   **ethylene/propyl acrylate copolymer**   **2004BEC**
Characterization: 7.0 mol% propyl acrylate, synthesized in the laboratory
Solvent (A):   **ethene**   **C$_2$H$_4$**   **74-85-1**

Type of data: cloud points

w_B 0.05   was kept constant

T/K	366.45	376.15	386.65	393.15	413.15	423.15	442.15	452.15	472.15
P/bar	1527	1476	1433	1408	1338	1310	1262	1244	1204

T/K	491.15	511.15	531.15
P/bar	1173	1141	1115

Polymer (B):   **ethylene/propyl acrylate copolymer**   **2004BEC**
Characterization: M_n/kg.mol^{-1} = 83, M_w/kg.mol^{-1} = 147,
     14 mol% propyl acrylate, synthesized in the laboratory
Solvent (A):   **ethene**   **C$_2$H$_4$**   **74-85-1**

Type of data: cloud points

w_B 0.05   was kept constant

T/K	353.15	363.15	373.65	397.85	415.45	434.15	452.15	473.15	492.15
P/bar	1380	1330	1297	1234	1192	1150	1127	1098	1074

T/K	509.15	532.15
P/bar	1056	1035

Polymer (B):   **ethylene/propyl acrylate copolymer**   **2004BEC**
Characterization: M_n/kg.mol^{-1} = 54.2, M_w/kg.mol^{-1} = 112,
     19 mol% propyl acrylate, synthesized in the laboratory
Solvent (A):   **ethene**   **C$_2$H$_4$**   **74-85-1**

Type of data: cloud points

w_B 0.05   was kept constant

T/K	367.15	373.15	393.15	412.15	431.15	451.15	468.15	491.55	512.15
P/bar	1250	1226	1183	1153	1124	1097	1077	1055	1031

T/K	529.85
P/bar	1018

Polymer (B):   **ethylene/propyl acrylate copolymer**   **2004BEC**
Characterization: M_n/kg.mol^{-1} = 58.1, M_w/kg.mol^{-1} = 127,
     26 mol% propyl acrylate, synthesized in the laboratory
Solvent (A):   **ethene**   **C$_2$H$_4$**   **74-85-1**

Type of data: cloud points

continued

continued

w_B	0.05	was kept constant						
T/K	354.15	374.15	392.15	413.15	433.15	451.15	471.15	491.15	512.15
P/bar	1208	1182	1147	1112	1083	1062	1038	1017	995

T/K	530.15
P/bar	978

Polymer (B):	**ethylene/propylene copolymer**	**1992CH1**
Characterization:	M_n/kg.mol^{-1} = 0.78, M_w/kg.mol^{-1} = 0.79, 50 mol% propene, alternating ethylene/propylene units from complete hydrogenation of polyisoprene	
Solvent (A):	**1-butene** C_4H_8	**106-98-9**

Type of data: cloud points

w_B	0.151	was kept constant						
T/K	355.05	373.15	398.15	421.65	424.05	426.35	428.15	432.95	444.45
P/bar	11.5	16.4	27.1	39.8	42.5	44.8	46.5	51.7	63.4

T/K	473.15
P/bar	90.8

Polymer (B):	**ethylene/propylene copolymer**	**1992CH1**
Characterization:	M_n/kg.mol^{-1} = 5.46, M_w/kg.mol^{-1} = 5.9, 50 mol% propene, alternating ethylene/propylene units from complete hydrogenation of polyisoprene	
Solvent (A):	**1-butene** C_4H_8	**106-98-9**

Type of data: cloud points

w_B	0.157	was kept constant				
T/K	376.15	378.85	382.65	397.95	423.15	447.95	473.15
P/bar	21.3	27.2	35.5	66.7	110.0	146.0	172.5

Comments: A lower critical endpoint is found at 373.15 K.

Polymer (B):	**ethylene/propylene copolymer**	**1992CH1**
Characterization:	M_n/kg.mol^{-1} = 25.2, M_w/kg.mol^{-1} = 26.0, 50 mol% propene, alternating ethylene/propylene units from complete hydrogenation of polyisoprene	
Solvent (A):	**1-butene** C_4H_8	**106-98-9**

Type of data: cloud points

w_B	0.160	was kept constant						
T/K	346.35	348.35	350.35	355.15	361.35	361.85	367.45	373.15	398.15
P/bar	12.3	15.2	20.5	29.0	40.2	38.2	50.3	61.2	108.0

continued

continued

T/K	416.35	423.15	448.35	473.35
P/bar	137.3	149.0	183.0	213.4

Comments: A lower critical endpoint is found at 345.15 K.

Polymer (B):	**ethylene/propylene copolymer**		**1992CH1**
Characterization:	$M_n/kg.mol^{-1} = 87.7$, $M_w/kg.mol^{-1} = 96.4$, 50 mol% propene,		
	alternating ethylene/propylene units from complete		
	hydrogenation of polyisoprene		
Solvent (A):	**1-butene** C_4H_8		**106-98-9**

Type of data: cloud points

w_B 0.152   was kept constant

T/K	334.45	336.05	339.55	343.75	355.15	373.15	398.15	423.75	448.15
P/bar	8.4	13.1	19.5	28.4	54.0	90.8	137.3	177.3	210.1

T/K	473.15
P/bar	240.5

Comments: A lower critical endpoint is found at 333.15 K.

Polymer (B):	**ethylene/propylene copolymer**		**2000VRI**
Characterization:	$M_n/kg.mol^{-1} = 51$, $M_w/kg.mol^{-1} = 120$, $M_z/kg.mol^{-1} = 210$,		
	42 mol% propene, 21 methyl groups/100 C-atoms		
Solvent (A):	**ethene** C_2H_4		**74-85-1**

Type of data: cloud points

w_B	0.0357	0.0357	0.0357	0.0357	0.0357	0.0357	0.0357	0.0498	0.0498
T/K	312.54	317.69	322.51	327.48	332.66	337.28	342.13	312.62	317.58
P/bar	2181	2077	1985	1897	1816	1756	1702	2163	2057

w_B	0.0498	0.0498	0.0498	0.0498	0.0498	0.0756	0.0756	0.0756	0.0756
T/K	322.53	327.10	332.09	337.40	342.30	312.76	317.91	322.27	327.47
P/bar	1961	1885	1809	1741	1684	2115	2006	1926	1843

w_B	0.0756	0.0756	0.0756	0.0989	0.0989	0.0989	0.0989	0.0989	0.0989
T/K	332.22	337.10	341.96	312.54	317.71	322.64	327.50	332.21	337.36
P/bar	1776	1715	1661	2083	1977	1890	1811	1749	1687

w_B	0.0989	0.1230	0.1230	0.1230	0.1230	0.1230	0.1230	0.1230	0.1491
T/K	342.25	312.38	317.41	322.40	327.20	332.31	337.13	342.06	312.62
P/bar	1634	2061	1960	1874	1797	1730	1673	1619	2011

w_B	0.1491	0.1491	0.1491	0.1491	0.1491	0.1491
T/K	317.54	322.33	327.28	332.16	337.03	341.77
P/bar	1917	1837	1765	1703	1647	1597

Polymer (B):   **ethylene/propylene copolymer**   **2000VRI**

Characterization: $M_n/\text{kg.mol}^{-1} = 8.7$, $M_w/\text{kg.mol}^{-1} = 24$, $M_z/\text{kg.mol}^{-1} = 47$,

                42 mol% propene, 22 methyl groups/100 C-atoms

Solvent (A):   **ethene** C_2H_4 **74-85-1**

Type of data: cloud points

w_B	0.0468	0.0468	0.0468	0.0468	0.0468	0.0468	0.0468	0.1168	0.1168
T/K	312.82	317.56	323.16	327.72	332.74	337.29	342.55	312.66	317.71
P/bar	1827	1748	1666	1610	1555	1512	1465	1759	1682

w_B	0.1668	0.1168	0.1168	0.1168	0.1168	0.1503	0.1503	0.1503	0.1503
T/K	322.51	327.38	332.26	337.28	42.01	312.75	317.60	322.28	327.76
P/bar	1617	1559	1505	1457	1413	1715	1646	1586	1524

w_B	0.1503	0.1503	0.1503	0.1983	0.1983	0.1983	0.1983	0.1983	0.1983
T/K	332.25	337.59	342.52	312.66	317.59	322.98	327.48	332.79	337.26
P/bar	1480	1432	1395	1615	1553	1492	1447	1399	1364

w_B	0.1983
T/K	342.69
P/bar	1325

Polymer (B):   **ethylene/propylene copolymer**   **1993GRE, 1994GR4**

Characterization: $M_n/\text{kg.mol}^{-1} = 0.78$, $M_w/\text{kg.mol}^{-1} = 0.79$, 50 mol% propene,

                alternating ethylene/propylene units from complete

                hydrogenation of polyisoprene

Solvent (A):   **ethene** C_2H_4 **74-85-1**

Type of data: cloud points

w_B	0.156	0.156	0.156	0.152	0.152
T/K	293.15	323.15	363.15	423.15	475.15
P/bar	571.5	555.0	550.0	600.9	689.6

Polymer (B):   **ethylene/propylene copolymer**   **1993GRE, 1994GR4**

Characterization: $M_n/\text{kg.mol}^{-1} = 5.46$, $M_w/\text{kg.mol}^{-1} = 5.90$, 50 mol% propene,

                alternating ethylene/propylene units from complete

                hydrogenation of polyisoprene

Solvent (A):   **ethene** C_2H_4 **74-85-1**

Type of data: cloud points

w_B	0.156	0.156	0.156	0.158	0.158	0.157	0.158	0.158	0.158
T/K	308.15	323.15	325.65	348.15	373.15	398.15	448.15	468.15	473.15
P/bar	1080.1	1018.6	1001.1	941.9	897.9	899.9	899.1	892.8	900.0

Polymer (B):   **ethylene/propylene copolymer**   **1993GRE, 1994GR4**

Characterization: $M_n/\text{kg.mol}^{-1} = 87.7$, $M_w/\text{kg.mol}^{-1} = 96.4$, 50 mol% propene, alternating ethylene/propylene units from complete hydrogenation of polyisoprene

Solvent (A):   **ethene**   **C$_2$H$_4$**   **74-85-1**

Type of data: cloud points

w_B	0.038	0.038	0.038	0.038	0.038
T/K	353.15	373.15	398.15	423.15	473.15
P/bar	1722.5	1569.0	1389.0	1235.5	1132.7

Polymer (B):   **ethylene/propylene copolymer**   **1997HAN**

Characterization: $M_n/\text{kg.mol}^{-1} = 2.17$, $M_w/\text{kg.mol}^{-1} = 2.60$, 50 mol% propene, $T_g/K = 209.2$, alternating ethylene/propylene units from complete hydrogenation (98%) of polyisoprene

Solvent (A):   **ethene**   **C$_2$H$_4$**   **74-85-1**

Type of data: cloud points

w_B	0.001	0.001	0.001	0.0067	0.0067	0.0067	0.0067	0.0095	0.0095
T/K	323.15	372.95	423.15	294.55	323.35	373.05	423.05	295.05	323.35
P/bar	381.2	396.7	410.5	490.4	484.3	483.1	491.0	539.4	523.0

w_B	0.0095	0.0095	0.026	0.026	0.026	0.026	0.034	0.034	0.034
T/K	373.15	423.15	298.45	323.35	373.05	423.05	323.35	372.95	423.55
P/bar	517.1	522.8	623.3	600.1	576.5	570.4	622.7	596.2	588.9

Type of data: coexistence data

$T/K = 423.15$

$P/$ bar	feed phase	w_B gel phase	sol phase
505	0.034	unknown	0.016

Polymer (B):   **ethylene/propylene copolymer**   **1992CH1, 1992CH2**

Characterization: $M_n/\text{kg.mol}^{-1} = 25.2$, $M_w/\text{kg.mol}^{-1} = 26.0$, 50 mol% propene, alternating ethylene/propylene units from complete hydrogenation of polyisoprene

Solvent (A):   **1-hexene**   **C$_6$H$_{12}$**   **592-41-6**

Type of data: cloud points

w_B	0.160	was kept constant							
T/K	427.15	447.95	467.85	473.15	477.95	478.65	473.15	477.95	478.55
P/bar	7.2	12.7	17.2	23.5	29.4	31.6	19.3	20.0	20.4
	(VLE	VLE	VLE	LLE	LLE	LLE	VLLE	VLLE	VLLE)

Comments: A lower critical endpoint is found at 472.15 K.

Polymer (B):   ethylene/propylene copolymer   **1992CH1**
Characterization: M_n/kg.mol^{-1} = 87.7, M_w/kg.mol^{-1} = 96.4, 50 mol% propene,
                          alternating ethylene/propylene units from complete
                          hydrogenation of polyisoprene
Solvent (A):   **1-hexene**   **C$_6$H$_{12}$**   **592-41-6**

Type of data: cloud points

w_B	0.150	was kept constant			

T/K	416.95	453.15	459.35	454.95	474.45	463.95
P/bar	8.3	14.0	18.3	28.5	43.3	17.3
	(VLE	VLE	LLE	LLE	LLE	VLLE)

Comments: A lower critical endpoint is found at 457.15 K.

Polymer (B):   ethylene/propylene copolymer   **1998HAN**
Characterization: M_n/kg.mol^{-1} = 32.3, M_w/kg.mol^{-1} = 71, 8.7 mol% propene,
                          T_m/K = 386.2, synthesized with metallocene catalyst
Solvent (A):   **propane**   **C$_3$H$_8$**   **74-98-6**

Type of data: cloud points

w_B	0.05	0.05	0.05	0.05	0.05
T/K	372.65	397.65	420.15	445.15	472.15
P/bar	568.7	559.3	559.6	558.9	560.9

Polymer (B):   ethylene/propylene copolymer   **1998HAN**
Characterization: M_n/kg.mol^{-1} = 34.5, M_w/kg.mol^{-1} = 72.5, 10.3 mol% propene,
                          T_m/K = 362.2, synthesized with metallocene catalyst
Solvent (A):   **propane**   **C$_3$H$_8$**   **74-98-6**

Type of data: cloud points

w_B	0.05	0.05	0.05	0.05	0.05
T/K	376.25	400.35	424.35	446.95	472.15
P/bar	565.4	550.3	551.2	554.0	557.8

Polymer (B):   ethylene/propylene copolymer   **1998HAN**
Characterization: M_n/kg.mol^{-1} = 27.3, M_w/kg.mol^{-1} = 62.8, 28.3 mol% propene,
                          T_m/K = 303.2, synthesized with metallocene catalyst
Solvent (A):   **propane**   **C$_3$H$_8$**   **74-98-6**

Type of data: cloud points

w_B	0.05	0.05	0.05	0.05	0.05	0.05
T/K	351.55	375.95	400.55	426.05	446.55	470.45
P/bar	471.2	474.4	482.9	493.7	502.3	511.5

Polymer (B):   **ethylene/propylene copolymer**   **1997HAN**
Characterization: M_n/kg.mol^{-1} = 2.17, M_w/kg.mol^{-1} = 2.60, 50 mol% propene,
                          T_g/K = 209.2, alternating ethylene/propylene units from
                          complete hydrogenation (98%) of polyisoprene
Solvent (A):   **propene**   **C$_3$H$_6$**   **115-07-1**

Type of data: cloud points

w_B	0.009	0.009	0.009	0.009	0.043	0.043	0.043	0.043	0.132
T/K	324.15	373.55	375.15	422.95	323.65	373.35	375.95	423.25	323.15
P/bar	43.7	139.4	142.5	198.6	53.7	152.0	156.3	224.2	70.4

w_B	0.132	0.132	0.343	0.343	0.343
T/K	372.95	423.45	323.05	373.25	423.35
P/bar	175.6	261.8	32.4	137.2	210.2

Polymer (B):   **ethylene/propylene copolymer**   **1997HAN**
Characterization: M_n/kg.mol^{-1} = 87.3, M_w/kg.mol^{-1} = 96.0, 50 mol% propene,
                          T_g/K = 220.2, alternating ethylene/propylene units from
                          complete hydrogenation of polyisoprene
Solvent (A):   **propene**   **C$_3$H$_6$**   **115-07-1**

Type of data: cloud points

w_B	0.045	0.045
T/K	373.15	423.15
P/bar	363.0	413.8

Polymer (B):   **ethylene/propylene copolymer**   **1997HAN**
Characterization: M_n/kg.mol^{-1} = 53.3, M_w/kg.mol^{-1} = 96.0, 50 mol% propene,
                          T_g/K = 220.2, alternating ethylene/propylene units from
                          complete hydrogenation of polyisoprene
Solvent (A):   **propene**   **C$_3$H$_6$**   **115-07-1**

Type of data: cloud points

w_B	0.048	0.048	0.048	0.099	0.099	0.099
T/K	323.45	372.95	422.85	323.25	373.15	423.05
P/bar	218.1	288.0	348.5	216.9	286.9	348.9

Polymer (B):   **ethylene/propylene copolymer**   **1997HAN**
Characterization: M_n/kg.mol^{-1} = 177, M_w/kg.mol^{-1} = 195, 50 mol% propene,
                          T_g/K = 225.2, alternating ethylene/propylene units from
                          complete hydrogenation (94%) of polyisoprene
Solvent (A):   **propene**   **C$_3$H$_6$**   **115-07-1**

Type of data: cloud points

w_B	0.008	0.008	0.008	0.050	0.050	0.050	0.100	0.100
T/K	323.35	372.85	423.25	322.95	372.95	423.45	323.35	423.35
P/bar	341.2	387.9	437.0	346.9	393.0	441.9	229.8	438.6

Polymer (B): **ethylene/propylene copolymer** **1992CH1**
Characterization: M_n/kg.mol^{-1} = 0.78, M_w/kg.mol^{-1} = 0.79, 50 mol% propene, alternating ethylene/propylene units from complete hydrogenation of polyisoprene
Solvent (A): **propene** **C$_3$H$_6$** **115-07-1**

Type of data: cloud points

w_B 0.172   was kept constant

T/K	319.45	343.35	350.55	352.15	355.05	373.35	397.95	423.15	447.95
P/bar	19.0	30.2	34.4	37.7	43.0	73.0	113.0	144.0	170.0
	(VLE	VLE	VLE	LLE	LLE	LLE	LLE	LLE	LLE)

T/K	473.35	352.15	353.75	355.05
P/bar	180.0	35.9	36.9	38.3
	(LLE	VLLE	VLLE	VLLE)

Comments: A lower critical endpoint is found at 351.1 K.

Polymer (B): **ethylene/propylene copolymer** **1992CH1**
Characterization: M_n/kg.mol^{-1} = 5.46, M_w/kg.mol^{-1} = 5.90, 50 mol% propene, alternating ethylene/propylene units from complete hydrogenation of polyisoprene
Solvent (A): **propene** **C$_3$H$_6$** **115-07-1**

Type of data: cloud points

w_B 0.155   was kept constant

T/K	272.65	298.15	325.15	356.15	373.15	398.95	423.35	447.95	472.75
P/bar	5.5	64.0	126.0	190.5	224.4	267.0	296.0	325.0	344.0

Comments: A lower critical endpoint is found at 272.6 K.

Polymer (B): **ethylene/propylene copolymer** **1993GRE, 1994GR4, 1994GR5**
Characterization: M_n/kg.mol^{-1} = 5.46, M_w/kg.mol^{-1} = 5.90, 50 mol% propene, alternating ethylene/propylene units from complete hydrogenation of polyisoprene
Solvent (A): **propene** **C$_3$H$_6$** **115-07-1**

Type of data: cloud points

w_B	0.018	0.073	0.150	0.216	0.281	0.392	0.018	0.073	0.150
T/K	373.15	373.75	373.15	373.35	373.45	373.15	423.15	423.15	423.15
P/bar	179.5	202.8	212.5	209.8	200.2	189.1	248.6	273.7	295.0

w_B	0.216	0.329
T/K	423.15	423.15
P/bar	293.0	267.1

Polymer (B):	**ethylene/propylene copolymer**				**1992CH1**

Characterization: M_n/kg.mol^{-1} = 25.2, M_w/kg.mol^{-1} = 26.0, 50 mol% propene, alternating ethylene/propylene units from complete hydrogenation of polyisoprene

Solvent (A):	**propene**	**C$_3$H$_6$**	**115-07-1**

Type of data: cloud points

w_B 0.151   was kept constant

T/K	323.55	355.55	373.55	398.05	423.55	448.15
P/bar	273.0	312.0	332.0	361.2	388.0	412.0

Polymer (B):	**ethylene/propylene copolymer**				**1992CH1**

Characterization: M_n/kg.mol^{-1} = 87.7, M_w/kg.mol^{-1} = 96.4, 50 mol% propene, alternating ethylene/propylene units from complete hydrogenation of polyisoprene

Solvent (A):	**propene**	**C$_3$H$_6$**	**115-07-1**

Type of data: cloud points

w_B 0.158   was kept constant

T/K	264.15	265.85	271.15	274.15	277.95	280.35	285.25	287.05	293.55
P/bar	354.0	347.0	337.0	331.0	325.5	322.5	316.7	315.0	311.0

T/K	303.05	317.15	322.55	331.15	354.65	373.05	397.75	423.35	447.95
P/bar	310.6	319.5	322.3	330.0	352.0	371.0	396.4	423.0	442.3

T/K	473.55
P/bar	461.0

Polymer (B):	**ethylene/propylene copolymer**		**1993GRE, 1994GR4, 1994GR5**

Characterization: M_n/kg.mol^{-1} = 87.7, M_w/kg.mol^{-1} = 96.4, 50 mol% propene, alternating ethylene/propylene units from complete hydrogenation of polyisoprene

Solvent (A):	**propene**	**C$_3$H$_6$**	**115-07-1**

Type of data: cloud points

w_B	0.008	0.045	0.088	0.158	0.231	0.008	0.045	0.088	0.158
T/K	373.15	373.15	373.15	373.15	373.15	423.15	423.15	423.15	423.15
P/bar	393.1	363.0	371.3	371.2	343.0	402.1	413.8	423.0	423.1

w_B	0.231
T/K	423.15
P/bar	393.0

Polymer (B):	**ethylene/propylene/isoprene terpolymer**	**1996ALB**

Characterization: M_n/kg.mol^{-1} = 7.65, M_w/kg.mol^{-1} = 13.6, 28.2 mol% propene, 43.6 mol% isoprene, alternating ethylene/propylene units from partial hydrogenation of polyisoprene

Solvent (A):	**dimethyl ether**	**C$_2$H$_6$O**	**115-10-6**

Type of data: cloud points

w_B = 0.0536　　　was kept constant

T/K	424.75	383.25	347.55	329.95
P/bar	216.6	133.0	46.2	12.2

Comments:　　A lower critical endpoint is found at 329.95 K.

Polymer (B):	**ethylene/propylene/isoprene terpolymer**	**1996ALB**

Characterization: M_n/kg.mol^{-1} = 7.69, M_w/kg.mol^{-1} = 13.7, 36.7 mol% propene, 26.6 mol% isoprene, alternating ethylene/propylene units from partial hydrogenation of polyisoprene

Solvent (A):	**dimethyl ether**	**C$_2$H$_6$O**	**115-10-6**

Type of data: cloud points

w_B = 0.0524　　　was kept constant

T/K	423.45	411.95	410.65	384.35	362.35	348.65	328.15
P/bar	208.9	186.1	184.4	132.9	85.5	53.5	11.9

Comments:　　A lower critical endpoint is found at 328.15 K.

Polymer (B):	**ethylene/propylene/isoprene terpolymer**	**1996ALB**

Characterization: M_n/kg.mol^{-1} = 7.69, M_w/kg.mol^{-1} = 13.7, 36.7 mol% propene, 26.6 mol% isoprene, alternating ethylene/propylene units from partial hydrogenation of polyisoprene

Solvent (A):	**ethane**	**C$_2$H$_6$**	**74-84-0**

Type of data: cloud points

w_B = 0.0722　　　was kept constant

T/K	423.65	414.15	393.75	392.85	383.05	382.25	373.75	364.85	362.25
P/bar	818.1	824.2	869.4	874.3	918.7	923.0	964.9	1005.4	1038.7

Polymer (B):	**ethylene/propylene/isoprene terpolymer**	**1996ALB**

Characterization: M_n/kg.mol^{-1} = 7.69, M_w/kg.mol^{-1} = 13.7, 36.7 mol% propene, 26.6 mol% isoprene, alternating ethylene/propylene units from partial hydrogenation of polyisoprene

Solvent (A):	**ethene**	**C$_2$H$_4$**	**74-85-1**

Type of data: cloud points

continued

continued

$w_B = 0.0843$ was kept constant

T/K	423.75	423.15	421.65	409.15	394.35
P/bar	1249.1	1250.5	1256.2	1293.2	1352.9

Polymer (B):   **ethylene/propylene/isoprene terpolymer**   **1996ALB**
Characterization: M_n/kg.mol^{-1} = 7.69, M_w/kg.mol^{-1} = 13.7, 36.7 mol% propene, 26.6 mol% isoprene, alternating ethylene/propylene units from partial hydrogenation of polyisoprene
Solvent (A):   **propane**   **C$_3$H$_8$**   **74-98-6**

Type of data: cloud points

$w_B = 0.0528$ was kept constant

T/K	422.75	422.05	390.05	389.85	388.75	388.45	363.95	347.75	333.65
P/bar	377.8	378.3	370.2	370.2	366.6	369.5	398.0	423.0	453.0

T/K	327.25
P/bar	472.3

Polymer (B):   **ethylene/propylene/isoprene terpolymer**   **1996ALB**
Characterization: M_n/kg.mol^{-1} = 7.69, M_w/kg.mol^{-1} = 13.7, 36.7 mol% propene, 26.6 mol% isoprene, alternating ethylene/propylene units from partial hydrogenation of polyisoprene
Solvent (A):   **propene**   **C$_3$H$_6$**   **115-07-1**

Type of data: cloud points

$w_B = 0.0478$ was kept constant

T/K	424.25	390.45	363.65	339.05	335.85	310.95
P/bar	357.5	320.7	301.1	304.5	305.0	337.5

Polymer (B):   **ethylene/vinyl acetate copolymer**   **1982RAE**
Characterization: M_n/kg.mol^{-1} = 20.4, M_w/kg.mol^{-1} = 145, 7.5 wt% vinyl acetate, Leuna-Werke, Germany
Solvent (A):   **ethene**   **C$_2$H$_4$**   **74-85-1**

Type of data: cloud points

T/K = 433.15

w_B	0.05	0.10	0.15	0.20	0.30
P/MPa	140.3	139.3	137.3	131.2	121.0

Polymer (B):   **ethylene/vinyl acetate copolymer**   **1984WAG**

Characterization: M_n/kg.mol^{-1} = 20.4, M_w/kg.mol^{-1} = 145, 7.5 wt% vinyl acetate, Leuna-Werke, Germany

Solvent (A):   **ethene**   **C_2H_4**   **74-85-1**

Type of data: cloud points

w_B	0.05	0.10	0.15	0.20	0.30	0.05	0.10	0.15	0.20
T/K	393.15	393.15	393.15	393.15	393.15	413.15	413.15	413.15	413.15
P/MPa	155.0	154.5	149.5	144.9	133.2	146.4	144.9	141.9	136.8

w_B	0.30	0.05	0.10	0.15	0.20	0.30	0.05	0.10	0.15
T/K	413.15	433.15	433.15	433.15	433.15	433.15	453.15	453.15	453.15
P/MPa	126.1	140.7	139.3	137.9	131.2	121.0	135.3	134.8	133.2

w_B	0.20	0.30	0.05	0.10	0.15	0.20	0.30
T/K	453.15	453.15	473.15	473.15	473.15	473.15	473.15
P/MPa	127.2	116.0	130.7	129.7	126.7	122.6	113.0

Polymer (B):   **ethylene/vinyl acetate copolymer**   **1991NIE**

Characterization: M_n/kg.mol^{-1} = 29.3, M_w/kg.mol^{-1} = 162, 10.9 wt% vinyl acetate, Exxon Chemical Company

Solvent (A):   **ethene**   **C_2H_4**   **74-85-1**

Type of data: coexistence data

T/K = 403.15

Total feed concentration of the copolymer in the homogeneous system: w_B = 0.151.

P/ MPa	w_A bottom phase	top phase
88.6	0.449	0.991
108.8	0.567	0.983
125.5	0.674	0.968
136.2	0.768	0.949
141.0	0.849	(cloud point)

Polymer (B):   **ethylene/vinyl acetate copolymer**   **1991NIE**

Characterization: M_n/kg.mol^{-1} = 29.3, M_w/kg.mol^{-1} = 162, 10.9 wt% vinyl acetate, Exxon Chemical Company

Solvent (A):   **ethene**   **C_2H_4**   **74-85-1**

Type of data: coexistence data

T/K = 433.15

continued

continued

Total feed concentration of the copolymer in the homogeneous system: $w_B = 0.155$.

$P/$ MPa	w_A bottom phase	top phase
94.8	0.501	0.992
104.9	0.568	0.985
115.2	0.652	0.975
128.2	0.782	0.952
131.0	0.845	(cloud point)

Polymer (B):	**ethylene/vinyl acetate copolymer**	**1982RAE**
Characterization:	$M_n/\text{kg.mol}^{-1} = 19.7$, $M_w/\text{kg.mol}^{-1} = 120$, 12.7 wt% vinyl acetate, Leuna-Werke, Germany	
Solvent (A):	**ethene** C_2H_4	**74-85-1**

Type of data: cloud points

$T/K = 433.15$

w_B	0.05	0.10	0.15	0.20	0.30
P/MPa	132.4	128.5	125.0	121.6	114.7

Polymer (B):	**ethylene/vinyl acetate copolymer**	**1984WAG**
Characterization:	$M_n/\text{kg.mol}^{-1} = 19.7$, $M_w/\text{kg.mol}^{-1} = 120$, 12.7 wt% vinyl acetate, Leuna-Werke, Germany	
Solvent (A):	**ethene** C_2H_4	**74-85-1**

Type of data: cloud points

w_B	0.05	0.075	0.10	0.15	0.20	0.30	0.05	0.075	0.10
T/K	393.15	393.15	393.15	393.15	393.15	393.15	413.15	413.15	413.15
P/MPa	146.6	145.1	143.3	138.9	133.9	125.5	137.8	137.3	134.8

w_B	0.15	0.20	0.30	0.05	0.075	0.10	0.15	0.20	0.30
T/K	413.15	413.15	413.15	433.15	433.15	433.15	433.15	433.15	433.15
P/MPa	130.4	127.0	119.6	132.4	130.4	128.5	125.0	121.6	114.7

w_B	0.05	0.075	0.10	0.15	0.20	0.30	0.05	0.075	0.10
T/K	453.15	453.15	453.15	453.15	453.15	453.15	473.15	473.15	473.15
P/MPa	127.5	125.5	124.0	120.6	117.7	109.8	122.6	121.1	119.6

w_B	0.15	0.20	0.30
T/K	473.15	473.15	473.15
P/MPa	116.7	113.8	105.5

Polymer (B):	**ethylene/vinyl acetate copolymer**	**1991NIE**
Characterization:	$M_n/\text{kg.mol}^{-1} = 33.4$, $M_w/\text{kg.mol}^{-1} = 138$, 17.5 wt% vinyl acetate, Exxon Chemical Company	
Solvent (A):	**ethene** $\quad$ **C_2H_4**	**74-85-1**

Type of data: cloud points

w_B	0.030	0.030	0.030	0.081	0.081	0.081	0.081	0.211	0.211
T/K	373.65	403.15	433.15	363.15	393.15	423.15	453.15	388.15	403.15
P/MPa	158.6	143.5	133.0	156.2	141.4	131.8	124.5	136.5	130.5

w_B	0.211	0.211	0.344	0.344	0.344	0.344	0.462	0.462	0.462
T/K	433.15	448.15	398.65	403.15	433.15	453.15	373.15	433.15	463.15
P/MPa	122.8	119.6	125.5	117.5	111.4	108.2	104.2	95.0	93.5

w_B	0.585	0.585	0.585
T/K	393.15	433.15	453.15
P/MPa	78.2	77.2	76.8

Type of data: coexistence data

$T/\text{K} = 403.15$

Total feed concentration of the copolymer in the homogeneous system: $w_B = 0.145$.

Demixing pressure	w_A bottom phase	top phase	Fractionation during demixing bottom phase		top phase		
$P/$ MPa			$M_n/$ kg/mol	$M_w/$ kg/mol	$M_n/$ kg/mol	$M_w/$ kg/mol	$M_z/$ kg/mol
48.7	0.276	0.9992					
100.1	0.543	0.989			7.8	15.3	30.6
118.7	0.680	0.980			11.7	23.8	42.4
134.1	0.855	(cloud point)	33.4	137.7	(feed sample)		

$T/\text{K} = 433.15$

Total feed concentration of the copolymer in the homogeneous system: $w_B = 0.133$.

$P/$ MPa	w_A bottom phase	top phase
43.1	0.280	
47.8		0.999
75.5	0.410	
77.5		0.995
89.7	0.497	0.993
103.5	0.584	0.984

continued

continued

$P/$ MPa	w_A bottom phase	top phase
107.9	0.623	0.985
113.8	0.679	0.972
121.0	0.757	0.948
123.8	0.792	0.940
126.1	0.867	(cloud point)

$T/K = 463.15$

Total feed concentration of the copolymer in the homogeneous system: $w_B = 0.138$.

$P/$ MPa	w_A bottom phase	top phase
48.1	0.252	0.9994
97.4	0.542	0.9905
108.7	0.650	0.9792
122.4	0.862	(cloud point)

Polymer (B):	**ethylene/vinyl acetate copolymer**							**1991NIE**
Characterization:	M_n/kg.mol^{-1} = 33.6, M_w/kg.mol^{-1} = 231, 26.0 wt% vinyl acetate,							
	Exxon Chemical Company							
Solvent (A):	**ethene**		$\mathbf{C_2H_4}$					**74-85-1**

Type of data: cloud points

w_B	0.040	0.040	0.040	0.067	0.067	0.101	0.101	0.101	0.157
T/K	373.15	403.15	463.15	393.15	433.15	393.15	433.15	453.15	373.15
P/MPa	146.6	139.8	125.5	139.7	129.0	138.5	128.0	123.8	143.5

w_B	0.157	0.157	0.157	0.262	0.262	0.262	0.262	0.360	0.360
T/K	393.15	433.15	453.15	393.15	413.15	433.15	453.15	403.15	418.15
P/MPa	136.5	126.2	122.5	130.2	125.4	121.2	117.0	115.5	113.0

w_B	0.360	0.408	0.408	0.408	0.408	0.480	0.480	0.480	0.480
T/K	453.15	393.15	413.15	433.15	453.15	373.15	393.15	433.15	449.15
P/MPa	108.2	110.6	107.5	105.2	102.8	99.2	96.5	94.0	93.8

Polymer (B):	**ethylene/vinyl acetate copolymer**		**1982RAE**
Characterization:	M_n/kg.mol^{-1} = 13.7, M_w/kg.mol^{-1} = 60.8, 27.3 wt% vinyl acetate,		
	Leuna-Werke, Germany		
Solvent (A):	**ethene**	$\mathbf{C_2H_4}$	**74-85-1**

continued

continued

Type of data: cloud points

$T/K = 433.15$

w_B	0.05	0.10	0.15	0.20	0.30
P/MPa	121.0	120.0	118.0	115.5	109.0

Polymer (B):   **ethylene/vinyl acetate copolymer**   **1984WAG**
Characterization: M_n/kg.mol^{-1} = 13.7, M_w/kg.mol^{-1} = 60.8, 27.3 wt% vinyl acetate,
Leuna-Werke, Germany
Solvent (A):   **ethene**   **C$_2$H$_4$**   **74-85-1**

Type of data: cloud points

w_B	0.05	0.10	0.15	0.20	0.30	0.05	0.10	0.15	0.20
T/K	393.15	393.15	393.15	393.15	393.15	413.15	413.15	413.15	413.15
P/MPa	128.5	127.5	125.5	123.0	116.5	125.5	124.0	122.0	119.0

w_B	0.30	0.05	0.10	0.15	0.20	0.30	0.05	0.10	0.15
T/K	413.15	433.15	433.15	433.15	433.15	433.15	453.15	453.15	453.15
P/MPa	112.0	121.0	120.0	118.0	115.5	109.0	118.5	117.0	115.0

w_B	0.20	0.30	0.05	0.10	0.15	0.20	0.30
T/K	453.15	453.15	473.15	473.15	473.15	473.15	473.15
P/MPa	112.0	105.0	115.5	113.5	111.5	109.0	102.5

Polymer (B):   **ethylene/vinyl acetate copolymer**   **1991NIE**
Characterization: M_n/kg.mol^{-1} = 35, M_w/.g.mol^{-1} = 126.5, 27.5 wt% vinyl acetate,
Exxon Chemical Company
Solvent (A):   **ethene**   **C$_2$H$_4$**   **74-85-1**

Type of data: coexistence data

$T/K = 433.15$

Total feed concentration of the copolymer in the homogeneous system: $w_B = 0.138$.

P/ MPa	w_A bottom phase	top phase
70.8		0.998
80.0		0.996
80.8	0.450	
92.8	0.551	0.987
104.0		0.981
104.2	0.600	
113.8	0.693	
114.2		0.964
119.0	0.745	0.945
120.5		0.936
123.8	0.792	
124.7	0.850	(cloud point)

Polymer (B):   **ethylene/vinyl acetate copolymer**   **1999KIN**
Characterization: M_n/kg.mol^{-1} = 61.9, M_w/kg.mol^{-1} = 167, 27.5 wt% vinyl acetate
Solvent (A):   **ethene**   **C$_2$H$_4$**   **74-85-1**

Type of data: cloud points

w_B	0.030	0.060	0.100	0.150	0.200	0.230	0.240	0.250	0.270
T/K	393.15	393.15	393.15	393.15	393.15	393.15	393.15	393.15	393.15
P/MPa	121.6	128.7	128.3	125.9	121.4	118.0	116.9	115.7	113.4

w_B	0.300	0.350	0.030	0.060	0.100	0.150	0.200	0.230	0.240
T/K	393.15	393.15	413.15	413.15	413.15	413.15	413.15	413.15	413.15
P/MPa	111.4	110.4	117.5	124.0	123.6	121.5	117.4	114.5	113.4

w_B	0.250	0.270	0.300	0.350	0.030	0.060	0.100	0.150	0.200
T/K	413.15	413.15	413.15	413.15	433.15	433.15	433.15	433.15	433.15
P/MPa	112.3	110.2	108.0	107.0	113.7	119.5	119.2	117.4	113.6

w_B	0.230	0.240	0.250	0.270	0.300	0.350	0.030	0.060	0.100
T/K	433.15	433.15	433.15	433.15	433.15	433.15	453.15	453.15	453.15
P/MPa	110.7	109.8	109.2	107.3	105.4	104.4	110.5	116.3	116.0

w_B	0.150	0.200	0.230	0.240	0.250	0.270	0.300	0.350	0.030
T/K	453.15	453.15	453.15	453.15	453.15	453.15	453.15	453.15	473.15
P/MPa	113.9	110.5	108.1	107.1	106.4	104.7	102.9	101.9	108.1

w_B	0.060	0.100	0.150	0.200	0.230	0.240	0.250	0.270	0.300
T/K	473.15	473.15	473.15	473.15	473.15	473.15	473.15	473.15	473.15
P/MPa	113.0	112.9	111.2	108.0	105.9	104.9	104.2	102.1	100.8

w_B	0.350	0.030	0.060	0.100	0.150	0.200	0.230	0.240	0.250
T/K	473.15	493.15	493.15	493.15	493.15	493.15	493.15	493.15	493.15
P/MPa	99.8	105.7	110.4	110.3	108.5	105.7	103.7	102.8	101.9

w_B	0.270	0.300	0.350
T/K	493.15	493.15	493.15
P/MPa	100.0	98.8	97.8

Polymer (B):   **ethylene/vinyl acetate copolymer**   **1983RAE**
Characterization: M_n/kg.mol^{-1} = 27.5, M_w/kg.mol^{-1} = 60.3, 31.4 wt% vinyl acetate,
                           Leuna-Werke, Germany
Solvent (A):   **ethene**   **C$_2$H$_4$**   **74-85-1**

Type of data: cloud points

w_B	0.014	0.04	0.07	0.10	0.20	0.25	0.30	0.05	0.10
T/K	393.15	393.15	393.15	393.15	393.15	393.15	393.15	433.15	433.15
P/MPa	122.6	124.0	123.1	121.6	118.7	115.2	114.2	117.9	115.1

w_B	0.20	0.30	0.014	0.04	0.07	0.10	0.20	0.25	0.30
T/K	433.15	433.15	473.15	473.15	473.15	473.15	473.15	473.15	473.15
P/MPa	112.0	108.3	111.3	113.3	110.3	107.4	106.4	103.9	102.5

Polymer (B):	ethylene/vinyl acetate copolymer	1991FIN

Characterization: M_n/kg.mol^{-1} = 2.2, M_w/kg.mol^{-1} = 5.7, 29.0 wt% vinyl acetate, Leuna-Werke, Germany

Solvent (A):	ethene	C_2H_4	74-85-1

Type of data: coexistence data

T/K = 443.15

Total feed concentration of the copolymer in the homogeneous system: w_B = 0.250.

$P/$ MPa	w_A bottom phase	top phase
48.06	0.2719	
52.08		0.9710
55.12	0.3199	
67.77	0.4045	0.9364
80.71	0.5468	0.8985
81.49	0.5211	0.8930
90.12	0.7401	0.7584

Polymer (B):	ethylene/vinyl acetate copolymer	1984WAG

Characterization: M_n/kg.mol^{-1} = 27.5, M_w/kg.mol^{-1} = 60.3, 31.4 wt% vinyl acetate, Leuna-Werke, Germany

Solvent (A):	ethene	C_2H_4	74-85-1

Type of data: cloud points

w_B	0.014	0.04	0.07	0.10	0.15	0.20	0.25	0.30	0.014
T/K	393.15	393.15	393.15	393.15	393.15	393.15	393.15	393.15	413.15
P/MPa	122.6	124.0	123.1	121.6	122.1	119.7	115.2	114.2	119.2

w_B	0.04	0.07	0.10	0.15	0.20	0.25	0.30	0.014	0.04
T/K	413.15	413.15	413.15	413.15	413.15	413.15	413.15	433.15	433.15
P/MPa	121.6	118.7	118.1	118.7	115.2	112.8	110.3	116.2	118.2

w_B	0.07	0.10	0.15	0.20	0.25	0.30	0.014	0.04	0.07
T/K	433.15	433.15	433.15	433.15	433.15	433.15	453.15	453.15	453.15
P/MPa	115.7	114.5	114.5	110.8	110.3	108.4	114.2	115.2	112.8

w_B	0.10	0.15	0.20	0.25	0.30	0.014	0.04	0.07	0.10
T/K	453.15	453.15	453.15	453.15	453.15	473.15	473.15	473.15	473.15
P/MPa	111.8	111.8	108.8	108.8	106.4	111.3	113.3	110.3	107.4

w_B	0.15	0.20	0.25	0.30
T/K	473.15	473.15	473.15	473.15
P/MPa	108.4	106.4	105.9	102.5

Polymer (B):   **ethylene/vinyl acetate copolymer**   **1982RAE**
Characterization: $M_n/\text{kg.mol}^{-1} = 16.2$, $M_w/\text{kg.mol}^{-1} = 48.5$, 31.8 wt% vinyl acetate,
                          Leuna-Werke, Germany
Solvent (A):   **ethene**   **C₂H₄**   **74-85-1**

Type of data: cloud points

$T/\text{K} = 433.15$

w_B	0.05	0.10	0.15	0.20	0.30
P/MPa	119.7	119.5	117.5	115.5	111.5

Polymer (B):   **ethylene/vinyl acetate copolymer**   **1984WAG**
Characterization: $M_n/\text{kg.mol}^{-1} = 16.2$, $M_w/\text{kg.mol}^{-1} = 48.5$, 31.8 wt% vinyl acetate,
                          Leuna-Werke, Germany
Solvent (A):   **ethene**   **C₂H₄**   **74-85-1**

Type of data: cloud points

w_B	0.05	0.10	0.15	0.20	0.30	0.05	0.10	0.15	0.20
T/K	393.15	393.15	393.15	393.15	393.15	413.15	413.15	413.15	413.15
P/MPa	126.8	126.0	125.8	124.5	119.5	123.0	122.2	121.0	120.0

w_B	0.30	0.05	0.10	0.15	0.20	0.30	0.05	0.10	0.15
T/K	413.15	433.15	433.15	433.15	433.15	433.15	453.15	453.15	453.15
P/MPa	115.5	119.7	119.5	117.5	115.5	111.5	116.0	115.7	114.0

w_B	0.20	0.30	0.05	0.10	0.15	0.20	0.30
T/K	453.15	453.15	473.15	473.15	473.15	473.15	473.15
P/MPa	112.0	108.0	113.5	112.5	111.0	109.2	105.5

Polymer (B):   **ethylene/vinyl acetate copolymer**   **1984WAG**
Characterization: $M_n/\text{kg.mol}^{-1} = 21.8$, $M_w/\text{kg.mol}^{-1} = 46.4$, 32.1 wt% vinyl acetate,
                          Leuna-Werke, Germany
Solvent (A):   **ethene**   **C₂H₄**   **74-85-1**

Type of data: cloud points

w_B	0.12	0.16	0.20	0.25	0.30	0.0075	0.12	0.16	0.20
T/K	393.15	393.15	393.15	393.15	393.15	413.15	413.15	413.15	413.15
P/MPa	121.1	120.6	118.7	116.2	112.3	114.2	117.2	117.2	114.7

w_B	0.25	0.30	0.0075	0.04	0.08	0.12	0.16	0.20	0.25
T/K	413.15	413.15	433.15	433.15	433.15	433.15	433.15	433.15	433.15
P/MPa	113.7	108.8	111.3	114.2	113.7	114.7	113.7	111.8	110.3

w_B	0.30	0.0075	0.04	0.08	0.12	0.16	0.20	0.25	0.30
T/K	433.15	453.15	453.15	453.15	453.15	453.15	453.15	453.15	453.15
P/MPa	106.9	110.9	112.3	110.8	111.8	110.8	109.8	107.4	103.9

w_B	0.0075	0.04	0.08	0.12	0.16	0.20	0.25	0.30
T/K	473.15	473.15	473.15	473.15	473.15	473.15	473.15	473.15
P/MPa	108.4	109.8	108.4	108.8	107.9	105.9	104.9	102.0

Polymer (B):   **ethylene/vinyl acetate copolymer**   **1983RAE**

Characterization: M_n/kg.mol^{-1} = 15.5, M_w/kg.mol^{-1} = 40.8, 32.3 wt% vinyl acetate, Leuna-Werke, Germany

Solvent (A):   **ethene** C_2H_4 **74-85-1**

Type of data: cloud points

w_B	0.009	0.04	0.08	0.12	0.20	0.25	0.30	0.009	0.04
T/K	393.15	393.15	393.15	393.15	393.15	393.15	393.15	473.15	473.15
P/MPa	118.7	120.1	118.2	117.4	114.9	111.8	105.9	107.9	109.8

w_B	0.08	0.12	0.20	0.25	0.30
T/K	473.15	473.15	473.15	473.15	473.15
P/MPa	107.9	106.9	105.9	102.7	99.1

Polymer (B):   **ethylene/vinyl acetate copolymer**   **1984WAG**

Characterization: M_n/kg.mol^{-1} = 15.5, M_w/kg.mol^{-1} = 40.8, 32.3 wt% vinyl acetate, Leuna-Werke, Germany

Solvent (A):   **ethene** C_2H_4 **74-85-1**

Type of data: cloud points

w_B	0.009	0.04	0.08	0.12	0.20	0.25	0.304	0.009	0.04
T/K	393.15	393.15	393.15	393.15	393.15	393.15	393.15	413.15	413.15
P/MPa	118.7	120.1	118.2	117.4	114.9	111.8	105.9	116.2	118.2

w_B	0.08	0.12	0.20	0.25	0.304	0.009	0.04	0.08	0.12
T/K	413.15	413.15	413.15	413.15	413.15	433.15	433.15	433.15	433.15
P/MPa	115.2	114.9	112.3	108.4	104.5	112.3	115.2	112.8	112.5

w_B	0.20	0.25	0.304	0.009	0.04	0.08	0.12	0.20	0.25
T/K	433.15	433.15	433.15	453.15	453.15	453.15	453.15	453.15	453.15
P/MPa	110.3	106.4	103.2	110.8	112.8	110.3	108.9	107.9	104.5

w_B	0.304	0.009	0.04	0.08	0.12	0.20	0.25	0.30
T/K	453.15	473.15	473.15	473.15	473.15	473.15	473.15	473.15
P/MPa	100.2	107.9	109.8	107.9	106.9	105.9	102.7	99.1

Polymer (B):   **ethylene/vinyl acetate copolymer**   **1991NIE**

Characterization: M_n/kg.mol^{-1} = 13.0, M_w/kg.mol^{-1} = 22.1, 32.5 wt% vinyl acetate, Exxon Chemical Company

Solvent (A):   **ethene** C_2H_4 **74-85-1**

Type of data: cloud points

w_B	0.068	0.068	0.068	0.068	0.157	0.157	0.178	0.178	0.178
T/K	383.15	393.15	433.15	453.15	383.15	433.15	373.15	403.15	453.15
P/MPa	118.2	115.8	108.5	105.7	116.5	107.0	116.0	110.0	103.2

w_B	0.300	0.300	0.300	0.393	0.393	0.393	0.393	0.393	0.439
T/K	393.15	403.15	433.15	353.15	373.15	403.15	433.15	453.15	373.15
P/MPa	107.5	106.0	102.3	107.4	103.5	99.2	96.4	95.0	98.0

w_B	0.439
T/K	433.15
P/MPa	92.2

Polymer (B):   **ethylene/vinyl acetate copolymer**   **1983RAE**
Characterization: $M_n/\text{kg.mol}^{-1} = 5.4$, $M_w/\text{kg.mol}^{-1} = 8.2$, 33.0 wt% vinyl acetate,
Leuna-Werke, Germany
Solvent (A):   **ethene** C_2H_4 **74-85-1**

Type of data: cloud points

w_B	0.014	0.04	0.08	0.12	0.20	0.25	0.32	0.014	0.04
T/K	393.15	393.15	393.15	393.15	393.15	393.15	393.15	473.15	473.15
P/MPa	92.7	98.6	100.0	99.0	98.6	96.6	94.3	86.8	93.6

w_B	0.08	0.12	0.20	0.25	0.32
T/K	473.15	473.15	473.15	473.15	473.15
P/MPa	93.2	91.2	90.7	88.7	86.8

Polymer (B):   **ethylene/vinyl acetate copolymer**   **1984WAG**
Characterization: $M_n/\text{kg.mol}^{-1} = 5.4$, $M_w/\text{kg.mol}^{-1} = 8.2$, 33.0 wt% vinyl acetate,
Leuna-Werke, Germany
Solvent (A):   **ethene** C_2H_4 **74-85-1**

Type of data: cloud points

w_B	0.014	0.04	0.08	0.12	0.16	0.20	0.25	0.32	0.014
T/K	393.15	393.15	393.15	393.15	393.15	393.15	393.15	393.15	413.15
P/MPa	92.7	98.6	100.0	99.0	98.6	99.0	96.6	94.2	90.2

w_B	0.04	0.08	0.12	0.16	0.20	0.25	0.32	0.014	0.04
T/K	413.15	413.15	413.15	413.15	413.15	413.15	413.15	433.15	433.15
P/MPa	97.1	98.1	96.8	97.1	96.9	95.1	93.2	88.7	96.1

w_B	0.08	0.12	0.16	0.20	0.25	0.32	0.014	0.04	0.08
T/K	433.15	433.15	433.15	433.15	433.15	433.15	453.15	453.15	453.15
P/MPa	96.1	94.6	95.1	94.3	92.7	90.2	87.8	95.1	94.6

w_B	0.12	0.16	0.20	0.25	0.32	0.014	0.04	0.08	0.12
T/K	453.15	453.15	453.15	453.15	453.15	473.15	473.15	473.15	473.15
P/MPa	92.7	93.2	92.7	90.4	88.7	86.8	93.6	93.2	91.2

w_B	0.16	0.20	0.25	0.32
T/K	473.15	473.15	473.15	473.15
P/MPa	91.2	90.7	88.7	80.8

Polymer (B):   **ethylene/vinyl acetate copolymer**   **1983RAE**
Characterization: $M_n/\text{kg.mol}^{-1} = 15.0$, $M_w/\text{kg.mol}^{-1} = 28.3$, 33.2 wt% vinyl acetate,
Leuna-Werke, Germany
Solvent (A):   **ethene** C_2H_4 **74-85-1**

Type of data: cloud points

w_B	0.05	0.10	0.20	0.30
T/K	433.15	433.15	433.15	433.15
P/MPa	111.2	111.0	108.4	104.4

Polymer (B):   **ethylene/vinyl acetate copolymer**   **1984WAG**
Characterization: M_n/kg.mol^{-1} = 15.0, M_w/kg.mol^{-1} = 28.3, 33.2 wt% vinyl acetate,
                      Leuna-Werke, Germany
Solvent (A):   **ethene**   **C$_2$H$_4$**   **74-85-1**

Type of data: cloud points

w_B	0.007	0.03	0.08	0.12	0.16	0.20	0.25	0.30	0.007
T/K	393.15	393.15	393.15	393.15	393.15	393.15	393.15	393.15	413.15
P/MPa	114.2	116.2	118.2	116.7	116.7	115.2	113.3	110.8	110.8

w_B	0.03	0.08	0.12	0.16	0.20	0.25	0.30	0.007	0.03
T/K	413.15	413.15	413.15	413.15	413.15	413.15	413.15	433.15	433.15
P/MPa	113.3	114.9	112.8	112.8	112.3	109.8	106.7	108.4	110.8

w_B	0.08	0.12	0.16	0.20	0.25	0.30	0.007	0.03	0.08
T/K	433.15	433.15	433.15	433.15	433.15	433.15	453.15	453.15	453.15
P/MPa	111.8	109.3	108.8	108.4	107.4	104.1	105.9	108.4	108.4

w_B	0.12	0.16	0.20	0.25	0.30	0.007	0.03	0.08	0.12
T/K	453.15	453.15	453.15	453.15	453.15	473.15	473.15	473.15	473.15
P/MPa	106.9	106.9	105.9	103.9	100.8	104.4	105.9	106.4	104.4

w_B	0.16	0.20	0.25	0.30
T/K	473.15	473.15	473.15	473.15
P/MPa	104.4	103.0	101.0	98.6

Polymer (B):   **ethylene/vinyl acetate copolymer**   **1984WAG**
Characterization: M_n/kg.mol^{-1} = 9.6, M_w/kg.mol^{-1} = 16.9, 33.4 wt% vinyl acetate,
                      Leuna-Werke, Germany
Solvent (A):   **ethene**   **C$_2$H$_4$**   **74-85-1**

Type of data: cloud points

w_B	0.0105	0.04	0.08	0.12	0.16	0.20	0.25	0.30	0.0105
T/K	393.15	393.15	393.15	393.15	393.15	393.15	393.15	393.15	413.15
P/MPa	114.2	114.7	110.8	110.3	110.3	107.8	107.8	106.4	111.0

w_B	0.04	0.08	0.12	0.16	0.20	0.25	0.30	0.0105	0.04
T/K	413.15	413.15	413.15	413.15	413.15	413.15	413.15	433.15	433.15
P/MPa	111.8	108.4	107.4	105.9	103.9	104.2	103.5	108.4	108.8

w_B	0.08	0.12	0.16	0.20	0.25	0.30	0.0105	0.04	0.08
T/K	433.15	433.15	433.15	433.15	433.15	433.15	453.15	453.15	453.15
P/MPa	105.4	104.4	102.5	102.0	102.0	100.5	106.9	106.4	103.0

w_B	0.12	0.16	0.20	0.25	0.30	0.0105	0.04	0.08	0.12
T/K	453.15	453.15	453.15	453.15	453.15	473.15	473.15	473.15	473.15
P/MPa	101.5	100.5	100.1	100.1	98.5	104.4	104.4	101.0	99.5

w_B	0.16	0.20	0.25	0.30
T/K	473.15	473.15	473.15	473.15
P/MPa	98.5	98.0	97.6	96.6

Polymer (B):   **ethylene/vinyl acetate copolymer**   **1983RAE, 1984WAG**
Characterization: M_n/kg.mol^{-1} = 19.8, M_w/kg.mol^{-1} = 44.0, 42.7 wt% vinyl acetate,
                          Leuna-Werke, Germany
Solvent (A):   **ethene**   **C$_2$H$_4$**   **74-85-1**

Type of data: cloud points

w_B	0.05	0.10	0.15	0.20	0.30	0.05	0.10	0.15	0.20
T/K	393.15	393.15	393.15	393.15	393.15	413.15	413.15	413.15	413.15
P/MPa	121.3	118.6	116.5	114.1	112.4	117.2	115.5	112.7	110.3

w_B	0.30	0.05	0.10	0.15	0.20	0.30	0.05	0.10	0.15
T/K	413.15	433.15	433.15	433.15	433.15	433.15	453.15	453.15	453.15
P/MPa	109.6	114.5	112.0	109.8	107.5	106.2	111.3	108.9	106.9

w_B	0.20	0.30	0.05	0.10	0.15	0.20	0.30
T/K	453.15	453.15	473.15	473.15	473.15	473.15	473.15
P/MPa	105.5	103.4	108.6	106.8	103.4	102.7	101.0

Polymer (B):   **ethylene/vinyl acetate copolymer**   **1991NIE**
Characterization: M_n/kg.mol^{-1} = 35, 58.0 wt% vinyl acetate,
                          Exxon Chemical Company
Solvent (A):   **ethene**   **C$_2$H$_4$**   **74-85-1**

Type of data: coexistence data

T/K = 463.15

Total feed concentration of the copolymer in the homogeneous system: w_B = 0.157.

P/ MPa	w_A bottom phase	top phase
93.5	0.376	
133.0		0.980
139.5	0.560	
151.6	0.602	
156.2		0.972
160.2	0.630	
164.5		0.964
170.0		0.942
171.5	0.718	
174.5		0.920
175.7	0.843	(cloud point)

Polymer (B): **DL-lactide/glycolide copolymer** **2001CON**
Characterization: M_n/kg.mol^{-1} = 61.7, M_w/kg.mol^{-1} = 95, 15 mol% glycolide,
T_g/K = 320.55, Alkermes, Inc., Cincinnati, OH
Solvent (A): **carbon dioxide** **CO$_2$** **124-38-9**

Type of data: cloud points

w_B	0.05	0.05	0.05	0.05	0.05
T/K	312.15	317.95	333.95	347.55	364.85
P/bar	1822	1815	1801	1784	1770

Polymer (B): **DL-lactide/glycolide copolymer** **2001CON**
Characterization: M_n/kg.mol^{-1} = 86.1, M_w/kg.mol^{-1} = 149, 15 mol% glycolide,
T_g/K = 323.85, Alkermes, Inc., Cincinnati, OH
Solvent (A): **carbon dioxide** **CO$_2$** **124-38-9**

Type of data: cloud points

w_B	0.05	0.05	0.05	0.05	0.05
T/K	309.85	315.75	329.45	345.75	358.65
P/bar	1918	1901	1881	1860	1843

Polymer (B): **DL-lactide/glycolide copolymer** **2001CON**
Characterization: M_n/kg.mol^{-1} = 77.7, M_w/kg.mol^{-1} = 130, 25 mol% glycolide,
T_g/K = 320.05, Alkermes, Inc., Cincinnati, OH
Solvent (A): **carbon dioxide** **CO$_2$** **124-38-9**

Type of data: cloud points

w_B	0.05	0.05	0.05	0.05	0.05
T/K	312.05	320.25	335.35	352.85	352.95
P/bar	2394	2373	2322	2249	2198

Polymer (B): **DL-lactide/glycolide copolymer** **2001CON**
Characterization: M_n/kg.mol^{-1} = 57.9, M_w/kg.mol^{-1} = 83, 35 mol% glycolide,
T_g/K = 317.65, Alkermes, Inc., Cincinnati, OH
Solvent (A): **carbon dioxide** **CO$_2$** **124-38-9**

Type of data: cloud points

w_B	0.05	0.05	0.05	0.05
T/K	314.65	328.75	343.95	359.85
P/bar	2999	2946	2877	2791

Polymer (B):   **DL-lactide/glycolide copolymer**   **2001CON**
Characterization: M_n/kg.mol^{-1} = 80.1, M_w/kg.mol^{-1} = 141, 35 mol% glycolide,
                      T_g/K = 321.35, Alkermes, Inc., Cincinnati, OH
Solvent (A):   **carbon dioxide**   **CO$_2$**   **124-38-9**

Type of data: cloud points

w_B	0.05	0.05	0.05	0.05
T/K	316.55	333.55	353.05	369.85
P/bar	3108	3022	2922	2822

Polymer (B):   **DL-lactide/glycolide copolymer**   **2000LE1**
Characterization: M_η/kg.mol^{-1} = 5.0, 10 wt% glycolide,
                      Polysciences, Inc., Warrington, PA
Solvent (A):   **chlorodifluoromethane**   **CHClF$_2$**   **75-45-6**

Type of data: cloud points

w_B	0.0370	0.0370	0.0370	0.0370	0.0370	0.0370
T/K	331.85	343.05	352.75	362.85	372.85	383.15
P/MPa	3.73	7.70	11.06	14.05	16.95	19.67

Polymer (B):   **DL-lactide/glycolide copolymer**   **2000LE1**
Characterization: M_η/kg.mol^{-1} = 5.0, 20 wt% glycolide,
                      Polysciences, Inc., Warrington, PA
Solvent (A):   **chlorodifluoromethane**   **CHClF$_2$**   **75-45-6**

Type of data: cloud points

w_B	0.0399	0.0399	0.0399	0.0399	0.0399	0.0399	0.0399
T/K	325.05	334.55	344.05	352.95	362.95	374.05	382.55
P/MPa	3.45	7.10	10.65	13.60	16.85	20.15	22.60

Polymer (B):   **DL-lactide/glycolide copolymer**   **2000LE1**
Characterization: M_η/kg.mol^{-1} = 10, 30 wt% glycolide,
                      Polysciences, Inc., Warrington, PA
Solvent (A):   **chlorodifluoromethane**   **CHClF$_2$**   **75-45-6**

Type of data: cloud points

w_B	0.0314	0.0314	0.0314	0.0314	0.0314	0.0314	0.0314	0.0314
T/K	314.65	324.85	334.75	344.55	352.45	362.85	372.75	382.55
P/MPa	3.80	8.35	12.50	16.50	19.45	23.25	26.62	29.65

Polymer (B):   DL-lactide/glycolide copolymer   **2000LE1**

Characterization: M_η/kg.mol^{-1} = 14.5, 45 wt% glycolide,

Resomer RG502, Boehringer Ingelheim Chemicals

Solvent (A):   **chlorodifluoromethane**   **CHClF$_2$**   **75-45-6**

Type of data: cloud points

w_B	0.0324	0.0324	0.0324	0.0324
T/K	313.65	335.25	353.15	373.15
P/MPa	7.55	17.35	24.75	31.45

Polymer (B):   DL-lactide/glycolide copolymer   **2001CON**

Characterization: M_n/kg.mol^{-1} = 77.7, M_w/kg.mol^{-1} = 130, 25 mol% glycolide,

T_g/K = 320.05, Alkermes, Inc., Cincinnati, OH

Solvent (A):   **chlorodifluoromethane**   **CHClF$_2$**   **75-45-6**

Type of data: cloud points

w_B	0.05	0.05	0.05	0.05
T/K	309.35	315.85	324.95	337.95
P/bar	14.5	25.8	63.8	118.9

Polymer (B):   DL-lactide/glycolide copolymer   **2001CON**

Characterization: M_n/kg.mol^{-1} = 50.8, M_w/kg.mol^{-1} = 69.6, 50 mol% glycolide,

T_g/K = 320.35, Alkermes, Inc., Cincinnati, OH

Solvent (A):   **chlorodifluoromethane**   **CHClF$_2$**   **75-45-6**

Type of data: cloud points

w_B	0.05	0.05	0.05
T/K	311.75	325.65	340.05
P/bar	105.1	174.1	241.0

Polymer (B):   DL-lactide/glycolide copolymer   **2001KUK**

Characterization: M_η/kg.mol^{-1} = 5.0, 10 wt% glycolide,

Polysciences, Inc., Warrington, PA

Solvent (A):   **dimethyl ether**   **C$_2$H$_6$O**   **115-10-6**

Type of data: cloud points

w_B	0.0314	0.0314	0.0314	0.0314	0.0314	0.0314	0.0314	0.0314
T/K	303.65	314.35	323.55	333.15	343.25	352.75	363.05	373.25
P/MPa	8.55	11.37	13.50	15.62	17.65	19.50	21.44	23.29

Polymer (B):   **DL-lactide/glycolide copolymer**   **2001KUK**
Characterization: M_η/kg.mol^{-1} = 5.0, 20 wt% glycolide,
                          Polysciences, Inc., Warrington, PA
Solvent (A):   **dimethyl ether   C$_2$H$_6$O**   **115-10-6**

Type of data: cloud points

w_B	0.0293	0.0293	0.0293	0.0293	0.0293	0.0293	0.0293	0.0293
T/K	303.35	312.85	323.75	333.75	343.15	353.45	363.45	373.05
P/MPa	21.65	23.55	25.10	26.50	28.00	29.75	31.25	32.65

Polymer (B):   **DL-lactide/glycolide copolymer**   **2001KUK**
Characterization: M_η/kg.mol^{-1} = 10, 30 wt% glycolide,
                          Polysciences, Inc., Warrington, PA
Solvent (A):   **dimethyl ether   C$_2$H$_6$O**   **115-10-6**

Type of data: cloud points

w_B	0.0275	0.0275	0.0275	0.0275	0.0275	0.0275	0.0275	0.0275
T/K	305.15	314.15	323.75	333.65	343.25	353.25	362.95	373.45
P/MPa	49.05	48.45	47.95	47.35	46.65	46.05	45.15	44.25

Polymer (B):   **DL-lactide/glycolide copolymer**   **2001KUK**
Characterization: M_η/kg.mol^{-1} = 14.5, 45 wt% glycolide,
                          Resomer RG502, Boehringer Ingelheim Chemicals
Solvent (A):   **dimethyl ether   C$_2$H$_6$O**   **115-10-6**

Type of data: cloud points

w_B	0.0309	0.0309	0.0309	0.0309	0.0309	0.0309
T/K	324.85	333.85	343.35	353.05	363.35	373.35
P/MPa	61.35	58.85	56.65	55.25	53.75	52.35

Polymer (B):   **DL-lactide/glycolide copolymer**   **2001CON**
Characterization: M_n/kg.mol^{-1} = 77.7, M_w/kg.mol^{-1} = 130, 25 mol% glycolide,
                          T_g/K = 320.05, Alkermes, Inc., Cincinnati, OH
Solvent (A):   **trifluoromethane   CHF$_3$**   **75-46-7**

Type of data: cloud points

w_B	0.05	0.05	0.05	0.05	0.05
T/K	301.75	310.55	310.65	339.45	354.25
P/bar	1341	1360	1394	1436	1474

Polymer (B):	polybutadiene, hydrogenated						2003TRU	

Characterization: $M_n/\text{kg.mol}^{-1} = 48$ (osmometry) or 43 (GPC),
$M_w/\text{kg.mol}^{-1} = 52$ (light scattering) or 49 (GPC),
$M_w/\text{kg.mol}^{-1} = 54$ (GPC), BD(CH$_3$/100 C) = 2.05,
PBD 50000, DSM, Geleen, The Netherlands

Solvent (A):	ethene	C$_2$H$_4$					74-85-1	

Type of data: cloud points

w_B	0.0050	0.0050	0.0050	0.0050	0.0050	0.0050	0.0050	0.0050	0.0050
T/K	399.45	403.35	407.65	412.15	416.95	421.55	427.35	432.15	437.45
P/MPa	160.2	157.7	155.0	152.4	149.7	147.2	144.4	142.1	139.8
w_B	0.0075	0.0075	0.0075	0.0075	0.0075	0.0075	0.0075	0.0075	0.0100
T/K	401.45	406.45	411.95	417.75	423.35	428.25	434.35	439.35	400.05
P/MPa	161.7	158.3	154.9	151.6	148.6	146.1	143.2	141.1	164.5
w_B	0.0100	0.0100	0.0100	0.0100	0.0100	0.0100	0.0100	0.0100	0.0125
T/K	403.95	408.65	412.85	417.65	423.65	428.55	433.75	438.45	401.05
P/MPa	161.8	158.6	155.9	153.1	150.0	147.4	144.9	142.9	165.1
w_B	0.0125	0.0125	0.0125	0.0125	0.0125	0.0125	0.0125	0.0150	0.0150
T/K	406.95	411.45	417.25	422.95	428.45	433.85	438.85	400.35	403.75
P/MPa	161.0	158.2	154.6	151.5	148.6	146.1	143.7	166.7	164.2
w_B	0.0150	0.0150	0.0150	0.0150	0.0150	0.0150	0.0150	0.0199	0.0199
T/K	408.25	413.25	417.85	423.15	427.95	432.95	438.65	399.95	403.55
P/MPa	161.1	157.8	155.1	152.3	149.7	147.2	144.8	168.5	165.9
w_B	0.0199	0.0199	0.0199	0.0199	0.0199	0.0199	0.0199	0.0249	0.0249
T/K	408.35	413.65	419.25	424.15	428.35	433.35	437.25	399.95	403.45
P/MPa	162.6	159.2	156.0	153.2	151.1	148.6	146.9	169.5	166.8
w_B	0.0249	0.0249	0.0249	0.0249	0.0249	0.0249	0.0249	0.0301	0.0301
T/K	408.15	413.15	418.65	423.75	428.95	433.45	437.85	398.95	403.95
P/MPa	163.5	160.3	157.0	154.2	151.4	149.2	147.2	170.7	167.0
w_B	0.0301	0.0301	0.0301	0.0301	0.0301	0.0301	0.0350	0.0350	0.0350
T/K	408.55	413.95	419.05	423.75	430.25	435.65	400.05	404.15	408.45
P/MPa	163.8	160.4	157.3	154.6	151.2	148.5	170.6	167.5	164.5
w_B	0.0350	0.0350	0.0350	0.0350	0.0350	0.0350	0.0402	0.0402	0.0402
T/K	413.25	418.15	423.45	429.05	434.05	437.65	399.15	403.15	407.15
P/MPa	161.4	158.3	155.3	152.3	150.0	148.1	171.8	168.6	165.8
w_B	0.0402	0.0402	0.0402	0.0402	0.0402	0.0402	0.0450	0.0450	0.0450
T/K	412.95	418.95	424.25	429.55	434.75	439.15	401.45	403.25	408.15
P/MPa	162.1	158.3	155.3	152.5	150.0	148.0	170.0	168.6	165.2
w_B	0.0450	0.0450	0.0450	0.0450	0.0450	0.0450	0.0500	0.0500	0.0500
T/K	413.65	418.55	423.95	428.55	434.05	438.25	398.95	401.85	406.85
P/MPa	161.7	158.6	155.5	153.1	150.4	148.3	172.2	169.6	166.3
w_B	0.0500	0.0500	0.0500	0.0500	0.0500	0.0500	0.0549	0.0549	0.0549
T/K	411.65	417.15	422.45	427.85	432.55	438.25	399.65	403.25	407.25
P/MPa	162.9	159.7	156.6	153.6	151.3	148.5	171.7	169.0	166.1

continued

continued

w_B	0.0549	0.0549	0.0549	0.0549	0.0549	0.0549	0.0600	0.0600	0.0600
T/K	412.55	417.65	422.75	427.35	432.55	437.35	398.85	402.25	407.85
P/MPa	162.6	159.4	156.5	153.9	151.3	149.0	172.5	169.7	165.8
w_B	0.0600	0.0600	0.0600	0.0600	0.0600	0.0600	0.0646	0.0646	0.0646
T/K	412.25	416.75	422.85	427.45	432.35	437.95	400.55	403.95	408.25
P/MPa	162.8	160.1	156.4	153.9	151.5	148.8	171.1	168.5	165.6
w_B	0.0646	0.0646	0.0646	0.0646	0.0646	0.0646	0.0701	0.0701	0.0701
T/K	413.25	417.85	422.15	427.45	431.55	437.45	400.55	405.15	408.45
P/MPa	162.4	159.5	157.1	154.1	152.2	149.2	171.2	167.8	165.4
w_B	0.0701	0.0701	0.0701	0.0701	0.0701	0.0701	0.0751	0.0751	0.0751
T/K	416.05	419.65	425.15	429.65	432.75	438.15	399.95	402.65	408.05
P/MPa	160.6	158.3	155.4	153.0	151.5	148.8	171.5	169.4	165.6
w_B	0.0751	0.0751	0.0751	0.0751	0.0751	0.0751	0.0795	0.0795	0.0795
T/K	412.15	417.15	422.45	427.15	432.55	437.45	399.35	403.75	407.45
P/MPa	162.8	159.8	156.7	154.1	151.4	149.1	172.0	168.7	166.0
w_B	0.0795	0.0795	0.0795	0.0795	0.0795	0.0795	0.0795	0.0852	0.0852
T/K	411.85	416.15	420.45	426.95	430.95	434.85	438.35	400.25	402.95
P/MPa	163.0	160.3	157.7	154.2	152.2	150.3	148.6	170.9	168.8
w_B	0.0852	0.0852	0.0852	0.0852	0.0852	0.0852	0.0852	0.0897	0.0897
T/K	408.05	412.15	418.15	421.95	426.75	432.35	438.25	400.65	404.35
P/MPa	165.2	162.6	158.8	156.6	153.9	151.1	148.3	170.4	167.8
w_B	0.0897	0.0897	0.0897	0.0897	0.0897	0.0995	0.0995	0.0995	0.0995
T/K	410.65	416.85	423.25	429.25	435.65	401.05	406.25	412.45	417.75
P/MPa	163.3	159.5	155.9	152.6	149.5	169.9	166.2	162.1	158.9
w_B	0.0995	0.0995	0.1098	0.1098	0.1098	0.1098	0.1098	0.1098	0.1197
T/K	427.75	432.75	400.65	407.35	415.35	423.65	431.15	436.95	400.65
P/MPa	153.3	150.8	170.2	165.2	160.1	155.3	151.4	148.5	169.4
w_B	0.1197	0.1197	0.1197	0.1197	0.1197	0.1197	0.1197	0.1303	0.1303
T/K	405.05	410.45	416.65	422.45	427.65	434.15	439.05	400.35	405.95
P/MPa	166.2	162.6	158.8	155.5	152.7	149.5	147.2	169.2	165.2
w_B	0.1303	0.1303	0.1303	0.1303	0.1303	0.1303	0.1396	0.1396	0.1396
T/K	411.25	417.05	422.65	427.55	433.25	438.75	400.05	405.95	411.35
P/MPa	161.8	158.1	155.0	152.4	149.6	147.2	168.8	164.8	161.3
w_B	0.1396	0.1396	0.1396	0.1396	0.1492	0.1492	0.1492	0.1492	0.1492
T/K	417.65	423.35	428.35	435.85	401.05	405.95	410.55	416.85	422.65
P/MPa	157.5	154.4	151.8	148.2	167.9	164.5	161.5	157.6	154.4
w_B	0.1492	0.1492	0.1492	0.1589	0.1589	0.1589	0.1589	0.1589	0.1589
T/K	427.45	432.55	438.65	400.95	406.35	411.75	416.65	422.75	428.85
P/MPa	152.0	149.6	146.7	167.5	163.7	160.2	157.4	154.0	150.9
w_B	0.1589	0.1589	0.1697	0.1697	0.1697	0.1697	0.1697	0.1697	0.1697
T/K	433.75	438.75	400.45	405.15	408.85	416.05	421.55	426.85	431.85
P/MPa	148.6	146.4	167.1	163.5	161.3	157.0	153.9	151.2	148.7
w_B	0.1697	0.1805	0.1805	0.1805	0.1805	0.1805	0.1805	0.1805	0.1805
T/K	437.55	400.55	405.55	410.65	416.35	421.55	426.75	432.35	437.35
P/MPa	146.2	165.9	162.4	159.2	155.8	153.0	150.4	147.7	145.5

Polymer (B):	poly(1-butene)							1999KOA

Characterization: $M_n/\text{kg.mol}^{-1} = 270$, $M_w/\text{kg.mol}^{-1} = 1280$

Solvent (A): 1-butene C_4H_8 106-98-9

Type of data: cloud points

w_B	0.0031	0.0031	0.0031	0.0031	0.0031	0.0031	0.0031	0.0031	0.0031
T/K	406.13	406.17	410.97	416.12	420.84	426.09	431.54	436.27	440.16
P/bar	101.6	101.8	110.0	118.6	126.7	135.2	144.0	151.4	156.9
w_B	0.0031	0.0054	0.0054	0.0054	0.0054	0.0054	0.0054	0.0054	0.0054
T/K	445.86	406.60	411.50	416.51	421.41	426.28	431.11	436.01	440.91
P/bar	164.5	103.6	113.0	120.3	128.4	136.2	143.9	151.4	158.3
w_B	0.0054	0.0067	0.0067	0.0067	0.0067	0.0067	0.0067	0.0067	0.0067
T/K	445.84	405.84	410.63	415.81	420.38	425.37	430.81	435.67	439.42
P/bar	165.4	102.0	110.8	118.8	127.1	135.2	143.3	151.4	154.7
w_B	0.0067	0.0067	0.0108	0.0108	0.0108	0.0108	0.0108	0.0108	0.0108
T/K	445.33	445.22	406.22	410.98	415.89	420.99	425.76	430.71	435.54
P/bar	165.4	165.1	103.1	111.9	120.2	128.1	136.0	143.1	150.7
w_B	0.0108	0.0108	0.0140	0.0140	0.0140	0.0140	0.0140	0.0140	0.0140
T/K	440.76	445.64	405.82	410.94	416.23	421.34	426.30	431.17	436.07
P/bar	158.4	165.4	103.0	111.9	120.5	128.7	136.9	144.6	151.9
w_B	0.0140	0.0140	0.0236	0.0236	0.0236	0.0236	0.0236	0.0236	0.0236
T/K	440.90	445.84	406.46	411.48	416.40	421.29	426.16	431.08	436.12
P/bar	159.0	166.1	103.0	111.6	119.8	128.3	136.2	144.1	151.4
w_B	0.0236	0.0236	0.0236	0.0337	0.0337	0.0337	0.0337	0.0337	0.0337
T/K	436.14	441.11	446.22	406.53	411.52	416.45	421.28	426.12	430.97
P/bar	151.7	159.1	166.4	102.2	110.8	119.2	127.6	135.2	143.0
w_B	0.0337	0.0337	0.0337	0.0434	0.0434	0.0434	0.0434	0.0434	0.0434
T/K	435.95	440.77	445.19	406.39	411.35	416.35	421.26	426.21	431.04
P/bar	150.9	158.1	164.7	101.6	110.6	118.7	126.9	135.0	142.6
w_B	0.0434	0.0434	0.0434	0.0549	0.0549	0.0549	0.0549	0.0549	0.0549
T/K	436.02	440.80	445.71	406.56	411.35	416.42	421.30	426.30	431.22
P/bar	150.1	157.7	164.3	100.6	108.9	117.9	126.3	134.2	142.0
w_B	0.0549	0.0549	0.0549	0.0628	0.0628	0.0628	0.0628	0.0628	0.0628
T/K	436.39	441.37	446.44	406.63	410.69	411.51	420.59	421.30	426.16
P/bar	149.8	157.3	164.1	101.1	107.9	109.5	124.9	126.4	133.8
w_B	0.0628	0.0628	0.0628	0.0628	0.0628	0.0690	0.0690	0.0690	0.0690
T/K	430.36	430.99	436.29	441.24	446.88	411.48	416.46	421.34	426.20
P/bar	140.5	141.1	149.7	156.8	165.1	108.8	116.9	125.4	133.1
w_B	0.0690	0.0690	0.0690	0.0690	0.0690	0.0770	0.0770	0.0770	0.0770
T/K	431.24	431.22	435.91	441.20	446.28	406.84	411.71	416.75	421.57
P/bar	141.1	140.8	148.4	156.1	163.6	100.1	108.9	117.1	125.4
w_B	0.0770	0.0770	0.0770	0.0770	0.0828	0.0828	0.0828	0.0828	0.0828
T/K	426.70	431.72	435.57	445.58	406.61	416.41	417.35	412.43	421.32
P/bar	134.3	141.6	147.8	161.2	98.2	116.5	118.1	109.6	124.5

continued

continued

w_B	0.0828	0.0828	0.0828	0.0828	0.0828	0.0862	0.0862	0.0862	0.0862
T/K	426.33	431.25	436.04	440.82	445.85	407.25	412.31	416.76	421.68
P/bar	132.7	140.1	147.9	154.9	161.8	99.6	108.9	116.6	124.7

w_B	0.0862	0.0862	0.0862	0.0862	0.0955	0.0955	0.0955	0.0955	0.0955
T/K	426.70	431.71	436.57	441.50	406.29	411.15	416.16	421.10	426.05
P/bar	132.5	140.6	148.0	155.3	98.7	106.8	115.1	123.9	132.0

w_B	0.0955	0.0955	0.0955	0.0955	0.0960	0.0960	0.0960	0.0960	0.0960
T/K	430.98	435.98	440.63	445.47	406.47	411.27	416.14	420.94	425.74
P/bar	139.5	147.4	153.9	161.2	98.9	107.1	115.4	123.5	131.1

w_B	0.0960	0.0960	0.0960	0.0960	0.1079	0.1079	0.1079	0.1079	0.1079
T/K	430.64	435.49	440.39	445.54	406.42	411.32	416.18	420.95	425.92
P/bar	138.7	146.2	153.8	160.9	98.0	105.9	114.1	122.3	130.1

w_B	0.1079	0.1079	0.1079	0.1079	0.1200	0.1200	0.1200	0.1200	0.1200
T/K	430.81	435.74	440.76	445.92	406.13	411.07	416.18	421.14	425.96
P/bar	137.7	145.4	153.2	160.2	95.7	104.8	112.8	121.7	129.6

w_B	0.1200	0.1200	0.1200	0.1200	0.1282	0.1282	0.1282	0.1282	0.1282
T/K	430.71	435.56	440.52	445.65	406.30	411.21	416.25	421.18	426.13
P/bar	136.7	144.2	151.9	159.2	95.7	104.8	112.8	121.0	129.1

w_B	0.1282	0.1282	0.1282	0.1282	0.1360	0.1360	0.1360	0.1360	0.1360
T/K	431.05	436.04	441.22	446.36	406.38	411.32	416.49	417.17	421.44
P/bar	136.9	144.6	151.9	159.8	95.2	103.7	112.8	113.8	120.8

w_B	0.1360	0.1360	0.1360	0.1360	0.1360	0.1518	0.1518	0.1518	0.1518
T/K	426.37	431.22	436.23	442.30	446.81	405.90	410.64	420.79	425.73
P/bar	128.9	136.4	144.2	153.2	159.8	93.3	101.6	118.6	126.7

w_B	0.1518	0.1518	0.1518	0.1518	0.1665	0.1665	0.1665	0.1665	0.1665
T/K	430.50	435.37	440.21	445.37	405.98	410.69	415.79	420.84	425.85
P/bar	134.4	141.9	149.0	156.6	91.9	100.4	108.9	117.3	125.7

w_B	0.1665	0.1665	0.1665	0.1665
T/K	430.86	435.82	440.79	445.47
P/bar	133.8	141.1	148.3	155.6

Polymer (B):	**poly(butyl acrylate)**		**1998ONE**
Characterization:	M_n/kg.mol^{-1} = 5.6, Air Products		
Solvent (A):	**carbon dioxide**	CO_2	**124-38-9**

Type of data: cloud points

w_B	0.00113	0.00113	0.00113	0.00113	0.00113
T/K	298.15	308.15	318.15	328.15	338.15
P/MPa	21.99	24.13	27.10	29.58	31.92

Polymer (B):	**poly(butyl methacrylate)**		**2000BY1, 2000KIM**
Characterization:	Aldrich Chem. Co., Inc., Milwaukee, WI		
Solvent (A):	**carbon dioxide**	CO_2	**124-38-9**

continued

continued

Type of data: cloud points

w_B	0.05	0.05	0.05	0.05	0.05
T/K	426.35	446.85	465.85	485.85	507.45
P/bar	2021.9	1716.8	1572.3	1484.1	1419.3
$\rho/(g/cm^3)$	1.085	1.020	0.973	0.932	0.895

Polymer (B):	**poly(butylene glycol)**	**1998ONE**
Characterization:	$M_n/kg.mol^{-1} = 0.8$, Air Products	
Solvent (A):	**carbon dioxide** **CO$_2$**	**124-38-9**

Type of data: cloud points

w_B	0.0050		T/K	298.15		P/MPa	12.27

Polymer (B):	**poly(butylene oxide)-b-poly(ethylene oxide)**	**1998ONE**
	diblock copolymer	
Characterization:	$M_n/kg.mol^{-1} = 1.62$ ($= 0.96 + 0.66$), 44.4 mol% ethylene oxide,	
	SAM 185, hydroxyl terminated	
Solvent (A):	**carbon dioxide** **CO$_2$**	**124-38-9**

Type of data: cloud points

w_B	0.00005	0.00005	0.00050	0.001
T/K	296.15	308.15	296.15	308.15
P/MPa	11.86	14.00	12.55	27.58

Polymer (B):	**poly(butyl methacrylate)**	**2004BEC**
Characterization:	$M_n/kg.mol^{-1} = 38.1$, $M_w/kg.mol^{-1} = 65.5$,	
	synthesized in the laboratory	
Solvent (A):	**ethene** **C$_2$H$_4$**	**74-85-1**

Type of data: cloud points

w_B	0.05	was kept constant

T/K	373.15	393.15	413.15	433.15	453.15	473.15	488.15	508.15
P/bar	1380	1320	1215	1161	1147	1111	1077	1031

Polymer (B):	**poly(εcaprolactone)**	**1993HA4**
Characterization:	$M_n/kg.mol^{-1} = 9.3$, $M_w/kg.mol^{-1} = 14.6$,	
	$T_g/K = 213$, Union Carbide Corp.	
Solvent (A):	**chlorodifluoromethane** **CHClF$_2$**	**75-45-6**

Type of data: cloud points

Comments: $50 < (T/K - 273.15) < 150$.

continued

continued

w_B = 0.00029 P/MPa = $-20.871 + 0.43397\,(T/K - 273.15) - 6.7838\ 10^{-4}\,(T/K - 273.15)^2$
w_B = 0.00288 P/MPa = $-19.636 + 0.42290\,(T/K - 273.15) - 4.4064\ 10^{-4}\,(T/K - 273.15)^2$
w_B = 0.00973 P/MPa = $-18.914 + 0.43503\,(T/K - 273.15) - 5.7375\ 10^{-4}\,(T/K - 273.15)^2$
w_B = 0.0206 P/MPa = $-20.629 + 0.47439\,(T/K - 273.15) - 7.6059\ 10^{-4}\,(T/K - 273.15)^2$
w_B = 0.0310 P/MPa = $-20.806 + 0.47302\,(T/K - 273.15) - 7.3343\ 10^{-4}\,(T/K - 273.15)^2$
w_B = 0.0406 P/MPa = $-20.118 + 0.46526\,(T/K - 273.15) - 7.1455\ 10^{-4}\,(T/K - 273.15)^2$
w_B = 0.0497 P/MPa = $-20.174 + 0.46303\,(T/K - 273.15) - 6.9886\ 10^{-4}\,(T/K - 273.15)^2$
w_B = 0.0725 P/MPa = $-20.729 + 0.46978\,(T/K - 273.15) - 7.2765\ 10^{-4}\,(T/K - 273.15)^2$
w_B = 0.1489 P/MPa = $-23.208 + 0.49300\,(T/K - 273.15) - 8.4772\ 10^{-4}\,(T/K - 273.15)^2$
w_B = 0.1887 P/MPa = $-22.928 + 0.44778\,(T/K - 273.15) - 6.0281\ 10^{-4}\,(T/K - 273.15)^2$

Polymer (B):   **poly(ɛ-caprolactone)**   1993HA4
Characterization: M_n/kg.mol^{-1} = 26.4, M_w/kg.mol^{-1} = 40.5,
                        T_g/K = 213, Union Carbide Corp.
Solvent (A):   **chlorodifluoromethane   CHClF$_2$**   75-45-6

Type of data: cloud points

Comments: $50 < (T/K - 273.15) < 150$.

w_B = 0.00031 P/MPa = $-20.927 + 0.48674\,(T/K - 273.15) - 7.8029\ 10^{-4}\,(T/K - 273.15)^2$
w_B = 0.00308 P/MPa = $-19.729 + 0.47761\,(T/K - 273.15) - 7.1376\ 10^{-4}\,(T/K - 273.15)^2$
w_B = 0.0100 P/MPa = $-20.096 + 0.49072\,(T/K - 273.15) - 7.5342\ 10^{-4}\,(T/K - 273.15)^2$
w_B = 0.0307 P/MPa = $-18.227 + 0.45775\,(T/K - 273.15) - 6.1003\ 10^{-4}\,(T/K - 273.15)^2$
w_B = 0.0505 P/MPa = $-19.114 + 0.47121\,(T/K - 273.15) - 6.5685\ 10^{-4}\,(T/K - 273.15)^2$
w_B = 0.0724 P/MPa = $-19.939 + 0.47596\,(T/K - 273.15) - 6.4877\ 10^{-4}\,(T/K - 273.15)^2$

Polymer (B):   **poly(1,1-dihydroperfluorooctyl acrylate)**   1996HAR
Characterization: M_w/kg.mol^{-1} = 100
Solvent (A):   **carbon dioxide   CO$_2$**   124-38-9

Type of data: cloud points

w_B	0.010	T/K	318.15	P/bar	177.9

Polymer (B):   **poly(dimethylsiloxane)**   1999LIJ
Characterization: M_n/kg.mol^{-1} = 31.3, M_n/kg.mol^{-1} = 93.7,
                        Scientific Polymer Products, Inc., Ontario, NY
Solvent (A):   **carbon dioxide   CO$_2$**   124-38-9

Type of data: cloud points

w_B	0.05	0.05	0.05	0.05	0.05	0.05	0.05
T/K	313.2	307.7	306.0	303.7	348.2	341.2	340.8
P/MPa	30.7	33.0	35.0	38.4	34.8	33.4	32.2

Polymer (B):	**poly(dimethylsiloxane)**							**1998ONE**
Characterization:	M_n/kg.mol^{-1} = 13, low polydispersity							
Solvent (A):	**carbon dioxide**		**CO$_2$**					**124-38-9**

Type of data: cloud points

w_B	0.00227	0.00505	0.01060	0.02420	0.00227	0.07160			
T/K	283.15	283.15	283.15	283.15	293.15	293.15			
P/MPa	7.58	9.58	13.24	18.62	10.20	21.65			

w_B	0.00227	0.00505	0.01060	0.01870	0.02420	0.03830	0.07160	0.09470	0.17300
T/K	298.15	298.15	298.15	298.15	298.15	298.15	298.15	298.15	298.15
P/MPa	11.72	13.10	14.96	15.72	16.41	19.03	20.34	21.03	20.89

w_B	0.00227	0.00505	0.01060	0.01870	0.02420	0.03830	0.07160	0.09470	0.17300
T/K	308.15	308.15	308.15	308.15	308.15	308.15	308.15	308.15	308.15
P/MPa	14.75	16.13	17.31	17.65	25.23	20.41	21.30	21.51	21.51

w_B	0.00227	0.00505	0.01060	0.01870	0.02420	0.03830	0.07160	0.09470	0.17300
T/K	323.15	323.15	323.15	323.15	323.15	323.15	323.15	323.15	323.15
P/MPa	18.41	20.34	21.44	21.86	22.13	23.72	24.34	24.68	24.48

w_B	0.00227	0.00505	0.01060	0.01870	0.02420	0.03830	0.07160	0.09470	0.17300
T/K	338.15	338.15	338.15	338.15	338.15	338.15	338.15	338.15	338.15
P/MPa	22.55	24.27	25.51	25.65	25.99	27.44	30.34	28.06	28.06

Polymer (B):	**poly(dimethylsiloxane)**		**1998ONE**
Characterization:	end-capped with $C_3H_6NH_2$, prepared in the laboratory		
Solvent (A):	**carbon dioxide** **CO$_2$**		**124-38-9**

Type of data: cloud points

w_B	0.0040	0.0040	0.0040	0.0040	0.0040
T/K	298.15	308.15	318.15	328.15	338.15
P/MPa	24.13	21.51	22.75	24.61	25.68

Polymer (B):	**poly(dimethylsiloxane) monomethacrylate**		**1998ONE**
Characterization:	M_n/kg.mol^{-1} = 10, Aldrich Chem. Co., Inc., Milwaukee, WI		
Solvent (A):	**carbon dioxide** **CO$_2$**		**124-38-9**

Type of data: cloud points

T/K = 298.15

w_B	0.00230	0.00507	0.01050	0.01490	0.02290
P/MPa	10.76	11.31	13.17	13.86	14.41

T/K = 303.15

w_B	0.00230	0.00507	0.01050	0.01490	0.02290
P/MPa	12.00	12.62	14.82	15.03	15.72

T/K = 313.15

w_B	0.00230	0.00507	0.01050	0.01490	0.02290
P/MPa	15.31	16.13	17.44	17.79	18.48

continued

continued

T/K = 323.15

w_B	0.00230	0.00507	0.01050	0.01490	0.02290
P/MPa	18.27	18.89	20.48	27.37	21.44

T/K = 333.15

w_B	0.00230	0.00507	0.01050	0.01490	0.02290
P/MPa	20.82	21.99	23.10	23.17	23.86

Polymer (B):	**poly(dimethylsiloxane)-g-poly(ethylene oxide) graft copolymer**	**1998ONE**
Characterization:	unknown composition and molar mass, DABCO DC5357, Dow Corning	
Solvent (A):	**carbon dioxide CO$_2$**	**124-38-9**

Type of data: cloud points

w_B	0.00110	0.00110	0.00110	0.00110	0.00110
T/K	298.15	303.15	313.15	323.15	333.15
P/MPa	22.96	24.89	28.61	31.44	35.03

Polymer (B):	**poly(dimethylsiloxane)-g-poly(ethylene oxide) graft copolymer**	**1998ONE**
Characterization:	unknown composition and molar mass, fractionated in the laboratory from DABCO DC5357, Dow Corning	
Solvent (A):	**carbon dioxide CO$_2$**	**124-38-9**

Type of data: cloud points

w_B	0.00110	0.00110	0.00110	0.00110	0.00110
T/K	298.15	303.15	313.15	323.15	333.15
P/MPa	7.79	9.72	12.55	15.72	18.13

Polymer (B):	**poly(dimethylsiloxane)-g-poly(ethylene oxide) graft copolymer**	**1998ONE**
Characterization:	unknown composition and molar mass, made in the laboratory from DABCO DC5384, Dow Corning	
Solvent (A):	**carbon dioxide CO$_2$**	**124-38-9**

Type of data: cloud points

w_B	0.00575	0.00789	0.00575	0.00789	0.01120	0.00575	0.00789	0.01120	0.00575
T/K	298.15	298.15	308.15	308.15	308.15	318.15	318.15	318.15	328.15
P/MPa	13.93	21.03	14.48	15.65	16.89	15.79	17.58	19.24	18.96

w_B	0.00789	0.01120	0.00575	0.00789	0.01120
T/K	328.15	328.15	338.15	338.15	338.15
P/MPa	20.96	23.86	20.82	23.37	25.37

| **Polymer (B):** | **poly(dimethylsiloxane)-g-poly(ethylene oxide) graft copolymer (trifluoroacetate end-capped)** | **1998ONE** |

Characterization: unknown composition and molar mass, made in the laboratory from DABCO DC5357, Dow Corning

| **Solvent (A):** | **carbon dioxide** | **CO$_2$** | **124-38-9** |

Type of data: cloud points

w_B	0.00150	0.00430	0.00575	0.00789	0.00150	0.00430	0.00575	0.00789	0.01120
T/K	298.15	298.15	298.15	298.15	303.15	303.15	308.15	308.15	308.15
P/MPa	13.51	17.03	13.93	21.03	16.20	19.72	14.48	15.65	16.89

w_B	0.00150	0.00430	0.00575	0.00789	0.01120	0.00150	0.00430	0.00575	0.00789
T/K	313.15	313.15	318.15	318.15	318.15	323.15	323.15	328.15	328.15
P/MPa	19.86	24.89	15.79	17.58	19.24	23.37	29.37	18.96	20.96

w_B	0.01120	0.00150	0.00430	0.00575	0.00789	0.01120
T/K	328.15	333.15	333.15	338.15	338.15	338.15
P/MPa	23.86	27.79	34.89	20.82	23.37	25.37

| **Polymer (B):** | **poly(dimethylsiloxane)-g-poly(ethylene oxide) graft copolymer (PDMS13-g-EO10)** | **1998ONE** |

Characterization: M_n/kg.mol^{-1} = 1.7, synthesized by Unilever Research, Edgewater, NJ

| **Solvent (A):** | **carbon dioxide** | **CO$_2$** | **124-38-9** |

Type of data: cloud points

w_B	0.00381	0.00381	0.00381
T/K	298.15	308.15	333.15
P/MPa	15.17	19.86	29.99

| **Polymer (B):** | **poly(dimethylsiloxane)-g-poly(ethylene oxide) graft copolymer (PDMS20-g-EO10-g-EO10)** | **1998ONE** |

Characterization: M_n/kg.mol^{-1} = 2.8, synthesized by Unilever Research, Edgewater, NJ

| **Solvent (A):** | **carbon dioxide** | **CO$_2$** | **124-38-9** |

Type of data: cloud points

w_B	0.00148	0.00148	0.00148
T/K	298.15	313.15	333.15
P/MPa	16.69	21.86	29.37

| **Polymer (B):** | **poly(dimethylsiloxane)-g-poly(ethylene oxide)-g-poly(propylene oxide) graft copolymer [PDMS22-g-(PEO/PPO = 75/25)]** | **1998ONE** |

Characterization: M_n/kg.mol^{-1} = 6.0, Abil B-8851 Surfactant, Goldschmidt

| **Solvent (A):** | **carbon dioxide** | **CO$_2$** | **124-38-9** |

continued

continued

Type of data: cloud points

w_B	0.00119	0.00119	0.00119	0.00119
T/K	298.15	313.15	323.15	333.15
P/MPa	28.75	32.61	33.03	34.13

Polymer (B):	**poly(dimethylsiloxane)-g-poly(ethylene oxide)-g-**	**1998ONE**
	poly(propylene oxide) graft copolymer	
	[PDMS73-g-(PEO/PPO = 75/25)]	
Characterization:	M_n/kg.mol^{-1} = 13, Abil B-88184 Surfactant, Goldschmidt	
Solvent (A):	**carbon dioxide** **CO$_2$**	**124-38-9**

Type of data: cloud points

w_B	0.00165	0.00165	0.00165	0.00165	0.00563	0.00563
T/K	298.15	313.15	323.15	333.15	313.15	333.15
P/MPa	28.61	32.82	35.30	37.23	37.23	40.95

Polymer (B):	**poly(di-1H,1H,2H,2H-perfluorodecyl diitaconate)**	**2003LID**
Characterization:	M_w/kg.mol^{-1} = 150 (this is an estimated value only),	
	synthesized in the laboratory	
Solvent (A):	**carbon dioxide** **CO$_2$**	**124-38-9**

Type of data: cloud points

w_B	0.0096	0.0096	0.0096	0.0096	0.0096	0.0096	0.0096	0.0096	0.0150
T/K	424.45	396.05	372.15	358.05	342.75	328.45	316.05	295.05	395.35
P/bar	398	359	321	299	260	238	218	118	364
ρ/(g/cm^3)	0.635	0.676	0.710	0.736	0.770	0.816	0.859	0.870	

w_B	0.0150	0.0150	0.0150	0.0150	0.0150	0.0150	0.0150	0.0670	0.0670
T/K	375.55	356.15	345.75	334.85	328.45	319.25	306.15	419.45	389.65
P/bar	336	326	309	291	276	258	221	396	355
ρ/(g/cm^3)								0.711	0.752

w_B	0.0670	0.0670	0.0670	0.0670	0.1100	0.1100	0.1100	0.1100	0.1100
T/K	355.55	334.65	314.85	296.95	385.65	352.85	337.15	325.25	310.65
P/bar	287	237	180	126	349	281	239	209	163
ρ/(g/cm^3)	0.802	0.838	0.872	0.907	0.749	0.798	0.826	0.847	0.877

Polymer (B):	**poly(di-1H,1H,2H,2H-perfluorododecyl diitaconate)**	**2003LID**
Characterization:	M_w/kg.mol^{-1} = 150 (this is an estimated value only),	
	synthesized in the laboratory	
Solvent (A):	**carbon dioxide** **CO$_2$**	**124-38-9**

Type of data: cloud points

w_B	0.0095	0.0095	0.0095	0.0095	0.0206	0.0206	0.0206	0.0206
T/K	386.35	349.95	324.85	297.25	377.45	345.75	312.45	297.85
P/bar	293.1	225.5	161.7	81.4	282.4	221.7	130.2	88.3

Polymer (B): **poly(di-1H,1H,2H,2H-perfluorohexyl diitaconate)** 2003LID
Characterization: $M_w/\text{kg.mol}^{-1} = 150$ (this is an estimated value only),
synthesized in the laboratory
Solvent (A): **carbon dioxide** CO_2 124-38-9

Type of data: cloud points

w_B 0.010   was kept constant

T/K	409.35	381.95	358.85	342.55	333.45	324.05	318.75	306.05	294.65
P/bar	388	343	294	252	228	215	182	137	95
$\rho/(\text{g/cm}^3)$	0.675	0.709	0.744	0.773	0.783	0.806	0.818	0.820	0.838

Polymer (B): **poly(di-1H,1H,2H,2H-perfluorooctyl diitaconate)** 2003LID
Characterization: $M_w/\text{kg.mol}^{-1} = 150$ (this is an estimated value only),
synthesized in the laboratory
Solvent (A): **carbon dioxide** CO_2 124-38-9

Type of data: cloud points

w_B 0.010   was kept constant

T/K	421.25	404.15	373.35	358.45	342.15	315.65	306.25	294.85
P/bar	402	379	327	313	256	177	146	106
$\rho/(\text{g/cm}^3)$	0.644	0.669	0.713	0.757	0.761	0.810	0.827	0.847

Polymer (B): **poly(di-1H,1H,2H,2H-perfluorodecyl monoitaconate)** 2003LID
Characterization: $M_w/\text{kg.mol}^{-1} = 150$ (this is an estimated value only),
synthesized in the laboratory
Solvent (A): **carbon dioxide** CO_2 124-38-9

Type of data: cloud points

w_B	0.0100	0.0100	0.0100	0.0100	0.0100	0.0100	0.0100	0.0305	0.0305
T/K	387.35	369.55	350.25	337.45	322.95	301.15	293.25	398.65	365.15
P/bar	335	318	298	275	244	183	142	351	306
$\rho/(\text{g/cm}^3)$	0.733	0.809	0.855	0.883	0.941	0.963	1.001	0.702	0.817

w_B	0.0305	0.0305	0.0305	0.0305	0.0305	0.0550	0.0550	0.0550	0.0550
T/K	346.45	337.55	323.45	306.55	296.15	374.25	349.25	343.35	325.45
P/bar	283	267	237	210	170	326	284	274	243
$\rho/(\text{g/cm}^3)$	0.870	0.887	0.930	0.950	0.978	0.744	0.835	0.844	0.882

w_B	0.0550	0.0550	0.0992	0.0992	0.0992	0.0992	0.0992	0.0992
T/K	314.95	296.15	379.75	356.75	340.85	331.55	322.55	309.95
P/bar	217	171	344	297	258	246	227	186
$\rho/(\text{g/cm}^3)$	0.935	0.972	0.752	0.822	0.897	0.909	0.919	0.928

Polymer (B): **poly(di-1H,1H,2H,2H-perfluorododecyl monoitaconate)** **2003LID**
Characterization: M_w/kg.mol^{-1} = 150 (this is an estimated value only),
 synthesized in the laboratory
Solvent (A): **carbon dioxide** CO_2 **124-38-9**

Type of data: cloud points

w_B 0.0129 was kept constant

T/K	402.05	369.65	359.25	348.65	342.35	326.05	319.55	300.15
P/bar	401	361	357	345	335	320	306	287
ρ/(g/cm^3)	0.804	0.884	0.923	0.961	0.983	1.047	1.069	1.147

Polymer (B): **poly(di-1H,1H,2H,2H-perfluorohexyl monoitaconate)** **2003LID**
Characterization: M_w/kg.mol^{-1} = 150 (this is an estimated value only),
 synthesized in the laboratory
Solvent (A): **carbon dioxide** CO_2 **124-38-9**

Type of data: cloud points

w_B 0.0098 was kept constant

T/K	400.95	382.45	366.15	352.45	335.75
P/bar	450	487	498	518	539
ρ/(g/cm^3)	0.850	0.927	0.982	1.030	1.114

Polymer (B): **poly(di-1H,1H,2H,2H-perfluorooctyl monoitaconate)** **2003LID**
Characterization: M_w/kg.mol^{-1} = 150 (this is an estimated value only),
 synthesized in the laboratory
Solvent (A): **carbon dioxide** CO_2 **124-38-9**

Type of data: cloud points

w_B 0.0103 was kept constant

T/K	388.55	370.55	359.65	347.55	336.05	321.75	306.05	293.45
P/bar	355	356	359	364	345	318	252	265
ρ/(g/cm^3)	0.766	0.854	0.907	0.971	1.009	1.057	1.120	1.154

Polymer (B): **poly(ethyl acrylate)** **2004BEC**
Characterization: M_n/kg.mol^{-1} = 85.4, M_w/kg.mol^{-1} = 154,
 Polysciences, Inc., Warrington, PA
Solvent (A): **ethene** C_2H_4 **74-85-1**

Type of data: cloud points

w_B 0.05 was kept constant

T/K	354.15	364.15	373.15	393.15	414.35	434.65	453.85	473.45	492.35
P/bar	1613	1541	1485	1376	1288	1217	1162	1117	1077

T/K	513.55	533.15
P/bar	1038	1010

Polymer (B):	**polyethylene**		**1992WOH**
Characterization:	M_n/kg.mol^{-1} = 1.94, M_w/kg.mol^{-1} = 5.37,		
	LDPE-wax, ρ = 0.9238 g/cm^3, Leuna-Werke, Germany		
Solvent (A):	**1-butene**	**C$_4$H$_8$**	**106-98-9**

Type of data:　　coexistence data

T/K = 493.15

Total feed concentrations of the homogeneous system before demixing: w_A = 0.75 and w_B = 0.25.

P/ MPa	w_A	
	liquid phase	gas phase
9.030	0.2985	
9.180		0.9936
9.220	0.3299	
9.313	0.3234	
9.317		0.9893
9.410		0.9866
9.601		0.9834
9.838	0.3441	
10.027		0.9875
12.062		0.9819
12.250	0.4023	
12.440	0.4093	0.9730
12.720	0.4161	0.9730
12.960		0.9663
13.010	0.4246	
13.860		0.9651
14.620	0.4739	
14.711		0.9528
15.280		0.9365
15.520	0.4942	
15.685		0.9396
15.940		0.9310
16.317	0.5054	
16.685	0.5196	
17.074		0.9146
18.305		0.8966
18.876	0.5837	
18.929	0.6014	
19.717		0.8440
20.400	0.7361	
22.263		0.7497
23.400	0.7369	

Comments:　　Vapor-liquid equilibrium data for this system are given in Chapter 2.

Polymer (B):   **polyethylene**   **1993HA3**
Characterization: M_n/kg.mol^{-1} = 37, M_w/kg.mol^{-1} = 49, 36% crystallinity, T_m/K = 371
Solvent (A):   **ethane**   **C$_2$H$_6$**   **74-84-0**

Type of data: cloud points

w_B 0.055   was kept constant

T/K	371.15	385.15	394.45	402.85	424.75
P/bar	1070.0	1060.0	1054.9	1034.2	996.3

Polymer (B):   **polyethylene**   **1993HA3**
Characterization: M_n/kg.mol^{-1} = 37, M_w/kg.mol^{-1} = 46, 57% crystallinity, T_m/K = 396
Solvent (A):   **ethane**   **C$_2$H$_6$**   **74-84-0**

Type of data: cloud points

w_B 0.055   was kept constant

T/K	388.95	389.45	398.45	408.25	424.55
P/bar	1187.0	1185.9	1158.3	1134.2	1096.2

Polymer (B):   **polyethylene**   **1993HA3**
Characterization: M_n/kg.mol^{-1} = 39, M_w/kg.mol^{-1} = 50, 40% crystallinity, T_m/K = 396
Solvent (A):   **ethane**   **C$_2$H$_6$**   **74-84-0**

Type of data: cloud points

w_B 0.055   was kept constant

T/K	382.05	388.45	393.75	403.65	413.15	423.35
P/bar	1148.0	1137.6	1123.8	1099.7	1079.0	1061.8

Polymer (B):   **polyethylene**   **1993HA3**
Characterization: M_n/kg.mol^{-1} = 41, M_w/kg.mol^{-1} = 53, 49% crystallinity, T_m/K = 381
Solvent (A):   **ethane**   **C$_2$H$_6$**   **74-84-0**

Type of data: cloud points

w_B 0.055   was kept constant

T/K	380.95	386.95	398.55	410.05	424.15
P/bar	1138.0	1130.7	1099.7	1072.1	1044.5

Polymer (B):   **polyethylene**   **1993HA3**
Characterization: M_n/kg.mol^{-1} = 21, M_w/kg.mol^{-1} = 106, 42% crystallinity, T_m/K = 396
Solvent (A):   **ethane**   **C$_2$H$_6$**   **74-84-0**

Type of data: cloud points

continued

continued

| w_B | 0.055 | was kept constant |

T/K	389.15	391.55	393.15	402.65	403.35	412.85	413.05	423.75	425.75
P/bar	1281.0	1280.7	1273.8	1246.2	1239.3	1206.6	1199.7	1158.3	1170.4

Polymer (B):	**polyethylene**		**1993HA3**

Characterization: M_n/kg.mol^{-1} = 76, M_w/kg.mol^{-1} = 108, 46% crystallinity, T_m/K = 394

Solvent (A):	**ethane**	**C$_2$H$_6$**	**74-84-0**

Type of data: cloud points

| w_B | 0.055 | was kept constant |

T/K	390.15	392.15	393.85	403.65	413.05	425.05
P/bar	1244.0	1242.8	1235.9	1206.6	1179.0	1144.5

Polymer (B):	**polyethylene**		**1996HEU**

Characterization: M_n/kg.mol^{-1} = 0.638, M_w/kg.mol^{-1} = 0.768, M_z/kg.mol^{-1} = 0.911, Exxon Polywax 655

Solvent (A):	**ethene**	**C$_2$H$_4$**	**74-85-1**

Type of data: cloud points

T/K = 413.15

x_B	0.0504	0.0434	0.0380	0.0309	0.0283	0.0265	0.0245	0.0230	0.0205
P/MPa	42.8	43.8	44.4	45.2	45.4	45.5	45.9	46.2	46.6

x_B	0.0160	0.0096	0.0070
P/MPa	46.8	45.8	45.1

critical concentration: $x_{B, crit}$ = 0.0265,   *critical pressure*: P_{crit}/MPa = 45.5

T/K = 433.15

x_B	0.0504	0.0434	0.0380	0.0309	0.0283	0.0265	0.0245	0.0230	0.0205
P/MPa	43.0	44.0	44.6	45.4	45.6	45.8	46.2	46.5	46.9

x_B	0.0160	0.0096	0.0070
P/MPa	47.0	46.1	45.4

critical concentration: $x_{B, crit}$ = 0.0265,   *critical pressure*: P_{crit}/MPa = 45.8

T/K = 453.15

x_B	0.0504	0.0434	0.0380	0.0309	0.0283	0.0265	0.0245	0.0230	0.0205
P/MPa	43.1	44.1	44.8	45.6	45.8	46.0	46.4	46.8	47.2

x_B	0.0160	0.0096	0.0070
P/MPa	47.3	46.4	45.6

critical concentration: $x_{B, crit}$ = 0.0265,   *critical pressure*: P_{crit}/MPa = 46.0

T/K = 473.15

continued

continued

x_B	0.0504	0.0434	0.0380	0.0309	0.0283	0.0265	0.0245	0.0230	0.0205
P/MPa	43.3	44.3	45.0	45.8	46.0	46.2	46.6	47.0	47.4

x_B	0.0160	0.0096	0.0070
P/MPa	47.6	46.7	45.9

critical concentration: $x_{B, crit}$ = 0.0265,   *critical pressure*: P_{crit}/MPa = 46.2

T/K = 493.15

x_B	0.0504	0.0434	0.0380	0.0309	0.0283	0.0265	0.0245	0.0230	0.0205
P/MPa	43.5	44.5	45.2	46.0	46.2	46.4	46.9	47.3	47.7

x_B	0.0160	0.0096	0.0070
P/MPa	47.8	46.9	46.2

critical concentration: $x_{B, crit}$ = 0.0265,   *critical pressure*: P_{crit}/MPa = 46.4

T/K = 513.15

x_B	0.0504	0.0434	0.0380	0.0309	0.0283	0.0265	0.0245	0.0230	0.0205
P/MPa	43.6	44.8	45.4	46.2	46.4	46.6	47.1	47.6	48.0

x_B	0.0160	0.0096	0.0070
P/MPa	48.1	47.3	46.5

critical concentration: $x_{B, crit}$ = 0.0265,   *critical pressure*: P_{crit}/MPa = 46.6

Polymer (B):	**polyethylene**		**1996HEU**

Characterization: M_n/kg.mol^{-1} = 1.122, M_w/kg.mol^{-1} = 1.311, M_z/kg.mol^{-1} = 1.538,
Exxon Polywax 1000

Solvent (A):	**ethene**	**C_2H_4**	**74-85-1**

Type of data: cloud points

T/K = 413.15

x_B	0.0320	0.0290	0.0270	0.0250	0.0230	0.0210	0.0180	0.0150	0.0132
P/MPa	54.1	55.4	56.1	56.8	57.4	58.1	58.8	59.4	59.7

x_B	0.0120	0.0110	0.0100	0.0080	0.0050	0.0030
P/MPa	59.8	60.3	60.8	61.5	61.2	59.2

critical concentration: $x_{B, crit}$ = 0.0120,   *critical pressure*: P_{crit}/MPa = 59.8

T/K = 433.15

x_B	0.0320	0.0290	0.0270	0.0250	0.0230	0.0210	0.0180	0.0150	0.0132
P/MPa	53.9	55.1	55.8	56.4	57.0	57.6	58.3	58.8	59.1

x_B	0.0120	0.0110	0.0100	0.0080	0.0050	0.0030
P/MPa	59.2	59.5	59.9	60.4	60.1	58.1

critical concentration: $x_{B, crit}$ = 0.0120,   *critical pressure*: P_{crit}/MPa = 59.2

T/K = 453.15

continued

continued

x_B	0.0320	0.0290	0.0270	0.0250	0.0230	0.0210	0.0180	0.0150	0.0132
P/MPa	53.7	54.8	55.5	56.0	56.7	57.3	57.8	58.2	58.5

x_B	0.0120	0.0110	0.0100	0.0080	0.0050	0.0030
P/MPa	58.6	59.0	59.4	59.7	59.2	57.4

critical concentration: $x_{B, crit} = 0.0120$,   *critical pressure*: P_{crit}/MPa = 58.6

T/K = 473.15

x_B	0.0320	0.0290	0.0270	0.0250	0.0230	0.0210	0.0180	0.0150	0.0132
P/MPa	53.5	54.6	55.2	55.7	56.2	56.7	57.3	57.7	57.9

x_B	0.0120	0.0110	0.0100	0.0080	0.0050	0.0030
P/MPa	58.0	58.3	58.7	59.1	58.5	56.4

critical concentration: $x_{B, crit} = 0.0120$,   *critical pressure*: P_{crit}/MPa = 58.0

T/K = 493.15

x_B	0.0320	0.0290	0.0270	0.0250	0.0230	0.0210	0.0180	0.0150	0.0132
P/MPa	53.2	54.2	54.7	55.2	55.6	56.0	56.5	56.9	57.1

x_B	0.0120	0.0110	0.0100	0.0080	0.0050	0.0030
P/MPa	57.2	57.7	58.1	58.5	57.6	55.4

critical concentration: $x_{B, crit} = 0.0120$,   *critical pressure*: P_{crit}/MPa = 57.2

T/K = 513.15

x_B	0.0320	0.0290	0.0270	0.0250	0.0230	0.0210	0.0180	0.0150	0.0132
P/MPa	52.9	53.7	54.2	54.6	55.0	55.4	55.8	56.2	56.3

x_B	0.0120	0.0110	0.0100	0.0080	0.0050	0.0030
P/MPa	56.4	57.0	57.5	58.0	57.0	54.7

critical concentration: $x_{B, crit} = 0.0120$,   *critical pressure*: P_{crit}/MPa = 56.4

Polymer (B):	**polyethylene**	**1996HEU**
Characterization:	M_n/kg.mol^{-1} = 1.160, M_w/kg.mol^{-1} = 1.600, M_z/kg.mol^{-1} = 3.154, Exxon Polywax H101	
Solvent (A):	**ethene** C_2H_4	**74-85-1**

Type of data: cloud points

T/K = 413.15

x_B	0.0197	0.0165	0.0134	0.0125	0.0118	0.0111	0.0105	0.0099	0.0093
P/MPa	56.7	60.6	63.0	63.6	63.8	64.0	64.7	65.4	66.0

x_B	0.0080	0.0060	0.0049	0.0036	0.0030	0.0024
P/MPa	67.6	69.8	70.6	70.8	70.3	69.7

critical concentration: $x_{B, crit} = 0.0111$,   *critical pressure*: P_{crit}/MPa = 64.0

T/K = 433.15

continued

continued

x_B	0.0197	0.0165	0.0134	0.0125	0.0118	0.0111	0.0105	0.0099	0.0093
P/MPa	56.3	60.1	62.3	62.8	63.1	63.2	63.9	64.4	65.0

x_B	0.0080	0.0060	0.0049	0.0036	0.0030	0.0024
P/MPa	66.5	68.6	69.5	69.5	69.0	68.3

critical concentration: $x_{B, crit} = 0.0111$,   *critical pressure*: P_{crit}/MPa = 63.2

T/K = 453.15

x_B	0.0197	0.0165	0.0134	0.0125	0.0118	0.0111	0.0105	0.0099	0.0093
P/MPa	55.9	59.5	61.7	62.1	62.3	62.4	62.9	63.5	64.1

x_B	0.0080	0.0060	0.0049	0.0036	0.0030	0.0024
P/MPa	65.6	67.4	68.2	68.4	67.8	67.1

critical concentration: $x_{B, crit} = 0.0111$,   *critical pressure*: P_{crit}/MPa = 64.1

T/K = 473.15

x_B	0.0197	0.0165	0.0134	0.0125	0.0118	0.0111	0.0105	0.0099	0.0093
P/MPa	55.6	59.1	60.9	61.3	61.4	61.5	62.1	62.7	63.3

x_B	0.0080	0.0060	0.0049	0.0036	0.0030	0.0024
P/MPa	64.6	66.3	66.9	67.1	66.6	65.7

critical concentration: $x_{B, crit} = 0.0111$,   *critical pressure*: P_{crit}/MPa = 61.5

T/K = 493.15

x_B	0.0197	0.0165	0.0134	0.0125	0.0118	0.0111	0.0105	0.0099	0.0093
P/MPa	55.2	58.6	60.3	60.6	60.7	60.8	61.3	61.9	62.5

x_B	0.0080	0.0060	0.0049	0.0036	0.0030	0.0024
P/MPa	63.7	65.3	65.8	65.8	65.1	64.1

critical concentration: $x_{B, crit} = 0.0111$,   *critical pressure*: P_{crit}/MPa = 60.8

T/K = 513.15

x_B	0.0197	0.0165	0.0134	0.0125	0.0118	0.0111	0.0105	0.0099	0.0093
P/MPa	54.8	57.9	59.4	59.6	59.7	59.9	60.3	60.9	61.4

x_B	0.0080	0.0060	0.0049	0.0036	0.0030	0.0024
P/MPa	62.7	64.3	64.7	64.4	63.8	62.8

critical concentration: $x_{B, crit} = 0.0111$,   *critical pressure*: P_{crit}/MPa = 59.9

Polymer (B):   **polyethylene**   **1981SP2, 1983SPA, 1986SPA**
Characterization: M_n/kg.mol^{-1} = 1.1, M_w/kg.mol^{-1} = 4.0, M_z/kg.mol^{-1} = 86,
                         LDPE-wax, ρ = 0.865 g/cm^3 at 295 K
Solvent (A):   **ethene**   **C$_2$H$_4$**   **74-85-1**

Type of data: cloud points

w_B	0.132	0.295	0.140	0.270	0.250	0.139	0.646
T/K	383.15	383.15	433.15	433.15	473.15	473.15	473.15
P/bar	57.5	60.9	59.2	61.1	65.4	64.0	50.9

continued

continued

Type of data: coexistence data

$T/K = 383.15$

Total feed concentrations of the homogeneous system before demixing: $w_A = 0.868$ and $w_B = 0.132$.

Demixing pressure	w_B gel phase	w_B sol phase	Fractionation during demixing gel phase		Fractionation during demixing sol phase	
$P/$ MPa			$M_n/$ kg/mol	$M_w/$ kg/mol	$M_n/$ kg/mol	$M_w/$ kg/mol
57.5		0.140			1.13	4.04
56.9		0.120			1.06	1.59
53.9	0.604		1.41	5.84		
53.0		0.077			0.88	1.13
43.2		0.032			0.85	1.06
42.2	0.700		1.22	4.77		

$T/K = 433.15$

Total feed concentrations of the homogeneous system before demixing: $w_A = 0.860$ and $w_B = 0.140$.

Demixing pressure	w_B gel phase	w_B sol phase	Fractionation during demixing gel phase		Fractionation during demixing sol phase	
$P/$ MPa			$M_n/$ kg/mol	$M_w/$ kg/mol	$M_n/$ kg/mol	$M_w/$ kg/mol
59.2		0.140			1.13	4.04
54.9	0.605		1.58	7.11		
54.3		0.102			0.94	1.32
46.6		0.056			0.91	1.21
43.7	0.701		1.31	5.52		

$T/K = 473.15$

Total feed concentrations of the homogeneous system before demixing: $w_A = 0.861$ and $w_B = 0.139$.

Demixing pressure	w_B gel phase	w_B sol phase	Fractionation during demixing gel phase		Fractionation during demixing sol phase	
$P/$ MPa			$M_n/$ kg/mol	$M_w/$ kg/mol	$M_n/$ kg/mol	$M_w/$ kg/mol
64.0		0.139			1.13	4.04
62.3		0.124			1.06	1.83

continued

continued

P/ MPa			M_n/ kg/mol	M_w/ kg/mol	M_n/ kg/mol	M_w/ kg/mol
61.3	0.624		1.96	9.63		
60.6		0.112			1.00	1.55
59.7	0.642		1.92	9.08		
54.3		0.072			0.95	1.33
51.4	0.692		1.61	7.32		
49.4	0.700		1.49	6.54		
38.4	0.768		1.28	5.54		

$T/K = 473.15$

Total feed concentrations of the homogeneous system before demixing: $w_A = 0.354$ and $w_B = 0.646$.

Demixing pressure	w_B gel phase	sol phase	Fractionation during demixing gel phase		sol phase	
P/ MPa			M_n/ kg/mol	M_w/ kg/mol	M_n/ kg/mol	M_w/ kg/mol
50.9	0.646		1.13	4.04		
49.8	0.654		1.22	6.23		
47.4		0.073			0.89	1.17
44.8	0.701		1.29	6.49		
38.6		0.028			0.79	1.02
37.7	0.767		1.26	5.54		
23.3	0.847		1.18	4.48		

Polymer (B):	**polyethylene**	**1981SP1**
Characterization:	M_n/kg.mol^{-1} = 1.2, M_w/kg.mol^{-1} = 4.4, M_z/kg.mol^{-1} = 86, LDPE-wax, $\rho = 0.865$ g/cm^3 at 295 K	
Solvent (A):	**ethene** $\quad$ **C$_2$H$_4$**	**74-85-1**

Type of data: cloud points

$T/K = 383.15$

w_B	0.132	0.247	0.316	0.370	0.435	0.610	0.641
P/MPa	57.5	60.4	60.7	58.5	57.4	48.6	46.0

critical concentration: $w_{B, crit} = 0.370$, $\qquad$ *critical pressure*: P_{crit}/MPa = 58.5

$T/K = 433.15$

w_B	0.114	0.140	0.186	0.280	0.365	0.381	0.425	0.484	0.515
P/MPa	58.3	59.2	60.3	61.0	59.7	58.6	58.1	56.9	55.7

continued

continued

w_B	0.548	0.589	0.719	0.819	0.855
P/MPa	54.4	52.1	40.5	26.0	19.8

critical concentration: $w_{B, crit} = 0.381$,   *critical pressure*: P_{crit}/MPa = 58.6

T/K = 473.15

w_B	0.139	0.268	0.392	0.430	0.455	0.646
P/MPa	64.0	65.4	63.2	61.0	59.1	50.9

Type of data: coexistence data

T/K = 383.15

Total feed concentrations of the homogeneous system before demixing: $w_A = 0.868$ and $w_B = 0.132$.

P/ MPa	w_B gel phase	sol phase
57.5	0.555	0.132 (cloud and shadow points)
56.9		0.120
55.4	0.584	
53.9	0.604	
53.0		0.077
52.1		0.066
43.2		0.032
42.2	0.700	
31.0		0.003

T/K = 383.15

Total feed concentrations of the homogeneous system before demixing: $w_A = 0.756$ and $w_B = 0.247$.

P/ MPa	w_B gel phase	sol phase
60.4	0.461	0.247 (cloud and shadow points)
57.5	0.515	
55.8	0.550	
54.8		0.124
49.5		0.073
47.9	0.635	
46.7		0.056
36.6	0.722	
35.8		0.020
23.0	0.812	

continued

continued

$T/K = 383.15$

Total feed concentrations of the homogeneous system before demixing: $w_A = 0.684$ and $w_B = 0.316$.

$P/$ MPa	w_B gel phase	sol phase
60.7	0.432	0.316 (cloud and shadow points)
58.7		0.229
57.5	0.493	
56.7		0.173
56.0	0.526	
48.7	0.622	
47.6		0.058
46.2		0.059
35.4		0.016
34.6	0.735	
33.7		0.021
20.6	0.813	

$T/K = 383.15$

Total feed concentrations of the homogeneous system before demixing: $w_A = 0.630$ and $w_B = 0.370$.

$P/$ MPa	w_B gel phase	sol phase
58.5	0.370	0.370 (cloud and shadow points at the critical concentration)
58.3	0.396	
57.8		0.306
56.7		0.228
56.0	0.470	
28.4		0.008
27.7	0.780	
25.3	0.793	
23.1	0.803	
20.6	0.827	
20.6	0.815	
12.7	0.875	

continued

continued

T/K = 383.15

Total feed concentrations of the homogeneous system before demixing: $w_A = 0.565$ and $w_B = 0.435$.

P/ MPa	w_B gel phase	sol phase
57.4	0.435	0.260 (cloud and shadow points)
56.3	0.454	
54.7		0.176
53.8	0.517	
51.3	0.575	
48.4		0.091
44.6	0.652	
43.7		0.061
39.0	0.699	
36.7		0.033
27.3	0.781	
26.5		0.003
22.7	0.810	
21.6	0.813	

T/K = 383.15

Total feed concentrations of the homogeneous system before demixing: $w_A = 0.390$ and $w_B = 0.610$.

P/ MPa	w_B gel phase	sol phase
48.6	0.610	0.096 (cloud and shadow points)
45.2	0.648	
44.1	0.656	
43.4		0.064
36.6	0.722	
33.6	0.745	
33.0		0.013
25.1		0.001
24.0		0.000
23.6	0.795	
22.7	0.800	
22.0		0.000
15.0	0.854	
14.5		0.000
14.2	0.852	
10.4	0.883	
7.5	0.918	
4.4	0.950	

continued

continued

$T/K = 433.15$

Total feed concentrations of the homogeneous system before demixing: $w_A = 0.933$ and $w_B = 0.067$.

$P/$ MPa	w_B gel phase	sol phase
54.4	0.644	0.067 (cloud and shadow points)
53.6		0.064
52.3		0.061
50.3		0.049
46.6		0.038
39.0	0.730	

$T/K = 433.15$

Total feed concentrations of the homogeneous system before demixing: $w_A = 0.886$ and $w_B = 0.114$.

$P/$ MPa	w_B gel phase	sol phase
58.3	0.578	0.114 (cloud and shadow points)
57.4		0.106
56.1		0.098
55.3	0.635	
52.7	0.643	
51.5		0.065
50.5	0.667	
48.8		0.060
46.0		0.041
44.2	0.695	
38.5		0.024
37.8	0.761	
26.4		0.005
17.2		0.000

$T/K = 433.15$

Total feed concentrations of the homogeneous system before demixing: $w_A = 0.860$ and $w_B = 0.140$.

$P/$ MPa	w_B gel phase	sol phase
59.2	0.557	0.140 (cloud and shadow points)
57.8		0.129
55.8		0.117

continued

continued

$P/$ MPa	w_B gel phase	sol phase
54.9	0.605	
54.3		0.102
53.3	0.619	
46.6		0.056
43.7	0.701	
43.3		0.039

$T/K = 433.15$

Total feed concentrations of the homogeneous system before demixing: $w_A = 0.814$ and $w_B = 0.186$.

$P/$ MPa	w_B gel phase	sol phase
60.3	0.523	0.186 (cloud and shadow points)
58.4		0.165
57.6	0.563	
56.1		0.136
55.1	0.593	
53.9		0.114
52.7	0.619	
46.2		0.061
45.2	0.686	
38.4	0.734	
37.4		0.028

$T/K = 433.15$

Total feed concentrations of the homogeneous system before demixing: $w_A = 0.720$ and $w_B = 0.280$.

$P/$ MPa	w_B gel phase	sol phase
61.0	0.453	0.280 (cloud and shadow points)
56.5	0.541	
56.2		0.177
55.4	0.563	
54.4		0.150
53.0	0.605	
51.4		0.105
49.5	0.643	
44.3	0.696	
40.6		0.041

continued

continued

$T/K = 433.15$

Total feed concentrations of the homogeneous system before demixing: $w_A = 0.635$ and $w_B = 0.365$.

$P/$ MPa	w_B gel phase	sol phase
59.7	0.396	0.365 (cloud and shadow points)
57.8	0.468	
56.2	0.512	
55.8		0.204
54.7	0.558	
51.9	0.611	
50.0		0.105
48.6	0.646	
46.7		0.075
45.7	0.682	

$T/K = 433.15$

Total feed concentrations of the homogeneous system before demixing: $w_A = 0.619$ and $w_B = 0.381$.

$P/$ MPa	w_B gel phase	sol phase
58.6	0.381	0.381 (cloud and shadow points at the critical concentration)
57.3		0.265
55.9	0.513	
55.2		0.187
54.4	0.543	

$T/K = 433.15$

Total feed concentrations of the homogeneous system before demixing: $w_A = 0.575$ and $w_B = 0.425$.

$P/$ MPa	w_B gel phase	sol phase
58.4	0.425	0.305 (cloud and shadow points)
56.4		0.221
55.6	0.517	
55.3		0.195
54.2		0.196
52.8	0.574	

continued

continued

$T/K = 433.15$

Total feed concentrations of the homogeneous system before demixing: $w_A = 0.516$ and $w_B = 0.484$.

$P/$ MPa	w_B gel phase	sol phase
56.9	0.484	0.245 (cloud and shadow points)
55.0	0.535	
53.1	0.582	
51.1		0.129
49.6	0.622	
47.7		0.083
34.6		0.018
34.1	0.772	

$T/K = 433.15$

Total feed concentrations of the homogeneous system before demixing: $w_A = 0.485$ and $w_B = 0.515$.

$P/$ MPa	w_B gel phase	sol phase
55.7	0.515	0.206 (cloud and shadow points)
55.3	0.521	
53.0		0.149
52.0		0.135

$T/K = 433.15$

Total feed concentrations of the homogeneous system before demixing: $w_A = 0.452$ and $w_B = 0.548$.

$P/$ MPa	w_B gel phase	sol phase
54.4	0.548	0.175 (cloud and shadow points)
53.9	0.558	
52.7		0.132
51.2		0.118
49.4	0.636	
47.9		0.090

continued

continued

$T/K = 433.15$

Total feed concentrations of the homogeneous system before demixing: $w_A = 0.411$ and $w_B = 0.589$.

$P/$ MPa	w_B gel phase	sol phase
52.1	0.589	0.136 (cloud and shadow points)
51.4	0.604	
48.6		0.093
47.0	0.657	
45.4		0.065
43.4	0.696	
41.8		0.043

$T/K = 433.15$

Total feed concentrations of the homogeneous system before demixing: $w_A = 0.281$ and $w_B = 0.719$.

$P/$ MPa	w_B gel phase	sol phase
40.5	0.719	0.038 (cloud and shadow points)
39.9	0.728	
38.9		0.026
37.6	0.739	

$T/K = 433.15$

Total feed concentrations of the homogeneous system before demixing: $w_A = 0.181$ and $w_B = 0.819$.

$P/$ MPa	w_B gel phase	sol phase
26.0	0.819	0.003 (cloud and shadow points)
25.3	0.822	
24.7		0.000
22.8	0.843	
21.7	0.844	

continued

continued

T/K = 433.15

Total feed concentrations of the homogeneous system before demixing: $w_A = 0.145$ and $w_B = 0.855$.

P/ MPa	w_B gel phase	sol phase
19.8	0.855	0.000 (cloud and shadow points)
19.5		0.000
17.9	0.869	
14.1	0.893	
10.5	0.920	

T/K = 473.15

Total feed concentrations of the homogeneous system before demixing: $w_A = 0.861$ and $w_B = 0.139$.

P/ MPa	w_B gel phase	sol phase
64.0	0.605	0.139 (cloud and shadow points)
62.3		0.124
61.3	0.624	
60.6		0.112
59.7	0.642	
53.7		0.072
52.2		0.061
51.4	0.692	
39.3		0.019
38.4	0.768	

T/K = 473.15

Total feed concentrations of the homogeneous system before demixing: $w_A = 0.732$ and $w_B = 0.268$.

P/ MPa	w_B gel phase	sol phase
65.4	0.533	0.268 (cloud and shadow points)
64.0		0.249
61.8		0.217
60.1	0.582	
59.5		0.203
58.6	0.590	
55.8		0.142
55.2	0.625	
50.8		0.085
49.6	0.671	

continued

continued

$T/K = 473.15$

Total feed concentrations of the homogeneous system before demixing: $w_A = 0.608$ and $w_B = 0.392$.

$P/$ MPa	w_B gel phase	sol phase
63.2	0.608	0.392 (cloud and shadow points)
62.1		0.368
60.3		0.305
59.1	0.565	
58.2	0.582	
52.4		0.099
51.6	0.652	
50.0	0.671	
48.9		0.065
40.2		0.029
39.4	0.757	
30.0		0.008
28.9	0.815	

$T/K = 473.15$

Total feed concentrations of the homogeneous system before demixing: $w_A = 0.570$ and $w_B = 0.430$.

$P/$ MPa	w_B gel phase	sol phase
61.0	0.453	0.430 (cloud and shadow points)
60.7	0.471	
59.9		0.376
58.1		0.276
56.4	0.580	

$T/K = 473.15$

Total feed concentrations of the homogeneous system before demixing: $w_A = 0.545$ and $w_B = 0.455$.

$P/$ MPa	w_B gel phase	sol phase
59.1	0.455	0.425 (cloud and shadow points)
58.1		0.315
57.6	0.545	
56.7		0.213
55.8	0.587	

continued

continued

$P/$ MPa	w_B gel phase	sol phase
55.0		0.176
53.9	0.609	
52.5		0.128
43.6	0.711	
42.2		0.033
28.2		0.006
25.8		0.005
23.9		0.001
17.2	0.885	
12.3	0.919	

$T/K = 473.15$

Total feed concentrations of the homogeneous system before demixing: $w_A = 0.354$ and $w_B = 0.646$.

$P/$ MPa	w_B gel phase	sol phase
50.9	0.646	0.105 (cloud and shadow points)
49.8	0.654	
48.6		0.077
47.4		0.073
44.8	0.701	
38.6		0.028
37.7	0.767	
37.2	0.767	
36.3		0.026
24.8		0.005
23.9		0.002
23.3	0.847	
23.0	0.860	
15.3	0.885	
15.2	0.889	
6.6	0.956	

Polymer (B):	**polyethylene**	**1980RAE**
Characterization:	M_n/kg.mol^{-1} = 2.4, M_w/kg.mol^{-1} = 6.2,	
	LDPE-wax, $\rho = 0.923$ g/cm^3, Leuna-Werke, Germany	
Solvent (A):	**ethene** C_2H_4	**74-85-1**

Type of data: cloud points

continued

continued

w_B	0.05	0.05	0.10	0.10	0.10	0.10	0.10	0.10	0.10
T/K	403.15	433.15	403.15	413.15	433.15	453.15	473.15	493.15	513.15
P/MPa	102.5	98.1	116.2	113.3	108.4	104.4	100.0	96.6	94.6

w_B	0.20	0.20	0.30	0.30
T/K	403.15	433.15	403.15	433.15
P/MPa	114.3	106.9	102.0	96.6

Polymer (B):   **polyethylene**   **1991NIE**

Characterization: M_n/kg.mol^{-1} = 2.41, M_w/kg.mol^{-1} = 4.88, HDPE,

                             Exxon Polymer Plant, Antwerp, fractionated in the laboratory

Solvent (A):   **ethene**   **C_2H_4**   **74-85-1**

Type of data: cloud points

w_B	0.072	0.072	0.115	0.115	0.195	0.195	0.195	0.271	0.271
T/K	403.65	443.15	423.15	463.65	407.65	418.15	453.15	413.15	443.15
P/MPa	107.8	100.2	103.0	97.1	104.8	102.6	97.5	98.0	93.2

w_B	0.271	0.360	0.360	0.360
T/K	462.15	406.15	433.15	462.15
P/MPa	91.3	91.2	87.0	84.2

Polymer (B):   **polyethylene**   **1996HEU**

Characterization: M_n/kg.mol^{-1} = 2.94, M_w/kg.mol^{-1} = 3.30, M_z/kg.mol^{-1} = 3.72,

                             Exxon Polywax 2000

Solvent (A):   **ethene**   **C_2H_4**   **74-85-1**

Type of data: cloud points

T/K = 413.15

x_B	0.0118	0.0102	0.0089	0.0068	0.0056	0.0045	0.0040	0.0036	0.0030
P/MPa	72.7	75.3	77.9	82.3	84.7	86.1	86.6	86.8	87.0

x_B	0.0027	0.0023	0.0019	0.0017	0.0012	0.0008
P/MPa	87.5	87.9	88.2	88.0	87.1	85.4

critical concentration: $x_{B, crit}$ = 0.0030,   *critical pressure:* P_{crit}/MPa = 87.0

T/K = 433.15

x_B	0.0118	0.0102	0.0089	0.0068	0.0056	0.0045	0.0040	0.0036	0.0030
P/MPa	71.5	73.9	76.2	80.0	82.0	83.2	83.6	83.8	84.0

x_B	0.0027	0.0023	0.0019	0.0017	0.0012	0.0008
P/MPa	84.4	84.8	85.0	84.9	84.1	82.6

critical concentration: $x_{B, crit}$ = 0.0030,   *critical pressure:* P_{crit}/MPa = 84.4

continued

continued

$T/K = 453.15$

x_B	0.0118	0.0102	0.0089	0.0068	0.0056	0.0045	0.0040	0.0036	0.0030
P/MPa	70.5	72.6	74.8	78.4	80.1	81.1	81.4	81.5	81.6

x_B	0.0027	0.0023	0.0019	0.0017	0.0012	0.0008
P/MPa	81.9	82.3	82.5	82.4	81.5	79.8

critical concentration: $x_{B, crit} = 0.0030$,   *critical pressure*: P_{crit}/MPa = 81.6

$T/K = 473.15$

x_B	0.0118	0.0102	0.0089	0.0068	0.0056	0.0045	0.0040	0.0036	0.0030
P/MPa	69.6	71.4	73.4	76.5	78.1	79.0	79.3	79.4	79.6

x_B	0.0027	0.0023	0.0019	0.0017	0.0012	0.0008
P/MPa	80.1	80.5	80.7	80.5	79.4	77.9

critical concentration: $x_{B, crit} = 0.0030$,   *critical pressure*: P_{crit}/MPa = 79.6

$T/K = 493.15$

x_B	0.0118	0.0102	0.0089	0.0068	0.0056	0.0045	0.0040	0.0036	0.0030
P/MPa	68.5	70.3	72.0	74.9	76.3	77.1	77.4	77.6	77.7

x_B	0.0027	0.0023	0.0019	0.0017	0.0012	0.0008
P/MPa	78.5	79.3	79.4	79.2	77.9	76.3

critical concentration: $x_{B, crit}$ – 0.0030,   *critical pressure*: P_{crit}/MPa = 77.7

$T/K = 513.15$

x_B	0.0118	0.0102	0.0089	0.0068	0.0056	0.0045	0.0040	0.0036	0.0030
P/MPa	67.5	69.3	70.9	73.5	74.8	75.6	75.8	76.0	76.2

x_B	0.0027	0.0023	0.0019	0.0017	0.0012	0.0008
P/MPa	77.1	77.8	78.0	77.8	76.4	74.4

critical concentration: $x_{B, crit} = 0.0030$,   *critical pressure*: P_{crit}/MPa = 76.2

Polymer (B):	**polyethylene**	**1981LOO**
Characterization:	M_n/kg.mol^{-1} = 3.0, M_w/kg.mol^{-1} = 3.7, M_z/kg.mol^{-1} = 4.3, linear, Central Lab. DSM, Geleen, The Netherlands	
Solvent (A):	**ethene** C_2H_4	**74-85-1**

Type of data: cloud points

w_B	0.1514	0.1514	0.1514	0.1514	0.1514	0.1514	0.1514	0.1514	0.1514
T/K	390.79	395.30	399.17	403.90	409.01	413.81	418.24	423.19	427.84
P/bar	1085	1072	1059	1045	1034	1020	1011	1000	991

w_B	0.1514	0.1514	0.1514	0.1596	0.1596	0.1596	0.1596	0.1596	0.1596
T/K	432.16	437.78	442.03	404.69	409.64	413.81	418.56	422.65	428.13
P/bar	982	972	964	1040	1024	1016	1005	996	986

w_B	0.1596	0.1596	0.1596	0.1711	0.1711	0.1711	0.1711	0.1711	0.1711
T/K	432.90	437.28	442.28	384.22	388.12	392.50	396.94	403.09	406.80
P/bar	977	969	960	1094	1081	1067	1054	1037	1028

continued

continued

w_B	0.1711	0.1711	0.1711	0.1711	0.1711	0.1711	0.1711	0.1711	0.1809
T/K	410.90	414.83	419.56	424.24	428.75	433.33	438.53	443.03	386.43
P/bar	1017	1007	996	988	979	970	960	953	1083
w_B	0.1809	0.1809	0.1809	0.1809	0.1809	0.1809	0.1809	0.1809	0.1809
T/K	391.63	396.28	401.19	405.29	410.15	414.96	418.81	423.31	428.17
P/bar	1067	1054	1040	1029	1016	1004	996	986	977
w_B	0.1809	0.1809	0.1809	0.1892	0.1892	0.1892	0.1892	0.1892	0.1892
T/K	432.86	437.61	442.41	385.89	390.01	394.34	399.64	403.43	407.90
P/bar	968	960	952	1093	1072	1059	1043	1033	1021
w_B	0.1892	0.1892	0.1892	0.1892	0.1892	0.1892	0.1892	0.2003	0.2003
T/K	412.30	416.51	421.20	426.25	430.93	435.89	439.12	385.82	390.59
P/bar	1011	1001	991	980	972	963	957	1081	1065
w_B	0.2003	0.2003	0.2003	0.2003	0.2003	0.2003	0.2003	0.2003	0.2003
T/K	394.94	399.19	403.26	408.79	413.12	417.74	422.74	427.28	431.14
P/bar	1051	1040	1029	1014	1004	993	984	975	967
w_B	0.2003	0.2003	0.2111	0.2111	0.2111	0.2111	0.2111	0.2111	0.2111
T/K	436.18	441.21	387.41	391.69	396.85	400.18	404.68	409.20	414.85
P/bar	958	949	1076	1062	1046	1037	1025	1013	1001
w_B	0.2111	0.2111	0.2111	0.2111	0.2111	0.2111	0.2194	0.2194	0.2194
T/K	419.03	424.17	428.31	432.99	438.43	443.28	387.21	392.26	396.71
P/bar	990	980	972	963	954	945	1075	1058	1045
w_B	0.2194	0.2194	0.2194	0.2194	0.2194	0.2194	0.2194	0.2194	0.2194
T/K	400.30	404.04	409.14	413.76	418.20	422.33	427.11	431.56	436.49
P/bar	1036	1025	1012	1001	991	983	974	965	956
w_B	0.2194	0.2303	0.2303	0.2303	0.2303	0.2303	0.2303	0.2303	0.2303
T/K	441.54	386.89	391.01	394.99	399.48	403.93	408.92	413.10	417.61
P/bar	948	1073	1060	1047	1035	1023	1010	1000	991
w_B	0.2303	0.2303	0.2303	0.2303	0.2303	0.2389	0.2389	0.2389	0.2389
T/K	421.51	426.70	431.47	436.80	439.63	386.20	391.75	396.39	400.63
P/bar	983	973	963	954	949	1074	1056	1042	1031
w_B	0.2389	0.2389	0.2389	0.2389	0.2389	0.2389	0.2389	0.2389	0.2389
T/K	405.11	409.98	414.30	418.18	422.63	428.35	432.85	436.89	441.80
P/bar	1018	1006	996	988	979	967	959	952	944
w_B	0.2502	0.2502	0.2502	0.2502	0.2502	0.2502	0.2502	0.2502	0.2502
T/K	387.26	391.60	395.90	400.58	405.07	409.48	413.31	417.58	422.23
P/bar	1068	1054	1041	1028	1017	1006	997	989	978
w_B	0.2502	0.2502	0.2502	0.2502	0.2575	0.2575	0.2575	0.2575	0.2575
T/K	426.27	431.03	435.65	441.30	385.84	390.30	395.60	400.18	404.28
P/bar	970	961	953	943	1071	1057	1040	1028	1017
w_B	0.2575	0.2575	0.2575	0.2575	0.2575	0.2575	0.2575	0.2575	0.2698
T/K	409.73	413.39	418.11	422.17	426.63	432.22	437.16	442.22	386.41
P/bar	1004	995	985	977	967	958	949	941	1066

continued

continued

w_B	0.2698	0.2698	0.2698	0.2698	0.2698	0.2698	0.2698	0.2698	0.2698
T/K	391.42	395.69	400.48	404.31	408.64	412.98	417.87	422.31	427.20
P/bar	1050	1038	1025	1014	1004	994	983	974	965

w_B	0.2698	0.2698	0.2698	0.2781	0.2781	0.2781	0.2781	0.2781	0.2781
T/K	431.74	436.77	441.49	386.07	391.37	394.94	399.29	403.49	408.15
P/bar	956	947	940	1062	1049	1038	1026	1015	1003

w_B	0.2781	0.2781	0.2781	0.2781	0.2781	0.2781	0.2781	0.2909	0.2909
T/K	413.43	417.65	421.34	426.67	431.77	436.78	441.58	386.75	391.18
P/bar	991	982	975	964	955	946	939	1062	1048

w_B	0.2909	0.2909	0.2909	0.2909	0.2909	0.2909	0.2909	0.2909	0.2909
T/K	394.69	399.77	404.02	409.15	411.89	417.03	421.20	426.52	431.09
P/bar	1035	1023	1012	1000	993	982	974	964	955

w_B	0.2909	0.2909	0.2954	0.2954	0.2954	0.2954	0.2954	0.2954	0.2954
T/K	435.87	440.35	386.33	390.42	394.71	398.67	403.67	407.91	413.24
P/bar	947	940	1063	1053	1038	1026	1013	1002	990

w_B	0.2954	0.2954	0.2954	0.2954	0.2954	0.2954
T/K	417.88	421.58	426.90	431.20	435.82	440.40
P/bar	980	973	963	955	947	939

Polymer (B):	**polyethylene**		**1991NIE**
Characterization:	M_n/kg.mol^{-1} = 5.2, M_w/kg.mol^{-1} = 8.6, HDPE,		
	Exxon Polymer Plant, Antwerp, fractionated in the laboratory		
Solvent (A):	**ethene**	C_2H_4	**74-85-1**

Type of data: cloud points

w_B	0.056	0.056	0.056	0.112	0.112	0.112	0.185	0.185	0.185
T/K	403.65	434.15	459.15	404.65	435.15	465.15	403.15	408.15	418.15
P/MPa	142.3	125.6	119.0	139.5	123.9	117.0	137.1	133.3	127.8

w_B	0.185	0.185	0.248	0.248	0.248
T/K	433.15	457.65	404.65	433.65	464.15
P/MPa	122.3	116.5	130.6	118.5	112.5

Polymer (B):	**polyethylene**		**1981LOO**
Characterization:	M_n/kg.mol^{-1} = 7.6, M_w/kg.mol^{-1} = 9.2, M_z/kg.mol^{-1} = 10.5,		
	linear, Central Lab. DSM, Geleen, The Netherlands		
Solvent (A):	**ethene**	C_2H_4	**74-85-1**

Type of data: cloud points

w_B	0.0100	0.0100	0.0100	0.0100	0.0100	0.0100	0.0100	0.0100	0.0143
T/K	414.55	420.04	425.88	429.99	433.83	437.08	440.26	443.92	414.68
P/bar	1219	1194	1180	1164	1154	1144	1139	1127	1248

w_B	0.0143	0.0143	0.0143	0.0143	0.0143	0.0143	0.0153	0.0153	0.0153
T/K	420.49	423.43	429.26	434.23	439.91	445.35	411.80	416.94	420.85
P/bar	1225	1216	1194	1180	1162	1147	1259	1238	1223

continued

continued

w_B	0.0153	0.0153	0.0153	0.0153	0.0153	0.0158	0.0158	0.0158	0.0158
T/K	425.66	430.02	434.29	438.90	443.31	414.56	415.72	420.29	425.27
P/bar	1205	1192	1178	1165	1152	1254	1250	1233	1215

w_B	0.0158	0.0158	0.0158	0.0158	0.0198	0.0198	0.0198	0.0198	0.0198
T/K	429.21	434.20	439.15	443.99	409.66	413.40	416.97	421.46	427.06
P/bar	1201	1185	1169	1156	1293	1273	1262	1243	1224

w_B	0.0198	0.0198	0.0198	0.0198	0.0230	0.0230	0.0230	0.0230	0.0230
T/K	431.47	435.57	440.13	441.32	409.48	414.04	419.34	423.26	427.07
P/bar	1209	1195	1182	1178	1298	1276	1257	1243	1228

w_B	0.0230	0.0230	0.0230	0.0230	0.0252	0.0252	0.0252	0.0252	0.0252
T/K	430.37	432.91	436.69	441.79	413.38	416.09	421.13	426.77	430.96
P/bar	1217	1208	1197	1181	1289	1279	1258	1237	1222

w_B	0.0252	0.0252	0.0252	0.0298	0.0298	0.0298	0.0298	0.0298	0.0298
T/K	435.04	439.34	443.17	410.53	415.61	419.46	424.87	428.67	433.11
P/bar	1209	1195	1184	1313	1291	1275	1253	1240	1224

w_B	0.0298	0.0298	0.0308	0.0308	0.0308	0.0308	0.0308	0.0308	0.0308
T/K	438.18	442.41	410.75	416.80	419.19	425.40	429.42	434.91	438.94
P/bar	1207	1194	1310	1282	1266	1250	1236	1217	1204

w_B	0.0308	0.0345	0.0345	0.0345	0.0345	0.0345	0.0345	0.0345	0.0345
T/K	444.13	410.55	414.74	418.50	423.22	428.02	432.49	436.83	441.23
P/bar	1189	1319	1299	1284	1266	1248	1232	1218	1204

w_B	0.0406	0.0406	0.0406	0.0406	0.0406	0.0406	0.0406	0.0406	0.0494
T/K	410.35	414.75	420.35	424.66	429.28	433.80	438.49	442.85	411.23
P/bar	1331	1312	1288	1271	1254	1238	1223	1209	1334

w_B	0.0494	0.0494	0.0494	0.0494	0.0494	0.0494	0.0494	0.0545	0.0545
T/K	414.90	419.72	425.27	428.85	434.49	437.80	442.03	410.10	414.87
P/bar	1317	1297	1275	1262	1243	1232	1217	1345	1324

w_B	0.0545	0.0545	0.0545	0.0545	0.0545	0.0545	0.0545	0.0587	0.0587
T/K	419.04	421.07	424.87	429.52	434.51	438.33	442.90	410.35	415.70
P/bar	1305	1297	1292	1264	1247	1234	1219	1345	1322

w_B	0.0587	0.0587	0.0587	0.0587	0.0587	0.0587	0.0636	0.0636	0.0636
T/K	419.19	424.40	428.05	432.33	436.93	441.75	410.40	415.23	419.06
P/bar	1306	1287	1272	1257	1241	1224	1347	1326	1309

w_B	0.0636	0.0636	0.0636	0.0636	0.0636	0.0700	0.0700	0.0700	0.0700
T/K	424.37	429.01	434.20	439.00	442.79	410.10	414.11	418.37	422.88
P/bar	1288	1270	1251	1237	1223	1349	1331	1312	1294

w_B	0.0700	0.0700	0.0700	0.0700	0.0831	0.0831	0.0831	0.0831	0.0831
T/K	427.26	431.78	436.87	441.22	412.06	417.79	421.70	426.39	431.00
P/bar	1278	1261	1243	1230	1344	1318	1302	1284	1267

w_B	0.0831	0.0831	0.0831	0.0895	0.0895	0.0895	0.0895	0.0895	0.0895
T/K	435.15	440.28	444.79	411.68	416.78	421.97	425.62	430.54	436.03
P/bar	1252	1236	1222	1345	1322	1300	1287	1268	1249

continued

continued

w_B	0.0895	0.0895	0.0996	0.0996	0.0996	0.0996	0.0996	0.0996	0.0996
T/K	440.13	444.60	412.60	415.74	419.39	423.42	428.54	431.97	436.19
P/bar	1237	1223	1339	1326	1310	1294	1275	1263	1248
w_B	0.0996	0.0996	0.1035	0.1035	0.1035	0.1035	0.1035	0.1035	0.1035
T/K	440.72	442.69	411.83	416.39	420.77	424.98	430.64	435.28	439.57
P/bar	1234	1227	1343	1323	1304	1289	1267	1251	1237
w_B	0.1035	0.1099	0.1099	0.1099	0.1099	0.1099	0.1099	0.1099	0.1099
T/K	444.18	413.05	417.06	421.48	427.23	430.96	435.71	439.81	445.96
P/bar	1223	1337	1319	1301	1279	1265	1249	1237	1217
w_B	0.1197	0.1197	0.1197	0.1197	0.1197	0.1197	0.1197	0.1197	0.1227
T/K	413.19	416.76	421.90	425.99	430.36	435.96	439.89	444.73	414.83
P/bar	1339	1319	1297	1282	1266	1246	1235	1220	1324
w_B	0.1227	0.1227	0.1227	0.1227	0.1227	0.1227	0.1227	0.1229	0.1229
T/K	417.89	421.28	425.56	430.96	437.21	439.29	443.51	413.93	419.11
P/bar	1312	1298	1282	1262	1241	1235	1222	1329	1307
w_B	0.1229	0.1229	0.1229	0.1229	0.1229	0.1229	0.1229	0.1298	0.1298
T/K	422.28	426.65	431.24	432.33	435.61	441.75	446.39	413.28	417.89
P/bar	1294	1278	1261	1258	1246	1227	1213	1331	1311
w_B	0.1298	0.1298	0.1298	0.1298	0.1298	0.1298	0.1310	0.1310	0.1310
T/K	421.89	426.64	431.01	435.86	440.60	445.97	415.33	418.42	420.96
P/bar	1295	1277	1261	1245	1230	1214	1321	1308	1297
w_B	0.1310	0.1310	0.1310	0.1310	0.1310	0.1379	0.1379	0.1379	0.1379
T/K	425.01	430.76	433.83	438.06	443.22	410.36	414.35	420.33	423.73
P/bar	1282	1261	1250	1237	1221	1342	1324	1298	1287
w_B	0.1379	0.1379	0.1379	0.1379	0.1406	0.1406	0.1406	0.1406	0.1406
T/K	429.78	433.41	438.03	442.86	412.97	417.14	421.64	427.02	431.40
P/bar	1263	1251	1236	1221	1330	1313	1294	1274	1258
w_B	0.1406	0.1406	0.1503	0.1503	0.1503	0.1503	0.1503	0.1503	0.1503
T/K	436.40	440.93	411.79	416.48	422.17	429.55	435.34	439.59	444.30
P/bar	1242	1228	1334	1313	1290	1263	1243	1230	1215
w_B	0.1532	0.1532	0.1532	0.1532	0.1532	0.1532	0.1532	0.1532	0.1616
T/K	411.26	416.97	420.92	426.83	431.82	436.24	440.46	445.83	412.59
P/bar	1334	1309	1293	1270	1253	1239	1225	1209	1328
w_B	0.1616	0.1616	0.1616	0.1616	0.1616	0.1616	0.1616	0.1702	0.1702
T/K	416.92	422.05	427.17	431.32	436.16	440.63	445.32	412.02	417.01
P/bar	1309	1289	1270	1255	1240	1225	1211	1329	1307
w_B	0.1702	0.1702	0.1702	0.1702	0.1702	0.1702	0.1752	0.1752	0.1752
T/K	421.63	426.01	430.61	435.39	440.57	445.06	413.91	417.86	422.77
P/bar	1289	1272	1256	1241	1224	1211	1318	1300	1282
w_B	0.1752	0.1752	0.1752	0.1752	0.1785	0.1785	0.1785	0.1785	0.1785
T/K	425.14	431.30	435.52	441.43	412.11	416.58	420.88	426.24	430.66
P/bar	1273	1250	1237	1218	1325	1306	1289	1269	1253

continued

continued

w_B	0.1785	0.1785	0.1785	0.1933	0.1933	0.1933	0.1933	0.1933	0.1933
T/K	435.25	440.00	445.07	414.15	419.27	424.33	427.21	429.72	432.54
P/bar	1238	1223	1208	1314	1306	1293	1273	1263	1254

w_B	0.1933	0.1933	0.1933	0.1972	0.1972	0.1972	0.1972	0.1972	0.1972
T/K	437.84	439.91	443.33	412.40	416.11	420.17	424.73	430.06	434.98
P/bar	1245	1222	1211	1320	1304	1289	1271	1252	1236

w_B	0.1972	0.1972	0.2138	0.2138	0.2138	0.2138	0.2138	0.2138	0.2138
T/K	439.15	443.73	412.65	416.24	421.78	425.56	431.16	434.82	439.52
P/bar	1222	1209	1318	1301	1279	1265	1245	1234	1219

w_B	0.2138	0.2313	0.2313	0.2313	0.2313	0.2313	0.2313	0.2313	0.2313
T/K	443.30	411.21	416.17	420.82	426.55	431.65	434.71	438.57	445.61
P/bar	1208	1318	1301	1279	1265	1245	1234	1219	1208

Polymer (B):	**polyethylene**	**1981LOO**

Characterization: M_n/kg.mol^{-1} = 5.0, M_w/kg.mol^{-1} = 21, M_z/kg.mol^{-1} = 39, linear, Central Lab. DSM, Geleen, The Netherlands

Solvent (A):	**ethene**	**C$_2$H$_4$**	**74-85-1**

Type of data: cloud points

w_B	0.1523	0.1523	0.1523	0.1523	0.1523	0.1523	0.1523	0.1523	0.1523
T/K	403.15	407.52	413.10	418.07	423.20	427.59	432.99	438.19	443.31
P/bar	1549	1523	1491	1465	1440	1420	1396	1376	1357

Polymer (B):	**polyethylene**	**1981LOO**

Characterization: M_n/kg.mol^{-1} = 7.0, M_w/kg.mol^{-1} = 37, M_z/kg.mol^{-1} = 100, linear, Central Lab. DSM, Geleen, The Netherlands

Solvent (A):	**ethene**	**C$_2$H$_4$**	**74-85-1**

Type of data: cloud points

w_B	0.1350	0.1350	0.1350	0.1350	0.1350	0.1350	0.1350
T/K	407.40	411.85	412.70	418.30	429.30	434.40	440.40
P/bar	1726	1663	1591	1557	1498	1475	1447

Polymer (B):	**polyethylene**	**1981LOO**

Characterization: M_n/kg.mol^{-1} = 8.0, M_w/kg.mol^{-1} = 55, M_z/kg.mol^{-1} = 300, linear, Central Lab. DSM, Geleen, The Netherlands

Solvent (A):	**ethene**	**C$_2$H$_4$**	**74-85-1**

Type of data: cloud points

w_B	0.0971	0.0971	0.0971	0.0971	0.0971	0.0971	0.1190	0.1190	0.1190
T/K	416.62	419.14	423.86	428.21	433.04	437.49	413.92	417.55	421.79
P/bar	1671	1653	1626	1603	1579	1557	1656	1638	1611

continued

continued

w_B	0.1190	0.1190	0.1190	0.1190	0.1302	0.1302	0.1302	0.1302	0.1302
T/K	426.24	430.97	435.34	440.34	414.61	419.82	424.60	428.98	433.31
P/bar	1588	1562	1541	1517	1626	1596	1567	1544	1522

w_B	0.1302	0.1406	0.1406	0.1406	0.1406	0.1406	0.1406	0.1406	0.1406
T/K	437.92	407.88	412.42	416.66	420.14	424.83	429.23	434.21	438.28
P/bar	1499	1656	1627	1600	1581	1554	1531	1507	1489

w_B	0.1531	0.1531	0.1531	0.1531	0.1531	0.1531	0.1531	0.1531
T/K	405.79	412.49	416.37	420.42	424.46	428.50	433.17	437.70
P/bar	1653	1609	1586	1563	1541	1520	1497	1477

Polymer (B):   **polyethylene**   **1995LOO**

Characterization: M_n/kg.mol^{-1} = 8.4, M_w/kg.mol^{-1} = 32, branched,

                          5.3 CH$_3$/100 C-atoms, DSM Research, Geleen, The Netherlands

Solvent (A):   **ethene**   **C$_2$H$_4$**   **74-85-1**

Type of data: cloud points

T/K = 393.15

w_B	0.01992	0.03010	0.03990	0.05008	0.06024	0.07985	0.08980	0.10021	0.11018
P/MPa	132.7	134.5	133.3	132.7	132.3	131.2	130.3	129.5	128.5

w_B	0.12022	0.13069	0.14108	0.15072	0.16028	0.17087	0.18041	0.20120	0.22374
P/MPa	127.9	126.9	126.3	125.8	125.2	124.6	124.1	123.2	122.1

critical concentration: $w_{B, crit}$ = 0.180,   *critical pressure*: P_{crit}/MPa = 124.1

T/K = 413.15

w_B	0.01992	0.03010	0.03990	0.05008	0.06024	0.07985	0.08980	0.10021	0.11018
P/MPa	124.5	125.4	125.1	124.7	124.2	123.3	122.9	122.2	121.0

w_B	0.12022	0.13069	0.14108	0.15072	0.16028	0.17087	0.18041	0.20120	0.22374
P/MPa	120.5	119.8	119.2	118.7	118.2	117.6	117.2	116.4	115.5

critical concentration: $w_{B, crit}$ = 0.180,   *critical pressure*: P_{crit}/MPa = 117.2

T/K = 433.15

w_B	0.01992	0.03010	0.03990	0.05008	0.06024	0.07985	0.08980	0.10021	0.11018
P/MPa	118.3	119.3	118.9	118.6	118.4	117.4	116.5	116.5	115.6

w_B	0.12022	0.13069	0.14108	0.15072	0.16028	0.17087	0.18041	0.20120	0.22374
P/MPa	114.8	114.3	113.8	113.2	112.8	112.3	111.9	111.3	110.4

critical concentration: $w_{B, crit}$ = 0.180,   *critical pressure*: P_{crit}/MPa = 111.9

Polymer (B):   **polyethylene**   **1991NIE**
Characterization: M_n/kg.mol^{-1} = 10.6, M_w/kg.mol^{-1} = 15.4, LDPE,
                          Exxon Polymer Plant, Antwerp, fractionated in the laboratory
Solvent (A):   **ethene**   **C$_2$H$_4$**   **74-85-1**

Type of data: cloud points

w_B	0.102	0.102	0.102	0.163	0.163	0.163	0.223	0.223	0.223
T/K	404.65	444.65	473.65	394.65	424.65	453.65	390.15	425.15	455.15
P/MPa	132.8	118.8	113.0	134.2	123.0	115.0	131.2	118.0	111.2

w_B	0.284	0.284	0.284	0.385	0.385	0.385	0.455	0.455	0.455
T/K	393.65	414.65	455.15	395.65	415.15	445.65	395.65	415.15	445.15
P/MPa	124.2	117.0	108.0	113.2	108.2	103.0	103.0	99.2	96.0

w_B	0.455	0.548	0.548	0.548	0.548	0.548
T/K	463.15	390.15	404.15	425.15	435.15	453.15
P/MPa	94.5	86.5	84.5	82.6	82.3	81.9

Polymer (B):   **polyethylene**   **1980RAE**
Characterization: M_n/kg.mol^{-1} = 11.2, M_w/kg.mol^{-1} = 23.2,
                          LDPE, ρ = 0.921 g/cm^3, Leuna-Werke, Germany
Solvent (A):   **ethene**   **C$_2$H$_4$**   **74-85-1**

Type of data: cloud points

w_B	0.05	0.05	0.05	0.05	0.05	0.05	0.10	0.10	0.10
T/K	413.15	433.15	453.15	473.15	493.15	513.15	393.15	413.15	433.15
P/MPa	123.6	115.7	110.8	105.4	103.9	102.5	141.2	134.4	125.5

w_B	0.10	0.20	0.20	0.20	0.20	0.20	0.20	0.30	0.30
T/K	453.15	393.15	413.15	433.15	453.15	473.15	493.15	403.15	413.15
P/MPa	119.6	132.4	125.5	118.7	114.3	109.8	105.9	115.5	112.1

w_B	0.30	0.30	0.30	0.30	0.30
T/K	433.15	453.15	473.15	493.15	513.15
P/MPa	106.3	103.5	99.8	96.6	94.1

Polymer (B):   **polyethylene**   **2004BEC**
Characterization: M_n/kg.mol^{-1} = 12, M_w/kg.mol^{-1} = 45.3, LDPE
Solvent (A):   **ethene**   **C$_2$H$_4$**   **74-85-1**

Type of data: cloud points

w_B	0.05	was kept constant

T/K	403.15	414.15	423.15	434.15	442.15	454.15	463.15	473.15	482.15
P/bar	1637	1585	1522	1468	1428	1389	1356	1323	1299

T/K	492.15	502.15	512.15	521.15
P/bar	1275	1249	1226	1204

Polymer (B):   **polyethylene**   **1981LOO**

Characterization: $M_n/\text{kg.mol}^{-1} = 12$, $M_w/\text{kg.mol}^{-1} = 53$, $M_z/\text{kg.mol}^{-1} = 150$, linear, Central Lab. DSM, Geleen, The Netherlands

Solvent (A):   **ethene**   **C$_2$H$_4$**   **74-85-1**

Type of data: cloud points

w_B	0.1224	0.1224	0.1224	0.1224	0.1224	0.1224	0.1224	0.1224	0.1224
T/K	399.71	404.53	409.75	414.34	419.00	424.72	430.57	435.84	441.65
P/bar	1743	1706	1667	1640	1610	1578	1547	1520	1494

Polymer (B):   **polyethylene**   **1995LOO**

Characterization: $M_n/\text{kg.mol}^{-1} = 14$, $M_w/\text{kg.mol}^{-1} = 70$, branched, 2.2 CH$_3$/100 C-atoms, DSM Research, Geleen, The Netherlands

Solvent (A):   **ethene**   **C$_2$H$_4$**   **74-85-1**

Type of data: cloud points

T/K = 393.15

w_B	0.05995	0.06995	0.08038	0.09086	0.10122	0.11012	0.12054	0.12983	0.14096
P/MPa	170.7	169.2	167.1	165.1	163.8	161.8	160.3	159.6	158.7

w_B	0.15070	0.15916	0.16967	0.17995	0.19177	0.20004
P/MPa	158.3	157.6	157.0	156.6	155.4	155.4

critical concentration: $w_{B,\,crit} = 0.133$,   *critical pressure*: $P_{crit}/\text{MPa} = 159.4$

T/K = 413.15

w_B	0.05995	0.06995	0.08038	0.09086	0.10122	0.11012	0.12054	0.12983	0.14096
P/MPa	156.6	155.6	154.2	152.3	150.8	149.5	148.3	147.3	146.5

w_B	0.15070	0.15916	0.16967	0.17995	0.19177	0.20004
P/MPa	146.3	145.9	145.3	144.9	144.1	143.7

critical concentration: $w_{B,\,crit} = 0.133$,   *critical pressure*: $P_{crit}/\text{MPa} = 147.2$

T/K = 433.15

w_B	0.05995	0.06995	0.08038	0.09086	0.10122	0.11012	0.12054	0.12983	0.14096
P/MPa	147.0	145.8	144.2	143.0	141.5	140.5	139.2	138.3	137.7

w_B	0.15070	0.15916	0.16967	0.17995	0.19177	0.20004
P/MPa	137.3	137.1	136.2	136.2	135.6	135.3

critical concentration: $w_{B,\,crit} = 0.133$,   *critical pressure*: $P_{crit}/\text{MPa} = 138.2$

Polymer (B):   **polyethylene**   **1995LOO**

Characterization: $M_n/\text{kg.mol}^{-1} = 18$, $M_w/\text{kg.mol}^{-1} = 157$, branched, 1.1 CH$_3$/100 C-atoms, DSM Research, Geleen, The Netherlands

Solvent (A):   **ethene**   **C$_2$H$_4$**   **74-85-1**

Type of data: cloud points

continued

continued

$T/K = 393.15$

w_B	0.00521	0.01007	0.01486	0.02010	0.02502	0.02990	0.03969	0.04998	0.06002
P/MPa	184.8	189.0	187.6	186.8	185.9	185.6	183.7	181.1	180.3

w_B	0.06991	0.07989	0.08987	0.09981	0.10985	0.12046	0.12931	0.14017	0.15189
P/MPa	178.2	176.8	175.2	174.5	173.2	172.3	171.0	169.7	168.8

w_B	0.16261	0.16737	0.17862	0.19022	0.20012
P/MPa	166.8	165.8	165.3	163.5	162.9

critical concentration: $w_{B, crit} = 0.110$,   *critical pressure*: P_{crit}/MPa = 173.3

$T/K = 413.15$

w_B	0.00521	0.01007	0.01486	0.02010	0.02502	0.02990	0.03969	0.04998	0.06002
P/MPa	169.7	170.3	171.2	171.3	170.2	169.7	168.0	166.3	165.0

w_B	0.06991	0.07989	0.08987	0.09981	0.10985	0.12046	0.12931	0.14017	0.15189
P/MPa	163.7	162.3	160.8	159.9	159.1	158.2	157.1	156.1	155.1

w_B	0.16261	0.16737	0.17862	0.19022	0.20012
P/MPa	154.1	153.5	152.8	151.6	150.5

critical concentration: $w_{B, crit} = 0.110$,   *critical pressure*: P_{crit}/MPa = 158.9

$T/K = 433.15$

w_B	0.00521	0.01007	0.01486	0.02010	0.02502	0.02990	0.03969	0.04998	0.06002
P/MPa	157.6	158.6	159.0	159.1	158.0	157.6	156.2	154.5	153.6

w_B	0.06991	0.07989	0.08987	0.09981	0.10985	0.12046	0.12931	0.14017	0.15189
P/MPa	152.5	151.2	150.1	149.3	148.6	147.8	146.9	146.0	145.1

w_B	0.16261	0.16737	0.17862	0.19022	0.20012
P/MPa	144.2	143.7	143.2	142.1	141.3

critical concentration: $w_{B, crit} = 0.110$,   *critical pressure*: P_{crit}/MPa = 148.5

Polymer (B):	**polyethylene**		**1995LOO**
Characterization:	M_n/kg.mol^{-1} = 19.1, M_w/kg.mol^{-1} = 229, branched,		
	1.2 CH$_3$/100 C-atoms, DSM Research, Geleen, The Netherlands		
Solvent (A):	**ethene**	**C$_2$H$_4$**	**74-85-1**

Type of data: cloud points

$T/K = 393.15$

w_B	0.02003	0.02990	0.04011	0.04997	0.05993	0.07050	0.08049	0.08976	0.09532
P/MPa	190.4	190.5	187.5	185.1	182.9	180.9	179.1	177.4	176.6

w_B	0.10090	0.10795	0.12307	0.11532	0.11961	0.13015	0.13933	0.14036	0.14915
P/MPa	176.0	175.1	174.6	174.2	174.0	172.7	171.7	171.6	170.4

w_B	0.16035	0.17063
P/MPa	169.2	168.2

critical concentration: $w_{B, crit} = 0.115$,   *critical pressure*: P_{crit}/MPa = 174.4

continued

continued

$T/K = 413.15$

w_B	0.02003	0.02990	0.04011	0.04997	0.05993	0.07050	0.08049	0.08976	0.09532
P/MPa	174.1	172.6	170.3	169.0	166.8	165.3	163.9	162.5	161.7

w_B	0.10090	0.10795	0.12307	0.11532	0.11961	0.13015	0.13933	0.14036	0.14915
P/MPa	161.1	160.5	160.1	159.8	159.5	158.4	157.6	157.5	156.6

w_B	0.16035	0.17063
P/MPa	155.5	154.8

critical concentration: $w_{B, crit} = 0.115$,   critical pressure: P_{crit}/MPa $= 159.8$

$T/K = 433.15$

w_B	0.02003	0.02990	0.04011	0.04997	0.05993	0.07050	0.08049	0.08976	0.09532
P/MPa	161.7	160.1	157.9	157.0	155.5	154.0	152.7	151.4	150.7

w_B	0.10090	0.10795	0.12307	0.11532	0.11961	0.13015	0.13933	0.14036	0.14915
P/MPa	150.2	149.6	149.3	149.1	148.8	147.8	147.2	147.1	146.3

w_B	0.16035	0.17063
P/MPa	145.4	144.9

critical concentration: $w_{B, crit} = 0.115$,   critical pressure: P_{crit}/MPa $= 149.0$

Polymer (B):	**polyethylene**		**1996MU1**
Characterization:	M_n/kg.mol^{-1} = 19.2, M_w/kg.mol^{-1} = 165,		
	LDPE, PE15280, BASF AG, Germany		
Solvent (A):	**ethene**	C_2H_4	**74-85-1**

Type of data: cloud points

w_B	0.010	0.010	0.010	0.010	0.010	0.074	0.074	0.074	0.074
T/K	523.35	498.35	473.35	449.25	423.45	498.55	473.55	448.45	424.65
P/MPa	124.8	130.7	137.3	145.9	156.5	127.2	133.9	142.4	153.1

w_B	0.075	0.075	0.075	0.075	0.075	0.100	0.100	0.100	0.100
T/K	523.75	498.15	473.75	448.05	423.65	524.65	498.85	473.55	449.35
P/MPa	121.7	127.1	133.6	142.4	152.5	118.8	124.1	130.5	138.5

w_B	0.100	0.114	0.114	0.114	0.114	0.114	0.128	0.128	0.128
T/K	423.65	523.45	499.45	473.65	449.05	424.15	522.55	498.95	472.85
P/MPa	149.0	119.9	124.9	131.0	139.1	150.1	118.3	123.2	129.2

w_B	0.128	0.128	0.144	0.144	0.144	0.144	0.144	0.200	0.200
T/K	448.45	424.35	524.25	498.75	474.15	449.05	422.65	523.35	499.65
P/MPa	136.8	146.9	115.4	119.7	125.6	133.3	143.5	110.9	115.0

w_B	0.200	0.200	0.200	0.261	0.261	0.261	0.261	0.261	0.362
T/K	474.15	449.75	423.05	523.45	498.35	474.45	447.55	424.45	523.55
P/MPa	120.0	125.1	135.1	108.9	111.8	116.5	122.2	129.4	89.1

w_B	0.362	0.362	0.362	0.362
T/K	499.15	473.05	447.25	425.05
P/MPa	91.4	92.9	95.9	99.1

Polymer (B):	**polyethylene**							**2002BEY**
Characterization:	M_n/kg.mol^{-1} = 19.2, M_w/kg.mol^{-1} = 165,							
	LDPE, PE15280, BASF AG, Germany							
Solvent (A):	**ethene**			**C₂H₄**				**74-85-1**

Type of data: cloud points

w_B	0.05	0.05	0.05	0.05	0.05	0.05	0.05	0.05	0.05
T/K	521.65	497.95	472.55	447.85	423.55	413.15	403.45	393.45	383.75
P/bar	1221	1272	1341	1421	1529	1591	1653	1725	1813

Polymer (B):	**polyethylene**				**1981SP1**
Characterization:	M_n/kg.mol^{-1} = 19.5, M_w/kg.mol^{-1} = 192, M_z/kg.mol^{-1} = 1450,				
	LDPE, ρ = 0.9225 g/cm^3 at 295 K				
Solvent (A):	**ethene**		**C₂H₄**		**74-85-1**

Type of data: cloud points

T/K = 433.15

w_B	0.060	0.077	0.100	0.122	0.239	0.265
P/MPa	152.0	153.2	150.9	146.3	133.8	129.5

T/K = 473.15

w_B	0.074	0.077	0.115	0.253	0.691
P/MPa	138.0	138.1	134.0	120.3	63.0

critical concentration: $w_{B, crit}$ = 0.115,   *critical pressure*: P_{crit}/MPa = 134.0

Type of data: coexistence data

T/K = 433.15

Total feed concentrations of the homogeneous system before demixing: w_A = 0.940 and w_B = 0.060.

$P/$ MPa	w_B gel phase	sol phase
152.0	0.167	0.060 (cloud and shadow points)
146.1	0.200	
143.3	0.211	
130.8		0.022
129.8	0.277	
124.0	0.317	
122.4	0.314	
119.5	0.338	
114.2		0.014
108.8	0.397	
94.3		0.005
84.8		0.003
74.4		0.003

continued

continued

T/K = 433.15

Total feed concentrations of the homogeneous system before demixing: w_A = 0.900 and w_B = 0.100.

P/ MPa	w_B gel phase	sol phase
150.9	0.124	0.100 (cloud and shadow points)
148.1		0.083
145.3	0.170	
143.9		0.076
140.5	0.204	
136.0		0.049
131.7		0.039
116.8		0.023
112.3	0.374	
106.4	0.431	
103.5		0.013
100.6	0.458	
99.5		0.012
67.0		0.004
64.7		0.003
54.2		0.002
48.2		0.002
42.5		0.001

T/K = 433.15

Total feed concentrations of the homogeneous system before demixing: w_A = 0.878 and w_B = 0.122.

P/ MPa	w_B gel phase	sol phase
146.3	0.122	0.113 (cloud and shadow points)
143.8	0.155	
142.1		0.105
139.1	0.205	
136.0		0.057
133.0	0.245	
130.2	0.253	
127.2		0.042
123.4		0.041
121.7	0.311	
107.8	0.422	
99.5		0.019
96.6	0.490	
87.8		0.007
59.8		0.003

continued

continued

T/K = 433.15

Total feed concentrations of the homogeneous system before demixing: w_A = 0.761 and w_B = 0.239.

P/ MPa	w_B gel phase	sol phase
133.8	0.239	0.074 (cloud and shadow points)
128.3		0.055
120.0		0.038
117.5	0.330	
111.3	0.377	
100.2		0.019
97.0	0.486	
85.7	0.555	
74.6		0.007
48.3		0.001

T/K = 433.15

Total feed concentrations of the homogeneous system before demixing: w_A = 0.735 and w_B = 0.265.

P/ MPa	w_B gel phase	sol phase
129.6	0.265	0.062 (cloud and shadow points)
127.8	0.274	
126.0	0.283	
122.1		0.050
119.3	0.316	
115.3		0.048
111.4	0.372	
106.1		0.032
103.4	0.445	
100.8		0.025
99.7	0.457	
91.0		0.018
89.4	0.533	
80.3		0.013
79.2	0.592	
65.0		0.006
63.6	0.675	
51.0	0.739	
49.5		0.004
46.0	0.764	

continued

continued

$T/K = 473.15$

Total feed concentrations of the homogeneous system before demixing: $w_A = 0.926$ and $w_B = 0.074$.

$P/$ MPa	w_B gel phase	sol phase
138.0	0.152	0.074 (cloud and shadow points)
136.0	0.172	
135.3		0.061
127.6		0.041
125.3	0.237	
123.0		0.035
121.5	0.261	
117.5	0.286	
112.2		0.024
100.4		0.012
99.2	0.395	
96.5	0.409	
88.5	0.491	
86.0	0.520	
81.8		0.004
48.2		0.002

$T/K = 473.15$

Total feed concentrations of the homogeneous system before demixing: $w_A = 0.885$ and $w_B = 0.115$.

$P/$ MPa	w_B gel phase	sol phase
134.0	0.115	0.115 (critical point)
131.7	0.166	
131.5		0.075
128.2	0.202	
124.2	0.226	
120.4		0.045
117.2	0.272	
112.3		0.028
111.5	0.305	
94.5		0.012
87.0	0.522	
72.3	0.628	
58.0		0.002
42.4		0.001

continued

continued

$T/K = 473.15$

Total feed concentrations of the homogeneous system before demixing: $w_A = 0.747$ and $w_B = 0.253$.

$P/$ MPa	w_B gel phase	sol phase
120.3	0.253	0.076 (cloud and shadow points)
117.0	0.275	
112.8		0.061
105.1	0.352	
101.2		0.036
95.8		0.034
93.6	0.453	
91.5	0.464	
85.3	0.509	
84.7	0.533	
82.3		0.018
75.2	0.601	
72.8		0.010
60.6		0.003
59.2	0.709	
51.7		0.005

$T/K = 473.15$

Total feed concentrations of the homogeneous system before demixing: $w_A = 0.309$ and $w_B = 0.691$.

$P/$ MPa	w_B gel phase	sol phase
63.0	0.691	0.008 (cloud and shadow points)
61.6		0.005
58.8	0.725	
53.3	0.742	
42.6	0.799	

Polymer (B):	**polyethylene**							**1981LOO**	

Characterization: $M_n/\text{kg.mol}^{-1} = 20$, $M_w/\text{kg.mol}^{-1} = 58$, $M_z/\text{kg.mol}^{-1} = 148$, linear, Central Lab. DSM, Geleen, The Netherlands

Solvent (A):	**ethene**	**C_2H_4**						**74-85-1**	

Type of data: cloud points

w_B	0.0691	0.0691	0.0691	0.0691	0.0691	0.0691	0.0691	0.0857	0.0857
T/K	408.99	415.94	419.96	423.18	427.38	431.04	436.91	414.66	420.17
P/bar	1831	1775	1749	1729	1701	1678	1644	1750	1712

continued

continued

w_B	0.0857	0.0857	0.0857	0.0857	0.0997	0.0997	0.0997	0.0997	0.0997
T/K	424.87	430.69	435.40	440.97	412.80	416.99	419.38	424.36	428.50
P/bar	1683	1649	1622	1595	1735	1707	1688	1659	1634

w_B	0.0997	0.0997	0.1129	0.1129	0.1129	0.1129	0.1129	0.1129	0.1129
T/K	432.23	439.10	408.53	415.11	419.51	420.81	423.57	427.38	432.65
P/bar	1611	1577	1739	1691	1662	1653	1639	1616	1588

w_B	0.1129	0.1129	0.1290	0.1290	0.1290	0.1290	0.1290	0.1290	0.1290
T/K	436.67	440.60	410.18	415.22	419.45	423.99	428.09	433.71	439.75
P/bar	1567	1547	1699	1663	1638	1609	1585	1554	1524

Polymer (B):	**polyethylene**	**1991NIE**
Characterization:	M_n/kg.mol^{-1} = 21.9, M_w/kg.mol^{-1} = 103.2, LDPE,	
	Exxon Polymer Plant, Antwerp	
Solvent (A):	**ethene** C_2H_4	**74-85-1**

Type of data: cloud points

w_B	0.072	0.095	0.095	0.095	0.174	0.174	0.174	0.243	0.243
T/K	427.15	399.15	428.65	468.15	399.15	428.15	466.65	400.65	430.15
P/MPa	148.8	156.2	141.0	127.6	148.2	134.8	123.0	140.6	128.2

w_B	0.243	0.346	0.346	0.346	0.445	0.445	0.445
T/K	460.15	419.65	451.15	471.15	397.65	427.65	462.15
P/MPa	120.0	122.5	114.5	111.0	111.4	104.8	99.5

Type of data: coexistence data

T/K = 403.15

Total feed concentrations of the homogeneous system before demixing: w_A = 0.844 and w_B = 0.156.

$P/$ MPa	w_A gel phase	sol phase
133.8	0.844	(cloud point)
123.3	0.693	0.961
107.0	0.567	0.987
89.2	0.456	0.989
72.6	0.369	0.995

Polymer (B):	**polyethylene**	**1998HOR**
Characterization:	M_n/kg.mol^{-1} = 23, M_w/kg.mol^{-1} = 160, LDPE,	
	Exxon Polymer Plant, Antwerp	
Solvent (A):	**ethene** C_2H_4	**74-85-1**

Type of data: cloud points

continued

continued

$T/K = 420.15$

w_B	0.184	0.184	0.184	0.184
P/bar	1396	1400	1405	1408
shear rate/s	0	500	1000	1500

$T/K = 422.15$

w_B	0.274	0.274	0.274	0.274
P/bar	1314	1319	1324	1328
shear rate/s	0	500	1000	1500

$T/K = 431.15$

w_B	0.184	0.184	0.184	0.184
P/bar	1347	1352	1358	1361
shear rate/s	0	500	1000	1500

$T/K = 439.15$

w_B	0.274	0.274	0.274	0.274
P/bar	1249	1254	1258	1262
shear rate/s	0	500	1000	1500

$T/K = 441.15$

w_B	0.225	0.225	0.225	0.225
P/bar	1281	1290	1300	1304
shear rate/s	0	500	1000	1500

$T/K = 453.15$

w_B	0.225	0.225	0.225	0.225
P/bar	1245	1255	1260	1263
shear rate/s	0	500	1000	1500

$T/K = 468.15$

w_B	0.225	0.225	0.225	0.225
P/bar	1203	1214	1219	1222
shear rate/s	0	500	1000	1500

Polymer (B):	**polyethylene**							**1980RAE**
Characterization:	M_n/kg.mol^{-1} = 27.4, M_w/kg.mol^{-1} = 374,							
	LDPE, ρ = 0.916 g/cm^3, Leuna-Werke, Germany							
Solvent (A):	**ethene**			**C$_2$H$_4$**				**74-85-1**

Type of data: cloud points

w_B	0.05	0.05	0.10	0.10	0.10	0.10	0.10	0.10	0.15
T/K	403.15	433.15	413.15	433.15	453.15	473.15	493.15	513.154	403.15
P/MPa	142.7	127.9	152.8	141.6	153.4	128.7	123.7	119.6	145.7

w_B	0.15
T/K	433.15
P/MPa	137.5

Polymer (B):	polyethylene							**2000CH4**
Characterization:	M_n/kg.mol^{-1} = 28.8, M_w/kg.mol^{-1} = 32.0, NIST SRM1483							
Solvent (A):	**ethene**		**C$_2$H$_4$**					**74-85-1**

Type of data: cloud points

w_B	0.050	0.050	0.050	0.050	0.050	0.050	0.075	0.075	0.075
T/K	453.15	443.15	433.15	423.15	413.15	403.15	453.15	443.15	433.15
P/bar	1406	1445	1488	1534	1590	1650	1431	1454	1500

w_B	0.075	0.075	0.075	0.100	0.100	0.100	0.100	0.100	0.100
T/K	423.15	413.15	403.15	453.15	443.15	433.15	423.15	413.15	403.15
P/bar	1536	1591	1654	1422	1452	1502	1536	1591	1651

w_B	0.124	0.124	0.124	0.124	0.124	0.124	0.150	0.150	0.150
T/K	453.15	443.15	433.15	423.15	413.15	403.15	453.15	443.15	433.15
P/bar	1445	1475	1515	1559	1610	1668	1425	1453	1495

w_B	0.150	0.150
T/K	423.15	413.15
P/bar	1527	1594

Polymer (B):	polyethylene							**1980RAE**
Characterization:	M_n/kg.mol^{-1} = 29.8, M_w/kg.mol^{-1} = 328,							
	LDPE, ρ = 0.918 g/cm^3, Leuna-Werke, Germany							
Solvent (A):	**ethene**		**C$_2$H$_4$**					**74-85-1**

Type of data: cloud points

w_B	0.05	0.05	0.05	0.05	0.05	0.10	0.10	0.10	0.15
T/K	433.15	453.15	473.15	493.15	513.154	473.15	493.15	513.15	493.15
P/MPa	152.0	144.2	138.3	132.4	126.5	144.2	134.4	124.1	128.0

w_B	0.15	0.20	0.20	0.20	0.20	0.20	0.30	0.30	0.30
T/K	513.15	443.15	453.15	473.15	493.15	513.15	473.15	493.15	513.15
P/MPa	121.1	128.5	125.0	119.6	115.7	111.8	97.6	92.2	89.7

Polymer (B):	polyethylene							**1995LOO**
Characterization:	M_n/kg.mol^{-1} = 30, M_w/kg.mol^{-1} = 223, branched,							
	2.0 CH$_3$/100 C-atoms, DSM Research, Geleen, The Netherlands							
Solvent (A):	**ethene**		**C$_2$H$_4$**					**74-85-1**

Type of data: cloud points

T/K = 393.15

w_B	0.01001	0.01500	0.02505	0.03011	0.04000	0.05014	0.06031	0.07409	0.07981
P/MPa	171.3	172.2	172.0	171.6	170.5	169.7	168.7	167.9	167.6

w_B	0.08537	0.09035	0.10035	0.11036	0.12001	0.12989	0.13512	0.14555	0.15929
P/MPa	167.2	166.6	166.4	165.6	165.0	164.6	164.0	163.1	162.1

continued

continued

w_B	0.17102	0.18313
P/MPa	161.2	160.1

critical concentration: $w_{B, crit}$ = 0.094,   *critical pressure*: P_{crit}/MPa = 166.6

T/K = 413.15

w_B	0.01001	0.01500	0.02505	0.03011	0.04000	0.05014	0.06031	0.07409	0.07981
P/MPa	157.6	158.1	158.0	157.8	156.8	156.1	155.4	154.5	154.3

w_B	0.08537	0.09035	0.10035	0.11036	0.12001	0.12989	0.13512	0.14555	0.15929
P/MPa	154.0	153.5	153.4	152.7	152.3	151.8	151.3	150.6	149.7

w_B	0.17102	0.18313
P/MPa	149.1	148.2

critical concentration: $w_{B, crit}$ = 0.094,   *critical pressure*: P_{crit}/MPa = 153.6

T/K = 433.15

w_B	0.01001	0.01500	0.02505	0.03011	0.04000	0.05014	0.06031	0.07409	0.07981
P/MPa	147.2	147.7	147.5	147.3	146.7	146.0	145.3	144.6	144.4

w_B	0.08537	0.09035	0.10035	0.11036	0.12001	0.12989	0.13512	0.14555	0.15929
P/MPa	144.1	143.7	143.7	143.1	142.6	142.3	141.9	141.2	140.5

w_B	0.17102	0.18313
P/MPa	139.9	139.2

critical concentration: $w_{B, crit}$ = 0.094,   *critical pressure*: P_{crit}/MPa = 143.8

Polymer (B):	**polyethylene**		**1995LOO**
Characterization:	M_n/kg.mol^{-1} = 31, M_w/kg.mol^{-1} = 2000, branched,		
	2.1 CH$_3$/100 C-atoms, DSM Research, Geleen, The Netherlands		
Solvent (A):	**ethene**	**C$_2$H$_4$**	**74-85-1**

Type of data: critical points (cloud points)

w_B	0.102	0.102	0.102
T/K	393.15	413.15	433.15
P/MPa	176.2	161.4	150.3

Polymer (B):	**polyethylene**		**2003BUB**
Characterization:	M_n/kg.mol^{-1} = 33, M_w/kg.mol^{-1} = 130,		
	synthesized in the laboratory by high-pressure polymerization		
Solvent (A):	**ethene**	**C$_2$H$_4$**	**74-85-1**

Type of data: cloud points

w_B	0.03	was kept constant

T/K	403.15	413.15	422.15	433.15	452.15	462.15	473.15	483.15	493.15
P/bar	1748	1666	1606	1542	1458	1418	1383	1351	1324

T/K	503.15	513.15	523.15	532.15
P/bar	1297	1274	1255	1239

Polymer (B):	polyethylene						1995LOO	
Characterization:	M_n/kg.mol^{-1} = 34.2, M_w/kg.mol^{-1} = 236, branched,							
	2.3 CH$_3$/100 C-atoms, DSM Research, Geleen, The Netherlands							
Solvent (A):	ethene			C_2H_4			74-85-1	

Type of data: cloud points

T/K = 393.15

w_B	0.05455	0.06532	0.07490	0.08566	0.09653	0.10998	0.12111	0.13189	0.13990
P/MPa	176.2	174.7	173.9	173.3	172.7	171.4	171.3	170.5	170.0

w_B	0.14871	0.15965	0.17140	0.17983	0.19003	0.20131	0.21076		
P/MPa	169.3	168.4	167.2	166.2	165.2	163.9	162.6		

T/K = 413.15

w_B	0.05455	0.06532	0.07490	0.08566	0.09653	0.10998	0.12111	0.13189	0.13990
P/MPa	161.7	160.5	159.5	159.1	158.6	157.5	157.4	156.8	156.5

w_B	0.14871	0.15965	0.17140	0.17983	0.19003	0.20131	0.21076		
P/MPa	155.8	155.1	154.2	153.6	152.5	151.5	150.5		

T/K = 433.15

w_B	0.05455	0.06532	0.07490	0.08566	0.09653	0.10998	0.12111	0.13189	0.13990
P/MPa	150.8	149.8	149.0	148.6	148.2	147.2	147.0	146.6	146.5

w_R	0.14871	0.15965	0.17140	0.17983	0.19003	0.20131	0.21076		
P/MPa	145.8	145.2	144.5	144.0	143.1	142.3	141.4		

Polymer (B):	polyethylene						1991NIE	
Characterization:	M_n/kg.mol^{-1} = 36.3, M_w/kg.mol^{-1} = 220.5, LDPE,							
	Exxon Polymer Plant, Antwerp							
Solvent (A):	ethene			C_2H_4			74-85-1	

Type of data: cloud points

w_B	0.029	0.029	0.029	0.029	0.029	0.042	0.042	0.042	0.047
T/K	413.15	427.15	437.15	467.15	470.15	407.15	437.15	467.15	407.15
P/MPa	169.8	161.0	155.6	142.8	141.8	172.7	154.0	141.7	170.7

w_B	0.047	0.047	0.047	0.061	0.061	0.061	0.061	0.084	0.084
T/K	417.15	437.15	467.15	407.15	417.15	437.15	467.15	422.65	457.15
P/MPa	163.5	152.5	140.6	169.0	162.5	151.5	139.5	157.2	141.5

w_B	0.100	0.100	0.100	0.100	0.100	0.100	0.138	0.138	0.180
T/K	402.65	407.65	427.65	447.65	457.65	473.65	426.15	446.15	407.15
P/MPa	167.0	163.5	152.0	143.2	139.5	134.5	149.0	140.3	155.0

w_B	0.180	0.180	0.180	0.242	0.242	0.242	0.283	0.283	0.283
T/K	427.15	447.15	476.65	425.15	450.15	476.15	407.15	427.15	447.15
P/MPa	144.5	136.7	128.0	138.0	129.6	123.0	139.5	132.0	126.0

continued

continued

w_B	0.283	0.409	0.409	0.409	0.409	0.434	0.434	0.434	0.572
T/K	472.15	409.65	429.15	447.65	466.15	424.15	460.65	476.15	416.65
P/MPa	120.3	119.2	114.0	109.7	106.5	111.0	104.2	102.0	89.0

w_B	0.572	0.572	0.572
T/K	426.65	446.65	466.15
P/MPa	87.5	85.3	83.5

Type of data: coexistence data

$T/K = 403.15$

Total feed concentrations of the homogeneous system before demixing: $w_A = 0.863$ and $w_B = 0.137$.

Demixing pressure	w_A gel phase	sol phase	Fractionation during demixing gel phase			sol phase		
$P/$ MPa			$M_n/$ kg/mol	$M_w/$ kg/mol	$M_z/$ kg/mol	$M_n/$ kg/mol	$M_w/$ kg/mol	$M_z/$ kg/mol
162.0	0.863		36.3	220.5	(feed sample, cloud point)			
154.8	0.767	0.956	49.9	209.3	505.4	14.7	35.9	65.4
138.4	0.674	0.981						
122.8	0.649	0.983				6.5	15.1	27.4
111.4	0.620							
92.6		0.993						
68.1		0.996						
46.3		0.997						

$T/K = 433.15$

Total feed concentrations of the homogeneous system before demixing: $w_A = 0.848$ and $w_B = 0.152$.

Demixing pressure	w_A gel phase	sol phase	Fractionation during demixing gel phase			sol phase		
$P/$ MPa			$M_n/$ kg/mol	$M_w/$ kg/mol	$M_z/$ kg/mol	$M_n/$ kg/mol	$M_w/$ kg/mol	$M_z/$ kg/mol
144.2	0.848		36.3	220.5	(feed sample, cloud point)			
138.9	0.778	0.963						
127.5	0.682	0.977						
110.8	0.571	0.985						
103.0	0.526	0.990						

continued

continued

$T/K = 463.15$

Total feed concentrations of the homogeneous system before demixing: $w_A = 0.841$ and $w_B = 0.159$.

Demixing pressure	w_A gel phase	sol phase	Fractionation during demixing gel phase			sol phase		
$P/$ MPa			$M_n/$ kg/mol	$M_w/$ kg/mol	$M_z/$ kg/mol	$M_n/$ kg/mol	$M_w/$ kg/mol	$M_z/$ kg/mol
135.2	0.841		36.3	220.5	(feed sample, cloud point)			
124.2	0.719	0.968	56.9	221.3	508.2	10.6	24.8	42.8
112.7	0.612	0.976						
97.8	0.504	0.984				5.9	12.3	29.2
78.7	0.440	0.988						
57.7	0.372	0.993						
37.0	0.289	0.998						

Polymer (B):	**polyethylene**		**1981LOO**
Characterization:	$M_n/\text{kg.mol}^{-1} = 43$, $M_w/\text{kg.mol}^{-1} = 118$, $M_z/\text{kg.mol}^{-1} = 231$, linear, Central Lab. DSM, Geleen, The Netherlands		
Solvent (A):	**ethene**	**C₂H₄**	**74-85-1**

Type of data: cloud points

w_B	0.0099	0.0099	0.0099	0.0099	0.0099	0.0099	0.0109	0.0109	0.0109
T/K	421.01	421.30	425.91	430.03	434.57	439.03	420.92	425.60	430.18
P/bar	1825	1821	1784	1760	1731	1698	1831	1796	1762
w_B	0.0109	0.0109	0.0149	0.0149	0.0149	0.0149	0.0149	0.0149	0.0149
T/K	434.96	439.31	412.54	416.83	420.89	425.05	429.91	434.62	439.25
P/bar	1733	1707	1900	1868	1835	1803	1772	1739	1709
w_B	0.0201	0.0201	0.0201	0.0201	0.0201	0.0201	0.0249	0.0249	0.0249
T/K	416.80	420.52	425.83	430.22	435.00	439.40	421.40	426.26	430.73
P/bar	1864	1834	1795	1762	1731	1705	1824	1786	1756
w_B	0.0249	0.0249	0.0297	0.0297	0.0297	0.0297	0.0297	0.0297	0.0297
T/K	435.46	440.20	414.88	420.10	424.22	428.39	433.26	438.21	442.35
P/bar	1727	1696	1870	1828	1794	1763	1733	1701	1679
w_B	0.0348	0.0348	0.0348	0.0348	0.0348	0.0373	0.0373	0.0373	0.0373
T/K	422.56	426.96	431.71	436.17	440.49	413.82	418.18	422.36	427.12
P/bar	1803	1772	1740	1713	1687	1869	1836	1804	1769
w_B	0.0373	0.0373	0.0373	0.0398	0.0398	0.0398	0.0398	0.0398	0.0398
T/K	431.74	436.36	440.92	405.17	409.72	414.40	418.47	422.83	427.09
P/bar	1739	1709	1681	1939	1898	1860	1827	1794	1762

continued

continued

w_B	0.0398	0.0398	0.0398	0.0450	0.0450	0.0450	0.0450	0.0450	0.0450
T/K	432.35	436.96	441.97	417.54	418.18	422.24	427.20	431.60	436.60
P/bar	1730	1700	1671	1830	1824	1787	1753	1720	1694
w_B	0.0450	0.0477	0.0477	0.0477	0.0477	0.0477	0.0477	0.0499	0.0499
T/K	441.03	418.58	422.95	427.29	431.79	436.84	441.17	409.01	414.15
P/bar	1667	1812	1778	1749	1719	1689	1665	1884	1845
w_B	0.0499	0.0499	0.0499	0.0499	0.0499	0.0499	0.0550	0.0550	0.0550
T/K	418.72	422.93	427.34	432.33	436.74	441.74	419.28	423.33	423.89
P/bar	1806	1777	1745	1712	1684	1657	1791	1766	1760
w_B	0.0550	0.0550	0.0550	0.0550	0.0573	0.0573	0.0573	0.0573	0.0573
T/K	427.55	432.58	436.93	441.80	404.74	409.86	414.62	414.79	419.02
P/bar	1737	1703	1678	1650	1913	1866	1826	1824	1795
w_B	0.0573	0.0573	0.0573	0.0573	0.0573	0.0580	0.0580	0.0580	0.0580
T/K	423.29	428.29	432.79	437.61	442.36	409.63	414.21	418.36	422.62
P/bar	1764	1729	1701	1671	1645	1865	1827	1797	1764
w_B	0.0580	0.0580	0.0580	0.0580	0.0587	0.0587	0.0587	0.0587	0.0587
T/K	427.70	432.28	436.79	441.54	418.80	423.13	427.81	432.38	437.47
P/bar	1728	1700	1673	1647	1793	1762	1731	1701	1670
w_B	0.0587	0.0605	0.0605	0.0605	0.0605	0.0605	0.0605	0.0605	0.0605
T/K	442.01	405.43	409.85	414.55	419.09	423.41	427.42	433.00	437.77
P/bar	1646	1897	1857	1820	1783	1752	1724	1690	1661
w_B	0.0605	0.0622	0.0622	0.0622	0.0622	0.0622	0.0622	0.0622	0.0652
T/K	441.88	418.78	422.44	426.93	427.23	431.84	436.68	441.03	405.09
P/bar	1639	1785	1758	1730	1726	1697	1666	1643	1895
w_B	0.0652	0.0652	0.0652	0.0652	0.0652	0.0652	0.0652	0.0652	0.0683
T/K	409.84	414.46	418.94	423.54	428.68	432.77	437.44	441.51	404.38
P/bar	1853	1814	1781	1748	1713	1689	1661	1637	1897
w_B	0.0683	0.0683	0.0683	0.0683	0.0683	0.0683	0.0683	0.0683	0.0700
T/K	409.13	413.65	418.21	422.36	427.61	432.09	436.74	441.18	405.26
P/bar	1854	1816	1781	1751	1717	1688	1659	1635	1890
w_B	0.0700	0.0700	0.0700	0.0700	0.0700	0.0700	0.0700	0.0700	0.0750
T/K	409.84	414.82	419.82	423.26	428.46	433.57	438.19	442.32	405.58
P/bar	1849	1807	1769	1742	1709	1677	1650	1627	1881
w_B	0.0750	0.0750	0.0750	0.0750	0.0750	0.0750	0.0750	0.0750	0.0797
T/K	410.21	414.77	418.96	423.64	428.31	432.86	437.32	441.95	404.76
P/bar	1841	1802	1771	1739	1707	1679	1652	1627	1886
w_B	0.0797	0.0797	0.0797	0.0797	0.0797	0.0797	0.0797	0.0797	0.0869
T/K	409.38	414.14	418.27	422.68	427.34	432.04	436.64	441.37	411.81
P/bar	1845	1805	1774	1743	1711	1681	1653	1627	1821
w_B	0.0869	0.0869	0.0869	0.0869	0.0869	0.0869	0.0869	0.0902	0.0902
T/K	417.19	418.00	422.67	426.84	431.92	436.11	442.35	405.18	409.84
P/bar	1779	1778	1740	1712	1679	1654	1619	1874	1834

continued

continued

w_B	0.0902	0.0902	0.0902	0.0902	0.0902	0.0902	0.0902	0.0953	0.0953
T/K	413.91	418.56	423.08	428.24	432.76	437.16	442.10	412.46	416.52
P/bar	1801	1765	1733	1699	1671	1645	1618	1808	1777
w_B	0.0953	0.0953	0.0953	0.0953	0.0953	0.0953	0.0998	0.0998	0.0998
T/K	420.57	425.44	429.77	434.03	438.53	443.12	411.33	415.38	419.16
P/bar	1747	1714	1686	1660	1635	1610	1815	1783	1754
w_B	0.0998	0.0998	0.0998	0.0998	0.0998	0.1048	0.1048	0.1048	0.1048
T/K	423.23	427.54	431.79	435.89	440.40	412.74	417.02	421.06	425.41
P/bar	1727	1699	1672	1648	1623	1799	1766	1738	1708
w_B	0.1048	0.1048	0.1048	0.1048	0.1100	0.1100	0.1100	0.1100	0.1100
T/K	429.62	434.32	438.32	443.01	405.24	409.84	414.34	418.62	422.83
P/bar	1682	1654	1631	1606	1858	1820	1783	1751	1722
w_B	0.1100	0.1100	0.1100	0.1100	0.1149	0.1149	0.1149	0.1149	0.1149
T/K	427.58	432.18	437.42	441.61	413.29	417.09	420.66	424.84	429.28
P/bar	1692	1663	1633	1610	1787	1758	1734	1706	1678
w_B	0.1149	0.1149	0.1149	0.1247	0.1247	0.1247	0.1247	0.1247	0.1247
T/K	433.87	438.27	442.59	411.99	416.88	420.82	424.73	429.14	433.38
P/bar	1650	1626	1603	1790	1753	1725	1699	1672	1647
w_B	0.1247	0.1247	0.1292	0.1292	0.1292	0.1292	0.1292	0.1292	0.1292
T/K	437.66	442.05	409.85	413.90	418.14	423.21	427.93	432.68	437.34
P/bar	1623	1599	1798	1767	1736	1701	1672	1644	1617
w_B	0.1292	0.1292	0.1292	0.1292	0.1292	0.1292	0.1292	0.1292	0.1292
T/K	441.75	408.07	413.16	416.59	421.01	425.90	430.63	436.33	441.63
P/bar	1594	1809	1769	1745	1713	1682	1653	1623	1593
w_B	0.1403	0.1403	0.1403	0.1403	0.1403	0.1403	0.1403	0.1403	0.1403
T/K	405.29	409.91	413.99	417.79	422.89	427.71	432.35	436.99	441.65
P/bar	1822	1784	1754	1725	1692	1661	1634	1608	1584
w_B	0.1440	0.1440	0.1440	0.1440	0.1440	0.1440	0.1440	0.1440	0.1497
T/K	410.14	415.10	419.19	423.59	428.02	433.13	438.04	443.61	410.64
P/bar	1778	1742	1712	1684	1656	1627	1600	1572	1767
w_B	0.1497	0.1497	0.1497	0.1497	0.1497	0.1497	0.1497	0.1546	0.1546
T/K	415.37	419.17	423.78	428.00	432.45	437.02	441.55	410.97	414.92
P/bar	1732	1705	1676	1650	1624	1599	1577	1762	1732
w_B	0.1546	0.1546	0.1546	0.1546	0.1546	0.1546	0.1604	0.1604	0.1604
T/K	419.85	424.41	429.17	433.69	438.86	442.65	407.49	411.78	416.21
P/bar	1699	1669	1641	1616	1588	1569	1799	1746	1716
w_B	0.1604	0.1604	0.1604	0.1604	0.1604	0.1604			
T/K	419.74	424.29	428.80	433.32	437.77	442.68			
P/bar	1691	1682	1635	1610	1586	1562			

Polymer (B):	**polyethylene**		**1995LOO**
Characterization:	M_n/kg.mol^{-1} = 47.3, M_w/kg.mol^{-1} = 52.0, branched,		
	2.0 CH$_3$/100 C-atoms, DSM Research, Geleen, The Netherlands		
Solvent (A):	**ethene**	**C$_2$H$_4$**	**74-85-1**

Type of data: cloud points

T/K = 393.15

w_B	0.00497	0.00753	0.01000	0.01249	0.01503	0.01994	0.02491	0.03005	0.03498
P/MPa	164.6	167.4	169.5	170.8	172.1	173.8	175.0	175.0	175.8

w_B	0.04018	0.04500	0.04995	0.05491	0.06004	0.06462	0.07007	0.07511	0.07950
P/MPa	176.3	176.2	176.4	176.7	176.8	176.6	177.3	176.7	177.1

w_B	0.08524	0.08967	0.09948	0.10978	0.11968	0.13029	0.13961	0.14923	0.15890
P/MPa	176.3	176.1	175.8	175.8	175.3	174.5	173.9	173.7	173.0

w_B	0.16967	0.18046
P/MPa	172.2	171.5

critical concentration: $w_{B, crit}$ = 0.082,   *critical pressure*: P_{crit}/MPa = 176.6

T/K = 413.15

w_B	0.00497	0.00753	0.01000	0.01249	0.01503	0.01994	0.02491	0.03005	0.03498
P/MPa	151.8	154.2	155.8	157.1	157.9	159.5	160.2	160.8	161.4

w_B	0.04018	0.04500	0.04995	0.05491	0.06004	0.06462	0.07007	0.07511	0.07950
P/MPa	161.8	161.9	162.1	162.2	162.2	162.4	162.3	162.2	162.1

w_B	0.08524	0.08967	0.09948	0.10978	0.11968	0.13029	0.13961	0.14923	0.15890
P/MPa	161.8	161.8	161.6	161.4	160.8	160.5	160.1	159.9	159.4

w_B	0.16967	0.18046
P/MPa	158.6	157.6

critical concentration: $w_{B, crit}$ = 0.082,   *critical pressure*: P_{crit}/MPa = 162.0

T/K = 433.15

w_B	0.00497	0.00753	0.01000	0.01249	0.01503	0.01994	0.02491	0.03005	0.03498
P/MPa	141.6	143.7	145.2	146.3	147.1	148.7	149.3	149.7	150.2

w_B	0.04018	0.04500	0.04995	0.05491	0.06004	0.06462	0.07007	0.07511	0.07950
P/MPa	150.7	150.7	150.9	150.9	151.0	151.2	151.2	151.0	151.0

w_B	0.08524	0.08967	0.09948	0.10978	0.11968	0.13029	0.13961	0.14923	0.15890
P/MPa	150.6	150.7	150.6	150.3	150.0	149.6	149.4	149.1	148.8

w_B	0.16967	0.18046
P/MPa	148.1	147.3

critical concentration: $w_{B, crit}$ = 0.082,   *critical pressure*: P_{crit}/MPa = 150.9

Polymer (B):	**polyethylene**		**1981SP1**

Characterization: $M_n/\text{kg.mol}^{-1} = 55$, $M_w/\text{kg.mol}^{-1} = 749$, $M_z/\text{kg.mol}^{-1} = 6775$, LDPE, $\rho = 0.9270$ g/cm^3 at 295 K

Solvent (A):	**ethene**	**C$_2$H$_4$**	**74-85-1**

Type of data: cloud points

$T/\text{K} = 433.15$

w_B	0.045	0.053	0.065	0.164
P/MPa	164.2	163.8	161.0	148.8

critical concentration: $w_{B,\,crit} = 0.065$,   *critical pressure*: $P_{crit}/\text{MPa} = 161.0$

$T/\text{K} = 473.15$

w_B	0.040	0.045	0.050	0.065	0.251
P/MPa	144.9	145.0	144.6	142.0	128.8

critical concentration: $w_{B,\,crit} = 0.065$,   *critical pressure*: $P_{crit}/\text{MPa} = 142.0$

Type of data: coexistence data

$T/\text{K} = 433.15$

Total feed concentrations of the homogeneous system before demixing: $w_A = 0.947$ and $w_B = 0.053$.

$P/$ MPa	w_B gel phase	sol phase
163.8	0.077	0.053 (cloud and shadow points)
160.5		0.040
159.1		0.038
157.9	0.122	
157.0	0.123	
155.5		0.031
153.1	0.153	
151.3	0.162	
145.0		0.025
141.7	0.200	
140.8	0.207	
138.5		0.022
123.6		0.014
121.2	0.350	
119.0		0.010
109.7		0.009
108.5	0.448	
105.5		0.008
100.0	0.509	
97.5		0.007
82.7		0.006
66.8		0.002
47.0		0.001

continued

continued

$T/K = 433.15$

Total feed concentrations of the homogeneous system before demixing: $w_A = 0.935$ and $w_B = 0.065$.

$P/$ MPa	w_B gel phase	sol phase
161.0	0.065	0.065 (critical point)
160.3		0.054
159.8	0.076	
158.7	0.097	
156.0		0.050
152.4	0.152	
150.0		0.039
147.3	0.171	
146.2	0.176	
143.0		0.030
139.5	0.206	
135.8		0.022
130.5	0.260	
118.8		0.014
117.1	0.388	
116.0	0.397	
113.6	0.418	
110.3		0.011
92.0		0.007
69.7		0.003
68.0		0.002

$T/K = 433.15$

Total feed concentrations of the homogeneous system before demixing: $w_A = 0.836$ and $w_B = 0.164$.

$P/$ MPa	w_B gel phase	sol phase
148.8	0.164	0.049 (cloud and shadow points)
147.1	0.173	
143.6		0.056
141.5	0.197	
140.0	0.206	
137.5		0.041
136.9	0.221	
135.1	0.234	
131.6		0.036
127.4	0.290	
126.4	0.294	

continued

continued

P/ MPa	w_B gel phase	sol phase
123.2		0.029
118.0		0.023
117.6	0.378	
114.8	0.409	
113.9	0.414	
105.0		0.017
101.0		0.016
98.3	0.459	
93.9	0.493	
91.7		0.013
89.9	0.572	
87.2		0.008
84.7	0.559	
80.5		0.004
73.5		0.003
52.5		0.002
47.5		0.001

T/K = 473.15

Total feed concentrations of the homogeneous system before demixing: $w_A = 0.960$ and $w_B = 0.040$.

P/ MPa	w_B gel phase	sol phase
144.9	0.095	0.040 (cloud and shadow points)
143.0		0.032
140.8		0.030
139.5	0.162	
137.2	0.190	
136.8		0.027
135.7	0.201	
127.6	0.261	
124.0		0.016
121.3		0.014
120.2	0.279	
114.1		0.011
107.0		0.008
105.5	0.365	
89.0		0.005
73.7		0.003
72.0		0.002

continued

continued

$T/K = 473.15$

Total feed concentrations of the homogeneous system before demixing: $w_A = 0.935$ and $w_B = 0.065$.

$P/$ MPa	w_B gel phase	sol phase
142.0	0.065	0.065 (critical point)
131.8	0.242	
129.6		0.032
128.7	0.253	
121.7		0.021
118.2	0.286	
108.4		0.011
104.2	0.380	
97.3		0.009
91.2	0.497	
88.2		0.007
72.0		0.004

$T/K = 473.15$

Total feed concentrations of the homogeneous system before demixing: $w_A = 0.749$ and $w_B = 0.251$.

$P/$ MPa	w_B gel phase	sol phase
128.8	0.251	0.051 (cloud and shadow points)
125.4	0.268	
116.5	0.292	
111.1		0.045
108.5	0.340	
104.8		0.038
101.3	0.416	
98.8		0.033
96.0	0.459	
88.0		0.016
86.6		0.015
84.0	0.554	
80.5		0.013
75.5		0.008
74.7	0.618	
73.3		0.009
65.8		0.005

Polymer (B):   **polyethylene**   **1999DIE**
Characterization: M_n/kg.mol^{-1} = 69.2, M_w/kg.mol^{-1} = 198, 54.0 % crystallinity
Solvent (A):   **ethene**   **C$_2$H$_4$**   **74-85-1**

Type of data: cloud points

w_B 0.03   was kept constant

T/K	384.15	388.15	393.15	403.15	413.15	423.15	433.15	443.15	453.15
P/bar	2200	2060	2000	1910	1840	1760	1673	1615	1562

T/K	463.15	473.15	483.15
P/bar	1519	1489	1452

Polymer (B):   **polyethylene**   **1981WO2**
Characterization: M_n/kg.mol^{-1} = 112, M_w/kg.mol^{-1} = 466,
     LDPE, ρ = 0.921 g/cm^3, Leuna-Werke, Germany
Solvent (A):   **ethene**   **C$_2$H$_4$**   **74-85-1**

Type of data: cloud points

w_B	0.05	0.05	0.05	0.05	0.05	0.10	0.10	0.10	0.20
T/K	473.15	483.15	493.15	503.15	513.15	493.15	503.15	513.154	503.15
P/MPa	128.7	125.5	122.6	121.2	118.2	113.7	112.1	110.0	100.7

w_B	0.20
T/K	513.15
P/MPa	99.4

Polymer (B):   **polyethylene**   **1991NIE**
Characterization: M_n/kg.mol^{-1} = 126, M_w/kg.mol^{-1} = 216, HDPE,
     Exxon Polymer Plant, Antwerp, fractionated in the laboratory
Solvent (A):   **ethene**   **C$_2$H$_4$**   **74-85-1**

Type of data: cloud points

w_B	0.061	0.061	0.061	0.101	0.101	0.101	0.101	0.124	0.124
T/K	403.75	423.35	460.55	402.65	422.15	443.25	462.65	403.65	423.15
P/MPa	202.5	182.0	161.2	199.2	179.0	164.9	155.2	193.1	175.5

w_B	0.124	0.166	0.166	0.166	0.166	0.315	0.315	0.315	0.315
T/K	453.15	403.15	418.15	433.35	452.95	405.65	423.15	452.95	465.45
P/MPa	157.8	185.2	173.1	163.3	155.0	156.3	148.0	136.3	133.5

w_B	0.394	0.394	0.394
T/K	406.15	432.15	458.15
P/MPa	139.0	130.4	124.5

Polymer (B):   **polyethylene**   **1991NIE**
Characterization: M_n/kg.mol^{-1} = 259, M_w/kg.mol^{-1} = 499, LDPE,
     Exxon Polymer Plant, Antwerp, fractionated in the laboratory
Solvent (A):   **ethene**   **C$_2$H$_4$**   **74-85-1**

Type of data: cloud points

w_B	0.042	0.042	0.042	0.074	0.074	0.120	0.120	0.120	0.120
T/K	418.15	433.15	453.15	433.15	463.15	404.15	418.15	433.15	463.15
P/MPa	171.0	160.5	150.7	159.7	146.0	180.2	168.2	158.1	144.5

w_B	0.216	0.216	0.216	0.216	0.310	0.310	0.310	0.371	0.371
T/K	403.15	433.15	458.15	473.15	403.15	433.15	453.15	403.15	433.15
P/MPa	166.0	148.8	138.8	134.0	145.8	133.7	128.2	133.6	123.5

w_B	0.371	0.431	0.431	0.431	0.431
T/K	463.15	393.15	413.15	433.15	452.65
P/MPa	116.0	122.8	116.5	111.8	108.0

Polymer (B):   **polyethylene**   **1965SWE**
Characterization: M_w/kg.mol^{-1} = 315, LDPE, Stamylan 3401, 2 CH$_3$/100 C-atoms,
     DSM, Geleen, The Netherlands
Solvent (A):   **ethene**   **C$_2$H$_4$**   **74-85-1**

Type of data: cloud points

w_B	0.0022	0.0022	0.0022	0.0022	0.0022	0.0022	0.0022	0.0035	0.0035
T/K	381.43	385.70	385.96	391.21	398.11	398.43	409.83	387.55	395.66
P/kg.cm^{-2}	1952	1916	1908	1840	1784	1788	1688	1904	1820

w_B	0.0035	0.0035	0.0045	0.0045	0.0045	0.0045	0.0077	0.0077	0.0077
T/K	401.24	401.78	396.72	400.24	408.44	408.63	383.34	388.93	394.62
P/kg.cm^{-2}	1762	1764	1820	1792	1726	1720	1960	1900	1846

w_B	0.0077	0.0077	0.0176	0.0176	0.0176	0.0176	0.0176	0.0441	0.0441
T/K	404.07	410.20	378.75	384.49	389.62	402.72	409.77	384.53	389.94
P/kg.cm^{-2}	1770	1722	1994	1928	1874	1760	1708	1890	1834

w_B	0.0441	0.0441	0.0493	0.0493	0.0493	0.0493	0.0493	0.0493	0.0493
T/K	390.15	403.20	380.43	383.93	388.98	393.50	400.13	407.90	408.01
P/kg.cm^{-2}	1832	1722	1924	1884	1834	1798	1738	1686	1684

w_B	0.0662	0.0662	0.0662	0.0662	0.0662	0.0662	0.0662	0.0662	0.0867
T/K	379.28	383.02	388.42	393.77	394.04	399.87	400.08	409.43	381.67
P/kg.cm^{-2}	1909	1867	1808	1769	1767	1720	1715	1652	1852

w_B	0.0867	0.0867	0.0867	0.0867	0.0867	0.1066	0.1066	0.1066	0.1066
T/K	389.65	400.00	404.73	407.37	407.62	378.99	383.99	388.64	393.32
P/kg.cm^{-2}	1781	1695	1661	1644	1642	1850	1804	1759	1720

w_B	0.1066	0.1066	0.1066	0.1066	0.1265	0.1265	0.1265	0.1265	0.1265
T/K	398.50	403.10	407.97	411.87	386.94	395.55	402.38	405.69	411.76
P/kg.cm^{-2}	1681	1644	1616	1593	1754	1684	1636	1614	1576

continued

continued

w_B	0.1445	0.1445	0.1445	0.1445	0.1445	0.1672	0.1672	0.1672	0.1672
T/K	379.71	382.44	390.82	395.95	409.30	387.92	394.83	398.48	406.71
P/kg.cm^{-2}	1804	1774	1704	1664	1574	1700	1644	1620	1568

w_B	0.1890	0.1890	0.1890	0.1890	0.2186	0.2186	0.2186	0.2186	0.2429
T/K	385.99	387.47	398.22	401.02	389.01	389.60	399.42	408.58	384.21
P/kg.cm^{-2}	1692	1680	1600	1584	1632	1630	1566	1514	1624

w_B	0.2429	0.2429	0.2429	0.2429	0.2988	0.2988	0.2988	0.2988	0.2988
T/K	391.53	397.50	401.90	409.38	382.49	392.94	398.30	399.85	404.35
P/kg.cm^{-2}	1568	1531	1508	1464	1544	1480	1434	1436	1402

Polymer (B):	**polyethylene**		**1972STE**
Characterization:	LDPE, synthesized by radical polymerization at high pressures		
Solvent (A):	**ethene**	**C₂H₄**	**74-85-1**

Type of data: cloud points

w_B	0.168	0.155	0.141	0.071	0.067
T/K	542.15	533.15	523.15	448.15	448.15
P/at	1190	1200	1230	1550	1530

Comments: The initiator was oxygen.

w_B	0.046	0.036	0.032	0.030	0.021	0.027	0.120	0.077
T/K	424.15	404.15	394.15	415.15	394.15	384.15	412.15	424.15
P/at	1520	1630	1680	1420	1550	1750	1720	1580

Comments: The initiator was 0.1 wt% peroxide.

w_B	0.045	0.040	0.050	0.027	0.039	0.139	0.028	0.043
T/K	422.15	409.15	437.15	414.15	409.15	430.15	408.15	429.15
P/at	1540	1600	1400	1450	1450	1610	1630	1550

Comments: The initiator was 5.0 wt% peroxide.

Polymer (B):	**polyethylene**		**1993HA3**
Characterization:	M_n/kg.mol^{-1} = 37, M_w/kg.mol^{-1} = 49, 36% crystallinity, T_m/K = 371		
Solvent (A):	**propane**	**C₃H₈**	**74-98-6**

Type of data: cloud points

w_B	0.055	0.055	0.055	0.055	0.055
T/K	366.15	389.05	398.35	408.55	423.45
P/bar	455.9	451.6	451.6	451.6	455.9

Polymer (B):	**polyethylene**		**1993HA3**
Characterization:	M_n/kg.mol^{-1} = 37, M_w/kg.mol^{-1} = 46, 57% crystallinity, T_m/K = 396		
Solvent (A):	**propane**	**C₃H₈**	**74-98-6**

Type of data: cloud points

w_B	0.055	0.055	0.055	0.055	0.055
T/K	380.35	381.85	394.55	408.05	422.85
P/bar	527.9	527.4	524.0	524.0	524.0

Polymer (B):   **polyethylene**   **1993HA3**
Characterization: M_n/kg.mol^{-1} = 39, M_w/kg.mol^{-1} = 50, 40% crystallinity, T_m/K = 396
Solvent (A):   **propane**   **C$_3$H$_8$**   **74-98-6**

Type of data: cloud points

w_B	0.055	0.055	0.055	0.055	0.055	0.055
T/K	377.15	380.15	384.15	393.05	407.85	424.05
P/bar	510.2	510.2	506.8	506.8	506.8	506.8

Polymer (B):   **polyethylene**   **1993HA3**
Characterization: M_n/kg.mol^{-1} = 41, M_w/kg.mol^{-1} = 53, 49% crystallinity, T_m/K = 381
Solvent (A):   **propane**   **C$_3$H$_8$**   **74-98-6**

Type of data: cloud points

w_B	0.055	0.055	0.055	0.055	0.055
T/K	375.15	380.95	392.55	408.95	424.05
P/bar	499.0	499.9	503.3	503.3	506.8

Polymer (B):   **polyethylene**   **1993HA3**
Characterization: M_n/kg.mol^{-1} = 21, M_w/kg.mol^{-1} = 106, 42% crystallinity, T_m/K = 396
Solvent (A):   **propane**   **C$_3$H$_8$**   **74-98-6**

Type of data: cloud points

w_B	0.055	0.055	0.055
T/K	380.65	392.45	425.45
P/bar	575.7	575.7	575.7

Polymer (B):   **polyethylene**   **1993HA3**
Characterization: M_n/kg.mol^{-1} = 76, M_w/kg.mol^{-1} = 108, 46% crystallinity, T_m/K = 394
Solvent (A):   **propane**   **C$_3$H$_8$**   **74-98-6**

Type of data: cloud points

w_B	0.055	0.055	0.055	0.055	0.055
T/K	380.15	385.35	394.55	409.45	424.65
P/bar	551.6	558.5	558.5	551.6	551.6

Polymer (B):   **polyethylene**   **1998HAN**
Characterization: M_n/kg.mol^{-1} = 26.75, M_w/kg.mol^{-1} = 32.1, linear, HDPE,
                        T_m/K = 407.2, NIST 1483
Solvent (A):   **propane**   **C$_3$H$_8$**   **74-98-6**

Type of data: cloud points

w_B	0.01	0.01	0.01	0.01	0.01
T/K	388.65	398.35	422.05	447.45	471.65
P/bar	571.6	566.7	566.1	560.5	560.8

Polymer (B): **poly(ethylene glycol)** **1999GOU**
Characterization: $M_n/kg.mol^{-1} = 0.2$, Sigma Chemical Co., Inc., St. Louis, MO
Solvent (A): **carbon dioxide** **CO_2** **124-38-9**

Type of data: coexistence data

| $P/$ | w_A | |
MPa	liquid phase	gas phase
$T/K = 313.15$		
3.87	0.0818	1.0000
7.75	0.1556	1.0000
9.72	0.1846	1.0000
12.05	0.1983	1.0000
15.83	0.2010	0.9965
18.41	0.2187	0.9834
24.55	0.2299	0.9759
$T/K = 333.15$		
4.26	0.0413	1.0000
7.38	0.1138	1.0000
10.90	0.1488	1.0000
14.37	0.1749	0.9962
17.80	0.1970	0.9895
20.96	0.2115	0.9819
24.45	0.2193	0.9830
$T/K = 348.15$		
5.94	0.0836	1.0000
8.31	0.1056	1.0000
11.27	0.1384	0.9972
14.32	0.1532	0.9991
17.85	0.1748	0.9931
21.38	0.1978	0.9844
24.87	0.2084	0.9852

Polymer (B): **poly(ethylene glycol)** **1990DAN**
Characterization: $M_n/kg.mol^{-1} = 0.4$, Aldrich Chem. Co., Inc., Milwaukee, WI
Solvent (A): **carbon dioxide** **CO_2** **124-38-9**

Type of data: coexistence data

continued

continued

$P/$ MPa	w_A liquid phase	gas phase
$T/K = 313.15$		
1.41	0.0259	1.0000
3.00	0.0578	1.0000
6.49	0.1402	1.0000
7.94	0.2860	1.0000
9.90	0.3225	0.99763
12.48	0.4023	0.99703
13.92	0.4245	0.99579
17.10	0.4317	0.99476
18.39	0.4338	0.99186
20.41	0.4606	0.98754
22.07	0.5306	0.98672
$T/K = 323.15$		
1.92	0.0164	1.0000
4.06	0.0348	1.0000
7.15	0.0707	1.0000
8.31	0.1081	1.0000
12.00	0.2501	0.99759
15.33	0.2781	0.99485
18.35	0.3506	0.99132
20.38	0.3649	0.99114
22.17	0.3658	0.98383
23.27	0.3704	0.98287
26.43	0.3975	0.97875

Polymer (B):	**poly(ethylene glycol)**	**1999GOU**
Characterization:	$M_n/kg.mol^{-1} = 0.4$, Aldrich Chem. Co., Inc., Milwaukee, WI	
Solvent (A):	**carbon dioxide** CO_2	**124-38-9**

Type of data: coexistence data

$P/$ MPa	w_A liquid phase	gas phase
$T/K = 313.15$		
5.23	0.1423	1.0000
10.62	0.2478	1.0000
16.11	0.2683	0.9933
19.74	0.2862	0.9928
24.01	0.3007	0.9897

continued

continued

P/ MPa	w_A liquid phase	gas phase
T/K = 333.15		
5.28	0.0949	1.0000
10.37	0.1801	1.0000
14.98	0.2227	0.9990
20.63	0.2534	0.9961
24.62	0.2616	0.9927
T/K = 348.15		
5.89	0.0888	1.0000
11.22	0.1522	1.0000
16.34	0.2078	0.9991
20.80	0.2235	0.9963
24.40	0.2390	0.9929

Polymer (B):	**poly(ethylene glycol)**	**1990DAN**
Characterization:	M_n/kg.mol^{-1} = 0.6, Aldrich Chem. Co., Inc., Milwaukee, WI	
Solvent (A):	**carbon dioxide** CO_2	**124-38-9**

Type of data: coexistence data

P/ MPa	w_A liquid phase	gas phase
T/K = 313.15		
1.13	0.0183	1.0000
2.37	0.0469	1.0000
3.61	0.0711	1.0000
5.84	0.1251	1.0000
7.13	0.1385	1.0000
8.02	0.2438	1.0000
10.49	0.4147	1.0000
13.68	0.4625	1.0000
14.86	0.5076	0.99892
15.86	0.5437	0.99851
18.80	0.5559	0.99798
20.40	0.5570	0.99746
21.44	0.5720	0.99607
22.75	0.6243	0.99560
25.16	0.6242	0.99445
28.42	0.6139	0.99118

continued

continued

$P/$ MPa	w_A liquid phase	gas phase
$T/K = 323.15$		
1.99-11.80	up to 0.2156	1.0000
13.88	0.4413	0.99990
16.79	0.4598	0.99980
18.03	0.4840	0.99959
20.36	0.5110	0.99921
22.71	0.5365	0.99873
24.33	0.5360	0.99825
27.45	0.5828	0.99817
29.00	0.5863	0.99757

Polymer (B):   **poly(ethylene glycol)**   **1999GOU**
Characterization: $M_n/kg.mol^{-1} = 0.6$, Aldrich Chem. Co., Inc., Milwaukee, WI
Solvent (A):   **carbon dioxide**   **CO_2**   **124-38-9**

Type of data: coexistence data

$P/$ MPa	w_A liquid phase	gas phase
$T/K = 313.15$		
5.71	0.0958	1.0000
10.16	0.2303	1.0000
15.41	0.2467	1.0000
21.34	0.2739	0.9995
26.34	0.2875	0.9946
$T/K = 333.15$		
5.63	0.0921	1.0000
10.07	0.1789	1.0000
16.37	0.2144	1.0000
21.24	0.2408	0.9991
26.50	0.2671	0.9979
$T/K = 348.15$		
5.63	0.0831	1.0000
10.31	0.1422	1.0000
14.82	0.1803	1.0000
19.13	0.2083	0.9998
24.78	0.2312	0.9993

Polymer (B): **poly(ethylene glycol)**   **1998ONE**
Characterization: $M_n/\text{kg.mol}^{-1} = 0.6$, Polysciences, Inc., Warrington, PA
Solvent (A): **carbon dioxide**   **CO_2**   **124-38-9**

Type of data: cloud points

$T/K = 298.15$

w_B	0.00296	0.00355	0.00394	0.00610	0.00708	0.00924
P/MPa	13.72	20.55	17.65	25.51	21.77	30.99

$T/K = 308.15$

w_B	0.00296	0.00355	0.00394	0.00610	0.00708	0.00924
P/MPa	17.03		20.89	28.41		33.41

$T/K = 313.15$

w_B	0.00296	0.00355	0.00394	0.00610	0.00708	0.00924
P/MPa	19.93	26.37	22.06	29.58	26.20	33.72

$T/K = 318.15$

w_B	0.00296	0.00355	0.00394	0.00610	0.00708	0.00924
P/MPa	21.51	27.44		30.44	27.51	

$T/K = 323.15$

w_B	0.00296	0.00355	0.00394	0.00610	0.00708	0.00924
P/MPa	23.03	28.61	24.61	30.72	28.48	34.68

$T/K = 328.15$

w_B	0.00296	0.00355	0.00394	0.00610	0.00708	0.00924
P/MPa	24.75			31.47		35.34

$T/K = 338.15$

w_B	0.00296	0.00355	0.00394	0.00610	0.00708	0.00924
P/MPa	27.96		28.75	32.89		36.58

Polymer (B): **poly(ethylene glycol)**   **2000WIE**
Characterization: $M_w/\text{kg.mol}^{-1} = 1.5$, Hoechst AG, Frankfurt/M., Germany
Solvent (A): **carbon dioxide**   **CO_2**   **124-38-9**

Type of data: gas solubilities and cloud points

$T/K = 338.15$

P/bar	41	54	78	118	157	203	240	271	274
w_B	0.910	0.883	0.831	0.800	0.785	0.737	0.702	0.700	0.713

$T/K = 353.15$

P/bar	45	74	106	138	200	286
w_B	0.920	0.857	0.813	0.785	0.735	0.740

$T/K = 373.15$

P/bar	41	61	79	98	130	162	202	244	270
w_B	0.950	0.919	0.892	0.863	0.844	0.821	0.794	0.792	0.755

Polymer (B):	**poly(ethylene glycol)**	**1990DAN**
Characterization:	$M_n/\text{kg.mol}^{-1} = 1.0$, Aldrich Chem. Co., Inc., Milwaukee, WI	
Solvent (A):	**carbon dioxide** CO_2	**124-38-9**

Type of data: coexistence data

| $P/$ | w_A | |
MPa	liquid phase	gas phase
T/K = 313.15		
1.75	0.0386	1.0000
3.59	0.0706	1.0000
5.75	0.1205	1.0000
8.47	0.1752	1.0000
10.20	0.1852	1.0000
11.90	0.2424	1.0000
15.07	0.4583	0.99984
18.01	0.5020	0.99980
19.32	0.4973	0.99975
21.65	0.5388	0.99956
23.01	0.5285	0.99934
24.51	0.5530	0.99928
26.31	0.5779	0.99925

Polymer (B):	**poly(ethylene glycol)**	**2000WIE**
Characterization:	$M_w/\text{kg.mol}^{-1} = 4.0$, Hoechst AG, Frankfurt/M., Germany	
Solvent (A):	**carbon dioxide** CO_2	**124-38-9**

Type of data: gas solubilities and cloud points

T/K = 328.15

P/bar	43	60	81	138	182	221	260	282
w_B	0.902	0.881	0.842	0.795	0.762	0.740	0.729	0.714

T/K = 333.15

P/bar	5	20	34	40	61	80	139	220	286
w_B	0.987	0.956	0.938	0.927	0.891	0.853	0.802	0.751	0.734

T/K = 353.15

P/bar	6	16	24	31	44	54	66	79	117
w_B	0.989	0.978	0.963	0.958	0.946	0.927	0.911	0.894	0.867

P/bar	140	181	219	265
w_B	0.843	0.807	0.785	0.766

T/K = 373.15

P/bar	11	20	32	40	44	53	58	81	101
w_B	0.989	0.985	0.968	0.958	0.953	0.946	0.938	0.913	0.898

P/bar	178	220	257
w_B	0.841	0.812	0.803

Polymer (B):   **poly(ethylene glycol)**   **2000WIE**
Characterization: $M_w/\text{kg.mol}^{-1}$ = 8.0, Hoechst AG, Frankfurt/M., Germany
Solvent (A):   **carbon dioxide**   **CO_2**   **124-38-9**

Type of data: gas solubilities and cloud points

T/K = 353.15

P/bar	40	63	81	98	140	178	218	256
w_B	0.947	0.912	0.891	0.874	0.831	0.802	0.783	0.768

Polymer (B):   **poly(ethylene glycol)**   **2000WIE**
Characterization: $M_w/\text{kg.mol}^{-1}$ = 0.2, Hoechst AG, Frankfurt/M., Germany
Solvent (A):   **propane**   **C_3H_8**   **74-98-6**

Type of data: gas solubilities and cloud points

T/K = 303.15

P/bar	8	14	14	24	56	74	132	205	246
w_B	0.9961	0.9821	0.9817	0.9799	0.9787	0.9793	0.9786	0.9797	0.9792

T/K = 323.15

P/bar	12	14	17	22	33	80	112	200	246
w_B	0.9883	0.9856	0.9808	0.9769	0.9768	0.9765	0.9759	0.9755	0.9736

T/K = 373.15

P/bar	15	22	28	34	54	102	157	194	240
w_B	0.9929	0.9860	0.9789	0.9699	0.9677	0.9644	0.9622	0.9629	0.9631

T/K = 393.15

P/bar	31	39	51	62	83	120	162	201	241
w_B	0.9797	0.9738	0.9670	0.9636	0.9606	0.9576	0.9540	0.9532	0.9528

Polymer (B):   **poly(ethylene glycol)**   **2000WIE**
Characterization: $M_w/\text{kg.mol}^{-1}$ = 1.5, Hoechst AG, Frankfurt/M., Germany
Solvent (A):   **propane**   **C_3H_8**   **74-98-6**

Type of data: gas solubilities and cloud points

T/K = 323.15

P/bar	10	20	31	38	49	59	140	160	180
w_B	0.9781	0.9570	0.9535	0.9530	0.9530	0.9536	0.9520	0.9520	0.9510

T/K = 353.15

P/bar	10	21	41	50	59	70	90	120	160
w_B	0.9842	0.9680	0.9483	0.9454	0.9446	0.9440	0.9440	0.9410	0.9410

T/K = 373.15

continued

continued

P/bar	10	20	30	40	51	61	90	110	170
w_B	0.9877	0.9738	0.9620	0.9469	0.9434	0.9428	0.9380	0.9370	0.9310

T/K = 393.15

P/bar	10	21	30	41	50	66
w_B	0.9881	0.9770	0.9678	0.9548	0.9479	0.9393

Polymer (B):	**poly(ethylene glycol)**		**2000WIE**
Characterization:	M_w/kg.mol^{-1} = 4.0, Hoechst AG, Frankfurt/M., Germany		
Solvent (A):	**propane**	**C$_3$H$_8$**	**74-98-6**

Type of data: gas solubilities and cloud points

T/K = 333.15

P/bar	20	23	31	50	51	120	160	190	220
w_B	0.9577	0.9499	0.9485	0.9470	0.9458	0.944	0.943	0.944	0.942

T/K = 353.15

P/bar	11	21	52	64	120	160	200	200	250
w_B	0.9798	0.9625	0.9409	0.9394	0.9390	0.938	0.935	0.935	0.936

T/K = 373.15

P/bar	9	20	30	40	53	90	150	210	250
w_B	0.9865	0.9744	0.9603	0.9473	0.9362	0.930	0.927	0.924	0.923

T/K = 393.15

P/bar	11	20	30	40	50	60	70	100	110
w_B	0.9899	0.9788	0.9661	0.9563	0.9462	0.9397	0.9350	0.9280	0.9240

P/bar	140	190	250
w_B	0.9230	0.9120	0.9060

Polymer (B):	**poly(ethylene glycol)**		**2000WIE**
Characterization:	M_w/kg.mol^{-1} = 8.0, Hoechst AG, Frankfurt/M., Germany		
Solvent (A):	**propane**	**C$_3$H$_8$**	**74-98-6**

Type of data: gas solubilities and cloud points

T/K = 353.15

P/bar	15	27	35	45	80	110	150	190	250
w_B	0.9750	0.9512	0.9435	0.9403	0.9360	0.9350	0.9340	0.9290	0.9220

T/K = 373.15

P/bar	12	25	36	47	55	80	130	170	210
w_B	0.9815	0.9650	0.9475	0.9376	0.9371	0.9340	0.9260	0.9240	0.9240

P/bar	250
w_B	0.9200

continued

continued

$T/K = 393.15$

P/bar	11	20	30	46	59	68	110	150	200
w_B	0.9874	0.9761	0.9632	0.9460	0.9380	0.9329	0.9250	0.9230	0.9130

P/bar	250
w_B	0.9090

Polymer (B):	poly(ethylene glycol) dimethyl ether	1998ONE
Characterization:	M_n/kg.mol^{-1} = 0.4, Polysciences, Inc., Warrington, PA	
Solvent (A):	carbon dioxide CO$_2$	124-38-9

Type of data: cloud points

w_B	0.00848	0.01390	0.02200	0.03510	0.00848	0.01390	0.02200	0.03510
T/K	298.15	298.15	298.15	298.15	313.15	313.15	313.15	313.15
P/MPa	8.76	9.79	10.48	10.34	13.03	14.75	15.93	15.17

w_B	0.00848	0.01390	0.02200	0.03510	0.00848	0.01390	0.02200	0.03510
T/K	323.15	323.15	323.15	323.15	333.15	333.15	333.15	333.15
P/MPa	16.13	17.10	17.37	16.89	18.48	19.86	20.27	19.65

Polymer (B):	poly(ethylene glycol) dimethyl ether	1998ONE
Characterization:	M_n/kg.mol^{-1} = 1.0, Polysciences, Inc., Warrington, PA	
Solvent (A):	carbon dioxide CO$_2$	124-38-9

Type of data: cloud points

w_B	0.00431	0.01010	0.01490	0.02390	0.00431	0.01010	0.01490	0.02390
T/K	298.15	298.15	298.15	298.15	303.15	303.15	303.15	303.15
P/MPa	17.03	19.65	21.79	27.30	17.93	21.72	22.41	28.54

w_B	0.00431	0.01010	0.01490	0.02390	0.00431	0.01010	0.01490	0.02390
T/K	313.15	313.15	313.15	313.15	323.15	323.15	323.15	323.15
P/MPa	22.20	25.30	26.75	31.72	24.96	28.89	29.23	36.40

w_B	0.00431	0.01010	0.01490	0.02390
T/K	333.15	333.15	333.15	333.15
P/MPa	28.75	31.30	32.75	40.27

Polymer (B):	poly(ethylene oxide)-b-poly(propylene oxide) diblock copolymer	1998ONE
Characterization:	M_n/kg.mol^{-1} = 2.0 (= 0.14 + 1.86), 8.6 mol% ethylene oxide, Jeffamine M2000, hydroxyl terminated EO, NH$_2$ terminated PO	
Solvent (A):	carbon dioxide CO$_2$	124-38-9

Type of data: cloud points

w_B	0.001	T/K	308.15	P/MPa	27.58

Polymer (B):   **poly(ethylene oxide)-b-poly(propylene oxide)**   **1998ONE**
                          diblock copolymer
Characterization: M_n/kg.mol^{-1} = 0.6 (= 0.04 + 0.56), 10.0 mol% ethylene oxide,
                          Jeffamine M600, hydroxyl terminated EO, NH$_2$ terminated PO
Solvent (A):   **carbon dioxide**   **CO$_2$**   **124-38-9**

Type of data: cloud points

w_B	0.00785	0.00785	0.00785	0.00785	0.00785
T/K	298.15	308.15	318.15	328.15	338.15
P/MPa	27.37	29.51	29.85	30.34	31.51

Polymer (B):   **poly(ethylene oxide)-b-poly(propylene oxide)-**   **1998ONE**
                          b-poly(ethylene oxide) triblock copolymer
Characterization: M_n/kg.mol^{-1} = 1.05 (= 0.05 + 0.95 + 0.05), 11.8 mol% ethylene oxide,
                          (1.1/16.5/1.1), Pluronic L31, hydroxyl terminated
Solvent (A):   **carbon dioxide**   **CO$_2$**   **124-38-9**

Type of data: cloud points

T/K = 298.15

w_B	0.00278	0.00558	0.00834	0.01110	0.01390	0.01670	0.01950	0.02500	0.03060
P/MPa	10.69	12.27	13.72	16.41	18.89	19.86	20.48	22.41	23.72

T/K = 303.15

w_B	0.00278	0.00558	0.00834	0.01110	0.01390	0.01670	0.01950	0.02500	0.03060
P/MPa	12.13	14.55	15.31	18.06	19.99	20.96	22.55	24.06	25.17

T/K = 313.15

w_B	0.00278	0.00558	0.00834	0.01110	0.01390	0.01670	0.01950	0.02500	0.03060
P/MPa	16.55	17.51	18.75	21.93	23.92	23.58	25.44	27.03	28.27

T/K = 323.15

w_B	0.00278	0.00558	0.00834	0.01110	0.01390	0.01670	0.01950	0.02500	0.03060
P/MPa	19.99	21.61	22.61	24.89	26.13	27.23	28.61	29.65	30.89

T/K = 333.15

w_B	0.00278	0.00558	0.00834	0.01110	0.01390	0.01670	0.01950	0.02500	0.03060
P/MPa	22.75	25.23	26.75	27.85	29.72	30.34	31.03	32.13	33.65

T/K = 338.15

w_B	0.00278	0.00558	0.00834	0.01110	0.01390	0.01670	0.01950	0.02500	0.03060
P/MPa	23.65		27.85	29.37	30.82	31.58	32.34	33.44	

Polymer (B):	poly(ethylene oxide)-b-poly(propylene oxide)-	1998ONE
	b-poly(ethylene oxide) triblock copolymer	

Characterization: M_n/kg.mol^{-1} = 2.95 (= 0.15 + 2.65 + 0.15), 12.6 mol% ethylene oxide, (3.3/45.7/3.3), Pluronic L81, hydroxyl terminated

Solvent (A):	carbon dioxide CO$_2$	124-38-9

Type of data: cloud points

w_B	0.00123	0.00123	0.00123	0.00123	0.00123	0.00123
T/K	298.15	303.15	313.15	323.15	333.15	338.15
P/MPa	19.65	21.17	24.27	29.30	33.03	33.78

Polymer (B):	poly(ethylene oxide)-b-poly(propylene oxide)-	1998ONE
	b-poly(ethylene oxide) triblock copolymer	

Characterization: M_n/kg.mol^{-1} = 2.21 (= 0.12 + 1.97 + 0.12), 13.7 mol% ethylene oxide, (2.7/34.0/2.7), Pluronic L61, hydroxyl terminated

Solvent (A):	carbon dioxide CO$_2$	124-38-9

Type of data: cloud points

T/K = 298.15

w_B	0.00278	0.00558	0.00834	0.01110	0.01340
P/MPa	16.62	20.13	25.37	27.58	30.82

T/K = 303.15

w_B	0.00278	0.00558	0.00834	0.01110	0.01340
P/MPa	16.96	21.72	26.34	28.96	31.37

T/K = 313.15

w_B	0.00278	0.00558	0.00834	0.01110	0.01340
P/MPa	20.62	26.13	27.65	31.51	33.44

T/K = 323.15

w_B	0.00278	0.00558	0.00834	0.01110	0.01340
P/MPa	23.30	27.99	30.61	34.13	35.51

T/K = 333.15

w_B	0.00278	0.00558	0.00834	0.01110	0.01340
P/MPa	25.92	32.13	34.75	36.34	37.85

T/K = 338.15

w_B	0.00278	0.00558	0.00834	0.01110	0.01340
P/MPa	27.51	32.54	35.44	37.16	39.09

Polymer (B):   **poly(ethylene oxide)-b-poly(propylene oxide)-**   **1998ONE**
                     b-poly(ethylene oxide) triblock copolymer

Characterization: M_n/kg.mol^{-1} = 2.5 (= 0.25 + 2.0 + 0.25), 24.8 mol% ethylene oxide,
                        (5.6/34.0/5.6), Pluronic L62, hydroxyl terminated

Solvent (A):   **carbon dioxide**   **CO$_2$**   **124-38-9**

Type of data: cloud points

w_B	0.00207	0.00207	0.00207	0.00207	0.00207	0.00207
T/K	298.15	303.15	313.15	323.15	333.15	338.15
P/MPa	28.61	29.51	31.03	35.23	38.27	33.78

Polymer (B):   **poly(ethylene oxide)-b-poly(propylene oxide)-**   **1998ONE**
                     b-poly(ethylene oxide) triblock copolymer

Characterization: M_n/kg.mol^{-1} = 3.75 (= 0.375 + 3.0 + 0.375), 24.8 mol% ethylene
                        oxide, (8.5/51.6/8.5), Pluronic L92, hydroxyl terminated

Solvent (A):   **carbon dioxide**   **CO$_2$**   **124-38-9**

Type of data: cloud points

w_B	0.00207	0.00207	0.00207
T/K	298.15	303.15	318.15
P/MPa	30.33	32.33	36.20

Polymer (B):   **poly(ethylene oxide)-b-poly(propylene oxide)-**   **1998ONE**
                     b-poly(ethylene oxide) triblock copolymer

Characterization: M_n/kg.mol^{-1} = 1.85 (= 0.28 + 1.29 + 0.28), 36.2 mol% ethylene oxide,
                        (6.3/22.2/6.3), Pluronic L43, hydroxyl terminated

Solvent (A):   **carbon dioxide**   **CO$_2$**   **124-38-9**

Type of data: cloud points

w_B	0.0028	0.003
T/K	298.15	308.15
P/MPa	25.51	27.58

Polymer (B):   **poly(fluoroalkoxyphosphazene)**   **1998ONE**
Characterization: molar mass unknown, EYPEL-F gum
Solvent (A):   **carbon dioxide**   **CO$_2$**   **124-38-9**

Type of data: cloud points

w_B	0.00386	0.00930	0.02080	0.04010	0.00386	0.00930	0.02080	0.04010
T/K	298.15	298.15	298.15	298.15	308.15	308.15	308.15	308.15
P/MPa	15.17	15.44	15.58	15.10	9.37	19.93	19.86	19.44

w_B	0.00386	0.00930	0.02080	0.04010	0.00386	0.00930	0.02080	0.04010
T/K	323.15	323.15	323.15	323.15	338.15	338.15	338.15	338.15
P/MPa	25.51	25.58	25.86	25.44	30.27	30.82	31.10	30.82

Polymer (B):	**poly(hexyl acrylate)**					2004BYU

Characterization: M_w/kg.mol^{-1} = 90,
Scientific Polymer Products, Inc., Ontario, NY

Solvent (A): **carbon dioxide** **CO_2** 124-38-9

Type of data: cloud points

w_B 0.049 was kept constant

T/K	398.45	399.35	403.45	405.65	414.85	435.05
P/bar	2553.4	1932.8	1743.1	1598.3	1450.0	1186.2

Polymer (B):	**poly(hexyl methacrylate)**					2004BYU

Characterization: M_w/kg.mol^{-1} = 400, Aldrich Chem. Co., Inc., Milwaukee, WI

Solvent (A): **carbon dioxide** **CO_2** 124-38-9

Type of data: cloud points

w_B 0.051 was kept constant

T/K	429.85	431.45	433.35	437.75	446.05	467.95	472.05
P/bar	2208.6	2070.7	1953.4	1805.2	1683.1	1462.1	1424.1

Polymer (B):	**poly(ββhydroxybutyrate)**					2003KHO

Characterization: M_w/kg.mol^{-1} = 800, Aldrich Chem. Co., Inc., Milwaukee, WI

Solvent (A): **carbon dioxide** **CO_2** 124-38-9

Type of data: polymer solubility

T/K = 308.15

P/bar	122	152	182	213	243	274	304	334	355
ρ_A/g.dm^{-3}	771	818	850	876	897	916	931	946	955
c_B/g.dm^{-3}	0.54	0.68	1.07	1.13	1.46	1.58	1.86	1.96	2.41

T/K = 318.15

P/bar	122	152	182	213	243	274	304	334	355
ρ_A/g.dm^{-3}	661	745	792	826	852	875	893	910	919
c_B/g.dm^{-3}	0.31	0.58	1.06	1.37	1.67	1.98	2.11	2.43	2.65

T/K = 328.15

P/bar	122	152	182	213	243	274	304	334	355
ρ_A/g.dm^{-3}	504	657	726	771	804	831	853	872	884
c_B/g.dm^{-3}	0.23	0.59	1.05	1.44	1.79	2.38	2.69	2.98	3.49

T/K = 338.15

P/bar	122	152	182	213	243	274	304	334	355
ρ_A/g.dm^{-3}	398	561	654	712	754	786	812	834	848
c_B/g.dm^{-3}	0.18	0.43	1.01	1.43	2.23	2.78	3.18	3.69	4.77

continued

continued

$T/K = 348.15$

P/bar	122	152	182	213	243	274	304	334	355
ρ_A/g.dm^{-3}	327	477	585	652	702	740	772	796	811
c_B/g.dm^{-3}	0.15	0.29	0.69	1.49	2.65	3.65	4.71	5.33	8.01

Polymer (B):   **polyisobutylene**   **1994GR1, 1994GR4**
Characterization: M_n/kg.mol^{-1} = 0.87, M_w/kg.mol^{-1} = 1.0,
                        synthesized in the laboratory by living carbocationic polymerization
Solvent (A):   **carbon dioxide**   **CO$_2$**   **124-38-9**

Type of data: cloud points

w_B 0.0482   was kept constant

T/K	459.25	459.05	456.85	423.15	418.95	413.25	407.45	398.25	390.15
P/bar	812.5	826.9	823.4	839.0	870.7	891.0	1009.0	1122.8	1343.1

T/K	386.45	383.95
P/bar	1680.2	1903.9

Polymer (B):   **polyisobutylene (monohydroxy)**   **1994GR1, 1994GR4**
Characterization: M_n/kg.mol^{-1} = 0.975, M_w/kg.mol^{-1} = 1.12,
                        synthesized in the laboratory by living carbocationic polymerization
Solvent (A):   **carbon dioxide**   **CO$_2$**   **124-38-9**

Type of data: cloud points

w_B 0.0477   was kept constant

T/K	448.15	438.15	433.15	428.15	423.15	413.15
P/bar	909.7	986.1	1044.0	1127.4	1190.0	1502.6

Polymer (B):   **polyisobutylene (dihydroxy)**   **1994GR1, 1994GR4**
Characterization: M_n/kg.mol^{-1} = 1.13, M_w/kg.mol^{-1} = 1.30,
                        synthesized in the laboratory by living carbocationic polymerization
Solvent (A):   **carbon dioxide**   **CO$_2$**   **124-38-9**

Type of data: cloud points

w_B 0.0480   was kept constant

T/K	423.15	418.15	413.15	407.45	400.45	398.15	388.15	386.15	385.15
P/bar	855.9	868.1	884.4	943.6	999.3	1028.2	1190.7	1232.1	1265.3

T/K	384.25	383.15	382.15
P/bar	1290.8	1323.3	1379.3

Polymer (B):	polyisobutylene						**1994GR1, 1994GR4**	

Characterization: $M_n/\text{kg.mol}^{-1} = 0.87$, $M_w/\text{kg.mol}^{-1} = 1.0$,
synthesized in the laboratory by living carbocationic polymerization

Solvent (A):	chlorodifluoromethane			CHClF$_2$			**75-45-6**	

Type of data: cloud points

w_B	0.0348	0.0348	0.0348	0.0400	0.0400	0.0458	0.0458	0.0458	0.0458
T/K	398.15	373.15	348.15	398.15	348.15	473.15	448.15	398.15	393.15
P/bar	242.7	236.2	267.7	243.7	287.1	318.1	295.2	251.2	247.3
w_B	0.0458	0.0458	0.0458	0.0458	0.0458	0.0459	0.0459	0.0459	0.0459
T/K	388.15	383.15	378.15	369.65	363.15	423.15	398.15	358.15	353.15
P/bar	245.8	243.1	243.2	243.1	252.2	268.5	250.4	262.0	279.1
w_B	0.0459	0.0459	0.0459	0.0459					
T/K	348.15	343.15	335.65	333.15					
P/bar	294.4	327.0	388.0	416.1					

Polymer (B):	polyisobutylene (monohydroxy)						**1994GR1, 1994GR4**	

Characterization: $M_n/\text{kg.mol}^{-1} = 0.975$, $M_w/\text{kg.mol}^{-1} = 1.12$,
synthesized in the laboratory by living carbocationic polymerization

Solvent (A):	chlorodifluoromethane			CHClF$_2$			**75-45-6**	

Type of data: cloud points

w_B	0.0414	0.0414	0.0414	0.0414	0.0414	0.0414	0.0414	0.0414	0.0414
T/K	398.15	393.15	388.15	383.15	378.15	373.15	368.15	363.15	358.15
P/bar	306.7	309.4	310.9	314.5	317.8	321.7	328.2	336.2	345.2
w_B	0.0414	0.0414	0.0414	0.0414	0.0441	0.0441	0.0441	0.0441	0.0441
T/K	353.15	348.15	342.45	338.15	448.05	423.15	398.15	393.15	388.15
P/bar	364.1	391.1	430.7	470.4	332.1	304.7	300.0	300.2	303.7
w_B	0.0441								
T/K	383.15								
P/bar	308.1								

Polymer (B):	polyisobutylene (dihydroxy)						**1994GR1, 1994GR4**	

Characterization: $M_n/\text{kg.mol}^{-1} = 1.13$, $M_w/\text{kg.mol}^{-1} = 1.30$,
synthesized in the laboratory by living carbocationic polymerization

Solvent (A):	chlorodifluoromethane			CHClF$_2$			**75-45-6**	

Type of data: cloud points

w_B	0.0449	0.0449	0.0449	0.0449	0.0458	0.0458	0.0458	0.0458	0.0458
T/K	472.65	448.15	423.15	398.15	393.15	388.15	383.15	373.15	368.15
P/bar	324.4	305.7	283.4	261.6	256.3	252.7	249.2	244.6	243.7
w_B	0.0458	0.0458	0.0458	0.0458	0.0458	0.0458	0.0458	0.0458	0.0458
T/K	363.15	358.15	353.15	348.15	343.15	338.15	333.15	328.15	323.15
P/bar	243.0	244.4	248.5	254.5	264.5	279.6	302.7	338.0	386.8

continued

continued

w_B	0.0462	0.0462	0.0462	0.0462	0.0462
T/K	393.15	387.75	383.15	373.15	348.25
P/bar	257.1	252.9	249.2	243.5	249.5

Polymer (B): **polyisobutylene** **1994GR3, 1994GR4**
Characterization: M_n/kg.mol^{-1} = 3.7, M_w/kg.mol^{-1} = 4.1 (1.37 per arm),
  synthesized in the laboratory
Solvent (A): **chlorodifluoromethane** **CHClF$_2$** **75-45-6**

Type of data: cloud points

w_B 0.015   was kept constant

T/K	448.65	433.15	423.15	413.25	403.15	401.15	399.15	397.15	395.05
P/bar	575.2	619.4	656.0	727.6	825.3	860.7	900.5	937.4	985.1

T/K	393.15	390.25	389.15	388.15
P/bar	1044.7	1139.2	1178.7	1233.2

Polymer (B): **polyisobutylene** **1994GR1, 1994GR4**
Characterization: M_n/kg.mol^{-1} = 0.87, M_w/kg.mol^{-1} = 1.0,
  synthesized in the laboratory by living carbocationic polymerization
Solvent (A): **dimethyl ether** **C$_2$H$_6$O** **115-10-6**

Type of data: cloud points

w_B	0.0552	0.0552	0.0552	0.0552	0.0552	0.0552	0.0754	0.0754	0.0754
T/K	465.15	461.45	448.15	408.15	390.15	390.15	423.15	397.75	388.35
P/bar	165.1	158.8	141.8	80.1	44.4	43.5	104.4	62.2	43.3

w_B	0.0754	0.0754
T/K	387.55	293.35
P/bar	42.7	5.7

lower critical end point: w_B = 0.0552, T/K = 390.15, P/bar = 43.5

Polymer (B): **polyisobutylene** **1994GR2, 1994GR4**
Characterization: M_n/kg.mol^{-1} = 9.5, M_w/kg.mol^{-1} = 10.9,
  synthesized in the laboratory by living carbocationic polymerization
Solvent (A): **dimethyl ether** **C$_2$H$_6$O** **115-10-6**

Type of data: cloud points

w_B 0.0510   was kept constant

T/K	423.15	412.25	401.45	394.05	383.15	373.15	368.45	363.15	358.15
P/bar	404.0	403.8	401.6	397.5	405.4	429.9	442.2	461.6	488.6

T/K	352.25	348.15	347.85	344.85	343.35
P/bar	523.0	557.2	557.0	584.9	605.4

Polymer (B):	polyisobutylene (monohydroxy)						**1994GR1, 1994GR4**	

Characterization: M_n/kg.mol^{-1} = 0.975, M_w/kg.mol^{-1} = 1.12, synthesized in the laboratory by living carbocationic polymerization

Solvent (A):   **dimethyl ether   C$_2$H$_6$O**   **115-10-6**

Type of data: cloud points

w_B 0.0572   was kept constant

T/K	423.15	413.15	403.15	386.15	378.15	371.85	371.65	348.15	323.15
P/bar	145.0	133.4	119.8	94.1	74.0	57.7	57.6	27.3	12.7

Polymer (B):	polyisobutylene (monohydroxy)						**1994GR2, 1994GR4**	

Characterization: M_n/kg.mol^{-1} = 9.6, M_w/kg.mol^{-1} = 11.0, synthesized in the laboratory by living carbocationic polymerization

Solvent (A):   **dimethyl ether   C$_2$H$_6$O**   **115-10-6**

Type of data: cloud points

w_B 0.0510   was kept constant

T/K	423.15	403.15	393.15	383.15	374.65	373.15	363.15	354.75	353.05
P/bar	354.4	348.4	345.3	345.9	351.7	351.1	364.2	385.0	393.8
T/K	348.15	343.15	339.15	334.35	332.35	333.15	328.15	323.15	320.65
P/bar	414.6	442.3	477.4	530.8	559.3	554.9	646.8	794.0	904.7

Polymer (B):	polyisobutylene (dihydroxy)						**1994GR1, 1994GR4**	

Characterization: M_n/kg.mol^{-1} = 1.13, M_w/kg.mol^{-1} = 1.30, synthesized in the laboratory by living carbocationic polymerization

Solvent (A):   **dimethyl ether   C$_2$H$_6$O**   **115-10-6**

Type of data: cloud points

w_B 0.0975   was kept constant

T/K	472.75	448.05	423.15	408.15	378.15	388.15	388.15	363.25	333.25
P/bar	183.5	150.4	110.4	88.1	66.7	43.2	42.2	22.4	14.5
T/K	328.15	293.45							
P/bar	11.2	5.5							

Polymer (B):	polyisobutylene (dihydroxy)						**1994GR2, 1994GR4**	

Characterization: M_n/kg.mol^{-1} = 9.6, M_w/kg.mol^{-1} = 11.0, synthesized in the laboratory by living carbocationic polymerization

Solvent (A):   **dimethyl ether   C$_2$H$_6$O**   **115-10-6**

Type of data: cloud points

w_B 0.0558   was kept constant

continued

continued

T/K	423.65	413.15	403.15	393.15	383.15	373.15	363.15	353.15	343.15
P/bar	376.9	373.7	367.9	366.0	366.0	375.8	389.2	420.9	466.8

T/K	333.05	328.15	323.15	320.55
P/bar	566.7	654.1	794.5	913.2

Polymer (B): **polyisobutylene** **1994GR3, 1994GR4**
Characterization: M_n/kg.mol^{-1} = 3.7, M_w/kg.mol^{-1} = 4.1 (1.37 per arm),
synthesized in the laboratory
Solvent (A): **dimethyl ether** **C$_2$H$_6$O** **115-10-6**

Type of data: cloud points

w_B 0.0635 was kept constant

T/K	423.65	409.85	403.15	391.95	383.15	372.65	367.15	351.15	343.15
P/bar	212.0	192.0	181.1	161.9	147.0	129.4	120.4	93.4	89.1

T/K	332.55	328.65	322.75	312.15	309.15	303.25	297.35	292.85	288.05
P/bar	79.5	77.1	75.7	84.3	91.1	117.1	148.9	191.8	261.1

T/K	283.15	282.85	282.15	280.95	280.55	280.45	280.25	279.25	
P/bar	387.2	421.7	456.3	506.9	529.8	545.9	592.9	652.0	

Polymer (B): **polyisobutylene** **1994GR1, 1994GR4**
Characterization: M_n/kg.mol^{-1} = 0.87, M_w/kg.mol^{-1} = 1.0,
synthesized in the laboratory by living carbocationic polymerization
Solvent (A): **ethane** **C$_2$H$_6$** **74-84-0**

Type of data: cloud points

w_B 0.1308 was kept constant

T/K	423.15	418.15	408.15	391.15	387.65	382.85	377.45	373.15	367.15
P/bar	389.2	385.5	378.3	366.5	363.2	357.6	353.5	348.0	341.0

T/K	358.15	348.15	338.15	327.15	323.45	293.85	292.85		
P/bar	328.7	313.4	300.8	283.6	275.9	229.9	37.7		

Polymer (B): **polyisobutylene** **1994GR2, 1994GR4**
Characterization: M_n/kg.mol^{-1} = 9.5, M_w/kg.mol^{-1} = 10.9,
synthesized in the laboratory by living carbocationic polymerization
Solvent (A): **ethane** **C$_2$H$_6$** **74-84-0**

Type of data: cloud points

w_B 0.0921 was kept constant

T/K	423.15	403.15	394.05	377.15	352.25	347.35	337.05	317.35	307.55
P/bar	1009.8	1014.9	1022.8	1028.0	1039.9	1046.2	1051.4	1085.6	1125.6

T/K	298.15
P/bar	1180.0

Polymer (B):   **polyisobutylene (monohydroxy)**   **1994GR1, 1994GR4**
Characterization: M_n/kg.mol^{-1} = 0.975, M_w/kg.mol^{-1} = 1.12,
                         synthesized in the laboratory by living carbocationic polymerization
Solvent (A):   **ethane**   **C$_2$H$_6$**   **74-84-0**

Type of data: cloud points

w_B	0.1061	0.1061	0.1061	0.1061	0.1061	0.1061	0.1061	0.1061	0.1061
T/K	423.05	413.15	407.65	403.15	397.65	393.15	383.05	363.05	353.15
P/bar	451.3	449.0	444.6	440.0	427.4	433.8	432.0	419.6	417.7

w_B	0.1061	0.1344	0.1344	0.1344	0.1344	0.1344	0.1344	0.1344	0.1344
T/K	340.55	423.15	413.15	406.25	401.35	397.85	393.15	388.15	382.95
P/bar	407.8	451.2	451.5	447.1	442.0	440.0	436.4	432.6	429.1

w_B	0.1344	0.1344	0.1344	0.1344
T/K	372.15	358.65	340.95	321.55
P/bar	431.5	430.4	415.4	412.2

Polymer (B):   **polyisobutylene (monohydroxy)**   **1994GR2, 1994GR4**
Characterization: M_n/kg.mol^{-1} = 9.6, M_w/kg.mol^{-1} = 11.0,
                         synthesized in the laboratory by living carbocationic polymerization
Solvent (A):   **ethane**   **C$_2$H$_6$**   **74-84-0**

Type of data: cloud points

w_B	0.0882	was kept constant

T/K	423.15	393.15	383.15	353.15	333.15	323.15	308.15
P/bar	983.8	1015.0	1036.7	1066.1	1111.1	1147.3	1205.4

Polymer (B):   **polyisobutylene (dihydroxy)**   **1994GR1, 1994GR4**
Characterization: M_n/kg.mol^{-1} = 1.13, M_w/kg.mol^{-1} = 1.30,
                         synthesized in the laboratory by living carbocationic polymerization
Solvent (A):   **ethane**   **C$_2$H$_6$**   **74-84-0**

Type of data: cloud points

w_B	0.1039	0.1039	0.1039	0.1039	0.1039	0.1039	0.1039	0.1039	0.1039
T/K	423.15	417.65	413.15	403.15	393.15	389.65	383.15	377.15	373.15
P/bar	464.7	465.1	465.0	465.6	467.0	468.0	470.9	475.3	479.1

w_B	0.1039	0.1039	0.1039	0.1039	0.1039	0.1039	0.1039	0.2437	0.2437
T/K	368.15	363.15	358.15	353.15	348.15	346.25	343.15	448.15	443.15
P/bar	485.5	493.7	505.0	519.4	538.2	546.3	562.2	521.6	521.6

Polymer (B):   **polyisobutylene (dihydroxy)**   **1994GR2, 1994GR4**
Characterization: M_n/kg.mol^{-1} = 9.6, M_w/kg.mol^{-1} = 11.0,
                        synthesized in the laboratory by living carbocationic polymerization
Solvent (A):   **ethane**   **C$_2$H$_6$**   **74-84-0**

Type of data: cloud points

w_B 0.0878   was kept constant

T/K	423.15	403.15	383.95	353.25	333.15
P/bar	985.8	1006.0	1018.0	1038.3	1092.7

Polymer (B):   **polyisobutylene**   **1994GR3, 1994GR4**
Characterization: M_n/kg.mol^{-1} = 3.7, M_w/kg.mol^{-1} = 4.1 (1.37 per arm),
                        synthesized in the laboratory
Solvent (A):   **ethane**   **C$_2$H$_6$**   **74-84-0**

Type of data: cloud points

w_B 0.104   was kept constant

T/K	423.05	403.15	392.35	383.55	371.65	363.25	362.55	353.05	343.15
P/bar	656.4	650.4	646.4	642.3	636.6	633.4	632.7	631.3	627.0

T/K	333.65	323.65	313.65	307.65	304.85	298.25	277.65	269.65	266.65
P/bar	621.4	617.9	613.5	611.8	610.2	608.8	605.9	606.5	607.3

T/K	258.65	252.65	243.05	237.95	234.45
P/bar	609.8	613.0	621.3	628.0	631.6

Polymer (B):   **polyisobutylene**   **1999CHE**
Characterization: M_n/kg.mol^{-1} = 1.0, M_w/kg.mol^{-1} = 1.12,
                        synthesized in the laboratory by living carbocationic polymerization
Solvent (A):   **ethene**   **C$_2$H$_4$**   **74-85-1**

Type of data: cloud points

w_B 0.000013   was kept constant

T/K	447.45	435.35	422.25	409.45	399.75	386.05	374.85	358.25	346.85
P/bar	249.2	277.0	269.5	244.4	240.4	227.7	215.6	195.2	175.7

T/K	332.15	323.45	312.85	302.75	295.25
P/bar	167.8	156.8	143.1	119.1	110.5

w_B 0.0025   was kept constant

T/K	429.35	413.55	408.15	398.95	381.55	366.25	351.75	339.25	325.85
P/bar	352.2	338.7	332.4	321.5	302.8	275.3	256.3	237.7	215.0

T/K	315.95	293.05	271.25	253.55	239.75	229.95
P/bar	194.7	165.2	122.3	86.6	62.0	60.2

Polymer (B):   **polyisobutylene**   **1994GR1, 1994GR4**

Characterization: M_n/kg.mol^{-1} = 0.87, M_w/kg.mol^{-1} = 1.0,

                  synthesized in the laboratory by living carbocationic polymerization

Solvent (A):   **propane**   **C$_3$H$_8$**   **74-98-6**

Type of data: cloud points

w_B	0.0636	0.0617	0.0636	0.0713	0.0713	0.0713	0.0636	0.0713
T/K	423.15	398.15	373.15	373.15	347.65	328.35	293.35	293.15
P/bar	151.5	118.5	72.1	76.9	26.1	19.7	8.3	8.3

lower critical end point: w_B = 0.0713, T/K = 347.65, P/bar = 26.1

Polymer (B):   **polyisobutylene**   **1994GR3, 1994GR4**

Characterization: M_n/kg.mol^{-1} = 3.7, M_w/kg.mol^{-1} = 4.1 (1.37 per arm),

                  synthesized in the laboratory

Solvent (A):   **propane**   **C$_3$H$_8$**   **74-98-6**

Type of data: cloud points

w_B 0.0657   was kept constant

T/K	423.65	398.15	373.15	368.15	362.65	353.15	328.15	323.15	317.05
P/bar	270.8	239.6	202.2	194.2	185.2	171.9	124.4	112.5	99.4

T/K	312.15	308.15	298.25	293.15	291.15	288.15	281.45
P/bar	89.6	79.0	56.7	46.1	41.6	33.7	17.2

Polymer (B):   **polyisobutylene**   **1994GR2, 1994GR4**

Characterization: M_n/kg.mol^{-1} = 9.5, M_w/kg.mol^{-1} = 10.9,

                  synthesized in the laboratory by living carbocationic polymerization

Solvent (A):   **propane**   **C$_3$H$_8$**   **74-98-6**

Type of data: cloud points

w_B 0.0684   was kept constant

T/K	423.15	398.15	373.45	372.55	348.65	337.05	323.05	312.75	303.15
P/bar	419.5	395.8	366.5	365.9	335.7	314.3	294.2	278.6	259.1

T/K	302.55
P/bar	10.4

Polymer (B):   **polyisobutylene (monohydroxy)**   **1994GR1, 1994GR4**

Characterization: M_n/kg.mol^{-1} = 0.975, M_w/kg.mol^{-1} = 1.12,

                  synthesized in the laboratory by living carbocationic polymerization

Solvent (A):   **propane**   **C$_3$H$_8$**   **74-98-6**

Type of data: cloud points

continued

continued

w_B	0.0732	0.0727	0.0697	0.0723	0.0717	0.0697	0.0697
T/K	423.15	398.15	373.15	348.15	328.35	325.05	295.35
P/bar	175.0	142.5	108.6	69.8	35.6	19.7	8.7

lower critical end point: w_B = 0.0697, T/K = 325.05, P/bar = 19.7

Polymer (B):	**polyisobutylene (monohydroxy)**	**1994GR2, 1994GR4**

Characterization: M_n/kg.mol^{-1} = 9.6, M_w/kg.mol^{-1} = 11.0,
synthesized in the laboratory by living carbocationic polymerization

Solvent (A):	**propane**	**C$_3$H$_8$**	**74-98-6**

Type of data: cloud points

w_B 0.0700   was kept constant

T/K	423.15	397.95	387.95	382.15	373.25	348.05	323.15	321.35
P/bar	413.3	388.1	376.7	369.6	358.5	328.8	291.0	16.4

Polymer (B):	**polyisobutylene (dihydroxy)**	**1994GR1, 1994GR4**

Characterization: M_n/kg.mol^{-1} = 1.13, M_w/kg.mol^{-1} = 1.30,
synthesized in the laboratory by living carbocationic polymerization

Solvent (A):	**propane**	**C$_3$H$_8$**	**74-98-6**

Type of data: cloud points

w_B	0.0688	0.0688	0.0688	0.0688	0.0688	0.0688	0.0688	0.0688
T/K	423.15	373.15	341.95	325.65	321.95	310.65	298.15	294.05
P/bar	193.0	136.3	115.1	105.0	114.8	160.8	208.1	214.2

Polymer (B):	**polyisobutylene (dihydroxy)**	**1994GR2, 1994GR4**

Characterization: M_n/kg.mol^{-1} = 9.6, M_w/kg.mol^{-1} = 11.0,
synthesized in the laboratory by living carbocationic polymerization

Solvent (A):	**propane**	**C$_3$H$_8$**	**74-98-6**

Type of data: cloud points

w_B 0.0681   was kept constant

T/K	423.15	418.15	413.15	402.85	391.85	380.55	360.15	352.85	342.85
P/bar	438.0	430.0	425.3	416.2	404.7	392.6	366.4	357.9	344.9

T/K	332.85	322.75	310.25	307.25
P/bar	322.4	318.1	296.3	295.6

Polymer (B):	**polyisoprene**	**1996ALB**

Characterization: M_n/kg.mol^{-1} = 7.63, M_w/kg.mol^{-1} = 13.4

Solvent (A):	**dimethyl ether**	**C$_2$H$_6$O**	**115-10-6**

continued

continued

Type of data: cloud points

$w_B = 0.0530$ was kept constant

T/K	424.25	422.25	410.75	383.75	382.65	363.25	348.25	347.95	332.75
P/bar	210.1	206.0	185.6	133.5	131.5	89.9	55.3	54.5	13.4

lower critical end point: $w_B = 0.0530$, T/K = 332.75, P/bar = 13.4

Polymer (B): **polyisoprene** 1996ALB
Characterization: M_n/kg.mol^{-1} = 7.63, M_w/kg.mol^{-1} = 13.4
Solvent (A): **ethane** **C_2H_6** 74-84-0

Type of data: cloud points

$w_B = 0.0723$ was kept constant

T/K	423.05	412.95	411.95	390.45	390.35	362.35	339.55	336.55	333.65
P/bar	787.5	783.1	799.2	811.3	815.3	846.6	932.3	964.9	985.1

T/K	333.25
P/bar	993.2

Polymer (B): **polyisoprene** 1996ALB
Characterization: M_n/kg.mol^{-1} = 7.63, M_w/kg.mol^{-1} = 13.4
Solvent (A): **ethene** **C_2H_4** 74-85-1

Type of data: cloud points

$w_B = 0.0841$ was kept constant

T/K	424.45	423.55	399.45	393.15	389.75	389.45	388.15	388.05
P/bar	972.3	972.8	998.0	1002.5	1003.6	1011.1	1018.5	1023.3

Polymer (B): **polyisoprene** 1996ALB
Characterization: M_n/kg.mol^{-1} = 7.63, M_w/kg.mol^{-1} = 13.4
Solvent (A): **propane** **C_3H_8** 74-98-6

Type of data: cloud points

$w_B = 0.0529$ was kept constant

T/K	423.15	390.05	389.35	363.35	348.25	336.45	336.15	326.25	324.85
P/bar	324.65								

T/K	383.7	375.3	376.0	377.1	387.0	420.7	416.6	455.5	467.3
P/bar	467.5								

Polymer (B):   **polyisoprene**   **1996ALB**
Characterization: M_n/kg.mol^{-1} = 7.63, M_w/kg.mol^{-1} = 13.4
Solvent (A):   **propene**   **C$_3$H$_6$**   **115-07-1**

Type of data: cloud points

w_B = 0.0482   was kept constant

T/K	423.35	390.75	363.05	362.85	336.55	323.25	309.75	299.05	292.05
P/bar	338.1	301.6	265.5	261.4	221.0	202.4	185.6	176.2	172.5

T/K	284.05	283.65	280.95
P/bar	175.3	173.9	174.8

Polymer (B):   **polyisoprene (hydrogenated)**   **1997HAN**
Characterization: M_n/kg.mol^{-1} = 2.4, M_w/kg.mol^{-1} = 2.6, 38% hydrogenation,
                 T_g/K = 208.2, synthesized in the laboratory
Solvent (A):   **propene**   **C$_3$H$_6$**   **115-07-1**

Type of data: cloud points

w_B	0.15	0.15	0.15	0.15	0.15	0.15	0.15	0.008	0.113
T/K	305.65	315.25	336.75	357.75	376.65	401.35	423.95	421.95	422.15
P/bar	14.2	40.4	91.7	134.7	169.3	208.2	238.4	156.8	230.4

w_B	0.289	0.360	0.477	0.623
T/K	423.05	422.35	423.25	422.95
P/bar	233.3	220.1	215.6	170.3

Type of data: coexistence data

T/K = 422.75

P/ bar	feed phase	w_B gel phase	sol phase
180	0.289	0.571	0.016
150	0.360	0.656	0.007

Polymer (B):   **poly(DL-lactide)**   **1996KIM**
Characterization: M_η/kg.mol^{-1} = 2.0, Boehringer Ingelheim Co., Germany
Solvent (A):   **carbon dioxide**   **CO$_2$**   **124-38-9**

Type of data: polymer solubility/cloud points

T/K = 333.15

P/bar	137.9	151.7	165.5	179.3	193.1
w_B	0.000018	0.000029	0.000040	0.000072	0.000118

Polymer (B):	**poly(L-lactide)**	**1991TOM**
Characterization:	M_n/kg.mol^{-1} = 2.75, M_w/kg.mol^{-1} = 5.5,	
	L104, Boehringer Ingelheim Chemicals	
Solvent (A):	**carbon dioxide** **CO_2**	**124-38-9**

Type of data: polymer solubility/cloud points

T/K = 318.15

P/bar	200	250	300
w_B	0.000128	0.000200	0.000427

T/K = 328.15

P/bar	200	250	300
w_B	0.000241	0.000260	0.000535

T/K = 338.15

P/bar	200	250	300
w_B	0.000280	0.000477	0.000738

Polymer (B):	**poly(DL-lactide)**	**2001CON**
Characterization:	M_n/kg.mol^{-1} = 60.6, M_w/kg.mol^{-1} = 84.5, T_g/K = 323.05	
Solvent (A):	**carbon dioxide** **CO_2**	**124-38-9**

Type of data: cloud points

w_B	0.05	0.05	0.05	0.05	0.05
T/K	317.55	329.65	336.85	349.45	363.75
P/bar	1312	1323	1332	1345	1360
ρ/(g/cm^3)	1.213	1.193	1.182	1.161	1.137

Polymer (B):	**poly(DL-lactide)**	**2001CON**
Characterization:	M_n/kg.mol^{-1} = 80.9, M_w/kg.mol^{-1} = 128.5, T_g/K = 323.35	
Solvent (A):	**carbon dioxide** **CO_2**	**124-38-9**

Type of data: cloud points

w_B	0.05	0.05	0.05	0.05	0.05
T/K	305.85	314.35	329.35	345.75	365.85
P/bar	1389	1398	1408	1418	1429
ρ/(g/cm^3)	1.192	1.179	1.154	1.130	1.099

Polymer (B):	**poly(DL-lactide)**	**2002KUK**
Characterization:	M_n/kg.mol^{-1} = 2.0, Resomer R104, Boehringer Ingelheim Chemicals	
Solvent (A):	**chlorodifluoromethane** **$CHClF_2$**	**75-45-6**

Type of data: cloud points

w_B	0.0274	0.0274	0.0274	0.0274	0.0274
T/K	342.85	353.15	362.35	372.95	383.05
P/MPa	3.61	6.43	8.89	11.50	13.84

Polymer (B):   **poly(DL-lactide)**   **2000LE1**
Characterization: M_η/kg.mol^{-1} = 2.0, Resomer R104, Boehringer Ingelheim Chemicals
Solvent (A):   **chlorodifluoromethane**   **CHClF$_2$**   **75-45-6**

Type of data: cloud points

w_B	0.0568	0.0568	0.0568	0.0568	0.0568	0.0568
T/K	346.95	355.15	364.75	373.05	382.45	392.75
P/MPa	4.25	6.40	8.85	11.05	13.43	15.72

Polymer (B):   **poly(L-lactide)**   **2000LE1**
Characterization: M_η/kg.mol^{-1} = 2.0, Polysciences, Inc., Warrington, PA
Solvent (A):   **chlorodifluoromethane**   **CHClF$_2$**   **75-45-6**

Type of data: cloud points

w_B	0.0009	0.0009	0.0009	0.0009	0.0009	0.0009	0.0106	0.0106	0.0106
T/K	346.95	354.65	364.55	373.35	383.75	393.55	344.35	353.55	363.45
P/MPa	3.85	6.05	8.65	10.80	13.45	15.20	3.35	5.80	8.55
w_B	0.0106	0.0106	0.0106	0.0288	0.0288	0.0288	0.0288	0.0288	0.0288
T/K	373.45	383.55	392.85	344.35	353.05	363.35	372.75	382.05	392.75
P/MPa	11.05	13.25	15.20	3.60	5.95	8.72	10.90	13.05	15.62
w_B	0.0371	0.0371	0.0371	0.0371	0.0371	0.0371	0.0577	0.0577	0.0577
T/K	345.85	352.85	363.35	372.85	382.95	392.45	347.35	353.75	363.25
P/MPa	4.10	5.96	8.66	11.00	13.35	15.42	4.55	6.23	8.85
w_B	0.0577	0.0577	0.0577	0.1029	0.1029	0.1029	0.1029	0.1029	0.1391
T/K	373.95	383.15	392.75	352.65	362.95	372.05	383.05	393.65	345.35
P/MPa	11.30	13.50	15.85	6.20	8.95	11.05	13.50	15.70	3.68
w_B	0.1391	0.1391	0.1391	0.1391	0.1391	0.1798	0.1798	0.1798	0.1798
T/K	355.85	364.55	373.95	383.65	393.95	346.55	355.25	366.45	374.55
P/MPa	6.60	8.94	11.25	13.53	15.87	3.40	5.77	8.87	10.95
w_B	0.1798	0.1798							
T/K	383.05	393.35							
P/MPa	12.93	15.22							

Polymer (B):   **poly(DL-lactide)**   **2002KUK**
Characterization: M_η/kg.mol^{-1} = 30, Resomer R203, Boehringer Ingelheim Chemicals
Solvent (A):   **chlorodifluoromethane**   **CHClF$_2$**   **75-45-6**

Type of data: cloud points

w_B	0.0050	0.0050	0.0050	0.0050	0.0050	0.0287	0.0287	0.0287	0.0287
T/K	343.05	352.85	362.85	373.05	382.55	338.25	343.65	352.75	362.75
P/MPa	4.23	7.13	9.81	12.45	14.85	3.35	5.03	7.71	10.55
w_B	0.0287	0.0287	0.0477	0.0477	0.0477	0.0477	0.0477	0.0477	0.0784
T/K	372.95	382.75	338.15	343.25	352.95	362.85	372.95	382.25	338.15
P/MPa	13.13	15.55	3.25	4.81	7.68	10.53	13.25	15.63	3.09

continued

continued

w_B	0.0784	0.0784	0.0784	0.0784	0.0784	0.1468	0.1468	0.1468	0.1468
T/K	342.75	352.85	363.35	372.85	382.05	339.95	343.25	352.85	363.35
P/MPa	4.50	7.45	10.55	13.17	15.45	3.26	4.29	7.20	10.25

w_B	0.1468	0.1468
T/K	373.15	382.75
P/MPa	12.84	15.30

Polymer (B):	**poly(L-lactide)**	**2000LE1**
Characterization:	M_η/kg.mol^{-1} = 50, Polysciences, Inc., Warrington, PA	
Solvent (A):	**chlorodifluoromethane** CHClF$_2$	**75-45-6**

Type of data: cloud points

w_B	0.0359	0.0359	0.0359	0.0359	0.0359	0.0359	0.0359
T/K	334.85	339.45	353.25	362.35	372.05	382.45	392.05
P/MPa	3.25	4.78	9.10	11.90	14.68	17.48	20.00

Polymer (B):	**poly(DL-lactide)**	**2001CON**
Characterization:	M_n/kg.mol^{-1} = 80.9, M_w/kg.mol^{-1} = 128.5, T_g/K = 323.35	
Solvent (A):	**chlorodifluoromethane** CHClF$_2$	**75-45-6**

Type of data: cloud points

w_B	0.05	0.05	0.05	0.05
T/K	323.45	328.15	333.65	345.05
P/bar	20.3	36.9	58.3	99.3
ρ/(g/cm^3)	1.106	1.101	1.096	1.084

Polymer (B):	**poly(L-lactide)**	**2000LE1**
Characterization:	M_η/kg.mol^{-1} = 100, Polysciences, Inc., Warrington, PA	
Solvent (A):	**chlorodifluoromethane** CHClF$_2$	**75-45-6**

Type of data: cloud points

w_B	0.0382	0.0382	0.0382	0.0382	0.0382	0.0382	0.0382
T/K	333.55	342.95	353.45	362.15	372.25	383.35	392.55
P/MPa	3.15	6.25	9.55	12.13	15.00	18.00	20.25

Polymer (B):	**poly(L-lactide)**	**2000LE1**
Characterization:	M_η/kg.mol^{-1} = 300, Polysciences, Inc., Warrington, PA	
Solvent (A):	**chlorodifluoromethane** CHClF$_2$	**75-45-6**

Type of data: cloud points

w_B	0.0568	0.0568	0.0568	0.0568
T/K	344.55	343.25	363.05	384.05
P/MPa	4.55	7.15	13.46	19.50

Polymer (B):   **poly(DL-lactide)**   **2002KUK**
Characterization: M_η/kg.mol^{-1} = 2.0, Resomer R104, Boehringer Ingelheim Chemicals
Solvent (A):   **difluoromethane   CH$_2$F$_2$**   **75-10-5**

Type of data: cloud points

w_B	0.0322	0.0322	0.0322	0.0322	0.0322	0.0322	0.0322	0.0322
T/K	304.45	315.55	322.75	333.25	342.95	353.15	362.85	372.75
P/MPa	33.00	33.00	33.45	34.55	36.00	37.65	39.25	40.90

Polymer (B):   **poly(DL-lactide)**   **2002KUK**
Characterization: M_η/kg.mol^{-1} = 30, Resomer R203, Boehringer Ingelheim Chemicals
Solvent (A):   **difluoromethane   CH$_2$F$_2$**   **75-10-5**

Type of data: cloud points

w_B	0.0047	0.0047	0.0047	0.0047	0.0047	0.0047	0.0047	0.0047	0.0164
T/K	305.05	311.95	323.85	333.75	343.55	352.85	362.95	372.95	303.55
P/MPa	49.65	48.25	47.35	48.05	48.55	49.25	49.95	50.75	54.75
w_B	0.0164	0.0164	0.0164	0.0164	0.0164	0.0164	0.0164	0.0278	0.0278
T/K	313.15	323.25	333.55	342.65	353.15	363.05	372.95	304.85	313.55
P/MPa	52.55	51.65	51.55	52.05	52.95	53.85	54.75	55.05	53.15
w_B	0.0278	0.0278	0.0278	0.0278	0.0278	0.0278	0.0480	0.0480	0.0480
T/K	323.85	334.45	343.35	352.95	363.45	373.05	304.15	313.25	323.75
P/MPa	52.05	52.15	52.55	53.45	54.55	55.45	55.55	53.05	52.05
w_B	0.0480	0.0480	0.0480	0.0480	0.0480	0.0908	0.0908	0.0908	0.0908
T/K	333.95	343.35	353.35	363.05	373.05	304.35	312.85	323.15	334.75
P/MPa	52.05	52.55	53.55	54.45	55.35	52.85	50.85	50.05	50.25
w_B	0.0908	0.0908	0.0908	0.0908	0.1500	0.1500	0.1500	0.1500	0.1500
T/K	342.85	352.85	363.35	375.55	303.45	313.45	323.15	333.45	342.85
P/MPa	50.75	51.65	52.85	54.25	49.35	47.65	47.25	47.75	48.55
w_B	0.1500	0.1500	0.1500
T/K	352.85	362.95	372.75
P/MPa	49.65	50.95	52.15

Polymer (B):   **poly(DL-lactide)**   **2001KUK**
Characterization: M_η/kg.mol^{-1} = 2.0, Resomer R104, Boehringer Ingelheim Chemicals
Solvent (A):   **dimethyl ether   C$_2$H$_6$O**   **115-10-6**

Type of data: cloud points

w_B	0.0297	0.0297	0.0297	0.0297
T/K	358.35	363.15	367.75	375.15
P/MPa	4.25	5.50	6.35	8.10

Polymer (B):	poly(DL-lactide)	2001KUK
Characterization:	M_η/kg.mol^{-1} = 30, Resomer R203, Boehringer Ingelheim Chemicals	
Solvent (A):	dimethyl ether C_2H_6O	115-10-6

Type of data: cloud points

w_B	0.0057	0.0057	0.0057	0.0057	0.0057	0.0057	0.0291	0.0291	0.0291
T/K	333.25	338.75	343.15	352.65	363.85	372.45	328.55	333.15	343.75
P/MPa	1.65	3.05	4.25	6.85	9.65	11.75	2.19	3.42	6.29

w_B	0.0291	0.0291	0.0291	0.0500	0.0500	0.0500	0.0500	0.0500	0.0500
T/K	352.75	362.85	373.35	328.65	333.05	343.05	354.15	363.35	372.95
P/MPa	8.53	11.04	13.45	2.47	3.75	6.55	9.27	11.45	13.73

w_B	0.0799	0.0799	0.0799	0.0799	0.0799	0.0799	0.1467	0.1467	0.1467
T/K	327.15	333.55	343.95	352.95	363.15	373.25	329.25	332.95	343.55
P/MPa	1.70	3.65	6.65	8.93	11.60	13.83	1.70	2.80	5.90

w_B	0.1467	0.1467	0.1467
T/K	353.05	363.05	372.45
P/MPa	8.50	11.01	13.23

Polymer (B):	poly(DL-lactide)	2002KUK
Characterization:	M_η/kg.mol^{-1} = 2.0, Resomer R104, Boehringer Ingelheim Chemicals	
Solvent (A):	1,1,1,2-tetrafluoroethane $C_2H_2F_4$	811-97-2

Type of data: cloud points

w_B	0.0292	0.0292	0.0292	0.0292	0.0292	0.0292	0.0292	0.0292
T/K	306.75	315.05	325.05	334.65	346.15	353.65	363.85	373.45
P/MPa	19.18	20.43	21.95	23.65	25.60	26.90	28.50	30.05

Polymer (B):	poly(DL-lactide)	2002KUK
Characterization:	M_η/kg.mol^{-1} = 30, Resomer R203, Boehringer Ingelheim Chemicals	
Solvent (A):	1,1,1,2-tetrafluoroethane $C_2H_2F_4$	811-97-2

Type of data: cloud points

w_B	0.0054	0.0054	0.0054	0.0054	0.0054	0.0054	0.0054	0.0054	0.0304
T/K	302.95	313.45	324.25	333.85	343.15	352.85	362.75	372.95	304.05
P/MPa	21.75	24.45	26.95	29.35	31.55	33.35	34.95	36.85	24.95

w_B	0.0304	0.0304	0.0304	0.0304	0.0304	0.0304	0.0304	0.0526	0.0526
T/K	314.75	323.75	333.25	343.15	353.25	363.35	372.95	303.65	313.25
P/MPa	27.37	29.55	31.75	34.03	36.15	38.38	40.15	25.24	27.30

w_B	0.0526	0.0526	0.0526	0.0526	0.0526	0.0526	0.0855	0.0855	0.0855
T/K	323.85	333.35	342.95	352.85	363.05	373.05	304.95	313.35	323.75
P/MPa	29.70	31.95	34.15	36.25	38.35	40.12	23.85	25.80	28.23

w_B	0.0855	0.0855	0.0855	0.0855	0.0855	0.1505	0.1505	0.1505	0.1505
T/K	333.35	343.35	352.75	363.15	373.05	303.75	313.85	323.85	333.45
P/MPa	30.50	32.85	34.95	37.09	38.91	21.63	24.25	26.81	29.13

continued

continued

w_B	0.1505	0.1505	0.1505	0.1505
T/K	343.15	352.95	363.25	373.25
P/MPa	31.41	33.60	35.71	37.60

Polymer (B):   **poly(DL-lactide)**   **2002KUK**
Characterization: $M_\eta/kg.mol^{-1} = 2.0$, Resomer R104, Boehringer Ingelheim Chemicals
Solvent (A):   **trifluoromethane**   **CHF$_3$**   **75-46-7**

Type of data: cloud points

w_B	0.0306	0.0306	0.0306	0.0306	0.0306	0.0306	0.0306	0.0306
T/K	306.05	313.25	323.35	332.75	342.65	352.55	363.05	373.85
P/MPa	46.05	48.75	52.35	55.75	59.05	62.05	64.85	67.35

Polymer (B):   **poly(DL-lactide)**   **2002KUK**
Characterization: $M_\eta/kg.mol^{-1} = 30$, Resomer R203, Boehringer Ingelheim Chemicals
Solvent (A):   **trifluoromethane**   **CHF$_3$**   **75-46-7**

Type of data: cloud points

w_B	0.0055	0.0055	0.0055	0.0055	0.0055	0.0055	0.0055	0.0055	0.0310
T/K	303.75	313.25	323.65	333.15	342.85	353.15	363.25	372.85	303.95
P/MPa	49.05	53.75	58.25	63.25	67.45	71.85	75.35	78.25	53.75

w_B	0.0310	0.0310	0.0310	0.0310	0.0310	0.0310	0.0310	0.0793	0.0793
T/K	313.25	323.25	333.15	343.15	353.55	363.35	372.85	304.05	313.25
P/MPa	58.55	63.35	67.95	72.35	76.55	80.25	83.45	55.45	60.05

w_B	0.0793	0.0793	0.0793	0.0793	0.0793	0.0793	0.1317	0.1317	0.1317
T/K	323.55	333.35	343.55	353.25	363.15	373.35	303.65	313.55	323.75
P/MPa	64.95	69.45	73.95	77.75	81.35	84.75	54.15	59.15	64.05

w_B	0.1317	0.1317	0.1317	0.1317	0.1317
T/K	333.15	343.45	353.05	363.15	372.95
P/MPa	68.25	72.65	76.55	80.15	83.45

Polymer (B):   **poly(DL-lactide)**   **2001CON**
Characterization: $M_n/kg.mol^{-1} = 80.9$, $M_w/kg.mol^{-1} = 128.5$, $T_g/K = 323.35$
Solvent (A):   **trifluoromethane**   **CHF$_3$**   **75-46-7**

Type of data: cloud points

w_B	0.05	0.05	0.05	0.05	0.05
T/K	300.55	306.35	317.25	330.95	344.15
P/bar	543	575	633	705	767
$\rho/(g/cm^3)$	1.334	1.332	1.321	1.311	1.302

Polymer (B):	poly(methyl acrylate)						2004BEC	

Characterization: M_n/kg.mol^{-1} = 76.5, M_w/kg.mol^{-1} = 187, synthesized in the laboratory

Solvent (A):	propene	C_3H_6					115-07-1	

Type of data: cloud points

w_B 0.05   was kept constant

T/K	473.15	480.15	481.15	482.15	483.15	491.15	493.75	501.15	510.15
P/bar	2750	2465	2418	2391	2361	2109	2018	1850	1691

T/K	512.75	521.95	523.45	525.45	526.45	531.65
P/bar	1641	1513	1487	1465	1457	1399

Polymer (B):	poly(methyl methacrylate)	1993HA4

Characterization: M_n/kg.mol^{-1} = 70, M_w/kg.mol^{-1} = 74, 57% syndiotactic, 6% isotactic, 37% atactic triads, T_g/K = 398, synthesized in the laboratory via group transfer polymerization by S. Smith, Procter & Gamble Co.

Solvent (A):	chlorodifluoromethane	CHClF$_2$	75-45-6

Type of data: cloud points

Comments: 65 < (T/K − 273.15) < 150.

w_B = 0.00030 P/MPa = −20.369 + 0.37989 (T/K − 273.15) − 4.2093 10^{-4} (T/K − 273.15)^2
w_B = 0.00311 P/MPa = −21.274 + 0.41400 (T/K − 273.15) − 5.4051 10^{-4} (T/K − 273.15)^2
w_B = 0.00983 P/MPa = −22.489 + 0.44372 (T/K − 273.15) − 6.4710 10^{-4} (T/K − 273.15)^2
w_B = 0.0311 P/MPa = −20.804 + 0.41613 (T/K − 273.15) − 5.0146 10^{-4} (T/K − 273.15)^2
w_B = 0.0507 P/MPa = −22.873 + 0.46489 (T/K − 273.15) − 7.6064 10^{-4} (T/K − 273.15)^2
w_B = 0.0826 P/MPa = −21.010 + 0.41800 (T/K − 273.15) − 4.9133 10^{-4} (T/K − 273.15)^2
w_B = 0.1542 P/MPa = −24.096 + 0.45924 (T/K − 273.15) − 6.8208 10^{-4} (T/K − 273.15)^2

Polymer (B):	poly(methyl methacrylate)						2004BEC	

Characterization: M_n/kg.mol^{-1} = 71.2, M_w/kg.mol^{-1} = 105, synthesized in the laboratory

Solvent (A):	propene	C_3H_6					115-07-1	

Type of data: cloud points

w_B 0.05   was kept constant

T/K	427.15	433.15	433.65	443.35	444.15	453.15	461.15	462.15	462.65
P/bar	2700	2379	2356	2060	2027	1891	1739	1728	1712

T/K	471.15	471.85	472.35	481.15	490.65	502.15	511.15	523.15	532.15
P/bar	1577	1573	1563	1458	1364	1265	1204	1122	1078

Polymer (B):	**poly(octadecyl acrylate)**	**2002BYU**

Characterization: M_w/kg.mol^{-1} = 93.3, T_g/K = 313,
Aldrich Chem. Co., Inc., Milwaukee, WI

Solvent (A):	**ethene**	**C$_2$H$_4$**	**74-85-1**

Type of data: cloud points

w_B 0.051   was kept constant

T/K	314.45	324.25	342.05	364.55	383.85	402.65	423.95
P/bar	750.0	739.7	730.7	726.6	723.8	724.5	725.9

Polymer (B):	**poly(propyl acrylate)**	**2004BEC**

Characterization: M_n/kg.mol^{-1} = 48, M_w/kg.mol^{-1} = 108,
Polysciences, Inc., Warrington, PA

Solvent (A):	**ethene**	**C$_2$H$_4$**	**74-85-1**

Type of data: cloud points

w_B 0.05   was kept constant

T/K	352.95	363.25	373.25	393.15	412.15	432.15	452.15	472.15	491.15
P/bar	1093	1059	1033	992	961	936	914	892	881

T/K	511.15	531.15
P/bar	877	874

Polymer (B):	**polypropylene**	**2000OLI**

Characterization: M_n/kg.mol^{-1} = 48.9, M_w/kg.mol^{-1} = 245,
92 % isotactic, Polibrasil Resinas S.A.

Solvent (A):	**n-butane**	**C$_4$H$_{10}$**	**106-97-8**

Type of data: cloud points

w_B	0.0030	0.0030	0.0030	0.0030	0.0030	0.0030	0.0030	0.0030	0.0030
T/K	423.15	413.15	413.15	403.15	403.15	393.15	393.15	383.15	383.15
P/bar	106.3	88.6	33.5	69.8	28.5	51.6	24.4	31.7	21.1
	LLE	LLE	VLLE	LLE	VLLE	LLE	VLLE	LLE	VLLE

w_B	0.0030	0.0030	0.0030	0.0030	0.0069	0.0069	0.0069	0.0069	0.0069
T/K	373.15	368.15	363.15	358.15	423.15	413.15	413.15	403.15	403.15
P/bar	16.7	15.2	13.9	12.4	112.4	95.0	33.6	77.4	28.4
	VLE	VLE	VLE	VLE	LLE	LLE	VLLE	LLE	VLLE

w_B	0.0069	0.0069	0.0069	0.0069	0.0069	0.0069	0.0069	0.0069	0.0069
T/K	393.15	393.15	383.15	383.15	373.15	373.15	368.15	363.15	358.15
P/bar	58.4	24.6	38.9	21.1	21.5	17.0	16.0	15.0	13.6
	LLE	VLLE	LLE	VLLE	LLE	VLLE	VLE	VLE	VLSE

w_B	0.0100	0.0100	0.0100	0.0100	0.0100	0.0100	0.0100	0.0100	0.0100
T/K	423.15	413.15	413.15	403.15	403.15	393.15	393.15	383.15	383.15
P/bar	113.4	95.7	33.4	77.4	28.4	58.9	24.4	40.9	21.4
	LLE	LLE	VLLE	LLE	VLLE	LLE	VLLE	LLE	VLLE

continued

continued

w_B	0.0100	0.0100	0.0100	0.0100	0.0100	0.0298	0.0298	0.0298	0.0298
T/K	373.15	373.15	368.15	363.15	358.15	423.15	423.15	413.15	413.15
P/bar	22.5	17.0	16.7	15.5	13.6	114.2	39.2	96.0	33.6
	LLE	VLLE	VLE	VLE	VLSE	LLE	VLLE	LLE	VLLE

w_B	0.0298	0.0298	0.0298	0.0298	0.0298	0.0298	0.0298	0.0298	0.0298
T/K	403.15	403.15	393.15	393.15	383.15	383.15	373.15	373.15	368.15
P/bar	77.9	29.2	60.5	25.3	42.2	21.5	22.0	17.2	17.0
	LLE	VLLE	LLE	VLLE	LLE	VLLE	LLE	VLLE	VLE

w_B	0.0298	0.0298	0.0496	0.0496	0.0496	0.0496	0.0496	0.0496	0.0496
T/K	363.15	358.15	423.15	423.15	413.15	413.15	403.15	403.15	393.15
P/bar	15.7	13.7	114.8	39.5	98.5	33.7	82.9	28.7	64.8
	VLE	VLSE	LLE	VLLE	LLE	VLLE	LLE	VLLE	LLE

w_B	0.0496	0.0496	0.0496	0.0496	0.0496	0.0496	0.0496	0.0496	0.0742
T/K	393.15	383.15	383.15	373.15	373.15	368.15	363.15	358.15	423.15
P/bar	24.8	45.1	21.0	24.5	17.1	17.2	15.6	13.9	112.1
	VLLE	LLE	VLLE	LLE	VLLE	VLE	VLE	VLSE	LLE

w_B	0.0742	0.0742	0.0742	0.0742	0.0742	0.0742	0.0742	0.0992	0.0992
T/K	413.15	403.15	393.15	383.15	373.15	363.15	358.15	423.15	413.15
P/bar	95.4	78.1	60.0	41.2	21.9	14.7	13.3	109.0	93.9
	LLE	LLE	LLE	LLE	LLE	VLE	VLSE	LLE	LLE

w_B	0.0992	0.0992	0.0992	0.0992	0.0992	0.0992	0.0992
T/K	403.15	393.15	383.15	373.15	368.15	363.15	358.15
P/bar	77.0	57.4	38.1	21.3	16.0	14.5	13.1
	LLE	LLE	LLE	LLE	VLE	VLE	VLSE

Polymer (B):	**polypropylene**		**2001DAR**
Characterization:	$M_n/kg.mol^{-1} = 101$, $M_w/kg.mol^{-1} = 476$,		
	87 % isotactic, Polibrasil Resinas S.A.		
Solvent (A):	**n-butane**	C_4H_{10}	**106-97-8**

Type of data: cloud points

w_B	0.00995	0.00995	0.00995	0.00995	0.00995	0.00995	0.00995	0.04975	0.04975
T/K	423.15	413.15	403.15	393.15	383.15	373.15	363.15	423.15	413.15
P/bar	111.9	95.9	80.2	60.0	40.6	22.7	15.4	113.2	98.0
	LLE	LLE	LLE	LLE	LLE	LLE	VLE	LLE	LLE

w_B	0.04975	0.04975	0.04975	0.04975	0.07318	0.07318	0.07318	0.07318	0.07318
T/K	403.15	383.15	373.15	363.15	423.15	403.15	393.15	383.15	373.15
P/bar	82.3	42.1	22.7	15.7	111.3	79.8	61.2	41.2	21.8
	LLE	LLE	LLE	VLE	LLE	LLE	LLE	LLE	LLE

w_B	0.07318
T/K	363.15
P/bar	15.7
	VLE

Polymer (B):   **polypropylene**   **2000OLI**
Characterization: M_n/kg.mol^{-1} = 48.9, M_w/kg.mol^{-1} = 245,
                           92 % isotactic, Polibrasil Resinas S.A.
Solvent (A):   **1-butene**   **C$_4$H$_8$**   **106-98-9**

Type of data: cloud points

w_B	0.0030	0.0030	0.0030	0.0030	0.0030	0.0030	0.0030	0.0069	0.0069
T/K	423.15	413.15	403.15	393.15	383.15	373.15	363.15	423.15	413.15
P/bar	113.7	100.7	91.1	73.9	53.1	35.8	15.9	127.3	110.4
	LLE	LLE	LLE	LLE	LLE	LLE	VLSE	LLE	LLE

w_B	0.0069	0.0069	0.0069	0.0069	0.0069	0.0099	0.0099	0.0099	0.0099
T/K	403.15	393.15	383.15	373.15	363.15	423.15	418.15	413.15	408.15
P/bar	96.6	78.2	58.9	38.6	17.6	129.2	120.2	114.5	106.7
	LLE	LLE	LLE	LLE	VLSE	LLE	LLE	LLE	LLE

w_B	0.0099	0.0099	0.0099	0.0099	0.0099	0.0099	0.0099	0.0099	0.0099
T/K	403.15	398.15	393.15	388.15	383.15	378.15	373.15	368.15	363.15
P/bar	99.2	90.2	80.7	71.4	61.4	51.2	41.1	29.8	18.4
	LLE	LLE	LLE	LLE	LLE	LLE	LLE	LLE	LLSE

w_B	0.0099	0.0996	0.0996	0.0996	0.0996	0.0996	0.0996	0.0996	0.0996
T/K	358.15	423.15	418.15	413.15	408.15	403.15	398.15	393.15	383.15
P/bar	14.7	129.1	120.3	112.8	106.1	97.2	88.2	77.5	58.8
	VLSE	LLE	LLE	LLE	LLE	LLE	LLE	LLE	LLE

w_B	0.0996	0.0996	0.0297	0.0297	0.0297	0.0297	0.0297	0.0297	0.0297
T/K	373.15	363.15	423.15	418.15	413.15	408.15	403.15	398.15	393.15
P/bar	38.9	18.9	132.3	124.1	116.2	106.7	98.3	89.4	80.1
	LLE	LLSE	LLE	LLE	LLE	LLE	LLE	LLE	LLE

w_B	0.0297	0.0297	0.0297	0.0297	0.0297	0.0297	0.0496	0.0496	0.0496
T/K	388.15	383.15	378.15	373.15	368.15	363.15	423.15	418.15	413.15
P/bar	69.4	60.8	50.1	40.1	29.0	18.4	132.2	124.7	116.1
	LLE	LLE	LLE	LLE	LLE	LLSE	LLE	LLE	LLE

w_B	0.0496	0.0496	0.0496	0.0496	0.0496	0.0496	0.0496	0.0742	0.0742
T/K	408.15	403.15	398.15	393.15	383.15	373.15	363.15	423.15	418.15
P/bar	107.5	99.8	89.5	80.9	61.1	40.5	18.4	133.1	125.9
	LLE	LLE	LLE	LLE	LLE	LLE	LLSE	LLE	LLE

w_B	0.0742	0.0742	0.0742	0.0742	0.0742	0.0742	0.0742	0.0742	
T/K	413.15	408.15	403.15	398.15	393.15	383.15	373.15	363.15	
P/bar	116.3	107.3	101.2	89.9	82.1	62.1	42.6	20.3	
	LLE	LLE	LLE	LLE	LLE	LLE	LLE	LLSE	

Polymer (B):   **polypropylene**   **2001DAR**
Characterization: M_n/kg.mol^{-1} = 77, M_w/kg.mol^{-1} = 200,
                           98 % isotactic, metallocene product
Solvent (A):   **1-butene**   **C$_4$H$_8$**   **106-98-9**

Type of data: cloud points

continued

continued

w_B	0.00285	0.00285	0.00285	0.00285	0.00285	0.00285	0.00285	0.00285	0.00285
T/K	423.15	413.15	403.15	393.15	383.15	373.15	368.15	363.15	358.15
P/bar	112.5	97.3	82.1	63.8	46.0	27.7	19.6	16.4	14.6
	LLE	LLE	LLE	LLE	LLE	LLE	LLE	VLE	VLE

w_B	0.00285	0.00285	0.00992	0.00992	0.00992	0.00992	0.00992	0.00992	0.00992
T/K	353.15	348.15	423.15	413.15	403.15	393.15	383.15	373.15	368.15
P/bar	13.2	11.9	127.3	111.3	93.9	75.7	56.4	36.7	26.5
	VLE	VLSE	LLE	LLE	LLE	LLE	LLE	LLE	LLE

w_B	0.00992	0.00992	0.00992	0.00992	0.04932	0.04932	0.04932	0.04932	0.04932
T/K	363.15	358.15	353.15	348.15	423.15	408.15	403.15	393.15	383.15
P/bar	17.7	15.7	13.8	12.1	129.2	104.4	95.7	78.1	59.1
	VLE	VLE	VLE	VLSE	LLE	LLE	LLE	LLE	LLE

w_B	0.04932	0.04932	0.04932	0.04932	0.04932	0.07434	0.07434	0.07434	0.07434
T/K	373.15	363.15	358.15	353.15	348.15	423.15	413.15	403.15	393.15
P/bar	37.8	19.8	16.9	15.5	14.4	124.5	108.0	90.6	73.1
	LLE	LLE	VLE	VLE	VLE	LLE	LLE	LLE	LLE

w_B	0.07434	0.07434	0.07434	0.07434	0.07434	0.07434	0.07434	0.09885	0.09885
T/K	383.15	373.15	368.15	363.15	358.15	353.15	348.15	423.15	413.15
P/bar	54.3	33.9	24.9	18.1	16.6	15.2	14.5	122.1	106.6
	LLE	LLE	LLE	VLE	VLE	VLE	VLSE	LLE	LLE

w_B	0.09885	0.09885	0.09885	0.09885	0.09885	0.09885	0.09885	0.09885	0.09885
T/K	403.15	393.15	383.15	373.15	368.15	363.15	358.15	353.15	348.15
P/bar	89.2	72.1	52.8	32.3	23.7	16.6	15.4	13.9	12.6
	LLE	LLE	LLE	LLE	LLE	VLE	VLE	VLE	VLSE

Polymer (B):	**polypropylene**		**2001DAR**
Characterization:	M_n/kg.mol^{-1} = 101, M_w/kg.mol^{-1} = 476,		
	87 % isotactic, Polibrasil Resinas S.A.		
Solvent (A):	**1-butene**	C_4H_8	**106-98-9**

Type of data: cloud points

w_B	0.00983	0.00983	0.00983	0.00983	0.00983	0.00983	0.00983	0.04985	0.04985
T/K	423.15	413.15	403.15	393.15	383.15	373.15	363.15	423.15	413.15
P/bar	131.0	116.7	98.6	80.5	61.6	41.9	20.3	132.3	116.9
	LLE	LLE	LLE	LLE	LLE	LLE	LLE	LLE	LLE

w_B	0.04985	0.04985	0.04985	0.04985	0.04985	0.07432	0.07432	0.07432	0.07432
T/K	403.15	393.15	383.15	373.15	363.15	423.15	413.15	403.15	393.15
P/bar	100.9	85.7	63.7	43.2	21.0	132.4	116.0	99.1	82.5
	LLE	LLE	LLE	LLE	LLE	LLE	LLE	LLE	LLE

w_B	0.07432	0.07432	0.07432
T/K	383.15	373.15	363.15
P/bar	62.8	41.9	21.0
	LLE	LLE	LLE

Polymer (B):   **polypropylene**   **1998HAN**
Characterization: M_n/kg.mol^{-1} = 45.5, M_w/kg.mol^{-1} = 91, isotactic,
                      T_m/K = 434.2, synthesized with metallocene catalyst
Solvent (A):   **propane**   **C$_3$H$_8$**   **74-98-6**

Type of data: cloud points

w_B	0.05	0.05	0.05
T/K	422.65	447.35	471.55
P/bar	436.6	444.9	456.0

Polymer (B):   **polypropylene**   **2000OLI**
Characterization: M_n/kg.mol^{-1} = 48.9, M_w/kg.mol^{-1} = 245,
                      92 % isotactic, Polibrasil Resinas S.A.
Solvent (A):   **propane**   **C$_3$H$_8$**   **74-98-6**

Type of data: cloud points

w_B	0.0030	0.0030	0.0030	0.0030	0.0030	0.0030	0.0030	0.0030	0.0100
T/K	403.15	398.15	393.15	388.15	383.15	378.15	373.15	368.15	403.15
P/bar	252.4	243.0	234.1	224.9	216.7	207.9	199.7	190.8	263.7
	LLE	LLE	LLE	LLE	LLE	LLE	LLE	LLSE	LLE

w_B	0.0100	0.0100	0.0100	0.0100	0.0100	0.0100	0.0100	0.0498	0.0498
T/K	398.15	393.15	388.15	383.15	378.15	373.15	368.15	403.15	398.15
P/bar	255.3	246.3	237.2	229.1	220.1	211.0	202.3	262.7	254.2
	LLE	LLE	LLE	LLE	LLE	LLE	LLSE	LLE	LLE

w_B	0.0498	0.0498	0.0498	0.0498	0.0498	0.0498	0.0748	0.0748	0.0748
T/K	393.15	388.15	383.15	378.15	373.15	368.15	403.15	398.15	393.15
P/bar	246.0	236.9	227.9	219.3	210.5	202.4	263.4	254.6	247.1
	LLE	LLE	LLE	LLE	LLE	LLSE	LLE	LLE	LLE

w_B	0.0748	0.0748	0.0748	0.0748	0.0748	0.1000	0.1000	0.1000	0.1000
T/K	388.15	383.15	378.15	373.15	368.15	403.15	398.15	393.15	388.15
P/bar	237.1	228.9	220.2	211.0	202.5	259.4	251.1	242.7	234.5
	LLE	LLE	LLE	LLE	LLSE	LLE	LLE	LLE	LLE

w_B	0.1000	0.1000	0.1000	0.1000
T/K	383.15	378.15	373.15	368.15
P/bar	225.0	216.7	207.1	198.8
	LLE	LLE	LLE	LLSE

Polymer (B):   **polypropylene**   **1995PRA**
Characterization: isotactic, Scientific Polymer Products, Inc., Ontario, NY
Solvent (A):   **propane**   **C$_3$H$_8$**   **74-98-6**

Type of data: cloud points

w_B	0.04	0.075	0.15	0.15	0.22
T/K	398.15	414.15	402.15	403.15	403.15
P/bar	297.2	297.2	295.1	291.0	259.1

Polymer (B):	polypropylene							**2000OLI**

Characterization: $M_n/\text{kg.mol}^{-1} = 48.9$, $M_w/\text{kg.mol}^{-1} = 245$,
92 % isotactic, Polibrasil Resinas S.A.

Solvent (A):	propene	C_3H_6						**115-07-1**

Type of data: cloud points

w_B	0.0030	0.0030	0.0030	0.0030	0.0030	0.0099	0.0099	0.0099	0.0099
T/K	388.15	383.15	378.15	373.15	368.15	383.15	378.15	373.15	368.15
P/bar	253.7	245.0	236.4	228.5	220.7	261.0	253.1	242.9	234.4
	LLE	LLE	LLE	LLE	LLSE	LLE	LLE	LLE	LLSE
w_B	0.0297	0.0297	0.0297	0.0297	0.0497	0.0497	0.0497	0.0497	0.0740
T/K	383.15	378.15	373.15	368.15	383.15	378.15	373.15	368.15	383.15
P/bar	267.2	258.7	249.4	240.7	266.4	257.6	248.5	240.6	259.1
	LLE	LLE	LLE	LLSE	LLE	LLE	LLE	LLSE	LLE
w_B	0.0740	0.0740	0.0740	0.0984	0.0984	0.0984	0.0984		
T/K	378.15	373.15	368.15	383.15	378.15	373.15	368.15		
P/bar	250.7	242.3	232.6	257.5	248.3	240.1	231.2		
	LLE	LLE	VLSE	LLE	LLE	LLE	VLSE		

Polymer (B):	polypropylene							**2001DAR**

Characterization: $M_n/\text{kg.mol}^{-1} = 77$, $M_w/\text{kg.mol}^{-1} = 200$,
98 % isotactic, metallocene product

Solvent (A):	propene	C_3H_6						**115-07-1**

Type of data: cloud points

w_B	0.00294	0.00294	0.00294	0.00294	0.00294	0.00294	0.00975	0.00975	0.00975
T/K	383.15	378.15	373.15	368.15	363.15	358.15	383.15	378.15	373.15
P/bar	241.2	232.8	223.5	215.2	207.0	198.5	249.7	241.1	233.5
	LLE	LLE	LLE	LLE	LLE	LLE	LLE	LLE	LLE
w_B	0.00975	0.00975	0.00975	0.03012	0.03012	0.03012	0.03012	0.03012	0.03012
T/K	368.15	363.15	358.15	383.15	378.15	373.15	368.15	363.15	358.15
P/bar	224.2	215.6	207.7	257.5	248.5	240.0	229.8	220.2	212.3
	LLE	LLE	LLE	LLE	LLE	LLE	LLE	LLE	LLE
w_B	0.04888	0.04888	0.04888	0.04888	0.04888	0.04888	0.07329	0.07329	0.07329
T/K	383.15	378.15	373.15	368.15	363.15	358.15	383.15	378.15	373.15
P/bar	257.9	248.6	240.2	230.9	221.0	212.8	249.4	241.0	232.6
	LLE	LLE	LLE	LLE	LLE	LLE	LLE	LLE	LLE
w_B	0.07329	0.07329	0.07329	0.09786	0.09786	0.09786	0.09786	0.09786	0.09786
T/K	368.15	363.15	358.15	383.15	378.15	373.15	368.15	363.15	358.15
P/bar	225.0	216.3	206.7	245.4	237.5	229.2	220.0	211.2	201.9
	LLE	LLE	LLE	LLE	LLE	LLE	LLE	LLE	LLE

Polymer (B):	**poly(propylene oxide)**						**1998ONE**

Characterization: M_n/kg.mol^{-1} = 0.4, Polysciences, Inc., Warrington, PA

Solvent (A):   **carbon dioxide**   **CO$_2$**   124-38-9

Type of data: cloud points

T/K = 303.15

w_B	0.014	0.017	0.024	0.031	0.037	0.047	0.082	0.095	
P/MPa	7.10	7.24	7.52	7.93	8.20	8.76	8.89	9.03	

T/K = 313.15

w_B	0.0035	0.007	0.010	0.014	0.017	0.024	0.031	0.037	0.047
P/MPa	8.96	9.38	9.65	9.86	10.14	10.62	11.10	11.38	12.00

w_B	0.082	0.095
P/MPa	12.27	12.13

T/K = 323.15

w_B	0.0035	0.007	0.010	0.014	0.017	0.024	0.031	0.037	0.047
P/MPa	10.76	11.38	11.72	12.27	12.69	13.24	13.72	14.20	14.82

w_B	0.095
P/MPa	15.51

Polymer (B):	**poly(propylene oxide)**						**1998ONE**

Characterization: M_n/kg.mol^{-1} = 1.025, Polysciences, Inc., Warrington, PA

Solvent (A):   **carbon dioxide**   **CO$_2$**   124-38-9

Type of data: cloud points

T/K = 303.15

w_B	0.005	0.016	0.027	0.037	0.047	0.057	0.067	0.086	0.104
P/MPa	8.07	9.86	11.31	12.41	13.58	14.34	15.31	17.93	20.41

T/K = 313.15

w_B	0.005	0.016	0.027	0.037	0.047	0.057	0.067	0.086	0.104
P/MPa	11.03	13.10	14.75	15.79	16.89	17.79	18.82	20.96	23.30

T/K = 323.15

w_B	0.005	0.016	0.027	0.037	0.047	0.057	0.067	0.086	0.104
P/MPa	14.13	16.20	17.79	18.96	20.13	20.89	21.99	23.72 2	

Polymer (B):	**poly(propylene oxide)**						**1998ONE**

Characterization: M_n/kg.mol^{-1} = 2.0, Polysciences, Inc., Warrington, PA

Solvent (A):   **carbon dioxide**   **CO$_2$**   124-38-9

Type of data: cloud points

T/K	298.15	298.15	298.15	308.15	308.15	308.15	323.15	323.15	323.15
w_B	0.00219	0.00495	0.01040	0.00219	0.00495	0.01040	0.00219	0.00495	0.01040
P/MPa	13.51	17.79	21.37	17.24	21.03	24.55	21.58	24.55	28.61

continued

continued

T/K	333.15	333.15	333.15
w_B	0.00219	0.00495	0.01040
P/MPa	23.58	28.13	31.37

Polymer (B):	**poly(propylene oxide)-b-poly(ethylene oxide)-b-poly(propylene oxide) triblock copolymer**	**1998ONE**
Characterization:	M_n/kg.mol^{-1} = 3.125 (= 1.25 + 0.625 + 1.25), 24.7 mol% ethylene oxide, (21.6/14.2/21.6), Pluronic 25R2, hydroxyl terminated	
Solvent (A):	**carbon dioxide CO$_2$**	**124-38-9**

Type of data: cloud points

w_B	0.00128	0.00128	0.00128	0.00128
T/K	298.15	308.15	318.15	328.15
P/MPa	24.06	28.96	30.27	33.03

Polymer (B):	**poly(propylene oxide)-b-poly(ethylene oxide)-b-poly(propylene oxide) triblock copolymer**	**1998ONE**
Characterization:	M_n/kg.mol^{-1} = 2.125 (= 0.85 + 0.425 + 0.85), 24.8 mol% ethylene oxide, (14.7/9.7/14.7), Pluronic 17R2, hydroxyl terminated	
Solvent (A):	**carbon dioxide CO$_2$**	**124-38-9**

Type of data: cloud points

$T/K = 298.15$

w_B	0.00202	0.00478	0.00753	0.01030	0.01570
P/MPa	15.93	22.89	25.92	28.27	33.72

$T/K = 303.15$

w_B	0.00202	0.00478	0.00753	0.01030	0.01570
P/MPa	17.79	24.48	27.92	29.92	35.37

$T/K = 313.15$

w_B	0.00202	0.00478	0.00753	0.01030	0.01570
P/MPa	21.65	26.89	30.06	32.75	38.13

$T/K = 323.15$

w_B	0.00202	0.00478	0.00753	0.01030	0.01570
P/MPa	25.51	29.85	33.72	35.78	40.27

$T/K = 333.15$

w_B	0.00202	0.00478	0.00753	0.01030
P/MPa	27.23	33.16	36.06	37.92

$T/K = 338.15$

w_B	0.00202	0.00478	0.00753	0.01030
P/MPa	28.96	34.34	37.78	39.58

Polymer (B): poly(propylene oxide)-b-poly(ethylene oxide)-   **1998ONE**
                 b-poly(propylene oxide) triblock copolymer
                 trifluoroacetate end-capped
Characterization: M_n/kg.mol^{-1} = 2.125 (= 0.85 + 0.425 + 0.85), 24.8 mol% ethylene
                     oxide, (14.7/9.7/14.7), Pluronic 17R2, trifluoroacetate end-capped
Solvent (A): **carbon dioxide**   **CO$_2$**   **124-38-9**

Type of data: cloud points

w_B	0.00346	0.00346	0.00346	0.00346	0.00548	0.00548	0.00548	0.00548
T/K	298.15	308.15	323.15	338.15	298.15	308.15	323.15	338.15
P/MPa	11.65	15.17	20.62	24.82	13.51	13.51	17.31	27.17

Polymer (B): poly(propylene oxide)-b-poly(ethylene oxide)-   **1998ONE**
                 b-poly(propylene oxide) triblock copolymer
Characterization: M_n/kg.mol^{-1} = 2.83 (= 0.85 + 1.13 + 0.85), 46.7 mol% ethylene oxide,
                     (14.7/25.8/14.7), Pluronic 17R4, hydroxyl terminated
Solvent (A): **carbon dioxide**   **CO$_2$**   **124-38-9**

Type of data: cloud points

w_B	0.00134	0.00134	0.00134	0.00134
T/K	298.15	303.15	318.15	333.15
P/MPa	26.06	29.92	33.30	37.44

Polymer (B): poly(propylene oxide)-b-poly(ethylene oxide)-   **1998ONE**
                 b-poly(propylene oxide) triblock copolymer
                 trifluoroacetate end-capped
Characterization: M_n/kg.mol^{-1} = 2.83 (= 0.85 + 1.13 + 0.85), 46.7 mol% ethylene oxide,
                     (14.7/25.8/14.7), Pluronic 17R4, trifluoroacetate end-capped
Solvent (A): **carbon dioxide**   **CO$_2$**   **124-38-9**

Type of data: cloud points

w_B	0.00127	0.00127	0.00127	0.00127	0.00127	0.00402	0.00402	0.00402	0.00402
T/K	298.15	308.15	318.15	328.15	338.15	298.15	308.15	318.15	328.15
P/MPa	16.06	19.24	23.24	26.06	28.13	18.48	22.89	26.54	29.85

w_B	0.00402	0.00961	0.00961	0.00961	0.00961
T/K	338.15	298.15	308.15	318.15	328.15
P/MPa	33.51	21.03	24.96	28.27	31.72

Polymer (B): poly(propylene oxide)-b-poly(ethylene oxide)-   **1998ONE**
                 b-poly(propylene oxide) triblock copolymer
Characterization: M_n/kg.mol^{-1} = 2.0 (= 0.5 + 1.0 + 0.5), 56.9 mol% ethylene oxide,
                     (8.6/22.7/8.6), Pluronic 10R5, hydroxyl terminated
Solvent (A): **carbon dioxide**   **CO$_2$**   **124-38-9**

continued

continued

Type of data: cloud points

w_B	0.00167	0.00167	0.00167	0.00167	0.00167	0.00444	0.00444	0.00444	0.00444
T/K	298.15	308.15	313.15	323.15	338.15	298.15	308.15	313.15	323.15
P/MPa	19.99	23.79	25.79	27.99	31.92	26.89	31.16	32.41	35.71

Polymer (B):	**poly(propylene oxide)-b-poly(ethylene oxide)-**	**1998ONE**
	b-poly(propylene oxide) triblock copolymer	
	trifluoroacetate end-capped	

Characterization: $M_n/kg.mol^{-1} = 2.0 (= 0.5 + 1.0 + 0.5)$, 56.9 mol% ethylene oxide, (8.6/22.7/8.6), Pluronic 10R5, trifluoroacetate end-capped in the laboratory

Solvent (A):	**carbon dioxide**	**CO_2**	**124-38-9**

Type of data: cloud points

w_B	0.00188	0.00426	0.01070	0.01720	0.00188	0.00426	0.01070	0.01720	0.00426
T/K	298.15	298.15	298.15	298.15	308.15	308.15	308.15	308.15	313.15
P/MPa	11.86	18.96	23.79	26.82	16.55	22.55	26.82	29.99	24.96

w_B	0.01070	0.01720	0.00188	0.00426	0.01070	0.01720	0.00426	0.01070	0.00426
T/K	313.15	313.15	323.15	323.15	323.15	323.15	333.15	333.15	338.15
P/MPa	29.03	32.34	19.58	26.96	31.92	35.51	27.85	34.89	31.30

Polymer (B):	**polystyrene**	**1981ALB**
Characterization:	$M_n/kg.mol^{-1} = 15.6$, $M_w/kg.mol^{-1} = 35.4$	
Solvent (A):	**sulfur dioxide** **SO_2**	**7446-09-5**

Type of data: coexistence data

Comments: PS/SO$_2$ weight ratio w_B/w_A in the total system was about 1/5. The equilibration time was 12 hours.

$T/K = 293.15$ $P_A/MPa = 0.326$

	upper phase			lower phase		
w_B	$M_n/$ kg/mol	$M_w/$ kg/mol	w_B	$M_n/$ kg/mol	$M_w/$ kg/mol	
0.315	16.6	35.8	0.0037	4.1	10.0	

Polymer (B):	**polystyrene**	**1981ALB**
Characterization:	$M_n/kg.mol^{-1} = 66.7$, $M_w/kg.mol^{-1} = 196$	
Solvent (A):	**sulfur dioxide** **SO_2**	**7446-09-5**

continued

continued

Type of data: coexistence data

Comments: PS/SO_2 weight ratio w_B/w_A in the total system was about 1/5. The equilibration time was 12 hours.

$T/K = 293.15$ $P_A/MPa = 0.326$

	upper phase			lower phase		
w_B	$M_n/$ kg/mol	$M_w/$ kg/mol	w_B	$M_n/$ kg/mol	$M_w/$ kg/mol	
0.343	63.8	194	0.0037	4.9	17.5	

Polymer (B):   **polystyrene**   **1981ALB**
Characterization: $M_n/\text{kg.mol}^{-1} = 136$, $M_w/\text{kg.mol}^{-1} = 402$
Solvent (A):   **sulfur dioxide**   **SO_2**   **7446-09-5**

Type of data: coexistence data

Comments: PS/SO_2 weight ratio w_B/w_A in the total system was about 1/5. The equilibration time was 12 hours.

$T/K = 293.15$ $P_A/MPa = 0.326$

	upper phase			lower phase		
w_B	$M_n/$ kg/mol	$M_w/$ kg/mol	w_B	$M_n/$ kg/mol	$M_w/$ kg/mol	
0.349	139	401	0.0023	4.01	9.81	

Polymer (B):   **poly(1,1,2,2-tetrahydroperfluorodecyl acrylate)**   **2002BLA**
Characterization: $M_n/\text{kg.mol}^{-1} = 86.1$, $M_w/\text{kg.mol}^{-1} = 254$
Solvent (A):   **carbon dioxide**   **CO_2**   **124-38-9**

Type of data: cloud points

$T/K = 298.15$

w_B	0.0591	0.0392	0.0307	0.0204	0.0050	0.0050	0.003	0.002	0.001
P/bar	114.05	115.80	117.93	116.16	110.82	110.03	110.05	110.14	109.22

w_B	0.001	0.001	0.001	0.0005	0.0005	0.0001	0.0001	0.0001	
P/bar	108.88	108.81	109.85	108.50	106.30	82.99	90.63	97.08	

continued

continued

T/K = 323.15

w_B	0.0591	0.0392	0.0307	0.0204	0.0050	0.0050	0.003	0.002	0.001
P/bar	201.62	203.74	204.16	202.72	197.46	195.76	198.02	196.23	195.27

w_B	0.001	0.001	0.001	0.0005	0.0001	0.0001
P/bar	197.10	195.57	195.78	193.55	176.55	179.58

T/K = 343.15

w_B	0.0591	0.0392	0.0307	0.0204	0.0050	0.0050	0.003	0.002	0.001
P/bar	259.58	262.09	262.09	260.84	257.62	255.15	255.89	254.01	254.02

w_B	0.001	0.001	0.0005	0.0001	0.0001
P/bar	253.73	253.84	250.34	237.60	237.85

T/K = 373.15

w_B	0.0591	0.0392	0.0307	0.0204	0.0050	0.003	0.002	0.001	0.001
P/bar	330.83	332.90	332.94	332.62	327.78	326.96	324.97	325.03	324.39

w_B	0.0005	0.0001	0.0001
P/bar	322.57	308.28	303.97

Type of data: liquid-liquid-vapor three phase equilibrium data

T/K = 298.15

w_B	0.01	0.05	0.1	0.2
P/bar	63.95	64.39	64.25	64.39

Polymer (B):	**poly[P-tris(trifluoroethoxy)-N-trimethylsilyl]**	**1998ONE**
	phosphazene	
Characterization:	M_n/kg.mol^{-1} = 62, M_w/kg.mol^{-1} = 96	
Solvent (A):	**carbon dioxide CO$_2$**	**124-38-9**

Type of data: cloud points

w_B	0.00516	0.0103	0.0187	0.0405	0.00516	0.0103	0.0187	0.0405
T/K	298.15	298.15	298.15	298.15	308.15	308.15	308.15	308.15
P/MPa	13.10	13.82	13.96	14.10	17.41	18.00	18.20	18.42

w_B	0.00516	0.01030	0.01870	0.04050	0.00516	0.01030	0.01870	0.04050
T/K	323.15	323.15	323.15	323.15	338.15	338.15	338.15	338.15
P/MPa	22.82	23.61	23.82	24.17	27.72	28.51	28.99	29.23

Polymer (B):	**poly(vinyl ethyl ether)**	**1992KIA, 1994KIA**
Characterization:	M_w/kg.mol^{-1} = 3.8, Scientific Polymer Products, Ontario, NY	
Solvent (A):	**carbon dioxide CO$_2$**	**124-38-9**

Type of data: cloud points

continued

continued

w_B	0.850	0.850	0.850	0.850	0.850	0.850	0.850	0.850	0.850
T/K	304.05	308.35	310.55	313.15	313.45	317.05	323.15	328.15	333.15
P/bar	37.34	39.01	40.84	43.34	45.02	46.17	52.63	56.64	60.52

w_B	0.850	0.850	0.800	0.800	0.800	0.800	0.800	0.800	0.800
T/K	338.05	343.05	303.75	308.15	313.15	318.15	323.15	328.15	333.15
P/bar	65.21	70.26	45.51	48.73	53.12	58.21	63.54	68.26	74.89

w_B	0.800	0.800	0.750	0.750	0.750	0.750	0.750	0.750	0.750
T/K	338.15	343.15	300.75	303.35	308.15	313.15	317.35	323.15	327.85
P/bar	81.23	87.56	49.09	50.48	54.74	61.23	68.66	75.81	81.89

w_B	0.750	0.750	0.750	0.700	0.700	0.700	0.700	0.700	0.700
T/K	332.45	337.45	342.95	302.95	308.15	313.35	318.25	323.15	328.15
P/bar	88.65	95.94	104.77	58.12	62.58	69.97	79.39	88.78	98.90

w_B	0.700	0.700	0.700	0.624	0.624	0.624	0.624	0.624	0.624
T/K	333.15	340.25	342.65	309.05	310.25	311.25	312.75	313.65	314.45
P/bar	110.45	126.51	131.86	72.56	75.10	76.40	79.23	81.79	83.70

w_B	0.624
T/K	315.95
P/bar	87.40

Polymer (B):	**tetraethylene glycol monododecyl ether**	**1996HAR**
Characterization:	$M/kg.mol^{-1} = 0.362$	
Solvent (A):	**carbon dioxide** CO_2	**124-38-9**

Type of data: cloud points

w_B	0.010		T/K	318.15		P/bar	149.6

Polymer (B):	**tetraethylene glycol monododecyl ether**	**2001LIU**
Characterization:	$M/kg.mol^{-1} = 0.362$, surfactant $C_{12}E_4$,	
	Aldrich Chem. Co., Inc., Milwaukee, WI	
Solvent (A):	**carbon dioxide** CO_2	**124-38-9**

Type of data: cloud points

T/K 313.15

w_B	0.00645	0.01127	0.01589	0.02017	0.02480	0.02920
P/MPa	10.20	13.02	15.15	16.87	18.35	19.65
$\rho/(g/cm^3)$	0.642	0.736	0.775	0.798	0.814	0.828

T/K 323.15

w_B	0.00392	0.00729	0.01166	0.01780	0.02107
P/MPa	11.25	13.27	15.86	18.18	19.37
$\rho/(g/cm^3)$	0.521	0.634	0.703	0.748	0.765

4.2. Table of binary systems where data were published only in graphical form as phase diagrams or related figures

Polymer (B)	Solvent (A)	Ref.
Cellulose acetate/cellulose butyrate copolymer	carbon dioxide	1997DIN
Dimethylsiloxane/(3-acetoxypropyl)methylsiloxane copolymer	carbon dioxide	2003KIL
Dimethylsiloxane/(4-acetylbutyl)methylsiloxane copolymer	carbon dioxide	2003KIL
Dimethylsiloxane/(4,4-dimethylpentyl)methylsiloxane copolymer	carbon dioxide	2003KIL
Dimethylsiloxane/(3-ethoxypropyl)methylsiloxane copolymer	carbon dioxide	2003KIL
Dimethylsiloxane/glycidyloxypropylsiloxane/ tris(trimethylsiloxy)silylpropylsiloxane terpolymer	carbon dioxide	1993HOE
Dimethylsiloxane/hydromethylsiloxane copolymer	carbon dioxide carbon dioxide	1993HOE 2003KIL
Dimethylsiloxane/(3-methoxycarbonylpropyl)methylsiloxane copolymer	carbon dioxide	2003KIL
Dimethylsiloxane/(3-trimethylsilylpropyl)methylsiloxane copolymer	carbon dioxide	2003KIL

Polymer (B)	Solvent (A)	Ref.
Ethylcellulose		
	carbon dioxide	2004LI2
	1-chloro-1,1-difluoroethane	2004LI2
	chlorodifluoromethane	2004LI2
	1,1-difluoroethane	2004LI2
	difluoromethane	2004LI2
	dimethylether	2004LI2
Ethylene/acrylic acid copolymer		
	n-butane	1994LE1
	n-butane	1996LE3
	1-butene	1993HA2
	1-butene	1994LE1
	1-butene	1994LE2
	1-butene	1996LE3
	dimethyl ether	1994LE1
	dimethyl ether	1996LE1
	dimethyl ether	1996LE3
	ethene	1992LUF
	ethene	1993HA2
	ethene	1997LEE
	propane	1994LE1
	propane	1996LE1
	propene	1993HA2
	propene	1994LE1
	propene	1996LE1
Ethylene/1-butene copolymer		
	1-butene	1999CHE
	dimethyl ether	2000BY3
	ethene	1999CHE
	propane	1995CHE
	propane	1998HAN
Ethylene/butyl acrylate copolymer		
	ethene	1996BYU
Ethylene/1-hexene copolymer		
	2-methylpropane	1999PAN
	propane	1998HAN
	propane	1999PAN

Polymer (B)	Solvent (A)	Ref.
Ethylene/methacrylic acid copolymer		
	n-butane	1997LEE
	1-butene	1997LEE
	dimethyl ether	1997LEE
	ethene	1997LEE
	propane	1997LEE
Ethylene/methyl acrylate copolymer		
	n-butane	1993PRA
	n-butane	1994LOS
	n-butane	1994LE2
	n-butane	1996LE1
	n-butane	1996LE3
	1-butene	1993PRA
	1-butene	1996LE1
	carbon dioxide	1999LO2
	chlorodifluoromethane	1991MEI
	chlorodifluoromethane	1993HA1
	chlorodifluoromethane	1993PRA
	dimethyl ether	1996LE3
	ethane	1992HAS
	ethane	1993HA2
	ethane	1994LOS
	ethane	1996LE3
	ethene	1992HAS
	ethene	1993HA2
	ethene	1996BYU
	ethene	1996LE3
	n-hexane	1994LOS
	methanol	1996LE3
	propane	1991MEI
	propane	1992HAS
	propane	1993HA1
	propane	1993HA2
	propane	1994LOS
	propene	1992HAS
	propene	1993HA2
	propene	1996LE3
Ethylene/1-octene copolymer		
	propane	1997WH2
	propane	1998HAN

Polymer (B)	Solvent (A)	Ref.
Ethylene/propylene copolymer		
	1-butene	1993CHE
	ethene	1997HAN
	ethene	2001KEM
	propane	1997WH2
	propane	1998HAN
	propene	1993CHE
	propene	1997HAN
Ethylene/propylene/diene terpolymer		
	ethene	2004PFO
Ethylene/vinyl acetate copolymer		
	ethene	1982WOH
	ethene	1996FOL
	ethene	2000KIN
Ethylene glycol/propylene glycol copolymer		
	carbon dioxide	2002DRO
Methyl methacrylate/2-hydroxyethyl methacrylate copolymer		
	carbon dioxide	1997LEP
Perfluoropolyether diamide		
	carbon dioxide	2001CHE
Polybutadiene, fluorinated		
	carbon dioxide	2002MC1
Poly(1-butene)		
	propane	1997WH2
Poly(butyl acrylate)		
	carbon dioxide	1996RIN
	ethene	1996BYU
	ethene	1999LO2
Poly(butyl methacrylate)		
	carbon dioxide	1996RIN
	carbon dioxide	1999LO2

Polymer (B)	Solvent (A)	Ref.
Poly(ε-caprolactone)		
	chlorodifluoromethane	1994LEL
Polycarbonate bisphenol-A		
	carbon dioxide	1998CHA
Poly(decyl methacrylate)		
	carbon dioxide	2002MC2
Poly(1,1-dihydroperfluorooctyl acrylate)		
	carbon dioxide	1998LUN
Poly(1,1-dihydroperfluorooctyl acrylate)-b-polystyrene diblock copolymer		
	carbon dioxide	2002TAY
Poly(1,1-dihydroperfluorooctyl acrylate)-b-poly(vinyl acetate) diblock copolymer		
	carbon dioxide	2002TAY
Poly(1,1-dihydroperfluorooctyl methacrylate)		
	carbon dioxide	2004DIC
Poly(dimethylsiloxane)		
	carbon dioxide	1995XI1
	carbon dioxide	1998CHA
	carbon dioxide	2000TEP
	carbon dioxide	2003SHE
Poly(ethyl acrylate)		
	carbon dioxide	1996RIN
	carbon dioxide	1999LO2
	ethene	1999LO2
Polyethylene		
	n-butane	1963EHR
	n-butane	1989KIR
	n-butane	1994KI2
	n-butane	1994LE1
	n-butane	1994XI2
	n-butane	1995HAS
	n-butane	1995XI2

Polymer (B)	Solvent (A)	Ref.
Polyethylene (*continued*)		
	1-butene	1992WOH
	1-butene	1994LE1
	1-butene	1995HAS
	carbon dioxide	1997DIN
	dimethyl ether	1994LE1
	dimethyl ether	1994LE2
	dimethyl ether	1995HAS
	ethane	1963EHR
	ethane	1992HAS
	ethane	1993HA2
	ethene	1965CE1
	ethene	1965CE2
	ethene	1965EHR
	ethene	1966KON
	ethene	1975LIN
	ethene	1976LUF
	ethene	1976STE
	ethene	1977LIN
	ethene	1979KOB
	ethene	1980RAE
	ethene	1981SP2
	ethene	1981WO1
	ethene	1981WO2
	ethene	1983LOO
	ethene	1983SPA
	ethene	1984FON
	ethene	1984KOB
	ethene	1992HAS
	ethene	1992WOH
	ethene	1993HA2
	ethene	1994CHE
	ethene	1995LOO
	ethene	1996BYU
	ethene	1998HEU
	fluorotrichloromethane	1988WAL
	4-methyl-1-pentene	1992WOH
	n-pentane	1963EHR
	n-pentane	1989KIR
	n-pentane	1994KI2
	n-pentane	1994XI1
	n-pentane	1994XI2
	n-pentane	1995XI2
	n-pentane	1997KIR
	n-pentane	1998ZHU
	n-pentane	1999MAR

Polymer (B)	Solvent (A)	Ref.
Polyethylene (*continued*)		
	propane	1963EHR
	propane	1991MEI
	propane	1991WAT
	propane	1992CO3
	propane	1992HAS
	propane	1993HA1
	propane	1994LE1
	propane	1995HAS
	propane	1995PRA
	propane	1997WH2
	propane	1998HAN
	propene	1992HAS
	propene	1993HA2
	propene	1994LE1
	propene	1995HAS
Poly(ethylene glycol)		
	carbon dioxide	1989DAN
	carbon dioxide	1997WEI
	carbon dioxide	2002DRO
Poly(ethylene glycol) dimethyl ether		
	carbon dioxide	2002DRO
Poly(ethylene glycol) monomethyl ether		
	carbon dioxide	2002DRO
Poly(ethylene glycol) nonylphenyl ether		
	carbon dioxide	1998DIM
Poly(ethylene oxide)-b-poly(propylene oxide)-b-poly(ethylene oxide) triblock copolymer		
	carbon dioxide	2002DRO
Poly(ethylene terephthalate)		
	carbon dioxide	1998CHA
Poly(ethylhexyl acrylate)		
	carbon dioxide	1996RIN
	carbon dioxide	1999LO2
	ethene	1999LO2

Polymer (B)	Solvent (A)	Ref.
Poly(ethyl methacrylate)		
	chlorodifluoromethane	1994LEL
Poly(ethyl vinyl ether)		
	carbon dioxide	2002DRO
Polyglycolide		
	carbon dioxide	1993DEB
Poly(hexyl methacrylate)		
	carbon dioxide	2002MC2
Polyisobutylene		
	n-butane	1983ALI
	carbon dioxide	1998CHA
	ethene	1999CHE
Polyisoprene, fluorinated		
	carbon dioxide	2002MC1
Poly(DL-lactide)		
	carbon dioxide	1993DEB
	carbon dioxide	2003SHE
	chlorodifluoromethane	1994LEL
Poly(methyl acrylate)		
	carbon dioxide	1996RIN
	carbon dioxide	1999LO2
	carbon dioxide	2003SHE
	chlorodifluoromethane	1991MEI
Poly(methyl methacrylate)		
	carbon dioxide	1976SC2
	carbon dioxide	1990BEC
	carbon dioxide	1998CHA
	chlorodifluoromethane	1994LEL
Poly(4-methyl-1-pentene		
	propane	1997WH2
Poly(methylpropenoxyalkylsiloxane)		
	carbon dioxide	2002MC2

Polymer (B)	Solvent (A)	Ref.
Poly(methylpropenoxyperfluoroalkylsiloxane)	carbon dioxide	2002MC2
Poly(octadecyl acrylate)	carbon dioxide	1996RIN
	carbon dioxide	1999LO2
	ethene	1999LO2
Poly(octyl methacrylate)	carbon dioxide	2002MC2
Poly(perfluoropropylene oxide) acid sorbitol ester	carbon dioxide	1997SIN
Poly(propyl acrylate)	carbon dioxide	1996RIN
Polypropylene	n-butane	2001NDI
	1-butene	2001NDI
	n-pentane	1998KI2
	n-pentane	1999MAR
	propane	1995CHE
	propane	1997WH1
	propene	1997WH1
Poly(propylene glycol)	carbon dioxide	1996PAR
	carbon dioxide	2002DRO
	ethane	2000MAR
Poly(propylene glycol) monobutyl ether	carbon dioxide	2002DRO
Poly(propylene glycol) monomethyl ether	carbon dioxide	2002DRO
Poly(propylene oxide)	carbon dioxide	2003SHE

Polymer (B)	Solvent (A)	Ref.
Poly(propylene oxide)-b-poly(ethylene oxide)- b-poly(propylene oxide) triblock copolymer		
	carbon dioxide	2002DRO
Polystyrene		
	n-butane	1983ALI
	n-butane	1988SA1
	n-butane	1988SA2
	carbon dioxide	1976SC2
	carbon dioxide	1987KUM
	carbon dioxide	1998CHA
	ethane	1987KUM
	ethane	1994PRA
	n-pentane	1988SA1
	n-pentane	1988SA2
	propane	1994PRA
Poly(1,1,2,2-tetrahydroperfluorodecyl acrylate)		
	carbon dioxide	1995MAW
	carbon dioxide	2002BLA
Poly(1,1,2,2-tetrahydroperfluorodecyl methacrylate)		
	carbon dioxide	2002MC2
Poly(1,1,2,2-tetrahydroperfluorohexyl methacrylate)		
	carbon dioxide	2002MC2
Poly(1,1,2,2-tetrahydroperfluorooctyl methacrylate)		
	carbon dioxide	2002MC2
Polyurethane and precursors		
	carbon dioxide	1996PAR
	carbon dioxide	1997DIN
Poly(vinyl acetate)		
	carbon dioxide	1988MAS
	carbon dioxide	1996RIN
	carbon dioxide	1999LO2
	carbon dioxide	2003SHE
Poly(vinyl acetate)-b-poly(1,1,2,2-tetrahydro- perfluorooctyl acrylate) diblock copolymer		
	carbon dioxide	2002COL

Polymer (B)	Solvent (A)	Ref.
Poly(vinyl chloride)		
	carbon dioxide	1976SC1
	carbon dioxide	1976SC2
Poly(vinyl ethyl ether)		
	carbon dioxide	1994KIA
Poly(vinyl formate)		
	carbon dioxide	2003SHE
Poly(vinylidene fluoride)		
	carbon dioxide	2000DIN
	1,1,1-chlorodifluoroethane	2000DIN
	chlorodifluoromethane	2000DIN
	difluoromethane	2000DIN
	1,1,1,2-tetrafluoroethane	2000DIN
	trifluoromethane	2000DIN
Tetrafluoroethylene/hexafluoropropylene copolymer		
	carbon dioxide	1995TUM
	carbon dioxide	1996MER
	carbon dioxide	1996RIN
	carbon dioxide	1998MC1
	carbon dioxide	1999MER
	chlorotrifluoromethane	1996MER
	hexafluoroethane	1996MER
	hexafluoropropene	1996MER
	octafluoropropane	1996MER
	sulfur hexafluoride	1996MER
	sulfur hexafluoride	1998MC1
	sulfur hexafluoride	1999MER
	tetrafluoromethane	1996MER
	tetrafluoromethane	1998MC1
	trifluoromethane	1999MER
Tetrafluoroethylene/2,2-bis(trifluoromethyl)-4,5-difluoro-1,3-dioxole copolymer		
	carbon dioxide	2000DIN
Vinylidene fluoride/chlorotrifluoroethylene copolymer		
	carbon dioxide	1997DIN

Polymer (B)	Solvent (A)	Ref.
Vinylidene fluoride/hexafluoropropylene copolymer		
	carbon dioxide	1996RIN
	carbon dioxide	1997DIN
	carbon dioxide	1997MER
	carbon dioxide	2000DIN
	1-chloro-1,1-difluoroethane	2000DIN
	chlorodifluoromethane	2000DIN
	chlorotrifluoromethane	1997MER
	chlorotrifluoromethane	2000DIN
	1,1-difluoroethane	2000DIN
	difluoromethane	2000DIN
	hexafluoropropene	1997MER
	pentafluoroethane	2000DIN
	1,1,1,2-tetrafluoroethane	2000DIN
	trifluoromethane	1997MER
	trifluoromethane	2000DIN

4.3. Cloud-point and/or coexistence curves of quasiternary and/or quasiquaternary solutions

Polymer (B):	ethylene/acrylic acid copolymer					1992WIN
Characterization:	M_n/kg.mol^{-1} = 37.5, M_w/kg.mol^{-1} = 183, 3.0 wt% acrylic acid					
Solvent (A):	ethene	C_2H_4				74-85-1
Solvent (C):	acrylic acid	$C_3H_4O_2$				79-10-7

Type of data: cloud points

T/K = 413.15

w_A	0.858	0.851	0.839	0.822	0.797	0.770
w_B	0.135	0.135	0.135	0.135	0.135	0.135
w_C	0.007	0.014	0.026	0.043	0.068	0.095
P/MPa	164.4	157.1	152.9	148.0	142.7	138.2

T/K = 433.15

w_A	0.858	0.851	0.839	0.822	0.797	0.770
w_B	0.135	0.135	0.135	0.135	0.135	0.135
w_C	0.007	0.014	0.026	0.043	0.068	0.095
P/MPa	150.6	146.2	143.2	139.4	134.1	130.5

T/K = 453.15

w_A	0.858	0.851	0.839	0.822	0.797	0.770
w_B	0.135	0.135	0.135	0.135	0.135	0.135
w_C	0.007	0.014	0.026	0.043	0.068	0.095
P/MPa	141.3	138.2	135.8	132.7	128.4	124.7

T/K = 473.15

w_A	0.858	0.851	0.839	0.822	0.797	0.770
w_B	0.135	0.135	0.135	0.135	0.135	0.135
w_C	0.007	0.014	0.026	0.043	0.068	0.095
P/MPa	133.8	131.7	129.7	127.5	123.7	120.5

Polymer (B):	ethylene/acrylic acid copolymer					1992WIN
Characterization:	M_n/kg.mol^{-1} = 30, M_w/kg.mol^{-1} = 150, 6.0 wt% acrylic acid					
Solvent (A):	ethene	C_2H_4				74-85-1
Solvent (C):	acrylic acid	$C_3H_4O_2$				79-10-7

Type of data: cloud points

T/K = 413.15

w_A	0.860	0.848	0.839	0.822	0.797	0.761
w_B	0.135	0.135	0.135	0.135	0.135	0.135
w_C	0.005	0.017	0.026	0.043	0.068	0.104
P/MPa	195.1	175.5	166.0	157.5	150.4	144.0

continued

continued

$T/K = 433.15$

w_A	0.860	0.851	0.848	0.839	0.822	0.797	0.761
w_B	0.135	0.135	0.135	0.135	0.135	0.135	0.135
w_C	0.005	0.014	0.017	0.026	0.043	0.068	0.104
P/MPa	173.4	163.2	169.4	154.5	147.2	141.2	135.1

$T/K = 453.15$

w_A	0.860	0.851	0.848	0.839	0.822	0.797	0.761
w_B	0.135	0.135	0.135	0.135	0.135	0.135	0.135
w_C	0.005	0.014	0.017	0.026	0.043	0.068	0.104
P/MPa	155.2	149.1	147.5	144.5	139.1	134.1	130.0

$T/K = 473.15$

w_A	0.860	0.851	0.848	0.839	0.822	0.797	0.761
w_B	0.135	0.135	0.135	0.135	0.135	0.135	0.135
w_C	0.005	0.014	0.017	0.026	0.043	0.068	0.104
P/MPa	143.1	140.5	139.2	137.5	133.3	128.5	124.6

Polymer (B): ethylene/acrylic acid copolymer   **1992WIN**
Characterization: $M_n/kg.mol^{-1} = 25$, $M_w/kg.mol^{-1} = 126$, 7.3 wt% acrylic acid
Solvent (A): ethene C_2H_4 **74-85-1**
Solvent (C): acrylic acid $C_3H_4O_2$ **79-10-7**

Type of data: cloud points

$T/K = 413.15$

w_A	0.858	0.851	0.842	0.814	0.791	0.752
w_B	0.135	0.135	0.135	0.135	0.135	0.135
w_C	0.007	0.014	0.023	0.051	0.074	0.113
P/MPa	210.3	189.5	174.4	157.6	152.8	150.3

$T/K = 433.15$

w_A	0.858	0.851	0.842	0.814	0.791	0.752
w_B	0.135	0.135	0.135	0.135	0.135	0.135
w_C	0.007	0.014	0.023	0.051	0.074	0.113
P/MPa	180.1	169.3	160.2	147.3	144.0	142.4

$T/K = 453.15$

w_A	0.858	0.851	0.842	0.814	0.791	0.752
w_B	0.135	0.135	0.135	0.135	0.135	0.135
w_C	0.007	0.014	0.023	0.051	0.074	0.113
P/MPa	162.4	155.3	149.0	139.5	137.3	136.5

$T/K = 473.15$

w_A	0.858	0.851	0.842	0.814	0.791	0.752
w_B	0.135	0.135	0.135	0.135	0.135	0.135
w_C	0.007	0.014	0.023	0.051	0.074	0.113
P/MPa	148.5	144.7	140.5	134.1	131.9	130.5

Polymer (B):	ethylene/acrylic acid copolymer							2002BEY	

Characterization: M_n/kg.mol^{-1} = 22.7, M_w/kg.mol^{-1} = 258, 6.0 wt% acrylic acid,
Exxon Co., Machelen, Belgium

Solvent (A):	ethene	C_2H_4						74-85-1	
Solvent (C):	n-decane	$C_{10}H_{22}$						124-18-5	

Type of data: cloud points

w_A	0.900	0.900	0.900	0.900	0.900	0.900	0.548	0.548	0.548
w_B	0.050	0.050	0.050	0.050	0.050	0.050	0.050	0.050	0.050
w_C	0.050	0.050	0.050	0.050	0.050	0.050	0.402	0.402	0.402
T/K	522.55	497.85	472.85	447.95	423.55	413.65	522.65	498.35	473.15
P/bar	1242	1330	1468	1691	2074	2306	799	844	917

w_A	0.548	0.548	0.548	0.548	0.548
w_B	0.050	0.050	0.050	0.050	0.050
w_C	0.402	0.402	0.402	0.402	0.402
T/K	448.55	423.75	413.55	403.85	393.35
P/bar	1041	1260	1406	1581	1818

Polymer (B):	ethylene/acrylic acid copolymer							2002BEY	

Characterization: M_n/kg.mol^{-1} = 22.7, M_w/kg.mol^{-1} = 258, 6.0 wt% acrylic acid,
Exxon Co., Machelen, Belgium

Solvent (A):	ethene	C_2H_4						74-85-1	
Solvent (C):	ethanol	C_2H_6O						64-17-5	

Type of data: cloud points

w_A	0.900	0.900	0.900	0.900	0.900	0.900	0.900	0.801	0.801
w_B	0.050	0.050	0.050	0.050	0.050	0.050	0.050	0.050	0.050
w_C	0.050	0.050	0.050	0.050	0.050	0.050	0.050	0.149	0.149
T/K	522.85	497.55	472.25	447.55	423.65	413.15	403.65	523.25	497.95
P/bar	1168	1212	1276	1371	1491	1591	1607	1036	1074

w_A	0.801	0.801	0.801	0.801	0.801	0.703	0.703	0.703	0.703
w_B	0.050	0.050	0.050	0.050	0.050	0.050	0.050	0.050	0.050
w_C	0.149	0.149	0.149	0.149	0.149	0.247	0.247	0.247	0.247
T/K	472.95	448.15	424.45	414.15	404.15	523.15	498.05	472.65	447.45
P/bar	1119	1161	1271	1301	1339	966	981	1016	1069

w_A	0.703	0.703	0.703	0.543	0.543	0.543	0.543	0.543	0.543
w_B	0.050	0.050	0.050	0.050	0.050	0.050	0.050	0.050	0.050
w_C	0.247	0.247	0.247	0.407	0.407	0.407	0.407	0.407	0.407
T/K	423.15	413.75	403.65	521.85	497.95	473.05	448.15	423.55	414.05
P/bar	1154	1209	1264	811	834	886	989	1256	1483

Polymer (B):	ethylene/acrylic acid copolymer							2002BEY	

Characterization: M_n/kg.mol^{-1} = 19.9, M_w/kg.mol^{-1} = 235, 7.5 wt% acrylic acid,
Exxon Co., Machelen, Belgium

continued

continued

Solvent (A):	ethene	C_2H_4	74-85-1
Solvent (C):	ethanol	C_2H_6O	64-17-5

Type of data: cloud points

w_A	0.893	0.893	0.893	0.893	0.893	0.893	0.893	0.548	0.548
w_B	0.050	0.050	0.050	0.050	0.050	0.050	0.050	0.050	0.050
w_C	0.057	0.057	0.057	0.057	0.057	0.057	0.057	0.402	0.402
T/K	523.35	497.75	472.55	447.95	423.75	414.15	403.85	523.15	497.65
P/bar	1189	1234	1310	1411	1546	1604	1673	766	774

w_A	0.548	0.548	0.548	0.548
w_B	0.050	0.050	0.050	0.050
w_C	0.402	0.402	0.402	0.402
T/K	472.75	448.25	433.25	423.35
P/bar	809	890	989	1079

Polymer (B):	ethylene/acrylic acid copolymer		2002BEY
Characterization:	M_n/kg.mol^{-1} = 23.4, M_w/kg.mol^{-1} = 227, 9.0 wt% acrylic acid,		
	Exxon Co., Machelen, Belgium		
Solvent (A):	ethene	C_2H_4	74-85-1
Solvent (C):	ethanol	C_2H_6O	64-17-5

Type of data: cloud points

w_A	0.900	0.900	0.900	0.900	0.900	0.900	0.900	0.556	0.556
w_B	0.050	0.050	0.050	0.050	0.050	0.050	0.050	0.050	0.050
w_C	0.050	0.050	0.050	0.050	0.050	0.050	0.050	0.394	0.394
T/K	523.35	497.95	473.05	448.05	423.85	413.85	403.65	524.05	498.15
P/bar	1200	1243	1331	1439	1601	1669	1746	795	804

w_A	0.556	0.556	0.556	0.556	0.556	0.556
w_B	0.050	0.050	0.050	0.050	0.050	0.050
w_C	0.394	0.394	0.394	0.394	0.394	0.394
T/K	472.65	447.95	423.55	413.65	403.65	393.95
P/bar	826	873	978	1051	1164	1334

Polymer (B):	ethylene/acrylic acid copolymer		2002BEY
Characterization:	M_n/kg.mol^{-1} = 22.7, M_w/kg.mol^{-1} = 258, 6.0 wt% acrylic acid,		
	Exxon Co., Machelen, Belgium		
Solvent (A):	ethene	C_2H_4	74-85-1
Solvent (C):	ethyl acetate	$C_4H_8O_2$	141-78-6

Type of data: cloud points

w_A	0.850	0.850	0.850	0.850	0.850	0.850	0.552	0.552	0.552
w_B	0.050	0.050	0.050	0.050	0.050	0.050	0.050	0.050	0.050
w_C	0.055	0.055	0.055	0.055	0.055	0.055	0.398	0.398	0.398
T/K	521.95	497.45	472.15	447.25	432.85	423.85	523.25	498.15	472.85
P/bar	1252	1339	1464	1657	1825	1954	936	974	1019

continued

continued

w_A	0.552	0.552	0.552	0.552	0.552	0.258	0.258	0.258	0.258
w_B	0.050	0.050	0.050	0.050	0.050	0.050	0.050	0.050	0.050
w_C	0.398	0.398	0.398	0.398	0.398	0.692	0.692	0.692	0.692
T/K	448.15	433.25	423.65	414.05	403.15	522.95	498.35	473.15	448.65
P/bar	1099	1168	1224	1298	1396	666	677	701	746

w_A	0.258	0.258	0.258	0.258
w_B	0.050	0.050	0.050	0.050
w_C	0.692	0.692	0.692	0.692
T/K	423.75	414.15	403.65	393.75
P/bar	842	910	1010	1156

Polymer (B):	**ethylene/acrylic acid copolymer**		**2002BEY**
Characterization:	M_n/kg.mol^{-1} = 22.7, M_w/kg.mol^{-1} = 258, 6.0 wt% acrylic acid,		
	Exxon Co., Machelen, Belgium		
Solvent (A):	**ethene**	C_2H_4	**74-85-1**
Solvent (C):	**n-heptane**	C_7H_{16}	**142-82-5**

Type of data: cloud points

w_A	0.900	0.900	0.900	0.900	0.900	0.900	0.552	0.552	0.552
w_B	0.050	0.050	0.050	0.050	0.050	0.050	0.050	0.050	0.050
w_C	0.050	0.050	0.050	0.050	0.050	0.050	0.398	0.398	0.398
T/K	522.45	497.95	472.45	447.75	423.35	413.35	522.85	498.05	473.05
P/bar	1257	1340	1476	1695	2056	2290	863	910	993

w_A	0.552	0.552	0.552	0.552	0.552
w_B	0.050	0.050	0.050	0.050	0.050
w_C	0.398	0.398	0.398	0.398	0.398
T/K	448.35	423.75	413.75	403.45	393.65
P/bar	1129	1366	1515	1709	1961

Polymer (B):	**ethylene/acrylic acid copolymer**		**2002BEY**
Characterization:	M_n/kg.mol^{-1} = 22.7, M_w/kg.mol^{-1} = 258, 6.0 wt% acrylic acid,		
	Exxon Co., Machelen, Belgium		
Solvent (A):	**ethene**	C_2H_4	**74-85-1**
Solvent (C):	**octanoic acid**	$C_8H_{16}O_2$	**124-07-2**

Type of data: cloud points

w_A	0.904	0.904	0.904	0.904	0.904	0.904	0.904	0.904	0.904
w_B	0.050	0.050	0.050	0.050	0.050	0.050	0.050	0.050	0.050
w_C	0.046	0.046	0.046	0.046	0.046	0.046	0.046	0.046	0.046
T/K	521.85	497.55	472.55	447.85	423.75	413.75	404.05	393.25	383.75
P/bar	1145	1180	1236	1308	1401	1454	1505	1565	1637

w_A	0.552	0.552	0.552	0.552	0.552	0.552	0.552	0.552	0.552
w_B	0.050	0.050	0.050	0.050	0.050	0.050	0.050	0.050	0.050
w_C	0.398	0.398	0.398	0.398	0.398	0.398	0.398	0.398	0.398
T/K	521.95	498.05	472.75	447.75	423.65	414.15	403.85	393.65	384.05
P/bar	697	691	696	706	719	728	738	749	764

continued

continued

w_A	0.292	0.292	0.292	0.292	0.292	0.292	0.292	0.292	0.292
w_B	0.050	0.050	0.050	0.050	0.050	0.050	0.050	0.050	0.050
w_C	0.658	0.658	0.658	0.658	0.658	0.658	0.658	0.658	0.658
T/K	522.65	497.95	472.65	448.65	424.05	414.25	404.25	393.55	384.15
P/bar	346	325	307	289	269	260	251	243	235

Polymer (B):	**ethylene/acrylic acid copolymer**		**2002BEY**

Characterization: M_n/kg.mol^{-1} = 22.7, M_w/kg.mol^{-1} = 258, 6.0 wt% acrylic acid, Exxon Co., Machelen, Belgium

Solvent (A):	**ethene**	**C$_2$H$_4$**	**74-85-1**
Solvent (C):	**2,2,4-trimethylpentane**	**C$_8$H$_{18}$**	**540-84-1**

Type of data: cloud points

w_A	0.900	0.900	0.900	0.900	0.900	0.900	0.551	0.551	0.551
w_B	0.050	0.050	0.050	0.050	0.050	0.050	0.050	0.050	0.050
w_C	0.050	0.050	0.050	0.050	0.050	0.050	0.399	0.399	0.399
T/K	522.45	497.95	472.55	447.95	423.65	413.55	522.95	498.05	472.25
P/bar	1233	1314	1445	1648	2005	2211	888	944	1036

w_A	0.551	0.551	0.551	0.551
w_B	0.050	0.050	0.050	0.050
w_C	0.399	0.399	0.399	0.399
T/K	447.95	423.75	413.25	403.55
P/bar	1181	1430	1586	1784

Polymer (B):	**ethylene/1-butene copolymer**		**1999CHE**

Characterization: M_n/kg.mol^{-1} = 7.57, M_w/kg.mol^{-1} = 7.87, 29.0 mol% 1-butene, completely hydrogenated polybutadiene, 29 mol% 1,2-units, Polymer Source, Inc., Dorval, Quebec

Solvent (A):	**ethene**	**C$_2$H$_4$**	**74-85-1**
Solvent (C):	**1-butene**	**C$_4$H$_8$**	**106-98-9**

Type of data: cloud points

w_A	0.49105		w_B	0.00085		w_C	0.50810		

T/K	431.85	430.45	418.95	410.95	401.25	399.75	381.75	373.35	368.95
P/bar	400.7	399.3	395.4	391.1	386.6	385.7	376.9	373.0	369.3

T/K	363.75	357.65	353.95	350.05	348.85	347.35
P/bar	365.9	361.4	353.8	407.4	555.0	983.2

Comments: The last three data points are temperature-induced phase transitions.

w_A	0.40287		w_B	0.00927		w_C	0.58786		

T/K	428.85	428.85	415.05	412.05	398.45	396.95	384.25	381.45	368.65
P/bar	698.7	692.7	676.2	674.2	649.6	645.1	606.2	607.0	544.7

T/K	367.95	353.65	352.75	352.75	352.65	352.55
P/bar	538.2	498.6	494.0	600.6	995.1	1426

Comments: The last three data points are temperature-induced phase transitions.

| **Polymer (B):** | **ethylene/butyl acrylate copolymer** | | | | | | | **1999DIE** |

Characterization: $M_n/\text{kg.mol}^{-1} = 16.4$, $M_w/\text{kg.mol}^{-1} = 294$,
1.0 mol% butyl acrylate, synthesized in the laboratory

| **Solvent (A):** | **ethene** | C_2H_4 | **74-85-1** |
| **Solvent (C):** | **butyl acrylate** | $C_7H_{12}O_2$ | **141-32-2** |

Type of data: cloud points

w_B 0.03   was kept constant

x_C	0.0044	0.0044	0.0044	0.0044	0.0044	0.0044	0.0108	0.0108	0.0108
T/K	383.15	388.15	393.15	403.15	413.15	423.15	383.15	388.15	393.15
P/bar	1838	1795	1735	1658	1600	1552	1783	1748	1699

x_C	0.0108	0.0108	0.0108	0.0154	0.0154	0.0154	0.0154	0.0154	0.0154
T/K	403.15	413.15	423.15	383.15	388.15	393.15	403.15	413.15	423.15
P/bar	1628	1570	1507	1770	1730	1684	1607	1545	1485

x_C	0.0268	0.0268	0.0268	0.0268	0.0268	0.0268	0.0396	0.0396	0.0396
T/K	383.15	388.15	393.15	403.15	413.15	423.15	383.15	388.15	393.15
P/bar	1690	1640	1610	1520	1470	1420	1609	1555	1515

x_C	0.0396	0.0396	0.0396
T/K	403.15	413.15	423.15
P/bar	1442	1390	1340

Comments: The mole fraction x_C is given for the monomer mixture, i.e., without the polymer.

| **Polymer (B):** | **ethylene/butyl acrylate copolymer** | | | | | | | **1999DIE** |

Characterization: $M_n/\text{kg.mol}^{-1} = 39.9$, $M_w/\text{kg.mol}^{-1} = 155$,
8.7 mol% butyl acrylate, synthesized in the laboratory

| **Solvent (A):** | **ethene** | C_2H_4 | **74-85-1** |
| **Solvent (C):** | **butyl acrylate** | $C_7H_{12}O_2$ | **141-32-2** |

Type of data: cloud points

w_B 0.03   was kept constant

x_C	0.0075	0.0075	0.0075	0.0075	0.0075	0.0075	0.0075	0.0075	0.0075
T/K	343.15	353.15	363.15	373.15	383.15	393.15	403.15	413.15	423.15
P/bar	1510	1420	1370	1345	1315	1290	1272	1250	1215

x_C	0.0094	0.0094	0.0094	0.0094	0.0094	0.0094	0.0094	0.0210	0.0210
T/K	363.15	373.15	383.15	393.15	403.15	413.15	423.15	343.15	353.15
P/bar	1330	1305	1270	1240	1240	1200	1200	1360	1320

x_C	0.0210	0.0210	0.0210	0.0210	0.0210	0.0210	0.0210	0.0242	0.0242
T/K	363.15	373.15	383.15	393.15	403.15	413.15	423.15	343.15	353.15
P/bar	1290	1240	1210	1190	1165	1160	1140	1320	1270

x_C	0.0242	0.0242	0.0242	0.0242	0.0242	0.0242	0.0242	0.0315	0.0315
T/K	363.15	373.15	383.15	393.15	403.15	413.15	423.15	343.15	353.15
P/bar	1250	1215	1195	1165	1145	1130	1110	1245	1220

continued

continued

x_C	0.0315	0.0315	0.0315	0.0315	0.0315	0.0315	0.0315	0.0382	0.0382
T/K	363.15	373.15	383.15	393.15	403.15	413.15	423.15	343.15	353.15
P/bar	1180	1155	1130	1110	1100	1070	1050	1215	1165

x_C	0.0382	0.0382	0.0382	0.0382	0.0382	0.0382
T/K	363.15	373.15	383.15	403.15	413.15	423.15
P/bar	1135	1120	1095	1060	1035	1010

Comments: The mole fraction x_C is given for the monomer mixture, i.e., without the polymer.

Polymer (B):	**ethylene/butyl acrylate copolymer**		**1999DIE**
Characterization:	22.2 mol% butyl acrylate, synthesized in the laboratory		
Solvent (A):	**ethene**	C_2H_4	**74-85-1**
Solvent (C):	**butyl acrylate**	$C_7H_{12}O_2$	**141-32-2**

Type of data: cloud points

w_B 0.03   was kept constant

x_C	0.0030	0.0030	0.0030	0.0030	0.0030	0.0030	0.0030	0.0030	0.0030
T/K	323.15	333.15	343.15	353.15	363.15	373.15	383.15	393.15	403.15
P/bar	1285	1260	1230	1195	1175	1155	1142	1130	1120

x_C	0.0030	0.0030	0.0130	0.0130	0.0130	0.0130	0.0130	0.0130	0.0130
T/K	413.15	423.15	323.15	333.15	343.15	353.15	363.15	373.15	383.15
P/bar	1115	1095	1200	1160	1145	1120	1110	1090	1080

x_C	0.0130	0.0130	0.0130	0.0130	0.0220	0.0220	0.0220	0.0220	0.0220
T/K	393.15	403.15	413.15	423.15	323.15	333.15	343.15	353.15	363.15
P/bar	1065	1050	1040	1035	1110	1090	1070	1060	1045

x_C	0.0220	0.0220	0.0220	0.0220	0.0220	0.0360	0.0360	0.0360	0.0360
T/K	383.15	393.15	403.15	413.15	423.15	323.15	333.15	343.15	353.15
P/bar	1015	1010	1005	997	990	975	970	955	950

x_C	0.0360	0.0360	0.0360	0.0360	0.0360	0.0360	0.0360
T/K	363.15	373.15	383.15	393.15	403.15	413.15	423.15
P/bar	940	930	925	920	915	910	890

Comments: The mole fraction x_C is given for the monomer mixture, i.e., without the polymer.

Polymer (B):	**ethylene/butyl acrylate copolymer**		**1999DIE**
Characterization:	$M_n/kg.mol^{-1} = 39.9$, $M_w/kg.mol^{-1} = 155$,		
	8.7 mol% butyl acrylate, synthesized in the laboratory		
Solvent (A):	**ethene**	C_2H_4	**74-85-1**
Solvent (C):	**methyl acrylate**	$C_4H_6O_2$	**96-33-3**

Type of data: cloud points

w_B 0.03   was kept constant

x_C	0.0213	0.0213	0.0213	0.0213	0.0213	0.0213	0.0213	0.0431	0.0431
T/K	353.15	363.15	373.15	383.15	393.15	403.15	423.15	353.15	363.15
P/bar	1390	1350	1310	1290	1260	1230	1200	1300	1280

continued

continued

x_C	0.0431	0.0431	0.0431	0.0431	0.0431	0.0431	0.0675	0.0675	0.0675
T/K	373.15	383.15	393.15	403.15	413.15	423.15	353.15	363.15	373.15
P/bar	1250	1220	1200	1170	1160	1130	1260	1200	1160

x_C	0.0675	0.0675	0.0675	0.0675	0.0675
T/K	383.15	393.15	403.15	413.15	423.15
P/bar	1150	1125	1110	1095	1065

Comments: The mole fraction x_C is given for the monomer mixture, i.e., without the polymer.

Polymer (B): **ethylene/2-ethylhexyl acrylate copolymer**   **1999DIE**
Characterization: M_n/kg.mol^{-1} = 34.9, M_w/kg.mol^{-1} = 107.7,
                     9.3 mol% 2-ethylhexyl acrylate, synthesized in the laboratory
Solvent (A): **ethene**   **C$_2$H$_4$**   **74-85-1**
Solvent (C): **2-ethylhexyl acrylate**   **C$_{11}$H$_{20}$O$_2$**   **103-11-7**

Type of data: cloud points

w_B 0.03   was kept constant

x_C	0.0000	0.0000	0.0000	0.0324	0.0324	0.0324	0.0368	0.0368	0.0368
T/K	403.15	423.15	443.15	403.15	423.15	443.15	403.15	423.15	443.15
P/bar	1440	1390	1335	1240	1200	1100	1170	1140	1080

Comments: The mole fraction x_C is given for the monomer mixture, i.e., without the polymer.

Polymer (B): **ethylene/1-hexene copolymer**   **1999KIN, 2001DOE**
Characterization: M_n/kg.mol^{-1} = 60, M_w/kg.mol^{-1} = 129, 16.1 wt% 1-hexene
Solvent (A): **ethene**   **C$_2$H$_4$**   **74-85-1**
Solvent (C): **n-butane**   **C$_4$H$_{10}$**   **106-97-8**

Type of data: cloud points

w_A	0.800	0.800	0.800	0.800	0.800	0.800	0.750	0.750	0.750
w_B	0.150	0.150	0.150	0.150	0.150	0.150	0.150	0.150	0.150
w_C	0.050	0.050	0.050	0.050	0.050	0.050	0.100	0.100	0.100
T/K	393.15	413.15	433.15	453.15	473.15	493.15	393.15	413.15	433.15
P/MPa	145.0	135.3	127.3	121.2	116.6	113.0	134.0	125.9	119.1

w_A	0.750	0.750	0.750	0.700	0.700	0.700	0.700	0.700	0.700
w_B	0.150	0.150	0.150	0.150	0.150	0.150	0.150	0.150	0.150
w_C	0.100	0.100	0.100	0.150	0.150	0.150	0.150	0.150	0.150
T/K	453.15	473.15	493.15	393.15	413.15	433.15	453.15	473.15	493.15
P/MPa	113.6	109.6	106.2	125.4	117.8	111.7	107.1	103.7	101.0

Polymer (B): **ethylene/1-hexene copolymer**   **2001DOE**
Characterization: M_n/kg.mol^{-1} = 60, M_w/kg.mol^{-1} = 129, 16.1 wt% 1-hexene
Solvent (A): **ethene**   **C$_2$H$_4$**   **74-85-1**
Solvent (C): **carbon dioxide**   **CO$_2$**   **124-38-9**

Type of data: cloud points

continued

continued

w_A	0.800	0.800	0.800	0.800	0.800	0.800	0.750	0.750	0.750
w_B	0.150	0.150	0.150	0.150	0.150	0.150	0.150	0.150	0.150
w_C	0.050	0.050	0.050	0.050	0.050	0.050	0.100	0.100	0.100
T/K	393.15	413.15	433.15	453.15	473.15	493.15	393.15	413.15	433.15
P/MPa	160.0	148.2	138.3	130.8	125.7	122.0	168.3	153.7	142.3

w_A	0.750	0.750	0.750
w_B	0.150	0.150	0.150
w_C	0.100	0.100	0.100
T/K	393.15	413.15	433.15
P/MPa	134.0	128.2	124.0

Polymer (B):	ethylene/1-hexene copolymer		1999KIN, 2001DOE
Characterization:	M_n/kg.mol^{-1} = 60, M_w/kg.mol^{-1} = 129, 16.1 wt% 1-hexene		
Solvent (A):	ethene	C_2H_4	74-85-1
Solvent (C):	ethane	C_2H_6	74-84-0

Type of data: cloud points

w_A	0.800	0.800	0.800	0.800	0.800	0.800	0.750	0.750	0.750
w_B	0.150	0.150	0.150	0.150	0.150	0.150	0.150	0.150	0.150
w_C	0.050	0.050	0.050	0.050	0.050	0.050	0.100	0.100	0.100
T/K	393.15	413.15	433.15	453.15	473.15	493.15	393.15	413.15	433.15
P/MPa	151.7	140.7	133.4	126.3	120.8	116.5	147.1	136.8	129.1

w_A	0.750	0.750	0.750	0.700	0.700	0.700	0.700	0.700	0.700
w_B	0.150	0.150	0.150	0.150	0.150	0.150	0.150	0.150	0.150
w_C	0.100	0.100	0.100	0.150	0.150	0.150	0.150	0.150	0.150
T/K	453.15	473.15	493.15	393.15	413.15	433.15	453.15	473.15	493.15
P/MPa	123.0	118.1	114.3	143.2	133.5	125.5	120.0	116.0	112.0

Polymer (B):	ethylene/1-hexene copolymer		1999KIN, 2001DOE
Characterization:	M_n/kg.mol^{-1} = 60, M_w/kg.mol^{-1} = 129, 16.1 wt% 1-hexene		
Solvent (A):	ethene	C_2H_4	74-85-1
Solvent (C):	helium	He	7440-59-7

Type of data: cloud points

w_A	0.840	0.840	0.840	0.840	0.840	0.840	0.830	0.830	0.830
w_B	0.150	0.150	0.150	0.150	0.150	0.150	0.150	0.150	0.150
w_C	0.010	0.010	0.010	0.010	0.010	0.010	0.020	0.020	0.020
T/K	393.15	413.15	433.15	453.15	473.15	493.15	433.15	453.15	473.15
P/MPa	195.0	178.6	165.8	155.4	148.2	143.0	214.6	200.7	188.3

w_A	0.830
w_B	0.150
w_C	0.020
T/K	493.15
P/MPa	176.8

| Polymer (B): | ethylene/1-hexene copolymer | | 2000CH3 |

Polymer (B): ethylene/1-hexene copolymer   **2000CH3**
Characterization: M_n/kg.mol^{-1} = 52.6, M_w/kg.mol^{-1} = 80, 10.6 wt% 1-hexene
Solvent (A): ethene C_2H_4 **74-85-1**
Solvent (C): 1-hexene C_6H_{12} **592-41-6**

Type of data: cloud points

w_A	0.4255	0.4255	0.4255	0.2755	0.2755	0.2755
w_B	0.1490	0.1490	0.1490	0.1390	0.1390	0.1390
w_C	0.4255	0.4255	0.4255	0.5855	0.5855	0.5855
T/K	393.15	423.15	453.35	393.25	423.15	453.25
P/MPa	65.4	66.5	65.1	44.2	45.1	44.6

Polymer (B): ethylene/1-hexene copolymer   **1999KIN, 2001DOE**
Characterization: M_n/kg.mol^{-1} = 60, M_w/kg.mol^{-1} = 129, 16.1 wt% 1-hexene
Solvent (A): ethene C_2H_4 **74-85-1**
Solvent (C): 1-hexene C_6H_{12} **592-41-6**

Type of data: cloud points

w_A	0.6375	0.6375	0.6375	0.6375	0.6375	0.6375	0.4250	0.4250	0.4250
w_B	0.1500	0.1500	0.1500	0.1500	0.1500	0.1500	0.1500	0.1500	0.1500
w_C	0.2125	0.2125	0.2125	0.2125	0.2125	0.2125	0.4250	0.4250	0.4250
T/K	393.15	413.15	433.15	453.15	473.15	493.15	393.15	413.15	433.15
P/MPa	108.2	103.2	99.2	95.7	93.3	91.4	59.0	59.0	59.1
w_A	0.4250	0.4250	0.4250	0.2125	0.2125	0.2125	0.2125	0.2125	0.2125
w_B	0.1500	0.1500	0.1500	0.1500	0.1500	0.1500	0.1500	0.1500	0.1500
w_C	0.4250	0.4250	0.4250	0.6375	0.6375	0.6375	0.6375	0.6375	0.6375
T/K	453.15	473.15	493.15	393.15	413.15	433.15	453.15	473.15	493.15
P/MPa	59.0	59.1	59.1	25.6	27.0	28.3	30.3	32.3	34.6

Polymer (B): ethylene/1-hexene copolymer   **2001DOE**
Characterization: M_n/kg.mol^{-1} = 60, M_w/kg.mol^{-1} = 129, 16.1 wt% 1-hexene
Solvent (A): ethene C_2H_4 **74-85-1**
Solvent (C): 1-hexene C_6H_{12} **592-41-6**

Type of data: cloud points

w_A	0.6800	0.6800	0.6800	0.6800	0.6800	0.6800	0.4250	0.4250	0.4250
w_B	0.1500	0.1500	0.1500	0.1500	0.1500	0.1500	0.1500	0.1500	0.1500
w_C	0.1700	0.1700	0.1700	0.1700	0.1700	0.1700	0.4250	0.4250	0.4250
T/K	393.15	413.15	433.15	453.15	473.15	493.15	393.15	413.15	433.15
P/MPa	118.5	112.0	106.4	102.5	99.6	96.9	63.7	63.6	63.1
w_A	0.4250	0.4250	0.4250	0.1275	0.1275	0.1275	0.1275	0.1275	0.1725
w_B	0.1500	0.1500	0.1500	0.1500	0.1500	0.1500	0.1500	0.1500	0.1500
w_C	0.4250	0.4250	0.4250	0.7225	0.7225	0.7225	0.7225	0.7225	0.7225
T/K	453.15	473.15	493.15	393.15	413.15	433.15	453.15	473.15	493.15
P/MPa	62.6	62.3	62.1	12.7	15.4	17.8	19.9	21.9	23.5

Polymer (B):	**ethylene/1-hexene copolymer**		**2000CH3**
Characterization:	$M_n/\text{kg.mol}^{-1} = 48.1$, $M_w/\text{kg.mol}^{-1} = 103$, 35.0 wt% 1-hexene		
Solvent (A):	ethene	C_2H_4	**74-85-1**
Solvent (C):	1-hexene	C_6H_{12}	**592-41-6**

Type of data: cloud points

w_A	0.4558	0.4558	0.4558	0.4558
w_B	0.1400	0.1400	0.1400	0.1400
w_C	0.4042	0.4042	0.4042	0.4042
T/K	393.15	413.15	433.15	453.15
P/MPa	55.7	55.7	56.2	57.7

Polymer (B):	**ethylene/1-hexene copolymer**		**1999KIN**
Characterization:	$M_n/\text{kg.mol}^{-1} = 60$, $M_w/\text{kg.mol}^{-1} = 129$, 16.1 wt% 1-hexene		
Solvent (A):	ethene	C_2H_4	**74-85-1**
Solvent (C):	1-hexene	C_6H_{12}	**592-41-6**
Solvent (D):	n-butane	C_4H_{10}	**106-97-8**

Type of data: cloud points

w_A	0.6000	0.6000	0.6000	0.6000	0.6000	0.6000	0.5625	0.5625	0.5625
w_B	0.1500	0.1500	0.1500	0.1500	0.1500	0.1500	0.1500	0.1500	0.1500
w_C	0.2000	0.2000	0.2000	0.2000	0.2000	0.2000	0.1875	0.1875	0.1875
w_D	0.0500	0.0500	0.0500	0.0500	0.0500	0.0500	0.1000	0.1000	0.1000
T/K	393.15	413.15	433.15	453.15	473.15	493.15	393.15	413.15	433.15
P/MPa	99.9	95.8	91.6	89.7	87.7	86.0	90.5	87.2	84.6
w_A	0.5625	0.5625	0.5625	0.5250	0.5250	0.5250	0.5250	0.5250	0.5250
w_B	0.1500	0.1500	0.1500	0.1500	0.1500	0.1500	0.1500	0.1500	0.1500
w_C	0.1875	0.1875	0.1875	0.1750	0.1750	0.1750	0.1750	0.1750	0.1750
w_D	0.1000	0.1000	0.1000	0.1500	0.1500	0.1500	0.1500	0.1500	0.1500
T/K	453.15	473.15	493.15	393.15	413.15	433.15	453.15	473.15	493.15
P/MPa	82.6	81.1	79.9	85.5	83.1	81.1	79.5	78.2	77.2
w_A	0.4000	0.4000	0.4000	0.4000	0.4000	0.4000	0.3750	0.3750	0.3750
w_B	0.1500	0.1500	0.1500	0.1500	0.1500	0.1500	0.1500	0.1500	0.1500
w_C	0.4000	0.4000	0.4000	0.4000	0.4000	0.4000	0.3750	0.3750	0.3750
w_D	0.0500	0.0500	0.0500	0.0500	0.0500	0.0500	0.1000	0.1000	0.1000
T/K	393.15	413.15	433.15	453.15	473.15	493.15	393.15	413.15	433.15
P/MPa	57.0	57.1	57.2	57.3	57.3	57.4	55.6	55.6	55.7
w_A	0.3750	0.3750	0.3750	0.3500	0.3500	0.3500	0.3500	0.3500	0.3500
w_B	0.1500	0.1500	0.1500	0.1500	0.1500	0.1500	0.1500	0.1500	0.1500
w_C	0.3750	0.3750	0.3750	0.3500	0.3500	0.3500	0.3500	0.3500	0.3500
w_D	0.1000	0.1000	0.1000	0.1500	0.1500	0.1500	0.1500	0.1500	0.1500
T/K	453.15	473.15	493.15	393.15	413.15	433.15	453.15	473.15	493.15
P/MPa	55.8	55.9	56.1	54.9	54.9	55.1	55.4	55.6	55.6

continued

continued

w_A	0.2000	0.2000	0.2000	0.2000	0.2000	0.2000	0.1875	0.1875	0.1785
w_B	0.1500	0.1500	0.1500	0.1500	0.1500	0.1500	0.1500	0.1500	0.1500
w_C	0.6000	0.6000	0.6000	0.6000	0.6000	0.6000	0.5625	0.5625	0.5625
w_D	0.0500	0.0500	0.0500	0.0500	0.0500	0.0500	0.1000	0.1000	0.1000
T/K	393.15	413.15	433.15	453.15	473.15	493.15	393.15	413.15	433.15
P/MPa	25.1	26.6	27.7	29.7	32.0	34.0	23.7	25.1	26.6

w_A	0.1875	0.1875	0.1875	0.1750	0.1750	0.1750	0.1750	0.1750	0.1750
w_B	0.1500	0.1500	0.1500	0.1500	0.1500	0.1500	0.1500	0.1500	0.1500
w_C	0.5625	0.5625	0.5625	0.5250	0.5250	0.5250	0.5250	0.5250	0.5250
w_D	0.1000	0.1000	0.1000	0.1500	0.1500	0.1500	0.1500	0.1500	0.1500
T/K	453.15	473.15	493.15	393.15	413.15	433.15	453.15	473.15	493.15
P/MPa	28.6	30.8	33.1	23.4	24.9	26.5	28.4	30.6	32.8

Polymer (B):	ethylene/1-hexene copolymer		1999KIN

Characterization: $M_n/\text{kg.mol}^{-1} = 60$, $M_w/\text{kg.mol}^{-1} = 129$, 16.1 wt% 1-hexene

Solvent (A):	ethene	C_2H_4	74-85-1
Solvent (C):	1-hexene	C_6H_{12}	592-41-6
Solvent (D):	helium	He	7440-59-7

Type of data: cloud points

w_A	0.6300	0.6300	0.6300	0.6300	0.6300	0.6300	0.6225	0.6225	0.6225
w_B	0.1500	0.1500	0.1500	0.1500	0.1500	0.1500	0.1500	0.1500	0.1500
w_C	0.2100	0.2100	0.2100	0.2100	0.2100	0.2100	0.2075	0.2075	0.2075
w_D	0.0100	0.0100	0.0100	0.0100	0.0100	0.0100	0.0200	0.0200	0.0200
T/K	393.15	413.15	433.15	453.15	473.15	493.15	393.15	413.15	433.15
P/MPa	150.3	139.9	131.5	125.1	119.9	115.5	200.5	185.5	172.6

w_A	0.6225	0.6225	0.6225	0.4200	0.4200	0.4200	0.4200	0.4200	0.4200
w_B	0.1500	0.1500	0.1500	0.1500	0.1500	0.1500	0.1500	0.1500	0.1500
w_C	0.2075	0.2075	0.2075	0.4200	0.4200	0.4200	0.4200	0.4200	0.4200
w_D	0.0200	0.0200	0.0200	0.0100	0.0100	0.0100	0.0100	0.0100	0.0100
T/K	453.15	473.15	493.15	393.15	413.15	433.15	453.15	473.15	493.15
P/MPa	161.4	152.5	147.7	100.0	95.1	91.2	88.2	86.1	84.4

w_A	0.4150	0.4150	0.4150	0.4150	0.4150	0.4150	0.4100	0.4100	0.4100
w_B	0.1500	0.1500	0.1500	0.1500	0.1500	0.1500	0.1500	0.1500	0.1500
w_C	0.4150	0.4150	0.4150	0.4150	0.4150	0.4150	0.4100	0.4100	0.4100
w_D	0.0200	0.0200	0.0200	0.0200	0.0200	0.0200	0.0300	0.0300	0.0300
T/K	393.15	413.15	433.15	453.15	473.15	493.15	393.15	413.15	433.15
P/MPa	149.7	139.0	130.5	123.6	118.2	113.6	209.9	189.9	172.9

w_A	0.4100	0.4100	0.4100	0.2100	0.2100	0.2100	0.2100	0.2100	0.2100
w_B	0.1500	0.1500	0.1500	0.1500	0.1500	0.1500	0.1500	0.1500	0.1500
w_C	0.4100	0.4100	0.4100	0.6300	0.6300	0.6300	0.6300	0.6300	0.6300
w_D	0.0300	0.0300	0.0300	0.0100	0.0100	0.0100	0.0100	0.0100	0.0100
T/K	453.15	473.15	493.15	393.15	413.15	433.15	453.15	473.15	493.15
P/MPa	161.6	153.8	148.4	54.4	54.6	54.5	54.3	54.0	54.0

continued

continued

w_A	0.2075	0.2075	0.2075	0.2075	0.2075	0.2075	0.2050	0.2050	0.2050
w_B	0.1500	0.1500	0.1500	0.1500	0.1500	0.1500	0.1500	0.1500	0.1500
w_C	0.6225	0.6225	0.6225	0.6225	0.6225	0.6225	0.6150	0.6150	0.6150
w_D	0.0200	0.0200	0.0200	0.0200	0.0200	0.0200	0.0300	0.0300	0.0300
T/K	393.15	413.15	433.15	453.15	473.15	493.15	393.15	413.15	433.15
P/MPa	93.3	88.8	85.1	82.1	79.8	78.0	134.5	127.1	121.7

w_A	0.2050	0.2050	0.2050
w_B	0.1500	0.1500	0.1500
w_C	0.6150	0.6150	0.6150
w_D	0.0300	0.0300	0.0300
T/K	453.15	473.15	493.15
P/MPa	117.1	113.3	110.2

Polymer (B):	**ethylene/1-hexene copolymer**		**1999KIN**
Characterization:	M_n/kg.mol^{-1} = 60, M_w/kg.mol^{-1} = 129, 16.1 wt% 1-hexene		
Solvent (A):	**ethene**	**C$_2$H$_4$**	**74-85-1**
Solvent (C):	**1-hexene**	**C$_6$H$_{12}$**	**592-41-6**
Solvent (D):	**methane**	**CH$_4$**	**74-82-8**

Type of data: cloud points

w_A	0.6000	0.6000	0.6000	0.6000	0.6000	0.6000	0.5625	0.5625	0.5625
w_B	0.1500	0.1500	0.1500	0.1500	0.1500	0.1500	0.1500	0.1500	0.1500
w_C	0.2000	0.2000	0.2000	0.2000	0.2000	0.2000	0.1875	0.1875	0.1875
w_D	0.0500	0.0500	0.0500	0.0500	0.0500	0.0500	0.1000	0.1000	0.1000
T/K	393.15	413.15	433.15	453.15	473.15	493.15	393.15	413.15	433.15
P/MPa	123.0	117.5	112.4	108.1	104.7	102.0	137.2	129.7	124.2

w_A	0.5625	0.5625	0.5625	0.5250	0.5250	0.5250	0.5250	0.5250	0.5250
w_B	0.1500	0.1500	0.1500	0.1500	0.1500	0.1500	0.1500	0.1500	0.1500
w_C	0.1875	0.1875	0.1875	0.1750	0.1750	0.1750	0.1750	0.1750	0.1750
w_D	0.1000	0.1000	0.1000	0.1500	0.1500	0.1500	0.1500	0.1500	0.1500
T/K	453.15	473.15	493.15	393.15	413.15	433.15	453.15	473.15	493.15
P/MPa	119.6	115.8	112.8	151.9	144.0	137.1	131.4	127.0	123.4

w_A	0.4000	0.4000	0.4000	0.4000	0.4000	0.4000	0.3750	0.3750	0.3750
w_B	0.1500	0.1500	0.1500	0.1500	0.1500	0.1500	0.1500	0.1500	0.1500
w_C	0.4000	0.4000	0.4000	0.4000	0.4000	0.4000	0.3750	0.3750	0.3750
w_D	0.0500	0.0500	0.0500	0.0500	0.0500	0.0500	0.1000	0.1000	0.1000
T/K	393.15	413.15	433.15	453.15	473.15	493.15	393.15	413.15	433.15
P/MPa	83.0	80.9	79.0	77.4	76.0	74.8	100.5	97.1	94.0

w_A	0.3750	0.3750	0.3750	0.3500	0.3500	0.3500	0.3500	0.3500	0.3500
w_B	0.1500	0.1500	0.1500	0.1500	0.1500	0.1500	0.1500	0.1500	0.1500
w_C	0.3750	0.3750	0.3750	0.3500	0.3500	0.3500	0.3500	0.3500	0.3500
w_D	0.1000	0.1000	0.1000	0.1500	0.1500	0.1500	0.1500	0.1500	0.1500
T/K	453.15	473.15	493.15	393.15	413.15	433.15	453.15	473.15	493.15
P/MPa	91.3	89.1	87.2	121.6	115.2	109.6	105.4	101.9	99.0

continued

continued

w_A	0.2000	0.2000	0.2000	0.2000	0.2000	0.2000	0.1875	0.1875	0.1785
w_B	0.1500	0.1500	0.1500	0.1500	0.1500	0.1500	0.1500	0.1500	0.1500
w_C	0.6000	0.6000	0.6000	0.6000	0.6000	0.6000	0.5625	0.5625	0.5625
w_D	0.0500	0.0500	0.0500	0.0500	0.0500	0.0500	0.1000	0.1000	0.1000
T/K	393.15	413.15	433.15	453.15	473.15	493.15	393.15	413.15	433.15
P/MPa	40.9	42.5	44.0	45.5	46.9	48.3	58.6	58.8	59.1

w_A	0.1875	0.1875	0.1875	0.1750	0.1750	0.1750	0.1750	0.1750	0.1750
w_B	0.1500	0.1500	0.1500	0.1500	0.1500	0.1500	0.1500	0.1500	0.1500
w_C	0.5625	0.5625	0.5625	0.5250	0.5250	0.5250	0.5250	0.5250	0.5250
w_D	0.1000	0.1000	0.1000	0.1500	0.1500	0.1500	0.1500	0.1500	0.1500
T/K	453.15	473.15	493.15	393.15	413.15	433.15	453.15	473.15	493.15
P/MPa	59.2	59.4	59.5	82.5	81.0	79.6	78.5	77.4	76.6

Polymer (B):	ethylene/1-hexene copolymer		2001DOE

Characterization: M_n/kg.mol^{-1} = 60, M_w/kg.mol^{-1} = 129, 16.1 wt% 1-hexene

Solvent (A):	ethene	C_2H_4	**74-85-1**
Solvent (C):	1-hexene	C_6H_{12}	**592-41-6**
Solvent (D):	nitrogen	N_2	**7727-37-9**

Type of data: cloud points

w_A	0.6400	0.6400	0.6400	0.6400	0.6400	0.6400	0.6000	0.6000	0.6000
w_B	0.1500	0.1500	0.1500	0.1500	0.1500	0.1500	0.1500	0.1500	0.1500
w_C	0.1600	0.1600	0.1600	0.1600	0.1600	0.1600	0.1500	0.1500	0.1500
w_D	0.0500	0.0500	0.0500	0.0500	0.0500	0.0500	0.1000	0.1000	0.1000
T/K	393.15	413.15	433.15	453.15	473.15	493.15	393.15	413.15	433.15
P/MPa	141.5	131.7	123.8	117.7	112.9	109.5	166.4	151.7	141.4

w_A	0.6000	0.6000	0.6000	0.4000	0.4000	0.4000	0.4000	0.4000	0.4000
w_B	0.1500	0.1500	0.1500	0.1500	0.1500	0.1500	0.1500	0.1500	0.1500
w_C	0.1500	0.1500	0.1500	0.4000	0.4000	0.4000	0.4000	0.4000	0.4000
w_D	0.1000	0.1000	0.1000	0.0500	0.0500	0.0500	0.0500	0.0500	0.0500
T/K	453.15	473.15	493.15	393.15	413.15	433.15	453.15	473.15	493.15
P/MPa	132.8	126.5	122.0	85.3	81.9	79.2	77.0	75.7	74.2

w_A	0.3750	0.3750	0.3750	0.3750	0.3750	0.3750	0.1200	0.1200	0.1200
w_B	0.1500	0.1500	0.1500	0.1500	0.1500	0.1500	0.1500	0.1500	0.1500
w_C	0.3750	0.3750	0.3750	0.3750	0.3750	0.3750	0.6800	0.6800	0.6800
w_D	0.1000	0.1000	0.1000	0.1000	0.1000	0.1000	0.0500	0.0500	0.0500
T/K	393.15	413.15	433.15	453.15	473.15	493.15	393.15	413.15	433.15
P/MPa	114.6	107.7	100.3	96.1	91.6	88.7	28.4	29.9	31.3

w_A	0.1200	0.1200	0.1200	0.1125	0.1125	0.1125	0.1125	0.1125	0.1125
w_B	0.1500	0.1500	0.1500	0.1500	0.1500	0.1500	0.1500	0.1500	0.1500
w_C	0.6800	0.6800	0.6800	0.6375	0.6375	0.6375	0.6375	0.6375	0.6375
w_D	0.0500	0.0500	0.0500	0.1000	0.1000	0.1000	0.1000	0.1000	0.1000
T/K	453.15	473.15	493.15	393.15	413.15	433.15	453.15	473.15	493.15
P/MPa	32.6	33.8	34.9	49.1	48.8	48.6	48.6	48.8	49.0

Polymer (B):	**ethylene/1-hexene copolymer**							**1999KIN**
Characterization:	M_n/kg.mol^{-1} = 60, M_w/kg.mol^{-1} = 129, 16.1 wt% 1-hexene							
Solvent (A):	**ethene**		**C$_2$H$_4$**					**74-85-1**
Solvent (C):	**methane**		**CH$_4$**					**74-82-8**

Type of data: cloud points

w_A	0.800	0.800	0.800	0.800	0.800	0.800	0.750	0.750	0.750
w_B	0.150	0.150	0.150	0.150	0.150	0.150	0.150	0.150	0.150
w_C	0.050	0.050	0.050	0.050	0.050	0.050	0.100	0.100	0.100
T/K	393.15	413.15	433.15	453.15	473.15	493.15	393.15	413.15	433.15
P/MPa	169.0	156.7	147.1	139.4	132.4	126.4	186.5	170.5	157.8

w_A	0.750	0.750	0.750	0.700	0.700	0.700	0.700	0.700	0.700
w_B	0.150	0.150	0.150	0.150	0.150	0.150	0.150	0.150	0.150
w_C	0.100	0.100	0.100	0.150	0.150	0.150	0.150	0.150	0.150
T/K	453.15	473.15	493.15	393.15	413.15	433.15	453.15	473.15	493.15
P/MPa	149.0	142.0	137.5	202.0	186.9	174.0	162.8	153.7	149.1

Polymer (B):	**ethylene/1-hexene copolymer**							**2001DOE**
Characterization:	M_n/kg.mol^{-1} = 60, M_w/kg.mol^{-1} = 129, 16.1 wt% 1-hexene							
Solvent (A):	**ethene**		**C$_2$H$_4$**					**74-85-1**
Solvent (C):	**nitrogen**		**N$_2$**					**7727-37-9**

Type of data: cloud points

w_A	0.800	0.800	0.800	0.800	0.800	0.800	0.750	0.750	0.750
w_B	0.150	0.150	0.150	0.150	0.150	0.150	0.150	0.150	0.150
w_C	0.050	0.050	0.050	0.050	0.050	0.050	0.100	0.100	0.100
T/K	393.15	413.15	433.15	453.15	473.15	493.15	393.15	413.15	433.15
P/MPa	179.5	166.8	153.3	144.8	138.0	134.4	206.5	188.6	173.2

w_A	0.750	0.750	0.750
w_B	0.150	0.150	0.150
w_C	0.100	0.100	0.100
T/K	453.15	473.15	493.15
P/MPa	161.8	152.8	145.6

Polymer (B):	**ethylene/1-hexene copolymer**						**1999KIN, 2001DOE**
Characterization:	M_n/kg.mol^{-1} = 60, M_w/kg.mol^{-1} = 129, 16.1 wt% 1-hexene						
Solvent (A):	**ethene**		**C$_2$H$_4$**				**74-85-1**
Solvent (C):	**propane**		**C$_3$H$_8$**				**74-98-6**

Type of data: cloud points

w_A	0.800	0.800	0.800	0.800	0.800	0.800	0.750	0.750	0.750
w_B	0.150	0.150	0.150	0.150	0.150	0.150	0.150	0.150	0.150
w_C	0.050	0.050	0.050	0.050	0.050	0.050	0.100	0.100	0.100
T/K	393.15	413.15	433.15	453.15	473.15	493.15	393.15	413.15	433.15
P/MPa	145.8	136.0	128.1	122.1	117.4	113.6	138.3	129.3	123.3

continued

continued

w_A	0.750	0.750	0.750	0.700	0.700	0.700	0.700	0.700	0.700
w_B	0.150	0.150	0.150	0.150	0.150	0.150	0.150	0.150	0.150
w_C	0.100	0.100	0.100	0.150	0.150	0.150	0.150	0.150	0.150
T/K	453.15	473.15	493.15	393.15	413.15	433.15	453.15	473.15	493.15
P/MPa	117.1	113.3	110.0	130.3	122.1	116.0	111.8	108.9	106.6

Polymer (B):   **ethylene/1-hexene copolymer**   **1999PAN**

Characterization: M_n/kg.mol^{-1} = 40.9, M_w/kg.mol^{-1} = 90, 13.6 wt% 1-hexene,
MI = 1.2, ρ = 0.904 g/cm^3, T_m/K = 406.25

Solvent (A):   **2-methylpropane**   **C$_4$H$_{10}$**   **75-28-5**

Solvent (C):   **1-hexene**   **C$_6$H$_{12}$**   **592-41-6**

Type of data: cloud points

w_A	0.892	0.892	0.892	0.892	0.892	0.892	0.892	0.892	0.892
w_B	0.108	0.108	0.108	0.108	0.108	0.108	0.108	0.108	0.108
w_C	0.000	0.000	0.000	0.000	0.000	0.000	0.000	0.000	0.000
T/K	378.45	386.75	393.05	394.05	400.05	407.75	415.55	421.95	422.75
P/bar	333.5	334.5	336.4	336.8	338.1	340.8	342.8	343.4 344.1	

w_A	0.84835	0.84835	0.84835	0.84835	0.84835	0.84835	0.84835	0.84835	0.84835
w_B	0.107	0.107	0.107	0.107	0.107	0.107	0.107	0.107	0.107
w_C	0.04465	0.04465	0.04465	0.04465	0.04465	0.04465	0.04465	0.04465	0.04465
T/K	368.35	373.55	378.85	385.75	393.45	400.05	408.45	415.45	423.75
P/bar	301.55	302.95	304.15	307.55	311.55	313.6	317.95	321.55	324.8

w_A	0.40275	0.40275	0.40275	0.40275	0.40275	0.40275
w_B	0.105	0.105	0.105	0.105	0.105	0.105
w_C	0.49225	0.49225	0.49225	0.49225	0.49225	0.49225
T/K	367.15	373.15	376.35	377.85	383.35	393.15
P/bar	32.5	41.5	46.2	48.9	56.4	71.1

Polymer (B):   **ethylene/methyl acrylate copolymer**   **1999DIE**

Characterization: M_n/kg.mol^{-1} = 52.5, M_w/kg.mol^{-1} = 165,
8.0 mol% methyl acrylate, synthesized in the laboratory

Solvent (A):   **ethene**   **C$_2$H$_4$**   **74-85-1**

Solvent (C):   **butyl acrylate**   **C$_7$H$_{12}$O$_2$**   **141-32-2**

Type of data: cloud points

w_B	0.03	was kept constant

x_C	0.0129	0.0129	0.0129	0.0129	0.0129	0.0129	0.0288	0.0288	0.0288
T/K	373.15	383.15	393.15	403.15	413.15	423.15	383.15	393.15	403.15
P/bar	1645	1570	1520	1485	1470	1440	1480	1435	1415

x_C	0.0288	0.0288	0.0305	0.0305	0.0305	0.0305	0.0305	0.0403	0.0403
T/K	413.15	423.15	383.15	393.15	403.15	413.15	423.15	373.15	383.15
P/bar	1390	1360	1450	1410	1390	1370	1370	1440	1420

x_C	0.0403	0.0403	0.0403	0.0403
T/K	393.15	403.15	413.15	423.15
P/bar	1375	1340	1340	1320

Comments: The mole fraction x_C is given for the solvent mixture, i.e., without the polymer.

Polymer (B):   **ethylene/methyl acrylate copolymer**   **1999DIE**
Characterization: $M_n/\text{kg.mol}^{-1} = 52.5$, $M_w/\text{kg.mol}^{-1} = 165$,
       8.0 mol% methyl acrylate, synthesized in the laboratory
Solvent (A):   **ethene** $\mathbf{C_2H_4}$ **74-85-1**
Solvent (C):   **2-ethylhexyl acrylate** $\mathbf{C_{11}H_{20}O_2}$ **103-11-7**

Type of data: cloud points

w_B 0.03   was kept constant

x_C	0.0120	0.0120	0.0120	0.0185	0.0185	0.0185	0.0194	0.0194	0.0194
T/K	403.15	423.15	443.15	403.15	423.15	443.15	403.15	423.15	443.15
P/bar	1390	1340	1290	1370	1250	1210	1290	1275	1260

x_C	0.0324	0.0324	0.0324	0.0339	0.0339	0.0339
T/K	403.15	423.15	443.15	403.15	423.15	443.15
P/bar	1240	1200	1140	1250	1170	1190

Comments: The mole fraction x_C is given for the solvent mixture, i.e., without the polymer.

Polymer (B):   **ethylene/methyl acrylate copolymer**   **1999DIE**
Characterization: $M_n/\text{kg.mol}^{-1} = 52.5$, $M_w/\text{kg.mol}^{-1} = 165$,
       8.0 mol% methyl acrylate, synthesized in the laboratory
Solvent (A):   **ethene** $\mathbf{C_2H_4}$ **74-85-1**
Solvent (C):   **n-heptane** $\mathbf{C_7H_{16}}$ **142-82-5**

Type of data: cloud points

w_B 0.03   was kept constant

x_C	0.0136	0.0136	0.0136	0.0136	0.0136	0.0261	0.0261	0.0261	0.0261
T/K	383.15	393.15	403.15	413.15	423.15	383.15	393.15	403.15	413.15
P/bar	1500	1460	1405	1380	1360	1420	1400	1370	1340

x_C	0.0261	0.0414	0.0414	0.0414	0.0414	0.0414	0.0599	0.0599	0.0599
T/K	423.15	383.15	393.15	403.15	413.15	423.15	383.15	393.15	403.15
P/bar	1300	1350	1315	1290	1255	1250	1240	1225	1210

x_C	0.0599	0.0599	0.0977	0.0977	0.0977	0.0977
T/K	413.15	423.15	383.15	403.15	413.15	423.15
P/bar	1165	1150	1000	975	970	950

Comments: The mole fraction x_C is given for the solvent mixture, i.e., without the polymer.

Polymer (B):   **ethylene/methyl acrylate copolymer**   **1999DIE**
Characterization: $M_n/\text{kg.mol}^{-1} = 52.5$, $M_w/\text{kg.mol}^{-1} = 165$,
       8.0 mol% methyl acrylate, synthesized in the laboratory
Solvent (A):   **ethene** $\mathbf{C_2H_4}$ **74-85-1**
Solvent (C):   **methyl acrylate** $\mathbf{C_4H_6O_2}$ **96-33-3**

Type of data: cloud points

continued

continued

w_B	0.03		was kept constant						
x_C	0.0144	0.0144	0.0144	0.0144	0.0144	0.0590	0.0590	0.0590	0.0590
T/K	393.15	403.15	423.15	433.15	443.15	383.15	393.15	403.15	413.15
P/bar	1550	1470	1440	1440	1410	1400	1390	1380	1365

x_C	0.0590	0.0855	0.0855	0.0855	0.0855	0.0855	0.0855	0.0855	0.0855
T/K	423.15	373.15	383.15	393.15	403.15	413.15	423.15	433.15	443.15
P/bar	1350	1300	1290	1300	1340	1370	1350	1380	1360

Comments: The mole fraction x_C is given for the monomer mixture, i.e., without the polymer.

Polymer (B):	**ethylene/propylene copolymer**		**1992CH2**
Characterization:	$M_n/kg.mol^{-1} = 25.2$, $M_w/kg.mol^{-1} = 26.0$, 50 mol% propene,		
	alternating ethylene/propylene units from complete		
	hydrogenation of polyisoprene		
Solvent (A):	**ethene**	C_2H_4	**74-85-1**
Solvent (C):	**1-butene**	C_4H_8	**106-98-9**

Type of data: cloud points

w_A	0.074	0.074	0.074	0.074	0.074	0.074	0.117	0.117	0.117
w_B	0.187	0.187	0.187	0.187	0.187	0.187	0.162	0.162	0.162
w_C	0.739	0.739	0.739	0.739	0.739	0.739	0.721	0.721	0.721
T/K	355.35	373.45	398.15	423.15	448.15	473.55	355.35	373.35	398.25
P/bar	144.0	177.0	215.3	250.3	278.0	303.0	200.4	232.0	266.4

w_A	0.117	0.117	0.117	0.234	0.234	0.234	0.234
w_B	0.162	0.162	0.162	0.154	0.154	0.154	0.154
w_C	0.721	0.721	0.721	0.612	0.612	0.612	0.612
T/K	423.15	447.95	473.45	355.35	373.55	398.15	423.25
P/bar	296.8	322.0	344.5	318.5	340.2	365.0	389.0

Polymer (B):	**ethylene/propylene copolymer**		**2000VRI**
Characterization:	$M_n/kg.mol^{-1} = 51$, $M_w/kg.mol^{-1} = 120$, $M_z/kg.mol^{-1} = 210$,		
	42 mol% propene, 21 methyl groups/100 C-atoms		
Solvent (A):	**ethene**	C_2H_4	**74-85-1**
Solvent (C):	**carbon dioxide**	CO_2	**124-38-9**

Type of data: cloud points

w_A	0.9025	0.9025	0.9025	0.9025	0.9025	0.9025	0.9025	0.8550	0.8550
w_B	0.0500	0.0500	0.0500	0.0500	0.0500	0.0500	0.0500	0.0500	0.0500
w_C	0.0475	0.0475	0.0475	0.0475	0.0475	0.0475	0.0475	0.0950	0.0950
T/K	312.59	317.69	322.45	327.31	332.20	336.74	342.05	312.44	317.45
P/bar	2342	2204	2094	1996	1910	1844	1764	2585	2413

w_A	0.8550	0.8550	0.8550	0.8550	0.8550	0.8075	0.8075	0.8075	0.8075
w_B	0.0500	0.0500	0.0500	0.0500	0.0500	0.0500	0.0500	0.0500	0.0500
w_C	0.0950	0.0950	0.0950	0.0950	0.0950	0.1425	0.1425	0.1425	0.1425
T/K	322.59	327.36	332.22	337.07	342.01	312.67	317.61	322.38	327.24
P/bar	2263	2144	2041	1949	1871	2959	2712	2519	2355

continued

continued

w_A	0.8075	0.8075	0.8075	0.7600	0.7600	0.7600	0.7600	0.7600	0.7600
w_B	0.0500	0.0500	0.0500	0.0500	0.0500	0.0500	0.0500	0.0500	0.0500
w_C	0.1425	0.1425	0.1425	0.1900	0.1900	0.1900	0.1900	0.1900	0.1900
T/K	332.15	337.04	341.86	312.43	317.52	322.42	327.33	332.19	337.15
P/bar	2218	2101	2001	3705	3240	2913	2666	2470	2311

w_A	0.7600	0.8550	0.8550	0.8550	0.8550	0.8550	0.8550	0.8550	0.8100
w_B	0.0500	0.1000	0.1000	0.1000	0.1000	0.1000	0.1000	0.1000	0.1000
w_C	0.1900	0.0450	0.0450	0.0450	0.0450	0.0450	0.0450	0.0450	0.0900
T/K	342.00	312.57	317.48	322.32	327.21	332.18	336.83	342.08	312.68
P/bar	2181	2265	2140	2032	1938	1858	1785	1718	2496

w_A	0.8100	0.8100	0.8100	0.8100	0.8100	0.8100	0.7650	0.7650	0.7650
w_B	0.1000	0.1000	0.1000	0.1000	0.1000	0.1000	0.1000	0.1000	0.1000
w_C	0.0900	0.0900	0.0900	0.0900	0.0900	0.0900	0.1350	0.1350	0.1350
T/K	2334	2193	2081	1981	1894	1818	2900	2654	2457

w_A	0.7650	0.7650	0.7650	0.7650	0.7200	0.7200	0.7200	0.7200	0.7200
w_B	0.1000	0.1000	0.1000	0.1000	0.1000	0.1000	0.1000	0.1000	0.1000
w_C	0.1350	0.1350	0.1350	0.1350	0.1800	0.1800	0.1800	0.1800	0.1800
T/K	327.41	332.17	336.99	342.23	312.50	317.48	322.35	327.28	332.16
P/bar	2290	2166	2055	1950	3665	3191	2867	2618	2419

w_A	0.7200	0.7200
w_B	0.1000	0.1000
w_C	0.1800	0.1800
T/K	337.30	342.12
P/bar	2258	2132

Polymer (B):	ethylene/propylene copolymer	1992CH2

Characterization: M_n/kg.mol^{-1} = 25.2, M_w/kg.mol^{-1} = 26.0, 50 mol% propene, alternating ethylene/propylene units from complete hydrogenation of polyisoprene

Solvent (A):	ethene	C_2H_4	74-85-1
Solvent (C):	1-hexene	C_6H_{12}	592-41-6

Type of data: cloud points

$w_A = 0.122$ $w_B = 0.158$ $w_C = 0.820$

T/K	299.65	323.15	353.45	373.65	384.05	397.15	405.75	413.15	422.15
P/bar	16.9	23.0	32.1	38.5	41.2	44.0	58.0	68.0	85.0
	(VLE	VLE	VLE	VLE	VLE	VLE	LLE	LLE	LLE)

T/K	424.75	433.45	447.65	463.65	468.55	473.35	424.75	433.45	447.65
P/bar	89.2	100.0	124.0	138.0	144.0	148.0	53.1	54.1	59.8
	(LLE	LLE	LLE	LLE	LLE	LLE	VLLE	VLLE	VLLE)

T/K	468.15
P/bar	63.2
	(VLLE)

continued

continued

$w_A = 0.243$ $\qquad$ $w_B = 0.150$ $\qquad$ $w_C = 0.607$

T/K	300.65	306.15	314.15	321.85	353.75	371.95	395.85	425.55	447.85
P/bar	28.0	31.7	43.0	56.5	114.0	144.0	179.0	222.0	246.0
	(VLE	VLE	LLE	LLE	LLE	LLE	LLE	LLE	LLE)

T/K	472.55	315.95	323.35	341.15	353.75	371.95	395.85	424.15	
P/bar	266.0	35.0	39.2	47.0	55.6	63.4	73.0	81.2	
	(LLE	VLLE	VLLE	VLLE	VLLE	VLLE	VLLE	VLLE)	

$w_A = 0.355$ $\qquad$ $w_B = 0.153$ $\qquad$ $w_C = 0.492$

T/K	251.15	264.65	271.15	282.75	295.45	311.65	324.65	355.45	373.45
P/bar	187.0	176.0	175.0	182.5	194.7	216.5	233.0	269.0	290.0
	(LLE	LLE	LLE	LLE	LLE	LLE	LLE	LLE	LLE)

T/K	398.15	423.95	449.15	473.45	258.85	282.75	295.45	311.65	324.65
P/bar	320.5	348.0	370.0	384.0	17.2	30.0	36.5	45.8	55.0
	(LLE	LLE	LLE	LLE	VLLE	VLLE	VLLE	VLLE	VLLE)

T/K	355.45	373.45	398.15
P/bar	75.5	85.0	91.8
	(VLLE	VLLE	VLLE)

$w_A = 0.504$ $\qquad$ $w_B = 0.149$ $\qquad$ $w_C = 0.347$

T/K	296.05	310.85	325.75	343.25	358.85	374.85	397.25	424.15	446.55
P/bar	488.0	464.0	456.0	452.0	459.0	468.0	479.0	492.0	502.0
	(LLE	LLE	LLE	LLE	LLE	LLE	LLE	LLE	LLE)

T/K	469.65	295.65	311.15	325.95	343.25	358.85	374.85		
P/bar	511.0	48.6	59.4	70.0	79.0	88.6	92.0		
	(LLE	VLLE	VLLE	VLLE	VLLE	VLLE	VLLE)		

Polymer (B):	**ethylene/propylene copolymer**	**1997HAN**
Characterization:	M_n/kg.mol^{-1} = 2.2, M_w/kg.mol^{-1} = 2.6, 50 mol% propene,	
	T_g/K = 209.2, alternating ethylene/propylene units from	
	complete hydrogenation (98%) of polyisoprene	
Solvent (A):	**propene** $\qquad$ **C$_3$H$_6$**	**115-07-1**
Polymer (C):	**ethylene/propylene copolymer**	
Characterization:	M_n/kg.mol^{-1} = 177, M_w/kg.mol^{-1} = 195, 50 mol% propene,	
	T_g/K = 225.2, alternating ethylene/propylene units from	
	complete hydrogenation (94%) of polyisoprene	

Type of data: cloud points

Comments: The weight ratio of w_B/w_C was kept constant at 75/25 = 3/1 and the resulting molar mass averages are M_n/kg.mol^{-1} = 3.47 and M_w/kg.mol^{-1} = 51.

$w_B + w_C$	0.012	0.012	0.012	0.050	0.050	0.050	0.100	0.100	0.100
T/K	322.95	372.85	423.15	323.45	373.05	423.05	323.15	372.95	423.35
P/bar	324.5	375.1	424.6	310.3	364.5	417.1	282.0	340.8	396.8

continued

continued

Type of data: coexistence data

$T/K = 423.15$

$P/$ bar	w_B feed phase	gel phase	sol phase
356	0.050	unknown	0.041

Polymer (B):	**ethylene/propylene copolymer**	**1997HAN**
Characterization:	$M_n/kg.mol^{-1} = 2.2$, $M_w/kg.mol^{-1} = 2.6$, 50 mol% propene, $T_g/K = 209.2$, alternating ethylene/propylene units from complete hydrogenation (98%) of polyisoprene	
Solvent (A):	**propene** C_3H_6	**115-07-1**
Polymer (C):	**ethylene/propylene copolymer**	
Characterization:	$M_n/kg.mol^{-1} = 177$, $M_w/kg.mol^{-1} = 195$, 50 mol% propene, $T_g/K = 225.2$, alternating ethylene/propylene units from complete hydrogenation (94%) of polyisoprene	
Solvent (D):	**n-dodecane** $C_{12}H_{26}$	**112-40-3**

Type of data: cloud points

Comments: The weight ratio of $w_B/w_C/w_D$ was kept constant at 3/1/1 and the resulting molar mass averages are $M_n/kg.mol^{-1} = 3.47$ and $M_w/kg.mol^{-1} = 51$.

w_A	0.05	0.05	0.05
T/K	323.15	372.95	423.05
P/bar	282.0	340.8	396.8

Polymer (B):	**ethylene/vinyl acetate copolymer**	**1999KIN**
Characterization:	$M_n/kg.mol^{-1} = 61.9$, $M_w/kg.mol^{-1} = 167$, 27.5 wt% vinyl acetate	
Solvent (A):	**ethene** C_2H_4	**74-85-1**
Solvent (C):	**carbon dioxide** CO_2	**124-38-9**

Type of data: cloud points

w_A	0.800	0.800	0.800	0.800	0.800	0.800	0.750	0.750	0.750
w_B	0.150	0.150	0.150	0.150	0.150	0.150	0.150	0.150	0.150
w_C	0.050	0.050	0.050	0.050	0.050	0.050	0.100	0.100	0.100
T/K	393.15	413.15	433.15	453.15	473.15	493.15	393.15	413.15	433.15
P/MPa	127.7	123.0	119.2	115.9	113.0	110.2	128.5	123.8	120.1

w_A	0.750	0.750	0.750	0.700	0.700	0.700	0.700	0.700	0.700
w_B	0.150	0.150	0.150	0.150	0.150	0.150	0.150	0.150	0.150
w_C	0.100	0.100	0.100	0.150	0.150	0.150	0.150	0.150	0.150
T/K	453.15	473.15	493.15	393.15	413.15	433.15	453.15	473.15	493.15
P/MPa	117.0	113.9	111.2	130.3	125.7	121.8	118.9	115.8	113.1

Polymer (B):	ethylene/vinyl acetate copolymer								1999KIN
Characterization:	M_n/kg.mol^{-1} = 61.9, M_w/kg.mol^{-1} = 167, 27.5 wt% vinyl acetate								
Solvent (A):	**ethene**		**C$_2$H$_4$**						**74-85-1**
Solvent (C):	**ethane**		**C$_2$H$_6$**						**74-84-0**

Type of data: cloud points

w_A	0.800	0.800	0.800	0.800	0.800	0.800	0.750	0.750	0.750
w_B	0.150	0.150	0.150	0.150	0.150	0.150	0.150	0.150	0.150
w_C	0.050	0.050	0.050	0.050	0.050	0.050	0.100	0.100	0.100
T/K	393.15	413.15	433.15	453.15	473.15	493.15	393.15	413.15	433.15
P/MPa	125.6	121.2	117.2	113.9	110.8	108.0	126.0	121.5	117.3
w_A	0.750	0.750	0.750	0.700	0.700	0.700	0.700	0.700	0.700
w_B	0.150	0.150	0.150	0.150	0.150	0.150	0.150	0.150	0.150
w_C	0.100	0.100	0.100	0.150	0.150	0.150	0.150	0.150	0.150
T/K	453.15	473.15	493.15	393.15	413.15	433.15	453.15	473.15	493.15
P/MPa	114.0	111.3	108.4	124.8	120.0	115.8	112.4	109.7	107.5

Polymer (B):	ethylene/vinyl acetate copolymer								1999KIN
Characterization:	M_n/kg.mol^{-1} = 61.9, M_w/kg.mol^{-1} = 167, 27.5 wt% vinyl acetate								
Solvent (A):	**ethene**		**C$_2$H$_4$**						**74-85-1**
Solvent (C):	**helium**		**He**						**7440-59-7**

Type of data: cloud points

w_A	0.955	0.955	0.955	0.955	0.955	0.955	0.920	0.920	0.920
w_B	0.035	0.035	0.035	0.035	0.035	0.035	0.070	0.070	0.070
w_C	0.010	0.010	0.010	0.010	0.010	0.010	0.010	0.010	0.010
T/K	393.15	413.15	433.15	453.15	473.15	493.15	393.15	413.15	433.15
P/MPa	163.6	156.1	149.2	142.8	137.5	132.7	168.1	159.3	151.5
w_A	0.920	0.920	0.920	0.870	0.870	0.870	0.870	0.870	0.870
w_B	0.070	0.070	0.070	0.120	0.120	0.120	0.120	0.120	0.120
w_C	0.010	0.010	0.010	0.010	0.010	0.010	0.010	0.010	0.010
T/K	453.15	473.15	493.15	393.15	413.15	433.15	453.15	473.15	493.15
P/MPa	145.4	139.9	135.3	165.2	157.6	150.7	144.6	139.2	134.5
w_A	0.840	0.840	0.840	0.840	0.840	0.840	0.790	0.790	0.790
w_B	0.150	0.150	0.150	0.150	0.150	0.150	0.200	0.200	0.200
w_C	0.010	0.010	0.010	0.010	0.010	0.010	0.010	0.010	0.010
T/K	393.15	413.15	433.15	453.15	473.15	493.15	393.15	413.15	433.15
P/MPa	163.8	156.2	149.4	143.3	137.9	133.3	163.2	155.7	148.8
w_A	0.790	0.790	0.790	0.640	0.640	0.640	0.640	0.640	0.640
w_B	0.200	0.200	0.200	0.350	0.350	0.350	0.350	0.350	0.350
w_C	0.010	0.010	0.010	0.010	0.010	0.010	0.010	0.010	0.010
T/K	453.15	473.15	493.15	393.15	413.15	433.15	453.15	473.15	493.15
P/MPa	142.5	137.5	132.6	158.7	150.1	142.7	137.4	132.5	128.4

Polymer (B):   **ethylene/vinyl acetate copolymer**   **1999KIN**

Characterization: $M_n/\text{kg.mol}^{-1} = 61.9$, $M_w/\text{kg.mol}^{-1} = 167$, 27.5 wt% vinyl acetate

Solvent (A):   **ethene**   **C$_2$H$_4$**   **74-85-1**

Solvent (C):   **nitrogen**   **N$_2$**   **7727-37-9**

Type of data: cloud points

w_A	0.800	0.800	0.800	0.800	0.800	0.800	0.750	0.750	0.750
w_B	0.150	0.150	0.150	0.150	0.150	0.150	0.150	0.150	0.150
w_C	0.050	0.050	0.050	0.050	0.050	0.050	0.100	0.100	0.100
T/K	393.15	413.15	433.15	453.15	473.15	493.15	393.15	413.15	433.15
P/MPa	146.5	140.5	135.8	130.5	125.4	120.5	172.5	164.7	155.6

w_A	0.750	0.750	0.750	0.700	0.700	0.700	0.700	0.700	0.700
w_B	0.150	0.150	0.150	0.150	0.150	0.150	0.150	0.150	0.150
w_C	0.100	0.100	0.100	0.150	0.150	0.150	0.150	0.150	0.150
T/K	453.15	473.15	493.15	393.15	413.15	433.15	453.15	473.15	493.15
P/MPa	148.8	142.5	137.5	193.0	184.8	176.8	168.9	161.4	154.4

Polymer (B):   **ethylene/vinyl acetate copolymer**   **1999KIN**

Characterization: $M_n/\text{kg.mol}^{-1} = 61.9$, $M_w/\text{kg.mol}^{-1} = 167$, 27.5 wt% vinyl acetate

Solvent (A):   **ethene**   **C$_2$H$_4$**   **74-85-1**

Solvent (C):   **propane**   **C$_3$H$_8$**   **74-98-6**

Type of data: cloud points

w_A	0.800	0.800	0.800	0.800	0.800	0.800	0.750	0.750	0.750
w_B	0.150	0.150	0.150	0.150	0.150	0.150	0.150	0.150	0.150
w_C	0.050	0.050	0.050	0.050	0.050	0.050	0.100	0.100	0.100
T/K	393.15	413.15	433.15	453.15	473.15	493.15	393.15	413.15	433.15
P/MPa	122.0	117.7	113.6	110.5	107.6	104.9	116.3	112.4	109.0

w_A	0.750	0.750	0.750	0.700	0.700	0.700	0.700	0.700	0.700
w_B	0.150	0.150	0.150	0.150	0.150	0.150	0.150	0.150	0.150
w_C	0.100	0.100	0.100	0.150	0.150	0.150	0.150	0.150	0.150
T/K	453.15	473.15	493.15	393.15	413.15	433.15	453.15	473.15	493.15
P/MPa	105.8	103.4	100.8	112.9	109.2	105.9	103.0	100.4	98.0

Polymer (B):   **ethylene/vinyl acetate copolymer**   **1991NIE**

Characterization: $M_n/\text{kg.mol}^{-1} = 29.3$, $M_w/\text{kg.mol}^{-1} = 162$, 10.9 wt% vinyl acetate,

                       Exxon Chemical Company

Solvent (A):   **ethene**   **C$_2$H$_4$**   **74-85-1**

Solvent (C):   **vinyl acetate**   **C$_4$H$_6$O$_2$**   **108-05-4**

Type of data: coexistence data

$T/\text{K} = 403.15$

Total feed concentrations of the homogeneous system before demixing:

$w_A = 0.590$, $w_B = 0.153$, $w_C = 0.257$.

continued

continued

$P/$ MPa	w_A	w_B	w_C	w_A	w_B	w_C
		bottom phase			top phase	
56.5	0.311	0.542	0.147			
75.5				0.683	0.016	0.301
87.8	0.450	0.352	0.198	0.678	0.023	0.299
100.2	0.546	0.212	0.242	0.660	0.048	0.292
104.1	0.590	0.153	0.257 (cloud point)			

$T/K = 403.15$

Total feed concentrations of the homogeneous system before demixing:

$w_A = 0.347$, $w_B = 0.157$, $w_C = 0.496$.

$P/$ MPa	w_A	w_B	w_C	w_A	w_B	w_C
		bottom phase			top phase	
58.5				0.387	0.037	0.576
68.0	0.322	0.215	0.463	0.380	0.056	0.564
72.7	0.347	0.157	0.496 (cloud point)			

$T/K = 433.15$

Total feed concentrations of the homogeneous system before demixing:

$w_A = 0.254$, $w_B = 0.155$, $w_C = 0.591$.

$P/$ MPa	w_A	w_B	w_C	w_A	w_B	w_C
		bottom phase			top phase	
48.7	0.231	0.230	0.539	0.277	0.050	0.673
53.4	0.254	0.155	0.591 (cloud point)			

Polymer (B):		**ethylene/vinyl acetate copolymer**							**1991NIE**
Characterization:		$M_n/kg.mol^{-1} = 33.4$, $M_w/kg.mol^{-1} = 138$, 17.5 wt% vinyl acetate,							
		Exxon Chemical Company							
Solvent (A):		**ethene**		C_2H_4					**74-85-1**
Solvent (C):		**vinyl acetate**		$C_4H_6O_2$					**108-05-4**

Type of data: cloud points

w_A	0.422	0.422	0.422	0.217	0.217	0.217	0.120	0.120	0.120
w_B	0.159	0.159	0.159	0.156	0.156	0.156	0.298	0.298	0.298
w_C	0.419	0.419	0.419	0.627	0.627	0.627	0.582	0.582	0.582
T/K	373.15	403.15	433.15	373.15	403.15	433.15	378.15	388.15	403.15
P/MPa	75.3	69.8	68.0	45.8	38.3	39.0	21.4	19.5	18.0

continued

continued

w_A	0.120	0.120	0.120
w_B	0298	0.298	0.298
w_C	0.582	0.582	0.582
T/K	423.15	433.15	443.15
P/MPa	19.4	20.5	22.2

Type of data: coexistence data

$T/K = 433.15$

Total feed concentrations of the homogeneous system before demixing:

$w_A = 0.217$, $w_B = 0.156$, $w_C = 0.627$.

$P/$ MPa	w_A	w_B	w_C	w_A	w_B	w_C
		bottom phase			top phase	
27.5	0.160	0.360	0.480	0.251	0.041	0.708
33.5	0.188	0.276	0.536	0.242	0.061	0.697
39.0	0.217	0.156	0.627 (cloud point)			

$T/K = 433.15$

Total feed concentrations of the homogeneous system before demixing:

$w_A = 0.343$, $w_B = 0.136$, $w_C = 0.521$.

$P/$ MPa	w_A	w_B	w_C	w_A	w_B	w_C
		bottom phase			top phase	
19.9				0.4042	0.0004	0.5954
43.0	0.256	0.361	0.383	0.3936	0.0127	0.5937
57.3	0.343	0.136	0.521 (cloud point)			

$T/K = 433.15$

Total feed concentrations of the homogeneous system before demixing:

$w_A = 0.422$, $w_B = 0.159$, $w_C = 0.419$.

$P/$ MPa	w_A	w_B	w_C	w_A	w_B	w_C
		bottom phase			top phase	
37.7	0.221	0.532	0.247	0.498	0.025	0.477
50.0	0.288	0.404	0.308	0.487	0.038	0.475
61.9	0.365	0.272	0.363	0.476	0.062	0.462
68.0	0.422	0.159	0.419 (cloud point)			

continued

continued

$T/K = 433.15$

Total feed concentrations of the homogeneous system before demixing:

$w_A = 0.645$, $w_B = 0.158$, $w_C = 0.197$.

$P/$ MPa	w_A	w_B	w_C	w_A	w_B	w_C
		bottom phase			top phase	
30.4	0.176	0.760	0.064	0.779	0.0008	0.221
43.7	0.244	0.670	0.086	0.766	0.0009	0.233
66.6	0.370	0.506	0.124	0.754	0.0130	0.233
86.0	0.513	0.325	0.162	0.736	0.0290	0.235
97.2	0.645	0.158	0.197 (cloud point)			

$T/K = 463.15$

Total feed concentrations of the homogeneous system before demixing:

$w_A = 0.460$, $w_B = 0.135$, $w_C = 0.405$.

$P/$ MPa	w_A	w_B	w_C	w_A	w_B	w_C
		bottom phase			top phase	
47.3	0.273	0.506	0.221	0.582	0.012	0.406
58.8	0.346	0.363	0.291	0.546	0.051	0.403
70.7	0.460	0.135	0.405 (cloud point)			

Fractionation during demixing

		molar mass averages				
Demixing pressure		bottom phase			top phase	
$P/$ MPa	$M_n/$ kg/mol	$M_w/$ kg/mol	$M_z/$ kg/mol	$M_n/$ kg/mol	$M_w/$ kg/mol	$M_z/$ kg/mol
47.3				5.61	13.63	27.45
58.8				11.96	26.06	44.20
70.7	33.4	137.7 (feed sample)				

Polymer (B):	ethylene/vinyl acetate copolymer	**1991NIE**
Characterization:	M_n/kg.mol^{-1} = 33.6, M_w/kg.mol^{-1} = 231, 26.0 wt% vinyl acetate, Exxon Chemical Company	
Solvent (A):	ethene C_2H_4	**74-85-1**
Solvent (C):	vinyl acetate $C_4H_6O_2$	**108-05-4**

continued

continued

Type of data: cloud points

w_A	0.317	0.317	0.317	0.317	0.317	0.317	0.340	0.340	0.413
w_B	0.360	0.360	0.360	0.360	0.360	0.360	0.322	0.322	0.207
w_C	0.323	0.323	0.323	0.323	0.323	0.323	0.338	0.338	0.380
T/K	363.15	373.15	393.15	413.15	438.15	453.15	383.15	433.15	403.15
P/MPa	52.3	53.1	54.0	55.2	57.2	58.5	62.4	64.5	75.4

w_A	0.413	0.505	0.505	0.505	0.505	0.505	0.631	0.631	0.631
w_B	0.207	0.257	0.257	0.257	0.257	0.257	0.187	0.187	0.187
w_C	0.380	0.238	0.238	0.238	0.238	0.238	0.182	0.182	0.182
T/K	433.15	363.15	383.15	403.15	438.15	453.15	363.15	383.15	403.15
P/MPa	75.1	88.3	86.1	83.7	82.0	81.8	106.8	102.8	100.0

w_A	0.631	0.795	0.795	0.795
w_B	0.187	0.105	0.105	0.105
w_C	0.182	0.100	0.100	0.100
T/K	443.15	383.15	403.15	433.15
P/MPa	96.5	122.0	118.5	113.0

Polymer (B):   **ethylene/vinyl acetate copolymer**   **1991NIE**
Characterization: $M_n/kg.mol^{-1} = 35$, $M_w/kg.mol^{-1} = 126.5$, 27.5 wt% vinyl acetate,
                        Exxon Chemical Company
Solvent (A):   **ethene**   **C_2H_4**   **74-85-1**
Solvent (C):   **vinyl acetate**   **$C_4H_6O_2$**   **108-05-4**

Type of data: cloud points

w_A	0.594	0.594	0.594	0.200	0.200	0.200
w_B	0.148	0.148	0.148	0.152	0.152	0.152
w_C	0.258	0.258	0.258	0.648	0.648	0.648
T/K	398.15	403.15	433.15	401.15	413.15	433.15
P/MPa	90.2	86.4	84.6	26.1	27.2	29.0

Type of data: coexistence data

$T/K = 403.15$

Total feed concentrations of the homogeneous system before demixing:

$w_A = 0.353$, $w_B = 0.159$, $w_C = 0.488$.

$P/$ MPa	w_A	w_B	w_C	w_A	w_B	w_C
		bottom phase			top phase	
30.7				0.415	0.016	0.569
31.2	0.220	0.446	0.334			
39.1				0.409	0.025	0.563
39.7	0.255	0.349	0.396			
46.7	0.319	0.224	0.457	0.403	0.044	0.553
49.5	0.353	0.159	0.488 (cloud point)			

continued

continued

$T/K = 433.15$

Total feed concentrations of the homogeneous system before demixing:

$w_A = 0.594$, $w_B = 0.148$, $w_C = 0.258$.

$P/$ MPa	w_A	w_B bottom phase	w_C	w_A	w_B top phase	w_C
35.5	0.204	0.702	0.094			
48.2	0.273	0.613	0.114	0.709	0.008	0.283
56.6	0.322	0.551	0.127	0.705	0.013	0.282
67.0	0.392	0.447	0.161	0.692	0.020	0.288
78.6	0.500	0.290	0.210	0.677	0.040	0.283
84.6	0.594	0.148	0.258 (cloud point)			

$T/K = 433.15$

Total feed concentrations of the homogeneous system before demixing:

$w_A = 0.349$, $w_B = 0.152$, $w_C = 0.499$.

$P/$ MPa	w_A	w_B bottom phase	w_C	w_A	w_B top phase	w_C
27.8	0.176	0.529	0.295	0.419	0.012	0.567
35.8	0.221	0.439	0.340	0.413	0.021	0.566
46.2	0.289	0.308	0.403	0.402	0.051	0.547
51.0	0.349	0.152	0.499 (cloud point)			

$T/K = 433.15$

Total feed concentrations of the homogeneous system before demixing:

$w_A = 0.200$, $w_B = 0.152$, $w_C = 0.648$.

$P/$ MPa	w_A	w_B bottom phase	w_C	w_A	w_B top phase	w_C
14.5	0.140	0.400	0.460	0.238	0.014	0.748
19.5	0.154	0.356	0.490	0.231	0.031	0.738
26.0	0.181	0.233	0.586	0.228	0.051	0.721
29.0	0.200	0.152	0.648 (cloud point)			

Polymer (B):	ethylene/vinyl acetate copolymer		1999KIN

Characterization: $M_n/\text{kg.mol}^{-1} = 61.9$, $M_w/\text{kg.mol}^{-1} = 167$, 27.5 wt% vinyl acetate

Solvent (A):	ethene	C_2H_4	74-85-1
Solvent (C):	vinyl acetate	$C_4H_6O_2$	108-05-4

Type of data: cloud points

w_A	0.6715	0.6715	0.6715	0.6715	0.6715	0.6715	0.6375	0.6375	0.6375
w_B	0.1500	0.1500	0.1500	0.1500	0.1500	0.1500	0.1500	0.1500	0.1500
w_C	0.1785	0.1785	0.1785	0.1785	0.1785	0.1785	0.2125	0.2125	0.2125
T/K	393.15	413.15	433.15	453.15	473.15	493.15	393.15	413.15	433.15
P/MPa	99.5	97.5	95.5	94.1	92.7	91.5	94.3	92.6	91.1

w_A	0.6375	0.6375	0.6375	0.4250	0.4250	0.4250	0.4250	0.4250	0.4250
w_B	0.1500	0.1500	0.1500	0.1500	0.1500	0.1500	0.1500	0.1500	0.1500
w_C	0.2125	0.2125	0.2125	0.4250	0.4250	0.4250	0.4250	0.4250	0.4250
T/K	453.15	473.15	493.15	393.15	413.15	433.15	453.15	473.15	493.15
P/MPa	89.8	88.6	87.5	60.7	61.1	61.6	62.0	62.4	62.7

w_A	0.2125	0.2125	0.2125	0.2125	0.2125	0.2125	0.0850	0.0850	0.0850
w_B	0.1500	0.1500	0.1500	0.1500	0.1500	0.1500	0.1500	0.1500	0.1500
w_C	0.6375	0.6375	0.6375	0.6375	0.6375	0.6375	0.7650	0.7650	0.7650
T/K	393.15	413.15	433.15	453.15	473.15	493.15	393.15	413.15	433.15
P/MPa	26.7	28.9	30.9	33.1	34.8	37.6	4.8	7.1	9.8

w_A	0.0850	0.0850	0.0850
w_B	0.1500	0.1500	0.1500
w_C	0.7650	0.7650	0.7650
T/K	453.15	473.15	493.15
P/MPa	13.2	17.5	19.5

w_A	0.4250	0.4250	0.4250	0.4250	0.4250	0.4250	0.4250	0.3060	0.3060
w_B	0.1500	0.1500	0.1500	0.1500	0.1500	0.1500	0.1500	0.1500	0.1500
w_C	0.4250	0.4250	0.4250	0.4250	0.4250	0.4250	0.4250	0.5440	0.5440
T/K	348.15	353.15	358.15	368.15	383.15	393.15	423.15	348.15	353.15
P/MPa	48.7	47.5	46.8	46.7	47.5	48.0	50.8	42.2	41.4

w_A	0.3060	0.3060	0.3060	0.3060	0.3060	0.3060	0.3060	0.3060	0.3060
w_B	0.1500	0.1500	0.1500	0.1500	0.1500	0.1500	0.1500	0.1500	0.1500
w_C	0.5440	0.5440	0.5440	0.5440	0.5440	0.5440	0.5440	0.5440	0.5440
T/K	358.15	368.15	373.15	383.15	393.15	403.15	423.15	443.15	463.15
P/MPa	41.0	40.9	41.0	41.1	41.5	42.0	43.9	45.3	46.9

w_A	0.0850	0.0850	0.0850	0.0850	0.0850	0.0850	0.0850	0.0850	0.0850
w_B	0.1500	0.1500	0.1500	0.1500	0.1500	0.1500	0.1500	0.1500	0.1500
w_C	0.7650	0.7650	0.7650	0.7650	0.7650	0.7650	0.7650	0.7650	0.7650
T/K	341.15	343.15	348.15	353.15	358.15	368.15	373.15	383.15	403.15
P/MPa	4.8	3.5	2.9	2.8	2.7	3.0	3.4	4.0	5.7

w_A	0.0850	0.0850	0.0850
w_B	0.1500	0.1500	0.1500
w_C	0.7650	0.7650	0.7650
T/K	423.15	443.15	463.15
P/MPa	8.6	11.0	15.5

continued

continued

w_A	0.4825	0.4825	0.4825	0.4825	0.4825	0.4825	0.4600	0.4600	0.4600
w_B	0.0350	0.0350	0.0350	0.0350	0.0350	0.0350	0.0800	0.0800	0.0800
w_C	0.4825	0.4825	0.4825	0.4825	0.4825	0.4825	0.4600	0.4600	0.4600
T/K	393.15	413.15	433.15	453.15	473.15	493.15	393.15	413.15	433.15
P/MPa	59.0	59.9	60.6	61.2	61.6	61.9	61.9	62.3	62.9

w_A	0.4600	0.4600	0.4600	0.4250	0.4250	0.4250	0.4250	0.4250	0.4250
w_B	0.0800	0.0800	0.0800	0.1500	0.1500	0.1500	0.1500	0.1500	0.1500
w_C	0.4600	0.4600	0.4600	0.4250	0.4250	0.4250	0.4250	0.4250	0.4250
T/K	453.15	473.15	493.15	393.15	413.15	433.15	453.15	473.15	493.15
P/MPa	63.3	63.8	64.1	60.7	61.1	61.6	62.0	62.4	62.7

w_A	0.3750	0.3750	0.3750	0.3750	0.3750	0.3750			
w_B	0.2500	0.2500	0.2500	0.2500	0.2500	0.2500			
w_C	0.3750	0.3750	0.3750	0.3750	0.3750	0.3750			
T/K	393.15	413.15	433.15	453.15	473.15	493.15			
P/MPa	53.8	54.9	56.0	56.9	58.0	58.9			

w_A	0.3450	0.3450	0.3450	0.3450	0.3450	0.3450	0.3400	0.3400	0.3400
w_B	0.3100	0.3100	0.3100	0.3100	0.3100	0.3100	0.3200	0.3200	0.3200
w_C	0.3450	0.3450	0.3450	0.3450	0.3450	0.3450	0.3400	0.3400	0.3400
T/K	393.15	413.15	433.15	453.15	473.15	493.15	393.15	413.15	433.15
P/MPa	47.5	49.2	50.7	52.1	53.5	54.7	49.0	50.4	51.8

w_A	0.3400	0.3400	0.3400	0.3250	0.3250	0.3250	0.3250	0.3250	0.3250
w_B	0.3200	0.3200	0.3200	0.3500	0.3500	0.3500	0.3500	0.3500	0.3500
w_C	0.3400	0.3400	0.3400	0.3250	0.3250	0.3250	0.3250	0.3250	0.3250
T/K	453.15	473.15	493.15	393.15	413.15	433.15	453.15	473.15	493.15
P/MPa	53.2	54.5	55.7	48.6	50.1	51.6	53.0	54.3	55.5

w_A	0.2750	0.2750	0.2750	0.2750	0.2750	0.2750			
w_B	0.4500	0.4500	0.4500	0.4500	0.4500	0.4500			
w_C	0.2750	0.2750	0.2750	0.2750	0.2750	0.2750			
T/K	393.15	413.15	433.15	453.15	473.15	493.15			
P/MPa	42.8	44.6	46.2	47.7	49.1	50.3			

Polymer (B):	**ethylene/vinyl acetate copolymer**	**1984WOH**

Characterization: $M_n/kg.mol^{-1} = 15.5$, $M_w/kg.mol^{-1} = 40.8$, 32.3 wt% vinyl acetate

Solvent (A):	**ethene**	C_2H_4	**74-85-1**
Solvent (C):	**vinyl acetate**	$C_4H_6O_2$	**108-05-4**

Type of data: cloud points

Comments: The total monomer weight fraction ratio $w_A/w_C = 64/36$ was kept constant.

w_B	0.05	0.10	0.15	0.20	0.30	0.05	0.10	0.15	0.20
T/K	393.15	393.15	393.15	393.15	393.15	413.15	413.15	413.15	413.15
P/MPa	80.4	77.8	75.8	73.3	71.3	78.3	76.8	75.0	72.3

w_B	0.30	0.05	0.10	0.15	0.20	0.30	0.05	0.10	0.15
T/K	413.15	433.15	433.15	433.15	433.15	433.15	453.15	453.15	453.15
P/MPa	70.3	76.8	76.0	74.0	70.3	68.8	76.3	74.8	73.3

w_B	0.20	0.30
T/K	453.15	453.15
P/MPa	68.8	67.8

Polymer (B):   **ethylene/vinyl acetate copolymer**   **1991NIE**
Characterization: $M_n/\text{kg.mol}^{-1} = 13.0$, $M_w/\text{kg.mol}^{-1} = 22.1$, 32.5 wt% vinyl acetate,
                      Exxon Chemical Company
Solvent (A):   **ethene**   **C₂H₄** → **C_2H_4**   **74-85-1**
Solvent (C):   **vinyl acetate**   **$C_4H_6O_2$**   **108-05-4**

Type of data: cloud points

w_A	0.761	0.761	0.761	0.761	0.761	0.657	0.657	0.657	0.657
w_B	0.073	0.073	0.073	0.073	0.073	0.136	0.136	0.136	0.136
w_C	0.166	0.166	0.166	0.166	0.166	0.207	0.207	0.207	0.207
T/K	348.15	373.15	403.15	433.15	453.15	373.15	403.15	433.15	453.15
P/MPa	65.8	65.4	65.8	66.8	67.8	84.9	82.9	81.4	80.6

w_A	0.505	0.505	0.505	0.361	0.361	0.361	0.430	0.430	0.521
w_B	0.184	0.184	0.184	0.246	0.246	0.246	0.273	0.273	0.084
w_C	0.311	0.311	0.311	0.393	0.393	0.393	0.297	0.297	0.395
T/K	358.15	383.15	438.15	373.15	403.15	433.15	373.15	433.15	373.15
P/MPa	98.1	94.6	90.2	53.4	55.6	58.5	63.1	64.3	61.7

w_A	0.521
w_B	0.084
w_C	0.395
T/K	433.15
P/MPa	64.1

Polymer (B):   **ethylene/vinyl acetate copolymer**   **1991FIN**
Characterization: $M_n/\text{kg.mol}^{-1} = 13.1$, $M_w/\text{kg.mol}^{-1} = 41.3$, 34.2 wt% vinyl acetate,
                      Leuna-Werke, Germany
Solvent (A):   **ethene**   **C_2H_4**   **74-85-1**
Solvent (C):   **vinyl acetate**   **$C_4H_6O_2$**   **108-05-4**

Type of data: coexistence data

$T/K = 473.15$

Total feed concentrations of the homogeneous system before demixing:

$w_A = 0.4395$, $w_B = 0.2500$, $w_C = 0.2565$.

$P/$ MPa	w_A	w_B	w_C	w_A	w_B	w_C
		bottom phase			top phase	
28.15	0.1554	0.7016	0.1489			
29.32	0.1640	0.6923	0.1515	0.6095	0.0700	0.3299
31.28	0.1720	0.7010	0.1498	0.6801	0.0090	0.3305
32.54	0.1863	0.6719	0.1600	0.6573	0.0307	0.3314
36.19	0.1968	0.6550	0.1507			
38.05	0.2118	0.6400	0.1557	0.6211	0.0314	0.3279
50.12	0.2594	0.6017	0.1591	0.6653	0.0381	0.3331

continued

continued

P/ MPa	w_A	w_B	w_C	w_A	w_B	w_C
		bottom phase			top phase	
51.40	0.2736	0.5415	0.1870	0.6590	0.0266	0.3318
52.54	0.3404	0.4725	0.2066	0.6541	0.0300	0.3441
59.01	0.3348	0.4848	0.2051	0.6361	0.0552	0.3302
64.90	0.3782	0.4188	0.2262	0.6550	0.0692	0.3210
66.94	0.4114	0.3811	0.2295	0.6427	0.0483	0.3296
71.00	0.4675	0.3011	0.2425	0.6232	0.0734	0.3230

$T/K = 473.15$

Total feed concentrations of the homogeneous system before demixing:

$w_A = 0.4125$, $w_B = 0.2500$, $w_C = 0.3375$.

P/ MPa	w_A	w_B	w_C	w_A	w_B	w_C
		bottom phase			top phase	
20.11	0.1116	0.7008	0.1876	0.5426	0.0392	0.4182
20.70	0.1193	0.6861	0.1947	0.5672	0.0000	0.4328
20.99	0.1171	0.6957	0.1872	0.5395	0.0169	0.4436
21.09	0.1166	0.6876	0.1959	0.5397	0.0303	0.4299
21.68	0.1191	0.6864	0.1944	0.5416	0.0299	0.4284
21.68	0.1204	0.6815	0.1981	0.5366	0.0214	0.4420
29.32	0.1650	0.6410	0.2009	0.5621	0.0232	0.4292
30.70	0.1742	0.6255	0.2047	0.5571	0.0330	0.4217
36.58	0.2132	0.5672	0.2132	0.5274	0.0362	0.4432
38.35	0.2273	0.5445	0.2405	0.5283	0.0414	0.4431
42.77	0.2487	0.6113	0.2400	0.5705	0.0215	0.4070
44.63	0.2630	0.4983	0.2387	0.5659	0.0281	0.4060
48.16	0.2846	0.4441	0.2714	0.5340	0.0547	0.4112
49.15	0.3009	0.4163	0.2828	0.5317	0.0330	0.4350
50.81	0.3188	0.3966	0.2846	0.5334	0.0512	0.4154
51.60	0.3308	0.3757	0.2935	0.4975	0.0870	0.4155
56.99	0.3618	0.3357	0.3025	0.5201	0.0696	0.4102
60.43	0.3881	0.2553	0.3567	0.5047	0.0974	0.3979

Polymer (B): **ethylene/vinyl acetate copolymer** **1991NIE**
Characterization: M_n/kg.mol^{-1} = 35, 58.0 wt% vinyl acetate, Exxon Chemical Company
Solvent (A): **ethene** **C$_2$H$_4$** **74-85-1**
Solvent (C): **vinyl acetate** **C$_4$H$_6$O$_2$** **108-05-4**

Type of data: cloud points

continued

continued

w_A	0.096	0.096	0.096	0.148	0.152	0.152	0.160	0.196	0.196
w_B	0.200	0.200	0.200	0.214	0.214	0.214	0.205	0.198	0.198
w_C	0.704	0.704	0.704	0.638	0.642	0.642	0.635	0.606	0.606
T/K	363.15	393.15	423.15	393.15	363.15	393.15	393.15	363.15	393.15
P/MPa	4.2	5.3	6.8	7.9	6.3	8.4	9.8	12.0	18.0

w_A	0.196	0.299	0.401	0.401	0.401
w_B	0.198	0.203	0.204	0.204	0.204
w_C	0.606	0.498	0.405	0.405	0.405
T/K	423.15	363.15	363.15	393.15	423.15
P/MPa	23.3	36.0	65.0	64.5	64.2

Type of data: coexistence data

$T/K = 393.15$

Total feed concentrations of the homogeneous system before demixing:

$w_A = 0.167$, $w_B = 0.219$, $w_C = 0.614$.

$P/$ MPa	w_A	w_B	w_C	w_A	w_B	w_C
		bottom phase			top phase	
8.3				0.231	0.013	0.756
8.65	0.128	0.350	0.522			
9.6	0.130	0.337	0.533	0.227	0.021	0.752
11.0	0.151	0.286	0.563			
11.65				0.217	0.050	0.733
12.1	0.167	0.219	0.614 (cloud point)			

$T/K = 393.15$

Total feed concentrations of the homogeneous system before demixing:

$w_A = 0.251$, $w_B = 0.200$, $w_C = 0.549$.

$P/$ MPa	w_A	w_B	w_C	w_A	w_B	w_C
		bottom phase			top phase	
9.85	0.112	0.489	0.399	0.330	0.004	0.666
13.9	0.132	0.487	0.381	0.346	0.004	0.645
17.4	0.150	0.441	0.409	0.343	0.007	0.650
20.0	0.164	0.415	0.421	0.336	0.009	0.650
24.9	0.200	0.328	0.472			
25.2				0.319	0.037	0.644
27.1	0.224	0.272	0.504			
27.7				0.311	0.061	0.628
28.1	0.251	0.200	0.549 (cloud point)			

Comments: Some three-phase demixing data below 10 MPa can be found in the original source.

Polymer (B):   **ethylene/vinyl acetate copolymer**   **1999KIN**

Characterization: $M_n/\text{kg.mol}^{-1} = 61.9$, $M_w/\text{kg.mol}^{-1} = 167$, 27.5 wt% vinyl acetate

Solvent (A):   **ethene** C_2H_4 **74-85-1**

Solvent (C):   **vinyl acetate** $C_4H_6O_2$ **108-05-4**

Solvent (D):   **carbon dioxide** CO_2 **124-38-9**

Type of data: cloud points

w_A	0.6000	0.6000	0.6000	0.6000	0.6000	0.6000	0.5625	0.5625	0.5625
w_B	0.1500	0.1500	0.1500	0.1500	0.1500	0.1500	0.1500	0.1500	0.1500
w_C	0.2000	0.2000	0.2000	0.2000	0.2000	0.2000	0.1875	0.1875	0.1875
w_D	0.0500	0.0500	0.0500	0.0500	0.0500	0.0500	0.1000	0.1000	0.1000
T/K	393.15	413.15	433.15	453.15	473.15	493.15	393.15	413.15	433.15
P/MPa	97.9	96.0	94.3	92.6	91.1	90.0	101.5	99.4	97.3

w_A	0.5625	0.5625	0.5625	0.5250	0.5250	0.5250	0.5250	0.5250	0.5250
w_B	0.1500	0.1500	0.1500	0.1500	0.1500	0.1500	0.1500	0.1500	0.1500
w_C	0.1875	0.1875	0.1875	0.1750	0.1750	0.1750	0.1750	0.1750	0.1750
w_D	0.1000	0.1000	0.1000	0.1500	0.1500	0.1500	0.1500	0.1500	0.1500
T/K	453.15	473.15	493.15	393.15	413.15	433.15	453.15	473.15	493.15
P/MPa	95.7	94.2	92.9	105.1	102.4	100.1	97.7	95.9	94.3

w_A	0.400	0.400	0.400	0.400	0.400	0.400	0.375	0.375	0.375
w_B	0.150	0.150	0.150	0.150	0.150	0.150	0.150	0.150	0.150
w_C	0.400	0.400	0.400	0.400	0.400	0.400	0.375	0.375	0.375
w_D	0.050	0.050	0.050	0.050	0.050	0.050	0.100	0.100	0.100
T/K	393.15	413.15	433.15	453.15	473.15	493.15	393.15	413.15	433.15
P/MPa	64.9	65.1	65.3	65.5	65.7	65.8	69.6	69.6	69.5

w_A	0.375	0.375	0.375	0.350	0.350	0.350	0.350	0.350	0.350
w_B	0.150	0.150	0.150	0.150	0.150	0.150	0.150	0.150	0.150
w_C	0.375	0.375	0.375	0.350	0.350	0.350	0.350	0.350	0.350
w_D	0.100	0.100	0.100	0.150	0.150	0.150	0.150	0.150	0.150
T/K	453.15	473.15	493.15	393.15	413.15	433.15	453.15	473.15	493.15
P/MPa	69.5	69.4	69.3	74.9	74.6	74.3	74.1	73.7	73.5

w_A	0.2000	0.2000	0.2000	0.2000	0.2000	0.2000	0.1875	0.1875	0.1875
w_B	0.1500	0.1500	0.1500	0.1500	0.1500	0.1500	0.1500	0.1500	0.1500
w_C	0.6000	0.6000	0.6000	0.6000	0.6000	0.6000	0.5625	0.5625	0.5625
w_D	0.0500	0.0500	0.0500	0.0500	0.0500	0.0500	0.1000	0.1000	0.1000
T/K	393.15	413.15	433.15	453.15	473.15	493.15	393.15	413.15	433.15
P/MPa	32.5	34.4	36.1	38.0	39.4	40.9	38.9	40.3	41.8

w_A	0.1875	0.1875	0.1875	0.1750	0.1750	0.1750	0.1750	0.1750	0.1750
w_B	0.1500	0.1500	0.1500	0.1500	0.1500	0.1500	0.1500	0.1500	0.1500
w_C	0.5625	0.5625	0.5625	0.5250	0.5250	0.5250	0.5250	0.5250	0.5250
w_D	0.1000	0.1000	0.1000	0.1500	0.1500	0.1500	0.1500	0.1500	0.1500
T/K	453.15	473.15	493.15	393.15	413.15	433.15	453.15	473.15	493.15
P/MPa	43.1	44.2	45.2	45.2	46.1	47.2	48.3	48.9	49.6

Polymer (B):	ethylene/vinyl acetate copolymer							1999KIN

Characterization: M_n/kg.mol^{-1} = 61.9, M_w/kg.mol^{-1} = 167, 27.5 wt% vinyl acetate

Solvent (A):	ethene	C_2H_4	74-85-1
Solvent (C):	vinyl acetate	$C_4H_6O_2$	108-05-4
Solvent (D):	helium	He	7440-59-7

Type of data: cloud points

w_A	0.8300	0.8300	0.8300	0.8300	0.8300	0.8300	0.6300	0.6300	0.6300
w_B	0.1500	0.1500	0.1500	0.1500	0.1500	0.1500	0.1500	0.1500	0.1500
w_C	0.0000	0.0000	0.0000	0.0000	0.0000	0.0000	0.2100	0.2100	0.2100
w_D	0.0200	0.0200	0.0200	0.0200	0.0200	0.0200	0.0100	0.0100	0.0100
T/K	393.15	413.15	433.15	453.15	473.15	493.15	393.15	413.15	433.15
P/MPa	209.4	197.5	185.4	174.8	165.9	158.5	139.1	133.4	127.4
w_A	0.6300	0.6300	0.6300	0.6225	0.6225	0.6225	0.6225	0.6225	0.6225
w_B	0.1500	0.1500	0.1500	0.1500	0.1500	0.1500	0.1500	0.1500	0.1500
w_C	0.2100	0.2100	0.2100	0.2075	0.2075	0.2075	0.2075	0.2075	0.0750
w_D	0.0100	0.0100	0.0100	0.0200	0.0200	0.0200	0.0200	0.0200	0.0200
T/K	453.15	473.15	493.15	393.15	413.15	433.15	453.15	473.15	493.15
P/MPa	122.6	118.7	116.3	186.5	176.6	166.4	157.7	151.3	147.1
w_A	0.420	0.420	0.420	0.420	0.420	0.420	0.415	0.415	0.415
w_B	0.150	0.150	0.150	0.150	0.150	0.150	0.150	0.150	0.150
w_C	0.420	0.420	0.420	0.420	0.420	0.420	0.415	0.415	0.415
w_D	0.010	0.010	0.010	0.010	0.010	0.010	0.020	0.020	0.020
T/K	393.15	413.15	433.15	453.15	473.15	493.15	393.15	413.15	433.15
P/MPa	99.2	96.8	95.0	93.0	91.3	89.9	148.0	139.6	132.0
w_A	0.415	0.415	0.415	0.410	0.410	0.410	0.410	0.410	0.410
w_B	0.150	0.150	0.150	0.150	0.150	0.150	0.150	0.150	0.150
w_C	0.415	0.415	0.415	0.410	0.410	0.410	0.410	0.410	0.410
w_D	0.020	0.020	0.020	0.030	0.030	0.030	0.030	0.030	0.030
T/K	453.15	473.15	493.15	393.15	413.15	433.15	453.15	473.15	493.15
P/MPa	126.8	121.7	117.3	215.6	201.1	185.4	173.2	163.1	153.5
w_A	0.2100	0.2100	0.2100	0.2100	0.2100	0.2100	0.2075	0.2075	0.2075
w_B	0.1500	0.1500	0.1500	0.1500	0.1500	0.1500	0.1500	0.1500	0.1500
w_C	0.6300	0.6300	0.6300	0.6300	0.6300	0.6300	0.6225	0.6225	0.6225
w_D	0.0100	0.0100	0.0100	0.0100	0.0100	0.0100	0.0200	0.0200	0.0200
T/K	393.15	413.15	433.15	453.15	473.15	493.15	393.15	413.15	433.15
P/MPa	69.4	67.1	65.1	63.7	62.9	62.8	117.3	109.4	102.8
w_A	0.2075	0.2075	0.2075	0.2050	0.2050	0.2050	0.2050	0.2050	0.2050
w_B	0.1500	0.1500	0.1500	0.1500	0.1500	0.1500	0.1500	0.1500	0.1500
w_C	0.6225	0.6225	0.6225	0.6150	0.6150	0.6150	0.6150	0.6150	0.6150
w_D	0.0200	0.0200	0.0200	0.0300	0.0300	0.0300	0.0300	0.0300	0.0300
T/K	453.15	473.15	493.15	393.15	413.15	433.15	453.15	473.15	493.15
P/MPa	97.6	94.0	91.8	165.0	151.9	140.6	131.5	124.9	120.6

| Polymer (B): | ethylene/vinyl acetate copolymer | | 1989WIL |

Polymer (B): ethylene/vinyl acetate copolymer   **1989WIL**
Characterization: $M_n/\text{kg.mol}^{-1} = 26$, $M_w/\text{kg.mol}^{-1} = 157$, 33.6 wt% vinyl acetate, Bayer AG, Leverkusen, Germany
Solvent (A): ethene C_2H_4 **74-85-1**
Solvent (C): vinyl acetate $C_4H_6O_2$ **108-05-4**
Solvent (D): 2-methyl-2-propanol $C_4H_{10}O$ **75-65-0**

Type of data: cloud points

w_A	0.431	0.431	0.431	0.413	0.413	0.413	0.322	0.322	0.322
w_B	0.057	0.057	0.057	0.112	0.112	0.112	0.229	0.229	0.229
w_C	0.173	0.173	0.173	0.148	0.148	0.148	0.113	0.113	0.113
w_D	0.339	0.339	0.339	0.327	0.327	0.327	0.336	0.336	0.336
T/K	333.15	353.15	373.15	333.15	353.15	373.15	333.15	353.15	373.15
P/bar	480	480	480	485	480	480	365	380	390

Polymer (B): ethylene/vinyl acetate copolymer   **1989WIL**
Characterization: $M_n/\text{kg.mol}^{-1} = 35$, $M_w/\text{kg.mol}^{-1} = 320$, 45.0 wt% vinyl acetate, Bayer AG, Leverkusen, Germany
Solvent (A): ethene C_2H_4 **74-85-1**
Solvent (C): vinyl acetate $C_4H_6O_2$ **108-05-4**
Solvent (D): 2-methyl-2-propanol $C_4H_{10}O$ **75-65-0**

Type of data: cloud points

w_A	0.413	0.413	0.413	0.413	0.368	0.368	0.368	0.368
w_B	0.044	0.044	0.044	0.044	0.093	0.093	0.093	0.093
w_C	0.184	0.184	0.184	0.184	0.168	0.168	0.168	0.168
w_D	0.358	0.358	0.358	0.358	0.372	0.372	0.372	0.372
T/K	313.15	333.15	353.15	373.15	313.15	333.15	353.15	373.15
P/bar	337	345	363	385	300	310	328	355

w_A	0.370	0.370	0.370	0.370	0.316	0.316	0.316	0.316
w_B	0.133	0.133	0.133	0.133	0.227	0.227	0.227	0.227
w_C	0.141	0.141	0.141	0.141	0.099	0.099	0.099	0.099
w_D	0.356	0.356	0.356	0.356	0.357	0.357	0.357	0.357
T/K	313.15	333.15	353.15	373.15	313.15	333.15	353.15	373.15
P/bar	318	326	345	370	295	316	322	344

w_A	0.041	0.041	0.041	0.349	0.349	0.349	0.349	0.403	0.403
w_B	0.077	0.077	0.077	0.024	0.024	0.024	0.024	0.048	0.048
w_C	0.295	0.295	0.295	0.219	0.219	0.219	0.219	0.184	0.184
w_D	0.587	0.587	0.587	0.407	0.407	0.407	0.407	0.365	0.365
T/K	313.15	333.15	353.15	313.15	333.15	353.15	373.15	313.15	333.15
P/bar	21	21	21	248	265	286	309	335	342

w_A	0.403	0.403	0.335	0.335	0.335	0.335	0.295	0.295	0.295
w_B	0.048	0.048	0.097	0.097	0.097	0.097	0.148	0.148	0.148
w_C	0.184	0.184	0.171	0.171	0.171	0.171	0.158	0.158	0.158
w_D	0.365	0.365	0.392	0.392	0.392	0.392	0.399	0.399	0.399
T/K	353.15	373.15	313.15	333.15	353.15	373.15	313.15	333.15	353.15
P/bar	360	382	269	276	294	318	170	195	222

continued

continued

w_A	0.295	0.247	0.247	0.247	0.247	0.205	0.205	0.205	0.205
w_B	0.148	0.203	0.203	0.203	0.203	0.258	0.258	0.258	0.258
w_C	0.158	0.140	0.140	0.140	0.140	0.119	0.119	0.119	0.119
w_D	0.399	0.410	0.410	0.410	0.410	0.419	0.419	0.419	0.419
T/K	373.15	313.15	333.15	353.15	373.15	313.15	333.15	353.15	373.15
P/bar	252	128	156	180	209	75	102	138	175

w_A	0.324	0.324	0.324	0.324	0.304	0.304	0.304	0.304	0.263
w_B	0.023	0.023	0.023	0.023	0.049	0.049	0.049	0.049	0.096
w_C	0.198	0.198	0.198	0.198	0.188	0.188	0.188	0.188	0.173
w_D	0.456	0.456	0.456	0.456	0.459	0.459	0.459	0.459	0.468
T/K	313.15	333.15	353.15	373.15	313.15	333.15	353.15	373.15	313.15
P/bar	187	204	226	256	170	186	210	243	120

w_A	0.263	0.263	0.241	0.241	0.241
w_B	0.096	0.096	0.089	0.089	0.089
w_C	0.173	0.173	0.160	0.160	0.160
w_D	0.468	0.468	0.510	0.510	0.510
T/K	333.15	353.15	313.15	333.15	353.15
P/bar	141	170	82	104	130

Polymer (B):	**ethylene/vinyl acetate copolymer**		**1989WIL**
Characterization:	M_n/kg.mol^{-1} = 51, M_w/kg.mol^{-1} = 360, 70.0 wt% vinyl acetate, Bayer AG, Leverkusen, Germany		
Solvent (A):	**ethene**	**C$_2$H$_4$**	**74-85-1**
Solvent (C):	**vinyl acetate**	**C$_4$H$_6$O$_2$**	**108-05-4**
Solvent (D):	**2-methyl-2-propanol**	**C$_4$H$_{10}$O**	**75-65-0**

Type of data: cloud points

w_A	0.145	0.145	0.217	0.217	0.217	0.217	0.371	0.371	0.371
w_B	0.054	0.054	0.062	0.062	0.062	0.062	0.050	0.050	0.050
w_C	0.379	0.379	0.212	0.212	0.212	0.212	0.194	0.194	0.194
w_D	0.422	0.422	0.479	0.479	0.479	0.479	0.385	0.385	0.385
T/K	323.15	353.15	313.15	333.15	353.15	373.15	313.15	333.15	353.15
P/bar	36	48	47	57	67	87	210	230	255

w_A	0.371
w_B	0.050
w_B	0.194
w_B	0.385
T/K	373.15
P/bar	286

Polymer (B):	**ethylene/vinyl acetate copolymer**		**1999KIN**
Characterization:	M_n/kg.mol^{-1} = 61.9, M_w/kg.mol^{-1} = 167, 27.5 wt% vinyl acetate		
Solvent (A):	**ethene**	**C$_2$H$_4$**	**74-85-1**
Solvent (C):	**vinyl acetate**	**C$_4$H$_6$O$_2$**	**108-05-4**
Solvent (D):	**nitrogen**	**N$_2$**	**7727-37-9**

continued

continued

Type of data: cloud points

w_A	0.6000	0.6000	0.6000	0.6000	0.6000	0.6000	0.5625	0.5625	0.5625
w_B	0.1500	0.1500	0.1500	0.1500	0.1500	0.1500	0.1500	0.1500	0.1500
w_C	0.2000	0.2000	0.2000	0.2000	0.2000	0.2000	0.1875	0.1875	0.1875
w_D	0.0500	0.0500	0.0500	0.0500	0.0500	0.0500	0.1000	0.1000	0.1000
T/K	393.15	413.15	433.15	453.15	473.15	493.15	393.15	413.15	433.15
P/MPa	115.4	112.5	110.0	108.0	105.4	103.0	142.2	136.6	131.1

w_A	0.5625	0.5625	0.5625	0.5250	0.5250	0.5250	0.5250	0.5250	0.5250
w_B	0.1500	0.1500	0.1500	0.1500	0.1500	0.1500	0.1500	0.1500	0.1500
w_C	0.1875	0.1875	0.1875	0.1750	0.1750	0.1750	0.1750	0.1750	0.1750
w_D	0.1000	0.1000	0.1000	0.1500	0.1500	0.1500	0.1500	0.1500	0.1500
T/K	453.15	473.15	493.15	393.15	413.15	433.15	453.15	473.15	493.15
P/MPa	125.5	120.6	116.1	170.0	160.9	153.2	146.2	139.8	133.9

w_A	0.400	0.400	0.400	0.400	0.400	0.400	0.375	0.375	0.375
w_B	0.150	0.150	0.150	0.150	0.150	0.150	0.150	0.150	0.150
w_C	0.400	0.400	0.400	0.400	0.400	0.400	0.375	0.375	0.375
w_D	0.050	0.050	0.050	0.050	0.050	0.050	0.100	0.100	0.100
T/K	393.15	413.15	433.15	453.15	473.15	493.15	393.15	413.15	433.15
P/MPa	84.1	82.5	81.3	79.7	78.5	77.3	107.0	104.0	101.1

w_A	0.375	0.375	0.375	0.350	0.350	0.350	0.350	0.350	0.350
w_B	0.150	0.150	0.150	0.150	0.150	0.150	0.150	0.150	0.150
w_C	0.375	0.375	0.375	0.350	0.350	0.350	0.350	0.350	0.350
w_D	0.100	0.100	0.100	0.150	0.150	0.150	0.150	0.150	0.150
T/K	453.15	473.15	493.15	393.15	413.15	433.15	453.15	473.15	493.15
P/MPa	98.4	96.1	93.9	130.8	126.1	122.0	118.6	115.5	112.5

w_A	0.2000	0.2000	0.2000	0.2000	0.2000	0.2000	0.1875	0.1875	0.1875
w_B	0.1500	0.1500	0.1500	0.1500	0.1500	0.1500	0.1500	0.1500	0.1500
w_C	0.6000	0.6000	0.6000	0.6000	0.6000	0.6000	0.5625	0.5625	0.5625
w_D	0.0500	0.0500	0.0500	0.0500	0.0500	0.0500	0.1000	0.1000	0.1000
T/K	393.15	413.15	433.15	453.15	473.15	493.15	393.15	413.15	433.15
P/MPa	48.4	49.0	49.4	49.6	50.2	50.7	73.6	72.2	70.9

w_A	0.1875	0.1875	0.1875	0.1750	0.1750	0.1750	0.1750	0.1750	0.1750
w_B	0.1500	0.1500	0.1500	0.1500	0.1500	0.1500	0.1500	0.1500	0.1500
w_C	0.5625	0.5625	0.5625	0.5250	0.5250	0.5250	0.5250	0.5250	0.5250
w_D	0.1000	0.1000	0.1000	0.1500	0.1500	0.1500	0.1500	0.1500	0.1500
T/K	453.15	473.15	493.15	393.15	413.15	433.15	453.15	473.15	493.15
P/MPa	70.0	69.3	68.7	99.8	97.7	95.8	94.1	92.6	91.3

Polymer (B):	nylon 6		**1994SUR**
Characterization:	Aldrich Chem. Co., Inc., Milwaukee, WI		
Solvent (A):	**carbon dioxide**	**CO_2**	**124-38-9**
Solvent (C):	**2,2,2-trifluoroethanol**	**$C_2H_3F_3O$**	**75-89-8**

Type of data: cloud points

$T/K = 373.15$

continued

continued

w_A	0.3632	0.3950	0.4195	0.4455	0.4635	0.3311	0.3596	0.3970	0.4229
w_B	0.0367	0.0349	0.0334	0.0320	0.0309	0.0694	0.0664	0.0626	0.0599
w_C	0.6001	0.5701	0.5471	0.5225	0.5056	0.5995	0.5740	0.5404	0.5172
P/MPa	14.48	19.31	23.45	29.65	33.10	14.48	18.62	25.17	30.00

w_A	0.4540	0.2651	0.3019	0.3297	0.3560	0.3797	0.3300	0.3626	0.3853
w_B	0.0566	0.1100	0.1045	0.1003	0.0964	0.0928	0.1129	0.1074	0.1035
w_C	0.4894	0.6249	0.5936	0.5700	0.5476	0.5275	0.5571	0.5300	0.5112
P/MPa	35.17	13.10	17.59	22.41	26.55	32.07	20.69	26.89	33.79

w_A	0.2825	0.3067	0.3324	0.3520
w_B	0.1394	0.1347	0.1297	0.1259
w_C	0.5781	0.5586	0.5379	0.5221
P/MPa	13.45	16.90	21.93	24.14

Polymer (B):	**polybutadiene**		**2002JOU**
Characterization:	M_w/kg.mol^{-1} = 420, Aldrich Chem. Co., Inc., Milwaukee, WI		
Solvent (A):	**carbon dioxide**	**CO$_2$**	**124-38-9**
Solvent (C):	**cyclohexane**	**C$_6$H$_{12}$**	**110-82-7**

Type of data: cloud points (LCST behavior)

w_A	0.172	0.172	0.172	0.172	0.172	0.172	0.240	0.240	0.240
w_B	0.050	0.050	0.050	0.050	0.050	0.050	0.050	0.050	0.050
w_C	0.778	0.778	0.778	0.778	0.778	0.778	0.710	0.710	0.710
T/K	423.15	433.15	443.15	453.15	463.15	473.15	403.15	413.15	423.15
P/MPa	9.0	10.8	12.4	14.1	15.6	17.1	9.7	10.8	12.8

w_A	0.240	0.240	0.240	0.240	0.240	0.300	0.300	0.300	0.300
w_B	0.050	0.050	0.050	0.050	0.050	0.050	0.050	0.050	0.050
w_C	0.710	0.710	0.710	0.710	0.710	0.650	0.650	0.650	0.650
T/K	433.15	443.15	453.15	463.15	473.15	363.15	373.15	383.15	393.15
P/MPa	14.7	16.8	18.3	19.9	21.5	8.6	11.5	13.4	16.1

w_A	0.300	0.300
w_B	0.050	0.050
w_C	0.650	0.650
T/K	403.15	413.15
P/MPa	18.2	20.0

Comments: Gas solubilities and VLE data for this system are given in Chapter 2.

Polymer (B):	**polybutadiene**		**1992KIA, 1994KIA**
Characterization:	M_w/kg.mol^{-1} = 420, 36% 1,4-*cis*, 55% 1,4-*trans*, 9% 1,2-*vinyl*, Scientific Polymer Products, Ontario, NY		
Solvent (A):	**carbon dioxide**	**CO$_2$**	**124-38-9**
Solvent (C):	**tetrahydrofuran**	**C$_4$H$_8$O**	**109-99-9**

Type of data: cloud points (VLE-type)

continued

continued

w_A	0.200	0.200	0.200	0.200	0.200	0.200	0.200	0.200	0.200
w_B	0.040	0.040	0.040	0.040	0.040	0.040	0.040	0.040	0.040
w_C	0.760	0.760	0.760	0.760	0.760	0.760	0.760	0.760	0.760
T/K	301.35	303.15	308.15	313.15	318.15	323.15	328.15	333.15	338.15
P/bar	14.71	14.74	15.96	17.34	18.72	20.42	22.10	23.81	25.77

w_A	0.200	0.250	0.250	0.250	0.250	0.250	0.250	0.250	0.250
w_B	0.040	0.0375	0.0375	0.0375	0.0375	0.0375	0.0375	0.0375	0.0375
w_C	0.760	0.7125	0.7125	0.7125	0.7125	0.7125	0.7125	0.7125	0.7125
T/K	339.65	306.05	308.15	313.95	318.15	323.15	328.15	333.15	338.15
P/bar	26.40	21.94	22.52	24.56	26.66	28.60	31.03	33.76	35.89

w_A	0.250	0.300	0.300	0.300	0.300	0.300	0.300	0.300	0.300
w_B	0.0375	0.035	0.035	0.035	0.035	0.035	0.035	0.035	0.035
w_C	0.7125	0.665	0.665	0.665	0.665	0.665	0.665	0.665	0.665
T/K	342.85	299.65	303.15	308.15	313.15	318.15	323.15	330.25	333.15
P/bar	38.81	23.38	24.13	26.14	28.34	31.54	34.38	37.76	40.06

w_A	0.300	0.350	0.350	0.350	0.350	0.350	0.350	0.350	0.350
w_B	0.035	0.0325	0.0325	0.0325	0.0325	0.0325	0.0325	0.0325	0.0325
w_C	0.665	0.6175	0.6175	0.6175	0.6175	0.6175	0.6175	0.6175	0.6175
T/K	342.75	301.15	303.15	308.15	313.15	318.15	323.15	328.15	333.15
P/bar	46.49	27.42	28.04	30.37	32.83	35.89	39.20	42.06	45.11

w_A	0.350	0.350	0.250	0.250	0.250	0.250	0.250	0.250	0.250
w_B	0.0325	0.0325	0.075	0.075	0.075	0.075	0.075	0.075	0.075
w_C	0.6175	0.6175	0.675	0.675	0.675	0.675	0.675	0.675	0.675
T/K	338.95	342.95	309.45	311.55	314.15	317.95	320.65	324.15	329.15
P/bar	49.25	52.17	24.98	25.63	26.76	28.23	29.87	31.52	33.81

w_A	0.250	0.250	0.250	0.250	0.250	0.300	0.300	0.300	0.300
w_B	0.075	0.075	0.075	0.075	0.075	0.070	0.070	0.070	0.070
w_C	0.675	0.675	0.675	0.675	0.675	0.630	0.630	0.630	0.630
T/K	331.35	333.25	337.45	340.55	342.95	303.75	308.15	313.15	318.15
P/bar	33.92	35.78	37.56	39.04	41.13	26.30	27.12	29.32	31.65

w_A	0.300	0.300	0.300	0.300	0.300	0.350	0.350	0.350	0.350
w_B	0.070	0.070	0.070	0.070	0.070	0.065	0.065	0.065	0.065
w_C	0.630	0.630	0.630	0.630	0.630	0.585	0.585	0.585	0.585
T/K	323.15	328.15	333.15	341.65	343.15	303.35	308.15	313.15	318.15
P/bar	34.05	36.70	39.43	44.95	46.82	28.83	30.01	32.43	35.06

w_A	0.350	0.350	0.350	0.250	0.250	0.250	0.250	0.250	0.250
w_B	0.065	0.065	0.065	0.1125	0.1125	0.1125	0.1125	0.1125	0.1125
w_C	0.585	0.585	0.585	0.6375	0.6375	0.6375	0.6375	0.6375	0.6375
T/K	323.15	328.95	333.15	310.45	313.25	322.65	324.35	329.55	332.25
P/bar	38.25	41.63	44.93	27.02	28.30	32.27	33.23	35.98	37.43

w_A	0.250	0.250	0.250	0.300	0.300	0.300	0.300	0.300	0.300
w_B	0.1125	0.1125	0.1125	0.105	0.105	0.105	0.105	0.105	0.105
w_C	0.6375	0.6375	0.6375	0.595	0.595	0.595	0.595	0.595	0.595
T/K	336.65	339.05	343.15	308.55	313.95	320.55	326.75	331.25	335.35
P/bar	39.73	41.31	42.75	30.21	33.00	37.49	41.01	43.77	46.26

continued

continued

w_A	0.300	0.300	0.350	0.350
w_B	0.105	0.105	0.0975	0.0975
w_C	0.595	0.595	0.5525	0.5525
T/K	339.05	343.15	322.95	327.25
P/bar	48.59	50.79	42.55	46.82

Polymer (B):	polybutadiene		1992KIA, 1994KIA
Characterization:	M_w/kg.mol^{-1} = 5, 80% (1,4-*cis* + 1,4-*trans*), 20% 1,2-*vinyl*,		
	Scientific Polymer Products, Ontario, NY		
Solvent (A):	**carbon dioxide**	**CO$_2$**	**124-38-9**
Solvent (C):	**toluene**	**C$_7$H$_8$**	**108-88-3**

Type of data: cloud points

w_A	0.150	0.150	0.150	0.150	0.150	0.150	0.150	0.201	0.201
w_B	0.510	0.510	0.510	0.510	0.510	0.510	0.510	0.4794	0.4794
w_C	0.340	0.340	0.340	0.340	0.340	0.340	0.340	0.3196	0.3196
T/K	301.45	321.15	323.15	328.05	333.15	337.45	343.15	301.45	303.15
P/bar	38.84	51.05	52.82	55.52	58.61	61.69	64.78	47.21	47.77
	VLE	VLE	VLE	VLE	VLE	VLE	VLE	VLE	VLE

w_A	0.201	0.201	0.201	0.201	0.201	0.201	0.201	0.201	0.250
w_B	0.4794	0.4794	0.4794	0.4794	0.4794	0.4794	0.4794	0.4794	0.450
w_C	0.3196	0.3196	0.3196	0.3196	0.3196	0.3196	0.3196	0.3196	0.300
T/K	308.15	313.25	318.15	323.15	328.55	333.15	338.15	343.05	301.35
P/bar	50.95	55.09	58.63	63.07	69.39	73.74	80.21	85.95	62.18
	VLE	VLE	VLE	VLE	VLE	VLE	VLE	VLE	LLE

w_A	0.250	0.250	0.250	0.250	0.250	0.250	0.250	0.250	0.250
w_B	0.450	0.450	0.450	0.450	0.450	0.450	0.450	0.450	0.450
w_C	0.300	0.300	0.300	0.300	0.300	0.300	0.300	0.300	0.300
T/K	303.15	308.35	313.45	318.25	323.35	328.15	333.15	338.15	339.35
P/bar	62.15	64.61	69.21	75.48	83.03	90.85	100.04	108.57	110.87
	LLE	LLE	LLE	LLE	LLE	LLE	LLE	LLE	LLE

w_A	0.266	0.266	0.266	0.266	0.266	0.266	0.266	0.266	0.266
w_B	0.4404	0.4404	0.4404	0.4404	0.4404	0.4404	0.4404	0.4404	0.4404
w_C	0.2936	0.2936	0.2936	0.2936	0.2936	0.2936	0.2936	0.2936	0.2936
T/K	301.15	302.65	303.15	304.15	305.15	306.15	307.15	308.15	309.15
P/bar	134.84	133.39	131.79	130.18	128.47	126.93	126.46	125.88	125.15
	LLE	LLE	LLE	LLE	LLE	LLE	LLE	LLE	LLE

w_A	0.266	0.266	0.266	0.266	0.266	0.266	0.266	0.266	0.266
w_B	0.4404	0.4404	0.4404	0.4404	0.4404	0.4404	0.4404	0.4404	0.4404
w_C	0.2936	0.2936	0.2936	0.2936	0.2936	0.2936	0.2936	0.2936	0.2936
T/K	310.15	311.35	313.15	315.65	317.15	318.15	321.15	324.65	325.65
P/bar	124.24	124.24	122.60	122.52	123.62	124.53	125.19	129.55	130.04
	LLE	LLE	LLE	LLE	LLE	LLE	LLE	LLE	LLE

continued

continued

w_A	0.266	0.266	0.266	0.266	0.266	0.266	0.150	0.150	0.150
w_B	0.4404	0.4404	0.4404	0.4404	0.4404	0.4404	0.6375	0.6375	0.6375
w_C	0.2936	0.2936	0.2936	0.2936	0.2936	0.2936	0.2125	0.2125	0.2125
T/K	327.15	328.25	329.65	330.65	332.15	333.15	304.65	308.45	309.45
P/bar	131.82	132.26	134.08	135.66	137.50	138.52	41.37	43.23	45.42
	LLE	LLE	LLE	LLE	LLE	LLE	VLE	VLE	VLE

w_A	0.150	0.150	0.150	0.150	0.150	0.150	0.150	0.150	0.150
w_B	0.6375	0.6375	0.6375	0.6375	0.6375	0.6375	0.6375	0.765	0.765
w_C	0.2125	0.2125	0.2125	0.2125	0.2125	0.2125	0.2125	0.085	0.085
T/K	313.15	318.45	323.45	328.05	333.35	338.05	343.25	303.85	307.95
P/bar	47.59	52.68	56.72	59.67	64.72	68.75	71.58	56.61	61.41
	VLE	VLE	VLE	VLE	VLE	VLE	VLE	VLE	VLE

w_A	0.150	0.150	0.150	0.150	0.150	0.150	0.150
w_B	0.765	0.765	0.765	0.765	0.765	0.765	0.765
w_C	0.085	0.085	0.085	0.085	0.085	0.085	0.085
T/K	313.85	317.65	323.05	328.25	333.35	338.35	343.15
P/bar	66.88	70.82	75.24	80.68	85.43	93.90	98.53
	VLE	VLE	VLE	VLE	VLE	VLE	VLE

Comments: VLE- or LLE-type could not always be distinguished in a clear manner.

Polymer (B):   **polybutadiene**   **1992KIA, 1994KIA**

Characterization: M_w/kg.mol^{-1} = 420, 36% 1,4-*cis*, 55% 1,4-*trans*, 9% 1,2-*vinyl*, Scientific Polymer Products, Ontario, NY

Solvent (A):   **carbon dioxide** CO_2 **124-38-9**

Solvent (C):   **toluene** C_7H_8 **108-88-3**

Type of data: cloud points (VLE-type)

w_A	0.150	0.150	0.150	0.150	0.150	0.150	0.150	0.200	0.200
w_B	0.085	0.085	0.085	0.085	0.085	0.085	0.085	0.080	0.080
w_C	0.765	0.765	0.765	0.765	0.765	0.765	0.765	0.720	0.720
T/K	314.25	318.75	323.25	328.15	333.15	338.05	343.15	311.45	313.15
P/bar	33.13	34.84	36.78	38.55	40.55	42.69	44.59	38.35	39.00

w_A	0.200	0.200	0.200	0.200	0.200	0.200	0.250	0.250	0.250
w_B	0.080	0.080	0.080	0.080	0.080	0.080	0.075	0.075	0.075
w_C	0.720	0.720	0.720	0.720	0.720	0.720	0.675	0.675	0.675
T/K	316.05	323.15	328.25	333.15	339.15	343.15	301.75	308.15	313.15
P/bar	41.18	45.21	48.20	50.64	54.31	56.67	37.92	39.76	42.69

w_A	0.250	0.250	0.250	0.250	0.250	0.250	0.300	0.300	0.300
w_B	0.075	0.075	0.075	0.075	0.075	0.075	0.070	0.070	0.070
w_C	0.675	0.675	0.675	0.675	0.675	0.675	0.630	0.630	0.630
T/K	318.15	324.05	327.95	332.75	338.15	342.75	304.65	305.25	314.25
P/bar	45.37	52.63	55.92	59.36	63.14	65.34	43.47	44.40	50.80

continued

continued

w_A	0.300	0.300	0.300	0.300	0.300	0.300	0.350	0.350	0.350
w_B	0.070	0.070	0.070	0.070	0.070	0.070	0.065	0.065	0.065
w_C	0.630	0.630	0.630	0.630	0.630	0.630	0.585	0.585	0.585
T/K	317.15	323.15	328.15	333.15	338.15	343.15	302.25	303.15	308.25
P/bar	53.42	58.18	61.73	65.93	70.12	74.53	44.62	44.72	47.74

w_A	0.350	0.350	0.350	0.350	0.350	0.350	0.350
w_B	0.065	0.065	0.065	0.065	0.065	0.065	0.065
w_C	0.585	0.585	0.585	0.585	0.585	0.585	0.585
T/K	313.25	318.15	323.35	328.25	333.15	338.25	342.05
P/bar	51.55	55.00	59.56	63.96	68.16	73.51	78.04

Polymer (B):	**polybutadiene**		**2002JOU**
Characterization:	M_w/kg.mol^{-1} = 420, Aldrich Chem. Co., Inc., Milwaukee, WI		
Solvent (A):	**carbon dioxide**	**CO$_2$**	**124-38-9**
Solvent (C):	**toluene**	**C$_7$H$_8$**	**108-88-3**

Type of data: cloud points (LCST behavior)

w_A	0.186	0.186	0.278	0.278	0.278	0.278	0.278	0.278	0.278
w_B	0.059	0.059	0.059	0.059	0.059	0.059	0.059	0.059	0.059
w_C	0.755	0.755	0.663	0.663	0.663	0.663	0.663	0.663	0.663
T/K	463.15	473.15	413.15	423.15	433.15	443.15	453.15	463.15	473.15
P/MPa	11.7	12.0	12.5	13.4	14.1	15.2	17.3	19.0	20.6

w_A	0.337	0.337	0.337	0.337	0.337	0.337	0.337
w_B	0.059	0.059	0.059	0.059	0.059	0.059	0.059
w_C	0.604	0.604	0.604	0.604	0.604	0.604	0.604
T/K	393.15	403.15	413.15	423.15	433.15	443.15	453.15
P/MPa	12.4	13.1	15.6	17.5	19.5	21.8	23.7

Comments: Gas solubilities and VLE data for this system are given in Chapter 2.

Polymer (B):	**poly(butyl methacrylate)**		**2000BY1, 2000KIM**
Characterization:	Aldrich Chem. Co., Inc., Milwaukee, WI		
Solvent (A):	**carbon dioxide**	**CO$_2$**	**124-38-9**
Solvent (C):	**butyl methacrylate**	**C$_8$H$_{14}$O$_2$**	**97-88-1**

Type of data: cloud points

w_A	0.870	0.870	0.870	0.870	0.870	0.870	0.785	0.785	0.785
w_B	0.051	0.051	0.051	0.051	0.051	0.051	0.053	0.053	0.053
w_C	0.079	0.079	0.079	0.079	0.079	0.079	0.162	0.162	0.162
T/K	360.55	371.25	388.65	409.35	428.45	449.75	324.05	330.15	338.35
P/bar	1923.9	1463.4	1270.3	1179.3	1138.0	1118.7	1739.2	1291.0	1115.9

w_A	0.785	0.785	0.785	0.785	0.785	0.650	0.650	0.650	0.650
w_B	0.053	0.053	0.053	0.053	0.053	0.052	0.052	0.052	0.052
w_C	0.162	0.162	0.162	0.162	0.162	0.298	0.298	0.298	0.298
T/K	347.35	365.15	383.45	405.35	424.45	313.15	323.25	333.75	353.05
P/bar	1029.0	955.9	923.9	908.4	895.3	550.5	553.3	565.4	591.2

continued

continued

w_A	0.650	0.650	0.541	0.541	0.541	0.541	0.541	0.541	0.401
w_B	0.052	0.052	0.052	0.052	0.052	0.052	0.052	0.052	0.049
w_C	0.298	0.298	0.407	0.407	0.407	0.407	0.407	0.407	0.550
T/K	372.45	391.35	309.15	316.75	325.25	339.15	359.65	379.55	453.45
P/bar	609.5	629.1	150.0	187.2	215.5	263.0	313.0	331.3	329.2

w_A	0.401	0.401	0.401	0.401	0.401	0.401	0.401	0.401	0.401
w_B	0.049	0.049	0.049	0.049	0.049	0.049	0.049	0.049	0.049
w_C	0.550	0.550	0.550	0.550	0.550	0.550	0.550	0.550	0.550
T/K	436.05	416.45	403.85	395.15	379.55	367.15	351.25	347.05	339.05
P/bar	301.8	269.6	243.7	222.7	187.2	156.2	115.5	95.5	94.5

w_A	0.401	0.401
w_B	0.049	0.049
w_C	0.550	0.550
T/K	323.95	312.95
P/bar	78.3	65.2

Comments: The last four data points belong to a vapor-liquid phase transition.

Polymer (B):	**polyester (hyperbranched, aliphatic)**	**2003SEI**

Characterization: M_n/kg.mol^{-1} = 1.6, M_w/kg.mol^{-1} = 2.1,
hydroxyl functional hyperbranched polyesters produced from poly-
alcohol cores and hydroxy acids, 16 OH groups per macromolecule,
hydroxyl no. = 490-520 mg KOH/g, acid no. = 5-9 mg KOH/g,
Boltorn H20, Perstorp Specialty Chemicals AB, Perstorp, Sweden

Solvent (A):	**carbon dioxide**	**CO_2**	**124-38-9**
Solvent (C):	**ethanol**	**C_2H_6O**	**64-17-5**

Type of data: cloud points (UCST behavior)

w_A	0.00	0.00	0.00	0.00	0.00	0.153	0.153	0.153
w_B	0.10	0.10	0.10	0.10	0.10	0.0847	0.0847	0.0847
w_C	0.90	0.90	0.90	0.90	0.90	0.7623	0.7623	0.7623
T/K	332.9	332.1	331.2	333.5	332.4	331.8	331.5	331.6
P/MPa	1.10	3.10	6.10	10.10	15.10	3.10	4.10	12.00

Type of data: cloud points (UCST/LCST merging behavior)

w_A	0.505	0.505	0.505	0.505	0.505	0.505	0.505	0.505	0.505
w_B	0.0495	0.0495	0.0495	0.0495	0.0495	0.0495	0.0495	0.0495	0.0495
w_C	0.4455	0.4455	0.4455	0.4455	0.4455	0.4455	0.4455	0.4455	0.4455
T/K	335.0	333.8	333.5	333.2	335.5	335.7	337.3	337.4	338.6
P/MPa	10.4	10.9	11.6	12.6	10.09	10.27	10.75	10.90	11.28

w_A	0.505	0.505	0.505	0.505	0.505	0.532	0.532	0.532	0.532
w_B	0.0495	0.0495	0.0495	0.0495	0.0495	0.0468	0.0468	0.0468	0.0468
w_C	0.4455	0.4455	0.4455	0.4455	0.4455	0.4212	0.4212	0.4212	0.4212
T/K	341.3	345.3	347.3	352.4	355.4	333.8	334.1	334.1	334.0
P/MPa	12.05	13.47	14.10	15.97	16.50	15.60	16.10	16.60	17.10

continued

continued

w_A	0.532	0.532	0.532	0.532	0.532	0.532	0.532
w_B	0.0468	0.0468	0.0468	0.0468	0.0468	0.0468	0.0468
w_C	0.4212	0.4212	0.4212	0.4212	0.4212	0.4212	0.4212
T/K	334.1	335.2	336.0	336.8	337.6	338.2	338.7
P/MPa	15.20	15.59	15.86	16.07	16.38	16.76	17.02

Type of data: cloud points (LCST behavior)

w_A	0.271	0.271	0.271	0.271	0.271	0.271	0.328	0.328	0.328
w_B	0.0729	0.0729	0.0729	0.0729	0.0729	0.0729	0.0672	0.0672	0.0672
w_C	0.6561	0.6561	0.6561	0.6561	0.6561	0.6561	0.6048	0.6048	0.6048
T/K	446.2	448.9	450.9	456.7	459.8	465.2	438.5	441.8	445.4
P/MPa	12.97	13.78	14.26	15.38	16.83	17.57	14.94	15.74	16.56

w_A	0.328	0.328	0.352	0.352	0.352	0.352	0.352	0.396	0.396
w_B	0.0672	0.0672	0.0648	0.0648	0.0648	0.0648	0.0648	0.0604	0.0604
w_C	0.6048	0.6048	0.5832	0.5832	0.5832	0.5832	0.5832	0.5436	0.5436
T/K	448.7	452.3	413.8	414.4	416.6	421.0	429.7	407.2	409.8
P/MPa	17.23	17.95	13.21	13.32	14.09	15.07	17.06	16.32	16.99

w_A	0.396	0.396	0.456	0.456	0.483	0.483	0.483	0.483	0.483
w_B	0.0604	0.0604	0.0544	0.0544	0.0517	0.0517	0.0517	0.0517	0.0517
w_C	0.5436	0.5436	0.4896	0.4896	0.4653	0.4653	0.4653	0.4653	0.4653
T/K	413.4	416.6	381.5	384.3	354.7	357.0	361.7	364.5	366.8
P/MPa	17.98	18.73	17.10	17.84	13.56	14.30	15.72	16.63	17.31

Polymer (B):	**polyester (hyperbranched, aliphatic)**		**2003SEI**
Characterization:	M_n/kg.mol^{-1} = 1.6, M_w/kg.mol^{-1} = 2.1,		

hydroxyl functional hyperbranched polyesters produced from poly-alcohol cores and hydroxy acids, 16 OH groups per macromolecule, hydroxyl no. = 490-520 mg KOH/g, acid no. = 5-9 mg KOH/g, Boltorn H20, Perstorp Specialty Chemicals AB, Perstorp, Sweden

Solvent (A):	**water**	**H_2O**	**7732-18-5**
Solvent (C):	**carbon dioxide**	**CO_2**	**124-38-9**

Type of data: cloud points (UCST behavior)

w_A	0.90	0.90	0.90	0.90	0.864	0.864	0.864	0.864
w_B	0.10	0.10	0.10	0.10	0.096	0.096	0.096	0.096
w_C	0.00	0.00	0.00	0.00	0.04	0.04	0.04	0.04
T/K	369.9	370.2	369.8	369.9	356.7	355.7	355.5	355.1
P/MPa	1.10	3.10	6.10	8.10	12.10	13.60	15.10	17.10

Polymer (B):	**polyester (hyperbranched, aliphatic)**		**2003SEI**
Characterization:	M_n/kg.mol^{-1} = 2.8, M_w/kg.mol^{-1} = 5.1,		

hydroxyl functional hyperbranched polyesters produced from poly-alcohol cores and hydroxy acids, 64 OH groups per macromolecule, hydroxyl no. = 470-500 mg KOH/g, acid no. = 7-11 mg KOH/g, Boltorn H40, Perstorp Specialty Chemicals AB, Perstorp, Sweden

continued

continued

| **Solvent (A):** | carbon dioxide | CO_2 | **124-38-9** |
| **Solvent (C):** | water | H_2O | **7732-18-5** |

Type of data: cloud points (UCST behavior)

w_A	0.00	0.00	0.00	0.00	0.00	0.00	0.02	0.02	0.02
w_B	0.10	0.10	0.10	0.10	0.10	0.10	0.098	0.098	0.098
w_C	0.90	0.90	0.90	0.90	0.90	0.90	0.882	0.882	0.882
T/K	414.7	415.0	414.0	413.8	412.8	414.7	407.1	405.0	405.3
P/MPa	1.10	3.10	7.10	10.10	12.10	16.10	4.40	10.10	13.10

w_A	0.02	0.04	0.04	0.04
w_B	0.098	0.096	0.096	0.096
w_C	0.882	0.864	0.864	0.864
T/K	405.2	395.6	394.7	393.9
P/MPa	15.90	11.93	14.10	16.60

Polymer (B):	polyethylene		**2000BEH**
Characterization:	$M_n/kg.mol^{-1} = 13$, $M_w/kg.mol^{-1} = 89$, $M_z/kg.mol^{-1} = 600$,		
	LDPE, Stamylan, DSM, Geleen, The Netherlands		
Solvent (A):	carbon dioxide	CO_2	**124-38-9**
Solvent (C):	cyclohexane	C_6H_{12}	**110-82-7**

Type of data: cloud points

w_A	0.0437	0.0437	0.0437	0.0437	0.0437	0.0956	0.0956	0.0956	0.0956
w_B	0.0527	0.0527	0.0527	0.0527	0.0527	0.0474	0.0474	0.0474	0.0474
w_C	0.9037	0.9037	0.9037	0.9037	0.9037	0.8571	0.8571	0.8571	0.8571
T/K	517.28	522.34	526.99	532.72	533.23	501.75	506.27	511.19	516.35
P/bar	49.7	55.9	61.4	68.1	68.6	76.2	81.9	87.5	92.1

w_A	0.0956	0.0956
w_B	0.0474	0.0474
w_C	0.8571	0.8571
T/K	527.11	537.22
P/bar	103.6	115.1

Polymer (B):	polyethylene		**2000BEH**
Characterization:	$M_n/kg.mol^{-1} = 93$, $M_w/kg.mol^{-1} = 101$, $M_z/kg.mol^{-1} = 112$		
Solvent (A):	carbon dioxide	CO_2	**124-38-9**
Solvent (C):	cyclohexane	C_6H_{12}	**110-82-7**

Type of data: cloud points

w_A	0.0667	0.0667	0.0667	0.0667	0.0667	0.0667	0.1346	0.1346	0.1346
w_B	0.1150	0.1150	0.1150	0.1150	0.1150	0.1150	0.1067	0.1067	0.1067
w_C	0.8183	0.8183	0.8183	0.8183	0.8183	0.8183	0.8587	0.8587	0.8587
T/K	469.93	479.59	489.52	499.09	509.41	519.73	423.92	434.17	444.30
P/bar	58.5	74.2	89.3	103.8	118.0	131.6	76.3	94.7	112.0

continued

continued

w_A	0.1346	0.1346	0.1346	0.1346	0.1346	0.1346	0.1346
w_B	0.1067	0.1067	0.1067	0.1067	0.1067	0.1067	0.1067
w_C	0.8587	0.8587	0.8587	0.8587	0.8587	0.8587	0.8587
T/K	454.37	464.03	473.96	484.04	493.70	504.30	514.25
P/bar	129.0	144.3	159.3	174.1	188.0	201.3	213.5

Polymer (B):	**polyethylene**		**2002JOU**
Characterization:	M_w/kg.mol^{-1} = 125, Aldrich Chem. Co., Inc., Milwaukee, WI		
Solvent (A):	**carbon dioxide**	**CO$_2$**	**124-38-9**
Solvent (C):	**n-heptane**	**C$_7$H$_{16}$**	**142-82-5**

Type of data: cloud points (LCST behavior)

w_A	0.125	0.125	0.125	0.125	0.125	0.125	0.125	0.125	0.125
w_B	0.024	0.024	0.024	0.024	0.024	0.024	0.024	0.024	0.024
w_C	0.851	0.851	0.851	0.851	0.851	0.851	0.851	0.851	0.851
T/K	383.15	393.15	403.15	413.15	423.15	433.15	443.15	453.15	463.15
P/MPa	6.2	8.0	9.3	10.8	12.0	13.2	14.3	15.4	16.5

w_A	0.125	0.273	0.273	0.273	0.273	0.273	0.273
w_B	0.024	0.024	0.024	0.024	0.024	0.024	0.024
w_C	0.851	0.703	0.703	0.703	0.703	0.703	0.703
T/K	473.15	383.15	393.15	403.15	413.15	423.15	433.15
P/MPa	17.7	28.0	29.7	30.8	32.0	32.9	33.5

Polymer (B):	**polyethylene**		**2002HOR**
Characterization:	M_n/kg.mol^{-1} = 13, M_w/kg.mol^{-1} = 89, M_z/kg.mol^{-1} = 600,		
	LDPE, Stamylan, DSM, Geleen, The Netherlands		
Solvent (A):	**cyclohexane**	**C$_6$H$_{12}$**	**110-82-7**
Solvent (C):	**nitrogen**	**N$_2$**	**7727-37-9**

Type of data: cloud points

w_A	0.856	0.856	0.856	0.856	0.856	0.856	0.846	0.846	0.846
w_B	0.104	0.104	0.104	0.104	0.104	0.104	0.103	0.103	0.103
w_C	0.040	0.040	0.040	0.040	0.040	0.040	0.050	0.050	0.050
T/K	505.43	506.23	507.19	508.15	508.92	509.83	504.55	505.48	506.53
P/bar	95.0	96.0	97.0	98.0	99.0	100.0	111.5	112.5	113.5

w_A	0.846	0.846	0.846	0.846	0.846	0.837	0.837	0.837	0.837
w_B	0.103	0.103	0.103	0.103	0.103	0.102	0.102	0.102	0.102
w_C	0.050	0.050	0.050	0.050	0.050	0.061	0.061	0.061	0.061
T/K	507.55	508.51	510.47	512.43	514.47	498.67	501.19	503.70	506.17
P/bar	114.9	115.9	117.9	119.9	121.9	133.7	136.1	138.5	140.9

w_A	0.837	0.837	0.837
w_B	0.102	0.102	0.102
w_C	0.061	0.061	0.061
T/K	508.65	511.16	513.64
P/bar	143.3	145.7	147.9

| Polymer (B): | polyethylene | | | | | | | **2002HOR** |

Characterization: $M_n/\text{kg.mol}^{-1} = 13$, $M_w/\text{kg.mol}^{-1} = 89$, $M_z/\text{kg.mol}^{-1} = 600$, LDPE, Stamylan, DSM, Geleen, The Netherlands

| Solvent (A): | ethane | C_2H_6 | **74-84-0** |
| Solvent (C): | cyclohexane | C_6H_{12} | **110-82-7** |

Type of data: cloud points

w_A	0.052	0.052	0.052	0.052	0.052	0.052	0.052	0.068	0.068
w_B	0.104	0.104	0.104	0.104	0.104	0.104	0.104	0.100	0.100
w_C	0.844	0.844	0.844	0.844	0.844	0.844	0.844	0.832	0.832
T/K	499.75	502.22	502.88	503.78	504.69	505.57	506.47	493.67	496.17
P/bar	48.5	52.0	53.0	54.0	55.5	56.5	58.0	54.9	58.3
w_A	0.068	0.068	0.068	0.068	0.068	0.068	0.101	0.101	0.101
w_B	0.100	0.100	0.100	0.100	0.100	0.100	0.098	0.098	0.098
w_C	0.832	0.832	0.832	0.832	0.832	0.832	0.801	0.801	0.801
T/K	498.64	501.11	503.59	506.06	508.59	511.13	473.64	473.70	476.15
P/bar	61.9	65.3	68.3	71.3	74.3	77.3	61.9	61.9	65.5
w_A	0.101	0.101	0.101	0.101	0.101	0.101	0.101	0.101	
w_B	0.098	0.098	0.098	0.098	0.098	0.098	0.098	0.098	
w_C	0.801	0.801	0.801	0.801	0.801	0.801	0.801	0.801	
T/K	478.66	478.71	483.68	488.66	493.62	498.64	503.64	508.62	
P/bar	68.9	68.9	75.9	82.5	89.3	95.9	101.9	107.9	

| Polymer (B): | polyethylene | | | | **1999DIE** |

Characterization: $M_n/\text{kg.mol}^{-1} = 69.2$, $M_w/\text{kg.mol}^{-1} = 198$, 54.0 % crystallinity

| Solvent (A): | ethene | C_2H_4 | **74-85-1** |
| Solvent (C): | butyl acrylate | $C_7H_{12}O_2$ | **141-32-2** |

Type of data: cloud points

| w_B | 0.03 | was kept constant |

Comments: The mole fraction x_C is given for the monomer mixture, i.e., without the polymer.

x_C	0.0000	0.0029	0.0053	0.0124	0.0202	0.0342	0.0489

$T/K = 393.15$

P/bar	2000	1990	1960	1900	1845	1765	1700

$T/K = 403.15$

P/bar	1910	1893	1861	1820	1750	1670	1610

$T/K = 413.15$

P/bar	1840	1818	1756	1765	1660	1600	1522

$T/K = 423.15$

P/bar	1760	1745	1708	1690	1610	1517	1490

Polymer (B):	polyethylene							**2004CHE**
Characterization:	M_n/kg.mol^{-1} = 14.4, M_w/kg.mol^{-1} = 15.2, PE standard, hydrogenated polybutadiene, Scientific Polymer Products, Inc., Ontario, NY							
Solvent (A):	ethene			C$_2$H$_4$				**74-85-1**
Solvent (C):	n-hexane			C$_6$H$_{14}$				**110-54-3**

Type of data: cloud points

w_A	0.0390	0.0390	0.0390	0.0390	0.0390	0.0390	0.0390	0.0390	0.0390
w_B	0.0874	0.0874	0.0874	0.0874	0.0874	0.0874	0.0874	0.0874	0.0874
w_C	0.8736	0.8736	0.8736	0.8736	0.8736	0.8736	0.8736	0.8736	0.8736
T/K	373.21	393.12	413.12	433.17	433.17	453.14	453.17	473.10	473.14
P/MPa	1.9	2.0	2.4	4.0	2.8	6.7	3.2	9.2	3.7
	VLE	VLE	VLE	LLE	VLLE	LLE	VLLE	LLE	VLLE

w_A	0.0605	0.0605	0.0605	0.0605	0.0605	0.0605	0.0605	0.0605	0.0605
w_B	0.0789	0.0789	0.0789	0.0789	0.0789	0.0789	0.0789	0.0789	0.0789
w_C	0.8606	0.8606	0.8606	0.8606	0.8606	0.8606	0.8606	0.8606	0.8606
T/K	373.20	393.09	413.03	413.14	433.07	433.23	453.12	453.21	473.13
P/MPa	2.2	2.7	3.6	3.1	6.5	3.6	9.2	4.0	4.5
	VLE	VLE	LLE	VLLE	LLE	VLLE	LLE	VLLE	VLLE

w_A	0.0605	0.1277	0.1277	0.1277	0.1277	0.1277	0.1277	0.1277	0.1277
w_B	0.0789	0.0733	0.0733	0.0733	0.0733	0.0733	0.0733	0.0733	0.0733
w_C	0.8606	0.7990	0.7990	0.7990	0.7990	0.7990	0.7990	0.7990	0.7990
T/K	473.15	373.15	373.15	393.09	393.11	413.08	413.11	433.06	433.10
P/MPa	11.6	8.8	4.5	11.6	5.1	14.4	5.7	16.9	6.3
	LLE	LLE	VLLE	LLE	VLLE	LLE	VLLE	LLE	VLLE

w_A	0.1277	0.1277	0.1277	0.1277
w_B	0.0733	0.0733	0.0733	0.0733
w_C	0.7990	0.7990	0.7990	0.7990
T/K	453.18	453.18	473.10	473.11
P/MPa	19.1	7.0	20.9	8.0
	LLE	VLLE	LLE	VLLE

Polymer (B):	polyethylene							**2004CHE**
Characterization:	M_n/kg.mol^{-1} = 23.3, M_w/kg.mol^{-1} = 60.4, M_z/kg.mol^{-1} = 100.7, metallocene LLDPE, unspecified comonomer, industrial source							
Solvent (A):	ethene			C$_2$H$_4$				**74-85-1**
Solvent (C):	n-hexane			C$_6$H$_{14}$				**110-54-3**

Type of data: cloud points

w_A	0.0217	0.0217	0.0217	0.0217	0.0217	0.0217	0.0217	0.0217	0.0217
w_B	0.0802	0.0802	0.0802	0.0802	0.0802	0.0802	0.0802	0.0802	0.0802
w_C	0.8981	0.8981	0.8981	0.8981	0.8981	0.8981	0.8981	0.8981	0.8981
T/K	393.13	413.20	413.20	433.15	433.13	453.16	453.17	473.16	473.15
P/MPa	1.5	2.1	1.8	5.2	2.2	7.9	2.6	10.4	3.1
	VLE	LLE	VLLE	LLE	VLLE	LLE	VLLE	LLE	VLLE

continued

continued

w_A	0.0416	0.0416	0.0416	0.0416	0.0416	0.0416	0.0416	0.0416	0.0416
w_B	0.0896	0.0896	0.0896	0.0896	0.0896	0.0896	0.0896	0.0896	0.0896
w_C	0.8688	0.8688	0.8688	0.8688	0.8688	0.8688	0.8688	0.8688	0.8688
T/K	393.10	413.11	413.12	433.12	433.19	453.16	453.16	473.13	473.13
P/MPa	2.2	4.6	2.5	7.6	2.9	10.3	3.4	12.7	3.8
	VLE	LLE	VLLE	LLE	VLLE	LLE	VLLE	LLE	VLLE

w_A	0.0636	0.0636	0.0636	0.0636	0.0636	0.0636	0.0636	0.0636	0.0636
w_B	0.0875	0.0875	0.0875	0.0875	0.0875	0.0875	0.0875	0.0875	0.0875
w_C	0.8489	0.8489	0.8489	0.8489	0.8489	0.8489	0.8489	0.8489	0.8489
T/K	393.10	393.16	413.20	413.16	433.17	433.16	453.18	453.17	473.14
P/MPa	5.2	3.0	8.3	3.4	11.1	3.9	13.4	4.3	15.7
	LLE	VLLE	LLE	VLLE	LLE	VLLE	LLE	VLLE	LLE

Polymer (B):	**polyethylene**		**2001TOR**
Characterization:	M_n/kg.mol^{-1} = 43, M_w/kg.mol^{-1} = 105, M_z/kg.mol^{-1} = 190,		
	HDPE, DSM, Geleen, The Netherlands		
Solvent (A):	**ethene**	**C$_2$H$_4$**	**74-85-1**
Solvent (C):	**n-hexane**	**C$_6$H$_{14}$**	**110-54-3**

Type of data: cloud points

w_A	0.0207	0.0207	0.0207	0.0207	0.0207	0.0207	0.0207	0.0207	0.0207
w_B	0.0997	0.0997	0.0997	0.0997	0.0997	0.0997	0.0997	0.0997	0.0997
w_C	0.8796	0.8796	0.8796	0.8796	0.8796	0.8796	0.8796	0.8796	0.8796
T/K	408.9	429.1	429.0	433.7	443.3	453.4	463.5	473.9	484.2
P/MPa	1.36	1.64	2.31	3.11	4.54	5.91	1.28	8.63	9.87
	VLE	VLLE	LLE	LLE	LLE	LLE	LLE	LLE	LLE

w_A	0.0207	0.0374	0.0374	0.0374	0.0374	0.0374	0.0374	0.0374	0.0374
w_B	0.0997	0.0969	0.0969	0.0969	0.0969	0.0969	0.0969	0.0969	0.0969
w_C	0.8796	0.8657	0.8657	0.8657	0.8657	0.8657	0.8657	0.8657	0.8657
T/K	493.2	404.5	413.3	423.2	433.5	423.3	433.5	443.5	453.4
P/MPa	10.85	1.74	1.94	2.09	2.31	3.40	4.96	6.43	7.67
	LLE	VLE	VLE	VLLE	VLLE	LLE	LLE	LLE	LLE

w_A	0.0374	0.0374	0.0374	0.0691	0.0691	0.0691	0.0691	0.0691	0.0691
w_B	0.0969	0.0969	0.0969	0.0937	0.0937	0.0937	0.0937	0.0937	0.0937
w_C	0.8657	0.8657	0.8657	0.8372	0.8372	0.8372	0.8372	0.8372	0.8372
T/K	463.5	473.7	482.7	382.0	392.8	403.3	413.4	392.7	403.3
P/MPa	8.98	10.23	11.30	2.60	2.97	3.20	3.38	3.03	4.70
	LLE	LLE	LLE	VLE	VLLE	VLLE	VLLE	LLE	LLE

w_A	0.0691	0.0691	0.0691	0.0691	0.0691	0.1188	0.1188	0.1188	0.1188
w_B	0.0937	0.0937	0.0937	0.0937	0.0937	0.0898	0.0898	0.0898	0.0898
w_C	0.8372	0.8372	0.8372	0.8372	0.8372	0.7914	0.7914	0.7914	0.7914
T/K	413.3	422.8	433.3	442.9	454.0	377.2	388.0	398.9	409.0
P/MPa	6.26	7.76	9.25	10.56	12.02	10.01	11.45	12.97	14.32
	LLE	LLE	LLE	LLE	LLE	LLE	LLE	LLE	LLE

continued

continued

w_A	0.1188
w_B	0.0898
w_C	0.7914
T/K	418.8
P/MPa	15.60
	LLE

Polymer (B):	polyethylene		2001TOR
Characterization:	M_n/kg.mol^{-1} = 43, M_w/kg.mol^{-1} = 105, M_z/kg.mol^{-1} = 190, HDPE, DSM, Geleen, The Netherlands		
Solvent (A):	**ethene**	**C_2H_4**	**74-85-1**
Solvent (C):	**n-hexane**	**C_6H_{14}**	**110-54-3**
Solvent (D):	**1-octene**	**C_8H_{16}**	**111-66-0**

Type of data: cloud points

w_A	0.0202	0.0202	0.0202	0.0202	0.0202	0.0202	0.0202	0.0202	0.0202
w_B	0.1856	0.1856	0.1856	0.1856	0.1856	0.1856	0.1856	0.1856	0.1856
w_C	0.6354	0.6354	0.6354	0.6354	0.6354	0.6354	0.6354	0.6354	0.6354
w_D	0.1588	0.1588	0.1588	0.1588	0.1588	0.1588	0.1588	0.1588	0.1588
T/K	423.2	433.8	443.7	454.2	454.2	462.4	472.8	483.1	493.2
P/MPa	1.59	1.80	2.02	2.21	3.34	4.45	5.86	7.16	8.38
	VLE	VLE	VLE	VLLE	LLE	LLE	LLE	LLE	LLE

w_A	0.0202	0.0202	0.0202	0.0548	0.0548	0.0548	0.0548	0.0548	0.0548
w_B	0.1856	0.1856	0.1856	0.1790	0.1790	0.1790	0.1790	0.1790	0.1790
w_C	0.6354	0.6354	0.6354	0.6130	0.6130	0.6130	0.6130	0.6130	0.6130
w_D	0.1588	0.1588	0.1588	0.1532	0.1532	0.1532	0.1532	0.1532	0.1532
T/K	503.2	513.4	524.1	404.7	413.3	423.2	433.7	423.2	433.6
P/MPa	9.49	10.56	11.59	2.91	3.06	3.26	3.49	4.10	5.60
	LLE	LLE	LLE	VLE	VLE	VLLE	VLLE	LLE	LLE

w_A	0.0548	0.0548	0.0548	0.0548	0.0548	0.0186	0.0186	0.0186	0.0186
w_B	0.1790	0.1790	0.1790	0.1790	0.1790	0.0921	0.0921	0.0921	0.0921
w_C	0.6130	0.6130	0.6130	0.6130	0.6130	0.7114	0.7114	0.7114	0.7114
w_D	0.1532	0.1532	0.1532	0.1532	0.1532	0.1779	0.1779	0.1779	0.1779
T/K	442.7	453.0	463.3	473.6	484.2	424.0	433.3	443.0	453.5
P/MPa	6.98	8.44	9.75	11.00	12.26	1.42	1.59	1.75	1.98
	LLE	LLE	LLE	LLE	LLE	VLE	VLE	VLLE	VLLE

w_A	0.0186	0.0186	0.0186	0.0186	0.0186	0.0186	0.0186	0.0186	0.0544
w_B	0.0921	0.0921	0.0921	0.0921	0.0921	0.0921	0.0921	0.0921	0.0887
w_C	0.7114	0.7114	0.7114	0.7114	0.7114	0.7114	0.7114	0.7114	0.6855
w_D	0.1779	0.1779	0.1779	0.1779	0.1779	0.1779	0.1779	0.1779	0.1714
T/K	443.0	453.4	463.7	473.9	484.0	493.4	503.2	513.4	403.3
P/MPa	2.20	3.73	5.17	6.49	7.76	8.85	9.93	10.98	2.67
	LLE	LLE	LLE	LLE	LLE	LLE	LLE	LLE	VLE

continued

continued

w_A	0.0544	0.0544	0.0544	0.0544	0.0544	0.0544	0.0544	0.0544	0.0544
w_B	0.0887	0.0887	0.0887	0.0887	0.0887	0.0887	0.0887	0.0887	0.0887
w_C	0.6855	0.6855	0.6855	0.6855	0.6855	0.6855	0.6855	0.6855	0.6855
w_D	0.1714	0.1714	0.1714	0.1714	0.1714	0.1714	0.1714	0.1714	0.1714
T/K	413.3	423.6	433.3	423.5	433.2	443.7	453.7	464.1	473.5
P/MPa	2.88	3.07	3.21	4.35	5.83	7.35	8.72	10.06	11.23
	VLE	VLLE	VLLE	LLE	LLE	LLE	LLE	LLE	LLE

w_A	0.0544
w_B	0.0887
w_C	0.6855
w_D	0.1714
T/K	483.8
P/MPa	12.41
	LLE

Polymer (B):   **polyethylene**   **2004CHE**
Characterization: $M_n/kg.mol^{-1} = 82$, $M_w/kg.mol^{-1} = 108$, PE standard, hydrogenated
      polybutadiene, Scientific Polymer Products, Inc., Ontario, NY
Solvent (A):   **ethene**   **C_2H_4**   **74-85-1**
Solvent (C):   **n-hexane**   **C_6H_{14}**   **110-54-3**

Type of data: cloud points

w_A	0.0280	0.0280	0.0280	0.0280	0.0280	0.0280	0.0280	0.0280	0.0280
w_B	0.0803	0.0803	0.0803	0.0803	0.0803	0.0803	0.0803	0.0803	0.0803
w_C	0.8917	0.8917	0.8917	0.8917	0.8917	0.8917	0.8917	0.8917	0.8917
T/K	373.20	393.14	413.12	413.15	433.13	433.14	453.11	453.14	473.12
P/MPa	1.3	1.6	3.5	1.9	6.5	2.3	9.3	2.7	11.8
	VLE	VLE	LLE	VLLE	LLE	VLLE	LLE	VLLE	LLE

w_A	0.0280	0.0512	0.0512	0.0512	0.0512	0.0512	0.0512	0.0512	0.0512
w_B	0.0803	0.0784	0.0784	0.0784	0.0784	0.0784	0.0784	0.0784	0.0784
w_C	0.8917	0.8704	0.8704	0.8704	0.8704	0.8704	0.8704	0.8704	0.8704
T/K	473.14	373.21	393.16	393.16	413.13	413.10	433.15	433.18	453.19
P/MPa	3.2	2.1	4.3	2.5	7.5	2.8	10.3	3.3	12.7
	VLLE	VLE	LLE	VLLE	LLE	VLLE	LLE	VLLE	LLE

w_A	0.0280	0.0512	0.0512	0.0668	0.0668	0.0668	0.0668	0.0668	0.0668
w_B	0.0803	0.0784	0.0784	0.0771	0.0771	0.0771	0.0771	0.0771	0.0771
w_C	0.8917	0.8704	0.8704	0.8561	0.8561	0.8561	0.8561	0.8561	0.8561
T/K	453.16	473.15	473.10	373.19	373.20	393.16	393.13	413.14	413.13
P/MPa	3.7	14.9	4.3	3.4	2.6	6.7	3.0	9.7	3.4
	VLLE	LLE	VLLE	LLE	VLLE	LLE	VLLE	LLE	VLLE

w_A	0.0668	0.0668	0.0668	0.0668	0.0668	0.0668
w_B	0.0771	0.0771	0.0771	0.0771	0.0771	0.0771
w_C	0.8561	0.8561	0.8561	0.8561	0.8561	0.8561
T/K	433.16	433.18	453.14	453.18	473.15	473.17
P/MPa	12.2	3.9	14.8	4.4	17.0	4.9
	LLE	VLLE	LLE	VLLE	LLE	VLLE

Polymer (B):	**polyethylene**		**2004CHE**

Characterization: M_n/kg.mol^{-1} = 14.4, M_w/kg.mol^{-1} = 15.2, PE standard, hydrogenated polybutadiene, Scientific Polymer Products, Inc., Ontario, NY

Solvent (A):	**ethene**	**C$_2$H$_4$**	**74-85-1**
Solvent (C):	**n-hexane**	**C$_6$H$_{14}$**	**110-54-3**
Polymer (D):	**polyethylene**		

Characterization: M_n/kg.mol^{-1} = 82, M_w/kg.mol^{-1} = 108, PE standard, hydrogenated polybutadiene, Scientific Polymer Products, Inc., Ontario, NY

Type of data: cloud points

Comments: w_B/w_D = 0.2501/0.2503 was kept constant.

w_A	0.0175	0.0175	0.0175	0.0175	0.0282	0.0282	0.0282	0.0282	0.0668
w_B	0.04853	0.04853	0.04853	0.04853	0.04798	0.04798	0.04798	0.04798	0.03853
w_C	0.8854	0.8854	0.8854	0.8854	0.8758	0.8758	0.8758	0.8758	0.8561
w_D	0.04857	0.04857	0.04857	0.04857	0.04802	0.04802	0.04802	0.04802	0.03857
T/K	443.19	453.18	463.16	473.14	443.19	453.20	463.18	473.18	443.19
P/MPa	5.6	7.0	8.3	9.5	8.1	9.4	10.6	11.8	11.2

w_A	0.0668	0.0668	0.0668
w_B	0.03853	0.03853	0.03853
w_C	0.8561	0.8561	0.8561
w_D	0.03857	0.03857	0.03857
T/K	453.19	463.12	473.09
P/MPa	12.5	13.7	14.8

Polymer (B):	**polyethylene**		**2000CH3**

Characterization: M_n/kg.mol^{-1} = 101, M_w/kg.mol^{-1} = 120,
DSC transition temperature range 392.2-407.2 K, NIST SRM1484

Solvent (A):	**ethene**	**C$_2$H$_4$**	**74-85-1**
Solvent (C):	**1-hexene**	**C$_6$H$_{12}$**	**592-41-6**

Type of data: cloud points

w_A	0.418	0.418	0.418	0.418	0.418	0.418
w_B	0.147	0.147	0.147	0.147	0.147	0.147
w_C	0.435	0.435	0.435	0.435	0.435	0.435
T/K	392.85	398.55	413.25	433.45	445.25	455.65
P/MPa	76.1	73.1	74.2	72.8	73.0	73.0

Polymer (B):	**polyethylene**		**1992WOH**

Characterization: M_n/kg.mol^{-1} = 1.94, M_w/kg.mol^{-1} = 5.37,
LDPE-wax, ρ = 0.9238 g/cm^3, Leuna-Werke, Germany

Solvent (A):	**ethene**	**C$_2$H$_4$**	**74-85-1**
Solvent (C):	**4-methyl-1-pentene**	**C$_6$H$_{12}$**	**691-37-2**

Type of data: coexistence data

continued

continued

$T/K = 493.15$ $P/MPa = 24.5$

w_A	w_B	w_C	w_A	w_B	w_C
	gel phase			sol phase	
0.1290	0.8710	0.0000	1.0000	0.0000	0.0000
0.1287	0.8219	0.0494	0.8847	0.0150	0.1003
0.1274	0.7840	0.0886	0.7905	0.0140	0.1955
0.1373	0.7754	0.0873	0.7430	0.0200	0.2370
0.1371	0.7626	0.1003	0.6449	0.0240	0.3311
0.1429	0.6750	0.1821	0.5780	0.0310	0.3910
0.1416	0.6380	0.2204	0.4765	0.0340	0.4895
0.1398	0.5787	0.2815	0.3880	0.0380	0.5740
0.1462	0.5365	0.3173	0.3380	0.0570	0.6050
0.1510	0.4662	0.3828	0.2699	0.1073	0.6228
0.1479	0.4498	0.4023	0.2417	0.1111	0.6472
0.1490	0.4037	0.4473	0.2609	0.1254	0.6137
0.1520	0.3981	0.4499	0.2317	0.1421	0.6262

Polymer (B):	**polyethylene**	**1996HEU**
Characterization:	$M_n/kg.mol^{-1} = 0.449$, $M_w/kg.mol^{-1} = 0.553$, $M_z/kg.mol^{-1} = 0.630$,	
	Exxon Polywax 500	
Solvent (A):	**ethene** C_2H_4	**74-85-1**
Polymer (C):	**polyethylene**	
Characterization:	$M_n/kg.mol^{-1} = 1.122$, $M_w/kg.mol^{-1} = 1.311$, $M_z/kg.mol^{-1} = 1.538$,	
	Exxon Polywax 1000	

Comments: The polymer mixture concentration is kept constant at $w_B/w_C = 0.50/0.50$ to simulate a bimodal polyethylene with $M_n/kg.mol^{-1} = 0.638$ and $M_w/M_n = 1.44$.

Type of data: cloud points

$T/K = 413.15$

x_{B+C}	0.0500	0.0429	0.0359	0.0306	0.0274	0.0257	0.0245	0.0234	0.0200
P/MPa	44.1	45.8	47.1	47.8	48.1	48.2	49.1	49.7	51.3

x_{B+C}	0.0165	0.0115	0.0085	0.0050
P/MPa	52.3	52.0	50.8	48.6

critical concentration: $x_{B+C, crit} = 0.0257$,   *critical pressure*: $P_{crit}/MPa = 48.2$

$T/K = 433.15$

x_{B+C}	0.0500	0.0429	0.0359	0.0306	0.0274	0.0257	0.0245	0.0234	0.0200
P/MPa	44.4	46.1	47.4	48.1	48.3	48.5	49.2	49.9	51.5

x_{B+C}	0.0165	0.0115	0.0085	0.0050
P/MPa	52.3	52.0	50.9	48.7

critical concentration: $x_{B, crit} = 0.0257$,   *critical pressure*: $P_{crit}/MPa = 48.5$

continued

continued

T/K = 453.15

x_{B+C}	0.0500	0.0429	0.0359	0.0306	0.0274	0.0257	0.0245	0.0234	0.0200
P/MPa	44.7	46.3	47.6	48.3	48.5	48.6	49.4	50.0	51.5

x_{B+C}	0.0165	0.0115	0.0085	0.0050
P/MPa	52.4	52.1	51.0	48.8

critical concentration: $x_{B, crit}$ = 0.0257,   *critical pressure*: P_{crit}/MPa = 48.6

T/K = 473.15

x_{B+C}	0.0500	0.0429	0.0359	0.0306	0.0274	0.0257	0.0245	0.0234	0.0200
P/MPa	44.9	46.5	47.8	48.5	48.8	48.9	49.6	50.1	51.6

x_{B+C}	0.0165	0.0115	0.0085	0.0050
P/MPa	52.4	52.2	51.1	49.0

critical concentration: $x_{B, crit}$ = 0.0257,   *critical pressure*: P_{crit}/MPa = 48.9

T/K = 493.15

x_{B+C}	0.0500	0.0429	0.0359	0.0306	0.0274	0.0257	0.0245	0.0234	0.0200
P/MPa	45.2	46.7	48.0	48.6	48.9	49.0	49.6	50.2	51.6

x_{B+C}	0.0165	0.0115	0.0085	0.0050
P/MPa	52.4	52.2	51.1	49.1

critical concentration: $x_{B, crit}$ = 0.0257,   *critical pressure*: P_{crit}/MPa = 49.0

T/K = 513.15

x_{B+C}	0.0500	0.0429	0.0359	0.0306	0.0274	0.0257	0.0245	0.0234	0.0200
P/MPa	45.4	46.9	48.1	48.8	49.1	49.2	49.7	50.3	51.7

x_{B+C}	0.0165	0.0115	0.0085	0.0050
P/MPa	52.5	52.2	51.2	49.2

critical concentration: $x_{B, crit}$ = 0.0257,   *critical pressure*: P_{crit}/MPa = 49.2

Polymer (B):	**polyethylene**		**1996HEU**
Characterization:	M_n/kg.mol^{-1} = 0.449, M_w/kg.mol^{-1} = 0.553, M_z/kg.mol^{-1} = 0.630, Exxon Polywax 500		
Solvent (A):	**ethene**	**C$_2$H$_4$**	**74-85-1**
Polymer (C):	**polyethylene**		
Characterization:	M_n/kg.mol^{-1} = 2.94, M_w/kg.mol^{-1} = 3.30, M_z/kg.mol^{-1} = 3.72, Exxon Polywax 2000		

Comments: The polymer mixture concentration is kept constant at w_B/w_C = 0.65/0.35 to simulate a bimodal polyethylene with M_n/kg.mol^{-1} = 0.638 and M_w/M_n = 2.36.

Type of data: cloud points

T/K = 413.15

x_{B+C}	0.0480	0.0334	0.0300	0.0264	0.0244	0.0226	0.0211	0.0200	0.0160
P/MPa	48.4	56.2	57.7	58.9	59.4	59.7	62.6	64.0	67.8

continued

continued

x_{B+C}	0.0118	0.0050
P/MPa	69.4	66.6

critical concentration: $x_{B+C, crit}$ = 0.0226, *critical pressure*: P_{crit}/MPa = 59.7

T/K = 433.15

x_{B+C}	0.0480	0.0334	0.0300	0.0264	0.0244	0.0226	0.0211	0.0200	0.0160
P/MPa	48.8	55.7	57.1	58.4	59.0	59.3	61.8	63.1	67.0

x_{B+C}	0.0118	0.0050
P/MPa	68.8	65.8

critical concentration: $x_{B+C, crit}$ = 0.0226, *critical pressure*: P_{crit}/MPa = 59.3

T/K = 453.15

x_{B+C}	0.0480	0.0334	0.0300	0.0264	0.0244	0.0226	0.0211	0.0200	0.0160
P/MPa	49.2	55.4	56.7	58.0	58.6	59.0	61.3	62.5	66.2

x_{B+C}	0.0118	0.0050
P/MPa	68.5	64.9

critical concentration: $x_{B+C, crit}$ = 0.0226, *critical pressure*: P_{crit}/MPa = 59.0

T/K = 473.15

x_{B+C}	0.0480	0.0334	0.0300	0.0264	0.0244	0.0226	0.0211	0.0200	0.0160
P/MPa	49.4	55.1	56.3	57.6	58.1	58.7	60.6	61.9	65.8

x_{B+C}	0.0118	0.0050
P/MPa	68.0	64.3

critical concentration: $x_{B+C, crit}$ = 0.0226, *critical pressure*: P_{crit}/MPa = 58.7

T/K = 493.15

x_{B+C}	0.0480	0.0334	0.0300	0.0264	0.0244	0.0226	0.0211	0.0200	0.0160
P/MPa	49.7	54.7	55.9	57.2	57.7	58.3	60.0	61.1	65.0

x_{B+C}	0.0118	0.0050
P/MPa	67.5	63.3

critical concentration: $x_{B+C, crit}$ = 0.0226, *critical pressure*: P_{crit}/MPa = 58.3

T/K = 513.15

x_{B+C}	0.0480	0.0334	0.0300	0.0264	0.0244	0.0226	0.0211	0.0200	0.0160
P/MPa	50.0	54.4	55.6	56.8	57.4	58.0	59.6	60.6	64.4

x_{B+C}	0.0118	0.0050
P/MPa	67.0	62.4

critical concentration: $x_{B+C, crit}$ = 0.0226, *critical pressure*: P_{crit}/MPa = 58.0

Polymer (B):	**polyethylene**							**1996HEU**

Characterization: $M_n/\text{kg.mol}^{-1} = 0.449$, $M_w/\text{kg.mol}^{-1} = 0.553$, $M_z/\text{kg.mol}^{-1} = 0.630$,
Exxon Polywax 500

Solvent (A):	**ethene**	**C₂H₄**	**74-85-1**

Polymer (C):   **polyethylene**

Characterization: $M_n/\text{kg.mol}^{-1} = 2.94$, $M_w/\text{kg.mol}^{-1} = 3.30$, $M_z/\text{kg.mol}^{-1} = 3.72$,
Exxon Polywax 2000

Comments: The polymer mixture concentration is kept constant at $w_B/w_C = 0.29/0.71$ to simulate a bimodal polyethylene with $M_n/\text{kg.mol}^{-1} = 1.122$ and $M_w/M_n = 2.22$.

Type of data: cloud points

$T/K = 413.15$

x_{B+C}	0.0280	0.0250	0.0204	0.0160	0.0121	0.0106	0.0089	0.0080	0.0071
P/MPa	62.1	65.2	69.2	72.7	75.3	76.2	76.7	77.0	78.2

x_{B+C}	0.0061
P/MPa	79.4

critical concentration: $x_{B+C,\,crit} = 0.0080$, *critical pressure:* $P_{crit}/\text{MPa} = 77.0$

$T/K = 433.15$

x_{B+C}	0.0280	0.0250	0.0204	0.0160	0.0121	0.0106	0.0089	0.0080	0.0071
P/MPa	61.4	64.6	67.6	70.1	73.2	74.6	74.4	75.9	75.2

x_{B+C}	0.0061
P/MPa	76.3

critical concentration: $x_{B+C,\,crit} = 0.0080$, *critical pressure:* $P_{crit}/\text{MPa} = 75.9$

$T/K = 453.15$

x_{B+C}	0.0280	0.0250	0.0204	0.0160	0.0121	0.0106	0.0089	0.0080	0.0071
P/MPa	61.2	63.3	66.4	69.3	71.7	72.6	73.4	73.9	75.0

x_{B+C}	0.0061
P/MPa	76.2

critical concentration: $x_{B+C,\,crit} = 0.0080$, *critical pressure:* $P_{crit}/\text{MPa} = 73.9$

$T/K = 473.15$

x_{B+C}	0.0280	0.0250	0.0204	0.0160	0.0121	0.0106	0.0089	0.0080	0.0071
P/MPa	60.8	62.6	65.6	68.3	70.5	71.3	72.2	72.5	73.7

x_{B+C}	0.0061
P/MPa	74.9

critical concentration: $x_{B+C,\,crit} = 0.0080$, *critical pressure:* $P_{crit}/\text{MPa} = 72.5$

$T/K = 493.15$

x_{B+C}	0.0280	0.0250	0.0204	0.0160	0.0121	0.0106	0.0089	0.0080	0.0071
P/MPa	60.5	62.3	64.8	67.1	69.9	70.2	71.2	71.1	72.9

continued

continued

x_{B+C}	0.0061
P/MPa	73.8

critical concentration: $x_{B+C, crit}$ = 0.0080,   *critical pressure*: P_{crit}/MPa = 71.1

T/K = 513.15

x_{B+C}	0.0280	0.0250	0.0204	0.0160	0.0121	0.0106	0.0089	0.0080	0.0071
P/MPa	59.2	61.7	63.9	66.1	68.9	69.1	70.0	70.7	71.7

x_{B+C}	0.0061
P/MPa	72.0

critical concentration: $x_{B+C, crit}$ = 0.0080,   *critical pressure*: P_{crit}/MPa = 70.7

Polymer (B):	**polyethylene**		**1996HEU**
Characterization:	M_n/kg.mol^{-1} = 0.449, M_w/kg.mol^{-1} = 0.553, M_z/kg.mol^{-1} = 0.630, Exxon Polywax 500		
Solvent (A):	**ethene**	**C$_2$H$_4$**	**74-85-1**
Polymer (C):	**polyethylene**		
Characterization:	M_n/kg.mol^{-1} = 4.40, M_w/kg.mol^{-1} = 5.07, M_z/kg.mol^{-1} = 5.83, Exxon Polywax 3000		

Comments: The polymer mixture concentration is kept constant at w_B/w_C = 0.667/0.333 to simulate a bimodal polyethylene with M_n/kg.mol^{-1} = 0.638 and M_w/M_n = 3.18.

Type of data: cloud points

T/K = 413.15

x_{B+C}	0.0490	0.0329	0.0270	0.0238	0.0220	0.0204	0.0190	0.0180	0.0160
P/MPa	50.5	61.8	65.8	67.6	68.5	69.7	71.4	72.6	75.0

x_{B+C}	0.0120	0.0085	0.0050
P/MPa	78.6	80.4	80.8

critical concentration: $x_{B+C, crit}$ = 0.0204,   *critical pressure*: P_{crit}/MPa = 69.7

T/K = 433.15

x_{B+C}	0.0490	0.0329	0.0270	0.0238	0.0220	0.0204	0.0190	0.0180	0.0160
P/MPa	50.6	61.3	64.9	66.8	67.5	68.6	70.4	71.8	74.0

x_{B+C}	0.0120	0.0085	0.0050
P/MPa	77.3	78.9	79.4

critical concentration: $x_{B+C, crit}$ = 0.0204,   *critical pressure*: P_{crit}/MPa = 68.6

T/K = 453.15

x_{B+C}	0.0490	0.0329	0.0270	0.0238	0.0220	0.0204	0.0190	0.0180	0.0160
P/MPa	50.8	60.6	64.2	66.1	66.9	67.7	69.6	70.8	72.9

x_{B+C}	0.0120	0.0085	0.0050
P/MPa	76.3	77.6	78.2

critical concentration: $x_{B+C, crit}$ = 0.0204,   *critical pressure*: P_{crit}/MPa = 67.7

continued

continued

$T/K = 473.15$

x_{B+C}	0.0490	0.0329	0.0270	0.0238	0.0220	0.0204	0.0190	0.0180	0.0160
P/MPa	51.9	60.2	63.7	65.5	66.2	66.9	68.7	70.0	72.3

x_{B+C}	0.0120	0.0085	0.0050
P/MPa	75.2	76.9	77.2

critical concentration: $x_{B+C, crit} = 0.0204$,   *critical pressure*: P_{crit}/MPa = 66.9

$T/K = 493.15$

x_{B+C}	0.0490	0.0329	0.0270	0.0238	0.0220	0.0204	0.0190	0.0180	0.0160
P/MPa	51.1	59.9	63.2	64.9	65.6	66.2	67.9	69.1	71.4

x_{B+C}	0.0120	0.0085	0.0050
P/MPa	74.7	76.2	76.6

critical concentration: $x_{B+C, crit} = 0.0204$,   *critical pressure*: P_{crit}/MPa = 66.2

$T/K = 513.15$

x_{B+C}	0.0490	0.0329	0.0270	0.0238	0.0220	0.0204	0.0190	0.0180	0.0160
P/MPa	51.3	59.7	62.9	64.2	64.8	65.4	67.0	68.3	70.5

x_{B+C}	0.0120	0.0085	0.0050
P/MPa	73.8	75.4	75.9

critical concentration: $x_{B+C, crit} = 0.0204$,   *critical pressure*: P_{crit}/MPa = 65.4

Polymer (B):	**polyethylene**	**1996HEU**
Characterization:	M_n/kg.mol^{-1} = 0.449, M_w/kg.mol^{-1} = 0.553, M_z/kg.mol^{-1} = 0.630, Exxon Polywax 500	
Solvent (A):	**ethene** $\quad$ **C$_2$H$_4$**	**74-85-1**
Polymer (C):	**polyethylene**	
Characterization:	M_n/kg.mol^{-1} = 4.40, M_w/kg.mol^{-1} = 5.07, M_z/kg.mol^{-1} = 5.83, Exxon Polywax 3000	

Comments: The polymer mixture concentration is kept constant at $w_B/w_C = 0.333/0.667$ to simulate a bimodal polyethylene with M_n/kg.mol^{-1} = 1.122 and M_w/M_n = 3.17.

Type of data: cloud points

$T/K = 413.15$

x_{B+C}	0.0278	0.0204	0.0161	0.0120	0.0083	0.0073	0.0065	0.0059	0.0053
P/MPa	66.1	74.2	79.2	84.0	88.5	89.7	90.8	91.5	92.4

x_{B+C}	0.0048	0.0042	0.0030
P/MPa	93.9	95.6	98.9

critical concentration: $x_{B+C, crit} = 0.0053$,   *critical pressure*: P_{crit}/MPa = 92.4

$T/K = 433.15$

x_{B+C}	0.0278	0.0204	0.0161	0.0120	0.0083	0.0073	0.0065	0.0059	0.0053
P/MPa	65.3	72.7	77.1	81.5	85.6	86.7	87.6	88.1	89.0

continued

continued

x_{B+C}	0.0048	0.0042	0.0030
P/MPa	90.6	92.3	95.1

critical concentration: $x_{B+C, crit}$ = 0.0053, *critical pressure*: P_{crit}/MPa = 89.0

T/K = 453.15

x_{B+C}	0.0278	0.0204	0.0161	0.0120	0.0083	0.0073	0.0065	0.0059	0.0053
P/MPa	64.6	71.3	75.4	79.5	83.3	84.4	85.4	86.1	87.0

x_{B+C}	0.0048	0.0042	0.0030
P/MPa	88.6	90.2	93.1

critical concentration: $x_{B+C, crit}$ = 0.0053, *critical pressure*: P_{crit}/MPa = 87.0

T/K = 473.15

x_{B+C}	0.0278	0.0204	0.0161	0.0120	0.0083	0.0073	0.0065	0.0059	0.0053
P/MPa	64.2	70.6	74.5	78.3	82.2	83.2	84.1	84.8	85.5

x_{B+C}	0.0048	0.0042	0.0030
P/MPa	87.1	88.6	91.5

critical concentration: $x_{B+C, crit}$ = 0.0053, *critical pressure*: P_{crit}/MPa = 85.5

T/K = 493.15

x_{B+C}	0.0278	0.0204	0.0161	0.0120	0.0083	0.0073	0.0065	0.0059	0.0053
P/MPa	63.5	69.5	73.2	76.8	80.7	81.6	82.7	83.2	84.1

x_{B+C}	0.0048	0.0042	0.0030
P/MPa	85.4	86.8	89.6

critical concentration: $x_{B+C, crit}$ = 0.0053, *critical pressure*: P_{crit}/MPa = 84.1

T/K = 513.15

x_{B+C}	0.0278	0.0204	0.0161	0.0120	0.0083	0.0073	0.0065	0.0059	0.0053
P/MPa	62.6	68.0	71.7	75.4	79.2	80.2	81.1	81.9	82.6

x_{B+C}	0.0048	0.0042	0.0030
P/MPa	83.9	85.3	88.0

critical concentration: $x_{B+C, crit}$ = 0.0053, *critical pressure*: P_{crit}/MPa = 82.6

Polymer (B):	**polyethylene**		**1996HEU**
Characterization:	M_n/kg.mol^{-1} = 0.638, M_w/kg.mol^{-1} = 0.768, M_z/kg.mol^{-1} = 0.911, Exxon Polywax 655		
Solvent (A):	**ethene**	**C$_2$H$_4$**	**74-85-1**
Polymer (C):	**polyethylene**		
Characterization:	M_n/kg.mol^{-1} = 2.94, M_w/kg.mol^{-1} = 3.30, M_z/kg.mol^{-1} = 3.72, Exxon Polywax 2000		

Comments: The polymer mixture concentration is kept constant at w_B/w_C = 0.45/0.55 to simulate a bimodal polyethylene with M_n/kg.mol^{-1} = 1.122 and M_w/M_n = 1.93.

Type of data: cloud points

continued

continued

$T/K = 413.15$

x_{B+C}	0.0280	0.0200	0.0160	0.0120	0.0098	0.0090	0.0084	0.0080	0.0060
P/MPa	58.8	65.8	68.9	71.2	72.3	72.7	73.6	74.2	76.6

x_{B+C}	0.0030	0.0010
P/MPa	76.8	74.8

critical concentration: $x_{B+C, crit} = 0.0090$,   *critical pressure*: P_{crit}/MPa = 72.7

$T/K = 433.15$

x_{B+C}	0.0280	0.0200	0.0160	0.0120	0.0098	0.0090	0.0084	0.0080	0.0060
P/MPa	58.4	64.6	67.6	70.1	71.2	71.6	72.4	72.9	75.2

x_{B+C}	0.0030	0.0010
P/MPa	75.3	73.2

critical concentration: $x_{B+C, crit} = 0.0090$,   *critical pressure*: P_{crit}/MPa = 71.6

$T/K = 453.15$

x_{B+C}	0.0280	0.0200	0.0160	0.0120	0.0098	0.0090	0.0084	0.0080	0.0060
P/MPa	58.0	63.7	66.6	69.0	70.0	70.2	71.0	71.7	73.8

x_{B+C}	0.0030	0.0010
P/MPa	73.7	71.4

critical concentration: $x_{B+C, crit} = 0.0090$,   *critical pressure*: P_{crit}/MPa = 70.2

$T/K = 473.15$

x_{B+C}	0.0280	0.0200	0.0160	0.0120	0.0098	0.0090	0.0084	0.0080	0.0060
P/MPa	57.8	63.0	65.6	67.9	68.7	69.0	70.1	70.8	72.7

x_{B+C}	0.0030	0.0010
P/MPa	72.2	69.7

critical concentration: $x_{B+C, crit} = 0.0090$,   *critical pressure*: P_{crit}/MPa = 69.0

$T/K = 493.15$

x_{B+C}	0.0280	0.0200	0.0160	0.0120	0.0098	0.0090	0.0084	0.0080	0.0060
P/MPa	57.5	62.3	64.8	67.1	67.9	68.2	68.2	69.1	69.9

x_{B+C}	0.0030	0.0010
P/MPa	71.8	71.2

critical concentration: $x_{B+C, crit} = 0.0090$,   *critical pressure*: P_{crit}/MPa = 68.2

$T/K = 513.15$

x_{B+C}	0.0280	0.0200	0.0160	0.0120	0.0098	0.0090	0.0084	0.0080	0.0060
P/MPa	57.2	61.7	63.9	66.1	66.9	67.1	68.0	68.7	70.7

x_{B+C}	0.0030	0.0010
P/MPa	70.0	67.4

critical concentration: $x_{B+C, crit} = 0.0090$,   *critical pressure*: P_{crit}/MPa = 67.1

Polymer (B):	polyethylene							**1996HEU**

Polymer (B): **polyethylene** **1996HEU**
Characterization: $M_n/\text{kg.mol}^{-1} = 0.638$, $M_w/\text{kg.mol}^{-1} = 0.768$, $M_z/\text{kg.mol}^{-1} = 0.911$,
Exxon Polywax 655

Solvent (A): **ethene** **C_2H_4** **74-85-1**
Polymer (C): **polyethylene**
Characterization: $M_n/\text{kg.mol}^{-1} = 4.40$, $M_w/\text{kg.mol}^{-1} = 5.07$, $M_z/\text{kg.mol}^{-1} = 5.83$,
Exxon Polywax 3000

Comments: The polymer mixture concentration is kept constant at $w_B/w_C = 0.495/0.505$ to simulate a bimodal polyethylene with $M_n/\text{kg.mol}^{-1} = 1.122$ and $M_w/M_n = 2.62$.

Type of data: cloud points

$T/\text{K} = 413.15$

x_{B+C}	0.0277	0.0204	0.0161	0.0121	0.0100	0.0086	0.0075	0.0068	0.0063
P/MPa	61.9	68.7	72.7	76.7	78.7	80.2	81.4	82.1	83.6

x_{B+C}	0.0053	0.0045
P/MPa	86.3	88.5

critical concentration: $x_{B+C, \text{crit}} = 0.0068$, critical pressure: $P_{\text{crit}}/\text{MPa} = 82.1$

$T/\text{K} = 433.15$

x_{B+C}	0.0277	0.0204	0.0161	0.0121	0.0100	0.0086	0.0075	0.0068	0.0063
P/MPa	61.4	67.3	71.3	74.9	76.9	78.5	79.7	80.4	81.7

x_{B+C}	0.0053	0.0045
P/MPa	84.6	86.8

critical concentration: $x_{B+C, \text{crit}} = 0.0068$, critical pressure: $P_{\text{crit}}/\text{MPa} = 80.4$

$T/\text{K} = 453.15$

x_{B+C}	0.0277	0.0204	0.0161	0.0121	0.0100	0.0086	0.0075	0.0068	0.0063
P/MPa	61.0	66.5	70.1	73.5	75.5	76.7	77.7	78.3	79.5

x_{B+C}	0.0053	0.0045
P/MPa	82.3	84.5

critical concentration: $x_{B+C, \text{crit}} = 0.0068$, critical pressure: $P_{\text{crit}}/\text{MPa} = 78.8$

$T/\text{K} = 473.15$

x_{B+C}	0.0277	0.0204	0.0161	0.0121	0.0100	0.0086	0.0075	0.0068	0.0063
P/MPa	60.4	65.7	69.0	72.4	74.3	75.7	76.8	77.2	78.4

x_{B+C}	0.0053	0.0045
P/MPa	81.2	83.4

critical concentration: $x_{B+C, \text{crit}} = 0.0068$, critical pressure: $P_{\text{crit}}/\text{MPa} = 77.2$

$T/\text{K} = 493.15$

x_{B+C}	0.0277	0.0204	0.0161	0.0121	0.0100	0.0086	0.0075	0.0068	0.0063
P/MPa	60.0	65.0	68.3	71.6	73.3	74.5	75.6	76.4	77.7

x_{B+C}	0.0053	0.0045
P/MPa	80.4	82.1

critical concentration: $x_{B+C, \text{crit}} = 0.0068$, critical pressure: $P_{\text{crit}}/\text{MPa} = 76.4$

continued

continued

$T/K = 513.15$

x_{B+C}	0.0277	0.0204	0.0161	0.0121	0.0100	0.0086	0.0075	0.0068	0.0063
P/MPa	59.5	64.0	67.2	70.6	72.4	73.8	74.7	75.6	76.8

x_{B+C}	0.0053	0.0045
P/MPa	78.9	81.0

critical concentration: $x_{B+C, crit} = 0.0068$,   *critical pressure*: P_{crit}/MPa = 75.6

Polymer (B):	**polyethylene**		**1996HEU**
Characterization:	M_n/kg.mol^{-1} = 0.638, M_w/kg.mol^{-1} = 0.768, M_z/kg.mol^{-1} = 0.911,		
	Exxon Polywax 655		
Solvent (A):	**ethene**	C_2H_4	**74-85-1**
Polymer (C):	**polyethylene**		
Characterization:	M_n/kg.mol^{-1} = 4.40, M_w/kg.mol^{-1} = 5.07, M_z/kg.mol^{-1} = 5.83,		
	Exxon Polywax 3000		

Comments: The polymer mixture concentration is kept constant at $w_B/w_C = 0.085/0.915$ to simulate a bimodal polyethylene with M_n/kg.mol^{-1} = 2.94 and M_w/M_n = 1.60.

Type of data: cloud points

$T/K = 413.15$

x_{B+C}	0.0106	0.0087	0.0079	0.0057	0.0046	0.0038	0.0030	0.0025	0.0022
P/MPa	80.0	84.7	88.8	91.4	93.8	95.5	97.3	98.6	99.4

x_{B+C}	0.0019	0.0017
P/MPa	101.3	103.0

critical concentration: $x_{B+C, crit} = 0.0022$,   *critical pressure*: P_{crit}/MPa = 99.4

$T/K = 433.15$

x_{B+C}	0.0106	0.0087	0.0079	0.0057	0.0046	0.0038	0.0030	0.0025	0.0022
P/MPa	78.2	82.8	86.8	89.4	91.8	93.5	95.3	96.3	97.2

x_{B+C}	0.0019	0.0017
P/MPa	98.8	100.4

critical concentration: $x_{B+C, crit} = 0.0022$,   *critical pressure*: P_{crit}/MPa = 97.2

$T/K = 453.15$

x_{B+C}	0.0106	0.0087	0.0079	0.0057	0.0046	0.0038	0.0030	0.0025	0.0022
P/MPa	76.5	80.7	84.6	87.2	89.4	91.1	92.7	93.9	94.5

x_{B+C}	0.0019	0.0017
P/MPa	96.3	97.9

critical concentration: $x_{B+C, crit} = 0.0022$,   *critical pressure*: P_{crit}/MPa = 94.5

$T/K = 473.15$

x_{B+C}	0.0106	0.0087	0.0079	0.0057	0.0046	0.0038	0.0030	0.0025	0.0022
P/MPa	74.8	78.9	82.9	85.2	87.5	89.0	90.7	91.6	92.2

continued

continued

x_{B+C}	0.0019	0.0017
P/MPa	93.3	94.4

critical concentration: $x_{B+C, crit}$ = 0.0022, *critical pressure*: P_{crit}/MPa = 92.2

T/K = 493.15

x_{B+C}	0.0106	0.0087	0.0079	0.0057	0.0046	0.0038	0.0030	0.0025	0.0022
P/MPa	73.3	77.0	80.8	83.0	85.2	86.7	88.5	89.7	90.2

x_{B+C}	0.0019	0.0017
P/MPa	91.6	92.8

critical concentration: $x_{B+C, crit}$ = 0.0022, *critical pressure*: P_{crit}/MPa = 90.2

T/K = 513.15

x_{B+C}	0.0106	0.0087	0.0079	0.0057	0.0046	0.0038	0.0030	0.0025	0.0022
P/MPa	71.6	75.3	78.7	81.0	83.2	84.9	86.5	87.5	88.3

x_{B+C}	0.0019	0.0017
P/MPa	89.8	91.1

critical concentration: $x_{B+C, crit}$ = 0.0022, *critical pressure*: P_{crit}/MPa = 88.3

Polymer (B):	**polyethylene**		**1996HEU**
Characterization:	M_n/kg.mol^{-1} = 1.122, M_w/kg.mol^{-1} = 1.311, M_z/kg.mol^{-1} = 1.538, Exxon Polywax 1000		
Solvent (A):	**ethene**	C_2H_4	**74-85-1**
Polymer (C):	**polyethylene**		
Characterization:	M_n/kg.mol^{-1} = 4.40, M_w/kg.mol^{-1} = 5.07, M_z/kg.mol^{-1} = 5.83, Exxon Polywax 3000		

Comments: The polymer mixture concentration is kept constant at w_B/w_C = 0.167/0.833 to simulate a bimodal polyethylene with M_n/kg.mol^{-1} = 2.94 and M_w/M_n = 1.51.

Type of data: cloud points

T/K = 413.15

x_{B+C}	0.0107	0.0096	0.0084	0.0067	0.0057	0.0045	0.0037	0.0031	0.0025
P/MPa	76.6	79.9	83.1	86.6	89.2	91.7	93.6	94.8	96.2

x_{B+C}	0.0023	0.0020	0.0018
P/MPa	96.7	98.5	99.8

critical concentration: $x_{B+C, crit}$ = 0.0023, *critical pressure*: P_{crit}/MPa = 96.7

T/K = 433.15

x_{B+C}	0.0107	0.0096	0.0084	0.0067	0.0057	0.0045	0.0037	0.0031	0.0025
P/MPa	75.0	77.9	80.7	84.1	86.7	89.2	91.0	92.2	93.5

x_{B+C}	0.0023	0.0020	0.0018
P/MPa	94.1	95.7	96.9

critical concentration: $x_{B+C, crit}$ = 0.0023, *critical pressure*: P_{crit}/MPa = 94.1

continued

continued

$T/K = 453.15$

x_{B+C}	0.0107	0.0096	0.0084	0.0067	0.0057	0.0045	0.0037	0.0031	0.0025
P/MPa	73.7	76.5	79.1	82.3	84.6	86.9	88.4	89.4	90.3

x_{B+C}	0.0023	0.0020	0.0018
P/MPa	90.9	92.7	93.9

critical concentration: $x_{B+C, crit} = 0.0023$, *critical pressure*: $P_{crit}/MPa = 90.9$

$T/K = 473.15$

x_{B+C}	0.0107	0.0096	0.0084	0.0067	0.0057	0.0045	0.0037	0.0031	0.0025
P/MPa	72.6	75.1	77.6	80.2	82.4	84.5	85.9	86.9	87.9

x_{B+C}	0.0023	0.0020	0.0018
P/MPa	88.4	90.2	91.4

critical concentration: $x_{B+C, crit} = 0.0023$, *critical pressure*: $P_{crit}/MPa = 88.4$

$T/K = 493.15$

x_{B+C}	0.0107	0.0096	0.0084	0.0067	0.0057	0.0045	0.0037	0.0031	0.0025
P/MPa	71.3	73.8	76.1	78.8	80.8	82.5	83.9	84.9	85.7

x_{B+C}	0.0023	0.0020	0.0018
P/MPa	86.2	87.0	89.3

critical concentration: $x_{B+C, crit} = 0.0023$, *critical pressure*: $P_{crit}/MPa = 86.2$

$T/K = 513.15$

x_{B+C}	0.0107	0.0096	0.0084	0.0067	0.0057	0.0045	0.0037	0.0031	0.0025
P/MPa	70.0	72.3	74.6	77.2	79.2	80.9	82.4	83.2	84.1

x_{B+C}	0.0023	0.0020	0.0018
P/MPa	84.5	86.0	87.1

critical concentration: $x_{B+C, crit} = 0.0023$, *critical pressure*: $P_{crit}/MPa = 84.5$

Polymer (B):	**polyethylene**		**1991NIE**
Characterization:	$M_n/kg.mol^{-1} = 2.41$, $M_w/kg.mol^{-1} = 4.88$, HDPE,		
	Exxon Polymer Plant, Antwerp, fractionated in the laboratory		
Solvent (A):	**ethene**	C_2H_4	**74-85-1**
Solvent (C):	**vinyl acetate**	$C_4H_6O_2$	**108-05-4**

Type of data: cloud points

w_A	0.651	0.651	0.510	0.510	0.510	0.379	0.379	0.379	0.230
w_B	0.086	0.086	0.114	0.114	0.114	0.159	0.159	0.159	0.230
w_C	0.263	0.263	0.376	0.376	0.376	0.462	0.462	0.462	0.540
T/K	413.15	433.65	404.15	413.15	447.65	413.15	428.15	457.55	413.15
P/MPa	82.4	79.8	69.8	68.0	62.9	55.5	53.0	49.1	39.5

w_A	0.230	0.230
w_B	0.230	0.230
w_C	0.540	0.540
T/K	442.65	453.15
P/MPa	36.0	35.5

Polymer (B):	polyethylene							**1991NIE**

Characterization: M_n/kg.mol^{-1} = 10.6, M_w/kg.mol^{-1} = 15.4, LDPE, Exxon Polymer Plant, Antwerp, fractionated in the laboratory

Solvent (A):	ethene	C_2H_4	**74-85-1**
Solvent (C):	vinyl acetate	$C_4H_6O_2$	**108-05-4**

Type of data: cloud points

w_A	0.834	0.834	0.834	0.834	0.834	0.574	0.574	0.574	0.574
w_B	0.048	0.048	0.048	0.048	0.048	0.077	0.077	0.077	0.077
w_C	0.118	0.118	0.118	0.118	0.118	0.349	0.349	0.349	0.349
T/K	390.15	395.15	425.15	444.65	463.65	388.15	403.15	433.65	453.65
P/MPa	142.6	138.6	120.2	111.4	106.5	112.0	102.2	89.4	84.3
w_A	0.574	0.408	0.678	0.678	0.678	0.678	0.573	0.573	0.573
w_B	0.077	0.280	0.156	0.156	0.156	0.156	0.255	0.255	0.255
w_C	0.349	0.312	0.166	0.166	0.166	0.166	0.172	0.172	0.172
T/K	463.65	393.15	389.15	403.15	423.65	443.65	388.15	403.15	423.15
P/MPa	82.4	82.4	118.8	112.4	104.9	100.2	107.6	102.5	95.8
w_A	0.573	0.573	0.408	0.408	0.408	0.482	0.482	0.482	0.248
w_B	0.255	0.255	0.280	0.280	0.280	0.353	0.353	0.353	0.381
w_C	0.172	0.172	0.312	0.312	0.312	0.163	0.163	0.163	0.371
T/K	443.15	458.65	413.65	448.65	459.15	393.15	423.15	448.65	393.15
P/MPa	91.5	88.9	74.1	71.0	70.6	94.5	86.6	82.5	56.2
w_A	0.248	0.248	0.248	0.248					
w_B	0.381	0.381	0.381	0.381					
w_C	0.371	0.371	0.371	0.371					
T/K	403.15	423.15	443.15	457.65					
P/MPa	52.7	49.5	48.3	48.0					

Polymer (B):	polyethylene							**1991NIE**

Characterization: M_n/kg.mol^{-1} = 36.3, M_w/kg.mol^{-1} = 220.5, LDPE, Exxon Polymer Plant, Antwerp

Solvent (A):	ethene	C_2H_4	**74-85-1**
Solvent (C):	vinyl acetate	$C_4H_6O_2$	**108-05-4**

Type of data: coexistence data

T/K = 403.15

Total feed concentrations of the homogeneous system before demixing:

w_A = 0.483, w_B = 0.138, w_C = 0.379.

P/ MPa	w_A	w_B bottom phase	w_C	w_A	w_B top phase	w_C	
131.4	0.483	0.138	0.379				(cloud point)
121.8	0.447	0.205	0.348	0.539	0.036	0.425	
99.0	0.423	0.254	0.322	0.546	0.029	0.425	
69.6	0.337	0.404	0.258	0.553	0.019	0.428	
42.4				0.556	0.002	0.442	

Polymer (B):	polyethylene		1990KEN
Characterization:	M_n/kg.mol^{-1} = 8, M_w/kg.mol^{-1} = 177, M_z/kg.mol^{-1} = 1000,		
	HDPE, DSM, Geleen, The Netherlands		
Solvent (A):	n-hexane	C$_6$H$_{14}$	110-54-3
Solvent (C):	nitrogen	N$_2$	7727-37-9

Type of data: cloud points

w_A	0.99218	0.99218	0.99218	0.99218	0.99218	0.99218	0.97801	0.97801	0.97801
w_B	0.00557	0.00557	0.00557	0.00557	0.00557	0.00557	0.01970	0.01970	0.01970
w_C	0.00225	0.00225	0.00225	0.00225	0.00225	0.00225	0.00229	0.00229	0.00229
T/K	404.66	408.64	414.59	418.61	422.55	428.53	407.57	410.03	412.54
P/bar	17.6	24.1	33.1	39.1	45.1	53.6	17.4	21.4	25.4
w_A	0.97801	0.97801	0.97801	0.97801	0.97801	0.95555	0.95555	0.95555	0.95555
w_B	0.01970	0.01970	0.01970	0.01970	0.01970	0.04220	0.04220	0.04220	0.04220
w_C	0.00229	0.00229	0.00229	0.00229	0.00229	0.00225	0.00225	0.00225	0.00225
T/K	415.04	417.56	422.48	427.46	432.45	410.03	412.52	415.02	417.51
P/bar	29.3	32.9	40.4	47.6	54.7	13.4	17.8	21.6	25.0
w_A	0.95555	0.93668	0.93668	0.93668	0.93668	0.93668	0.93668	0.93668	0.93668
w_B	0.04220	0.06110	0.06110	0.06110	0.06110	0.06110	0.06110	0.06110	0.06110
w_C	0.00225	0.00222	0.00222	0.00222	0.00222	0.00222	0.00222	0.00222	0.00222
T/K	420.00	417.56	420.09	422.56	425.03	427.51	429.99	432.47	437.47
P/bar	29.0	18.2	23.2	27.2	30.6	35.0	38.3	41.8	49.4
w_A	0.93668	0.91600	0.91600	0.91600	0.91600	0.91600	0.91600	0.91600	0.91600
w_B	0.06110	0.08180	0.08180	0.08180	0.08180	0.08180	0.08180	0.08180	0.08180
w_C	0.00222	0.00220	0.00220	0.00220	0.00220	0.00220	0.00220	0.00220	0.00220
T/K	442.34	420.01	419.95	422.53	422.43	424.98	424.93	427.48	427.51
P/bar	56.4	14.6	14.8	18.4	18.6	22.4	22.6	26.4	26.4
w_A	0.91600	0.91600	0.91600	0.91600	0.91600	0.91600	0.87440	0.87440	0.87440
w_B	0.08180	0.08180	0.08180	0.08180	0.08180	0.08180	0.12350	0.12350	0.12350
w_C	0.00220	0.00220	0.00220	0.00220	0.00220	0.00220	0.00210	0.00210	0.00210
T/K	429.96	429.96	432.47	437.44	442.41	447.35	427.49	432.42	437.36
P/bar	30.1	30.2	34.0	41.1	48.3	55.6	18.0	25.4	32.8
w_A	0.87440	0.87440	0.87440	0.87440	0.86626	0.86626	0.86626	0.86626	0.86626
w_B	0.12350	0.12350	0.12350	0.12350	0.13160	0.13160	0.13160	0.13160	0.13160
w_C	0.00210	0.00210	0.00210	0.00210	0.00214	0.00214	0.00214	0.00214	0.00214
T/K	442.36	447.33	452.31	457.26	427.49	429.96	432.45	434.93	437.41
P/bar	40.1	47.3	54.4	61.4	15.4	19.3	23.2	27.0	30.8
w_A	0.86626	0.86626	0.86626	0.86626	0.98317	0.98317	0.98317	0.98317	0.98317
w_B	0.13160	0.13160	0.13160	0.13160	0.00443	0.00443	0.00443	0.00443	0.00443
w_C	0.00214	0.00214	0.00214	0.00214	0.01240	0.01240	0.01240	0.01240	0.01240
T/K	442.36	447.33	452.31	457.26	392.62	397.32	402.57	407.59	412.57
P/bar	37.9	45.0	52.4	58.8	34.5	41.6	49.6	57.0	64.3
w_A	0.98317	0.98317	0.98317	0.97797	0.97797	0.97797	0.97797	0.97797	0.97797
w_B	0.00443	0.00443	0.00443	0.00973	0.00973	0.00973	0.00973	0.00973	0.00973
w_C	0.01240	0.01240	0.01240	0.01230	0.01230	0.01230	0.01230	0.01230	0.01230
T/K	417.56	422.53	427.48	396.66	402.38	407.59	412.29	421.90	427.46
P/bar	71.4	78.4	85.0	38.2	46.8	54.6	61.6	75.2	82.3

continued

continued

w_A	0.97637	0.97637	0.97637	0.97637	0.97637	0.97637	0.97637	0.97637	0.96690
w_B	0.01093	0.01093	0.01093	0.01093	0.01093	0.01093	0.01093	0.01093	0.02060
w_C	0.01270	0.01270	0.01270	0.01270	0.01270	0.01270	0.01270	0.01270	0.01250
T/K	396.64	400.63	401.99	407.57	412.52	417.56	422.43	427.40	392.62
P/bar	39.4	45.5	47.5	55.8	63.2	70.4	77.2	84.0	29.8

w_A	0.96690	0.96690	0.96690	0.96690	0.96690	0.96690	0.96690	0.96600	0.96600
w_B	0.02060	0.02060	0.02060	0.02060	0.02060	0.02060	0.02060	0.02130	0.02130
w_C	0.01250	0.01250	0.01250	0.01250	0.01250	0.01250	0.01250	0.01270	0.01270
T/K	396.66	401.97	407.59	411.89	417.56	422.53	427.48	396.66	402.62
P/bar	36.0	44.2	52.0	59.0	67.2	74.2	80.7	37.2	46.2

w_A	0.96600	0.96600	0.96600	0.96600	0.96600	0.95800	0.95800	0.95800	0.95800
w_B	0.02130	0.02130	0.02130	0.02130	0.02130	0.02950	0.02950	0.02950	0.02950
w_C	0.01270	0.01270	0.01270	0.01270	0.01270	0.01250	0.01250	0.01250	0.01250
T/K	407.59	412.57	417.56	422.53	427.46	397.03	401.99	407.59	416.93
P/bar	53.7	61.1	68.3	75.0	81.4	33.2	40.8	49.1	63.0

w_A	0.95800	0.95800	0.95800	0.94630	0.94630	0.94630	0.94630	0.94630	0.94630
w_B	0.02950	0.02950	0.02950	0.04120	0.04120	0.04120	0.04120	0.04120	0.04120
w_C	0.01250	0.01250	0.01250	0.01250	0.01250	0.01250	0.01250	0.01250	0.01250
T/K	421.92	426.79	431.76	402.41	404.63	406.62	408.61	412.57	417.56
P/bar	70.1	76.5	82.9	37.0	40.5	43.6	46.5	52.3	59.6

w_A	0.94630	0.94630	0.94630	0.92590	0.92590	0.92590	0.92590	0.92590	0.92590
w_B	0.04120	0.04120	0.04120	0.06140	0.06140	0.06140	0.06140	0.06140	0.06140
w_C	0.01250	0.01250	0.01250	0.01270	0.01270	0.01270	0.01270	0.01270	0.01270
T/K	422.53	427.46	432.42	402.22	407.20	412.18	417.17	422.16	427.15
P/bar	66.9	73.6	80.4	30.6	37.9	45.4	52.8	60.1	67.2

w_A	0.92590	0.92590	0.90640	0.90640	0.90640	0.90640	0.90640	0.90640	0.90640
w_B	0.06140	0.06140	0.08120	0.08120	0.08120	0.08120	0.08120	0.08120	0.08120
w_C	0.01270	0.01270	0.01240	0.01240	0.01240	0.01240	0.01240	0.01240	0.01240
T/K	432.11	437.07	407.59	411.89	412.57	416.93	421.80	421.92	426.79
P/bar	74.1	80.5	32.9	39.2	40.3	46.8	54.0	54.2	61.2

w_A	0.90640	0.90640	0.90640	0.88390	0.88390	0.88390	0.88390	0.88390	0.88390
w_B	0.08120	0.08120	0.08120	0.10390	0.10390	0.10390	0.10390	0.10390	0.10390
w_C	0.01240	0.01240	0.01240	0.01220	0.01220	0.01220	0.01220	0.01220	0.01220
T/K	431.76	436.70	441.62	412.12	414.57	416.56	422.06	427.48	431.97
P/bar	68.2	74.9	81.2	32.9	36.5	39.5	47.8	55.8	62.2

w_A	0.88390	0.92975	0.92975	0.92975	0.92975	0.92975	0.92975	0.93350	0.93350
w_B	0.10390	0.05980	0.05980	0.05980	0.05980	0.05980	0.05980	0.06040	0.06040
w_C	0.01220	0.01045	0.01045	0.01045	0.01045	0.01045	0.01045	0.00610	0.00610
T/K	437.39	411.81	416.56	421.71	426.73	431.62	441.54	411.86	417.54
P/bar	69.8	38.5	45.6	53.3	50.6	57.6	31.0	23.4	32.3

w_A	0.93350	0.93350	0.93350	0.93350	0.93350	0.89241	0.89241	0.89241	0.89241
w_B	0.06040	0.06040	0.06040	0.06040	0.06040	0.09860	0.09860	0.09860	0.09860
w_C	0.00610	0.00610	0.00610	0.00610	0.00610	0.00899	0.00899	0.00899	0.00899
T/K	421.79	427.49	431.68	437.41	442.36	396.61	401.99	402.17	407.15
P/bar	38.8	47.5	53.7	62.3	69.1	26.2	34.8	35.0	43.0

continued

continued

w_A	0.89241	0.89241	0.89241	0.89241	0.89158	0.89158	0.89158	0.89158	0.89158
w_B	0.09860	0.09860	0.09860	0.09860	0.10340	0.10340	0.10340	0.10340	0.10340
w_C	0.00899	0.00899	0.00899	0.00899	0.00502	0.00502	0.00502	0.00502	0.00502
T/K	411.89	412.07	416.56	426.73	417.87	421.75	426.73	431.63	436.59
P/bar	50.2	50.6	57.2	72.2	17.7	23.7	31.3	38.6	45.8

w_A	0.89158	0.89158	0.88122	0.88122	0.88122	0.88122	0.88122	0.88122	0.88122
w_B	0.10340	0.10340	0.11110	0.11110	0.11110	0.11110	0.11110	0.11110	0.11110
w_C	0.00502	0.00502	0.00768	0.00768	0.00768	0.00768	0.00768	0.00768	0.00768
T/K	441.57	449.43	417.54	422.48	427.49	432.42	437.41	442.36	447.33
P/bar	53.0	64.0	24.2	31.5	38.9	46.2	53.5	60.3	67.1

Type of data: liquid-liquid-vapor three phase equilibrium data

w_A	0.99215	0.97801	0.95555	0.93668	0.91660	0.87440	0.86616	0.98320	0.97800
w_B	0.00560	0.01970	0.04220	0.06110	0.08180	0.12350	0.13160	0.00440	0.00970
w_C	0.00225	0.00229	0.00225	0.00222	0.00160	0.00210	0.00224	0.01240	0.01230
T/K	399.15	402.45	408.15	412.35	417.85	423.85	425.65	387.05	388.75
P/bar	8.8	9.3	10.1	10.8	11.1	12.3	12.6	25.8	26.0

w_A	0.97640	0.96690	0.96600	0.95800	0.94630	0.92590	0.90640	0.88390	0.92975
w_B	0.01090	0.02060	0.02130	0.02950	0.04120	0.06140	0.08120	0.10390	0.05980
w_C	0.01270	0.01250	0.01270	0.01250	0.01250	0.01270	0.01240	0.01220	0.01045
T/K	388.35	390.75	390.05	393.15	395.95	400.55	404.95	409.85	402.95
P/bar	26.5	26.7	27.0	27.1	27.2	27.9	28.9	29.4	24.8

w_A	0.93350	0.98111	0.89158	0.88122
w_B	0.06040	0.00990	0.10340	0.11110
w_C	0.00610	0.00899	0.00502	0.00768
T/K	407.85	392.95	417.15	415.75
P/bar	17.0	20.4	16.6	21.4

Polymer (B):	**polyethylene**		**2002HOR**
Characterization:	$M_n/\text{kg.mol}^{-1} = 13$, $M_w/\text{kg.mol}^{-1} = 89$, $M_z/\text{kg.mol}^{-1} = 600$,		
	LDPE, Stamylan, DSM, Geleen, The Netherlands		
Solvent (A):	**propane**	C_3H_8	**74-98-6**
Solvent (C):	**cyclohexane**	C_6H_{12}	**110-82-7**

Type of data: cloud points

w_A	0.100	0.100	0.100	0.100	0.100	0.100	0.100	0.100	0.100
w_B	0.097	0.097	0.097	0.097	0.097	0.097	0.097	0.097	0.097
w_C	0.803	0.803	0.803	0.803	0.803	0.803	0.803	0.803	0.803
T/K	493.13	495.21	495.42	497.45	497.48	499.62	499.75	502.06	504.69
P/bar	40.0	43.0	43.5	46.5	46.5	50.0	50.0	53.0	56.0

w_A	0.100	0.154	0.154	0.154	0.154	0.154	0.154	0.154	0.154
w_B	0.097	0.094	0.094	0.094	0.094	0.094	0.094	0.094	0.094
w_C	0.803	0.752	0.752	0.752	0.752	0.752	0.752	0.752	0.752
T/K	506.72	473.48	473.72	475.99	478.47	478.58	480.95	483.60	488.50
P/bar	59.0	41.0	40.9	44.5	48.0	47.9	51.5	54.9	61.9

continued

continued

w_A	0.154	0.154	0.154	0.154	0.200	0.200	0.200	0.200	0.200
w_B	0.094	0.094	0.094	0.094	0.086	0.086	0.086	0.086	0.086
w_C	0.752	0.752	0.752	0.752	0.714	0.714	0.714	0.714	0.714
T/K	493.51	498.50	503.48	508.46	455.40	460.29	465.25	470.02	474.75
P/bar	68.9	75.5	81.9	87.9	44.5	52.5	59.5	66.5	74.0

w_A	0.200	0.200	0.200
w_B	0.086	0.086	0.086
w_C	0.714	0.714	0.714
T/K	484.44	493.89	503.10
P/bar	87.5	100.5	113.0

Polymer (B):	poly(ethylene glycol)		1992KIA
Characterization:	M_w/kg.mol^{-1} = 8.0, Aldrich Chem. Co., Inc., Milwaukee, WI		
Solvent (A):	**carbon dioxide**	**CO$_2$**	**124-38-9**
Solvent (C):	**methanol**	**CH$_4$O**	**67-56-1**

Type of data: cloud points

w_A	0.196	0.196	0.196	0.196	0.196	0.196	0.196	0.351	0.351
w_B	0.1206	0.1206	0.1206	0.1206	0.1206	0.1206	0.1206	0.09735	0.09735
w_C	0.6834	0.6834	0.6834	0.6834	0.6834	0.6834	0.6834	0.55165	0.55165
T/K	304.35	308.55	313.45	318.45	323.35	333.35	343.15	304.75	323.15
P/bar	35.89	39.17	42.75	46.46	50.37	58.74	67.01	49.21	65.44
	VLE	VLE	VLE	VLE	VLE	VLE	VLE	VLE	VLE

w_A	0.351	0.351	0.505	0.505	0.505	0.505	0.505	0.200	0.200
w_B	0.09735	0.09735	0.07425	0.07425	0.07425	0.07425	0.07425	0.240	0.240
w_C	0.55165	0.55165	0.42075	0.42075	0.42075	0.42075	0.42075	0.560	0.560
T/K	333.25	343.25	304.95	313.55	323.35	333.05	338.45	308.45	313.45
P/bar	76.17	87.76	57.13	68.46	81.36	96.13	103.62	34.05	36.77
	VLE	VLE	VLE	VLE	VLE	VLE	VLE	VLE	VLE

w_A	0.200	0.200	0.200	0.200	0.295	0.295	0.295	0.295	0.399
w_B	0.240	0.240	0.240	0.240	0.2115	0.2115	0.2115	0.2115	0.1803
w_C	0.560	0.560	0.560	0.560	0.4935	0.4935	0.4935	0.4935	0.4207
T/K	318.25	323.25	333.15	343.15	313.15	324.05	333.15	343.15	304.85
P/bar	40.22	43.33	50.90	59.72	48.36	59.13	65.04	76.40	51.58
	VLE	VLF	VLE	VLE	VLE	VLE	VLE	VLE	VLE

w_A	0.399	0.399	0.399	0.399	0.498	0.498	0.498	0.498	0.498
w_B	0.1803	0.1803	0.1803	0.1803	0.1506	0.1506	0.1506	0.1506	0.1506
w_C	0.4207	0.4207	0.4207	0.4207	0.3514	0.3514	0.3514	0.3514	0.3514
T/K	312.95	323.25	333.15	343.15	303.85	308.45	313.15	318.15	323.15
P/bar	60.60	73.77	87.40	101.58	56.17	62.31	67.60	76.01	81.88
	VLE	VLE	VLE	VLE	VLE	VLE	VLE	VLE	VLE

w_A	0.498	0.498	0.498	0.200	0.200	0.200	0.300	0.300	0.300
w_B	0.1506	0.1506	0.1506	0.400	0.400	0.400	0.350	0.350	0.350
w_C	0.3514	0.3514	0.3514	0.400	0.400	0.400	0.350	0.350	0.350
T/K	328.15	333.15	343.15	313.15	323.15	333.15	313.15	323.15	333.15
P/bar	91.83	97.88	133.23	30.96	36.54	42.39	54.11	60.35	66.49
	VLE	VLE	LLE	VLE	VLE	VLE	VLE	VLE	VLE

continued

continued

w_A	0.400	0.400	0.400	0.500	0.500	0.200	0.200	0.200	0.200
w_B	0.300	0.300	0.300	0.250	0.250	0.560	0.560	0.560	0.560
w_C	0.300	0.300	0.300	0.250	0.250	0.240	0.240	0.240	0.240
T/K	313.15	323.15	333.15	313.15	323.15	304.45	308.65	313.15	318.25
P/bar	62.48	77.59	95.68	92.46	137.37	39.82	43.83	49.63	52.20
	VLE	VLE	LLE	LLE	LLE	VLE	VLE	VLE	VLE

w_A	0.200	0.200	0.200	0.250	0.250	0.250	0.250	0.250	0.250
w_B	0.560	0.560	0.560	0.525	0.525	0.525	0.525	0.525	0.525
w_C	0.240	0.240	0.240	0.225	0.225	0.225	0.225	0.225	0.225
T/K	323.15	333.15	343.15	305.05	308.55	313.15	318.15	323.15	328.15
P/bar	59.32	65.69	77.48	44.42	48.16	49.15	53.19	62.08	65.47
	VLE	VLE	VLE	VLE	VLE	VLE	VLE	VLE	VLE

w_A	0.250	0.250	0.250	0.250	0.250	0.250	0.250	0.250	0.300
w_B	0.525	0.525	0.525	0.525	0.525	0.525	0.525	0.525	0.490
w_C	0.225	0.225	0.225	0.225	0.225	0.225	0.225	0.225	0.210
T/K	333.15	343.15	313.15	318.15	323.15	328.15	333.15	343.15	313.15
P/bar	66.84	78.92	75.08	74.12	84.41	101.71	129.69	168.23	57.19
	VLE	VLE	LLE	LLE	LLE	LLE	LLE	LLE	VLE

w_A	0.300	0.300	0.300	0.300	0.300	0.300	0.300	0.300	0.300
w_B	0.490	0.490	0.490	0.490	0.490	0.490	0.490	0.490	0.490
w_C	0.210	0.210	0.210	0.210	0.210	0.210	0.210	0.210	0.210
T/K	318.15	323.15	333.15	343.15	313.15	318.15	323.15	333.15	343.15
P/bar	63.72	68.52	84.70	98.40	80.73	79.43	92.12	100.36	158.58
	VLE	VLE	VLE	VLE	LLE	LLE	LLE	LLE	LLE

Polymer (B):	poly(ethyl methacrylate)						2000BY1, 2000BY2
Characterization:	Aldrich Chem. Co., Inc., Milwaukee, WI						
Solvent (A):	**carbon dioxide**	CO_2					**124-38-9**
Solvent (C):	**ethyl methacrylate**	$C_6H_{10}O_2$					**97-63-2**

Type of data: cloud points

w_A	0.825	0.825	0.825	0.825	0.825	0.697	0.697	0.697	0.697
w_B	0.05	0.05	0.05	0.05	0.05	0.05	0.05	0.05	0.05
w_C	0.125	0.125	0.125	0.125	0.125	0.253	0.253	0.253	0.253
T/K	436.75	450.45	458.65	472.75	490.45	316.05	336.15	355.65	374.25
P/bar	1201	929	940	957	975	546	605	657	702

w_A	0.697	0.697	0.697	0.697	0.568	0.568	0.568	0.568	0.568
w_B	0.05	0.05	0.05	0.05	0.05	0.05	0.05	0.05	0.05
w_C	0.253	0.253	0.253	0.253	0.382	0.382	0.382	0.382	0.382
T/K	393.95	415.65	438.05	466.75	314.85	333.55	354.95	374.55	395.55
P/bar	742	781	815	853	246	321	396	448	476

w_A	0.493	0.493	0.493	0.493	0.493	0.493	0.493	0.493	0.493
w_B	0.05	0.05	0.05	0.05	0.05	0.05	0.05	0.05	0.05
w_C	0.457	0.457	0.457	0.457	0.457	0.457	0.457	0.457	0.457
T/K	319.15	324.25	333.75	343.45	363.55	383.55	403.45	423.55	
P/bar	67	85	122	163	228	284	328	353	

Polymer (B):	poly(hexyl acrylate)							**2004BYU**

Characterization: $M_w/\text{kg.mol}^{-1} = 90$,
Scientific Polymer Products, Inc., Ontario, NY

Solvent (A):	carbon dioxide	CO_2						**124-38-9**
Solvent (C):	hexyl acrylate	$C_9H_{16}O_2$						**2499-95-8**

Type of data: cloud points

w_A	0.899	0.899	0.899	0.899	0.899	0.822	0.822	0.822	0.822
w_B	0.051	0.051	0.051	0.051	0.051	0.050	0.050	0.050	0.050
w_C	0.050	0.050	0.050	0.050	0.050	0.128	0.128	0.128	0.128
T/K	358.25	362.45	383.85	402.85	428.65	317.45	318.35	319.05	328.95
P/bar	2246.6	1343.1	1077.6	983.5	902.4	1408.6	1115.5	1022.4	805.5

w_A	0.822	0.822	0.822	0.822	0.822	0.739	0.739	0.739	0.739
w_B	0.050	0.050	0.050	0.050	0.050	0.054	0.054	0.054	0.054
w_C	0.128	0.128	0.128	0.128	0.128	0.207	0.207	0.207	0.207
T/K	345.05	365.05	385.35	404.75	424.55	312.25	327.85	345.05	365.05
P/bar	705.9	679.3	681.0	691.4	702.1	316.6	338.6	375.5	417.2

w_A	0.739	0.739	0.739	0.611	0.611	0.611	0.611	0.611	0.611
w_B	0.054	0.054	0.054	0.050	0.050	0.050	0.050	0.050	0.050
w_C	0.207	0.207	0.207	0.339	0.339	0.339	0.339	0.339	0.339
T/K	385.75	409.25	428.75	303.25	307.45	313.85	323.85	332.55	323.75
P/bar	460.4	496.6	516.6	67.2	71.0	80.4	90.0	99.8	110.4
				VLE	VLE	VLE	VLLE	VLLE	LLE

w_A	0.611	0.611	0.611	0.611
w_B	0.050	0.050	0.050	0.050
w_C	0.339	0.339	0.339	0.339
T/K	335.45	356.05	377.35	396.75
P/bar	149.7	213.5	268.3	306.6

Polymer (B):	poly(hexyl methacrylate)							**2004BYU**

Characterization: $M_w/\text{kg.mol}^{-1} = 400$, Aldrich Chem. Co., Inc., Milwaukee, WI

Solvent (A):	carbon dioxide	CO_2						**124-38-9**
Solvent (C):	hexyl methacrylate	$C_{10}H_{18}O_2$						**142-09-6**

Type of data: cloud points

w_A	0.852	0.852	0.852	0.852	0.852	0.852	0.811	0.811	0.811
w_B	0.056	0.056	0.056	0.056	0.056	0.056	0.045	0.045	0.045
w_C	0.092	0.092	0.092	0.092	0.092	0.092	0.144	0.144	0.144
T/K	384.75	387.85	390.65	404.25	424.55	444.65	355.45	357.25	361.25
P/bar	1794.8	1612.1	1508.6	1227.6	1058.6	1022.4	1825.9	1541.7	1336.2

w_A	0.811	0.811	0.811	0.811	0.811	0.626	0.626	0.626	0.626
w_B	0.045	0.045	0.045	0.045	0.045	0.050	0.050	0.050	0.050
w_C	0.144	0.144	0.144	0.144	0.144	0.324	0.324	0.324	0.324
T/K	366.05	386.45	405.35	425.95	449.35	334.15	354.75	373.15	393.25
P/bar	1205.2	976.6	901.0	871.4	859.3	501.7	507.6	519.3	524.5

continued

continued

w_A	0.626	0.626	0.525	0.525	0.525	0.525	0.525	0.525	0.525
w_B	0.050	0.050	0.047	0.047	0.047	0.047	0.047	0.047	0.047
w_C	0.324	0.324	0.428	0.428	0.428	0.428	0.428	0.428	0.428
T/K	415.15	434.55	313.05	323.95	333.35	351.65	355.95	373.15	353.55
P/bar	528.3	493.8	67.2	81.7	92.1	112.2	119.2	139.8	131.0
			VLE	VLE	VLE	VLLE	VLLE	VLLE	LLE

w_A	0.525	0.525
w_B	0.047	0.047
w_C	0.428	0.428
T/K	374.45	395.05
P/bar	192.1	241.7

Polymer (B):	**polyisobutylene**		**2002JOU**
Characterization:	M_n/kg.mol^{-1} = 600, M_w/kg.mol^{-1} = 1000,		
	Aldrich Chem. Co., Inc., Milwaukee, WI		
Solvent (A):	**carbon dioxide**	**CO$_2$**	**124-38-9**
Solvent (C):	**n-heptane**	**C$_7$H$_{16}$**	**142-82-5**

Type of data: cloud points

w_A	0.088	0.088	0.088	0.088	0.088	0.088	0.088	0.088	0.260
w_B	0.025	0.025	0.025	0.025	0.025	0.025	0.025	0.025	0.025
w_C	0.887	0.887	0.887	0.887	0.887	0.887	0.887	0.887	0.715
T/K	403.15	413.15	423.15	433.15	443.15	453.15	463.15	473.15	323.15
P/MPa	3.8	5.7	7.6	9.1	10.2	12.3	13.9	15.3	10.1

w_A	0.260	0.260	0.260	0.260	0.260	0.260	0.260	0.260
w_B	0.025	0.025	0.025	0.025	0.025	0.025	0.025	0.025
w_C	0.715	0.715	0.715	0.715	0.715	0.715	0.715	0.715
T/K	333.15	343.15	353.15	363.15	373.15	383.15	393.15	403.15
P/MPa	11.6	13.6	15.2	17.6	19.5	21.2	22.9	23.7

Comments: Gas solubilities and VLE data for this system are given in Chapter 2.

Polymer (B):	**poly(L-lactide)**		**2000LE2, 2000LE3**
Characterization:	M_n/kg.mol^{-1} = 2.0, Polysciences, Inc., Warrington, PA		
Solvent (A):	**carbon dioxide**	**CO$_2$**	**124-38-9**
Solvent (C):	**chlorodifluoromethane**	**CHClF$_2$**	**75-45-6**

Type of data: cloud points

w_A	0.0000	0.0000	0.0000	0.0000	0.0000	0.0000	0.0722	0.0722	0.0722
w_B	0.0288	0.0288	0.0288	0.0288	0.0288	0.0288	0.0278	0.0278	0.0278
w_C	0.9712	0.9712	0.9712	0.9712	0.9712	0.9712	0.9000	0.9000	0.9000
T/K	344.35	353.05	363.35	372.75	382.05	392.75	305.45	314.45	323.95
P/MPa	3.60	5.95	8.72	10.90	13.05	15.62	1.76	2.15	2.69
	LLE	LLE	LLE	LLE	LLE	LLE	VLE	VLE	VLE

continued

continued

w_A	0.0722	0.0722	0.0722	0.0722	0.0722	0.0722	0.0722	0.0722	0.0722
w_B	0.0278	0.0278	0.0278	0.0278	0.0278	0.0278	0.0278	0.0278	0.0278
w_C	0.9000	0.9000	0.9000	0.9000	0.9000	0.9000	0.9000	0.9000	0.9000
T/K	328.75	333.85	342.65	353.15	357.45	360.55	361.55	363.05	333.85
P/MPa	3.00	3.30	3.97	4.80	5.10	5.33	5.40	5.50	3.98
	VLE	VLLE	VLLE	VLLE	VLLE	VLLE	VLLE	CP	LLE

w_A	0.0722	0.0722	0.0722	0.0722	0.0722	0.0722	0.1470	0.1470	0.1470
w_B	0.0278	0.0278	0.0278	0.0278	0.0278	0.0278	0.0310	0.0310	0.0310
w_C	0.9000	0.9000	0.9000	0.9000	0.9000	0.9000	0.8220	0.8220	0.8220
T/K	342.45	352.95	362.85	373.35	382.35	392.55	309.05	314.25	325.55
P/MPa	6.53	9.70	12.53	15.20	17.37	19.70	2.58	2.94	3.60
	LLE	LLE	LLE	LLE	LLE	LLE	VLE	VLE	VLLE

w_A	0.1470	0.1470	0.1470	0.1470	0.1470	0.1470	0.1470	0.1470	0.1470
w_B	0.0310	0.0310	0.0310	0.0310	0.0310	0.0310	0.0310	0.0310	0.0310
w_C	0.8220	0.8220	0.8220	0.8220	0.8220	0.8220	0.8220	0.8220	0.8220
T/K	333.95	344.35	354.35	356.35	325.35	333.75	344.45	355.55	364.55
P/MPa	4.27	5.03	5.95	6.10	3.97	6.70	10.03	13.20	16.10
	VLLE	VLLE	VLLE	CP	LLE	LLE	LLE	LLE	LLE

w_A	0.1470	0.1470	0.1470	0.3066	0.3066	0.3066	0.3066	0.3066	0.3066
w_B	0.0310	0.0310	0.0310	0.0316	0.0316	0.0316	0.0316	0.0316	0.0316
w_C	0.8220	0.8220	0.8220	0.6618	0.6618	0.6618	0.6618	0.6618	0.6618
T/K	372.65	382.65	391.85	304.05	310.25	316.35	323.75	333.25	336.35
P/MPa	18.10	20.60	22.75	3.45	3.72	4.15	4.75	5.65	5.95
	LLE	LLE	LLE	VLE	VLLE	VLLE	VLLE	VLLE	VLLE

w_A	0.3066	0.3066	0.3066	0.3066	0.3066	0.3066	0.3066	0.3066	0.3066
w_B	0.0316	0.0316	0.0316	0.0316	0.0316	0.0316	0.0316	0.0316	0.0316
w_C	0.6618	0.6618	0.6618	0.6618	0.6618	0.6618	0.6618	0.6618	0.6618
T/K	337.45	338.65	339.25	340.95	342.05	342.95	343.85	344.85	309.75
P/MPa	6.05	6.15	6.22	6.35	6.45	6.55	6.60	6.63	5.07
	VLLE	VLLE	VLLE	VLLE	VLLE	VLLE	VLLE	CP	LLE

w_A	0.3066	0.3066	0.3066	0.3066	0.3066	0.3066	0.3066	0.3066	0.3066
w_B	0.0316	0.0316	0.0316	0.0316	0.0316	0.0316	0.0316	0.0316	0.0316
w_C	0.6618	0.6618	0.6618	0.6618	0.6618	0.6618	0.6618	0.6618	0.6618
T/K	316.75	324.05	333.95	344.75	353.85	362.25	372.85	382.15	392.35
P/MPa	8.10	10.93	14.50	18.05	21.07	23.50	26.42	28.97	31.35
	LLE	LLE	LLE	LLE	LLE	LLE	LLE	LLE	LLE

w_A	0.4257	0.4257	0.4257	0.4257	0.4257	0.4257	0.4257	0.4257	0.4257
w_B	0.0326	0.0326	0.0326	0.0326	0.0326	0.0326	0.0326	0.0326	0.0326
w_C	0.5417	0.5417	0.5417	0.5417	0.5417	0.5417	0.5417	0.5417	0.5417
T/K	305.05	315.35	323.75	329.05	333.05	333.75	334.15	334.85	335.55
P/MPa	4.47	5.20	6.00	6.51	6.93	6.98	7.02	7.06	7.10
	VLLE	VLLE	VLLE	VLLE	VLLE	VLLE	VLLE	VLLE	VLLE

continued

continued

w_A	0.4257	0.4257	0.4257	0.4257	0.4257	0.4257	0.4257	0.4257	0.4257
w_B	0.0326	0.0326	0.0326	0.0326	0.0326	0.0326	0.0326	0.0326	0.0326
w_C	0.5417	0.5417	0.5417	0.5417	0.5417	0.5417	0.5417	0.5417	0.5417
T/K	336.25	337.15	305.05	315.35	324.35	328.05	338.15	350.85	362.45
P/MPa	7.13	7.15	9.25	13.90	17.40	18.95	22.40	26.97	30.70
	VLLE	CP	LLE	LLE	LLE	LLE	LLE	LLE	LLE

w_A	0.4257	0.4257	0.4257	0.5014	0.5014	0.5014	0.5014	0.5014	0.5014
w_B	0.0326	0.0326	0.0326	0.0281	0.0281	0.0281	0.0281	0.0281	0.0281
w_C	0.5417	0.5417	0.5417	0.4705	0.4705	0.4705	0.4705	0.4705	0.4705
T/K	372.85	383.05	393.45	302.45	315.85	325.05	333.75	342.55	354.05
P/MPa	33.60	36.30	38.15	12.62	18.22	22.55	25.72	28.90	33.02
	LLE	LLE	LLE	LLE	LLE	LLE	LLE	LLE	LLE

w_A	0.5014	0.5014	0.5014	0.6470	0.6470	0.6470	0.6470	0.6470	0.6470
w_B	0.0281	0.0281	0.0281	0.0260	0.0260	0.0260	0.0260	0.0260	0.0260
w_C	0.4705	0.4705	0.4705	0.3270	0.3270	0.3270	0.3270	0.3270	0.3270
T/K	363.65	373.35	383.55	315.35	324.55	335.15	343.85	352.95	362.95
P/MPa	35.87	38.65	41.22	30.25	34.42	38.18	41.50	44.45	47.75
	LLE	LLE	LLE	LLE	LLE	LLE	LLE	LLE	LLE

w_A	0.6470	0.6470	0.7996	0.7996	0.7996	0.7996	0.7996	0.7996
w_B	0.0260	0.0260	0.0282	0.0282	0.0282	0.0282	0.0282	0.0282
w_C	0.3270	0.3270	0.1722	0.1722	0.1722	0.1722	0.1722	0.1722
T/K	372.85	382.65	323.95	334.35	343.45	353.25	363.35	372.05
P/MPa	50.65	53.40	56.55	60.65	63.75	66.55	69.15	71.47
	LLE	LLE	LLE	LLE	LLE	LLE	LLE	LLE

Comments: CP is a critical point.

Polymer (B):	**poly(L-lactide)**		**2000LE3**
Characterization:	M_η/kg.mol^{-1} = 50, Polysciences, Inc., Warrington, PA		
Solvent (A):	**carbon dioxide**	**CO$_2$**	**124-38-9**
Solvent (C):	**chlorodifluoromethane**	**CHClF$_2$**	**75-45-6**

Type of data: cloud points

w_A	0.0000	0.0000	0.0000	0.0000	0.1166	0.1166	0.1166	0.1166	0.1166
w_B	0.0303	0.0303	0.0303	0.0303	0.0276	0.0276	0.0276	0.0276	0.0276
w_C	0.9697	0.9697	0.9697	0.9697	0.8558	0.8558	0.8558	0.8558	0.8558
T/K	336.95	350.95	360.35	373.85	323.85	333.55	343.15	353.05	363.25
P/MPa	4.08	8.60	11.38	15.03	5.75	9.45	12.85	16.30	19.65

w_A	0.1166	0.2978	0.2978	0.2978	0.2978	0.2978	0.2978	0.2978	0.2978
w_B	0.0276	0.0266	0.0266	0.0266	0.0266	0.0266	0.0266	0.0266	0.0266
w_C	0.8558	0.6756	0.6756	0.6756	0.6756	0.6756	0.6756	0.6756	0.6756
T/K	373.05	303.55	313.25	323.75	333.45	343.45	353.05	363.15	373.25
P/MPa	22.70	6.70	11.30	16.00	20.20	24.20	27.90	31.55	34.95

continued

continued

w_A	0.3783	0.3783	0.3783	0.3783	0.3783	0.3783	0.3783	0.3783	0.5102
w_B	0.0278	0.0278	0.0278	0.0278	0.0278	0.0278	0.0278	0.0278	0.0281
w_C	0.5939	0.5939	0.5939	0.5939	0.5939	0.5939	0.5939	0.5939	0.4617
T/K	303.15	313.35	323.95	333.35	343.35	353.05	362.95	372.95	323.45
P/MPa	11.50	16.50	21.60	25.80	30.05	33.95	37.70	41.15	32.80
w_A	0.5102	0.5102	0.5102	0.5102	0.5102	0.5890	0.5890	0.5890	0.5890
w_B	0.0281	0.0281	0.0281	0.0281	0.0281	0.0254	0.0254	0.0254	0.0254
w_C	0.4617	0.4617	0.4617	0.4617	0.4617	0.3856	0.3856	0.3856	0.3856
T/K	333.55	343.05	353.15	363.15	373.15	323.65	333.75	343.05	353.15
P/MPa	37.65	41.90	46.15	50.10	53.50	43.20	48.15	52.30	56.65
w_A	0.5890	0.5890	0.1853	0.1853	0.1853	0.1853	0.1853	0.1853	0.1853
w_B	0.0254	0.0254	0.0307	0.0307	0.0307	0.0307	0.0307	0.0307	0.0307
w_C	0.3856	0.3856	0.7840	0.7840	0.7840	0.7840	0.7840	0.7840	0.7840
T/K	363.25	373.25	313.55	323.45	333.55	343.35	353.55	363.05	373.25
P/MPa	60.65	64.05	5.50	9.65	13.60	17.30	21.00	24.25	27.57

Polymer (B):	poly(L-lactide)	2000LE3

Characterization: M_η/kg.mol^{-1} = 100, Polysciences, Inc., Warrington, PA

Solvent (A):	**carbon dioxide**	**CO$_2$**	**124-38-9**
Solvent (C):	**chlorodifluoromethane**	**CHClF$_2$**	**75-45-6**

Type of data: cloud points

w_A	0.0000	0.0000	0.0000	0.0000	0.0000	0.1006	0.1006	0.1006	0.1006
w_B	0.0291	0.0291	0.0291	0.0291	0.0291	0.0284	0.0284	0.0284	0.0284
w_C	0.9709	0.9709	0.9709	0.9709	0.9709	0.8710	0.8710	0.8710	0.8710
T/K	334.25	342.55	353.15	364.45	373.95	323.85	333.85	343.25	352.95
P/MPa	3.60	6.30	9.63	12.86	15.76	5.57	9.40	12.70	16.05
w_A	0.1006	0.1006	0.2016	0.2016	0.2016	0.2016	0.2016	0.2016	0.2016
w_B	0.0284	0.0284	0.0296	0.0296	0.0296	0.0296	0.0296	0.0296	0.0296
w_C	0.8710	0.8710	0.7688	0.7688	0.7688	0.7688	0.7688	0.7688	0.7688
T/K	362.75	373.35	312.85	323.95	333.65	343.65	353.65	362.95	372.85
P/MPa	19.30	22.58	6.95	11.50	15.55	19.35	23.15	26.45	29.70
w_A	0.2855	0.2855	0.2855	0.2855	0.2855	0.2855	0.2855	0.2855	0.3734
w_B	0.0302	0.0302	0.0302	0.0302	0.0302	0.0302	0.0302	0.0302	0.0311
w_C	0.6843	0.6843	0.6843	0.6843	0.6843	0.6843	0.6843	0.6843	0.5955
T/K	303.15	313.55	323.95	333.55	343.65	353.15	363.25	373.85	303.15
P/MPa	6.85	11.65	16.30	20.50	24.70	28.35	32.05	35.75	12.55
w_A	0.3734	0.3734	0.3734	0.3734	0.3734	0.3734	0.3734	0.6058	0.6058
w_B	0.0311	0.0311	0.0311	0.0311	0.0311	0.0311	0.0311	0.0289	0.0289
w_C	0.5955	0.5955	0.5955	0.5955	0.5955	0.5955	0.5955	0.3653	0.3653
T/K	313.15	323.65	333.25	343.05	353.25	362.95	373.35	323.85	333.95
P/MPa	17.65	22.60	26.95	31.20	35.35	39.12	42.85	47.85	52.68

continued

continued

w_A	0.6058	0.6058	0.6058	0.6058	0.4849	0.4849	0.4849	0.4849	0.4849
w_B	0.0289	0.0289	0.0289	0.0289	0.0310	0.0310	0.0310	0.0310	0.0310
w_C	0.3653	0.3653	0.3653	0.3653	0.4841	0.4841	0.4841	0.4841	0.4841
T/K	343.15	353.05	363.25	373.25	313.15	323.75	333.25	343.15	353.05
P/MPa	56.90	61.00	65.00	68.50	27.70	33.15	37.75	42.28	46.45

w_A	0.4849	0.4849
w_B	0.0310	0.0310
w_C	0.4841	0.4841
T/K	363.15	373.15
P/MPa	50.55	54.15

Polymer (B):	**poly(L-lactide)**		**2002LEE**
Characterization:	$M_\eta/\text{kg.mol}^{-1} = 2.0$, Polysciences, Inc., Warrington, PA		
Solvent (A):	**carbon dioxide**	**CO$_2$**	**124-38-9**
Solvent (C):	**dichloromethane**	**CH$_2$Cl$_2$**	**75-09-2**

Type of data: cloud points

w_A	0.8733	0.8733	0.8733	0.8733	0.8733	0.8733	0.8366	0.8366	0.8366
w_B	0.0506	0.0506	0.0506	0.0506	0.0506	0.0506	0.0477	0.0477	0.0477
w_C	0.0761	0.0761	0.0761	0.0761	0.0761	0.0761	0.1157	0.1157	0.1157
T/K	322.65	333.45	343.15	353.15	363.15	373.35	324.55	333.65	343.45
P/MPa	76.65	78.45	80.45	82.05	83.35	84.75	66.05	68.25	70.35

w_A	0.8366	0.8366	0.8366	0.8054	0.8054	0.8054	0.8054	0.8054	0.8054
w_B	0.0477	0.0477	0.0477	0.0495	0.0495	0.0495	0.0495	0.0495	0.0495
w_C	0.1157	0.1157	0.1157	0.1451	0.1451	0.1451	0.1451	0.1451	0.1451
T/K	353.55	363.25	372.95	323.65	333.35	343.25	353.25	363.55	375.15
P/MPa	72.15	74.15	75.65	56.75	59.65	62.35	64.75	67.05	69.35

w_A	0.7444	0.7444	0.7444	0.7444	0.7444	0.7444	0.6725	0.6725	0.6725
w_B	0.0515	0.0515	0.0515	0.0515	0.0515	0.0515	0.0492	0.0492	0.0492
w_C	0.2041	0.2041	0.2041	0.2041	0.2041	0.2041	0.2783	0.2783	0.2783
T/K	323.15	333.25	343.25	353.15	363.25	373.25	323.25	333.35	343.35
P/MPa	42.55	46.05	49.25	52.05	54.65	57.05	29.75	33.65	37.15

w_A	0.6725	0.6725	0.6725	0.5690	0.5690	0.5690	0.5690	0.5690	0.5690
w_B	0.0492	0.0492	0.0492	0.0498	0.0498	0.0498	0.0498	0.0498	0.0498
w_C	0.2783	0.2783	0.2783	0.3812	0.3812	0.3812	0.3812	0.3812	0.3812
T/K	353.15	362.75	373.15	323.75	333.25	343.15	353.15	363.45	373.15
P/MPa	40.35	43.05	45.95	14.85	18.55	22.25	25.55	28.75	31.25

w_A	0.4448	0.4448	0.4448	0.4448	0.4448	0.4448	0.4448	0.4448
w_B	0.0522	0.0522	0.0522	0.0522	0.0522	0.0522	0.0522	0.0522
w_C	0.5030	0.5030	0.5030	0.5030	0.5030	0.5030	0.5030	0.5030
T/K	304.05	313.95	323.75	333.85	343.25	353.25	363.25	372.65
P/MPa	4.82	5.65	6.55	7.58	8.60	11.37	14.45	17.05

Polymer (B):	**poly(L-lactide)**							**2002LEE**
Characterization:	$M_\eta/\mathrm{kg.mol}^{-1} = 50$, Polysciences, Inc., Warrington, PA							
Solvent (A):	**carbon dioxide**		**CO$_2$**					**124-38-9**
Solvent (C):	**dichloromethane**		**CH$_2$Cl$_2$**					**75-09-2**

Type of data: cloud points

w_A	0.7398	0.7398	0.7398	0.7398	0.7398	0.7398	0.7398	0.6666	0.6666
w_B	0.0508	0.0508	0.0508	0.0508	0.0508	0.0508	0.0508	0.0484	0.0484
w_C	0.2094	0.2094	0.2094	0.2094	0.2094	0.2094	0.2094	0.2850	0.2850
T/K	318.15	326.15	334.35	343.25	352.75	362.95	372.85	318.05	322.65
P/MPa	52.75	55.93	58.78	62.00	65.07	68.14	70.85	35.30	37.28

w_A	0.6666	0.6666	0.6666	0.6666	0.6666	0.5648	0.5648	0.5648	0.5648
w_B	0.0484	0.0484	0.0484	0.0484	0.0484	0.0507	0.0507	0.0507	0.0507
w_C	0.2850	0.2850	0.2850	0.2850	0.2850	0.3845	0.3845	0.3845	0.3845
T/K	333.25	343.55	352.45	363.35	373.25	313.95	323.05	333.35	343.45
P/MPa	41.78	45.82	49.06	52.75	55.67	15.07	19.30	23.87	27.84

w_A	0.5648	0.5648	0.5648	0.4408	0.4408	0.4408	0.4408	0.4408	0.4408
w_B	0.0507	0.0507	0.0507	0.0508	0.0508	0.0508	0.0508	0.0508	0.0508
w_C	0.3845	0.3845	0.3845	0.5084	0.5084	0.5084	0.5084	0.5084	0.5084
T/K	352.75	363.75	372.65	312.35	322.35	333.55	342.45	353.15	362.55
P/MPa	31.44	35.27	38.15	5.26	6.14	8.67	12.29	16.23	19.50

w_A	0.4408	0.3865	0.3865	0.3865	0.3865	0.3865	0.3865	0.3865
w_B	0.0508	0.0509	0.0509	0.0509	0.0509	0.0509	0.0509	0.0509
w_C	0.5084	0.5626	0.5626	0.5626	0.5626	0.5626	0.5626	0.5626
T/K	372.45	311.65	322.75	332.45	345.05	352.35	361.75	373.15
P/MPa	22.75	4.82	5.75	6.76	7.97	10.73	13.55	17.43

Polymer (B):	**poly(L-lactide)**							**2002LEE**
Characterization:	$M_\eta/\mathrm{kg.mol}^{-1} = 100$, Polysciences, Inc., Warrington, PA							
Solvent (A):	**carbon dioxide**		**CO$_2$**					**124-38-9**
Solvent (C):	**dichloromethane**		**CH$_2$Cl$_2$**					**75-09-2**

Type of data: cloud points

w_A	0.7479	0.7479	0.7479	0.7479	0.7479	0.7479	0.7479	0.6549	0.6549
w_B	0.0509	0.0509	0.0509	0.0509	0.0509	0.0509	0.0509	0.0531	0.0531
w_C	0.2012	0.2012	0.2012	0.2012	0.2012	0.2012	0.2012	0.2920	0.2920
T/K	318.45	324.75	333.45	343.25	352.75	362.35	372.25	316.85	324.45
P/MPa	58.20	60.55	63.67	67.05	69.97	72.85	75.52	34.94	38.55

w_A	0.6549	0.6549	0.6549	0.6549	0.6549	0.6023	0.6023	0.6023	0.6023
w_B	0.0531	0.0531	0.0531	0.0531	0.0531	0.0499	0.0499	0.0499	0.0499
w_C	0.2920	0.2920	0.2920	0.2920	0.2920	0.3478	0.3478	0.3478	0.3478
T/K	333.75	343.55	352.85	363.05	373.05	317.95	324.15	334.75	345.75
P/MPa	42.37	46.18	49.64	53.07	56.15	20.47	23.40	27.95	32.55

continued

continued

w_A	0.6023	0.6023	0.6023	0.4590	0.4590	0.4590	0.4590	0.4590	0.4590
w_B	0.0499	0.0499	0.0499	0.0513	0.0513	0.0513	0.0513	0.0513	0.0513
w_C	0.3478	0.3478	0.3478	0.4897	0.4897	0.4897	0.4897	0.4897	0.4897
T/K	352.05	362.75	372.65	314.35	322.85	333.15	343.55	352.95	364.25
P/MPa	34.85	38.70	42.05	5.45	7.37	11.72	15.00	19.50	23.56

w_A	0.4590
w_B	0.0513
w_C	0.4897
T/K	372.95
P/MPa	26.05

Polymer (B):	**poly(DL-lactide)**		**2001KUK**
Characterization:	M_η/kg.mol^{-1} = 30, Resomer R203, Boehringer Ingelheim Chemicals		
Solvent (A):	**carbon dioxide**	**CO$_2$**	**124-38-9**
Solvent (C):	**dimethyl ether**	**C$_2$H$_6$O**	**115-10-6**

Type of data: cloud points

w_A	0.0000	0.0000	0.0000	0.0000	0.0000	0.0000	0.1216	0.1216	0.1216
w_B	0.0500	0.0500	0.0500	0.0500	0.0500	0.0500	0.0521	0.0521	0.0521
w_C	0.9500	0.9500	0.9500	0.9500	0.9500	0.9500	0.8263	0.8263	0.8263
T/K	328.65	333.05	343.05	354.15	363.35	372.95	303.45	313.95	323.15
P/MPa	2.47	3.75	6.55	9.27	11.45	13.73	2.52	5.73	8.55

w_A	0.1216	0.1216	0.1216	0.1216	0.1216	0.2034	0.2034	0.2034	0.2034
w_B	0.0521	0.0521	0.0521	0.0521	0.0521	0.0439	0.0439	0.0439	0.0439
w_C	0.8263	0.8263	0.8263	0.8263	0.8263	0.7527	0.7527	0.7527	0.7527
T/K	333.55	342.85	353.55	362.75	373.35	303.05	314.25	325.95	333.65
P/MPa	11.50	13.75	16.12	18.25	20.65	8.55	11.95	15.15	17.39

w_A	0.2034	0.2034	0.2034	0.2034	0.3126	0.3126	0.3126	0.3126	0.3126
w_B	0.0439	0.0439	0.0439	0.0439	0.0468	0.0468	0.0468	0.0468	0.0468
w_C	0.7527	0.7527	0.7527	0.7527	0.6406	0.6406	0.6406	0.6406	0.6406
T/K	342.85	353.15	363.05	372.95	303.25	313.25	323.45	333.65	342.95
P/MPa	19.65	21.70	23.82	25.95	18.70	21.35	23.99	26.42	28.80

w_A	0.3126	0.3126	0.3126	0.4017	0.4017	0.4017	0.4017	0.4017	0.4017
w_B	0.0468	0.0468	0.0468	0.0485	0.0485	0.0485	0.0485	0.0485	0.0485
w_C	0.6406	0.6406	0.6406	0.5498	0.5498	0.5498	0.5498	0.5498	0.5498
T/K	353.75	362.85	372.95	304.55	313.05	322.95	333.85	342.95	353.45
P/MPa	30.82	32.35	34.28	27.12	29.33	31.72	34.12	36.20	37.79

w_A	0.4017	0.4017	0.4729	0.4729	0.4729	0.4729	0.4729	0.4729	0.4729
w_B	0.0485	0.0485	0.0456	0.0456	0.0456	0.0456	0.0456	0.0456	0.0456
w_C	0.5498	0.5498	0.4815	0.4815	0.4815	0.4815	0.4815	0.4815	0.4815
T/K	362.75	372.75	304.05	313.45	323.55	333.55	343.05	353.05	362.75
P/MPa	39.49	41.40	33.90	35.99	38.48	40.79	42.95	44.95	46.15

continued

continued

w_A	0.4729	0.6040	0.6040	0.6040	0.6040	0.6040	0.6040	0.6040	0.6040
w_B	0.0456	0.0428	0.0428	0.0428	0.0428	0.0428	0.0428	0.0428	0.0428
w_C	0.4815	0.3532	0.3532	0.3532	0.3532	0.3532	0.3532	0.3532	0.3532
T/K	373.15	303.65	313.75	323.25	333.05	343.35	353.25	363.25	373.25
P/MPa	47.75	45.95	48.95	51.35	53.75	56.25	57.75	58.65	60.25

w_A	0.6908	0.6908	0.6908	0.6908	0.6908	0.6908	0.6908	0.6908
w_B	0.0510	0.0510	0.0510	0.0510	0.0510	0.0510	0.0510	0.0510
w_C	0.2582	0.2582	0.2582	0.2582	0.2582	0.2582	0.2582	0.2582
T/K	304.15	313.25	324.55	334.15	342.95	353.05	363.45	373.35
P/MPa	56.25	59.35	62.75	65.45	67.85	69.75	70.85	72.45

Polymer (B):	**poly(L-lactide)**		**1991TOM**
Characterization:	$M_n/kg.mol^{-1} = 2.75$, $M_w/kg.mol^{-1} = 5.5$,		
	L104, Boehringer Ingelheim Chemicals		
Solvent (A):	**carbon dioxide**	**CO_2**	**124-38-9**
Solvent (C):	**2-propanone**	**C_3H_6O**	**67-64-1**

Type of data: polymer solubility/cloud points

Comments: Solubility measurements were carried out using a mixture of 99 wt% CO_2 and 1 wt% C_3H_6O.

$T/K = 318.15$

P/bar	150	200	250	300
w_B	0.000485	0.000554	0.000839	0.00157

$T/K = 328.15$

P/bar	150	200	250	300
w_B	0.000561	0.00127	0.00123	0.00254

$T/K = 338.15$

P/bar	200	250	300
w_B	0.00214	0.00220	0.00368

Polymer (B):	**poly(L-lactide)**		**2002BOT**
Characterization:	$M_n/kg.mol^{-1} = 44$, $M_w/kg.mol^{-1} = 100$, semicrystalline,		
	Polysciences, Inc., Warrington, PA		
Solvent (A):	**carbon dioxide**	**CO_2**	**124-38-9**
Solvent (C):	**trichloromethane**	**$CHCl_3$**	**67-66-3**

Type of data: cloud points

$T/K = 318.15$

c_B/(mg/ml $CHCl_3$)	3	8
P/bar	62.8–65.5	63.2–67.6

Comments: Cloud points were obtained at different CO_2 pressurization rates.

Polymer (B):   **poly(DL-lactide)**   **2002BOT**

Characterization: $M_n/\text{kg.mol}^{-1} = 75$, $M_w/\text{kg.mol}^{-1} = 106$,
Sigma Chemical Co., Inc., St. Louis, MO

Solvent (A):   **carbon dioxide**   **CO_2**   **124-38-9**

Solvent (C):   **trichloromethane**   **$CHCl_3$**   **67-66-3**

Type of data: cloud points

$T/K = 318.15$

$c_B/(\text{mg/ml } CHCl_3)$	3	8
P/bar	64.8–65.1	64.3–67.2

Comments: Cloud points were obtained at different CO_2 pressurization rates.

Polymer (B):   **poly(methyl methacrylate)**   **1992KIA**

Characterization: $M_w/\text{kg.mol}^{-1} = 120$, Aldrich Chem. Co., Inc., Milwaukee, WI

Solvent (A):   **carbon dioxide**   **CO_2**   **124-38-9**

Solvent (C):   **2-butanone**   **C_4H_8O**   **78-93-3**

Type of data: cloud points

w_A	0.150	0.150	0.150	0.150	0.150	0.150	0.150	0.150	0.150
w_B	0.1275	0.1275	0.1275	0.1275	0.1275	0.1275	0.1275	0.1275	0.1275
w_C	0.7225	0.7225	0.7225	0.7225	0.7225	0.7225	0.7225	0.7225	0.7225
T/K	298.15	303.15	308.15	313.15	318.15	323.15	328.15	333.15	338.15
P/bar	13.79	14.99	16.32	17.96	20.10	21.33	23.26	24.76	26.79
	VLE	VLE	VLE	VLE	VLE	VLE	VLE	VLE	VLE
w_A	0.150	0.200	0.200	0.200	0.200	0.200	0.200	0.200	0.200
w_B	0.1275	0.120	0.120	0.120	0.120	0.120	0.120	0.120	0.120
w_C	0.7225	0.680	0.680	0.680	0.680	0.680	0.680	0.680	0.680
T/K	343.15	299.35	303.15	308.15	314.15	318.15	323.15	328.15	333.15
P/bar	27.71	19.86	21.04	22.99	25.37	27.45	29.72	32.54	34.81
	VLE	VLE	VLE	VLE	VLE	VLE	VLE	VLE	VLE
w_A	0.200	0.200	0.250	0.250	0.250	0.250	0.250	0.250	0.250
w_B	0.120	0.120	0.1125	0.1125	0.1125	0.1125	0.1125	0.1125	0.1125
w_C	0.680	0.680	0.6375	0.6375	0.6375	0.6375	0.6375	0.6375	0.6375
T/K	338.15	343.05	300.65	308.25	313.35	315.25	323.25	328.15	333.35
P/bar	37.30	40.68	23.94	27.74	30.57	32.18	54.77	69.97	84.45
	VLE	VLE	VLE	VLE	VLE	VLE	LLE	LLE	LLE
w_A	0.250	0.250	0.264	0.264	0.264	0.264	0.264	0.264	0.264
w_B	0.1125	0.1125	0.1104	0.1104	0.1104	0.1104	0.1104	0.1104	0.1104
w_C	0.6375	0.6375	0.6256	0.6256	0.6256	0.6256	0.6256	0.6256	0.6256
T/K	337.65	343.15	300.55	303.85	308.35	313.15	318.15	323.15	328.15
P/bar	95.48	114.21	27.55	40.65	49.06	62.16	77.78	93.18	107.46
	LLE	LLE	LLE	LLE	LLE	LLE	LLE	LLE	LLE

continued

continued

w_A	0.264	0.150	0.150	0.150	0.150	0.150	0.150	0.150	0.150
w_B	0.1104	0.170	0.170	0.170	0.170	0.170	0.170	0.170	0.170
w_C	0.6256	0.680	0.680	0.680	0.680	0.680	0.680	0.680	0.680
T/K	332.65	303.35	308.15	313.15	318.15	323.15	328.15	333.15	338.15
P/bar	120.10	17.66	18.62	20.46	21.90	23.37	25.21	27.25	29.05
	LLE	VLE	VLE	VLE	VLE	VLE	VLE	VLE	VLE

w_A	0.150	0.200	0.200	0.200	0.200	0.200	0.200	0.200	0.200
w_B	0.170	0.160	0.160	0.160	0.160	0.160	0.160	0.160	0.160
w_C	0.680	0.640	0.640	0.640	0.640	0.640	0.640	0.640	0.640
T/K	342.15	302.65	315.75	318.35	323.15	328.15	332.95	337.95	343.05
P/bar	31.06	24.76	28.79	30.01	31.92	34.27	36.41	39.17	41.99
	VLE	VLE	VLE	VLE	VLE	VLE	VLE	VLE	VLE

w_A	0.223	0.223	0.223	0.223	0.223	0.223	0.223	0.223	0.223
w_B	0.1554	0.1554	0.1554	0.1554	0.1554	0.1554	0.1554	0.1554	0.1554
w_C	0.6216	0.6216	0.6216	0.6216	0.6216	0.6216	0.6216	0.6216	0.6216
T/K	302.45	303.15	308.15	313.15	318.15	324.15	329.15	333.15	338.15
P/bar	25.74	25.84	27.81	29.98	32.60	35.72	38.54	40.81	43.53
	VLE	VLE	VLE	VLE	VLE	VLE	VLE	VLE	VLE

w_A	0.223	0.247	0.247	0.247	0.247	0.247	0.247	0.247	0.247
w_B	0.1554	0.1506	0.1506	0.1506	0.1506	0.1506	0.1506	0.1506	0.1506
w_C	0.6216	0.6024	0.6024	0.6024	0.6024	0.6024	0.6024	0.6024	0.6024
T/K	342.45	301.45	303.15	308.15	313.15	318.15	323.15	325.65	328.15
P/bar	46.19	27.19	27.68	29.87	33.16	35.46	38.05	41.60	50.37
	VLE	VLE	VLE	VLE	VLE	VLE	VLE	VLE	LLE

w_A	0.247	0.247	0.247	0.264	0.264	0.264	0.264	0.264	0.264
w_B	0.1506	0.1506	0.1506	0.1472	0.1472	0.1472	0.1472	0.1472	0.1472
w_C	0.6024	0.6024	0.6024	0.5888	0.5888	0.5888	0.5888	0.5888	0.5888
T/K	333.15	338.35	343.05	303.05	308.15	313.35	318.25	323.25	328.15
P/bar	56.41	69.47	82.64	33.45	37.33	57.39	72.56	84.71	100.07
	LLE	LLE	LLE	VLE	LLE	LLE	LLE	LLE	LLE

w_A	0.264	0.150	0.150	0.150	0.150	0.150	0.150	0.150	0.150
w_B	0.1472	0.2125	0.2125	0.2125	0.2125	0.2125	0.2125	0.2125	0.2125
w_C	0.5888	0.6375	0.6375	0.6375	0.6375	0.6375	0.6375	0.6375	0.6375
T/K	333.15	303.15	308.15	313.15	318.15	323.15	328.15	333.15	338.15
P/bar	116.26	18.88	20.13	21.93	23.80	25.77	27.77	29.68	32.34
	LLE	VLE	VLE	VLE	VLE	VLE	VLE	VLE	VLE

w_A	0.200	0.200	0.200	0.200	0.200	0.200	0.200	0.200	0.200
w_B	0.200	0.200	0.200	0.200	0.200	0.200	0.200	0.200	0.200
w_C	0.600	0.600	0.600	0.600	0.600	0.600	0.600	0.600	0.600
T/K	304.45	308.15	313.15	318.15	323.15	328.15	333.25	338.15	342.45
P/bar	26.13	27.08	28.76	31.85	34.01	37.89	40.02	42.75	46.13
	VLE	VLE	VLE	VLE	VLE	VLE	VLE	VLE	VLE

continued

continued

w_A	0.229	0.229	0.229	0.229	0.229	0.229	0.229	0.229	0.229
w_B	0.19275	0.19275	0.19275	0.19275	0.19275	0.19275	0.19275	0.19275	0.19275
w_C	0.57825	0.57825	0.57825	0.57825	0.57825	0.57825	0.57825	0.57825	0.57825
T/K	304.75	310.35	318.95	319.95	323.45	326.15	330.75	338.25	343.05
P/bar	29.55	32.47	37.56	37.85	40.05	42.71	54.04	75.06	92.58
	VLE	VLE	VLE	VLE	VLE	VLE	LLE	LLE	LLE

w_A	0.245	0.245	0.245	0.245	0.245	0.245
w_B	0.18875	0.18875	0.18875	0.18875	0.18875	0.18875
w_C	0.56625	0.56625	0.56625	0.56625	0.56625	0.56625
T/K	302.75	308.65	313.55	318.45	323.15	328.25
P/bar	33.55	49.77	65.67	80.14	97.02	114.06
	LLE	LLE	LLE	LLE	LLE	LLE

Polymer (B):	**poly(methyl methacrylate)**		**2003DOM**
Characterization:	M_w/kg.mol^{-1} = 540, synthesized in the laboratory		
Solvent (A):	**carbon dioxide**	**CO$_2$**	**124-38-9**
Solvent (C):	**dichloromethane**	**CH$_2$Cl$_2$**	**75-09-2**

Type of data: polymer solubility/cloud points

T/K = 313 P/MPa = 18

w_A	0.99999	0.842	0.780	0.617	0.564
w_B	0.00001	0.00002	0.00008	0.00020	0.00025
w_C	0.000	0.158	0.220	0.383	0.436

Polymer (B):	**poly(methyl methacrylate)**		**2003DOM**
Characterization:	M_w/kg.mol^{-1} = 540, synthesized in the laboratory		
Solvent (A):	**carbon dioxide**	**CO$_2$**	**124-38-9**
Solvent (C):	**ethanol**	**C$_2$H$_6$O**	**64-17-5**

Type of data: polymer solubility/cloud points

T/K = 313 P/MPa = 18

w_A	0.99999	0.923	0.870	0.731	0.685
w_B	0.00001	0.00001	0.00001	0.00002	0.00001
w_C	0.000	0.077	0.130	0.269	0.315

Polymer (B):	**poly(methyl methacrylate)**		**1999LO1**
Characterization:	M_n/kg.mol^{-1} = 46.4, M_w/kg.mol^{-1} = 93.3,		
	Aldrich Chem. Co., Inc., Milwaukee, WI		
Solvent (A):	**carbon dioxide**	**CO$_2$**	**124-38-9**
Solvent (C):	**methyl methacrylate**	**C$_5$H$_8$O$_2$**	**82-62-6**

Type of data: cloud points

continued

continued

w_A	0.846	0.846	0.846	0.846	0.846	0.846	0.660	0.660	0.660
w_B	0.050	0.050	0.050	0.050	0.050	0.050	0.051	0.051	0.051
w_C	0.104	0.104	0.104	0.104	0.104	0.104	0.289	0.289	0.289
T/K	380.15	389.25	398.45	410.05	423.05	445.55	300.15	319.65	337.95
P/bar	2482.1	2302.8	2163.2	2030.5	1925.3	1801.2	887.3	906.6	919.4

w_A	0.660	0.660	0.660	0.660	0.660	0.465	0.465	0.465	0.465
w_B	0.051	0.051	0.051	0.051	0.051	0.051	0.051	0.051	0.051
w_C	0.289	0.289	0.289	0.289	0.289	0.484	0.484	0.484	0.484
T/K	358.75	378.35	398.15	419.25	443.65	300.15	318.25	332.45	348.85
P/bar	930.8	939.7	948.0	956.3	965.9	108.2	183.4	237.2	290.9

w_A	0.465	0.465	0.465	0.465
w_B	0.051	0.051	0.051	0.051
w_C	0.484	0.484	0.484	0.484
T/K	362.85	377.75	397.75	422.25
P/bar	329.9	367.5	412.3	463.3

Polymer (B):	poly(methyl methacrylate)		1994MUR
Characterization:	M_n/kg.mol^{-1} = 30, M_w/kg.mol^{-1} = 120,		
	Aldrich Chem. Co., Inc., Milwaukee, WI		
Solvent (A):	**carbon dioxide**	**CO$_2$**	**124-38-9**
Solvent (C):	**2-propanone**	**C$_3$H$_6$O**	**67-64-1**

Type of data: cloud points   (a) - visually   (b) - from *p-V* data

w_A	0.29	0.29	0.26	0.26
w_B	0.11	0.11	0.16	0.16
w_C	0.60	0.60	0.58	0.58
T/K	338.15	338.15	343.15	343.15
P/bar	61.63	71.91	53.91	76.81
	(a)	(b)	(a)	(b)

Polymer (B):	poly(methyl methacrylate)		1992KIA
Characterization:	M_w/kg.mol^{-1} = 120, Aldrich Chem. Co., Inc., Milwaukee, WI		
Solvent (A):	**carbon dioxide**	**CO$_2$**	**124-38-9**
Solvent (C):	**2-propanone**	**C$_3$H$_6$O**	**67-64-1**

Type of data: cloud points

w_A	0.150	0.150	0.150	0.150	0.150	0.150	0.200	0.200	0.200
w_B	0.1275	0.1275	0.1275	0.1275	0.1275	0.1275	0.120	0.120	0.120
w_C	0.7225	0.7225	0.7225	0.7225	0.7225	0.7225	0.680	0.680	0.680
T/K	308.25	313.15	318.15	323.15	328.15	333.05	297.45	308.15	313.15
P/bar	11.87	12.97	14.09	15.27	16.49	17.22	13.95	15.22	16.56
	VLE	VLE	VLE	VLE	VLE	VLE	VLE	VLE	VLE

continued

continued

w_A	0.200	0.200	0.200	0.200	0.200	0.200	0.245	0.250	0.250
w_B	0.120	0.120	0.120	0.120	0.120	0.120	0.11325	0.1125	0.1125
w_C	0.680	0.680	0.680	0.680	0.680	0.680	0.64175	0.6375	0.6375
T/K	318.15	323.15	328.15	333.15	338.15	342.35	342.65	303.65	318.35
P/bar	18.13	19.87	21.53	23.84	25.31	27.24	36.60	19.77	25.38
	VLE	VLE	VLE	VLE	VLE	VLE	LLE	VLE	VLE

w_A	0.250	0.250	0.250	0.250	0.250	0.250	0.250	0.250	0.250
w_B	0.1125	0.1125	0.1125	0.1125	0.1125	0.1125	0.1125	0.1125	0.1125
w_C	0.6375	0.6375	0.6375	0.6375	0.6375	0.6375	0.6375	0.6375	0.6375
T/K	323.05	325.65	328.15	331.85	332.45	333.45	334.25	334.55	335.25
P/bar	28.14	29.54	30.63	32.30	32.34	32.70	33.23	33.92	34.87
	VLE	VLE	VLE	VLE	VLE	VLE	VLE	VLE	VLE

w_A	0.250	0.250	0.250	0.250	0.250	0.250	0.250	0.250	0.273
w_B	0.1125	0.1125	0.1125	0.1125	0.1125	0.1125	0.1125	0.1125	0.10905
w_C	0.6375	0.6375	0.6375	0.6375	0.6375	0.6375	0.6375	0.6375	0.61795
T/K	336.35	337.15	337.55	338.45	340.35	341.15	342.75	342.95	303.45
P/bar	38.02	39.27	40.25	41.89	46.54	49.37	51.94	53.54	21.24
	LLE	LLE	LLE	LLE	LLE	LLE	LLE	LLE	VLE

w_A	0.273	0.273	0.273	0.273	0.273	0.273	0.273	0.273	0.273
w_B	0.10905	0.10905	0.10905	0.10905	0.10905	0.10905	0.10905	0.10905	0.10905
w_C	0.61795	0.61795	0.61795	0.61795	0.61795	0.61795	0.61795	0.61795	0.61795
T/K	308.15	313.15	317.65	317.95	318.15	318.45	319.15	320.05	320.55
P/bar	22.85	25.18	29.25	31.29	31.42	31.88	33.03	35.81	36.11
	VLE	VLE	VLE	VLE	VLE	LLE	LLE	LLE	LLE

w_A	0.273	0.273	0.273	0.273	0.273	0.273	0.273	0.273	0.273
w_B	0.10905	0.10905	0.10905	0.10905	0.10905	0.10905	0.10905	0.10905	0.10905
w_C	0.61795	0.61795	0.61795	0.61795	0.61795	0.61795	0.61795	0.61795	0.61795
T/K	320.95	321.35	322.45	324.15	325.15	326.15	327.15	328.15	330.75
P/bar	36.84	38.80	40.78	44.75	48.85	50.13	53.06	56.30	61.63
	LLE	LLE	LLE	LLE	LLE	LLE	LLE	LLE	LLE

w_A	0.273	0.273	0.303	0.303	0.303	0.303	0.303	0.303	0.303
w_B	0.10905	0.10905	0.10455	0.10455	0.10455	0.10455	0.10455	0.10455	0.10455
w_C	0.61795	0.61795	0.59245	0.59245	0.59245	0.59245	0.59245	0.59245	0.59245
T/K	332.95	333.75	298.85	299.05	299.15	299.25	300.15	301.55	302.35
P/bar	67.63	70.75	41.73	43.07	43.57	43.39	44.00	49.74	49.48
	LLE	LLE	LLE	LLE	LLE	LLE	LLE	LLE	LLE

w_A	0.303	0.303	0.303	0.303	0.303	0.303	0.303	0.303	0.303
w_B	0.10455	0.10455	0.10455	0.10455	0.10455	0.10455	0.10455	0.10455	0.10455
w_C	0.59245	0.59245	0.59245	0.59245	0.59245	0.59245	0.59245	0.59245	0.59245
T/K	303.85	305.55	306.15	307.35	308.75	309.15	310.25	315.85	317.85
P/bar	53.59	56.86	57.94	59.52	62.94	64.09	66.75	79.01	83.50
	LLE	LLE	LLE	LLE	LLE	LLE	LLE	LLE	LLE

continued

continued

w_A	0.303	0.303	0.303	0.303	0.303	0.303	0.303	0.303	0.303
w_B	0.10455	0.10455	0.10455	0.10455	0.10455	0.10455	0.10455	0.10455	0.10455
w_C	0.59245	0.59245	0.59245	0.59245	0.59245	0.59245	0.59245	0.59245	0.59245
T/K	320.45	321.05	322.25	323.15	323.85	324.65	326.85	329.15	330.85
P/bar	88.25	89.99	93.64	95.08	96.89	98.73	103.06	108.64	114.85
	LLE	LLE	LLE	LLE	LLE	LLE	LLE	LLE	LLE

w_A	0.303	0.150	0.150	0.150	0.150	0.150	0.150	0.150	0.150
w_B	0.10455	0.170	0.170	0.170	0.170	0.170	0.170	0.170	0.170
w_C	0.59245	0.680	0.680	0.680	0.680	0.680	0.680	0.680	0.680
T/K	331.45	305.85	308.15	313.15	318.15	323.55	328.45	333.15	338.15
P/bar	115.70	14.38	14.61	15.40	16.93	18.51	20.15	21.88	23.66
	LLE	VLE	VLE	VLE	VLE	VLE	VLE	VLE	VLE

w_A	0.200	0.200	0.200	0.200	0.200	0.200	0.250	0.250	0.250
w_B	0.160	0.160	0.160	0.160	0.160	0.160	0.150	0.150	0.150
w_C	0.640	0.640	0.640	0.640	0.640	0.640	0.600	0.600	0.600
T/K	318.15	323.15	327.55	333.15	338.15	343.15	307.15	312.85	315.45
P/bar	23.15	25.21	27.18	28.90	31.38	33.88	25.90	27.06	28.27
	VLE	VLE	VLE	VLE	VLE	VLE	VLE	VLE	VLE

w_A	0.250	0.250	0.250	0.250	0.250	0.250	0.250	0.250	0.250
w_B	0.150	0.150	0.150	0.150	0.150	0.150	0.150	0.150	0.150
w_C	0.600	0.600	0.600	0.600	0.600	0.600	0.600	0.600	0.600
T/K	315.85	316.95	317.65	318.75	319.85	320.05	320.85	321.15	321.55
P/bar	28.33	29.35	29.84	30.83	31.49	32.37	33.65	33.23	34.27
	VLE	VLE	VLE	VLE	VLE	VLE	LLE	LLE	LLE

w_A	0.250	0.250	0.250	0.250	0.250	0.250	0.250	0.250	0.250
w_B	0.150	0.150	0.150	0.150	0.150	0.150	0.150	0.150	0.150
w_C	0.600	0.600	0.600	0.600	0.600	0.600	0.600	0.600	0.600
T/K	321.95	322.45	322.85	323.25	323.65	324.75	325.35	326.45	327.85
P/bar	35.26	36.11	36.80	38.11	39.69	41.01	43.44	46.13	49.15
	LLE	LLE	LLE	LLE	LLE	LLE	LLE	LLE	LLE

w_A	0.250	0.250	0.250	0.250	0.250	0.250	0.250	0.250	0.250
w_B	0.150	0.150	0.150	0.150	0.150	0.150	0.150	0.150	0.150
w_C	0.600	0.600	0.600	0.600	0.600	0.600	0.600	0.600	0.600
T/K	328.15	329.35	330.25	330.95	332.05	332.95	335.25	336.15	336.75
P/bar	49.41	53.45	55.65	56.93	60.41	61.56	68.49	71.05	72.59
	LLE	LLE	LLE	LLE	LLE	LLE	LLE	LLE	LLE

w_A	0.250	0.250	0.250	0.250	0.250	0.250	0.250	0.250	0.250
w_B	0.150	0.150	0.150	0.150	0.150	0.150	0.150	0.150	0.150
w_C	0.600	0.600	0.600	0.600	0.600	0.600	0.600	0.600	0.600
T/K	337.65	337.95	338.75	339.05	339.55	340.05	340.15	340.35	340.75
P/bar	75.48	76.46	78.66	78.83	79.91	81.55	82.57	83.30	83.65
	LLE	LLE	LLE	LLE	LLE	LLE	LLE	LLE	LLE

continued

continued

w_A	0.250	0.250	0.250	0.250	0.250	0.268	0.268	0.268	0.268
w_B	0.150	0.150	0.150	0.150	0.150	0.1464	0.1464	0.1464	0.1464
w_C	0.600	0.600	0.600	0.600	0.600	0.5856	0.5856	0.5856	0.5856
T/K	341.45	341.65	342.25	342.65	342.85	307.65	311.95	312.35	312.65
P/bar	87.20	86.84	89.30	88.98	89.40	35.39	40.71	43.86	44.78
	LLE	LLE	LLE	LLE	LLE	LLE	LLE	LLE	LLE

w_A	0.268	0.268	0.268	0.268	0.268	0.268	0.268	0.268	0.268
w_B	0.1464	0.1464	0.1464	0.1464	0.1464	0.1464	0.1464	0.1464	0.1464
w_C	0.5856	0.5856	0.5856	0.5856	0.5856	0.5856	0.5856	0.5856	0.5856
T/K	315.85	316.25	316.45	316.65	316.95	317.25	317.55	320.45	320.75
P/bar	50.99	54.47	54.17	55.68	55.88	56.11	57.49	64.48	64.61
	LLE	LLE	LLE	LLE	LLE	LLE	LLE	LLE	LLE

w_A	0.268	0.268	0.268	0.268	0.268	0.268	0.268	0.268	0.268
w_B	0.1464	0.1464	0.1464	0.1464	0.1464	0.1464	0.1464	0.1464	0.1464
w_C	0.5856	0.5856	0.5856	0.5856	0.5856	0.5856	0.5856	0.5856	0.5856
T/K	320.95	321.25	324.25	324.45	324.75	325.05	328.75	328.95	329.15
P/bar	65.17	66.16	73.28	73.57	74.33	74.95	84.38	85.66	86.38
	LLE	LLE	LLE	LLE	LLE	LLE	LLE	LLE	LLE

w_A	0.268	0.268	0.268	0.268	0.268	0.268	0.268	0.268	0.268
w_B	0.1464	0.1464	0.1464	0.1464	0.1464	0.1464	0.1464	0.1464	0.1464
w_C	0.5856	0.5856	0.5856	0.5856	0.5856	0.5856	0.5856	0.5856	0.5856
T/K	329.85	330.15	333.85	334.05	334.55	334.65	334.85	335.65	336.15
P/bar	87.49	89.20	98.20	98.93	100.76	100.04	101.88	102.17	102.86
	LLE	LLE	LLE	LLE	LLE	LLE	LLE	LLE	LLE

w_A	0.268	0.268	0.268	0.268	0.268	0.283	0.283	0.283	0.283
w_B	0.1464	0.1464	0.1464	0.1464	0.1464	0.1434	0.1434	0.1434	0.1434
w_C	0.5856	0.5856	0.5856	0.5856	0.5856	0.5736	0.5736	0.5736	0.5736
T/K	336.25	336.45	336.55	336.75	337.85	311.75	311.95	312.05	312.35
P/bar	104.01	104.15	104.83	106.34	108.41	68.65	69.08	70.62	70.72
	LLE	LLE	LLE	LLE	LLE	LLE	LLE	LLE	LLE

w_A	0.283	0.283	0.283	0.283	0.283	0.283	0.283	0.283	0.283
w_B	0.1434	0.1434	0.1434	0.1434	0.1434	0.1434	0.1434	0.1434	0.1434
w_C	0.5736	0.5736	0.5736	0.5736	0.5736	0.5736	0.5736	0.5736	0.5736
T/K	314.65	314.95	318.25	318.55	318.65	318.95	321.85	322.15	322.35
P/bar	74.63	75.38	83.30	85.03	85.07	85.83	91.67	93.27	93.62
	LLE	LLE	LLE	LLE	LLE	LLE	LLE	LLE	LLE

w_A	0.283	0.283	0.283	0.283	0.283	0.283	0.283	0.283	0.283
w_B	0.1434	0.1434	0.1434	0.1434	0.1434	0.1434	0.1434	0.1434	0.1434
w_C	0.5736	0.5736	0.5736	0.5736	0.5736	0.5736	0.5736	0.5736	0.5736
T/K	322.55	322.75	322.95	323.25	326.05	326.45	326.55	329.45	329.95
P/bar	93.67	94.42	95.44	95.84	103.08	103.68	104.48	111.99	112.32
	LLE	LLE	LLE	LLE	LLE	LLE	LLE	LLE	LLE

continued

continued

w_A	0.283	0.283	0.283	0.283	0.150	0.150	0.150	0.150	0.150
w_B	0.1434	0.1434	0.1434	0.1434	0.2125	0.2125	0.2125	0.2125	0.2125
w_C	0.5736	0.5736	0.5736	0.5736	0.6375	0.6375	0.6375	0.6375	0.6375
T/K	330.15	330.65	331.05	331.35	324.95	328.35	333.35	338.15	343.15
P/bar	113.47	114.75	115.18	115.60	23.72	24.83	27.31	29.38	31.72
	LLE	LLE	LLE	LLE	VLE	VLE	VLE	VLE	VLE

w_A	0.200	0.200	0.200	0.200	0.200	0.200	0.200	0.200	0.200
w_B	0.200	0.200	0.200	0.200	0.200	0.200	0.200	0.200	0.200
w_C	0.600	0.600	0.600	0.600	0.600	0.600	0.600	0.600	0.600
T/K	300.75	303.15	308.15	313.15	318.15	323.65	329.75	333.15	338.15
P/bar	19.72	20.04	21.48	23.59	25.67	28.08	31.22	32.63	35.68
	VLE	VLE	VLE	VLE	VLE	VLE	VLE	VLE	VLE

w_A	0.200	0.200	0.200	0.200	0.200	0.200	0.200	0.200	0.200
w_B	0.200	0.200	0.200	0.200	0.200	0.200	0.200	0.200	0.200
w_C	0.600	0.600	0.600	0.600	0.600	0.600	0.600	0.600	0.600
T/K	342.15	342.45	342.55	342.65	342.75	342.85	342.95	343.05	343.15
P/bar	38.94	41.77	41.27	43.17	42.91	43.50	44.92	44.42	44.49
	LLE	LLE	LLE	LLE	LLE	LLE	LLE	LLE	LLE

w_A	0.222	0.222	0.222	0.222	0.222	0.222	0.222	0.222	0.222
w_B	0.1945	0.1945	0.1945	0.1945	0.1945	0.1945	0.1945	0.1945	0.1945
w_C	0.5835	0.5835	0.5835	0.5835	0.5835	0.5835	0.5835	0.5835	0.5835
T/K	312.05	318.15	323.15	324.75	326.15	327.75	328.85	329.45	329.65
P/bar	25.75	27.59	29.47	31.52	32.21	33.09	37.69	39.57	39.46
	VLE	VLE	VLE	VLE	VLE	VLE	VLE	VLE	VLE

w_A	0.222	0.222	0.222	0.222	0.222	0.222	0.222	0.222	0.222
w_B	0.1945	0.1945	0.1945	0.1945	0.1945	0.1945	0.1945	0.1945	0.1945
w_C	0.5835	0.5835	0.5835	0.5835	0.5835	0.5835	0.5835	0.5835	0.5835
T/K	329.85	330.65	330.85	331.05	331.35	331.85	332.15	332.45	332.65
P/bar	40.61	42.26	41.37	44.06	43.51	44.36	44.88	46.73	47.42
	LLE	LLE	LLE	LLE	LLE	LLE	LLE	LLE	LLE

w_A	0.222	0.222	0.222	0.222	0.222	0.222	0.222	0.222	0.222
w_B	0.1945	0.1945	0.1945	0.1945	0.1945	0.1945	0.1945	0.1945	0.1945
w_C	0.5835	0.5835	0.5835	0.5835	0.5835	0.5835	0.5835	0.5835	0.5835
T/K	332.85	333.95	334.35	334.55	334.75	334.85	334.95	334.95	335.25
P/bar	47.18	49.15	51.22	51.06	48.37	52.47	49.12	52.27	49.94
	LLE	LLE	LLE	LLE	LLE	LLE	LLE	LLE	LLE

w_A	0.222	0.222	0.222	0.222	0.222	0.222	0.222	0.222	0.222
w_B	0.1945	0.1945	0.1945	0.1945	0.1945	0.1945	0.1945	0.1945	0.1945
w_C	0.5835	0.5835	0.5835	0.5835	0.5835	0.5835	0.5835	0.5835	0.5835
T/K	335.25	335.45	335.45	335.75	335.85	335.95	336.15	336.35	336.45
P/bar	52.59	50.17	53.68	53.54	51.95	54.83	52.17	54.74	52.47
	LLE	LLE	LLE	LLE	LLE	LLE	LLE	LLE	LLE

continued

continued

w_A	0.222	0.222	0.222	0.222	0.222	0.222	0.222	0.222	0.222
w_B	0.1945	0.1945	0.1945	0.1945	0.1945	0.1945	0.1945	0.1945	0.1945
w_C	0.5835	0.5835	0.5835	0.5835	0.5835	0.5835	0.5835	0.5835	0.5835
T/K	336.75	336.95	337.05	337.15	337.35	337.45	337.55	337.65	337.75
P/bar	56.05	56.74	54.41	54.54	55.19	55.30	57.89	55.49	55.26
	LLE	LLE	LLE	LLE	LLE	LLE	LLE	LLE	LLE

w_A	0.222	0.222	0.222	0.222	0.222	0.222	0.222	0.222	0.222
w_B	0.1945	0.1945	0.1945	0.1945	0.1945	0.1945	0.1945	0.1945	0.1945
w_C	0.5835	0.5835	0.5835	0.5835	0.5835	0.5835	0.5835	0.5835	0.5835
T/K	337.95	338.45	339.25	339.45	339.55	339.65	339.75	339.85	340.05
P/bar	59.20	61.23	60.71	61.23	63.30	61.20	61.52	62.45	62.74
	LLE	LLE	LLE	LLE	LLE	LLE	LLE	LLE	LLE

w_A	0.222	0.222	0.222	0.222	0.222	0.222	0.222	0.222	0.222
w_B	0.1945	0.1945	0.1945	0.1945	0.1945	0.1945	0.1945	0.1945	0.1945
w_C	0.5835	0.5835	0.5835	0.5835	0.5835	0.5835	0.5835	0.5835	0.5835
T/K	340.15	341.35	341.45	341.65	341.85	342.05	342.35	342.45	343.05
P/bar	64.55	66.69	67.67	67.60	68.49	68.53	72.95	73.99	70.91
	LLE	LLE	LLE	LLE	LLE	LLE	LLE	LLE	LLE

w_A	0.241	0.241	0.241	0.241	0.241	0.241	0.241	0.241	0.241
w_B	0.18975	0.18975	0.18975	0.18975	0.18975	0.18975	0.18975	0.18975	0.18975
w_C	0.56925	0.56925	0.56925	0.56925	0.56925	0.56925	0.56925	0.56925	0.56925
T/K	297.45	303.15	308.15	309.05	310.45	311.05	311.65	311.95	312.25
P/bar	21.80	24.00	26.39	27.09	27.61	26.99	30.07	29.68	32.18
	VLE	VLE	VLE	VLE	VLE	VLE	VLE	VLE	LLE

w_A	0.241	0.241	0.241	0.241	0.241	0.241	0.241	0.241	0.241
w_B	0.18975	0.18975	0.18975	0.18975	0.18975	0.18975	0.18975	0.18975	0.18975
w_C	0.56925	0.56925	0.56925	0.56925	0.56925	0.56925	0.56925	0.56925	0.56925
T/K	312.55	312.85	313.25	314.55	314.85	315.75	316.05	316.35	316.65
P/bar	33.78	33.95	35.16	39.14	39.67	42.09	43.90	43.64	44.68
	LLE	LLE	LLE	LLE	LLE	LLE	LLE	LLE	LLE

w_A	0.241	0.241	0.241	0.241	0.241	0.241	0.241	0.241	0.241
w_B	0.18975	0.18975	0.18975	0.18975	0.18975	0.18975	0.18975	0.18975	0.18975
w_C	0.56925	0.56925	0.56925	0.56925	0.56925	0.56925	0.56925	0.56925	0.56925
T/K	317.55	317.75	318.05	318.25	318.55	318.85	319.15	319.45	320.25
P/bar	46.00	47.48	48.79	50.07	49.44	50.40	51.81	51.85	52.47
	LLE	LLE	LLE	LLE	LLE	LLE	LLE	LLE	LLE

w_A	0.241	0.241	0.241	0.241	0.241	0.241	0.241	0.241	0.241
w_B	0.18975	0.18975	0.18975	0.18975	0.18975	0.18975	0.18975	0.18975	0.18975
w_C	0.56925	0.56925	0.56925	0.56925	0.56925	0.56925	0.56925	0.56925	0.56925
T/K	320.45	320.95	321.15	323.15	323.65	324.05	324.45	324.75	325.25
P/bar	52.77	54.90	54.80	58.58	62.85	64.11	64.51	64.55	65.90
	LLE	LLE	LLE	LLE	LLE	LLE	LLE	LLE	LLE

continued

continued

w_A	0.241	0.241	0.241	0.241	0.241	0.241	0.241	0.241	0.241
w_B	0.18975	0.18975	0.18975	0.18975	0.18975	0.18975	0.18975	0.18975	0.18975
w_C	0.56925	0.56925	0.56925	0.56925	0.56925	0.56925	0.56925	0.56925	0.56925
T/K	326.65	327.95	329.05	329.35	329.85	330.55	331.05	331.25	331.55
P/bar	68.29	72.50	75.10	76.01	77.26	78.47	80.90	80.34	82.21
	LLE	LLE	LLE	LLE	LLE	LLE	LLE	LLE	LLE

w_A	0.241	0.241	0.241	0.241	0.241	0.241	0.241	0.241	0.241
w_B	0.18975	0.18975	0.18975	0.18975	0.18975	0.18975	0.18975	0.18975	0.18975
w_C	0.56925	0.56925	0.56925	0.56925	0.56925	0.56925	0.56925	0.56925	0.56925
T/K	331.65	332.65	332.95	333.35	333.35	333.55	333.65	333.65	334.25
P/bar	82.41	82.94	83.92	77.22	84.32	77.85	78.34	85.86	87.72
	LLE	LLE	LLE	LLE	LLE	LLE	LLE	LLE	LLE

w_A	0.241	0.241	0.241	0.241	0.241	0.241	0.241	0.241	0.241
w_B	0.18975	0.18975	0.18975	0.18975	0.18975	0.18975	0.18975	0.18975	0.18975
w_C	0.56925	0.56925	0.56925	0.56925	0.56925	0.56925	0.56925	0.56925	0.56925
T/K	335.55	336.35	336.45	336.55	337.25	337.35	337.45	338.15	338.25
P/bar	82.58	91.67	91.90	92.72	94.32	95.55	96.24	90.19	98.50
	LLE	LLE	LLE	LLE	LLE	LLE	LLE	LLE	LLE

w_A	0.241	0.241	0.241	0.241	0.241	0.241	0.241	0.241	0.241
w_B	0.18975	0.18975	0.18975	0.18975	0.18975	0.18975	0.18975	0.18975	0.18975
w_C	0.56925	0.56925	0.56925	0.56925	0.56925	0.56925	0.56925	0.56925	0.56925
T/K	338.45	339.15	339.25	340.45	340.85	341.75	342.35	342.85	343.35
P/bar	99.42	93.48	93.35	96.99	97.35	100.52	101.98	102.91	103.85
	LLE	LLE	LLE	LLE	LLE	LLE	LLE	LLE	LLE

w_A	0.241	0.241	0.241
w_B	0.18975	0.18975	0.18975
w_C	0.56925	0.56925	0.56925
T/K	343.45	343.55	343.65
P/bar	104.08	104.70	104.43
	LLE	LLE	LLE

Polymer (B):	**poly(methyl methacrylate)**		**2003DOM**
Characterization:	M_w/kg.mol^{-1} = 540, synthesized in the laboratory		
Solvent (A):	**carbon dioxide**	**CO$_2$**	**124-38-9**
Solvent (C):	**2-propanone**	**C$_3$H$_6$O**	**67-64-1**

Type of data: polymer solubility/cloud points

T/K = 313 P/MPa = 18

w_A	0.99999	0.930	0.855	0.734	0.682
w_B	0.00001	0.00016	0.00022	0.00007	0.00003
w_C	0.000	0.070	0.145	0.266	0.318

Polymer (B):	poly(methyl methacrylate)							**1992KIA, 1994KIA**
Characterization:	M_w/kg.mol^{-1} = 120, Aldrich Chem. Co., Inc., Milwaukee, WI							
Solvent (A):	carbon dioxide		CO_2					**124-38-9**
Solvent (C):	tetrahydrofuran		C_4H_8O					**109-99-9**

Type of data: cloud points

w_A	0.150	0.150	0.150	0.150	0.150	0.150	0.150	0.150	0.150
w_B	0.1275	0.1275	0.1275	0.1275	0.1275	0.1275	0.1275	0.1275	0.1275
w_C	0.7225	0.7225	0.7225	0.7225	0.7225	0.7225	0.7225	0.7225	0.7225
T/K	304.35	308.15	313.15	318.15	323.25	328.15	333.15	338.35	341.65
P/bar	16.07	17.20	18.44	19.68	22.30	23.73	25.64	27.71	29.60
	VLE	VLE	VLE	VLE	VLE	VLE	VLE	VLE	VLE
w_A	0.150	0.200	0.200	0.200	0.200	0.200	0.200	0.200	0.250
w_B	0.1275	0.120	0.120	0.120	0.120	0.120	0.120	0.120	0.1125
w_C	0.7225	0.680	0.680	0.680	0.680	0.680	0.680	0.680	0.6375
T/K	342.65	315.45	319.55	325.55	328.25	333.45	337.55	342.95	306.45
P/bar	29.98	25.21	27.40	30.07	31.84	34.14	36.47	39.10	25.45
	VLE	VLE	VLE	VLE	VLE	VLE	VLE	VLE	VLE
w_A	0.250	0.250	0.250	0.250	0.250	0.250	0.300	0.300	0.300
w_B	0.1125	0.1125	0.1125	0.1125	0.1125	0.1125	0.105	0.105	0.105
w_C	0.6375	0.6375	0.6375	0.6375	0.6375	0.6375	0.595	0.595	0.595
T/K	313.15	319.55	324.75	328.15	333.35	337.15	301.35	316.25	320.95
P/bar	27.72	31.17	33.65	36.12	39.02	41.27	26.17	34.76	37.82
	VLE	VLE	VLE	VLE	VLE	VLE	VLE	VLE	VLE
w_A	0.300	0.300	0.300	0.300	0.323	0.323	0.323	0.323	0.323
w_B	0.105	0.105	0.105	0.105	0.10155	0.10155	0.10155	0.10155	0.10155
w_C	0.595	0.595	0.595	0.595	0.57545	0.57545	0.57545	0.57545	0.57545
T/K	327.75	332.75	337.85	342.75	300.25	303.15	313.15	318.15	323.15
P/bar	41.67	44.53	48.08	51.77	27.32	28.40	34.87	37.31	41.47
	VLE	VLE	VLE	VLE	VLE	VLE	VLE	VLE	VLE
w_A	0.323	0.323	0.323	0.323	0.323	0.323	0.323	0.323	0.323
w_B	0.10155	0.10155	0.10155	0.10155	0.10155	0.10155	0.10155	0.10155	0.10155
w_C	0.57545	0.57545	0.57545	0.57545	0.57545	0.57545	0.57545	0.57545	0.57545
T/K	325.75	326.35	326.75	327.05	328.55	328.95	329.35	329.75	330.45
P/bar	43.61	44.42	45.44	45.57	46.53	47.04	52.40	53.09	55.49
	VLE	VLE	VLE	VLE	VLE	VLE	LLE	LLE	LLE
w_A	0.323	0.323	0.323	0.323	0.323	0.323	0.323	0.323	0.323
w_B	0.10155	0.10155	0.10155	0.10155	0.10155	0.10155	0.10155	0.10155	0.10155
w_C	0.57545	0.57545	0.57545	0.57545	0.57545	0.57545	0.57545	0.57545	0.57545
T/K	330.85	331.25	331.85	332.35	332.95	333.95	334.35	334.85	336.95
P/bar	53.29	55.19	57.69	58.44	60.35	65.18	65.83	65.83	72.73
	LLE	LLE	LLE	LLE	LLE	LLE	LLE	LLE	LLE
w_A	0.323	0.323	0.323	0.323	0.323	0.323	0.323	0.323	0.323
w_B	0.10155	0.10155	0.10155	0.10155	0.10155	0.10155	0.10155	0.10155	0.10155
w_C	0.57545	0.57545	0.57545	0.57545	0.57545	0.57545	0.57545	0.57545	0.57545
T/K	337.35	337.75	338.45	338.65	339.05	339.25	341.15	341.45	341.55
P/bar	73.51	73.94	81.72	78.11	84.01	77.61	89.07	87.17	89.50
	LLE	LLE	LLE	LLE	LLE	LLE	LLE	LLE	LLE

continued

continued

w_A	0.323	0.323	0.323	0.337	0.337	0.337	0.337	0.337	0.337
w_B	0.10155	0.10155	0.10155	0.09945	0.09945	0.09945	0.09945	0.09945	0.09945
w_C	0.57545	0.57545	0.57545	0.56355	0.56355	0.56355	0.56355	0.56355	0.56355
T/K	341.65	342.55	342.65	296.85	297.25	303.15	308.15	313.15	318.15
P/bar	89.50	95.53	91.98	28.76	35.96	45.57	54.97	72.10	85.66
	LLE	LLE	LLE	LLE	LLE	LLE	LLE	LLE	LLE

w_A	0.337	0.337	0.200	0.200	0.200	0.200	0.200	0.200	0.200
w_B	0.09945	0.09945	0.160	0.160	0.160	0.160	0.160	0.160	0.160
w_C	0.56355	0.56355	0.640	0.640	0.640	0.640	0.640	0.640	0.640
T/K	322.95	327.85	303.95	308.15	313.25	318.15	325.25	328.15	342.75
P/bar	98.40	110.32	18.12	19.27	21.20	22.62	25.12	26.63	32.73
	LLE	LLE	VLE	VLE	VLE	VLE	VLE	VLE	VLE

w_A	0.250	0.250	0.250	0.250	0.250	0.250	0.250	0.300	0.300
w_B	0.150	0.150	0.150	0.150	0.150	0.150	0.150	0.140	0.140
w_C	0.600	0.600	0.600	0.600	0.600	0.600	0.600	0.560	0.560
T/K	303.35	318.95	323.15	328.15	333.15	338.15	340.15	303.75	308.15
P/bar	23.50	30.17	32.54	35.07	37.76	40.91	42.45	28.79	30.23
	VLE	VLE	VLE	VLE	VLE	VLE	VLE	VLE	VLE

w_A	0.300	0.300	0.300	0.300	0.300	0.300	0.300	0.317	0.317
w_B	0.140	0.140	0.140	0.140	0.140	0.140	0.140	0.1366	0.1366
w_C	0.560	0.560	0.560	0.560	0.560	0.560	0.560	0.5464	0.5464
T/K	313.15	318.45	325.55	328.15	333.15	338.45	343.15	298.65	315.05
P/bar	33.03	35.85	40.12	42.09	46.09	49.48	53.10	28.07	35.95
	VLE	VLE	VLE	VLE	VLE	VLE	VLE	VLE	VLE

w_A	0.317	0.317	0.317	0.317	0.317	0.317	0.330	0.330	0.330
w_B	0.1366	0.1366	0.1366	0.1366	0.1366	0.1366	0.134	0.134	0.134
w_C	0.5464	0.5464	0.5464	0.5464	0.5464	0.5464	0.536	0.536	0.536
T/K	318.45	323.25	326.55	333.15	336.65	343.15	303.15	313.15	318.15
P/bar	37.53	40.74	42.79	62.58	73.90	92.56	38.84	62.63	77.52
	VLE	VLE	VLE	LLE	LLE	LLE	LLE	LLE	LLE

w_A	0.330	0.330	0.330	0.150	0.150	0.150	0.150	0.150	0.150
w_B	0.134	0.134	0.134	0.2125	0.2125	0.2125	0.2125	0.2125	0.2125
w_C	0.536	0.536	0.536	0.6375	0.6375	0.6375	0.6375	0.6375	0.6375
T/K	323.15	328.15	333.15	313.15	318.15	323.15	328.15	333.15	338.15
P/bar	93.80	105.12	118.22	18.95	20.85	22.46	24.68	25.77	27.35
	LLE	LLE	LLE	VLE	VLE	VLE	VLE	VLE	VLE

w_A	0.200	0.200	0.200	0.200	0.200	0.200	0.200	0.200	0.200
w_B	0.200	0.200	0.200	0.200	0.200	0.200	0.200	0.200	0.200
w_C	0.600	0.600	0.600	0.600	0.600	0.600	0.600	0.600	0.600
T/K	297.75	310.05	313.75	318.15	328.45	332.75	338.15	342.85	343.05
P/bar	26.79	26.92	28.92	30.96	38.51	39.76	43.31	45.54	46.03
	VLE	VLE	VLE	VLE	VLE	VLE	VLE	VLE	VLE

continued

continued

w_A	0.250	0.250	0.250	0.250	0.250	0.250	0.250	0.250	0.250
w_B	0.1875	0.1875	0.1875	0.1875	0.1875	0.1875	0.1875	0.1875	0.1875
w_C	0.5625	0.5625	0.5625	0.5625	0.5625	0.5625	0.5625	0.5625	0.5625
T/K	303.15	308.15	313.15	318.15	323.15	330.65	333.15	338.15	343.15
P/bar	28.73	31.58	34.64	37.56	40.55	64.06	74.00	84.12	101.15
	VLE	VLE	VLE	VLE	VLE	LLE	LLE	LLE	LLE

w_A	0.271	0.271	0.271	0.271	0.271	0.271	0.271	0.271	0.291
w_B	0.18225	0.18225	0.18225	0.18225	0.18225	0.18225	0.18225	0.18225	0.17725
w_C	0.54675	0.54675	0.54675	0.54675	0.54675	0.54675	0.54675	0.54675	0.53175
T/K	297.85	303.45	308.25	313.25	318.15	323.25	328.25	333.15	300.85
P/bar	27.35	33.56	44.95	58.51	71.35	84.28	95.48	107.63	68.59
	VLE	LLE	LLE	LLE	LLE	LLE	LLE	LLE	LLE

w_A	0.291	0.291	0.291
w_B	0.17725	0.17725	0.17725
w_C	0.53175	0.53175	0.53175
T/K	308.25	313.15	318.15
P/bar	90.62	107.69	119.97
	LLE	LLE	LLE

Polymer (B):   **poly(methyl methacrylate)**   **1994MUR**
Characterization: M_n/kg.mol^{-1} = 30, M_w/kg.mol^{-1} = 120,
                      Aldrich Chem. Co., Inc., Milwaukee, WI
Solvent (A):   **carbon dioxide**   **CO$_2$**   **124-38-9**
Solvent (C):   **toluene**   **C$_7$H$_8$**   **108-88-3**

Type of data: cloud points   (a) - visually   (b) - from *p-V* data

w_A	0.33	0.33
w_B	0.15	0.15
w_C	0.52	0.52
T/K	318.15	318.15
P/bar	82.94	92.26
	(a)	(b)

Polymer (B):   **poly(methyl methacrylate)**   **1992KIA, 1994KIA**
Characterization: M_w/kg.mol^{-1} = 120, Aldrich Chem. Co., Inc., Milwaukee, WI
Solvent (A):   **carbon dioxide**   **CO$_2$**   **124-38-9**
Solvent (C):   **toluene**   **C$_7$H$_8$**   **108-88-3**

Type of data: cloud points

w_A	0.150	0.150	0.150	0.150	0.150	0.150	0.150	0.150	0.150
w_B	0.1275	0.1275	0.1275	0.1275	0.1275	0.1275	0.1275	0.1275	0.1275
w_C	0.7225	0.7225	0.7225	0.7225	0.7225	0.7225	0.7225	0.7225	0.7225
T/K	303.35	304.25	308.15	313.15	318.15	323.15	328.15	333.15	338.15
P/bar	24.79	25.32	26.66	28.33	30.30	32.50	33.94	36.22	37.96
	VLE	VLE	VLE	VLE	VLE	VLE	VLE	VLE	VLE

continued

continued

w_A	0.200	0.200	0.200	0.250	0.250	0.250	0.250	0.300	0.300
w_B	0.120	0.120	0.120	0.1125	0.1125	0.1125	0.1125	0.105	0.105
w_C	0.680	0.680	0.680	0.6375	0.6375	0.6375	0.6375	0.595	0.595
T/K	333.15	338.15	343.15	330.35	334.35	338.55	344.35	303.95	308.15
P/bar	48.35	51.66	54.84	52.64	56.89	59.92	63.76	39.20	41.61
	VLE	VLE	VLE	VLE	VLE	VLE	VLE	VLE	VLE
w_A	0.300	0.300	0.300	0.300	0.300	0.300	0.300	0.358	0.358
w_B	0.105	0.105	0.105	0.105	0.105	0.105	0.105	0.0963	0.0963
w_C	0.595	0.595	0.595	0.595	0.595	0.595	0.595	0.5457	0.5457
T/K	313.95	318.15	323.15	328.15	333.15	339.05	343.15	307.15	308.15
P/bar	45.91	48.61	53.06	56.96	59.79	64.88	69.11	43.84	44.52
	VLE	VLE	VLE	VLE	VLE	VLE	VLE	VLE	VLE
w_A	0.358	0.358	0.358	0.358	0.358	0.358	0.358	0.358	0.358
w_B	0.0963	0.0963	0.0963	0.0963	0.0963	0.0963	0.0963	0.0963	0.0963
w_C	0.5457	0.5457	0.5457	0.5457	0.5457	0.5457	0.5457	0.5457	0.5457
T/K	313.15	318.15	320.65	321.15	322.15	323.15	323.95	324.95	326.55
P/bar	48.62	53.08	54.05	55.69	56.34	57.72	58.79	59.54	62.06
	VLE	VLE	VLE	VLE	VLE	VLE	VLE	VLE	VLE
w_A	0.358	0.358	0.358	0.358	0.358	0.358	0.358	0.358	0.358
w_B	0.0963	0.0963	0.0963	0.0963	0.0963	0.0963	0.0963	0.0963	0.0963
w_C	0.5457	0.5457	0.5457	0.5457	0.5457	0.5457	0.5457	0.5457	0.5457
T/K	326.85	327.25	327.65	328.75	326.75	327.75	327.95	328.15	328.45
P/bar	60.01	62.89	62.71	61.63	67.09	75.25	75.37	76.81	78.48
	VLE	VLE	VLE	VLE	LLE	LLE	LLE	LLE	LLE
w_A	0.358	0.358	0.358	0.358	0.358	0.358	0.358	0.358	0.358
w_B	0.0963	0.0963	0.0963	0.0963	0.0963	0.0963	0.0963	0.0963	0.0963
w_C	0.5457	0.5457	0.5457	0.5457	0.5457	0.5457	0.5457	0.5457	0.5457
T/K	328.95	329.55	329.85	330.05	330.75	331.25	331.45	331.95	332.05
P/bar	81.59	82.19	82.28	82.15	84.98	88.52	91.55	90.53	93.56
	LLE	LLE	LLE	LLE	LLE	LLE	LLE	LLE	LLE
w_A	0.358	0.358	0.358	0.358	0.358	0.358	0.374	0.374	0.374
w_B	0.0963	0.0963	0.0963	0.0963	0.0963	0.0963	0.0939	0.0939	0.0939
w_C	0.5457	0.5457	0.5457	0.5457	0.5457	0.5457	0.5321	0.5321	0.5321
T/K	332.55	333.55	333.85	333.95	337.45	337.65	301.55	304.25	308.15
P/bar	92.51	94.02	96.11	97.57	104.96	107.50	43.15	44.72	47.84
	LLE	LLE	LLE	LLE	LLE	LLE	VLE	VLE	VLE
w_A	0.374	0.374	0.374	0.374	0.374	0.374	0.374	0.374	0.374
w_B	0.0939	0.0939	0.0939	0.0939	0.0939	0.0939	0.0939	0.0939	0.0939
w_C	0.5321	0.5321	0.5321	0.5321	0.5321	0.5321	0.5321	0.5321	0.5321
T/K	313.15	314.45	315.85	317.05	317.95	320.05	321.95	324.55	325.35
P/bar	50.89	53.17	54.97	55.49	56.83	58.56	65.11	77.61	80.56
	VLE	VLE	VLE	VLE	VLE	VLE	LLE	LLE	LLE

continued

continued

w_A	0.374	0.374	0.399	0.399	0.399	0.399	0.399	0.399	0.399
w_B	0.0939	0.0939	0.09015	0.09015	0.09015	0.09015	0.09015	0.09015	0.09015
w_C	0.5321	0.5321	0.51085	0.51085	0.51085	0.51085	0.51085	0.51085	0.51085
T/K	327.95	328.45	298.55	303.15	307.95	309.75	312.35	312.65	313.25
P/bar	94.68	96.41	42.14	45.37	50.43	51.24	57.07	65.30	67.04
	LLE	LLE	VLE	VLE	VLE	VLE	LLE	LLE	LLE

w_A	0.399	0.399	0.399	0.413	0.413	0.413	0.413	0.413	0.413
w_B	0.09015	0.09015	0.09015	0.08805	0.08805	0.08805	0.08805	0.08805	0.08805
w_C	0.51085	0.51085	0.51085	0.49895	0.49895	0.49895	0.49895	0.49895	0.49895
T/K	316.25	316.65	322.65	301.25	303.15	305.85	306.25	307.45	307.85
P/bar	81.64	84.97	103.92	43.67	44.82	48.26	48.15	49.64	52.67
	LLE	LLE	LLE	VLE	VLE	VLE	VLE	VLE	VLE

w_A	0.413	0.413	0.413	0.413	0.216	0.216	0.216	0.216	0.216
w_B	0.08805	0.08805	0.08805	0.08805	0.1568	0.1568	0.1568	0.1568	0.1568
w_C	0.49895	0.49895	0.49895	0.49895	0.6272	0.6272	0.6272	0.6272	0.6272
T/K	308.05	309.45	310.25	310.45	298.65	303.15	308.15	313.15	318.15
P/bar	54.67	59.66	63.22	65.19	33.87	34.88	37.41	41.24	43.35
	LLE	LLE	LLE	LLE	VLE	VLE	VLE	VLE	VLE

w_A	0.216	0.216	0.216	0.216	0.216	0.250	0.250	0.250	0.250
w_B	0.1568	0.1568	0.1568	0.1568	0.1568	0.150	0.150	0.150	0.150
w_C	0.6272	0.6272	0.6272	0.6272	0.6272	0.600	0.600	0.600	0.600
T/K	323.15	328.45	336.45	338.15	343.15	323.15	328.15	333.15	337.65
P/bar	47.31	50.37	56.60	57.42	61.24	52.25	55.97	59.23	62.49
	VLE	VLE	VLE	VLE	VLE	VLE	VLE	VLE	VLE

w_A	0.250	0.300	0.300	0.300	0.300	0.300	0.300	0.300	0.300
w_B	0.150	0.140	0.140	0.140	0.140	0.140	0.140	0.140	0.140
w_C	0.600	0.560	0.560	0.560	0.560	0.560	0.560	0.560	0.560
T/K	342.75	297.15	303.15	303.45	308.15	313.15	318.15	323.15	328.15
P/bar	65.67	38.98	41.63	42.32	46.02	49.71	53.65	58.12	61.22
	VLE	VLE	VLE	LLE	LLE	LLE	LLE	VLE	VLE

w_A	0.300	0.300	0.300	0.300	0.300	0.300	0.300	0.300	0.300
w_B	0.140	0.140	0.140	0.140	0.140	0.140	0.140	0.140	0.140
w_C	0.560	0.560	0.560	0.560	0.560	0.560	0.560	0.560	0.560
T/K	328.65	333.15	336.55	336.75	337.05	338.05	338.25	338.45	339.25
P/bar	62.61	67.73	69.02	70.40	71.08	71.22	72.68	73.28	73.08
	VLE	VLE	VLE	VLE	VLE	VLE	VLE	VLE	VLE

w_A	0.300	0.300	0.300	0.300	0.300	0.300	0.300	0.300	0.300
w_B	0.140	0.140	0.140	0.140	0.140	0.140	0.140	0.140	0.140
w_C	0.560	0.560	0.560	0.560	0.560	0.560	0.560	0.560	0.560
T/K	341.05	341.85	342.15	342.35	336.55	336.65	336.95	337.05	337.65
P/bar	75.41	76.19	75.29	75.65	69.44	72.71	73.99	74.26	75.85
	VLE	VLE	VLE	VLE	LLE	LLE	LLE	LLE	LLE

continued

continued

w_A	0.300	0.300	0.300	0.300	0.300	0.300	0.300	0.300	0.300
w_B	0.140	0.140	0.140	0.140	0.140	0.140	0.140	0.140	0.140
w_C	0.560	0.560	0.560	0.560	0.560	0.560	0.560	0.560	0.560
T/K	338.15	338.25	338.45	338.95	339.05	339.35	339.95	340.15	341.35
P/bar	79.85	81.81	81.94	86.46	86.60	89.56	89.98	91.63	93.66
	LLE	LLE	LLE	LLE	LLE	LLE	LLE	LLE	LLE
w_A	0.300	0.300	0.300	0.300	0.316	0.316	0.316	0.316	0.316
w_B	0.140	0.140	0.140	0.140	0.1368	0.1368	0.1368	0.1368	0.1368
w_C	0.560	0.560	0.560	0.560	0.5472	0.5472	0.5472	0.5472	0.5472
T/K	342.15	342.25	342.35	342.55	297.45	323.75	324.75	324.95	329.25
P/bar	100.39	98.13	98.73	101.66	40.12	60.07	60.09	60.37	63.80
	LLE	LLE	LLE	LLE	VLE	VLE	VLE	VLE	VLE
w_A	0.316	0.316	0.316	0.316	0.316	0.316	0.316	0.316	0.316
w_B	0.1368	0.1368	0.1368	0.1368	0.1368	0.1368	0.1368	0.1368	0.1368
w_C	0.5472	0.5472	0.5472	0.5472	0.5472	0.5472	0.5472	0.5472	0.5472
T/K	329.45	331.85	332.25	333.75	333.95	334.15	334.35	336.25	338.15
P/bar	64.55	64.98	68.60	74.66	75.29	75.15	76.22	82.61	90.36
	VLE	VLE	VLE	LLE	LLE	LLE	LLE	LLE	LLE
w_A	0.316	0.316	0.333	0.333	0.333	0.333	0.333	0.333	0.333
w_B	0.1368	0.1368	0.1333	0.1333	0.1333	0.1333	0.1333	0.1333	0.1333
w_C	0.5472	0.5472	0.5333	0.5333	0.5333	0.5333	0.5333	0.5333	0.5333
T/K	341.65	342.65	297.35	303.15	308.15	313.15	316.95	317.15	317.35
P/bar	104.70	107.68	40.84	44.52	48.43	53.06	55.98	56.18	56.63
	LLE	LLE	VLE	VLE	VLE	VLE	VLE	VLE	VLE
w_A	0.333	0.333	0.333	0.333	0.333	0.333	0.333	0.333	0.333
w_B	0.1333	0.1333	0.1333	0.1333	0.1333	0.1333	0.1333	0.1333	0.1333
w_C	0.5333	0.5333	0.5333	0.5333	0.5333	0.5333	0.5333	0.5333	0.5333
T/K	319.55	320.55	319.15	319.45	319.55	319.75	320.35	320.65	320.85
P/bar	59.78	58.34	67.87	74.70	74.04	70.40	71.54	76.66	79.81
	VLE	VLE	LLE	LLE	LLE	LLE	LLE	LLE	LLE
w_A	0.333	0.333	0.333	0.333	0.333	0.333	0.333	0.333	0.333
w_B	0.1333	0.1333	0.1333	0.1333	0.1333	0.1333	0.1333	0.1333	0.1333
w_C	0.5333	0.5333	0.5333	0.5333	0.5333	0.5333	0.5333	0.5333	0.5333
T/K	321.05	321.35	321.45	321.55	322.25	323.35	323.65	323.95	324.35
P/bar	73.64	78.17	80.37	82.21	84.50	92.98	92.56	94.40	99.88
	LLE	LLE	LLE	LLE	LLE	LLE	LLE	LLE	LLE
w_A	0.333	0.333	0.333	0.333	0.349	0.349	0.349	0.349	0.349
w_B	0.1333	0.1333	0.1333	0.1333	0.1302	0.1302	0.1302	0.1302	0.1302
w_C	0.5333	0.5333	0.5333	0.5333	0.5208	0.5208	0.5208	0.5208	0.5208
T/K	324.45	324.65	324.85	325.05	310.05	312.85	314.35	315.05	315.25
P/bar	96.25	95.11	96.40	101.02	50.07	54.07	53.91	55.29	56.11
	LLE	LLE	LLE	LLE	VLE	VLE	VLE	VLE	VLE

continued

continued

w_A	0.349	0.349	0.349	0.349	0.349	0.349	0.349	0.363	0.363
w_B	0.1302	0.1302	0.1302	0.1302	0.1302	0.1302	0.1302	0.1274	0.1274
w_C	0.5208	0.5208	0.5208	0.5208	0.5208	0.5208	0.5208	0.5096	0.5096
T/K	315.95	317.15	317.45	320.95	321.05	322.45	322.55	301.75	301.95
P/bar	57.30	62.05	64.65	78.28	77.12	85.20	85.56	44.55	44.97
	VLE	VLE	VLE	LLE	LLE	LLE	LLE	VLE	VLE

w_A	0.363	0.363	0.363	0.363	0.363	0.363	0.363	0.363	0.363
w_B	0.1274	0.1274	0.1274	0.1274	0.1274	0.1274	0.1274	0.1274	0.1274
w_C	0.5096	0.5096	0.5096	0.5096	0.5096	0.5096	0.5096	0.5096	0.5096
T/K	304.45	304.95	305.65	307.05	307.35	307.45	307.55	307.85	308.35
P/bar	46.04	46.82	48.03	48.00	49.54	55.54	49.77	53.14	56.84
	VLE	VLE	VLE	VLE	VLE	VLE	VLE	VLE	VLE

w_A	0.363	0.363	0.363	0.363	0.363	0.363	0.363	0.363	0.363
w_B	0.1274	0.1274	0.1274	0.1274	0.1274	0.1274	0.1274	0.1274	0.1274
w_C	0.5096	0.5096	0.5096	0.5096	0.5096	0.5096	0.5096	0.5096	0.5096
T/K	309.65	311.85	312.15	312.55	312.65	312.85	313.25	313.95	314.85
P/bar	58.84	74.53	74.65	81.01	80.24	80.67	84.34	85.53	90.03
	VLE	LLE	LLE	LLE	LLE	LLE	LLE	LLE	LLE

w_A	0.363	0.363	0.385	0.385	0.385	0.385	0.385	0.200	0.200
w_B	0.1274	0.1274	0.123	0.123	0.123	0.123	0.123	0.2664	0.2664
w_C	0.5096	0.5096	0.492	0.492	0.492	0.492	0.492	0.5336	0.5336
T/K	315.75	315.95	306.75	307.45	307.35	307.75	308.25	316.45	316.95
P/bar	93.91	95.28	67.67	72.66	71.77	72.92	75.57	38.41	38.51
	LLE	LLE	LLE	LLE	LLE	LLE	LLE	VLE	VLE

w_A	0.200	0.200	0.200	0.200	0.200	0.200	0.200	0.200	0.200
w_B	0.2664	0.2664	0.2664	0.2664	0.2664	0.2664	0.2664	0.2664	0.2664
w_C	0.5336	0.5336	0.5336	0.5336	0.5336	0.5336	0.5336	0.5336	0.5336
T/K	317.05	321.05	322.25	322.55	324.75	328.85	329.15	329.65	332.95
P/bar	40.09	39.16	40.48	40.37	41.76	43.72	44.62	45.25	46.58
	VLE	VLE	VLE	VLE	VLE	VLE	VLE	VLE	VLE

w_A	0.200	0.200	0.242	0.242	0.300	0.300	0.300	0.300	0.300
w_B	0.2664	0.2664	0.2524	0.2524	0.2331	0.2331	0.2331	0.2331	0.2331
w_C	0.5336	0.5336	0.5056	0.5056	0.4669	0.4669	0.4669	0.4669	0.4669
T/K	342.75	343.15	342.75	342.85	312.95	314.45	314.75	314.85	316.05
P/bar	53.72	55.12	63.81	64.51	49.72	50.27	50.17	50.62	50.86
	VLE	VLE	LLE	LLE	VLE	VLE	VLE	VLE	VLE

w_A	0.300	0.300	0.300	0.300	0.300	0.300	0.300	0.300	0.300
w_B	0.2331	0.2331	0.2331	0.2331	0.2331	0.2331	0.2331	0.2331	0.2331
w_C	0.4669	0.4669	0.4669	0.4669	0.4669	0.4669	0.4669	0.4669	0.4669
T/K	316.15	316.65	317.55	317.75	317.95	318.95	319.85	320.25	318.35
P/bar	50.86	50.92	51.81	51.51	51.90	52.99	54.17	53.97	54.86
	VLE	VLE	VLE	VLE	VLE	VLE	VLE	VLE	VLE

continued

continued

w_A	0.300	0.300	0.300	0.300	0.300	0.300	0.300	0.300	0.300
w_B	0.2331	0.2331	0.2331	0.2331	0.2331	0.2331	0.2331	0.2331	0.2331
w_C	0.4669	0.4669	0.4669	0.4669	0.4669	0.4669	0.4669	0.4669	0.4669
T/K	322.75	323.55	325.35	326.45	327.95	330.35	333.05	333.45	333.85
P/bar	56.04	56.24	57.98	59.03	59.49	62.71	64.27	64.50	64.84
	VLE	VLE	VLE	VLE	VLE	VLE	VLE	VLE	VLE

w_A	0.300	0.300	0.300	0.300	0.300	0.300	0.300	0.300	0.300
w_B	0.2331	0.2331	0.2331	0.2331	0.2331	0.2331	0.2331	0.2331	0.2331
w_C	0.4669	0.4669	0.4669	0.4669	0.4669	0.4669	0.4669	0.4669	0.4669
T/K	334.35	335.95	336.45	337.25	337.75	338.15	338.65	339.05	339.65
P/bar	65.79	66.61	67.67	70.35	76.63	74.59	81.52	75.84	82.87
	VLE	VLE	LLE	LLE	LLE	LLE	LLE	LLE	LLE

w_A	0.300	0.300	0.300	0.313	0.313	0.313	0.313	0.313	
w_B	0.2331	0.2331	0.2331	0.2288	0.2288	0.2288	0.2288	0.2288	
w_C	0.4669	0.4669	0.4669	0.4582	0.4582	0.4582	0.4582	0.4582	
T/K	339.95	340.75	341.55	301.55	305.35	307.95	312.55	313.35	
P/bar	81.06	90.22	92.71	44.74	46.42	48.79	58.69	74.43	
	LLE	LLE	LLE	VLE	VLE	VLE	LLE	LLE	

Polymer (B):	**poly(methyl methacrylate)**		**1992KIA**
Characterization:	M_w/kg.mol^{-1} = 120, Aldrich Chem. Co., Inc., Milwaukee, WI		
Solvent (A):	**carbon dioxide**	**CO_2**	**124-38-9**
Solvent (C):	**trichloromethane**	**$CHCl_3$**	**67-66-3**

Type of data: cloud points

w_A	0.100	0.100	0.100	0.100	0.100	0.200	0.200	0.200	0.200
w_B	0.1503	0.1503	0.1503	0.1503	0.1503	0.1336	0.1336	0.1336	0.1336
w_C	0.7497	0.7497	0.7497	0.7497	0.7497	0.6664	0.6664	0.6664	0.6664
T/K	308.15	313.15	323.15	333.15	343.15	313.15	323.15	333.15	343.15
P/bar	22.48	24.10	27.45	30.11	33.92	37.82	44.24	51.51	58.01
	VLE	VLE	VLE	VLE	VLE	VLE	VLE	VLE	VLE

w_A	0.274	0.274	0.274	0.274	0.274	0.323	0.323	0.323	0.323
w_B	0.1212	0.1212	0.1212	0.1212	0.1212	0.1131	0.1131	0.1131	0.1131
w_C	0.6048	0.6048	0.6048	0.6048	0.6048	0.5639	0.5639	0.5639	0.5639
T/K	309.45	313.15	323.15	333.15	343.15	313.15	323.15	328.15	329.15
P/bar	41.89	45.00	51.84	60.32	69.08	48.00	56.37	60.94	61.14
	VLE	VLE	VLE	VLE	VLE	LLE	LLE	LLE	LLE

w_A	0.323	0.323	0.323	0.323	0.201	0.201	0.201	0.201	0.201
w_B	0.1131	0.1131	0.1131	0.1131	0.1598	0.1598	0.1598	0.1598	0.1598
w_C	0.5639	0.5639	0.5639	0.5639	0.6392	0.6392	0.6392	0.6392	0.6392
T/K	330.65	331.15	333.15	343.15	307.65	313.45	323.15	328.25	333.15
P/bar	63.63	64.32	66.98	103.68	35.48	39.62	46.16	49.84	53.19
	LLE	LLE	LLE	LLE	VLE	VLE	VLE	VLE	VLE

continued

continued

w_A	0.201	0.201	0.280	0.280	0.280	0.280	0.280	0.280	0.280
w_B	0.1598	0.1598	0.144	0.144	0.144	0.144	0.144	0.144	0.144
w_C	0.6392	0.6392	0.576	0.576	0.576	0.576	0.576	0.576	0.576
T/K	339.95	343.15	314.65	318.55	323.25	328.65	332.25	338.55	343.15
P/bar	57.03	60.67	47.70	50.66	54.10	59.52	63.04	68.26	73.50
	VLE	VLE	VLE	VLE	VLE	VLE	VLE	VLE	VLE
w_A	0.341	0.341	0.341	0.341	0.341	0.341	0.341	0.341	0.341
w_B	0.1318	0.1318	0.1318	0.1318	0.1318	0.1318	0.1318	0.1318	0.1318
w_C	0.5272	0.5272	0.5272	0.5272	0.5272	0.5272	0.5272	0.5272	0.5272
T/K	296.55	299.65	301.15	302.15	303.45	304.05	304.45	308.65	309.65
P/bar	37.00	39.23	40.78	41.47	41.82	42.22	43.53	51.41	56.93
	VLE	VLE	VLE	VLE	VLE	VLE	LLE	LLE	LLE
w_A	0.341	0.341	0.341	0.350	0.350	0.350	0.350	0.350	0.350
w_B	0.1318	0.1318	0.1318	0.130	0.130	0.130	0.130	0.130	0.130
w_C	0.5272	0.5272	0.5272	0.520	0.520	0.520	0.520	0.520	0.520
T/K	313.15	318.15	323.15	295.85	296.25	296.45	296.85	297.45	298.05
P/bar	69.27	89.70	111.10	37.09	37.28	37.62	38.45	39.09	39.53
	LLE	LLE	LLE	VLE	VLE	VLE	VLE	VLE	VLE
w_A	0.350	0.350	0.350	0.350	0.350	0.350	0.362	0.362	0.362
w_B	0.130	0.130	0.130	0.130	0.130	0.130	0.1276	0.1276	0.1276
w_C	0.520	0.520	0.520	0.520	0.520	0.520	0.5104	0.5104	0.5104
T/K	299.15	300.05	304.65	308.45	313.15	318.15	302.25	304.45	305.35
P/bar	43.76	48.85	67.60	84.01	99.84	121.42	86.54	101.03	100.20
	LLE	LLE	LLE	LLE	LLE	LLE	LLE	LLE	LLE
w_A	0.362	0.362	0.362	0.362	0.362	0.362	0.154	0.154	0.154
w_B	0.1276	0.1276	0.1276	0.1276	0.1276	0.1276	0.2115	0.2115	0.2115
w_C	0.5104	0.5104	0.5104	0.5104	0.5104	0.5104	0.6345	0.6345	0.6345
T/K	306.25	308.45	309.85	310.55	316.25	318.35	313.45	318.75	322.15
P/bar	102.75	109.37	121.91	119.41	140.20	150.31	30.47	32.82	34.80
	LLE	LLE	LLE	LLE	LLE	LLE	VLE	VLE	VLE
w_A	0.154	0.154	0.238	0.238	0.238	0.238	0.238	0.238	0.307
w_B	0.2115	0.2115	0.1905	0.1905	0.1905	0.1905	0.1905	0.1905	0.17325
w_C	0.6345	0.6345	0.5715	0.5715	0.5715	0.5715	0.5715	0.5715	0.51975
T/K	333.15	343.15	308.15	313.15	323.15	333.15	338.15	343.15	312.95
P/bar	40.91	45.71	37.82	40.98	49.09	58.51	61.50	67.11	48.59
	VLE	VLE	VLE	VLE	VLE	VLE	VLE	VLE	LLE
w_A	0.307	0.307	0.307	0.307	0.307	0.307	0.307	0.307	0.307
w_B	0.17325	0.17325	0.17325	0.17325	0.17325	0.17325	0.17325	0.17325	0.17325
w_C	0.51975	0.51975	0.51975	0.51975	0.51975	0.51975	0.51975	0.51975	0.51975
T/K	322.05	328.85	331.05	331.65	332.25	333.35	334.45	336.25	338.15
P/bar	60.31	62.25	68.20	70.79	73.94	76.99	81.99	89.11	96.24
	LLE	LLE	LLE	LLE	LLE	LLE	LLE	LLE	LLE
w_A	0.307								
w_B	0.17325								
w_C	0.51975								
T/K	343.15								
P/bar	109.08								
	LLE								

Polymer (B):   **poly(octadecyl acrylate)**   **2002BYU**
Characterization: M_w/kg.mol^{-1} = 93.3, T_g/K = 313,
                           Aldrich Chem. Co., Inc., Milwaukee, WI
Solvent (A):   **ethene**   **C$_2$H$_4$**   **74-85-1**
Solvent (C):   **octadecyl acrylate**   **C$_{21}$H$_{40}$O$_2$**   **4813-57-4**

Type of data: cloud points

w_A	0.836	0.836	0.836	0.836	0.836	0.836	0.836	0.799	0.799
w_B	0.049	0.049	0.049	0.049	0.049	0.049	0.049	0.050	0.050
w_C	0.119	0.119	0.119	0.119	0.119	0.119	0.119	0.151	0.151
T/K	416.65	417.45	418.25	420.45	422.25	443.75	466.95	314.05	324.15
P/bar	2100.0	1752.1	1631.4	1408.6	1309.3	1012.4	922.8	515.9	522.1

w_A	0.799	0.799	0.799	0.799	0.799	0.690	0.690	0.690	0.690
w_B	0.050	0.050	0.050	0.050	0.050	0.052	0.052	0.052	0.052
w_C	0.151	0.151	0.151	0.151	0.151	0.258	0.258	0.258	0.258
T/K	343.45	363.65	384.05	403.85	427.45	352.45	354.25	358.45	381.45
P/bar	539.3	545.9	563.1	574.8	584.8	1560.0	870.0	744.1	594.8

w_A	0.690	0.690	0.665	0.665	0.665	0.665	0.665	0.665	0.665
w_B	0.052	0.052	0.048	0.048	0.048	0.048	0.048	0.048	0.048
w_C	0.258	0.258	0.287	0.287	0.287	0.287	0.287	0.287	0.287
T/K	401.75	423.55	333.15	334.35	338.45	343.45	345.95	363.15	384.45
P/bar	558.3	550.0	1656.9	1119.0	698.3	622.4	594.8	504.1	474.5

w_A	0.665	0.665	0.603	0.603	0.603	0.603	0.559	0.559	0.559
w_B	0.048	0.048	0.047	0.047	0.047	0.047	0.041	0.041	0.041
w_C	0.287	0.287	0.350	0.350	0.350	0.350	0.400	0.400	0.400
T/K	407.65	428.75	313.65	333.25	352.95	372.55	309.45	313.95	319.75
P/bar	482.8	488.3	332.8	281.0	286.6	312.1	158.6	167.2	179.0

w_A	0.559	0.559	0.559	0.559	0.559	0.559	0.559	0.559	0.559
w_B	0.041	0.041	0.041	0.041	0.041	0.041	0.041	0.041	0.041
w_C	0.400	0.400	0.400	0.400	0.400	0.400	0.400	0.400	0.400
T/K	325.25	342.75	354.75	364.05	372.85	381.25	391.55	410.95	432.05
P/bar	190.0	223.5	247.9	264.1	277.9	291.4	305.2	326.2	346.6

w_A	0.501	0.501	0.501	0.501	0.501	0.501	0.501	0.501	0.501
w_B	0.049	0.049	0.049	0.049	0.049	0.049	0.049	0.049	0.049
w_C	0.450	0.450	0.450	0.450	0.450	0.450	0.450	0.450	0.450
T/K	314.95	323.85	342.75	362.55	383.25	403.65	308.55	312.45	322.25
P/bar	127.9	150.0	187.9	224.8	258.3	284.1	115.3	119.1	132.2

w_A	0.501	0.491	0.491	0.491	0.491	0.491	0.491	0.491	
w_B	0.049	0.057	0.057	0.057	0.057	0.057	0.057	0.057	
w_C	0.450	0.452	0.452	0.452	0.452	0.452	0.452	0.452	
T/K	327.55	308.35	313.95	334.15	353.15	374.05	392.25	411.15	
P/bar	137.0	181.7	189.7	223.6	248.3	271.7	293.1	312.8	

Comments: The two data pairs at 308.55 K and 312.45 K for w_C = 0.450 correspond to a liquid-vapor transition, and the two data pairs at 322.25 K and 327.55 K for w_C = 0.450 correspond to a liquid-liquid-vapor transition.

Polymer (B):   **poly(octadecyl methacrylate)**   **2001BYU**

Characterization: $M_w/\text{kg.mol}^{-1} = 170$,

     Scientific Polymer Products, Inc., Ontario, NY

Solvent (A):   **ethene**   **C_2H_4**   **74-85-1**

Solvent (C):   **octadecyl methacrylate**   **$C_{22}H_{42}O_2$**   **32360-05-7**

Type of data: cloud points

w_A	0.842	0.842	0.842	0.842	0.842	0.842	0.842	0.842	0.842
w_B	0.056	0.056	0.056	0.056	0.056	0.056	0.056	0.056	0.056
w_C	0.102	0.102	0.102	0.102	0.102	0.102	0.102	0.102	0.102
T/K	452.65	452.95	453.85	455.15	459.55	470.15	487.05	508.75	527.75
P/bar	1946.6	1887.9	1791.4	1687.9	1587.9	1426.2	1236.2	1109.7	1059.3
w_A	0.719	0.719	0.719	0.719	0.719	0.719	0.719	0.719	0.719
w_B	0.049	0.049	0.049	0.049	0.049	0.049	0.049	0.049	0.049
w_C	0.232	0.232	0.232	0.232	0.232	0.232	0.232	0.232	0.232
T/K	384.85	385.05	385.25	385.65	387.25	393.25	413.85	433.95	453.65
P/bar	1794.8	1660.3	1463.1	1387.9	1254.8	1081.7	870.6	781.7	748.3
w_A	0.719	0.663	0.663	0.663	0.663	0.663	0.663	0.663	0.557
w_B	0.049	0.040	0.040	0.040	0.040	0.040	0.040	0.040	0.050
w_C	0.232	0.297	0.297	0.297	0.297	0.297	0.297	0.297	0.393
T/K	472.95	359.35	360.15	361.85	363.55	384.05	403.15	424.75	312.65
P/bar	735.9	1374.1	1159.0	1019.0	925.2	652.8	611.7	561.0	499.7
w_A	0.557	0.557	0.557	0.557	0.557	0.445	0.445	0.445	0.445
w_B	0.050	0.050	0.050	0.050	0.050	0.058	0.058	0.058	0.058
w_C	0.393	0.393	0.393	0.393	0.393	0.497	0.497	0.497	0.497
T/K	318.05	333.15	352.75	373.95	392.95	308.15	312.85	333.05	352.85
P/bar	395.5	322.8	311.4	325.5	352.8	113.5	120.7	163.5	199.0
w_A	0.445	0.445	0.437	0.437	0.437	0.437	0.437	0.437	0.437
w_B	0.058	0.058	0.046	0.046	0.046	0.046	0.046	0.046	0.046
w_C	0.497	0.497	0.517	0.517	0.517	0.517	0.517	0.517	0.517
T/K	374.35	393.65	333.65	354.35	378.25	298.15	304.65	313.85	323.65
P/bar	238.3	263.1	164.8	205.2	249.3	92.4	104.5	118.7	131.7
w_A	0.437								
w_B	0.046								
w_C	0.517								
T/K	334.45								
P/bar	142.3								

Comments: The three data pairs at at 298.15 K, 304.65 and 313.85 K for $w_C = 0.517$ correspond to a liquid-vapor transition, and the two data pairs at 323.65 K and 334.45 K for $w_C = 0.517$ correspond to a liquid-liquid-vapor transition.

Polymer (B):   **polystyrene**   **1992KIA**

Characterization: $M_w/\text{kg.mol}^{-1} = 235$,

     Scientific Polymer Products, Inc., Ontario, NY

Solvent (A):   **carbon dioxide**   **CO_2**   **124-38-9**

continued

continued

Solvent (C):	2-butanone	C_4H_8O	78-93-3

Type of data: cloud points

w_A	0.150	0.150	0.150	0.150	0.150	0.150	0.150	0.150	0.150
w_B	0.1275	0.1275	0.1275	0.1275	0.1275	0.1275	0.1275	0.1275	0.1275
w_C	0.7225	0.7225	0.7225	0.7225	0.7225	0.7225	0.7225	0.7225	0.7225
T/K	303.25	304.65	305.65	308.15	310.65	313.65	315.65	318.15	320.65
P/bar	117.87	116.63	115.53	113.37	108.61	107.23	105.36	104.90	103.17
	LLE	LLE	LLE	LLE	LLE	LLE	LLE	LLE	LLE

w_A	0.150	0.150	0.150	0.150	0.150	0.150	0.150	0.150	0.150
w_B	0.1275	0.1275	0.1275	0.1275	0.1275	0.1275	0.1275	0.1275	0.1275
w_C	0.7225	0.7225	0.7225	0.7225	0.7225	0.7225	0.7225	0.7225	0.7225
T/K	323.35	325.65	328.15	330.65	332.85	333.25	336.15	338.15	340.65
P/bar	103.78	102.37	103.29	102.86	103.95	104.54	105.69	105.98	107.17
	LLE	LLE	LLE	LLE	LLE	LLE	LLE	LLE	LLE

w_A	0.150	0.100	0.100	0.100	0.100	0.100	0.100	0.100	0.100
w_B	0.1275	0.180	0.180	0.180	0.180	0.180	0.180	0.180	0.180
w_C	0.7225	0.720	0.720	0.720	0.720	0.720	0.720	0.720	0.720
T/K	343.15	316.45	318.75	321.45	325.65	331.25	336.55	339.25	342.95
P/bar	108.41	14.87	14.94	15.46	16.65	18.35	24.02	27.52	32.57
	LLE	VLE	VLE	VLE	VLE	VLE	LLE	LLE	LLE

Polymer (B):	polystyrene		1997BUN, 1999BUN
Characterization:	M_n/kg.mol^{-1} = 93, M_w/kg.mol^{-1} = 101, M_z/kg.mol^{-1} = 112		
Solvent (A):	carbon dioxide	CO_2	124-38-9
Solvent (C):	cyclohexane	C_6H_{12}	110-82-7

Type of data: cloud points

w_A	0.050	0.050	0.050	0.050	0.050	0.050	0.139	0.139	0.139
w_B	0.094	0.094	0.094	0.094	0.094	0.094	0.093	0.093	0.093
w_C	0.856	0.856	0.856	0.856	0.856	0.856	0.768	0.768	0.768
T/K	471.8	476.41	482.93	486.92	492.39	501.91	418.75	442.45	467.75
P/bar	42.6	50.5	59.5	66.5	75.0	90.5	66.5	73.0	77.0

w_A	0.139	0.139	0.139	0.139	0.139	0.165	0.165	0.165	0.1947
w_B	0.093	0.093	0.093	0.093	0.093	0.092	0.092	0.092	0.0800
w_C	0.768	0.768	0.768	0.768	0.768	0.743	0.743	0.743	0.7253
T/K	418.75	426.45	442.45	443.76	467.75	423.75	433.50	442.65	293.96
P/bar	66.5	81.7	107.2	107.7	150.2	117.7	135.0	149.5	46.0

w_A	0.1947	0.1947	0.1947	0.1947	0.1947	0.1947	0.1947	0.1947	0.1947
w_B	0.0800	0.0800	0.0800	0.0800	0.0800	0.0800	0.0800	0.0800	0.0800
w_C	0.7253	0.7253	0.7253	0.7253	0.7253	0.7253	0.7253	0.7253	0.7253
T/K	294.90	295.35	374.18	375.80	379.29	384.02	384.15	399.55	408.53
P/bar	40.5	37.0	76.3	78.0	84.3	95.1	101.0	122.2	135.7

continued

continued

w_A	0.1947	0.1947	0.1947	0.1947	0.1947	0.1947	0.1947	0.1947	0.1947
w_B	0.0800	0.0800	0.0800	0.0800	0.0800	0.0800	0.0800	0.0800	0.0800
w_C	0.7253	0.7253	0.7253	0.7253	0.7253	0.7253	0.7253	0.7253	0.7253
T/K	423.16	293.96	294.90	295.35	297.34	374.18	384.15	408.53	423.16
P/bar	161.8	35.5	35.0	35.7	36.7	70.2	77.0	84.5	89.7

w_A	0.2125	0.2125	0.2125	0.2125	0.2125	0.2125	0.2125	0.2125	0.2125
w_B	0.0815	0.0815	0.0815	0.0815	0.0815	0.0815	0.0815	0.0815	0.0815
w_C	0.7060	0.7060	0.7060	0.7060	0.7060	0.7060	0.7060	0.7060	0.7060
T/K	307.08	311.80	316.60	322.22	326.34	330.75	333.86	335.83	335.89
P/bar	60.9	55.1	53.1	52.0	51.7	54.1	57.3	60.7	59.5

w_A	0.2125	0.2125	0.2125	0.2125	0.2125	0.2125	0.2125	0.2125	0.2125
w_B	0.0815	0.0815	0.0815	0.0815	0.0815	0.0815	0.0815	0.0815	0.0815
w_C	0.7060	0.7060	0.7060	0.7060	0.7060	0.7060	0.7060	0.7060	0.7060
T/K	341.92	345.96	350.14	365.05	389.51	307.08	311.80	316.60	322.22
P/bar	65.0	70.3	76.0	97.9	140.3	42.1	43.45	46.9	50.0

w_A	0.2125	0.2125	0.2125	0.2125	0.2125	0.2125	0.2125	0.2125	0.2125
w_B	0.0815	0.0815	0.0815	0.0815	0.0815	0.0815	0.0815	0.0815	0.0815
w_C	0.7060	0.7060	0.7060	0.7060	0.7060	0.7060	0.7060	0.7060	0.7060
T/K	326.34	330.75	331.65	333.86	335.83	335.89	341.92	345.96	350.14
P/bar	52.0	71.9	57.6	63.5	53.9	52.5	60.2	55.4	55.5

w_A	0.2236	0.2236	0.2236	0.2236	0.2236	0.2236	0.2236	0.2236	0.2236
w_B	0.0795	0.0795	0.0795	0.6969	0.6969	0.6969	0.6969	0.6969	0.6969
w_C	0.6969	0.6969	0.6969	0.0795	0.0795	0.0795	0.0795	0.0795	0.0795
T/K	305.15	311.76	316.55	321.15	328.85	335.55	353.25	369.85	384.45
P/bar	40.5	43.3	46.2	48.1	51.6	56.5	66.7	74.8	83.2

w_A	0.2236	0.2236	0.2236	0.2236	0.2236	0.2236	0.2236	0.2236	0.2236
w_C	0.6969	0.6969	0.6969	0.0795	0.0795	0.0795	0.0795	0.0795	0.0795
w_B	0.0795	0.0795	0.0795	0.6969	0.6969	0.6969	0.6969	0.6969	0.6969
T/K	305.15	311.76	316.55	321.15	328.85	335.55	353.25	369.85	384.45
P/bar	98.1	90.7	86.3	84.6	89.8	95.6	110.0	133.9	158.7

Type of data: coexistence data

$T/K = 443.75$

Total feed concentrations of the homogeneous system before demixing:

$w_A = 0.165$, $w_B = 0.092$, $w_C = 0.743$.

$P/$ bar	Liquid phase I			Liquid phase II			
	w_A	w_B	w_C	w_A	w_B	w_C	
149.5				0.165	0.092	0.743	(cloud point)
139.7	0.113	0.262	0.625	0.180	0.024	0.796	
132.8	0.083	0.276	0.641	0.170	0.011	0.819	
105.4	0.053	0.329	0.618	0.190	0.004	0.806	
86.6				0.196	0.006	0.798	

continued

continued

$T/K = 443.75$

Total feed concentrations of the homogeneous system before demixing:

$w_A = 0.139$, $w_B = 0.093$, $w_C = 0.768$.

$P/$ bar	Liquid phase I			Liquid phase II		
	w_A	w_B	w_C	w_A	w_B	w_C
109.0				0.139	0.093	0.768 (cloud point)
99.1	0.077	0.248	0.675	0.140	0.019	0.841
92.4	0.083	0.269	0.648			
73.0	0.035	0.333	0.632	0.149	0.007	0.844

$T/K = 448.15$

Total feed concentrations of the homogeneous system before demixing:

$w_A = 0.1554$, $w_B = 0.0935$, $w_C = 0.7511$.

$P/$ bar	Liquid phase I			Liquid phase II		
	w_A	w_B	w_C	w_A	w_B	w_C
145.0				0.1554	0.0935	0.7511 (cloud point)
124.1	0.121	0.303	0.576	0.1630	0.0270	0.8100
102.6	0.139	0.343	0.518	0.1788	0.0028	0.8184
86.3	0.061	0.365	0.574	0.1900	0.0017	0.8083

$T/K = 468.15$

Total feed concentrations of the homogeneous system before demixing:

$w_A = 0.141$, $w_B = 0.093$, $w_C = 0.766$.

$P/$ bar	Liquid phase I			Liquid phase II		
	w_A	w_B	w_C	w_A	w_B	w_C
150.2				0.141	0.093	0.766 (cloud point)
136.5	0.065	0.357	0.578	0.142	0.0306	0.8274
121.1				0.162	0.010	0.828
105.7	0.072	0.366	0.562	0.149	0.002	0.849
87.1	0.038	0.3564	0.6056	0.155	0.011	0.883

continued

continued

$T/K = 467.85$

Total feed concentrations of the homogeneous system before demixing:

$w_A = 0.1018$, $w_B = 0.0980$, $w_C = 0.8002$.

$P/$ bar	Liquid phase I			Liquid phase II		
	w_A	w_B	w_C	w_A	w_B	w_C
89.1	0.0383	0.2350	0.7260	0.1311	0.0300	0.8389
78.3	0.0367	0.3260	0.6373	0.1268	0.0130	0.8602
67.0	0.0314	0.3410	0.6276	0.1310	0.0050	0.8640

$T/K = 477.85$

Total feed concentrations of the homogeneous system before demixing:

$w_A = 0.0893$, $w_B = 0.0950$, $w_C = 0.8157$.

$P/$ bar	Liquid phase I			Liquid phase II		
	w_A	w_B	w_C	w_A	w_B	w_C
101.5				0.0893	0.0950	0.8157 (cloud point)
93.5	0.0643	0.2940	0.6417	0.0989	0.0370	0.8641
85.6	0.0296	0.3330	0.6374	0.1093	0.0100	0.8807
74.2	0.0281	0.3830	0.5889	0.1071	0.0080	0.8849
62.7	0.0269	0.3860	0.5871	0.1109	0.0060	0.8831

Type of data: liquid-liquid-vapor three-phase equilibrium

$T/K = 443.15$

Liquid phase I			Liquid phase II			Vapor phase		
w_A	w_B	w_C	w_A	w_B	w_C	w_A	w_B	w_C
P/bar = 71.2								
0.096	0.319	0.585	0.151	0.008	0.841	0.650	0.000	0.350
P/bar = 88.0								
0.100	0.225	0.675	0.1575	0.0060	0.8365			
0.109	0.302	0.589	0.1880	0.0036	0.8084			
0.092	0.380	0.528	0.2050	0.0040	0.7910			
0.102	0.397	0.501	0.2310	0.0013	0.7677	0.678	0.000	0.322

continued

continued

P/bar = 101

0.038	0.248	0.714	0.139	0.019	0.842			
0.064	0.278	0.658	0.152	0.022	0.826			
0.087	0.270	0.643	0.182	0.010	0.808			
0.073	0.339	0.588	0.217	0.007	0.776			
0.091	0.447	0.462	0.251	0.015	0.734	0.652	0.000	0.348
0.092	0.415	0.493				0.669	0.000	0.331
0.068	0.611	0.321				0.739	0.000	0.261
0.080	0.681	0.239				0.803	0.000	0.197

P/bar = 135

0.142	0.291	0.567	0.2030	0.0095	0.7875			
0.138	0.3954	0.4666	0.2370	0.0110	0.7520			
0.159	0.458	0.383	0.3100	0.0033	0.6867			
0.148	0.503	0.349	0.3360	0.0037	0.6603			
0.189	0.582	0.229	0.3890	0.0024	0.6086	0.638	0.000	0.362

Polymer (B):	**polystyrene**	**1997BUN, 1999BUN**
Characterization:	$M_n/\text{kg.mol}^{-1} = 128$, $M_w/\text{kg.mol}^{-1} = 306$, $M_z/\text{kg.mol}^{-1} = 595$	
Solvent (A):	**carbon dioxide** **CO_2**	**124-38-9**
Solvent (C):	**cyclohcxane** **C_6H_{12}**	**110-82-7**

Type of data: coexistence data

T/K = 461.85

Total feed concentrations of the homogeneous system before demixing:

w_A = 0.1286, w_B = 0.0341, w_C = 0.8373.

P/	Liquid phase I			Liquid phase II		
bar	w_A	w_B	w_C	w_A	w_B	w_C
125.7				0.1286	0.0341	0.8373 (cloud point)
111.3	0.0819	0.2274	0.6907	0.1444	0.0150	0.8406
97.5	0.0875	0.2833	0.6292	0.1469	0.0122	0.8409
79.7	0.0991	0.3138	0.5871	0.1431	0.0098	0.8471
66.2	0.0977	0.3332	0.5691	0.1210	0.0042	0.8748

Polymer (B):	**polystyrene**	**1997BUN, 1999BUN**
Characterization:	$M_n/\text{kg.mol}^{-1} = 37.8$, $M_w/\text{kg.mol}^{-1} = 42.7$, $M_z/\text{kg.mol}^{-1} = 67.7$	
Solvent (A):	**carbon dioxide** **CO_2**	**124-38-9**
Solvent (C):	**cyclohexane** **C_6H_{12}**	**110-82-7**
Polymer (D):	**polystyrene**	
Characterization:	$M_n/\text{kg.mol}^{-1} = 148$, $M_w/\text{kg.mol}^{-1} = 160$, $M_z/\text{kg.mol}^{-1} = 186$	

continued

continued

Comments: The initial polymer mixture was made 50/50 by weight.

Type of data: coexistence data

$T/K = 443.15$

Total feed concentrations of the homogeneous system before demixing:

$w_A = 0.134$, $w_{B+D} = 0.094$, $w_C = 0.772$.

$P/$ bar	Liquid phase I			Liquid phase II		
	w_A	w_{B+D}	w_C	w_A	w_{B+D}	w_C
114.0				0.134	0.094	0.772 (cloud point)
101.5	0.084	0.263	0.653	0.134	0.038	0.828
86.2	0.084	0.266	0.650	0.151	0.021	0.828
74.1	0.101	0.321	0.578	0.174	0.008	0.818

Polymer (B):	**polystyrene**		**1992KIA**
Characterization:	$M_w/\text{kg.mol}^{-1} = 235$,		
	Scientific Polymer Products, Inc., Ontario, NY		
Solvent (A):	**carbon dioxide**	**CO$_2$**	**124-38-9**
Solvent (C):	**cyclohexane**	**C$_6$H$_{12}$**	**110-82-7**

Type of data: cloud points

w_A	0.108	0.108	0.108	0.108	0.108	0.200	0.200	0.200
w_B	0.1784	0.1784	0.1784	0.1784	0.1784	0.160	0.160	0.160
w_C	0.7136	0.7136	0.7136	0.7136	0.7136	0.640	0.640	0.640
T/K	306.05	313.65	323.55	333.25	343.25	323.15	333.15	343.15
$P/$bar	23.84	26.59	29.97	31.35	35.36	83.70	76.86	85.63
	VLE	VLE	VLE	VLE	VLE	LLE	LLE	LLE

Polymer (B):	**polystyrene**		**1998KOA**
Characterization:	$M_n/\text{kg.mol}^{-1} = 29.1$, $M_w/\text{kg.mol}^{-1} = 31.6$,		
	Aldrich Chem. Co., Inc., Milwaukee, WI		
Solvent (A):	**carbon dioxide**	**CO$_2$**	**124-38-9**
Solvent (C):	**methylcyclohexane**	**C$_7$H$_{14}$**	**108-87-2**

Type of data: cloud points

w_A	0.0377	0.0377	0.0377	0.0377	0.0377	0.0377	0.0377	0.0377	0.0377
w_B	0.0495	0.0495	0.0495	0.0495	0.0495	0.0495	0.0495	0.0495	0.0495
w_C	0.9128	0.9128	0.9128	0.9128	0.9128	0.9128	0.9128	0.9128	0.9128
T/K	298.40	298.34	297.77	297.55	297.16	296.86	296.73	296.40	295.91
$P/$MPa	2.40	2.85	4.65	5.40	7.25	8.025	8.50	9.75	12.05

continued

continued

w_A	0.0775	0.0775	0.0775	0.0775	0.0775	0.0775	0.0775	0.0775	0.1122
w_B	0.0494	0.0494	0.0494	0.0494	0.0494	0.0494	0.0494	0.0494	0.0494
w_C	0.8731	0.8731	0.8731	0.8731	0.8731	0.8731	0.8731	0.8731	0.8384
T/K	298.28	297.38	296.40	295.88	295.36	294.84	294.40	294.00	299.37
P/MPa	2.95	4.45	6.60	7.75	9.05	10.40	11.55	12.90	3.00
w_A	0.1122	0.1122	0.1122	0.1122	0.1122	0.1122	0.1221	0.1221	0.1221
w_B	0.0494	0.0494	0.0494	0.0494	0.0494	0.0494	0.0496	0.0496	0.0496
w_C	0.8384	0.8384	0.8384	0.8384	0.8384	0.8384	0.8283	0.8283	0.8283
T/K	298.28	296.73	295.36	294.38	293.52	293.24	299.99	298.16	296.85
P/MPa	4.00	6.60	8.80	10.60	12.35	12.90	3.30	5.45	7.15
w_A	0.1221	0.1221	0.1221	0.1221	0.1221	0.1479	0.1479	0.1479	0.1479
w_B	0.0496	0.0496	0.0496	0.0496	0.0496	0.0491	0.0491	0.0491	0.0491
w_C	0.8283	0.8283	0.8283	0.8283	0.8283	0.8030	0.8030	0.8030	0.8030
T/K	295.73	294.96	294.14	293.59	293.07	303.92	302.89	301.48	299.33
P/MPa	8.80	10.05	11.50	12.45	13.55	3.25	3.95	5.20	6.95
w_A	0.1479	0.1479	0.1479	0.1479	0.1812	0.1812	0.1812	0.1812	0.1812
w_B	0.0491	0.0491	0.0491	0.0491	0.0494	0.0494	0.0494	0.0494	0.0494
w_C	0.8030	0.8030	0.8030	0.8030	0.7694	0.7694	0.7694	0.7694	0.7694
T/K	298.40	296.32	295.36	294.43	315.80	314.29	311.86	308.28	305.83
P/MPa	8.00	10.30	11.45	12.80	4.60	5.10	6.05	7.70	9.05
w_A	0.1812	0.1812	0.1812	0.2095	0.2095	0.2095	0.2095	0.2095	0.2095
w_B	0.0494	0.0494	0.0494	0.0479	0.0479	0.0479	0.0479	0.0479	0.0479
w_C	0.7694	0.7694	0.7694	0.7426	0.7426	0.7426	0.7426	0.7426	0.7426
T/K	304.29	302.28	300.64	368.48	363.49	358.95	354.03	348.36	328.91
P/MPa	10.05	11.50	12.80	8.55	8.10	7.80	7.50	7.30	7.95
w_A	0.2095	0.2095	0.2095	0.2095	0.2114	0.2114	0.2114	0.2114	0.2114
w_B	0.0479	0.0479	0.0479	0.0479	0.0495	0.0495	0.0495	0.0495	0.0495
w_C	0.7426	0.7426	0.7426	0.7426	0.7391	0.7391	0.7391	0.7391	0.7391
T/K	323.61	319.01	316.29	313.30	367.47	359.02	352.44	347.40	344.34
P/MPa	8.80	9.80	10.60	11.65	9.15	8.50	8.15	8.00	7.95
w_A	0.2114	0.2114	0.2114	0.2114	0.2114	0.2114	0.2114	0.2114	0.2114
w_B	0.0495	0.0495	0.0495	0.0495	0.0495	0.0495	0.0495	0.0495	0.0495
w_C	0.7391	0.7391	0.7391	0.7391	0.7391	0.7391	0.7391	0.7391	0.7391
T/K	341.34	338.37	336.34	333.33	330.35	328.32	326.28	323.33	318.33
P/MPa	8.00	8.05	8.15	8.35	8.60	8.90	9.10	9.65	10.85
w_A	0.2114	0.2114	0.0242	0.0242	0.0242	0.0242	0.0242	0.0242	0.0428
w_B	0.0495	0.0495	0.1099	0.1099	0.1099	0.1099	0.1099	0.1099	0.1090
w_C	0.7391	0.7391	0.8659	0.8659	0.8659	0.8659	0.8659	0.8659	0.8482
T/K	315.33	313.30	301.62	301.06	300.09	299.52	299.22	298.98	300.92
P/MPa	11.80	12.55	2.25	4.10	7.65	10.10	11.45	12.65	2.05
w_A	0.0428	0.0428	0.0428	0.0428	0.0428	0.0428	0.0508	0.0508	0.0508
w_B	0.1090	0.1090	0.1090	0.1090	0.1090	0.1090	0.1097	0.1097	0.1097
w_C	0.8482	0.8482	0.8482	0.8482	0.8482	0.8482	0.8395	0.8395	0.8395
T/K	300.19	299.56	298.81	298.35	297.85	297.50	301.04	300.41	299.66
P/MPa	4.00	5.75	8.10	9.65	11.50	12.95	1.95	3.45	5.25

continued

continued

w_A	0.0508	0.0508	0.0508	0.0508	0.0742	0.0742	0.0742	0.0742	0.0742
w_B	0.1097	0.1097	0.1097	0.1097	0.1091	0.1091	0.1091	0.1091	0.1091
w_C	0.8395	0.8395	0.8395	0.8395	0.8167	0.8167	0.8167	0.8167	0.8167
T/K	298.80	298.31	297.68	297.12	301.05	301.00	300.30	299.44	298.56
P/MPa	7.60	9.10	11.15	13.05	1.85	1.95	3.10	4.65	6.40

w_A	0.0742	0.0742	0.0742	0.1037	0.1037	0.1037	0.1037	0.1037	0.1037
w_B	0.1091	0.1091	0.1091	0.1098	0.1098	0.1098	0.1098	0.1098	0.1098
w_C	0.8167	0.8167	0.8167	0.7865	0.7865	0.7865	0.7865	0.7865	0.7865
T/K	297.44	296.56	295.89	302.29	301.38	300.25	299.14	298.18	297.10
P/MPa	8.85	11.05	12.85	2.85	4.00	5.40	7.00	8.50	10.30

w_A	0.1037	0.1167	0.1167	0.1167	0.1167	0.1167	0.1167	0.1457	0.1457
w_B	0.1098	0.1095	0.1095	0.1095	0.1095	0.1095	0.1095	0.1095	0.1095
w_C	0.7865	0.7738	0.7738	0.7738	0.7738	0.7738	0.7738	0.7448	0.7448
T/K	296.18	303.57	303.17	301.87	300.72	299.33	297.62	310.79	310.28
P/MPa	12.10	3.10	3.50	4.80	6.20	7.95	10.45	3.825	4.075

w_A	0.1457	0.1457	0.1457	0.1457	0.1457	0.1457	0.1457	0.1657	0.1657
w_B	0.1095	0.1095	0.1095	0.1095	0.1095	0.1095	0.1095	0.1096	0.1096
w_C	0.7448	0.7448	0.7448	0.7448	0.7448	0.7448	0.7448	0.7247	0.7247
T/K	309.28	308.35	305.77	303.74	302.83	301.73	299.37	320.65	317.06
P/MPa	4.625	5.175	6.875	8.425	9.20	10.25	12.70	4.325	5.325

w_A	0.1657	0.1657	0.1657	0.1657	0.1657
w_B	0.1096	0.1096	0.1096	0.1096	0.1096
w_C	0.7247	0.7247	0.7247	0.7247	0.7247
T/K	312.30	309.34	307.24	305.58	303.32
P/MPa	7.175	8.625	9.925	11.00	12.75

Polymer (B):	**polystyrene**		**1994MUR**
Characterization:	M_n/kg.mol^{-1} = 83, M_w/kg.mol^{-1} = 190,		
	Scientific Polymer Products, Inc., Ontario, NY		
Solvent (A):	**carbon dioxide**	**CO_2**	**124-38-9**
Solvent (C):	**tetrahydrofuran**	**C_4H_8O**	**109-99-9**

Type of data: cloud points   (a) - visually   (b) - from *p-V* data

w_A	0.29	0.29
w_B	0.11	0.11
w_C	0.60	0.60
T/K	343.15	343.15
P/bar	62.81	75.98
	(a)	(b)

Polymer (B):	polystyrene		1992KIA, 1994KIA

Characterization: $M_w/\text{kg.mol}^{-1} = 235$,
Scientific Polymer Products, Inc., Ontario, NY

Solvent (A):	**carbon dioxide**	**CO$_2$**	**124-38-9**
Solvent (C):	**tetrahydrofuran**	**C$_4$H$_8$O**	**109-99-9**

Type of data: cloud points

w_A	0.200	0.200	0.200	0.200	0.200	0.200	0.200	0.200	0.200
w_B	0.120	0.120	0.120	0.120	0.120	0.120	0.120	0.120	0.120
w_C	0.680	0.680	0.680	0.680	0.680	0.680	0.680	0.680	0.680
T/K	299.15	303.15	308.25	313.15	318.15	323.15	328.15	333.15	337.15
P/bar	17.69	18.35	20.82	22.36	24.30	26.43	28.37	31.29	32.24
	VLE	VLE	VLE	VLE	VLE	VLE	VLE	VLE	VLE

w_A	0.250	0.250	0.250	0.250	0.250	0.250	0.250	0.250	0.250
w_B	0.1125	0.1125	0.1125	0.1125	0.1125	0.1125	0.1125	0.1125	0.1125
w_C	0.6375	0.6375	0.6375	0.6375	0.6375	0.6375	0.6375	0.6375	0.6375
T/K	302.35	303.15	308.35	313.15	318.15	323.15	328.55	333.15	336.65
P/bar	25.70	26.13	28.43	30.50	33.16	36.34	39.49	42.19	49.31
	VLE	VLE	VLE	VLE	VLE	VLE	VLE	LLE	LLE

w_A	0.250	0.250	0.250	0.260	0.260	0.260	0.260	0.260	0.260
w_B	0.1125	0.1125	0.1125	0.111	0.111	0.111	0.111	0.111	0.111
w_C	0.6375	0.6375	0.6375	0.629	0.629	0.629	0.629	0.629	0.629
T/K	338.25	340.65	342.85	297.95	303.15	308.15	313.15	315.65	318.25
P/bar	54.04	61.33	68.15	24.19	26.43	28.76	31.45	36.38	43.14
	LLE	LLE	LLE	VLE	VLE	VLE	VLE	LLE	LLE

w_A	0.260	0.260	0.260	0.260	0.260	0.260	0.260	0.260	0.260
w_B	0.111	0.111	0.111	0.111	0.111	0.111	0.111	0.111	0.111
w_C	0.629	0.629	0.629	0.629	0.629	0.629	0.629	0.629	0.629
T/K	320.65	323.25	325.65	328.65	330.65	333.15	335.65	338.15	340.15
P/bar	49.08	55.68	62.28	70.00	75.74	83.03	89.96	96.03	101.94
	LLE	LLE	LLE	LLE	LLE	LLE	LLE	LLE	LLE

w_A	0.260	0.150	0.150	0.150	0.150	0.150	0.150	0.150	0.150
w_B	0.111	0.170	0.170	0.170	0.170	0.170	0.170	0.170	0.170
w_C	0.629	0.680	0.680	0.680	0.680	0.680	0.680	0.680	0.680
T/K	341.95	298.85	303.25	308.15	313.15	318.15	323.15	328.15	333.15
P/bar	106.97	16.84	17.83	19.37	21.17	23.08	24.62	26.66	29.12
	LLE	VLE	VLE	VLE	VLE	VLE	VLE	VLE	VLE

w_A	0.150	0.150	0.200	0.200	0.200	0.200	0.200	0.200	0.200
w_B	0.170	0.170	0.160	0.160	0.160	0.160	0.160	0.160	0.160
w_C	0.680	0.680	0.640	0.640	0.640	0.640	0.640	0.640	0.640
T/K	338.15	341.55	298.15	305.15	307.65	313.55	318.15	323.15	328.15
P/bar	30.96	32.83	21.84	25.12	25.90	28.30	30.41	32.41	35.55
	VLE	VLE	VLE	VLE	VLE	VLE	VLE	VLE	VLE

continued

continued

w_A	0.200	0.200	0.200	0.210	0.210	0.210	0.210	0.210	0.210
w_B	0.160	0.160	0.160	0.158	0.158	0.158	0.158	0.158	0.158
w_C	0.640	0.640	0.640	0.632	0.632	0.632	0.632	0.632	0.632
T/K	333.15	338.15	342.55	298.45	303.15	308.15	313.15	318.15	323.15
P/bar	38.12	41.04	44.26	23.54	25.12	27.84	30.60	32.67	35.56
	VLE	VLE	VLE	VLE	VLE	VLE	VLE	VLE	VLE
w_A	0.210	0.210	0.210	0.210	0.222	0.222	0.222	0.222	0.222
w_B	0.158	0.158	0.158	0.158	0.1556	0.1556	0.1556	0.1556	0.1556
w_C	0.632	0.632	0.632	0.632	0.6224	0.6224	0.6224	0.6224	0.6224
T/K	328.15	333.15	338.15	342.95	300.65	303.15	308.15	313.15	318.15
P/bar	38.81	41.70	44.78	48.39	26.27	26.89	29.25	31.68	34.27
	VLE	VLE	VLE	VLE	VLE	VLE	VLE	VLE	VLE
w_A	0.222	0.222	0.222	0.222	0.222	0.222	0.222	0.222	0.231
w_B	0.1556	0.1556	0.1556	0.1556	0.1556	0.1556	0.1556	0.1556	0.1538
w_C	0.6224	0.6224	0.6224	0.6224	0.6224	0.6224	0.6224	0.6224	0.6152
T/K	323.15	328.15	330.15	333.15	335.65	338.15	340.15	343.15	300.95
P/bar	37.30	40.32	41.78	47.31	52.99	61.40	68.20	75.97	27.41
	VLE	VLE	VLE	VLE	LLE	LLE	LLE	LLE	VLE
w_A	0.231	0.231	0.231	0.231	0.231	0.231	0.231	0.231	0.231
w_B	0.1538	0.1538	0.1538	0.1538	0.1538	0.1538	0.1538	0.1538	0.1538
w_C	0.6152	0.6152	0.6152	0.6152	0.6152	0.6152	0.6152	0.6152	0.6152
T/K	303.25	305.65	308.45	310.65	313.15	315.65	318.15	320.65	323.15
P/bar	28.07	30.50	32.63	35.98	41.50	47.22	53.03	60.34	66.12
	VLE	VLE	VLE	LLE	LLE	LLE	LLE	LLE	LLE
w_A	0.231	0.231	0.231	0.231	0.231	0.231	0.231	0.231	0.236
w_B	0.1538	0.1538	0.1538	0.1538	0.1538	0.1538	0.1538	0.1538	0.1528
w_C	0.6152	0.6152	0.6152	0.6152	0.6152	0.6152	0.6152	0.6152	0.6112
T/K	325.65	328.15	330.65	333.15	335.85	338.15	340.65	342.75	302.85
P/bar	73.05	79.22	85.63	93.18	100.66	106.72	112.38	119.49	87.43
	LLE	LLE	LLE	LLE	LLE	LLE	LLE	LLE	LLE
w_A	0.236	0.236	0.236	0.236	0.236	0.236	0.236	0.236	0.236
w_B	0.1528	0.1528	0.1528	0.1528	0.1528	0.1528	0.1528	0.1528	0.1528
w_C	0.6112	0.6112	0.6112	0.6112	0.6112	0.6112	0.6112	0.6112	0.6112
T/K	303.15	304.15	305.65	308.15	310.65	313.15	315.65	318.15	320.65
P/bar	87.73	88.22	89.98	93.64	98.79	101.64	106.81	111.96	117.11
	LLE	LLE	LLE	LLE	LLE	LLE	LLE	LLE	LLE
w_A	0.236	0.150	0.150	0.150	0.150	0.150	0.150	0.150	0.150
w_B	0.1528	0.2125	0.2125	0.2125	0.2125	0.2125	0.2125	0.2125	0.2125
w_C	0.6112	0.6375	0.6375	0.6375	0.6375	0.6375	0.6375	0.6375	0.6375
T/K	323.35	301.35	303.15	308.25	313.15	318.15	323.15	328.15	333.15
P/bar	121.58	18.22	18.38	19.66	21.11	22.88	24.79	26.79	28.50
	LLE	VLE	VLE	VLE	VLE	VLE	VLE	VLE	VLE

continued

continued

w_A	0.150	0.200	0.200	0.200	0.200	0.200	0.200	0.200	0.200
w_B	0.2125	0.200	0.200	0.200	0.200	0.200	0.200	0.200	0.200
w_C	0.6375	0.600	0.600	0.600	0.600	0.600	0.600	0.600	0.600
T/K	338.15	299.55	315.35	317.45	323.25	328.15	333.15	338.35	342.65
P/bar	30.73	22.62	29.48	30.27	33.23	35.60	38.02	41.11	43.86
	VLE	VLE	VLE	VLE	VLE	VLE	VLE	VLE	VLE

w_A	0.216	0.216	0.216	0.216	0.216	0.216	0.216	0.216	0.216
w_B	0.196	0.196	0.196	0.196	0.196	0.196	0.196	0.196	0.196
w_C	0.588	0.588	0.588	0.588	0.588	0.588	0.588	0.588	0.588
T/K	299.65	303.15	308.15	313.15	318.15	323.15	328.15	333.15	337.45
P/bar	25.18	26.17	28.23	30.53	34.05	35.52	38.28	41.43	44.88
	VLE	VLE	VLE	VLE	VLE	VLE	VLE	VLE	VLE

w_A	0.216	0.231	0.231	0.231	0.231	0.231	0.231	0.231	0.231
w_B	0.196	0.19225	0.19225	0.19225	0.19225	0.19225	0.19225	0.19225	0.19225
w_C	0.588	0.57675	0.57675	0.57675	0.57675	0.57675	0.57675	0.57675	0.57675
T/K	342.15	298.65	303.65	308.15	313.15	318.15	323.15	328.15	333.15
P/bar	48.49	25.51	28.43	29.61	32.01	35.23	38.22	42.35	54.57
	VLE	VLE	VLE	VLE	VLE	VLE	VLE	LLE	LLE

w_A	0.231	0.231	0.231	0.231	0.240	0.240	0.240	0.240	0.240
w_B	0.19225	0.19225	0.19225	0.19225	0.190	0.190	0.190	0.190	0.190
w_C	0.57675	0.57675	0.57675	0.57675	0.570	0.570	0.570	0.570	0.570
T/K	338.55	340.65	342.15	342.65	298.85	300.65	303.15	305.65	308.15
P/bar	69.31	76.08	80.17	82.14	50.76	52.56	54.63	59.16	63.33
	LLE	LLE	LLE	LLE	LLE	LLE	LLE	LLE	LLE

w_A	0.240	0.240	0.240	0.240	0.240	0.240	0.240	0.240	0.240
w_B	0.190	0.190	0.190	0.190	0.190	0.190	0.190	0.190	0.190
w_C	0.570	0.570	0.570	0.570	0.570	0.570	0.570	0.570	0.570
T/K	310.65	313.15	315.65	318.15	320.65	323.15	325.65	328.15	330.65
P/bar	68.69	72.49	77.68	83.13	89.34	94.69	101.35	105.94	111.72
	LLE	LLE	LLE	LLE	LLE	LLE	LLE	LLE	LLE

w_A	0.240	0.240	0.240	0.150	0.150	0.150	0.150	0.150	0.150
w_B	0.190	0.190	0.190	0.283	0.283	0.283	0.283	0.283	0.283
w_C	0.570	0.570	0.570	0.567	0.567	0.567	0.567	0.567	0.567
T/K	333.15	335.15	335.65	305.05	305.65	308.15	313.15	322.65	333.15
P/bar	118.03	121.64	123.97	21.21	21.44	21.74	23.41	28.56	33.42
	LLE	LLE	LLE	VLE	VLE	VLE	VLE	VLE	VLE

w_A	0.150	0.150	0.200	0.200	0.200	0.200	0.200	0.200	0.200
w_B	0.283	0.283	0.2664	0.2664	0.2664	0.2664	0.2664	0.2664	0.2664
w_C	0.567	0.567	0.5336	0.5336	0.5336	0.5336	0.5336	0.5336	0.5336
T/K	338.15	343.15	302.45	308.15	313.65	318.15	323.15	330.65	333.15
P/bar	35.46	37.66	25.71	27.12	29.65	32.07	34.47	38.98	40.36
	VLE	VLE	VLE	VLE	VLE	VLE	VLE	VLE	VLE

w_A	0.200	0.200
w_B	0.2664	0.2664
w_C	0.5336	0.5336
T/K	340.65	343.15
P/bar	45.08	46.82
	VLE	VLE

Polymer (B): **polystyrene** **2001LI2**
Characterization: M_n/kg.mol^{-1} = 71, M_w/kg.mol^{-1} = 78, State Key Laboratory of
Polymer Science, Chinese Academy of Sciences
Solvent (A): **carbon dioxide** **CO$_2$** **124-38-9**
Solvent (C): **toluene** **C$_7$H$_8$** **108-88-3**

Type of data: cloud points

Comments: Cloud points were determined for a given concentration of polystyrene in the CO_2-free toluene solution, c_{0B}, and the given equilibrium CO_2-pressure of the ternary system. Volume expansion coefficients of the ternary system are given in Chapter 6.

c_{0B}/g.ml^{-3}	0.00046	0.00046	0.00046	0.00046	0.00046	0.00046	0.00046	0.00069	0.00069
T/K	298.15	303.15	308.15	313.15	318.15	323.15	328.15	298.15	303.15
P/MPa	4.70	4.74	4.78	4.85	4.91	4.96	4.99	4.65	4.71

c_{0B}/g.ml^{-3}	0.00069	0.00069	0.00069	0.00069	0.00069	0.00107	0.00107	0.00107	0.00107
T/K	308.15	313.15	318.15	323.15	328.15	298.15	303.15	308.15	313.15
P/MPa	4.74	4.78	4.86	4.91	4.93	4.61	4.67	4.70	4.74

c_{0B}/g.ml^{-3}	0.00107	0.00107	0.00107	0.0032	0.0032	0.0032	0.0032	0.0032	0.0032
T/K	318.15	323.15	328.15	298.15	303.15	308.15	313.15	318.15	323.15
P/MPa	4.82	4.86	4.89	4.59	4.61	4.64	4.70	4.75	4.82

c_{0B}/g.ml^{-3}	0.0032	0.0092	0.0092	0.0092	0.0092	0.0092	0.0092	0.0092
T/K	328.15	298.15	303.15	308.15	313.15	318.15	323.15	328.15
P/MPa	4.84	4.56	4.58	4.61	4.65	4.71	4.78	4.81

Polymer (B): **polystyrene** **1993DIX**
Characterization: M_n/kg.mol^{-1} = 191, M_w/kg.mol^{-1} = 200,
Pressure Chemical Company, Pittsburgh, PA
Solvent (A): **carbon dioxide** **CO$_2$** **124-38-9**
Solvent (C): **toluene** **C$_7$H$_8$** **108-88-3**

Type of data: cloud points

T/K = 395.15 P/bar = 60.6

w_A	0.4000	0.3500	0.3400
w_B	0.0060	0.0379	0.0845
w_C	0.5940	0.6120	0.5750

Polymer (B): **polystyrene** **1992KIA, 1994KIA**
Characterization: M_w/kg.mol^{-1} = 235,
Scientific Polymer Products, Inc., Ontario, NY
Solvent (A): **carbon dioxide** **CO$_2$** **124-38-9**
Solvent (C): **toluene** **C$_7$H$_8$** **108-88-3**

Type of data: cloud points

continued

continued

w_A	0.146	0.146	0.146	0.146	0.146	0.250	0.250	0.250	0.250
w_B	0.2844	0.2844	0.2844	0.2844	0.2844	0.250	0.250	0.250	0.250
w_C	0.5696	0.5696	0.5696	0.5696	0.5696	0.500	0.500	0.500	0.500
T/K	307.55	313.15	323.15	333.15	343.15	313.15	320.15	333.15	343.15
P/bar	30.54	32.94	38.38	43.11	48.73	44.95	48.53	58.93	67.01
	VLE	VLE	VLE	VLE	VLE	LLE	LLE	LLE	LLE
w_A	0.150	0.150	0.150	0.150	0.150	0.150	0.150	0.150	0.150
w_B	0.1275	0.1275	0.1275	0.1275	0.1275	0.1275	0.1275	0.1275	0.1275
w_C	0.7225	0.7225	0.7225	0.7225	0.7225	0.7225	0.7225	0.7225	0.7225
T/K	299.25	303.15	308.15	313.15	318.15	323.15	333.15	338.15	342.85
P/bar	30.23	31.52	33.75	36.21	38.91	42.02	46.75	49.39	53.32
	VLE	VLE	VLE	VLE	VLE	VLE	VLE	VLE	VLE
w_A	0.200	0.200	0.200	0.200	0.200	0.200	0.200	0.200	0.200
w_B	0.120	0.120	0.120	0.120	0.120	0.120	0.120	0.120	0.120
w_C	0.680	0.680	0.680	0.680	0.680	0.680	0.680	0.680	0.680
T/K	299.65	303.15	309.15	313.15	318.15	323.15	330.65	333.15	338.15
P/bar	34.96	36.34	39.56	41.95	45.77	48.66	53.68	55.09	59.04
	VLE	VLE	VLE	VLE	VLE	VLE	VLE	VLE	VLE
w_A	0.200	0.250	0.250	0.250	0.250	0.250	0.250	0.250	0.250
w_B	0.120	0.1125	0.1125	0.1125	0.1125	0.1125	0.1125	0.1125	0.1125
w_C	0.680	0.6375	0.6375	0.6375	0.6375	0.6375	0.6375	0.6375	0.6375
T/K	342.65	300.75	303.15	308.25	313.45	319.25	323.15	328.15	334.65
P/bar	62.41	40.74	41.56	44.78	47.90	53.19	55.85	60.44	65.27
	VLE	VLE	VLE	VLE	VLE	VLE	VLE	VLE	VLE
w_A	0.250	0.250	0.276	0.276	0.276	0.276	0.276	0.276	0.276
w_B	0.1125	0.1125	0.1086	0.1086	0.1086	0.1086	0.1086	0.1086	0.1086
w_C	0.6375	0.6375	0.6154	0.6154	0.6154	0.6154	0.6154	0.6154	0.6154
T/K	338.15	343.15	301.85	304.15	305.15	306.65	307.45	308.15	309.15
P/bar	68.13	73.67	110.58	112.35	114.22	115.24	116.09	117.21	118.06
	VLE	VLE	LLE	LLE	LLE	LLE	LLE	LLE	LLE
w_A	0.276	0.276	0.276	0.200	0.200	0.200	0.200	0.200	0.200
w_B	0.1086	0.1086	0.1086	0.160	0.160	0.160	0.160	0.160	0.160
w_C	0.6154	0.6154	0.6154	0.640	0.640	0.640	0.640	0.640	0.640
T/K	310.15	311.15	312.15	308.15	313.15	318.15	323.15	328.15	333.15
P/bar	119.90	120.79	121.49	40.38	42.88	45.96	49.38	52.63	55.98
	LLE	LLE	LLE	VLE	VLE	VLE	VLE	VLE	VLE
w_A	0.200	0.200	0.250	0.250	0.250	0.250	0.250	0.250	0.250
w_B	0.160	0.160	0.150	0.150	0.150	0.150	0.150	0.150	0.150
w_C	0.640	0.640	0.600	0.600	0.600	0.600	0.600	0.600	0.600
T/K	338.15	342.65	305.15	308.15	313.15	318.15	323.15	328.15	333.15
P/bar	59.59	63.53	43.47	44.55	48.23	51.72	55.79	59.56	64.22
	VLE	VLE	VLE	VLE	VLE	VLE	VLE	VLE	VLE

continued

continued

w_A	0.250	0.250	0.259	0.259	0.259	0.259	0.259	0.259	0.259
w_B	0.150	0.150	0.1482	0.1482	0.1482	0.1482	0.1482	0.1482	0.1482
w_C	0.600	0.600	0.5928	0.5928	0.5928	0.5928	0.5928	0.5928	0.5928
T/K	338.15	342.95	301.25	303.25	308.15	313.15	318.15	323.15	328.35
P/bar	68.82	72.79	41.99	42.95	46.10	49.41	54.01	61.30	73.64
	VLE	VLE	VLE	VLE	VLE	VLE	VLE	VLE	LLE

w_A	0.259	0.259	0.259	0.259	0.259	0.273	0.273	0.273	0.273
w_B	0.1482	0.1482	0.1482	0.1482	0.1482	0.1454	0.1454	0.1454	0.1454
w_C	0.5928	0.5928	0.5928	0.5928	0.5928	0.5816	0.5816	0.5816	0.5816
T/K	333.15	335.55	336.95	338.75	342.95	305.85	308.15	313.15	315.15
P/bar	84.45	90.19	93.51	98.07	109.99	45.51	47.18	50.89	52.30
	LLE	LLE	LLE	LLE	LLE	VLE	VLE	VLE	VLE

w_A	0.273	0.273	0.273	0.273	0.273	0.273	0.273	0.273	0.273
w_B	0.1454	0.1454	0.1454	0.1454	0.1454	0.1454	0.1454	0.1454	0.1454
w_C	0.5816	0.5816	0.5816	0.5816	0.5816	0.5816	0.5816	0.5816	0.5816
T/K	318.15	319.55	321.25	323.05	326.75	329.45	332.95	338.95	340.15
P/bar	54.97	55.94	57.88	61.00	68.07	74.17	82.64	96.33	99.19
	VLE	VLE	VLE	VLE	LLE	LLE	LLE	LLE	LLE

w_A	0.273	0.273	0.273	0.281	0.281	0.281	0.281	0.281	0.281
w_B	0.1454	0.1454	0.1454	0.1438	0.1438	0.1438	0.1438	0.1438	0.1438
w_C	0.5816	0.5816	0.5816	0.5752	0.5752	0.5752	0.5752	0.5752	0.5752
T/K	340.65	341.75	343.55	304.45	306.15	308.45	310.15	311.65	313.25
P/bar	99.86	102.99	106.64	98.17	100.86	103.03	106.70	109.46	111.21
	LLE	LLE	LLE	LLE	LLE	LLE	LLE	LLE	LLE

w_A	0.281	0.281	0.281	0.281	0.281	0.175	0.175	0.175	0.175
w_B	0.1438	0.1438	0.1438	0.1438	0.1438	0.2005	0.2005	0.2005	0.2005
w_C	0.5752	0.5752	0.5752	0.5752	0.5752	0.6245	0.6245	0.6245	0.6245
T/K	314.65	316.65	318.15	320.15	321.65	328.65	332.95	337.95	343.15
P/bar	112.42	115.64	119.02	121.94	125.13	54.96	57.36	61.07	64.09
	LLE	LLE	LLE	LLE	LLE	VLE	VLE	VLE	VLE

w_A	0.200	0.200	0.200	0.200	0.200	0.200	0.200	0.200	0.225
w_B	0.1944	0.1944	0.1944	0.1944	0.1944	0.1944	0.1944	0.1944	0.1883
w_C	0.6056	0.6056	0.6056	0.6056	0.6056	0.6056	0.6056	0.6056	0.5867
T/K	307.35	313.15	318.15	323.15	328.15	333.75	338.65	343.15	325.15
P/bar	44.09	47.97	49.61	54.70	57.32	60.96	64.09	67.93	60.64
	VLE	VLE	VLE	VLE	VLE	VLE	VLE	VLE	VLE

w_A	0.225	0.225	0.225	0.225	0.243	0.243	0.243	0.243	0.243
w_B	0.1883	0.1883	0.1883	0.1883	0.184	0.184	0.184	0.184	0.184
w_C	0.5867	0.5867	0.5867	0.5867	0.573	0.573	0.573	0.573	0.573
T/K	328.45	333.35	336.45	343.15	309.15	315.35	317.15	319.45	323.05
P/bar	63.60	68.43	70.69	77.12	51.84	55.45	56.34	58.50	62.18
	VLE	VLE	VLE	VLE	VLE	VLE	VLE	VLE	VLE

continued

continued

w_A	0.243	0.243	0.243	0.243	0.243	0.243	0.243	0.150	0.150
w_B	0.184	0.184	0.184	0.184	0.184	0.184	0.184	0.2125	0.2125
w_C	0.573	0.573	0.573	0.573	0.573	0.573	0.573	0.6375	0.6375
T/K	326.45	329.65	331.75	335.15	337.75	340.15	343.05	303.85	308.25
P/bar	64.48	67.53	71.08	78.89	84.44	90.61	95.43	31.35	32.14
	LLE	LLE	LLE	LLE	LLE	LLE	LLE	VLE	VLE

w_A	0.150	0.150	0.150	0.150	0.150	0.150	0.150	0.200	0.200
w_B	0.2125	0.2125	0.2125	0.2125	0.2125	0.2125	0.2125	0.200	0.200
w_C	0.6375	0.6375	0.6375	0.6375	0.6375	0.6375	0.6375	0.600	0.600
T/K	313.15	318.15	323.15	328.95	333.15	338.15	342.15	307.85	313.25
P/bar	34.44	36.71	39.56	42.06	43.86	46.09	49.01	41.20	43.20
	VLE	VLE	VLE	VLE	VLE	VLE	VLE	VLE	VLE

w_A	0.200	0.200	0.200	0.200	0.200	0.200	0.250	0.250	0.250
w_B	0.200	0.200	0.200	0.200	0.200	0.200	0.1875	0.1875	0.1875
w_C	0.600	0.600	0.600	0.600	0.600	0.600	0.5625	0.5625	0.5625
T/K	318.15	323.15	328.15	333.15	338.15	343.05	304.95	308.15	313.15
P/bar	46.03	49.81	52.83	56.08	60.18	64.65	43.60	44.78	48.39
	VLE	VLE	VLE	VLE	VLE	VLE	VLE	VLE	VLE

w_A	0.250	0.250	0.250	0.250	0.250	0.250	0.261	0.261	0.261
w_B	0.1875	0.1875	0.1875	0.1875	0.1875	0.1875	0.18475	0.18475	0.18475
w_C	0.5625	0.5625	0.5625	0.5625	0.5625	0.5625	0.55425	0.55425	0.55425
T/K	318.25	323.15	328.15	333.15	338.15	342.55	308.15	312.85	318.15
P/bar	51.84	55.94	59.39	64.48	68.78	72.92	48.16	51.84	55.65
	VLE	VLE	VLE	VLE	VLE	VLE	VLE	VLE	VLE

w_A	0.261	0.261	0.261	0.261	0.261	0.261	0.261	0.269	0.269
w_B	0.18475	0.18475	0.18475	0.18475	0.18475	0.18475	0.18475	0.18275	0.18275
w_C	0.55425	0.55425	0.55425	0.55425	0.55425	0.55425	0.55425	0.54825	0.54825
T/K	323.85	328.15	333.15	335.65	338.05	340.05	342.85	305.95	308.35
P/bar	61.20	63.96	68.42	70.69	73.18	77.75	84.51	84.25	100.53
	VLE	VLE	VLE	VLE	VLE	LLE	LLE	LLE	LLE

w_A	0.269	0.269	0.269	0.269	0.269	0.269	0.269	0.269	0.269
w_B	0.18275	0.18275	0.18275	0.18275	0.18275	0.18275	0.18275	0.18275	0.18275
w_C	0.54825	0.54825	0.54825	0.54825	0.54825	0.54825	0.54825	0.54825	0.54825
T/K	309.95	313.35	314.65	315.65	316.65	318.15	319.65	320.65	321.65
P/bar	102.60	107.02	108.70	109.46	111.27	113.60	116.36	117.31	119.31
	LLE	LLE	LLE	LLE	LLE	LLE	LLE	LLE	LLE

w_A	0.269	0.269
w_B	0.18275	0.18275
w_C	0.54825	0.54825
T/K	323.25	324.25
P/bar	121.84	123.91
	LLE	LLE

Polymer (B):	polystyrene		1992KIA
Characterization:	$M_w/\text{kg.mol}^{-1} = 235$,		
	Scientific Polymer Products, Inc., Ontario, NY		
Solvent (A):	carbon dioxide	CO_2	124-38-9
Solvent (C):	trichloromethane	$CHCl_3$	67-66-3

Type of data: cloud points

w_A	0.150	0.150	0.150	0.150	0.150	0.150	0.150	0.150	0.150
w_B	0.085	0.085	0.085	0.085	0.085	0.085	0.085	0.085	0.085
w_C	0.765	0.765	0.765	0.765	0.765	0.765	0.765	0.765	0.765
T/K	299.75	303.15	308.15	313.15	318.15	323.15	328.25	333.15	338.15
P/bar	22.69	23.70	25.58	27.97	29.65	31.58	34.31	36.77	39.27
	VLE	VLE	VLE	VLE	VLE	VLE	VLE	VLE	VLE

w_A	0.150	0.200	0.200	0.200	0.200	0.200	0.200	0.200	0.200
w_B	0.085	0.080	0.080	0.080	0.080	0.080	0.080	0.080	0.080
w_C	0.765	0.720	0.720	0.720	0.720	0.720	0.720	0.720	0.720
T/K	343.05	306.75	308.15	313.15	318.15	323.15	328.15	332.95	338.15
P/bar	42.09	33.69	34.41	36.54	39.07	43.64	46.39	49.77	53.22
	VLE	VLE	VLE	VLE	VLE	VLE	VLE	VLE	VLE

w_A	0.200	0.224	0.224	0.224	0.224	0.224	0.224	0.224	0.224
w_B	0.080	0.0776	0.0776	0.0776	0.0776	0.0776	0.0776	0.0776	0.0776
w_C	0.720	0.6984	0.6984	0.6984	0.6984	0.6984	0.6984	0.6984	0.6984
T/K	342.95	298.65	303.15	308.15	313.15	318.15	323.15	328.15	332.85
P/bar	55.29	31.32	33.72	35.98	39.56	42.22	46.71	49.77	53.16
	VLE	VLE	VLE	VLE	VLE	VLE	VLE	VLE	VLE

w_A	0.224	0.224	0.234	0.234	0.234	0.234	0.234	0.234	0.234
w_B	0.0776	0.0776	0.0766	0.0766	0.0766	0.0766	0.0766	0.0766	0.0766
w_C	0.6984	0.6984	0.6894	0.6894	0.6894	0.6894	0.6894	0.6894	0.6894
T/K	337.25	343.15	300.65	305.35	308.45	310.15	312.05	314.85	317.55
P/bar	59.29	80.93	34.57	37.27	42.88	47.64	53.85	63.17	72.33
	LLE	LLE	VLE	VLE	LLE	LLE	LLE	LLE	LLE

w_A	0.234	0.234	0.234	0.234	0.234	0.234	0.234	0.150	0.150
w_B	0.0766	0.0766	0.0766	0.0766	0.0766	0.0766	0.0766	0.1275	0.1275
w_C	0.6894	0.6894	0.6894	0.6894	0.6894	0.6894	0.6894	0.7225	0.7225
T/K	320.65	322.75	325.85	328.15	330.25	333.15	333.75	302.75	308.15
P/bar	82.41	88.94	98.73	107.26	113.50	123.51	125.48	29.06	30.57
	LLE	LLE	LLE	LLE	LLE	LLE	LLE	VLE	VLE

w_A	0.150	0.150	0.150	0.150	0.150	0.150	0.150	0.200	0.200
w_B	0.1275	0.1275	0.1275	0.1275	0.1275	0.1275	0.1275	0.120	0.120
w_C	0.7225	0.7225	0.7225	0.7225	0.7225	0.7225	0.7225	0.680	0.680
T/K	313.15	318.15	323.15	328.15	333.15	338.45	342.95	308.15	318.15
P/bar	32.71	35.49	38.40	40.38	42.95	46.62	49.38	37.47	43.73
	VLE	VLE	VLE	VLE	VLE	VLE	VLE	VLE	VLE

continued

continued

w_A	0.200	0.200	0.200	0.200	0.200	0.200	0.238	0.238	0.238
w_B	0.120	0.120	0.120	0.120	0.120	0.120	0.1143	0.1143	0.1143
w_C	0.680	0.680	0.680	0.680	0.680	0.680	0.6477	0.6477	0.6477
T/K	323.15	328.15	333.15	338.15	342.35	343.15	303.15	306.25	309.15
P/bar	46.92	49.84	53.09	57.06	59.89	60.67	73.77	81.56	90.09
	VLE	VLE	VLE	VLE	VLE	VLE	LLE	LLE	LLE

w_A	0.238	0.238	0.238	0.238	0.238	0.238	0.238	0.150	0.150
w_B	0.1143	0.1143	0.1143	0.1143	0.1143	0.1143	0.1143	0.170	0.170
w_C	0.6477	0.6477	0.6477	0.6477	0.6477	0.6477	0.6477	0.680	0.680
T/K	310.25	311.25	312.55	314.45	318.15	320.35	322.95	300.85	303.15
P/bar	92.65	95.09	98.04	103.85	114.38	121.32	129.49	27.54	27.77
	LLE	LLE	LLE	LLE	LLE	LLE	LLE	VLE	VLE

w_A	0.150	0.150	0.150	0.150	0.150	0.150	0.150	0.150	0.200
w_B	0.170	0.170	0.170	0.170	0.170	0.170	0.170	0.170	0.160
w_C	0.680	0.680	0.680	0.680	0.680	0.680	0.680	0.680	0.640
T/K	308.15	313.15	318.15	323.15	328.15	334.65	338.15	343.05	303.25
P/bar	29.65	32.34	34.37	37.13	39.59	44.15	45.42	49.57	34.96
	VLE	VLE	VLE	VLE	VLE	VLE	VLE	VLE	VLE

w_A	0.200	0.200	0.200	0.200	0.200	0.200	0.200	0.200	0.200
w_B	0.160	0.160	0.160	0.160	0.160	0.160	0.160	0.160	0.160
w_C	0.640	0.640	0.640	0.640	0.640	0.640	0.640	0.640	0.640
T/K	308.15	313.15	315.85	319.45	322.35	326.05	330.55	333.55	336.35
P/bar	37.23	41.47	49.08	60.29	68.91	81.03	94.35	104.17	113.01
	VLE	VLE	LLE	LLE	LLE	LLE	LLE	LLE	LLE

w_A	0.200	0.200	0.213	0.213	0.213	0.213	0.213	0.213	0.213
w_B	0.160	0.160	0.1574	0.1574	0.1574	0.1574	0.1574	0.1574	0.1574
w_C	0.640	0.640	0.6296	0.6296	0.6296	0.6296	0.6296	0.6296	0.6296
T/K	338.05	339.75	304.25	307.45	311.05	312.65	316.25	320.95	323.05
P/bar	117.78	121.77	91.63	100.30	108.31	112.12	122.24	132.67	135.83
	LLE	LLE	LLE	LLE	LLE	LLE	LLE	LLE	LLE

w_A	0.213
w_B	0.1574
w_C	0.6296
T/K	325.15
P/bar	141.07
	LLE

Polymer (B):	**polystyrene**		**2002HOR**
Characterization:	$M_n/kg.mol^{-1} = 93$, $M_w/kg.mol^{-1} = 101.4$, $M_z/kg.mol^{-1} = 111.9$,		
	BASF AG, Germany		
Solvent (A):	**cyclohexane**	**C$_6$H$_{12}$**	**110-82-7**
Solvent (C):	**ethane**	**C$_2$H$_6$**	**74-84-0**

Type of data: cloud points

continued

continued

w_A	0.846	0.846	0.846	0.846	0.846	0.846	0.846	0.846	0.846
w_B	0.104	0.104	0.104	0.104	0.104	0.104	0.104	0.104	0.104
w_C	0.050	0.050	0.050	0.050	0.050	0.050	0.050	0.050	0.050
T/K	453.68	456.17	458.64	463.63	468.63	473.61	478.60	483.57	493.59
P/bar	40	44	47	54	61	66	72	79	93

w_A	0.846	0.846	0.826	0.826	0.826	0.826	0.826	0.826	0.826
w_B	0.104	0.104	0.100	0.100	0.100	0.100	0.100	0.100	0.100
w_C	0.050	0.050	0.074	0.074	0.074	0.074	0.074	0.074	0.074
T/K	503.59	513.58	431.34	433.74	433.79	436.27	438.75	438.75	443.65
P/bar	107	121	39	43	43	47	50	51	59

w_A	0.826	0.826	0.826	0.826	0.826	0.826	0.826	0.826	0.826
w_B	0.100	0.100	0.100	0.100	0.100	0.100	0.100	0.100	0.100
w_C	0.074	0.074	0.074	0.074	0.074	0.074	0.074	0.074	0.074
T/K	448.68	453.65	458.67	463.69	468.66	473.67	483.68	493.65	503.64
P/bar	67	74	81	89	96	103	117	131	144

w_A	0.804	0.804	0.804	0.804	0.804	0.804	0.804	0.804	0.804
w_B	0.097	0.097	0.097	0.097	0.097	0.097	0.097	0.097	0.097
w_C	0.099	0.099	0.099	0.099	0.099	0.099	0.099	0.099	0.099
T/K	388.59	391.08	393.61	396.10	398.58	403.56	413.54	423.53	433.55
P/bar	34	38	42	46	48	55	65	76	92

w_A	0.804	0.804
w_B	0.097	0.097
w_C	0.099	0.099
T/K	443.55	453.52
P/bar	106	120

Polymer (B):	**polystyrene**		**2002HOR**
Characterization:	$M_n/kg.mol^{-1} = 93$, $M_w/kg.mol^{-1} = 101.4$, $M_z/kg.mol^{-1} = 111.9$, BASF AG, Germany		
Solvent (A):	**cyclohexane**	C_6H_{12}	**110-82-7**
Solvent (C):	**nitrogen**	N_2	**7727-37-9**

Type of data: cloud points

w_A	0.866	0.866	0.866	0.866	0.866	0.866	0.866	0.866	0.866
w_B	0.108	0.108	0.108	0.108	0.108	0.108	0.108	0.108	0.108
w_C	0.026	0.026	0.026	0.026	0.026	0.026	0.026	0.026	0.026
T/K	478.71	478.77	481.25	483.68	483.74	486.23	488.66	493.65	498.58
P/bar	75	75	78	82	82	85	88	95	101

w_A	0.866	0.866	0.866	0.857	0.857	0.857	0.857	0.857	0.857
w_B	0.108	0.108	0.108	0.096	0.096	0.096	0.096	0.096	0.096
w_C	0.026	0.026	0.026	0.047	0.047	0.047	0.047	0.047	0.047
T/K	502.27	508.62	513.64	456.17	458.67	458.69	461.19	461.22	463.69
P/bar	108	114	121	115	118	118	122	121	125

continued

continued

w_A	0.857	0.857	0.857	0.857	0.857	0.857	0.857	0.838	0.838
w_B	0.096	0.096	0.096	0.096	0.096	0.096	0.096	0.101	0.101
w_C	0.047	0.047	0.047	0.047	0.047	0.047	0.047	0.060	0.060
T/K	463.71	466.21	468.74	471.19	473.72	476.20	478.74	437.40	438.42
P/bar	125	127	131	134	137	140	143	153	154

w_A	0.838	0.838	0.838	0.838	0.838
w_B	0.101	0.101	0.101	0.101	0.101
w_C	0.060	0.060	0.060	0.060	0.060
T/K	439.41	440.35	441.46	442.42	443.37
P/bar	155.5	156.5	158	159	160

Polymer (B):	polystyrene		2002HOR
Characterization:	M_n/kg.mol^{-1} = 93, M_w/kg.mol^{-1} = 101.4, M_z/kg.mol^{-1} = 111.9, BASF AG, Germany		
Solvent (A):	**cyclohexane**	**C$_6$H$_{12}$**	**110-82-7**
Solvent (C):	**propane**	**C$_3$H$_8$**	**74-98-6**

Type of data: cloud points

w_A	0.839	0.839	0.839	0.839	0.839	0.839	0.839	0.839	0.839
w_B	0.108	0.108	0.108	0.108	0.108	0.108	0.108	0.108	0.108
w_C	0.053	0.053	0.053	0.053	0.053	0.053	0.053	0.053	0.053
T/K	471.05	473.61	476.07	478.58	483.57	488.58	493.56	503.56	513.53
P/bar	25	29	33	37	44	52	59	73	87

w_A	0.828	0.828	0.828	0.828	0.828	0.828	0.828	0.828	0.828
w_B	0.098	0.098	0.098	0.098	0.098	0.098	0.098	0.098	0.098
w_C	0.074	0.074	0.074	0.074	0.074	0.074	0.074	0.074	0.074
T/K	456.20	458.69	458.86	461.19	463.69	463.85	468.80	473.81	478.77
P/bar	23	27	27	31	35	35	43	50	58

w_A	0.828	0.828	0.828	0.828	0.828	0.828	0.799	0.799	0.799
w_B	0.098	0.098	0.098	0.098	0.098	0.098	0.099	0.099	0.099
w_C	0.074	0.074	0.074	0.074	0.074	0.074	0.102	0.102	0.102
T/K	483.76	488.74	493.76	498.64	503.67	508.70	438.58	441.09	443.57
P/bar	65	72	79	86	92	99	23	27	32

w_A	0.799	0.799	0.799	0.799	0.799	0.799	0.799	0.799	0.799
w_B	0.099	0.099	0.099	0.099	0.099	0.099	0.099	0.099	0.099
w_C	0.102	0.102	0.102	0.102	0.102	0.102	0.102	0.102	0.102
T/K	443.65	448.62	453.63	458.59	463.61	468.58	473.61	478.63	483.57
P/bar	30	37	45	53	61	68	75	83	90

w_A	0.799	0.799	0.799	0.799	0.799
w_B	0.099	0.099	0.099	0.099	0.099
w_C	0.102	0.102	0.102	0.102	0.102
T/K	488.55	493.59	498.56	503.56	508.57
P/bar	97	104	111	117	124

Polymer (B): **polystyrene** **1985MC2**
Characterization: $M_n/\text{kg.mol}^{-1} = 211$, $M_w/\text{kg.mol}^{-1} = 239$,
Dow Chemical Company
Solvent (A): **ethane** C_2H_6 **74-84-0**
Solvent (C): **toluene** C_7H_8 **108-88-3**

Type of data: coexistence data

$T/\text{K} = 311.1$

Total feed concentrations of the homogeneous system before demixing:

$w_A = 0.296$, $w_B = 0.050$, $w_C = 0.654$.

$P/$ MPa	w_A	w_B gel phase	w_C	w_A	w_B sol phase	w_C
9.1	0.258	0.003	0.739	0.199	0.158	0.643

Polymer (B): **polysulfone** **2003ZH1**
Characterization: $M_n/\text{kg.mol}^{-1} = 36.1$, $M_w/\text{kg.mol}^{-1} = 66.5$, by GPC using PS
standards, Polymer Products, Inc., Ontario, NY
Solvent (A): **carbon dioxide** CO_2 **124-38-9**
Solvent (C): **tetrahydrofuran** C_4H_8O **109-99-9**

Type of data: cloud points

w_A	0.8865	0.8865	0.8865	0.8865	0.8865	0.8865	0.8822	0.8822	0.8822
w_B	0.0150	0.0150	0.0150	0.0150	0.0150	0.0150	0.0200	0.0200	0.0200
w_C	0.0985	0.0985	0.0985	0.0985	0.0985	0.0985	0.0978	0.0978	0.0978
T/K	299.2	323.3	348.5	373.3	398.4	424.5	297.1	323.5	347.2
P/MPa	23.93	29.34	34.60	39.21	44.02	47.95	13.89	20.64	26.23
w_A	0.8822	0.8822	0.8822	0.8780	0.8780	0.8780	0.8780	0.8780	0.8780
w_B	0.0200	0.0200	0.0200	0.0246	0.0246	0.0246	0.0246	0.0246	0.0246
w_C	0.0978	0.0978	0.0978	0.0974	0.0974	0.0974	0.0974	0.0974	0.0974
T/K	372.7	399.4	425.4	301.2	323.2	345.6	372.4	399.2	425.0
P/MPa	32.25	37.93	42.66	26.39	30.78	35.73	40.57	45.07	49.06
w_A	0.8732	0.8732	0.8732	0.8732	0.8732	0.8732	0.8710	0.8710	0.8710
w_B	0.0301	0.0301	0.0301	0.0301	0.0301	0.0301	0.0321	0.0321	0.0321
w_C	0.0967	0.0967	0.0967	0.0967	0.0967	0.0967	0.0969	0.0969	0.0969
T/K	303.1	320.0	347.8	372.1	397.0	424.3	296.4	324.8	349.4
P/MPa	38.63	38.64	43.79	47.52	50.90	53.85	36.29	40.41	43.71
w_A	0.8710	0.8710	0.8710	0.8668	0.8668	0.8668	0.8668	0.8668	0.8668
w_B	0.0321	0.0321	0.0321	0.0366	0.0366	0.0366	0.0366	0.0366	0.0366
w_C	0.0969	0.0969	0.0969	0.0966	0.0966	0.0966	0.0966	0.0966	0.0966
T/K	375.5	400.0	426.3	298.6	323.6	348.6	373.8	398.8	423.7
P/MPa	47.35	50.68	53.63	13.79	20.94	27.45	33.14	38.42	42.95

continued

continued

w_A	0.8608	0.8608	0.8608	0.8608	0.8608	0.8608	0.8550	0.8550	0.8550
w_B	0.0444	0.0444	0.0444	0.0444	0.0444	0.0444	0.0496	0.0496	0.0496
w_C	0.0948	0.0948	0.0948	0.0948	0.0948	0.0948	0.0954	0.0954	0.0954
T/K	308.3	321.7	346.9	372.2	398.6	426.0	299.2	323.2	348.9
P/MPa	22.97	26.31	31.99	37.18	42.43	46.84	40.37	42.96	46.06

w_A	0.8550	0.8550	0.8550
w_B	0.0496	0.0496	0.0496
w_C	0.0954	0.0954	0.0954
T/K	374.8	400.8	425.3
P/MPa	49.24	52.00	54.91

Comments: Densities are given in Chapter 6.

Polymer (B):	**poly(vinyl ethyl ether)**		**1992KIA, 1994KIA**
Characterization:	M_w/kg.mol^{-1} = 3.8, Scientific Polymer Products, Ontario, NY		
Solvent (A):	**carbon dioxide**	**CO$_2$**	**124-38-9**
Solvent (C):	**toluene**	**C$_7$H$_8$**	**108-88-3**

Type of data: cloud points

w_A	0.150	0.150	0.150	0.150	0.150	0.150	0.150	0.150	0.150
w_B	0.680	0.680	0.680	0.680	0.680	0.680	0.680	0.680	0.680
w_C	0.170	0.170	0.170	0.170	0.170	0.170	0.170	0.170	0.170
T/K	300.85	304.15	308.15	313.15	318.15	323.15	327.25	333.15	338.15
P/bar	36.44	37.95	40.15	43.80	47.73	51.74	57.12	61.56	64.54

w_A	0.150	0.200	0.200	0.200	0.200	0.200	0.200	0.200	0.250
w_B	0.680	0.640	0.640	0.640	0.640	0.640	0.640	0.640	0.600
w_C	0.170	0.160	0.160	0.160	0.160	0.160	0.160	0.160	0.150
T/K	343.15	313.15	318.65	323.15	328.15	333.15	338.15	342.75	299.95
P/bar	68.29	52.37	56.36	60.27	66.06	71.24	77.23	82.76	48.94

w_A	0.250	0.250	0.250	0.250	0.250	0.250	0.250	0.250	0.250
w_B	0.600	0.600	0.600	0.600	0.600	0.600	0.600	0.600	0.600
w_C	0.150	0.150	0.150	0.150	0.150	0.150	0.150	0.150	0.150
T/K	303.15	308.15	313.25	318.45	323.15	328.15	333.15	338.15	342.65
P/bar	51.97	57.29	62.80	69.80	76.30	83.13	89.96	98.23	105.12

w_A	0.300	0.300	0.300	0.300	0.300	0.300	0.300	0.300	0.300
w_B	0.560	0.560	0.560	0.560	0.560	0.560	0.560	0.560	0.560
w_C	0.140	0.140	0.140	0.140	0.140	0.140	0.140	0.140	0.140
T/K	299.15	303.15	308.25	313.15	318.15	323.15	328.15	333.15	338.15
P/bar	52.40	56.37	62.51	69.37	76.43	84.14	92.71	101.22	112.52

w_A	0.300	0.350	0.350	0.350	0.350	0.350	0.350	0.350	0.350
w_B	0.560	0.520	0.520	0.520	0.520	0.520	0.520	0.520	0.520
w_C	0.140	0.130	0.130	0.130	0.130	0.130	0.130	0.130	0.130
T/K	342.95	302.65	308.65	316.85	318.15	324.35	328.15	332.15	333.65
P/bar	122.06	58.01	66.64	78.60	81.32	95.84	104.43	116.62	120.69

continued

continued

w_A	0.350	0.350	0.350	0.400	0.150	0.150	0.150	0.150	0.150
w_B	0.520	0.520	0.520	0.480	0.765	0.765	0.765	0.765	0.765
w_C	0.130	0.130	0.130	0.120	0.085	0.085	0.085	0.085	0.085
T/K	335.65	338.15	340.15	300.35	307.45	313.15	318.05	323.55	327.95
P/bar	124.60	132.08	136.91	57.65	36.64	41.20	44.32	49.74	51.39

w_A	0.150	0.150	0.150	0.200	0.200	0.200	0.200	0.200	0.200
w_B	0.765	0.765	0.765	0.720	0.720	0.720	0.720	0.720	0.720
w_C	0.085	0.085	0.085	0.080	0.080	0.080	0.080	0.080	0.080
T/K	331.55	338.15	343.25	301.45	303.35	308.15	313.15	319.45	323.25
P/bar	54.23	60.64	64.05	41.33	41.86	44.78	49.74	55.72	58.93

w_A	0.200	0.200	0.200	0.200	0.250	0.250	0.250	0.250	0.250
w_B	0.720	0.720	0.720	0.720	0.675	0.675	0.675	0.675	0.675
w_C	0.080	0.080	0.080	0.080	0.075	0.075	0.075	0.075	0.075
T/K	328.15	333.15	338.15	341.65	299.85	303.25	308.15	312.85	318.35
P/bar	64.16	70.49	76.63	79.91	47.41	50.13	55.72	60.90	67.93

w_A	0.250	0.250	0.250	0.250	0.250	0.300	0.300	0.300	0.300
w_B	0.675	0.675	0.675	0.675	0.675	0.630	0.630	0.630	0.630
w_C	0.075	0.075	0.075	0.075	0.075	0.070	0.070	0.070	0.070
T/K	323.15	331.85	333.25	337.15	343.05	300.65	303.25	308.35	314.95
P/bar	74.04	88.71	90.65	96.95	108.12	53.45	55.61	63.14	73.51

w_A	0.300	0.300	0.300	0.300	0.300	0.350	0.350	0.350	0.350
w_B	0.630	0.630	0.630	0.630	0.630	0.585	0.585	0.585	0.585
w_C	0.070	0.070	0.070	0.070	0.070	0.065	0.065	0.065	0.065
T/K	319.45	323.15	328.15	333.15	338.15	299.45	303.55	308.55	313.15
P/bar	81.32	89.66	101.02	113.34	124.93	54.47	60.25	68.02	80.83

w_A	0.350	0.350	0.350	0.350
w_B	0.585	0.585	0.585	0.585
w_C	0.065	0.065	0.065	0.065
T/K	318.15	323.15	328.15	331.85
P/bar	95.71	110.90	125.35	137.04

4.4. Table of ternary or quaternary systems where data were published only in graphical form as phase diagrams or related figures

Polymer (B)	Second/third/fourth component	Ref.
Cellulose acetate		
	carbon dioxide and 2-propanone	1998KI1
Cellulose acetate butyrate		
	carbon dioxide and ethanol	1998KI1
	carbon dioxide and 2-propanone	1998KI1
Cellulose propionate		
	carbon dioxide and ethanol	1998KI1
	carbon dioxide and 2-propanone	1998KI1
Cellulose triacetate		
	carbon dioxide and 2-propanone	1998KI1
Ethylcellulose		
	carbon dioxide and ethanol	1998KI1
	carbon dioxide and ethanol	2004LI2
	carbon dioxide and methanol	2004LI2
	carbon dioxide and 2-propanone	1998KI1
Ethylene/acrylic acid copolymer		
	n-butane and dimethyl ether	1996LE2
	n-butane and ethanol	1996LE1
	n-butane and ethanol	1996LE2
Ethylene/1-butene copolymer		
	1-butene and ethene	1999CHE
Ethylene/ethyl acrylate copolymer		
	1-butene and polyethylene	1998LEE
Ethylene/2-ethylhexyl acrylate copolymer		
	ethene and 2-ethylhexyl acrylate	1996BUB

Polymer (B)	Second/third/fourth component	Ref.
Ethylene/1-hexene copolymer		
	ethene and n-butane	2000KIN
	ethene and carbon dioxide	2000KIN
	ethene and 1-hexene	2000KIN
	ethene and nitrogen	2000KIN
	2-methylpropane and 1-hexene	1999PAN
Ethylene/methacrylic acid copolymer		
	n-butane and dimethyl ether	1997LEE
	n-butane and ethanol	1997LEE
Ethylene/methyl acrylate copolymer		
	n-butane and ethylene/methyl acrylate copolymer	1998LEE
	n-butane and ethylene/vinyl alcohol copolymer	1998LEE
	1-butene and ethylene/methyl acrylate copolymer	1998LEE
	chlorodifluoromethane and ethanol	1992MEI
	chlorodifluoromethane and ethanol	1993HA1
	chlorodifluoromethane and 2-propanone	1992MEI
	chlorodifluoromethane and 2-propanone	1993HA1
	ethene and methyl acrylate	1987LUF
Ethylene/methyl acrylate copolymer		
	propane and 1-butanol	1994LOS
	propane and ethanol	1992MEI
	propane and ethanol	1993HA1
	propane and ethanol	1993HA2
	propane and ethanol	1994LOS
	propane and n-hexane	1994LOS
	propane and 1-hexene	1994LOS
	propane and methanol	1994LOS
	propane and 1-propanol	1994LOS
	propane and 2-propanone	1992MEI
	propane and 2-propanone	1993HA1
	propane and 2-propanone	1993HA2
Ethylene/propylene copolymer		
	carbon dioxide and ethene	2001KEM
	carbon dioxide and C6-fraction solvent	1985MCH
	ethene and C6-fraction solvent	1985MCH
	ethene and 1-butene	1993CHE
	ethene and n-hexane	1985MC1
	ethene and 1-hexene	1993CHE
	methane and C6-fraction solvent	1985MCH
	propene and C6-fraction solvent	1985MCH
	propene and C6-fraction solvent	1986IRA
	propene and 2-methylbutane	1986IRA

Polymer (B)	Second/third/fourth component	Ref.
Ethylene/propylene/diene terpolymer		
	n-hexane and propene	2001VLI
Ethylene/vinyl acetate copolymer		
	ethene and n-butane	2000KIN
	ethene and carbon dioxide	2000KIN
	ethene and nitrogen	2000KIN
	ethene and vinyl acetate	1980RAE
	ethene and vinyl acetate	1983WOH
	ethene and vinyl acetate	1992FIN
	ethene and vinyl acetate	1996FOL
	ethene and vinyl acetate	2000KIN
Ethylene/vinyl alcohol copolymer		
	n-butane and ethylene/methyl acrylate copolymer	1998LEE
Polyamide 8		
	carbon dioxide and dimethylsulfoxide	1993YEO
Polybutadiene		
	carbon dioxide and polyisoprene	2003RAM
	carbon dioxide and tetrahydrofuran	1994KIA
	carbon dioxide and toluene	1994KIA
Poly(butyl acrylate)		
	carbon dioxide and butyl acrylate	1998MC2
Polycarbonate bisphenol-A		
	carbon dioxide and tetrahydrofuran	2001LI3
Polycarbosilane		
	carbon dioxide and toluene	1998KIM
Poly(dimethylsiloxane)		
	carbon dioxide and poly(ethylmethylsiloxane)	2004WAL
Polyethylene		
	n-butane and carbon dioxide	1994XI2
	carbon dioxide and cyclohexane	1993KI1
	carbon dioxide and n-pentane	1993KI1
	carbon dioxide and n-pentane	1993KI2
	carbon dioxide and n-pentane	1994XI1
	carbon dioxide and n-pentane	1997KIR

Polymer (B)	Second/third/fourth component	Ref.
Polyethylene (*continued*)		
	carbon dioxide and n-pentane	2003ZH2
	carbon dioxide and toluene	1993KI1
	ethanol and propane	1993HA1
	ethene and 4-methyl-1-pentene	1992WOH
	n-hexane and nitrogen	1988KLE
	propane and 2-propanone	1993HA2
Poly(ethylene oxide)		
	carbon dioxide and acetonitrile	2003STR
	carbon dioxide and acetonitrile/water	2003STR
	carbon dioxide and ethyl acetate	2003STR
	carbon dioxide and acetonitrile/ethyl acetate	2003STR
Poly(ethylene glycol)		
	carbon dioxide and ethanol	1998MIS
	carbon dioxide and ethanol	1999MIS
Poly(ethylene terephthalate)		
	carbon dioxide and styrene	2001LI1
Poly(2-ethylhexyl acrylate)		
	carbon dioxide and 2-ethylhexyl acrylate	1998MC2
Poly(ethylmethylsiloxane)		
	carbon dioxide and poly(dimethylsiloxane)	2004WAL
Polyisoprene		
	carbon dioxide and polybutadiene	2003RAM
	carbon dioxide and polystyrene	1999WAL
Poly(methyl acrylate)		
	chlorodifluoromethane and ethanol	1993HA1
	chlorodifluoromethane and 2-propanone	1993HA1
Poly(methyl methacrylate)		
	carbon dioxide and tetrahydrofuran	1994KIA
	carbon dioxide and toluene	1994KIA
Poly(perfluoropropylene oxide) acid sorbitol ester		
	carbon dioxide and water	1997SIN

Polymer (B)	Second/third/fourth component	Ref.
Polypropylene		
	carbon dioxide and n-pentane	1998KI2
	carbon dioxide and n-pentane	1999MAR
	propane and 1-butanol	1997WH2
	propane and ethanol	1997WH2
	propane and 1-propanol	1997WH2
Poly(propylene glycol)		
	ethane and tetrachloromethane	2000MAR
	ethane and trichloromethane	2000MAR
Polystyrene		
	benzene and carbon dioxide	1989SAS
	benzene and carbon dioxide	1990SAS
	carbon dioxide and cyclohexane	2001TOR
	carbon dioxide and cyclohexane	2002LID
	carbon dioxide and polyisoprene	1999WAL
	carbon dioxide and poly(vinyl methyl ether)	2000RAO
	carbon dioxide and tetrahydrofuran	1994KIA
	carbon dioxide and styrene	2002WU1
	carbon dioxide and styrene	2002WU2
	carbon dioxide and toluene	1985MC2
	carbon dioxide and toluene	1994KIA
	carbon dioxide and toluene	1998KIM
	carbon dioxide and toluene	2001LI2
	chlorodifluoromethane and toluene	1998TAN
	ethane and propane	1987KUM
	ethane and toluene	1985MC2
	ethane and toluene	1988SEC
Polysulfone		
	carbon dioxide and tetrahydrofuran	2002ZHA
	dimethyl ether and tetrahydrofuran	2004LI1
	dimethyl ether and N,N-dimethylformamide	2004LI1
Poly(vinyl acetate)		
	benzene and carbon dioxide	1988MAS
	benzene and carbon dioxide	1989SAS
	benzene and carbon dioxide	1990SAS
Poly(vinyl ethyl ether)		
	carbon dioxide and toluene	1994KIA

Polymer (B)	Second/third/fourth component	Ref.
Poly(vinyl methyl ether)		
	carbon dioxide and polystyrene	2000RAO
Tetrafluoroethylene/hexafluoropropylene copolymer		
	carbon dioxide and sulfur hexafluoride	1999MER
	carbon dioxide and trifluoromethane	1999MER
	trifluoromethane and sulfur hexafluoride	1999MER

4.5. References

1963EHR   Ehrlich, P. and Kurpen, J.J., Phase equilibria of polymer-solvent systems at high pressure near their critical loci. Polyethylene with n-alkanes, *J. Polym. Sci.: Part A*, 1, 3217, 1963.

1965CE1   Cernia, E.M. and Mancini, C., A thermodynamic approach to phase equilibria investigation of polyethylene-ethylene system at high pressure, *Kobunshi Kagaku*, 22, 797, 1965.

1965CE2   Cernia, E.M. and Mancini, C., A new method for determination of the solubility parameters of the polyethylene-ethylene pair at high temperature and pressure, *Polym. Lett.*, 3, 1093, 1965.

1965EHR   Ehrlich, P., Phase equilibria of polymer-solvent systems at high pressures near their critical loci. II. Polyethylene-ethylene, *J. Polym. Sci.: Part A*, 3, 131, 1965.

1965SWE   Swelheim, T., De Swaan Arons, J., and Diepen, G.A.M., Fluid phase equilibria in the system polyethylene-ethene, *Recueil*, 84, 261, 1965.

1966KON   Koningsveld, R., Diepen, G.A.M., and Chermin, H.A.G., Fluid phase equilibria in the system polyethylene-ethylene II, *Recueil*, 85, 504, 1966.

1972STE   Steiner, R. and Horle, K., Phasenverhalten von Ethylen/Polyethylen-Gemischen unter hohem Druck, *Chemie-Ing. Techn.*, 44, 1010, 1972.

1975LIN   Lindner, A., Untersuchungen von Phasengleichgewichten bei Monomer-Polymer-Systemen, *Dissertation*, TH Darmstadt, 1975.

1976LUF   Luft, G. and Lindner, A., Zum Einfluss des Polymermolekulargewichts auf das Phasenverhalten von Gas-Polymer-Systemen unter Hochdruck, *Angew. Makromol. Chemie*, 56, 99, 1976.

1976SC1   Schröder, E. and Arndt, K.-F., Löslichkeitsverhalten von Makromolekülen in komprimierten Gasen I. Einführung und Messtechnik, *Faserforsch. Textiltechn.*, 27, 135, 1976.

1976SC2   Schröder, E. and Arndt, K.-F., Löslichkeitsverhalten von Makromolekülen in komprimierten Gasen II. Experimentelle Ergebnisse. Über das Druckverhalten von PVC, PMMA und PS in fluidem CO_2, *Faserforsch. Textiltechn.*, 27, 141, 1976.

1976STE   Steiner, R, Entmischungsvorgänge bei Hochdruckreaktionen, *Chem.-Ing. Techn.*, 48, 533, 1976.

1977LIN   Lindner, A. and Luft, G., The influence of the polymer mean molecular weight on the phase behavior of gas-polymer mixtures under high pressure, *High Temp.-High Press.*, 9, 563, 1977.

1979KOB   Kobyakov, V.M., Kogan, V.B., Rätzsch, M., and Zernov, V.S., Fazovye ravnovesiya v smesyakh etilena s nizkomolekulyarnym polietilenom, *Plast. Massy*, 8, 24, 1979.

1980RAE   Rätzsch, M., Findeisen, R., and Sernov, V.S., Untersuchungen zum Phasenverhalten von Monomer-Polymer-Systemen unter hohem Druck (experimental data by R. Findeisen), *Z. Phys. Chemie, Leipzig*, 261, 995, 1980.

1981LOO   Loos, Th.W. de, Evenwichten tussen fluide fasen in systemen van lineair polyetheen en etheen, *Proefschrift*, TH Delft, 1981.

1981SP1   Spahl, R., Entmischungsverhalten von Ethylen und niedermolekularem Polyethylen unter hohem Druck, *Dissertation*, TH Darmstadt, 1981.

1981SP2   Spahl, R. and Luft, G., Entmischungsverhalten von Ethylen und niedermolekularem Polyethylen, *Ber. Bunsenges. Phys. Chem.*, 85, 379, 1981.

1981WO1   Wohlfarth, C. and Rätzsch, M.T., Berechnungen zum Hochdruckphasengleichgewicht in Mischungen aus Ethylen und Polyethylen I, *Acta Polym.*, 32, 733, 1981.

1981WO2   Wohlfarth, C., Rätzsch, M.T., and Weber, K., Berechnungen zum Hochdruckphasengleichgewicht in Mischungen aus Ethylen und Polyethylen II (experimental data for PE2 by R. Findeisen and P.Wagner), *Acta Polym.*, 32, 740, 1981.

1982RAE   Rätzsch, M.T., Wagner, P., Wohlfarth, C., and Heise, D., High-pressure phase equilibrium studies in mixtures of ethylene and (ethylene-vinyl acetate) copolymers. Part I. Dependence on vinyl acetate content of copolymers (Ger.), *Acta Polym.*, 33, 463, 1982.

1982WOH   Wohlfarth, C., Wagner, P., Rätzsch, M.T., and Westmeier, S., High-pressure phase equilibrium studies in mixtures of ethylene and (ethylene-vinyl acetate) copolymers. Part II. Temperature effect (Ger.), *Acta Polym.*, 33, 468, 1982.

1983ALI   Ali, S., Thermodynamic properties of polymer solutions in compressed gases, *Z. Phys. Chem., N. F.*, 137, 13, 1983.

1983LOO   Loos, Th.W. de, Poot, W., and Diepen, G.A.M., Fluid phase equilibriums in the system polyethylene + ethylene. 1. Systems of linear polyethylene + ethylene at high pressure, *Macromolecules*, 16, 111, 1983.

1983RAE   Rätzsch, M.T., Wagner, P., Wohlfarth, C., and Gleditzsch, S., Studies of phase equilibriums in mixtures of ethylene and ethylene-vinyl acetate copolymers at high pressures. Part III. Dependence on the molecular weight distribution (Ger.), *Acta Polym.*, 34, 340, 1983.

1983SPA   Spahl, R. and Luft, G., Fraktionierungserscheinungen bei der Entmischung von Ethylen-Polyethylen-Gemischen, *Angew. Makromol. Chemie*, 115, 87, 1983.

1983WOH   Wohlfarth, C. and Rätzsch, M.T., Calculation of high-pressure phase equilibria in mixtures of ethylene, vinyl acetate, and ethylene-vinyl acetate copolymers, *Acta Polym.*, 34, 255, 1983.

1984FON   Fonin, M.F., Saltanova, V.B., Anishchuk, V.V., Evdokimova, T.N., Shvyd'ko, T.I., and Gordeev, V.K., Solubility of ethylene and its diffusion in freshly prepared high-pressure polyethylene (Russ.), *Plast.Massy*, 7, 16, 1984.

1984KOB   Kobyakov, V.M. and Zernov, V.S., Solubility of ethylene in polyethylene at high pressures and temperatures (Russ.), in *Sintez, Svoistva, Pererab. Poliolefinov*, 1984, 60.

1984WAG   Wagner, P., Zum Phasengleichgewicht im System Ethylen + Ethylen-Vinylacetat-Copolymer unter hohem Druck, *Dissertation*, TH Leuna-Merseburg, 1984.

1984WOH   Wohlfarth, Ch., Wagner, P., Glindemann, D., Völkner, M., and Rätzsch, M.T., High-pressure phase equilibrium in the system ethylene + vinyl acetate + (ethylene-vinyl acetate) copolymers (Ger.), *Acta Polym.*, 35, 498, 1984.

1985MC1   McClellan, A.K. and McHugh, M.A., Separating polymer solutions using high pressure lower critical solution temperature (LCST) phenomena, *Polym. Eng. Sci.*, 25, 1088, 1985.

1985MC2   McClellan, A.K., Bauman, E.G., and McHugh, M.A., Polymer solution-supercritical fluid phase behavior, in *Supercritical Fluid Technology*, Elsevier Sci. Publ., Amsterdam, 1985, 162.

1985MCH   McHugh, M.A. and Guckes, T.L., Separating polymer solutions with supercritical fluids, *Macromolecules*, 18, 674, 1985.

1986IRA   Irani, C.A. and Cozewith, C., Lower critical solution temperature behavior of ethylene propylene copolymers in multicomponent solvents, *J. Appl. Polym. Sci.*, 31, 1879, 1986.

1986SPA   Spahl, R. and Luft, G., Einfluss von Molmasse und Molekülverzweigungen auf das Entmischungsverhalten von Ethylen/Polyethylen-Systemen unter Hochdruck, *Ber. Bunsenges. Phys. Chem.*, 86, 621, 1986.

1987KUM   Kumar, S.K., Chhabria, S.P., Reid, R.C., and Suter, U.W., Solubility of polystyrene in supercritical fluids, *Macromolecules*, 20, 2550, 1987.

1987LUF   Luft, G. and Subramanian, N.S., Phase behavior of mixtures of ethylene, methyl acrylate, and copolymers under high pressures, *Ind. Eng. Chem. Res.*, 26, 750, 1987.

1988KLE   Kleintjens, L.A., van der Haegen, R., van Opstal, L., and Koningsveld, R., Mean-field lattice-gas modelling of supercritical phase behavior, *J. Supercrit. Fluids*, 1, 23, 1988.

1988MAS   Masuoka, H., Takishima, S., and Wang, N.-H., Supercritical fluid extraction of high-boiling materials from polymers, *Rep. Asahi Glass Found. Ind. Technol.*, 52, 275, 1988.

1988SA1   Saraf, V.P. and Kiran, E., Solubility of polystyrenes in supercritical fluids, *J. Supercrit. Fluids*, 1, 37, 1988.

1988SA2   Saraf, V.P. and Kiran, E., Supercritical fluid-polymer interactions. Phase equilibrium data for solutions of polystyrenes in n-butane and n-pentane, *Polymer*, 29, 2061, 1988.

1988SEC   Seckner, A.J., McClellan, A.K., and McHugh, M.A., High-pressure solution behavior of the polystyrene-toluene-ethane system, *AIChE-J.*, 34, 9, 1988.

1988WAL   Walsh, D.J. and Dee, G.T., Calculations of the phase diagrams of polyethylene dissolved in supercritical solvents, *Polymer*, 29, 656, 1988.

1989DAN   Daneshwar, M. and Gulari, E., Partition coefficients of poly(ethylene glycol) in supercritical carbon dioxide, *ACS Symp. Ser.*, 406, 72, 1989.

1989KIR   Kiran, E., Saraf, V.P., and Sen, Y.L., Solubility of polymers in supercritical fluids, *Int. J. Thermophys.*, 10, 437, 1989.

1989SAS   Sasaki, M., Takishima, S., and Masuoka, H., Supercritical carbon dioxide extraction of benzene in poly(vinyl acetate) and polystyrene, *Sekiyu Gakkaishi*, 32, 67, 1989.

1989WIL   Will, B., Löslichkeit von Ethylen-Vinylacetat-Copolymeren in ethylenhaltigen Mischlösungsmitteln, *Dissertation*, Johannes-Gutenberg Universität Mainz, 1989.

1990BEC   Beckmann, E.J., Koningsveld, R., and Porter, R.S., Mean-field lattice equations of state. 3. Modeling the phase behavior of supercritical gas-polymer mixtures, *Macromolecules*, 23, 2321, 1990.

1990DAN   Daneshvar, M., Kim, S., and Gulari, E., High-pressure phase equilibria of poly(ethylene glycol)-carbon dioxide systems, *J. Phys. Chem.*, 94, 2124, 1990.

1990KEN   Kennis, H.A.J., Loos, Th.W. de, DeSwaan Arons, J., Van der Haegen, R., and Kleintjens, L.A., The influence of nitrogen on the liquid-liquid phase behaviour of the system n-hexane-polyethylene: experimental results and predictions with the mean-field lattice-gas model (experimental data by Th.W. de Loos from M.Sc. Thesis by H.A.J. Kennis, TU Delft 1987), *Chem. Eng. Sci.*, 45, 1875, 1990.

1990SAS   Sasaki, M., Takishima, S., and Masuoka, H., Supercritical carbon dioxide extraction of benzene in poly(vinyl acetate) and polystyrene. Part 2, *Sekiyu Gakkaishi*, 33, 304, 1990.

1991FIN   Finck, U., Phasengleichgewichte in Monomer + Copolymer - Systemen, *Diploma paper*, TH Leuna-Merseburg, 1991.

1991MEI   Meilchen, M.A., Hasch, B.M., and McHugh, M.A., Effect of copolymer composition on the phase behavior of mixtures of poly(ethylene-*co*-methyl acrylate) with propane and chlorodifluoromethane, *Macromolecules*, 24, 4874, 1991.

1991NIE   Nieszporek, B., Untersuchungen zum Phasenverhalten quasibinärer und quasiternärer Mischungen aus Polyethylen, Ethylen-Vinylacetat-Copolymeren, Ethylen und Vinylacetat unter Druck, *Dissertation*, TH Darmstadt, 1991.

1991TOM   Tom, J.W. and Debenedetti, P.G., Formation of bioerodible polymeric microspheres and microparticles by rapid expansion of supercritical solutions, *Biotechnol.Prog.*, 7, 403, 1991.

1991WAT   Watkins, J.J., Krukonis, V.J., Condo, P.D., Pradhan, D., and Ehrlich, P., Fractionation of high density polyethylene in propane by isothermal pressure profiling and isobaric temperature profiling, *J. Supercrit. Fluids*, 4, 24, 1991.

1992CH1   Chen, S.-J. and Radosz, M., Density-tuned polyolefin phase equilibria 1, *Macromolecules*, 25, 3089, 1992.

1992CH2   Chen, S.-J., Economou, I.G., Radosz, M., Density-tuned polyolefin phase equilibria 2, *Macromolecules*, 25, 4987, 1992.

1992CO3   Condo, P.D., Colman, E.J., and Ehrlich, P., Phase equilibria in linear polyethylene with supercritical propane, *Macromolecules*, 25, 750, 1992.

1992FIN   Finck, U., Wohlfarth, Ch., and Heuer, T., Calculation of high pressure phase equilibria of mixtures of ethylene, vinyl acetate and an (ethylene-vinyl acetate) copolymer, *Ber. Bunsenges. Phys. Chem.*, 96, 179, 1992.

1992HAS   Hasch, B.A., Meilchen, M.A., Lee, S.-H., and McHugh, M.A., High-pressure phase behavior of mixtures of poly(ethylene-*co*-methyl acrylate) with low-molecular weight hydrocarbons, *J. Polym. Sci., Part B: Polym. Phys.*, 30, 1365, 1992.

1992KIA   Kiamos, A.A., *M.Sc. Thesis*, High-pressure phase-equilibrium studies of polymer-solvent-supercritical fluid mixtures, John Hopkins University, 1992.

1992LUF   Luft, G. and Wind, R.W., Phasenverhalten von Mischungen aus Ethylen und Ethylen-Acrylsäure-Copolymeren unter hohem Druck, *Chem.-Ing.-Techn.*, 64, 1114, 1992.

1992MEI   Meilchen, M.A., Hasch, B.M., Lee, S.-H., and McHugh, M.A., Poly(ethylene-*co*-methyl acrylate)-solvent-cosolvent phase behaviour at high pressures, *Polymer*, 33, 1922, 1992.

1992WIN   Wind, R.W., Untersuchungen zum Phasenverhalten von Mischungen aus Ethylen, Acrylsäure und Ethylen-Acrylsäure-Copolymer unter hohem Druck, *Dissertation*, TH Darmstadt, 1992.

1992WOH   Wohlfarth, C., Finck, U., Schultz, R., and Heuer, T., Investigation of phase equilibria in mixtures composed of ethene, 1-butene, 4-methyl-1-pentene and a polyethylene wax, *Angew. Makromol. Chem.*, 198, 91, 1992.

1993CHE   Chen, S.-J., Economou, I.G., and Radosz, M., Phase behavior of LCST and UCST solutions of branchy copolymers, *Fluid Phase Equil.*, 83, 391, 1993.

1993DEB   Debenedetti, P.G., Tom, J.W., Yeo, S.-D., and Lim, G.-B., Application of supercritical fluids for the production of sustained delivery devices, *J. Controlled Release*, 24, 27, 1993.

1993DIX   Dixon, D.J. and Johnston, K.P., Formation of microporous polymer fibers and fibrils by precipitation with a compressed fluid antisolvent, *J. Appl. Polym. Sci.*, 50, 1929, 1993.

1993GRE   Gregg, C.J., Chen, S.-J., Stein, F.P., and Radosz, M., Phase behavior of binary ethylene-propylene copolymer solutions in sub- and supercritical ethylene and propylene, *Fluid Phase Equil.*, 83, 375, 1993.

1993HA1   Hasch, B.M., Meilchen, M.A., Lee, S.-H., and McHugh, M.A., Cosolvency effects on copolymer solutions at high pressure, *J. Polym. Sci., Part B: Polym. Phys.*, 31, 429, 1993.

1993HA2   Hasch, B.M., Lee, S.-H., and McHugh, M.A., The effect of copolymer architecture on solution behavior, *Fluid Phase Equil.*, 83, 341, 1993.

1993HA3   Hasch, B.M., Lee, S.-H., McHugh, M.A., Watkins, J.J., and Krukonis, V.J., The effect of backbone structure on the cloud point behaviour of polyethylene-ethane and polyethylene-propane mixtures (experimental data by M.A. McHugh), *Polymer*, 34, 2554, 1993.

1993HA4   Haschets, C.W. and Shine, A.D., Phase behavior of polymer-supercritical chlorodifluoro-methane solutions, *Macromolecules*, 26, 5052, 1993.

1993HOE   Hoefling, T.A., Newman, D.A., Enick, R.M., and Beckman, E.J., Effect of structure on the cloud-point curves of silicone-based amphiphiles in supercritical carbon dioxide, *J. Supercrit. Fluids*, 6, 165, 1993.

1993KI1   Kiran, E., Zhuang, W., and Sen, Y.L., Solubility and demixing of polyethylene in super-critical binary fluid mixtures: Carbon dioxide-cyclohexane, carbon dioxide-toluene, carbon dioxide-pentane, *J. Appl. Polym. Sci.*, 47, 895, 1993.

1993KI2   Kiran, E., Xiong, Y., and Zhunag, W., Modeling polyethylene solutions in near and super-critical fluids using the Sanchez-Lacombe model, *J. Supercrit. Fluids*, 6, 193, 1993.

1993PRA   Pratt, J.A., Lee, S.-H., and McHugh, M.A., Supercritical fluid fractionation of copolymers based on chemical composition and molecular weight, *J. Appl. Polym. Sci.*, 49, 953, 1993.

1993YEO   Yeo, S.-D., Debenedetti, P.G., Radosz, M., and Schmidt, H.-W., Supercritical antisolvent process for substituted para-linked aromatic polyamines: phase equilibrium and morphology study, *Macromolecules*, 26, 6207, 1993.

1994CHE   Chen, C.-K., Duran, M.A., and Radosz, M., Supercritical antisolvent fractionation of polyethylene simulated with multistage algorithm and SAFT equation of state: staging leads to high selectivity enhancements for light fractions, *Ind. Eng. Chem. Res.*, 33, 306, 1994.

1994GR1   Gregg, C.J., Stein, P.S., and Radosz, M., Phase behavior of telechelic polyisobutylene (PIB) in subcritical and supercritical fluids I., *Macromolecules*, 27, 4972, 1994.

1994GR2   Gregg, C.J., Stein, P.S., and Radosz, M., Phase behavior of telechelic polyisobutylene (PIB) in subcritical and supercritical fluids II., *Macromolecules*, 27, 4981, 1994.

1994GR3   Gregg, C.J., Stein, F.P., and Radosz, M., Phase behavior of telechelic polyisobutylene in subcritical and supercritical fluids 3., *J. Phys. Chem.*, 98, 10634, 1994.

1994GR4   Gregg, C.J., Phase equilibria of supercritical fluid solutions of associating polymers in nonpolar and polar fluids, *Ph.D. Thesis*, Lehigh University, Bethlehem, 1994.

1994GR5   Gregg, C.J., Stein, F.P., Morgan, C.K., and Radosz, M., A variable-volume optical pressure-volume-temperature cell for high-pressure cloud points, densities, and IR-spectra, applicable to supercritical fluid solutions of polymers up to 2 kbar, *J. Chem. Eng. Data*, 39, 219, 1994.

1994KIA   Kiamos, A.A. and Donohue, M.D., The effect of supercritical carbon dioxide on polymer-solvent mixtures, *Macromolecules*, 27, 357, 1994.

1994KI2   Kiran, E., Xiong, Y., and Zhuang, W., Effect of polydispersity on the demixing pressures of polyethylene in near- or supercritical alkanes, *J. Supercrit. Fluids*, 7, 283, 1994.

1994LE1   Lee, S.-H., LoStracco, M.A., Hasch, B.M., and McHugh, M.A., Solubility of poly(ethylene-*co*-acrylic acid) in low molecular weight hydrocarbons, *J. Phys. Chem.*, 98, 4055, 1994.

1994LE2   Lee, S.-H., LoStracco, A., and McHugh, M.A., High pressure, molecular weight-dependent behavior of (co)polymer-solvent mixtures, *Macromolecules*, 27, 4652, 1994.

1994LEL   Lele, A.K. and Shine, A.D., Effect of RESS dynamics on polymer morphology, *Ind. Eng. Chem. Res.*, 33, 1476, 1994.

1994LOS   LoStracco, M.A., Lee, S.-H., and McHugh, M.A., Comparison of the effect of density and hydrogen bonding on the cloud point behavior of poly(ethylene-*co*-methyl acrylate)-propane-cosolvent mixtures, *Polymer*, 35, 3272, 1994.

1994MUR   Muralidharan, V., Donohue, M.D., Argyropoulos, J., and Nielsen, K.A., Determination of phase boundaries from pressure-volume data for polymer-solvent-carbon dioxide mixtures, *J. Supercrit. Fluids*, 7, 275, 1994.

1994PRA   Pradhan, D., Chen, C., and Radosz, M., Fractionation of polystyrene with supercritical propane and ethane: characterisation, semibatch solubility experiments, and SAFT simulations, *Ind. Eng. Chem. Res.*, 33, 1984, 1994.

1994SUR   Suresh, S.J., Enick, R.M., and Beckman, E.J., Phase behavior of nylon-6/trifluoroethanol/carbon dioxide mixtures, *Macromolecules*, 27, 348, 1994.

1994XI1   Xiong, Y. and Kiran, E., Prediction of high-pressure phase behaviour in polyethylene/n-pentane/carbon dioxide ternary system with the Sanchez-Lacombe model, *Polymer*, 35, 4408, 1994.

1994XI2   Xiong, Y. and Kiran, E., High pressure phase behavior in polyethylene/n-butane binary and polyethylene/n-butane/CO_2 ternary systems, *J. Appl. Polym. Sci.*, 53, 1179, 1994.

1995CHE   Chen, S., Banaszak, M., and Radosz, M., Phase behavior of poly(ethylene-1-butene) in subcritical and supercritical propane, *Macromolecules*, 28, 1812, 1995.

1995HAS   Hasch, B.M. and McHugh, M.A., Calculating poly(ethylene-*co*-acrylic acid)-solvent phase behavior with the SAFT equation of state, *J. Polym. Sci.: Part B: Polym. Phys.*, 33, 715, 1995.

1995LOO   Loos, Th.W. de, Poot, W., and Lichtenthaler, R.N., The influence of branching on high-pressure vapor-liquid equilibria in systems of ethylene and polyethylene (experimental data by Th.W. de Loos), *J. Supercrit. Fluids*, 8, 282, 1995.

1995MAW   Mawson, S., Johnston, K.P., Combes, J.R., and DeSimone, J.M., Formation of poly(1,1,2,2-tetrahydroperfluorodecyl acrylate) submicron fibers and particles from supercritical carbon dioxide solutions, *Macromolecules*, 28, 3182, 1995.

1995PRA   Pradhan, D. and Ehrlich, P., Morphologies of microporous polyethylene and polypropylene crystallized from solution in supercritical propane, *J. Polym. Sci.: Part B: Polym. Phys.*, 33, 1053, 1995.

1995TUM   Tuminello, W.H., Dee, G.T., and McHugh, M.A., Dissolving perfluoropolymers in supercritical carbon dioxide, *Macromolecules*, 28, 1506, 1995.

1995XI1   Xiong, Y. and Kiran, E., Miscibility, density and viscosity of poly(dimethylsiloxane) in supercritical carbon dioxide, *Polymer*, 36, 4817, 1995.

1995XI2   Xiong, Y. and Kiran, E., Comparison of Sanchez-Lacombe and SAFT model in predicting solubility of polyethylene in high-pressure fluids, *J. Appl. Polym. Sci.*, 55, 1805, 1995.

1996ALB   Albrecht, K.L., Stein, F.P., Han, S.J., Gregg, C.J., and Radosz, M., Phase equilibria of saturated and unsaturated polyisoprene in sub- and supercritical ethane, ethylene, propane, propylene, and dimethyl ether, *Fluid Phase Equil.*, 117, 84, 1996.

1996BUB   Buback, M., Busch, M., Dietzsch, H., Dröge, T., and Lovis, K., Cloud-point curves in ethylene-acrylate-poly(ethylene-*co*-acrylate) systems, in *High Pressure Chemical Engineering*, von Rohr, Ph.R., Trepp, Ch., Eds., Elsevier Sci. B.V., 1996, 175.

1996BYU   Byun, H.-S., Hasch, B.M., McHugh, M.A., Mähling, F.-O., Busch, M., and Buback, M., Poly(ethylene-*co*-butyl acrylate). Phase behavior in ethylene compared to the poly(ethylene-*co*-methyl acrylate)-ethylene system and aspects of copolymerization kinetics at high pressures, *Macromolecules*, 29, 1625, 1996.

1996FOL   Folie, B., Gregg, C., Luft, G., and Radosz, M., Phase equilibria of poly(ethylene-*co*-vinyl acetate) copolymers in subcritical and supercritical ethylene and ethylene-vinyl acetate mixtures, *Fluid Phase Equil.*, 120, 11, 1996.

1996HAR   Harrison, K.L., Johnston, K.P., and Sanchez, I.C., Effect of surfactants on the interfacial tension between supercritical carbon dioxide and polyethylene glycol, *Langmuir*, 12, 2637, 1996.

1996HEU   Heukelbach, D., Einfluss der Polydispersität von Polyethylen auf das Phasenverhalten mit überkritischem Ethen, *Dissertation*, TH Darmstadt, 1996.

1996KIM   Kim, J.-H., Paxton, T.E., and Tomasko, D.L., Microencapsulation of Naproxen using rapid expansion of supercritical solutions, *Biotechnol. Progr.*, 12, 650, 1996.

1996LE1   Lee, S.-H., Hasch, B.M., and McHugh, M.A., Calculating copolymer solution behavior with statistical associating fluid theory, *Fluid Phase Equil.*, 117, 61, 1996.

1996LE2   Lee, S.-H., LoStracco, M.A., and McHugh, M.A., Cosolvent effect on the phase behavior of poly(ethylene-*co*-acrylic acid)-butane mixtures, *Macromolecules*, 29, 1349, 1996.

1996LE3   Lee, S.-H. and McHugh, M.A., Influence of chain architecture on high-pressure co-polymer solution behavior: Experiments and modeling, in *High Pressure Chemical Engineering*, von Rohr, Ph.R., Trepp, Ch., Eds., Elsevier Sci. B.V., 1996, 11.

1996MER   Mertdogan, C.A., Byun, H.-S., McHugh, M.A., and Tuminello, W.H., Solubility of poly (tetrafluoroethylene-*co*-19 mol% hexafluoropropylene) in supercritical CO_2 and halogenated supercritical solvents, *Macromolecules*, 29, 6548, 1996.

1996MU1   Müller, C., Untersuchungen zum Phasenverhalten von quasibinären Gemischen aus Ethylen und Ethylen-Copolymeren, *Dissertation*, Univ. Karlsruhe (TH), 1996.

1996MU2   Müller, C. and Oellrich, L.R., The influence of different inhibitors and ethylenes of different origin on the location of cloud points due to thermal polymerization of ethylene, *Acta Polym.*, 47, 404, 1996.

1996PAR   Parks, K.L. and Beckman, E.J., Generation of microcellular polyurethane foams via polymerization in carbon dioxide. I. Phase behavior of polyurethane precursors, *Polym. Eng. Sci.*, 36, 2404, 1996.

1996RIN   Rindfleisch, F., DiNoia, T.P., and McHugh, M.A., Solubility of polymers and copolymers in supercritical CO_2, *J. Phys. Chem.*, 100, 15581, 1996.

1997BUN   Bungert, B., Sadowski, G., and Arlt, W., Supercritical antisolvent fractionation: measurements in the systems monodisperse and bidisperse polystyrene-cyclohexane-carbon dioxide, *Fluid Phase Equil.*, 139, 349, 1997.

1997DIN   DiNoia, T.P., McHugh, M.A., Cocchiaro, J.E., and Morris, J.B., Solubility and phase behavior of PEP binders in supercritical carbon dioxide, *Waste Managment*, 17, 151, 1997.

1997HAN   Han, S.J., Gregg, C.J., and Radosz, M., How the solute polydispersity affects the cloud-point and coexistence pressures in propylene and ethylene solutions of alternating poly(ethylene-*co*-propylene), *Ind. Eng. Chem. Res.*, 36, 5520, 1997.

1997KIR   Kiran, E. and Zhuang, W., Miscibility and phase separation of polymers in near- and super-critical fluids, *ACS Symp. Ser.*, 670, 2, 1997.

1997LEE   Lee, S.-H. and McHugh, M.A., Phase behaviour studies with poly(ethylene-*co*-methacrylic acid) at high pressures, *Polymer*, 38, 1317, 1997.

1997LEP   Lepilleur, C., Beckman, E.J., Schonemann, H., and Krukonis, V.J., Effect of molecular architecture on the phase behavior of fluoroether-functional graft copolymers in supercritical CO_2, *Fluid Phase Equil.*, 134, 285, 1997.

1997MER   Mertdogan, C.A., DiNoia, T.P., and McHugh, M.A., Impact of backbone architecture on the solubility of fluoropolymers in supercritical CO_2 and halogenated supercritical solvents, *Macromolecules*, 30, 7511, 1997.

1997SIN   Singley, E.J., Liu, W., and Beckman, E.J., Phase behavior and emulsion formation of novel fluoroether amphiphiles in CO_2, *Fluid Phase Equil.*, 128, 199, 1997.

1997WEI   Weidner, E., Wiesmet, V., Knez, Z., and Skerget, M., Phase equilibrium (solid-liquid-gas) in polyethyleneglycol-carbon dioxide systems, *J. Supercrit. Fluids*, 10, 139, 1997.

1997WH1   Whaley, P.D., Winter, H.H., and Ehrlich, P., Phase equilibria of polypropylene with compressed propane and related systems. 1. Isotactic and atactic polypropylene with propane and propylene, *Macromolecules*, 30, 4882, 1997.

1997WH2   Whaley, P.D., Winter, H.H., and Ehrlich, P., Phase equilibria of polypropylene with compressed propane and related systems. 2. Fluid phase equilibria of polypropylene with propane containing alcohols as cosolvents and of some other branched polyolefins with propane, *Macromolecules*, 30, 4887, 1997.

1998CHA   Chang, S.-H., Park, S.-C., and Shim, J.-J., Phase equilibria of supercritical fluid-polymer systems, *J. Supercrit. Fluids*, 13, 113, 1998.

1998DIM   Dimitrov, K., Boyadzhiev, L., Tufeu, R., Cansell, F., and Barth, D., Solubility of poly(ethylene glycol) nonylphenyl ether in supercritical carbon dioxide, *J. Supercrit. Fluids*, 14, 41, 1998.

1998HAN   Han, S.J., Lohse, D.J., Radosz, M., and Sperling, L.H., Short chain branching effect on the cloud-point pressures of ethylene copolymers in subcritical and supercritical propane, *Macromolecules*, 31, 2533, 1998.

1998HEU   Heukelbach, D. and Luft, G., Critical points of mixtures of ethylene and polyethylene wax under high pressure, *Fluid Phase Equil.*, 146, 187, 1998.

1998HOR   Horst, R., Wolf, B.A., Kinzl, M., Luft, G., and Folie, B., Shear influences on the solubility of LDPE in ethene, *J. Supercrit. Fluids*, 14, 49, 1998.

1998KIM   Kim, S., Kim, Y.-S., and Lee, S.-B., Phase behaviors and fractionation of polymer solutions in supercritical carbon dioxide, *J. Supercrit. Fluids*, 13, 99, 1998.

1998KI1   Kiran, E. and Pöhler, H., Alternative solvents for cellulose derivatives. Miscibility and density of cellulosic polymers in carbon dioxide + acetone and carbon dioxide + ethanol binary fluid mixtures, *J. Supercrit. Fluids*, 13, 135, 1998.

1998KI2   Kiran, E. and Xiong, Y., Miscibility of isotactic polypropylene in n-pentane and n-pentane + carbon dioxide mixtures at high pressures, *J. Supercrit. Fluids*, 11, 173, 1998.

1998KOA   Koak, N., Loos, Th.W. de, and Heidemann, R.A., Upper-critical-solution-temperature behavior of the system polystyrene + methylcyclohexane. Influence of CO_2 on the liquid-liquid equilibria, *Fluid Phase Equil.*, 145, 311, 1998.

1998LEE   Lee, S.-H. and McHugh, M.A., Phase behavior of copolymer-copolymer-solvent mixtures at high pressures, *Polymer*, 39, 5447, 1998.

1998LUN   Luna-Barcenas, G., Mawson, S., Takishima, S., DeSimone, J.M., Sanchez, I.C., and Johnston, K.P., Phase behavior of poly(1,1-dihydroperfluorooctyl acrylate) in supercritical carbon dioxide, *Fluid Phase Equil.*, 146, 325, 1998.

1998MC1   McHugh, M.A., Mertdogan, C.A., DiNoia, T.P., Anolick, C., Tuminello, W.H., and Wheland, R., Impact of melting temperature on poly(tetrafluoroethylene-*co*-hexafluoro-propylene) solubility in supercritical fluid solvents, *Macromolecules*, 31, 2252, 1998.

1998MC2   McHugh, M.A., Rindfleisch, F., Kuntz, P.T., Schmaltz, C., and Buback, M., Cosolvent effect of alkyl acrylates on the phase behaviour of poly(alkyl acrylates)-supercritical CO_2 mixtures, *Polymer*, 39, 6045, 1998.

1998MIS   Mishima, K., Tokuyasu, T., Matsuyama, K., Komorita, N., Enjoji, T., and Nagatani, M., Solubility of polymer in the mixtures containing supercritical carbon dioxide and antisolvent, *Fluid Phase Equil.*, 144, 299, 1998.

1998ONE   O'Neill, M.L., Cao, Q., Fang, C.M., Johnston, K.P., Wilkinson, S.P., Smith, C.D., Kersch-
            ner, J.L., and Jureller, S.H., Solubility of homopolymers and copolymers in carbon dioxide,
            Ind. Eng. Chem. Res., 37, 3067, 1998.

1998TAN   Tan, C.-S. and Chang, W.-W., Precipitation of polystyrene from toluene with HFC-134a by
            the GAS process, *Ind. Eng. Chem. Res.*, 37, 1821, 1998.

1998ZHU   Zhuang, W. and Kiran, E., Kinetics of pressure-induced phase separation (PIPS) from
            polymer solutions by time resolved light scattering. Polyethylene + n-pentane, *Polymer*, 39,
            2903, 1998.

1999BUN   Bungert, B., Komplexe Phasengleichgewichte von Polymerlösungen, *Dissertation*, TU
            Berlin, 1999.

1999CHE   Chen, A.-Q. and Radosz, M., Phase equilibria of dilute poly(ethylene-*co*-1-butene)
            solutions in ethylene, 1-butene, and 1-butene+ethylene, *J. Chem. Eng. Data*, 44, 854, 1999.

1999DIE   Dietzsch, H., Hochdruck-Copolymerisation von Ethen und (Meth)Acrylsäureestern,
            Dissertation, University Goettingen, 1999.

1999GOU   Gourgouillon, D. and Nunes da Ponte, M., High pressure phase equilibria for poly(ethylene
            glycol) + CO_2: experimental results and modeling, *Phys. Chem. Chem. Phys.*, 1, 5369, 1999.

1999KIN   Kinzl, M., Einfluss der Zugabe von Inertkomponenten auf die Phasengleichgewichte von
            Copolymerlösungen in überkritischem Ethen, *Dissertation*, TH Darmstadt, 1999.

1999KOA   Koak, N., Visser, R.M., and Loos, Th.W. de, High-pressure phase behavior of the systems
            polyethylene + ethylene and polybutene + 1-butene, *Fluid Phase Equil.*, 158-160, 835,
            1999.

1999LIJ   Li, J., Zhang, M., and Kiran, E., Dynamics of pressure-induced phase separation in polymer
            solutions. The dependence of the demixing pressure on the rate of pressure quench in
            solutions of poly(dimethylsiloxane) in supercritical carbon dioxide, *Ind. Eng. Chem. Res.*,
            38, 4486, 1999.

1999LO1   Lora, M. and McHugh, M.A., Phase behavior and modeling of the poly(methyl
            methacrylate)-CO_2-methyl methacrylate system, *Fluid Phase Equil.*, 157, 285, 1999.

1999LO2   Lora, M., Rindfleisch, F., and McHugh, M.A., Influence of the alkyl tail on the solubility of
            poly(alkyl acrylates) in ethylene and CO_2 at high pressures: experiments and modeling, *J.
            Appl. Polym. Sci.*, 73, 1979, 1999.

1999MAR   Martin, T.M., Lateef, A.A., and Roberts, C.B., Measurement and modeling of cloud point
            behavior for polypropylene/n-pentane and polypropylene/n-pentane/carbon dioxide
            mixtures at high pressures, *Fluid Phase Equil.*, 154, 241, 1999.

1999MER   Mertdogan, C.A., McHugh, M.A., and Tuminello, W.H., Cosolvency effect of SF_6 on the
            solubility of poly(tetrafluoroethylene-*co*-19 mol% hexafluoropropylene) in supercritical
            CO_2 and CHF_3, *J. Appl. Polym. Sci.*, 74, 2039, 1999.

1999MIS   Mishima, K., Matsuyama, K., and Nagatani, M., Solubilities of poly(ethylene glycol)s in the
            mixtures of supercritical carbon dioxide and cosolvent, *Fluid Phase Equil.*, 161, 315, 1999.

1999PAN   Pan, C. and Radosz, M., Phase behavior of poly(ethylene-*co*-hexene-1) solutions in
            isobutane and propane, *Ind. Eng. Chem. Res.*, 38, 2842, 1999.

1999WAL   Walker, T.A., Raghavan, S.R., Royer, J.R., Smith, S.D., Wignall, G.D., Melnichenko, Y.,
            Khan, S.A., and Spontak, R.J., Enhanced miscibility of low-molecular-weight
            polystyrene/polyisoprene blends in supercritical CO_2, *J. Phys. Chem. B*, 103, 5472, 1999.

2000BAY   Bayraktar, Z. and Kiran, E., Miscibility, phase separation, and volumetric properties in
            solutions of poly(dimethylsiloxane) in supercritical carbon dioxide, *J. Appl. Polym. Sci.*, 75,
            1397, 2000.

2000BEH   Behme, S., Thermodynamik von Polymersystemen bei hohen Drucken, *Dissertation*, TU
            Berlin, 2000.

2000BY1   Byun, H.-S. and McHugh, M.A., Impact of 'free' monomer concentration on the phase
            behavior of supercritical carbon dioxide-polymer mixtures, *Ind. Eng. Chem. Res.*, 39, 4658,
            2000.

2000BY2   Byun, H.-S. and Choi, T.-H., Effect of monomer comcentration on the phase behavior of supercritical carbon dioxide-poly(ethyl methacrylate) mixture, *J. Korean Ind. Eng. Chem.*, 11, 396, 2000.

2000BY3   Byun, H.-S., Kim, K., and Lee, H.-S., High-pressure phase behavior and mixture density of binary poly(ethylene-*co*-butene)-dimethyl ether system, *Hwahak Konghak*, 38, 826, 2000.

2000CH1   Chan, A.K.C., Adidharma, H., and Radosz, M., Fluid-liquid and fluid-solid transitions of poly(ethylene-*co*-octene-1) in sub- and supercritical propane solutions, *Ind. Eng. Chem. Res.*, 39, 3069, 2000.

2000CH2   Chan, A.K.C., Russo, P.S., and Radosz, M., Fluid-liquid equilibria in poly(ethylene-*co*-hexene-1) + propane: a light-scattering probe of cloud-point pressure and critical polymer concentration, *Fluid Phase Equil.*, 173, 149, 2000.

2000CH3   Chan, A.K.C. and Radosz, M., Fluid-liquid and fluid-solid phase behavior of poly(ethylene-*co*-hexene-1) solutions in sub- and supercritical propane, ethylene, and ethylene + hexene-1, *Macromolecules*, 33, 6800, 2000.

2000CH4   Chan, A.K.C., Adidharma, H., and Radosz, M., Fluid-liquid transitions of poly(ethylene-*co*-octene-1) in supercritical ethylene solutions, *Ind. Eng. Chem. Res.*, 39, 4370, 2000.

2000DIN   DiNoia, T.P., Conway, S.E., Lim, J.S., McHugh, M.A., Solubility of vinylidene fluoride polymers in supercritical CO_2 and halogenated solvents, *J. Polym. Sci., Part B: Polym. Phys.*, 38, 2832, 2000.

2000KIM   Kim, K. and Byun, H.-S., Cosolvent effect on phase behavior of poly(butyl methacrylate)-CO_2-butyl methacrylate system at high pressure, *Hwahak Konghak*, 38, 479, 2000.

2000KIN   Kinzl, M., Luft, G., Adidharma, H., and Radosz, M., SAFT modeling of inert-gas effects on the cloud-point pressures in ethylene copolymerization systems: Poly(ethylene-*co*-vinyl acetate) + vinyl acetate + ethylene and poly(ethylene-*co*-hex-1-ene) + 1-hexene + ethylene with carbon dioxide, nitrogen, or butane, *Ind. Eng. Chem. Res.*, 39, 541, 2000.

2000LE1   Lee, J.M., Lee, B.-C., and Lee, S.-H., Cloud points of biodegradable polymers in compressed liquid and supercritical chlorodifluoromethane, *J. Chem. Eng. Data*, 45, 851, 2000.

2000LE2   Lee, J.M., Lee, B.-C., and Hwang, S.-J., Phase behavior of poly(L-lactide) in supercritical mixtures of CO_2 and chlorodifluoromethane, *J. Chem. Eng. Data*, 45, 1162, 2000.

2000MAR   Martin, T.M., Gupta, R.B., and Roberts, C.B., Measurements and modeling of cloud point behavior for poly(propylene glycol) in ethane and in ethane + cosolvent mixtures at high pressure, *Ind. Eng. Chem. Res.*, 39, 185, 2000.

2000OLI   Oliveira, J.V., Dariva, C., and Pinto, J.C., High-pressure phase equilibria for polypropylene-hydrocarbon systems, *Ind. Eng. Chem. Res.*, 39, 4627, 2000.

2000RAO   Rao, V.S.R. and Watkins, J.J., Phase separation in polystyrene-poly(vinyl methyl ether) blends dilated with compressed carbon dioxide, *Macromolecules*, 33, 5143, 2000.

2000TEP   Tepper, G. and Levit, N., Polymer deposition from supercritical solutions for sensing applications, *Ind. Eng. Chem. Res.*, 39, 4445, 2000.

2000VRI   Vries, T.J. de, Somers, P.J.A., Loos, Th.W. de, Vorstman, M.A.G., and Keurentjes, J.T.F., Phase behavior of poly(ethylene-co-propylene) in ethylene and carbon dioxide: experimental results and modeling with the SAFT equation of state, *Ind. Eng. Chem. Res.*, 39, 4510, 2000.

2000WIE   Wiesmet, V., Weidner, E., Behme, S., Sadowski, G., and Arlt, W., Measurement and modelling of high-pressure phase equilibria in the systems polyethyleneglycol (PEG)-propane, PEG-nitrogen and PEG-carbon dioxide, *J. Supercrit. Fluids*, 17, 1, 2000.

2001BYU   Byun, H.-S., Kim, K., Kim, N.H., and Kwak, C., Cosolvent effect and phase behavior of poly(octadecyl methacrylate)-CO_2 mixtures at high pressure, *J. Korean Ind. Eng. Chem.*, 12, 212, 2001.

2001CHE   Chernyak, Y., Henon, F., Harris, R.B., Gould, R.D., Franklin, R.K., Edwards, J.R., De Simone, J.M., and Carbonell, R.G., Formation of perfluoropolyether coatings by the rapid expansion of supercritical solutions (RESS) process. Part 1: Experimental results, *Ind. Eng. Chem. Res.*, 40, 6118, 2001.

2001CON   Conway, S.E., Byun, H.-S., McHugh, M.A., Wang, J.D., and Mandel, F.S., Poly(lactide-*co*-glycolide) solution behavior in supercritical CO_2, CHF_3, and $CHClF_2$, *J. Appl. Polym. Sci.*, 80, 1155, 2001.

2001DAR   Dariva, C., Oliveira, J.V., Tavares, F.W., and Pinto, J.C., Phase equilibria of polypropylene samples with hydrocarbon solvents at high pressures, *J. Appl. Polym. Sci.*, 81, 3044, 2001.

2001DOE   Doerr, H., Kinzl, M., and Luft, G., The influence of inert gases on the high-pressure phase equilibria of EH-copolymer/1-hexene/ethylene-mixtures, *Fluid Phase Equil.*, 178, 191, 2001.

2001KEM   Kemmere, M., de Vries, T.J., Vorstman, M., and Keurentjens J., A novel process for the catalytic polymerization of olefins in supercritical carbon dioxide, *Chem. Eng. Sci.*, 56, 4197, 2001.

2001KUK   Kuk, Y.-M., Lee, B.-C., Lee, Y.W., and Lim, J.S., Phase behavior of biodegradable polymers in dimethyl ether and dimethyl ether + carbon dioxide, *J. Chem. Eng. Data*, 46, 1344, 2001.

2001LI1   Li, D., Han, B., Liu, Z., and Zhao, D., Phase behavior of supercritical CO_2/styrene/poly(ethylene terephthalate) (PET) system and preparation of polystyrene/PET composites, *Polymer*, 42, 2331, 2001.

2001LI2   Li, D., Han, B., Liu, Z., Liu, J., Zhang, X., Wang, S., and Zhang, X., Effect of gas antisolvent on conformation of polystyrene in toluene: Viscosity and small-angle X-ray scattering study (experimental data by B. Han), *Macromolecules*, 34, 2195, 2001.

2001LI3   Li, D., Han, B., Huo, Q., Wang, J., and Dong, B., Small-angle X-ray scattering by dilute solutions of bisphenol-A polycarbonate during adding antisolvent CO_2, *Macromolecules*, 34, 4874, 2001.

2001LIU   Liu, J., Han, B., Li, G., Liu, Z., He, J., and Yang, G., Solubility of the non-ionic surfactant tetraethylene glycol n-lauryl ether in supercritical CO_2 with n-pentanol, *Fluid Phase Equil.* 187-188, 247, 2001.

2001NDI   Ndiaye, P.M., Dariva, C., Oliveira, J.V., and Tavares, F.W., Phase behavior of isotactic polypropylene/C_4-solvents at high pressure. Experimental data and SAFT modeling, *J. Supercrit. Fluids*, 21, 93, 2001.

2001TOR   Tork, T., Measurement and calculation of phase equilibria in polyolefin/solvent systems (Ger.), *Dissertation*, TU Berlin, 2001.

2001VLI   Vliet, R.E. van, Tiemersma, T.P., Krooshof, G.J., and Iedema, P.D., The use of liquid-liquid extraction in the EPDM solution polymerization process, *Ind. Eng. Chem. Res.*, 40, 4586, 2001.

2002BEY   Beyer, C. and Oellrich, L.R., Cosolvent studies with the system ethylene/poly(ethylene-*co*-acrylic acid): Effects of solvent, density, polarity, hydrogen bonding, and copolymer composition, *Helv. Chim. Acta*, 85, 659, 2002.

2002BLA   Blasig, A., Shi, C., Enick, R.M., and Thies, M.C., Effect of concentration and degree of saturation on RESS of a CO_2-soluble fluoropolymer (experimental data by A. Blasig and M.C. Thies), *Ind. Eng. Chem. Res.*, 41, 4976, 2002.

2002BOT   Bothum, G.D., White, K.L., and Knutson, B.L., Gas antisolvent fractionation on semicrystalline and amorphous poly(lactic acid) using compressed CO_2, *Polymer*, 43, 4445, 2002.

2002BYU   Byun, H.-S. and Choi, T.-H., Effect of the octadecyl acrylate concentration on the phase behavior of poly(octadecyl acrylate)/supercritical CO_2 and C_2H_4 at high pressures, *J. Appl. Polym. Sci.*, 86, 372, 2002.

2002COL   Colina, C.M., Hall, C.K., and Gubbins, K.E., Phase behavior of PVAC-PTAN block copolymer in supercritical carbon dioxide using SAFT, *Fluid Phase Equil.*, 194-197, 553, 2002.

2002DRO   Drohmann, C. and Beckman, E.J., Phase behavior of polymers containing ether groups in carbon dioxide, *J. Supercrit. Fluids*, 22, 103, 2002.

2002HOR   Horst, M.H. ter, Behme, S., Sadowski, G., and Loos, Th.W. de, The influence of supercritical gases on the phase behavior of polystyrene-cyclohexane and polyethylene-cyclohexane systems: experimental results and modeling with the SAFT equation of state, *J. Supercrit. Fluids*, 23, 181, 2002.

2002JOU   Joung, S.N., Park, J.-U., Kim, S.Y., and Yoo, K.-P., High-pressure phase behavior of polymer-solvent systems with addition of supercritical CO_2 at temperatures from 323.15 K to 503.15 K, *J. Chem. Eng. Data*, 47, 270, 2002.

2002KUK   Kuk, Y.-M., Lee, B.-C., Lee, Y.-W., and Lim, J. S., High-pressure phase behavior of poly(D,L-lactide) in chlorodifluoromethane, difluoromethane, trifluoromethane, and 1,1,1,2-tetrafluoroethane, *J. Chem. Eng. Data*, 47, 575, 2002.

2002LEE   Lee, B.-C. and Kuk, Y.-M., Phase behavior of poly(L-lactide) in supercritical mixtures of dichloromethane and carbon dioxide, *J. Chem. Eng. Data*, 47, 367, 2002.

2002LID   Li, D., Liu, Z., Han, N., Yang, G., Song, L., Wang, J., and Dong, B., Thermodynamic nature of monodisperse polystyrene in mixed solvent of cyclohexane and antisolvent carbon dioxide using synchrotron small-angle X-ray scattering, *Macromolecules*, 35, 10114, 2002.

2002MC1   McHugh, M.A., Park, I.-H., Reisinger, J.J., Ren, Y., Lodge, T.P., and Hillmyer, M.A., Solubility of CF_2-modified polybutadiene and polyisoprene in supercritical carbon dioxide, *Macromolecules*, 35, 4653, 2002.

2002MC2   McHugh, M.A., Garach-Domech, A., Park, I.-H., Li, D., Barbu, E., Graham, P., Tsibouklis, J., Impact of fluorination and side-chain length on poly(methylpropenoxyalkylsiloxane) and poly(alkyl methacrylate) solubility in supercritical carbon dioxide, *Macromolecules*, 35, 6479, 2002.

2002TAY   Taylor, D.K., Keiper, J.S., and DeSimone, J.M., Polymer self-assembly in carbon dioxide, *Ind. Eng. Chem. Res.*, 41, 4451, 2002.

2002WU1   Wu, J., Pan, Q., and Rempel, G.L., High-pressure phase equilibria for a styrene/CO_2/polystyrene ternary system, *J. Appl. Polym. Sci.*, 85, 1938, 2002.

2002WU2   Wu, J., Pan, Q., and Rempel, G.L., Prediction of phase behavior for styrene/CO_2/polystyrene mixtures, *Chin. J. Chem. Eng.*, 10, 706, 2002.

2002ZHA   Zhang, W. and Kiran, E., Phase behavior and density of polysulfone in binary fluid mixtures of tetrahydrofuran and carbon dioxide under high pressure: miscibility windows, *J. Appl. Polym. Sci.*, 86, 2357, 2002.

2003BUB   Buback, M. and Latz, H., Cloud-point pressure curves of ethene/poly[ethylene-*co*-((meth)acrylic acid)] mixtures, *Macromol. Chem. Phys.*, 204, 638, 2003.

2003DOM   Domingo, C., Vega, A., Fanovich, M.A., Elvira, C., and Subra, P., Behavior of poly(methyl methacrylate)-based systems in supercritical CO_2 and CO_2 plus cosolvent: solubility measurements and process assessment, *J. Appl. Polym. Sci.*, 90, 3652, 2003.

2003KIL   Kilic, S., Michalik, S., Wang, Y., Johnson, J.K., Enick, R.M., and Beckman, E.J., Effect of grafted Lewis base groups on the phase behavior of model poly(dimethyl siloxanes) in CO_2, *Ind. Eng. Chem. Res.*, 42, 6415, 2003.

2003KHO   Khosravi-Darani, K., Vasheghani-Farahani, E., Yamini, Y., and Bahramifar, N., Solubility of poly(beta-hydroxybutyrate) in supercritical carbon dioxide, *J. Chem. Eng. Data*, 48, 860, 2003.

2003LEE   Lee, B.-C., Lim, J.S., and Lee, Y.-W., Effect of solvent composition and polymer molecular weight on cloud points of poly(L-lactide) in chlorodifluoromethane + carbon dioxide, *J. Chem. Eng. Data*, 48, 774, 2003.

2003LID   Li, D., Shen, Z., McHugh, M.A., Tsibouklis, J., and Barbu, E., Solubility of poly(perfluoromonoitaconates) and poly(perfluorodiitaconates) in supercritical CO_2, *Ind. Eng. Chem. Res.*, 42, 6499, 2003.

2003RAM   Ramachandrarao, V.S., Vogt, B.D., Gupta, R.R., and Watkins, J.J., Effect of carbon dioxide sorption on the phase behavior of weakly interacting polymer mixtures: solvent-induced segregation in deuterated polybutadiene/polyisoprene blends, *J. Polym. Sci.: Part B: Polym. Phys.*, 41, 3114, 2003.

2003SEI   Seiler, M., Rolker, J., and Arlt, W., Phase behavior and thermodynamic phenomena of hyperbranched polymer solutions, *Macromolecules*, 36, 2085, 2003.

2003SHE   Shen, Z., McHugh, M.A., Xu, J., Belardi, J., Kilic, S., Mesiano, A., Bane, S., Karnikas, C., Beckman, E., and Enick, R., CO_2 solubility of oligomers and polymers that contain the carbonyl group, *Polymer*, 44, 1491, 2003.

2003STR   Striolo, A., Elvassore, N., Parton, T., and Bertucco, A., Relationship between volume expansion, solvent power, and precipitation in GAS processes, *AIChE-J.*, 49, 2671, 2003.

2003TRU   Trumpi, H., Loos, Th.W. de, Krenz, R.A., and Heidemann, R.A., High pressure phase equilibria in the system linear low density polyethylene + ethylene: experimental results and modelling, *J. Supercrit. Fluids*, 27, 205, 2003.

2003ZH1   Zhang, W. and Kiran, E., (p, V, T) Behaviour and miscibility of (polysulfone + THF + carbon dioxide) at high pressures, *J. Chem. Thermodyn.*, 35, 605, 2003.

2003ZH2   Zhang, W., Dindar, C., Bayraktar, Z., and Kiran, E., Phase behavior, density, and crystallization of polyethylene in n-pentane and in n-pentane/CO_2 at high pressures, *J. Appl. Polym. Sci.*, 89, 2201, -2003.

2004BEC   Becker, F., Buback, M., Latz, H., Sadowski, G., and Tumakaka, F., Cloud-point curves of ethylene-(meth)acrylate copolymers in fluid ethene up to high pressures and temperatures: experimental study and PC-SAFT modeling, *Fluid Phase Equil.*, 215, 263, 2004.

2004BYU   Byun, H.-S., Kim, J.-G., and Yang, J.-S., Phase behavior of the poly[hexyl (meth)acrylate]-supercritical solvents-monomer mixtures at high pressures, *Ind. Eng. Chem. Res.* 43 (2004) 1543-1552

2004CHE   Chen, X., Yasuda, K., Sato, Y., Takishima, S., and Masuoka, H., Measurement and correlation of phase equilibria of ethylene + n-hexane + metallocene polyethylene at temperatures between 373 and 473 K and at pressures up to 20 MPa, *Fluid Phase Equil.*, 215, 105, 2004.

2004DIC   Dickson, J.L., Ortiz-Estrada, C., Alvarado, J.F.J., Hwang, H.S., Sanchez, I.C., Luna-Barcenas, G., Lim, K.T., and Johnston, K.P., Critical flocculation density of dilute water-in-CO_2 emulsions stabilized with block copolymers, *J. Colloid Interface Sci.*, 272, 444, 2004.

2004LI1   Li, D. and McHugh, M.A., Limited polysulfone solubility in supercritical dimethyl ether with THF and DMF cosolvents, *J. Supercrit. Fluids*, 28, 79, 2004.

2004LI2   Li, D. and McHugh, M.A., Solubility behavior of ethyl cellulose in supercritical fluid solvents, *J. Supercrit. Fluids*, 28, 225, 2004.

2004PFO   Pfohl, O. and Dohrn, R., Provision of thermodynamic properties of polymer systems for industrial applications, *Fluid Phase Equil.*, 217, 189, 2004.

2004WAL   Walker, T.A., Colina, C.M., Gubbins, K.E., and Spontak, R.J., Thermodynamics of poly(dimethylsiloxane)/poly(ethylmethylsiloxane) (PDMS/PEMS) blends in the presence of high-pressure CO_2, *Macromolecules*, 37, 2588, 2004.

5. ENTHALPY CHANGES IN POLYMER SOLUTIONS AT ELEVATED PRESSURES

5.1. Enthalpies of mixing or intermediary enthalpies of dilution

Polymer (B):	poly(ethylene glycol) dimethyl ether	1999LOP
Characterization:	M_n/kg.mol^{-1} = 0.280, PEGDME 250, a mixture of oligomers of n = 3 to 9, Aldrich Chem. Co., Inc., Milwaukee, WI	
Solvent (A):	methanol CH$_4$O	67-56-1

T/K = 323.15 P/MPa = 8.0

x_A	0.1464	0.2180	0.2992	0.3981	0.4999	0.5002	0.5951	0.7016
$\Delta_M H$/J.mol^{-1}	115.8	221.8	378.4	515.4	591.7	604.7	612.3	571.3

x_A	0.7972	0.8559	0.8987
$\Delta_M H$/J.mol^{-1}	470.5	363.3	267.7

T/K = 373.15 P/MPa = 8.0

x_A	0.0829	0.1202	0.1549	0.1875	0.2181	0.2605	0.2994	0.3572
$\Delta_M H$/J.mol^{-1}	214.6	296.2	367.4	419.8	474.4	547.3	603.6	684.8

x_A	0.3983	0.4264	0.5000	0.5415	0.5950	0.6537	0.7016	0.7590
$\Delta_M H$/J.mol^{-1}	732.2	765.6	823.4	836.5	840.7	788.1	754.8	674.5

x_A	0.7972	0.8558	0.8987	0.9315
$\Delta_M H$/J.mol^{-1}	608.6	476.8	349.0	236.6

T/K = 423.15 P/MPa = 8.0

x_A	0.0829	0.1202	0.1550	0.1875	0.2181	0.2606	0.2994	0.3573
$\Delta_M H$/J.mol^{-1}	184.7	271.1	349.5	423.8	482.0	563.2	630.3	716.7

x_A	0.3984	0.4526	0.4999	0.5415	0.5950	0.6400	0.7016	0.7590
$\Delta_M H$/J.mol^{-1}	775.6	864.9	905.4	922.7	895.8	882.7	828.0	745.7

x_A	0.7972	0.8558	0.8987	0.9314
$\Delta_M H$/J.mol^{-1}	672.3	516.3	366.2	245.8

Polymer (B):	**tetra(ethylene glycol) dimethyl ether**						**1999LOP**
Characterization:	M_n/kg.mol^{-1} = 0.222, Aldrich Chem. Co., Inc., Milwaukee, WI						
Solvent (A):	**methanol**		**CH$_4$O**				**67-56-1**

T/K = 323.15 P/MPa = 8.0

x_A	0.0686	0.1000	0.1441	0.1849	0.2227	0.2907	0.3500	0.4022
$\Delta_M H$/J.mol^{-1}	134.9	193.3	273.5	336.1	392.5	478.7	538.0	588.1

x_A	0.4023	0.4699	0.4700	0.5091	0.5608	0.6056	0.6567	0.7000
$\Delta_M H$/J.mol^{-1}	577.6	605.6	611.3	616.5	614.0	604.6	586.0	548.7

x_A	0.7538	0.7977	0.8550	0.8988	0.9333			
$\Delta_M H$/J.mol^{-1}	481.6	418.8	312.2	220.4	141.3			

T/K = 373.15 P/MPa = 8.0

x_A	0.0686	0.1001	0.1442	0.1850	0.2229	0.2909	0.3502	0.4024
$\Delta_M H$/J.mol^{-1}	163.3	242.5	348.6	431.3	499.3	611.0	690.9	734.0

x_A	0.4025	0.4701	0.5093	0.5610	0.6058	0.6569	0.7002	0.7540
$\Delta_M H$/J.mol^{-1}	744.3	779.5	791.8	797.9	787.0	755.7	715.5	643.2

x_A	0.7978	0.8551	0.8989	0.9334				
$\Delta_M H$/J.mol^{-1}	564.8	430.5	310.8	205.7				

T/K = 423.15 P/MPa = 8.0

x_A	0.0686	0.1001	0.1442	0.1850	0.2229	0.2909	0.3502	0.4024
$\Delta_M H$/J.mol^{-1}	196.2	279.1	382.6	468.9	545.3	668.0	755.4	806.2

x_A	0.4024	0.4700	0.5092	0.5610	0.6057	0.6568	0.6991	0.7001
$\Delta_M H$/J.mol^{-1}	812.3	861.7	577.9	899.4	894.8	871.2	820.0	837.0

x_A	0.7002	0.7539	0.7977	0.8449	0.8989	0.9334		
$\Delta_M H$/J.mol^{-1}	811.8	707.0	616.3	485.2	322.7	210.5		

5.2. References

1999LOP   Lopez, E.R., Coxam, J.-Y., Fernandez, J., and Grolier, J.-P.E., Pressure and temperature dependence of excess enthalpies of methanol + tetraethylene glycol dimethyl ether and methanol + polyethylene glycol dimethyl ether 250, *J. Chem. Eng. Data*, 44, 1409, 1999.

6. PVT DATA OF POLYMERS AND SOLUTIONS

6.1. PVT data of selected polymers

Polymer (B):	ethylene/acrylic acid copolymer	1992WIN

Characterization: M_n/kg.mol^{-1} = 26, M_w/kg.mol^{-1} = 130, 12.4 wt% acrylic acid

T/K	393.2		413.2		433.2		453.2		473.2	

P/ MPa	V_{spez}/ cm^{3}g^{-1}	P/ MPa	V_{spez}/ cm^{3}g^{-1}	P/ MPa	V_{spez}/ cm^{3}g^{-1}	P/ MPa	V_{spez}/ cm^{3}g^{-1}	P/ MPa	V_{spez}/ cm^{3}g^{-1}
8.6	1.1749	3.4	1.1930	6.1	1.2047	5.6	1.2208	5.3	1.2355
19.6	1.1653	7.6	1.1884	10.7	1.1994	9.8	1.2157	10.2	1.2288
30.4	1.1568	14.7	1.1814	19.2	1.1905	17.9	1.2056	18.2	1.2187
39.7	1.1498	27.9	1.1698	31.6	1.1788	31.1	1.1923	32.2	1.2035
49.6	1.1430	43.1	1.1575	44.2	1.1683	45.5	1.1792	44.4	1.1916
62.8	1.1348	62.8	1.1434	63.4	1.1538	63.5	1.1652	64.1	1.1748
78.9	1.1249	82.0	1.1316	83.3	1.1409	82.8	1.1518	83.1	1.1607
94.1	1.1167	99.0	1.1227	103.1	1.1296	102.7	1.1397	102.8	1.1481
99.3	1.1128	102.2	1.1205	122.3	1.1196	122.1	1.1287	121.4	1.1369
112.3	1.1076	121.0	1.1114	141.5	1.1105	141.2	1.1189	141.9	1.1263
129.6	1.0995	141.7	1.1020	160.4	1.1022	159.8	1.1101	161.0	1.1169
		159.9	1.0942	179.3	1.0946	178.6	1.1021	180.0	1.1085
		178.3	1.0863	198.3	1.0873	197.5	1.0944	200.1	1.1005

Comments: More data are given in the original source 1992WIN or can also be found in 2001WOH.

Polymer (B):	ethylene/1-butene copolymer	1999MAI

Characterization: M_n/kg.mol^{-1} = 82.8, M_w/kg.mol^{-1} = 199, 52.2 wt% 1-butene, ρ_B = 0.8662 g/cm^3, synthesized in the laboratory

P/MPa	424.48	433.74	443.82	454.18	464.52	474.15	484.04	493.79	504.43
				T/K V_{spez}/cm^{3}g^{-1}					
0.1	1.2558	1.2645	1.2734	1.2828	1.2924	1.3025	1.3119	1.3210	1.3314
10	1.2443	1.2522	1.2604	1.2692	1.2778	1.2868	1.2953	1.3038	1.3133
20	1.2337	1.2410	1.2487	1.2569	1.2647	1.2728	1.2804	1.2887	1.2974
30	1.2243	1.2313	1.2384	1.2462	1.2534	1.2610	1.2681	1.2756	1.2839
40	1.2158	1.2224	1.2290	1.2365	1.2433	1.2504	1.2571	1.2642	1.2716
50	1.2080	1.2142	1.2205	1.2276	1.2342	1.2409	1.2472	1.2539	1.2611

continued

continued

60	1.2007	1.2066	1.2127	1.2193	1.2257	1.2322	1.2382	1.2446	1.2514
70	1.1939	1.1995	1.2053	1.2118	1.2179	1.2240	1.2300	1.2360	1.2425
80	1.1874	1.1929	1.1985	1.2046	1.2107	1.2166	1.2222	1.2281	1.2342
90	1.1815	1.1867	1.1922	1.1980	1.2039	1.2096	1.2150	1.2207	1.2265
100	1.1757	1.1807	1.1861	1.1917	1.1974	1.2029	1.2081	1.2137	1.2193
110	1.1703	1.1752	1.1803	1.1857	1.1913	1.1966	1.2018	1.2072	1.2125
120	1.1651	1.1698	1.1748	1.1802	1.1856	1.1906	1.1957	1.2009	1.2061
130	1.1602	1.1646	1.1696	1.1748	1.1800	1.1850	1.1899	1.1949	1.2000
140	1.1554	1.1597	1.1646	1.1697	1.1747	1.1796	1.1844	1.1893	1.1942
150	1.1508	1.1551	1.1598	1.1647	1.1697	1.1745	1.1791	1.1839	1.1887
160	1.1464	1.1505	1.1550	1.1600	1.1648	1.1695	1.1740	1.1787	1.1834
170	1.1422	1.1462	1.1507	1.1554	1.1601	1.1648	1.1691	1.1738	1.1782
180	1.1382	1.1419	1.1463	1.1510	1.1556	1.1601	1.1645	1.1690	1.1733
200	1.1304	1.1340	1.1383	1.1428	1.1472	1.1516	1.1558	1.1601	1.1642

Comments: More data can be found in 2001WOH.

Polymer (B):	**ethylene/1-octene copolymer**	**1999MAI**
Characterization:	M_n/kg.mol^{-1} = 41.4, M_w/kg.mol^{-1} = 86.9, 25.0 wt% 1-octene, ρ_B = 0.8750 g/cm^3, Dow Chemical	

P/MPa				*T*/K					
	423.10	433.01	443.28	453.42	463.80	473.36	483.63	493.64	503.31
				V_{spez}/cm^{3}g^{-1}					
0.1	1.2825	1.2930	1.3026	1.3122	1.3227	1.3324	1.3431	1.3533	1.3639
10	1.2704	1.2798	1.2887	1.2977	1.3072	1.3163	1.3256	1.3349	1.3444
20	1.2592	1.2677	1.2763	1.2845	1.2933	1.3019	1.3101	1.3188	1.3275
30	1.2494	1.2574	1.2650	1.2733	1.2814	1.2895	1.2971	1.3050	1.3133
40	1.2403	1.2479	1.2550	1.2625	1.2706	1.2781	1.2853	1.2926	1.3004
50	1.2320	1.2392	1.2459	1.2531	1.2603	1.2679	1.2745	1.2815	1.2888
60	1.2242	1.2311	1.2375	1.2443	1.2514	1.2582	1.2648	1.2714	1.2783
70	1.2169	1.2236	1.2296	1.2363	1.2430	1.2496	1.2556	1.2623	1.2687
80	1.2101	1.2166	1.2224	1.2286	1.2352	1.2416	1.2474	1.2533	1.2600
90	1.2038	1.2099	1.2155	1.2216	1.2279	1.2341	1.2397	1.2454	1.2515
100	1.1977	1.2038	1.2091	1.2150	1.2209	1.2270	1.2324	1.2380	1.2439
110	1.1919	1.1978	1.2030	1.2086	1.2145	1.2203	1.2256	1.2310	1.2366
120	1.1863	1.1922	1.1971	1.2025	1.2083	1.2138	1.2190	1.2243	1.2298
130	1.1811	1.1866	1.1915	1.1968	1.2023	1.2078	1.2127	1.2179	1.2233
140	1.1759	1.1814	1.1861	1.1913	1.1966	1.2019	1.2069	1.2118	1.2172
150	1.1710	1.1765	1.1809	1.1860	1.1912	1.1964	1.2012	1.2061	1.2112
160	1.1662	1.1716	1.1760	1.1808	1.1859	1.1911	1.1958	1.2005	1.2055
170	1.1617	1.1670	1.1712	1.1759	1.1809	1.1860	1.1907	1.1951	1.2001
180	1.1575	1.1624	1.1666	1.1713	1.1761	1.1811	1.1857	1.1902	1.1949
200	1.1492	1.1541	1.1581	1.1624	1.1672	1.1719	1.1764	1.1806	1.1853

Comments: More data can be found in 2001WOH.

Polymer (B): **ethylene/propylene copolymer** **1999MAI**

Characterization: M_n/kg.mol^{-1} = 104.5, M_w/kg.mol^{-1} = 178, 31.8 wt% propene,
ρ_B = 0.8871 g/cm^3, synthesized in the laboratory

P/MPa				T/K					
	422.35	432.50	442.35	452.05	462.27	472.37	482.18	492.51	502.28
				V_{spez}/cm^{3}g^{-1}					
0.1	1.2740	1.2836	1.2938	1.3028	1.3122	1.3220	1.3317	1.3422	1.3523
10	1.2624	1.2711	1.2804	1.2886	1.2975	1.3066	1.3155	1.3249	1.3339
20	1.2517	1.2598	1.2681	1.2761	1.2843	1.2929	1.3010	1.3095	1.3178
30	1.2423	1.2498	1.2578	1.2648	1.2731	1.2810	1.2886	1.2966	1.3042
40	1.2337	1.2408	1.2484	1.2550	1.2624	1.2702	1.2776	1.2851	1.2922
50	1.2259	1.2324	1.2399	1.2460	1.2531	1.2601	1.2675	1.2746	1.2814
60	1.2186	1.2247	1.2319	1.2377	1.2445	1.2512	1.2578	1.2651	1.2716
70	1.2118	1.2177	1.2245	1.2301	1.2366	1.2431	1.2494	1.2558	1.2626
80	1.2053	1.2110	1.2176	1.2229	1.2292	1.2354	1.2415	1.2477	1.2536
90	1.1993	1.2047	1.2110	1.2163	1.2222	1.2282	1.2342	1.2400	1.2458
100	1.1935	1.1987	1.2048	1.2100	1.2157	1.2214	1.2272	1.2329	1.2385
110	1.1881	1.1930	1.1990	1.2040	1.2095	1.2151	1.2206	1.2262	1.2315
120	1.1828	1.1876	1.1932	1.1984	1.2036	1.2090	1.2144	1.2197	1.2250
130	1.1777	1.1825	1.1880	1.1930	1.1980	1.2033	1.2086	1.2136	1.2187
140	1.1729	1.1775	1.1828	1.1877	1.1926	1.1978	1.2028	1.2079	1.2128
150	1.1683	1.1726	1.1777	1.1827	1.1875	1.1925	1.1976	1.2023	1.2071
160	1.1639	1.1680	1.1730	1.1778	1.1826	1.1876	1.1923	1.1972	1.2016
170	1.1595	1.1635	1.1684	1.1733	1.1779	1.1828	1.1873	1.1921	1.1966
180	1.1553	1.1593	1.1640	1.1687	1.1733	1.1781	1.1826	1.1872	1.1916
190	1.1513	1.1553	1.1597	1.1645	1.1689	1.1737	1.1780	1.1826	1.1868
200	1.1475	1.1513	1.1557	1.1604	1.1647	1.1694	1.1736	1.1783	1.1822

Comments: More data can be found in 2001WOH.

Polymer (B): **ethylene/vinyl acetate copolymer** **1995ZOL**

Characterization: Scientific Polymer Products, Inc., Ontario, NY

P/MPa				T/K					
	393.85	401.35	409.25	417.75	425.85	432.75	441.65	449.95	457.15
				V_{spez}/cm^{3}g^{-1}					
0.1	1.1796	1.1866	1.1938	1.2014	1.2089	1.2158	1.2238	1.2314	1.2385
20	1.1622	1.1687	1.1752	1.1816	1.1883	1.1947	1.2015	1.2080	1.2140
40	1.1474	1.1532	1.1590	1.1648	1.1711	1.1764	1.1827	1.1885	1.1939
60	1.1341	1.1395	1.1449	1.1503	1.1557	1.1609	1.1666	1.1719	1.1768
80	1.1220	1.1274	1.1324	1.1373	1.1423	1.1472	1.1524	1.1574	1.1619

continued

continued

100	1.1119	1.1170	1.1216	1.1262	1.1309	1.1356	1.1404	1.1451	1.1493
120	1.1023	1.1070	1.1114	1.1158	1.1203	1.1247	1.1293	1.1338	1.1377
140	1.0935	1.0980	1.1022	1.1064	1.1108	1.1150	1.1191	1.1233	1.1270
160	1.0855	1.0898	1.0937	1.0976	1.1019	1.1058	1.1099	1.1137	1.1174
180	1.0778	1.0819	1.0858	1.0895	1.0934	1.0976	1.1014	1.1052	1.1085
200	1.0707	1.0747	1.0785	1.0821	1.0857	1.0896	1.0933	1.0970	1.1002

Comments: More data can be found in 1995ZOL.

Polymer (B):	**poly(1-butene)**	**1999MAI**
Characterization:	M_n/kg.mol^{-1} = 46.7, M_w/kg.mol^{-1} = 131, isotactic,
synthesized in the laboratory by MBI/MAO metallocene catalysts

P/MPa	T/K								
	454.40	464.75	474.70	484.50	494.55	504.50	514.25	524.63	534.60
				V_{spez}/cm^{3}g^{-1}					
0.1	1.2796	1.2892	1.2980	1.3053	1.3144	1.3237	1.3333	1.3436	1.3534
10	1.2647	1.2735	1.2814	1.2878	1.2958	1.3042	1.3127	1.3217	1.3302
20	1.2515	1.2595	1.2666	1.2725	1.2794	1.2873	1.2949	1.3030	1.3104
30	1.2397	1.2476	1.2542	1.2596	1.2661	1.2731	1.2802	1.2875	1.2943
40	1.2295	1.2364	1.2431	1.2482	1.2541	1.2607	1.2672	1.2740	1.2803
50	1.2203	1.2269	1.2326	1.2378	1.2435	1.2496	1.2558	1.2623	1.2681
60	1.2119	1.2182	1.2236	1.2280	1.2339	1.2396	1.2456	1.2515	1.2571
70	1.2039	1.2100	1.2152	1.2194	1.2246	1.2305	1.2363	1.2420	1.2473
80	1.1967	1.2026	1.2075	1.2115	1.2165	1.2218	1.2276	1.2331	1.2382
90	1.1899	1.1956	1.2002	1.2041	1.2089	1.2140	1.2192	1.2250	1.2297
100	1.1835	1.1890	1.1935	1.1972	1.2019	1.2067	1.2118	1.2169	1.2220
110	1.1774	1.1828	1.1871	1.1906	1.1951	1.1999	1.2048	1.2099	1.2142
120	1.1716	1.1769	1.1811	1.1845	1.1889	1.1934	1.1981	1.2030	1.2074
130	1.1662	1.1713	1.1753	1.1786	1.1829	1.1873	1.1919	1.1967	1.2008
140	1.1609	1.1658	1.1698	1.1731	1.1772	1.1815	1.1859	1.1907	1.1946
150	1.1558	1.1608	1.1646	1.1678	1.1718	1.1760	1.1803	1.1850	1.1888
160	1.1511	1.1558	1.1595	1.1626	1.1666	1.1706	1.1750	1.1795	1.1831
170	1.1465	1.1511	1.1547	1.1578	1.1616	1.1656	1.1697	1.1742	1.1777
180	1.1421	1.1466	1.1501	1.1531	1.1569	1.1607	1.1650	1.1693	1.1725
190	1.1379	1.1423	1.1456	1.1486	1.1523	1.1561	1.1602	1.1643	1.1677
200	1.1339	1.1382	1.1414	1.1443	1.1478	1.1516	1.1556	1.1599	1.1629

Polymer (B):	**poly(butylene succinate)**	**2000SAT**
Characterization:	M_n/kg.mol^{-1} = 29, M_w/kg.mol^{-1} = 140, T_g/K = 243,
T_m/K = 388, 35 wt% crystallinity, branched

continued

continued

P/MPa	T/K				
	413.9	433.8	453.6	473.5	493.4
			V_{spez}/cm^{3}g^{-1}		
0.1	0.9012	0.9132	0.9251	0.9375	0.9508
10	0.8954	0.9070	0.9185	0.9302	0.9426
20	0.8902	0.9011	0.9115	0.9235	0.9351
50	0.8763	0.8864	0.8961	0.9062	0.9164
100	0.8565	0.8650	0.8737	0.8824	0.8910
150	0.8416	0.8493	0.8567	0.8647	0.8720
200	0.8274	0.8358	0.8425	0.8498	0.8564

Polymer (B): **poly(butylene terephthalate)** **1991FAK**
Characterization: M_n/kg.mol^{-1} = 23, M_w/kg.mol^{-1} = 70, T_g/K = 342,
T_m/K = 508, PBT Valox Grade 295, General Electric Plastics

P/MPa	T/K					
	514.55	529.05	545.05	559.85	575.95	590.85
			V_{spez}/cm^{3}g^{-1}			
0.1	0.8966	0.9048	0.9139	0.9236	0.9372	0.9543
10	0.8899	0.8978	0.9063	0.9155	0.9284	0.9447
20	0.8835	0.8910	0.8991	0.9077	0.9203	0.9360
30	0.8780	0.8851	0.8928	0.9013	0.9131	0.9282
40	0.8726	0.8798	0.8872	0.8953	0.9066	0.9214
50	0.8676	0.8746	0.8817	0.8894	0.9005	0.9148
60	0.8631	0.8697	0.8766	0.8846	0.8950	0.9089
70	0.8586	0.8653	0.8719	0.8794	0.8897	0.9032
80	0.8545	0.8611	0.8675	0.8749	0.8848	0.8979
90	0.8506	0.8570	0.8632	0.8703	0.8801	0.8930
100	0.8469	0.8531	0.8593	0.8662	0.8757	0.8882
110	0.8414	0.8493	0.8554	0.8621	0.8716	0.8837
120	0.8362	0.8457	0.8515	0.8584	0.8674	0.8793
130	0.8291	0.8423	0.8479	0.8546	0.8636	0.8752
140	0.8139	0.8389	0.8444	0.8509	0.8598	0.8711
150	0.8058	0.8356	0.8412	0.8476	0.8561	0.8673
160	0.7985	0.8324	0.8378	0.8441	0.8526	0.8634
170	0.7924	0.8271	0.8347	0.8408	0.8492	0.8597
180	0.7897	0.8223	0.8315	0.8377	0.8458	0.8560
190	0.7872	0.8175	0.8286	0.8347	0.8426	0.8562
200	0.7846	0.8101	0.8256	0.8316	0.8393	0.8491

Polymer (B):	**poly(butyl acrylate)**					**1999CHU**
Characterization:	M_n/kg.mol^{-1} = 38, M_w/kg.mol^{-1} = 101, Aldrich Chem. Co., Inc., Milwaukee, WI					

P/MPa			T/K				
	303.15	333.15	363.15	393.15	423.15	453.15	483.15
			V_{spez}/cm^{3}g^{-1}				
10	0.9137	0.9356	0.9546	0.9745	0.9976	1.0192	1.0415
20	0.9076	0.9285	0.9468	0.9654	0.9866	1.0067	1.0266
30	0.9026	0.9227	0.9406	0.9581	0.9786		1.0161
40	0.8980	0.9177	0.9347	0.9515	0.9710	0.9888	1.0066
50	0.8939	0.9127	0.9292	0.9453	0.9641	0.9813	0.9982
60	0.8899	0.9081	0.9242	0.9396	0.9576	0.9743	0.9906
70	0.8861	0.9038	0.9194	0.9343	0.9516	0.9680	0.9834
80	0.8825	0.8992	0.9149	0.9296	0.9455	0.9614	0.9765
90	0.8788	0.8955	0.9106	0.9246	0.9408	0.9556	0.9704
100	0.8753	0.8918	0.9064	0.9204	0.9358	0.9502	0.9648
110	0.8722	0.8885	0.9026	0.9160	0.9309	0.9452	0.9593
120	0.8691	0.8850	0.8987	0.9118	0.9266	0.9404	0.9539
130	0.8663	0.8817	0.8950	0.9080	0.9221	0.9359	0.9488
140	0.8633	0.8789	0.8917	0.9040	0.9182	0.9311	0.9441
150	0.8604	0.8755	0.8884	0.9004	0.9140	0.9271	0.9394
160	0.8577	0.8724	0.8853	0.8966	0.9102	0.9230	0.9352
170	0.8553	0.8698	0.8822	0.8934	0.9065	0.9189	0.9309
180	0.8524	0.8669	0.8788	0.8900	0.9030	0.9150	0.9266
190	0.8500	0.8641	0.8758	0.8866	0.8993	0.9113	0.9228
200	0.8476		0.8729	0.8835	0.8959	0.9077	0.9187

Polymer (B):	**poly(εcaprolactone)**				**1993ROG**
Characterization:	M_w/kg.mol^{-1} = 32, Scientific Polymer Products, Inc., Ontario, NY				

P/MPa			T/K			
	373.75	384.05	393.35	403.25	412.55	421.35
			V_{spez}/cm^{3}g^{-1}			
0.1	0.9647	0.9719	0.9773	0.9833	0.9890	0.9950
10	0.9584	0.9651	0.9702	0.9760	0.9814	0.9871
20	0.9522	0.9571	0.9628	0.9687	0.9737	0.9789
30	0.9467	0.9523	0.9566	0.9622	0.9678	0.9728
40	0.9415	0.9465	0.9512	0.9566	0.9618	0.9665
50	0.9369	0.9418	0.9461	0.9514	0.9563	0.9607
60	0.9324	0.9369	0.9413	0.9463	0.9507	0.9553
70	0.9280	0.9326	0.9367	0.9417	0.9452	0.9504

continued

continued

80	0.9240	0.9284	0.9323	0.9369	0.9409	0.9455
90	0.9209	0.9243	0.9284	0.9326	0.9367	0.9410
100	0.9164	0.9206	0.9248	0.9287	0.9326	0.9371
110	0.9130	0.9171	0.9212	0.9247	0.9285	0.9329
120	0.9095	0.9138	0.9170	0.9210	0.9247	0.9289
130	0.9064	0.9104	0.9135	0.9176	0.9212	0.9253
140	0.9033	0.9072	0.9103	0.9142	0.9174	0.9216
150	0.9005	0.9041	0.9069	0.9107	0.9138	0.9181
160	0.8974	0.9010	0.9040	0.9075	0.9106	0.9145
170	0.8948	0.8981	0.9012	0.9044	0.9081	0.9111
180	0.8922	0.8953	0.8985	0.9013	0.9046	0.9078
190	0.8894	0.8925	0.8955	0.8986	0.9021	0.9044
200	0.8870	0.8898	0.8928	0.8961	0.8987	0.9013

Polymer (B):   **polycarbonate bisphenol-A**   **1997SAT**

Characterization: M_w/kg.mol^{-1} = 60, Scientific Polymer Products, Inc., Ontario, NY

P/MPa	T/K								
	443.7	463.6	483.5	503.3	523.6	543.5	563.6	583.5	603.4
				V_{spez}/cm^{3}g^{-1}					
0.1	0.8738	0.8841	0.8934	0.9061	0.9175	0.9281	0.9402	0.9517	0.9643
10	0.8678	0.8777	0.8872	0.8994	0.9102	0.9203	0.9315	0.9423	0.9538
20	0.8627	0.8721	0.8817	0.8930	0.9030	0.9130	0.9231	0.9334	0.9439
50	0.8497	0.8583	0.8670	0.8770	0.8859	0.8946	0.9034	0.9123	0.9213
100			0.8469	0.8556	0.8630	0.8705	0.8781	0.8854	0.8928
150				0.8387	0.8453	0.8518	0.8584	0.8649	0.8716
200					0.8306	0.8365	0.8424	0.8483	0.8542

Polymer (B):   **polycarbonate tetramethyl bisphenol-A**   **1992KIM**

Characterization: M_w/kg.mol^{-1} = 33, Bayer AG, Leverkusen, Germany

P/MPa	T/K								
	491.55	497.75	506.65	514.75	524.15	535.75	544.65	553.75	562.75
				V_{spez}/cm^{3}g^{-1}					
0.1	0.9787	0.9837	0.9912	0.9961	1.0016	1.0091	1.0149	1.0212	1.0306
10	0.9700	0.9746	0.9811	0.9858	0.9911	0.9978	1.0033	1.0090	1.0175
20	0.9612	0.9657	0.9709	0.9753	0.9811	0.9870	0.9923	0.9978	1.0050

continued

continued

30	0.9545	0.9588	0.9637	0.9677	0.9727	0.9782	0.9837	0.9885	0.9951
40	0.9481	0.9522	0.9546	0.9604	0.9658	0.9706	0.9753	0.9802	0.9868
50	0.9423	0.9460	0.9500	0.9542	0.9588	0.9636	0.9681	0.9726	0.9791
60	0.9372	0.9402	0.9441	0.9479	0.9523	0.9570	0.9613	0.9655	0.9717
70		0.9354	0.9385	0.9423	0.9467	0.9509	0.9550	0.9590	0.9647
80					0.9411	0.9454	0.9492	0.9531	0.9584
90					0.9359	0.9402	0.9439	0.9480	0.9527
100						0.9352	0.9390	0.9427	0.9475
110							0.9342	0.9377	0.9424
120							0.9298	0.9330	0.9373
130								0.9285	0.9326
140								0.9244	0.9282
150								0.9200	0.9238
160								0.9159	0.9193

Polymer (B):	**poly(1,1-dihydroxyperfluorooctyl acrylate)**	**1998LUN**
Characterization:	M_w/kg.mol^{-1} = 1200-1600, prepared in the laboratory by solution-free radical polymerization in supercritical CO_2	

P/MPa	T/K						
	303.15	313.15	323.15	333.15	343.15	353.15	363.15
			V_{spez}/cm^{3}g^{-1}				
10	0.5428	0.5483	0.5537	0.5589	0.5685	0.5721	0.5735
20	0.5377	0.5426	0.5477		0.5570	0.5614	0.5652
30	0.5335	0.5386	0.5428	0.5475	0.5520	0.5556	0.5599
40	0.5301	0.5350	0.5389	0.5433	0.5477	0.5511	0.5556
50	0.5269	0.5316	0.5353	0.5395	0.5438	0.5469	0.5517
60	0.5239	0.5286	0.5319	0.5363	0.5404	0.5433	0.5481
70	0.5212	0.5258	0.5286	0.5332		0.5400	0.5445
80	0.5184	0.5230		0.5303		0.5368	0.5413
90	0.5159	0.5205	0.5238	0.5276	0.5314	0.5337	0.5379
100	0.5134	0.5180	0.5212	0.5250	0.5286	0.5310	0.5347
110	0.5115	0.5154	0.5188	0.5226	0.5236	0.5284	0.5312
120	0.5091	0.5131	0.5167	0.5203	0.5212	0.5261	0.5292
130	0.5074	0.5108	0.5145	0.5183	0.5188	0.5235	0.5266
140	0.5057	0.5089	0.5123	0.5161	0.5166	0.5207	0.5245
150	0.5039	0.5069	0.5100	0.5141	0.5145	0.5194	0.5224
160	0.5023	0.5051	0.5079	0.5122	0.5126	0.5174	0.5203
170	0.5006	0.5035	0.5059	0.5103	0.5109	0.5155	0.5183
180	0.4991	0.5019	0.5041	0.5082	0.5090	0.5137	0.5162
190	0.4976	0.5004	0.5025	0.5064	0.5073	0.5117	0.5144
200	0.4963	0.4988	0.5009	0.5043		0.5100	0.5125

Polymer (B):   poly(dimethylsiloxane)   **1998SAC**

Characterization: M_n/kg.mol^{-1} = 5.8, M_w/kg.mol^{-1} = 9

P/MPa	291.25	298.15	304.45	307.45	338.15	365.15	394.15	423.05
				T/K				
			V_{spez}/cm^{3}g^{-1}					
0.1	1.0267	1.0325	1.0407	1.0418	1.0692	1.0957	1.1245	1.1529
22.5	1.0043	1.0092	1.0158	1.0170	1.0396	1.0607	1.0825	1.1034
32.5	0.9959	1.0004	1.0069	1.0078	1.0287	1.0482	1.0686	1.0874
42.5	0.9882	0.9925	0.9987	0.9993	1.0190	1.0376	1.0562	1.0731
52.5	0.9812	0.9853	0.9909	0.9917	1.0104	1.0279	1.0452	1.0607
62.5	0.9746	0.9785	0.9841	0.9844	1.0027	1.0187	1.0353	1.0498
72.5	0.9686	0.9722	0.9774	0.9779	0.9950	1.0105	1.0261	1.0400
82.5	0.9628	0.9661	0.9713	0.9717	0.9881	1.0032	1.0180	1.0312
92.5	0.9574	0.9606	0.9655	0.9659	0.9816	0.9962	1.0103	1.0230
102.5	0.9522	0.9554	0.9600	0.9604	0.9759	0.9895	1.0033	1.0151
112.5	0.9473	0.9505	0.9550	0.9553	0.9700	0.9834	0.9965	1.0082
122.5	0.9427	0.9458	0.9500	0.9504	0.9646	0.9776	0.9904	1.0015
132.5	0.9383	0.9412	0.9453	0.9458	0.9598	0.9724	0.9845	0.9953
142.5	0.9342	0.9370	0.9409	0.9414	0.9550	0.9671	0.9789	0.9891
152.5	0.9301	0.9329	0.9368	0.9372	0.9503	0.9625	0.9740	0.9836
162.5	0.9264	0.9291	0.9328	0.9332	0.9461	0.9578	0.9688	0.9783
172.5	0.9228	0.9252	0.9290	0.9293	0.9417	0.9533	0.9641	0.9730
182.5	0.9192	0.9217	0.9252	0.9257	0.9377	0.9488	0.9595	0.9683
202.5	0.9124	0.9148	0.9182	0.9186	0.9301	0.9409	0.9510	0.9591

Polymer (B):   poly(ether sulfone)   **2004KIM**

Characterization: M_n/kg.mol^{-1} = 36.5, M_w/kg.mol^{-1} = 70, GPC against PS standards, Ultrason-E, ICI America, Inc.

P/MPa	497.05	504.25	513.95	522.95	532.65	542.05	551.05	560.25	569.35
				T/K					
			V_{spez}/cm^{3}g^{-1}						
0.1	0.7712	0.7758	0.7798	0.7841	0.7880	0.7938	0.7992	0.8047	0.8153
10	0.7663	0.7708	0.7744	0.7785	0.7820	0.7874	0.7924	0.7979	0.8049
20	0.7616	0.7659	0.7692	0.7731	0.7763	0.7811	0.7862	0.7913	0.7945
30	0.7576	0.7621	0.7650	0.7688	0.7718	0.7764	0.7808	0.7863	0.7892
40	0.7540	0.7584	0.7610	0.7646	0.7676	0.7721	0.7763	0.7812	0.7843
50	0.7506	0.7550	0.7573	0.7609	0.7638	0.7680	0.7720	0.7766	0.7793
60	0.7475	0.7517	0.7537	0.7572	0.7600	0.7641	0.7681	0.7725	0.7750
70	0.7445	0.7484	0.7504	0.7537	0.7565	0.7605	0.7644	0.7683	0.7710
80	0.7417	0.7453	0.7472	0.7503	0.7532	0.7571	0.7607	0.7645	0.7669
90	0.7390	0.7424	0.7441	0.7470	0.7500	0.7537	0.7573	0.7607	0.7632
100	0.7366	0.7395	0.7412	0.7438	0.7469	0.7505	0.7539	0.7571	0.7595

| Polymer (B): | polyethylene | | | | | | | 1999MAI |

Characterization: M_n/kg.mol^{-1} = 274, M_w/kg.mol^{-1} = 712, commercial sample, HDPE, linear

| P/MPa | T/K | | | | | | | | |
| | 452.10 | 462.80 | 472.75 | 482.75 | 492.80 | 502.95 | 512.85 | 522.85 | 532.85 |
				V_{spez}/cm^{3}g^{-1}					
0.1	1.3038	1.3142	1.3239	1.3340	1.3443	1.3552	1.3661	1.3770	1.3887
10	1.2906	1.3002	1.3091	1.3184	1.3276	1.3374	1.3470	1.3567	1.3668
20	1.2787	1.2876	1.2957	1.3044	1.3127	1.3216	1.3303	1.3389	1.3479
30	1.2682	1.2765	1.2843	1.2923	1.3000	1.3083	1.3163	1.3244	1.3326
40	1.2586	1.2665	1.2739	1.2813	1.2886	1.2963	1.3039	1.3115	1.3190
50	1.2498	1.2572	1.2643	1.2714	1.2782	1.2856	1.2928	1.3000	1.3070
60	1.2416	1.2487	1.2555	1.2622	1.2688	1.2757	1.2826	1.2896	1.2962
70	1.2340	1.2408	1.2474	1.2537	1.2600	1.2667	1.2734	1.2799	1.2863
80	1.2269	1.2335	1.2398	1.2459	1.2520	1.2583	1.2647	1.2710	1.2772
90	1.2199	1.2265	1.2328	1.2385	1.2443	1.2504	1.2567	1.2628	1.2688
100	1.2136	1.2200	1.2260	1.2317	1.2372	1.2431	1.2492	1.2550	1.2607
110	1.2076	1.2134	1.2197	1.2251	1.2305	1.2361	1.2420	1.2478	1.2533
120	1.2018	1.2075	1.2132	1.2190	1.2243	1.2296	1.2352	1.2409	1.2462
130	1.1965	1.2019	1.2074	1.2130	1.2182	1.2234	1.2289	1.2342	1.2395
140	1.1913	1.1965	1.2019	1.2070	1.2125	1.2174	1.2228	1.2281	1.2330
150	1.1863	1.1913	1.1966	1.2015	1.2065	1.2118	1.2171	1.2222	1.2269
160	1.1814	1.1863	1.1916	1.1964	1.2012	1.2060	1.2116	1.2165	1.2211
170	1.1767	1.1816	1.1868	1.1915	1.1962	1.2007	1.2062	1.2111	1.2156
180	1.1723	1.1770	1.1821	1.1866	1.1913	1.1959	1.2008	1.2060	1.2103
190	1.1680	1.1727	1.1775	1.1821	1.1866	1.1910	1.1958	1.2005	1.2052
200	1.1639	1.1685	1.1732	1.1777	1.1821	1.1864	1.1912	1.1958	1.1998

| Polymer (B): | poly(ethylene oxide) | | | | 1987VE1, 1987VE2 |

Characterization: M_w/kg.mol^{-1} = 25–30, M_w/M_n < 1.1

| P/MPa | T/K | | | | |
| | 353.15 | 363.15 | 373.85 | 388.75 | 393.45 |
			V_{spez}/cm^{3}g^{-1}		
0.1	0.92902	0.93589	0.94322	0.95012	0.95657
5.0	0.92644	0.93327	0.94056	0.94742	0.95374
10.0	0.92413	0.93075	0.93791	0.94455	0.95066
15.0	0.92191	0.92842	0.93537	0.94188	0.94787
20.0	0.91946	0.92592	0.93275	0.93914	0.94491
25.0	0.91709	0.92336	0.93015	0.93633	0.94206
30.0	0.91474	0.92090	0.92747	0.93362	0.93923
35.0	0.91232	0.91844	0.92498	0.93101	0.93650
40.0	0.90984	0.91592	0.92242	0.92842	0.93388

Polymer (B):   **poly(DL-lactide)**   **2000SAT**

Characterization: M_n/kg.mol^{-1} = 39, M_w/kg.mol^{-1} = 110, T_g/K = 337,
T_m/K = 441, Lacea H-100E, Mitsumi Chem., Inc., Tokyo, Japan

P/MPa	T/K		
	453.4	473.3	493.3
	V_{spez}/cm^{3}g^{-1}		
0.1	0.8968	0.9107	0.9251
10	0.8888	0.9019	0.9153
20	0.8814	0.8935	0.9065
50	0.8626	0.8732	0.8844
100	0.8392	0.8481	0.8576
150	0.8209	0.8288	0.8369
200	0.8058	0.8126	0.8202

Polymer (B):   **poly(αmethylstyrene)**   **1996MAI**

Characterization: M_n/kg.mol^{-1} = 58.6, M_w/kg.mol^{-1} = 61,
synthesized in the laboratory

P/MPa	T/K				
	481.65	491.25	501.65	511.95	522.30
	V_{spez}/cm^{3}g^{-1}				
0.1	0.9812	0.9867	0.9928	0.9996	1.0073
10	0.9743	0.9796	0.9852	0.9917	0.9990
20	0.9676	0.9728	0.9779	0.9841	0.9908
30	0.9623	0.9674	0.9721	0.9781	0.9847
40	0.9573	0.9621	0.9668	0.9724	0.9786
50	0.9527	0.9573	0.9617	0.9672	0.9730
60	0.9482	0.9526	0.9570	0.9622	0.9680
70	0.9441	0.9483	0.9526	0.9576	0.9630
80		0.9443	0.9484	0.9533	0.9587
90		0.9405	0.9445	0.9492	0.9545
100			0.9407	0.9454	0.9504
110			0.9371	0.9417	0.9466
120			0.9336	0.9381	0.9428
130				0.9347	0.9395
140				0.9313	0.9359
150					0.9327
160					0.9296

Polymer (B):   **poly(1-octene)**   **1999MAI**

Characterization: $M_n/\text{kg.mol}^{-1} = 35$, $M_w/\text{kg.mol}^{-1} = 84$, isotactic,
                        synthesized in the laboratory by MBI/MAO metallocene catalysts

P/MPa				T/K					
	455.90	465.85	476.10	485.95	496.20	505.75	516.50	526.15	536.75
				$V_{spez}/\text{cm}^3\text{g}^{-1}$					
0.1	1.2993	1.3095	1.3195	1.3308	1.3410	1.3511	1.3624	1.3731	1.3842
10	1.2831	1.2923	1.3014	1.3110	1.3202	1.3295	1.3394	1.3487	1.3583
20	1.2687	1.2772	1.2853	1.2935	1.3024	1.3111	1.3196	1.3279	1.3365
30	1.2566	1.2644	1.2719	1.2795	1.2874	1.2952	1.3038	1.3113	1.3191
40	1.2457	1.2532	1.2602	1.2672	1.2745	1.2818	1.2894	1.2968	1.3040
50	1.2358	1.2429	1.2497	1.2565	1.2631	1.2700	1.2771	1.2837	1.2909
60	1.2269	1.2336	1.2401	1.2465	1.2527	1.2593	1.2661	1.2723	1.2788
70	1.2186	1.2250	1.2312	1.2376	1.2436	1.2497	1.2561	1.2620	1.2683
80	1.2110	1.2172	1.2231	1.2292	1.2349	1.2410	1.2469	1.2526	1.2586
90	1.2040	1.2100	1.2156	1.2216	1.2271	1.2328	1.2386	1.2440	1.2497
100	1.1973	1.2031	1.2085	1.2143	1.2195	1.2252	1.2307	1.2359	1.2413
110	1.1910	1.1967	1.2020	1.2075	1.2126	1.2181	1.2233	1.2284	1.2337
120	1.1849	1.1906	1.1957	1.2012	1.2060	1.2114	1.2163	1.2213	1.2263
130	1.1793	1.1847	1.1897	1.1950	1.1999	1.2050	1.2098	1.2145	1.2194
140	1.1740	1.1791	1.1840	1.1892	1.1939	1.1990	1.2036	1.2082	1.2130
150	1.1687	1.1739	1.1785	1.1837	1.1883	1.1932	1.1977	1.2022	1.2067
160	1.1638	1.1687	1.1732	1.1784	1.1830	1.1878	1.1920	1.1964	1.2008
170	1.1591	1.1637	1.1683	1.1734	1.1779	1.1826	1.1867	1.1910	1.1951
180	1.1545	1.1591	1.1636	1.1684	1.1729	1.1776	1.1815	1.1857	1.1898
190	1.1501	1.1545	1.1589	1.1637	1.1681	1.1727	1.1767	1.1807	1.1846
200	1.1461	1.1502	1.1546	1.1591	1.1636	1.1681	1.1718	1.1758	1.1798

Polymer (B):   **poly(phenylmethylsiloxane)**   **1993JAN**

Characterization: $M_w/\text{kg.mol}^{-1} = 2.5$

P/MPa			T/K		
	339.6	350.5	361.7	371.5	397.3
			$V_{spez}/\text{cm}^3\text{g}^{-1}$		
10	0.8931	0.9009	0.9086	0.9142	0.9269
15	0.8899	0.8977	0.9052	0.9104	0.9227
20	0.8874	0.8951	0.9022	0.9076	0.9196
25	0.8849	0.8926	0.8996	0.9047	0.9164
30	0.8826	0.8901	0.8970	0.9022	0.9135
35	0.8804	0.8880	0.8948	0.8998	0.9110
40	0.8783	0.8858	0.8925	0.8972	0.9082
45	0.8763	0.8836	0.8901	0.8950	0.9057

continued

continued

50	0.8744	0.8816	0.8881	0.8928	0.9034
55	0.8723	0.8802	0.8860	0.8905	0.9010
60	0.8705	0.8775	0.8837	0.8884	0.8987
65	0.8687	0.8758	0.8819	0.8864	0.8966
70	0.8669	0.8737	0.8799	0.8843	0.8944
75	0.8650	0.8719	0.8779	0.8823	0.8923
80	0.8634	0.8703	0.8763	0.8805	0.8903
85	0.8617	0.8685	0.8744	0.8785	0.8882
90	0.8599	0.8666	0.8726	0.8766	0.8864
95	0.8585	0.8651	0.8710	0.8749	0.8846
100	0.8569	0.8634	0.8699	0.8732	0.8826

Polymer (B):   **poly(propylene)**   **1993ROG**

Characterization: $M_w/\text{kg.mol}^{-1} = 30$, atactic,
Scientific Polymer Products, Inc., Ontario, NY

P/MPa			T/K		
	353.15	363.15	373.15	383.15	393.15
			$V_{spez}/\text{cm}^3\text{g}^{-1}$		
0.1	1.2095	1.2164	1.2261	1.2366	1.2468
10	1.1979	1.2137	1.2137	1.2233	1.2326
20	1.1863	1.1939	1.2102	1.2102	1.2187
30	1.1773	1.1844	1.1922	1.1997	1.2081
40	1.1693	1.1761	1.1832	1.1903	1.1983
50	1.1620	1.1685	1.1752	1.1820	1.1897
60	1.1550	1.1615	1.1675	1.1740	1.1815
70	1.1486	1.1548	1.1604	1.1669	1.1736
80	1.1424	1.1481	1.1539	1.1601	1.1666
90	1.1367	1.1428	1.1478	1.1539	1.1599
100	1.1312	1.1369	1.1419	1.1479	1.1536

Polymer (B):   **polypropylene**   **1999MAI**

Characterization: $M_n/\text{kg.mol}^{-1} = 140$, $M_w/\text{kg.mol}^{-1} = 280$, isotactic,
synthesized in the laboratory by MBI/MAO metallocene catalysts

P/MPa				T/K					
	455.25	465.60	475.80	485.75	495.55	505.70	515.80	525.50	535.50
				$V_{spez}/\text{cm}^3\text{g}^{-1}$					
0.1	1.2993	1.3087	1.3185	1.3281	1.3379	1.3480	1.3589	1.3691	1.3782
10	1.2830	1.2915	1.3003	1.3089	1.3178	1.3266	1.3361	1.3450	1.3526

continued

continued

20	1.2685	1.2762	1.2843	1.2921	1.3002	1.3080	1.3165	1.3245	1.3310
30	1.2565	1.2635	1.2712	1.2783	1.2861	1.2930	1.3010	1.3081	1.3140
40	1.2453	1.2524	1.2595	1.2662	1.2735	1.2801	1.2874	1.2941	1.2993
50	1.2355	1.2418	1.2490	1.2555	1.2622	1.2683	1.2753	1.2815	1.2864
60	1.2267	1.2327	1.2392	1.2457	1.2521	1.2580	1.2644	1.2705	1.2749
70	1.2185	1.2243	1.2305	1.2364	1.2428	1.2485	1.2546	1.2603	1.2646
80	1.2109	1.2166	1.2226	1.2281	1.2340	1.2398	1.2455	1.2511	1.2550
90	1.2038	1.2093	1.2150	1.2206	1.2260	1.2312	1.2372	1.2423	1.2463
100	1.1973	1.2025	1.2081	1.2134	1.2186	1.2236	1.2289	1.2344	1.2381
110	1.1910	1.1961	1.2015	1.2066	1.2117	1.2166	1.2217	1.2265	1.2305
120	1.1851	1.1900	1.1953	1.2002	1.2051	1.2098	1.2149	1.2193	1.2228
130	1.1795	1.1842	1.1893	1.1942	1.1991	1.2035	1.2083	1.2127	1.2161
140	1.1741	1.1786	1.1838	1.1884	1.1931	1.1975	1.2021	1.2063	1.2097
150	1.1689	1.1734	1.1783	1.1831	1.1875	1.1918	1.1963	1.2003	1.2036
160	1.1640	1.1684	1.1732	1.1777	1.1821	1.1863	1.1907	1.1946	1.1978
170	1.1594	1.1636	1.1682	1.1728	1.1771	1.1810	1.1853	1.1891	1.1923
180	1.1548	1.1589	1.1634	1.1679	1.1721	1.1760	1.1803	1.1839	1.1870
190	1.1505	1.1545	1.1589	1.1635	1.1674	1.1712	1.1754	1.1788	1.1820
200	1.1463	1.1502	1.1546	1.1589	1.1629	1.1667	1.1708	1.1741	1.1771

Polymer (B): **poly(propylene glycol)** **2004BOL**
Characterization: $M_n/\text{kg.mol}^{-1} = 0.425$

P/MPa			*T*/K						
	293.15	303.15	313.15	323.15	333.15	343.15	353.15	363.15	373.15
			$V_{\text{spez}}/\text{cm}^3\text{g}^{-1}$						
0.1	0.9804	0.9881	0.9950	1.0040	1.0121	1.0215	1.0309	1.0395	1.0493
10	0.9747	0.9823	0.9891	0.9970	1.0050	1.0142	1.0225	1.0320	1.0406
20	0.9699	0.9766	0.9843	0.9911	0.9990	1.0070	1.0152	1.0235	1.0331
30	0.9653	0.9718	0.9785	0.9852	0.9921	1.0000	1.0081	1.0163	1.0246
40	0.9606	0.9671	0.9737	0.9794	0.9872	0.9940	1.0020	1.0101	1.0183
50	0.9560	0.9625	0.9690	0.9747	0.9814	0.9881	0.9960	1.0040	1.0121
60	0.9524	0.9588	0.9643	0.9699	0.9766	0.9833	0.9911	0.9990	1.0060
70	0.9488	0.9551	0.9606	0.9662	0.9728	0.9794	0.9862	0.9940	1.0010
80	0.9452	0.9515	0.9569	0.9625	0.9690	0.9756	0.9823	0.9891	0.9960
90	0.9425	0.9479	0.9533	0.9588	0.9653	0.9718	0.9785	0.9852	0.9921
100	0.9398	0.9452	0.9506	0.9560	0.9615	0.9681	0.9747	0.9814	0.9881

Polymer (B): **polystyrene** **1989OUG**
Characterization: $M_n/\text{kg.mol}^{-1} = 240$, $M_w/\text{kg.mol}^{-1} = 253$,
Scientific Polymer Products, Inc., Ontario, NY

continued

continued

P/MPa	T/K								
	391.45	413.75	435.55	455.75	478.55	501.05	534.95	556.25	557.25
				V_{spez}/cm^{3}g^{-1}					
0.1	0.9858	0.9981	1.0111	1.0231	1.0365	1.0500	1.0711	1.0855	1.1003
10	0.9804	0.9921	1.0044	1.0158	1.0285	1.0411	1.0607	1.0739	1.0875
20	0.9752	0.9865	0.9981	1.0090	1.0209	1.0326	1.0511	1.0632	1.0756
30	0.9704	0.9812	0.9925	1.0028	1.0142	1.0254	1.0428	1.0542	1.0657
40	0.9659	0.9762	0.9870	0.9970	1.0080	1.0186	1.0351	1.0457	1.0567
50	0.9617	0.9715	0.9821	0.9916	1.0021	1.0124	1.0282	1.0383	1.0488
60	0.9580	0.9672	0.9772	0.9866	0.9966	1.0064	1.0215	1.0311	1.0411
70	0.9546	0.9630	0.9728	0.9819	0.9916	1.0010	1.0154	1.0246	1.0342
80	0.9516	0.9589	0.9685	0.9773	0.9867	0.9959	1.0098	1.0187	1.0279
90	0.9486	0.9551	0.9644	0.9731	0.9822	0.9909	1.0045	1.0130	1.0218
100	0.9459	0.9515	0.9605	0.9689	0.9778	0.9863	0.9994	1.0076	1.0163
110	0.9435	0.9480	0.9568	0.9650	0.9737	0.9820	0.9945	1.0026	1.0109
120	0.9409	0.9448	0.9533	0.9613	0.9697	0.9778	0.9902	0.9978	1.0059
130	0.9386	0.9417	0.9498	0.9576	0.9658	0.9737	0.9857	0.9932	1.0016
140	0.9363	0.9388	0.9464	0.9541	0.9623	0.9699	0.9816	0.9890	0.9966
150	0.9342	0.9362	0.9432	0.9507	0.9587	0.9662	0.9775	0.9847	0.9920
160	0.9320	0.9336	0.9401	0.9475	0.9552	0.9625	0.9736	0.9806	0.9878
170	0.9299	0.9312	0.9371	0.9443	0.9518	0.9590	0.9698	0.9767	0.9837
180	0.9279	0.9289	0.9341	0.9412	0.9486	0.9556	0.9662	0.9729	0.9798
190	0.9260	0.9266	0.9313	0.9381	0.9454	0.9523	0.9627	0.9691	0.9759
200	0.9240	0.9246	0.9285	0.9353	0.9424	0.9490	0.9592	0.9656	0.9723

Polymer (B): **poly(vinyl chloride)** **1993ROG**

Characterization: M_w/kg.mol^{-1} = 63, Vista Chemical

P/MPa	T/K					
	373.35	383.25	393.25	403.25	413.25	423.25
			V_{spez}/cm^{3}g^{-1}			
0.1	0.7401	0.7432	0.7476	0.7519	0.7561	0.7610
10	0.7363	0.7394	0.7436	0.7474	0.7512	0.7561
20	0.7322	0.7358	0.7397	0.7425	0.7456	0.7509
30	0.7289	0.7321	0.7361	0.7391	0.7421	0.7470
40	0.7260	0.7291	0.7329	0.7355	0.7385	0.7431
50		0.7263	0.7296	0.7322	0.7351	0.7396
60		0.7234	0.7266	0.7291	0.7320	0.7362
70		0.7207	0.7238	0.7263	0.7287	0.7330
80		0.7182	0.7211	0.7235	0.7259	0.7301
90			0.7188	0.7208	0.7232	0.7273
100			0.7162	0.7184	0.7208	0.7244

continued

continued

110		0.7139	0.7158	0.7184	0.7219
120		0.7114	0.7136	0.7158	0.7194
130			0.7114	0.7136	0.7171
140			0.7092	0.7113	0.7145
150			0.7071	0.7092	0.7122
160			0.7052	0.7071	0.7099
170			0.7033	0.7048	0.7080

Polymer (B): **poly(vinylidene fluoride)**   **2001MEK**
Characterization: $M_n/\text{kg.mol}^{-1} = 95$, $M_w/\text{kg.mol}^{-1} = 197$,
ATOFINA Chemicals, Inc., Philadelphia, PA

P/MPa				T/K					
	453.15	458.15	463.15	468.15	473.15	478.15	483.15	488.15	493.15
				$V_{\text{spez}}/\text{cm}^3\text{g}^{-1}$					
0.1	0.6716	0.6742	0.6766	0.6782	0.6801	0.6813	0.6829	0.6849	0.6866
20	0.6615	0.6640	0.6662	0.6680	0.6697	0.6712	0.6733	0.6752	0.6768
40	0.6514	0.6534	0.6553	0.6574	0.6593	0.6615	0.6636	0.6654	0.6670
80	0.6368	0.6384	0.6398	0.6415	0.6430	0.6446	0.6463	0.6476	0.6490
120	0.6238	0.6267	0.6281	0.6296	0.6310	0.6323	0.6338	0.6352	0.6363

Polymer (B): **poly(vinyl methyl ether)**   **1993JAN**
Characterization: $M_w/\text{kg.mol}^{-1} = 64$

P/MPa		T/K				
	311.5	340.3	353.5	374.5	394.5	415.5
			$V_{\text{spez}}/\text{cm}^3\text{g}^{-1}$			
0.1	0.9671	0.9861	0.9935	1.0053	1.0189	1.0344
5.0	0.9646	0.9830	0.9902	1.0016	1.0149	1.0299
10.0	0.9621	0.9799	0.9870	0.9981	1.0110	1.0257
20.0	0.9574	0.9742	0.9809	0.9913	1.0038	1.0174
30.0	0.9528	0.9687	0.9754	0.9853	0.9973	1.0105
40.0	0.9485	0.9637	0.9703	0.9795	0.9909	1.0034
50.0	0.9443	0.9589	0.9654	0.9742	0.9850	0.9971
60.0	0.9404	0.9545	0.9608	0.9692	0.9796	0.9915
70.0	0.9366	0.9501	0.9564	0.9645	0.9747	0.9859
80.0	0.9329	0.9462	0.9523	0.9600	0.9699	0.9808
90.0	0.9296	0.9423	0.9484	0.9558	0.9653	0.9760
100.0	0.9263	0.9387	0.9447	0.9518	0.9609	0.9714
110.0	0.9232	0.9351	0.9412	0.9480	0.9571	0.9671
120.0	0.9202	0.9318	0.9379	0.9443	0.9531	0.9630

6.2. Excess volumes and/or densities at elevated pressures
6.2.1. Binary polymer solutions

Polymer (B):	**ethylene/1-butene copolymer**								**2000BY2**

Characterization: M_n/kg.mol^{-1} = 175, M_w/kg.mol^{-1} = 177, 94 mol% 1-butene, Exxon Research and Engineering Co.

Solvent (A):	**dimethyl ether**		**C$_2$H$_6$O**						**115-10-6**

T/K = 383.15 w_B 0.05

P/bar	346.6	415.5	519.0	691.4	1036.2	1381.0	1725.9	2070.7	2415.5
ρ/g cm^{-3}	0.669	0.683	0.701	0.725	0.763	0.793	0.818	0.840	0.860

T/K = 393.15 w_B 0.05

P/bar	346.6	415.5	519.0	691.4	1036.2	1381.0	1725.9	2070.7	2415.5
ρ/g cm^{-3}	0.660	0.675	0.694	0.719	0.759	0.790	0.816	0.839	0.859

T/K = 403.15 w_B 0.05

P/bar	346.6	415.5	519.0	691.4	1036.2	1381.0	1725.9	2070.7	2415.5
ρ/g cm^{-3}	0.643	0.659	0.678	0.705	0.746	0.777	0.803	0.826	0.846

T/K = 423.15 w_B 0.05

P/bar	346.6	415.5	519.0	691.4	1036.2	1381.0	1725.9	2070.7	2415.5
ρ/g cm^{-3}	0.612	0.632	0.655	0.684	0.727	0.760	0.787	0.811	0.831

Polymer (B):	**ethylene/1-butene copolymer**								**2000BY2**

Characterization: M_n/kg.mol^{-1} = 230, M_w/kg.mol^{-1} = 232, 20.2 mol% 1-butene, Exxon Research and Engineering Co.

Solvent (A):	**dimethyl ether-d6**		**C$_2$D$_6$O**						**17222-37-6**

T/K = 383.15 w_B 0.05

P/bar	1001.7	1036.2	1105.2	1174.1	1280.6	1381.0	1484.5	1725.9	1932.8
ρ/g cm^{-3}	0.850	0.857	0.860	0.868	0.872	0.889	0.898	0.921	0.938

P/bar	2070.7	2312.1	2401.7
ρ/g cm^{-3}	0.949	0.969	0.980

T/K = 393.15 w_B 0.05

P/bar	760.3	829.3	1036.2	1381.0	1725.9	2070.7	2415.5
ρ/g cm^{-3}	0.813	0.823	0.848	0.883	0.915	0.945	0.959

T/K = 403.15 w_B 0.05

P/bar	650.0	670.7	691.4	829.3	1036.2	1381.0	1725.9	2139.7	2415.5
ρ/g cm^{-3}	0.778	0.781	0.785	0.804	0.830	0.866	0.896	0.929	0.959

T/K = 423.15 w_B 0.05

P/bar	622.4	691.4	829.3	1036.2	1381.0	1725.9	2070.7	2415.5
ρ/g cm^{-3}	0.748	0.760	0.782	0.809	0.846	0.878	0.906	0.933

Polymer (B):   **ethylene/1-butene copolymer**   **2000BY2**
Characterization: M_n/kg.mol^{-1} = 175, M_w/kg.mol^{-1} = 177, 94 mol% 1-butene,
                          Exxon Research and Engineering Co.
Solvent (A):   **dimethyl ether-d6**   **C$_2$D$_6$O**   **17222-37-6**

T/K = 383.15 w_B 0.05

P/bar	346.6	415.5	519.0	691.4	1036.2	1381.0	1725.9	2070.7	2415.5
ρ/g cm^{-3}	0.685	0.698	0.715	0.739	0.776	0.804	0.828	0.849	0.867

T/K = 393.15 w_B 0.05

P/bar	346.6	415.5	519.0	691.4	1036.2	1381.0	1725.9	2070.7	2415.5
ρ/g cm^{-3}	0.676	0.690	0.708	0.733	0.772	0.801	0.826	0.847	0.866

T/K = 403.15 w_B 0.05

P/bar	346.6	415.5	519.0	691.4	1036.2	1381.0	1725.9	2070.7	2415.5
ρ/g cm^{-3}	0.659	0.675	0.694	0.720	0.759	0.789	0.814	0.835	0.854

T/K = 423.15 w_B 0.05

P/bar	346.6	415.5	519.0	691.4	1036.2	1381.0	1725.9	2070.7	2415.5
ρ/g cm^{-3}	0.630	0.649	0.671	0.699	0.740	0.772	0.798	0.821	0.840

Polymer (B):   **ethylene/1-butene copolymer (deuterated)**   **2000BY1**
Characterization: M_n/kg.mol^{-1} = 155, M_w/kg.mol^{-1} = 169, 3 mol% 1-butene,
                          synthesized from polymerization of deuterated 1,3-butadiene with
                          controlled 1,2-addition, Polymer Source, Inc., Dorval, Quebec
Solvent (A):   **ethane**   **C$_2$H$_6$**   **74-84-0**

T/K = 393.15 w_B 0.024

P/bar	1242.8	1311.7	1380.7	1518.5	1725.4	1897.8	2070.1	2242.5	2414.9
ρ/g cm^{-3}	0.511	0.518	0.523	0.532	0.546	0.554	0.563	0.571	0.580

T/K = 403.15 w_B 0.024

P/bar	1242.8	1311.7	1380.7	1518.5	1725.4	1897.8	2070.1	2242.5	2414.9
ρ/g cm^{-3}	0.506	0.513	0.518	0.528	0.541	0.551	0.561	0.570	0.578

T/K = 423.15 w_B 0.024

P/bar	1242.8	1311.7	1380.7	1518.5	1725.4	1897.8	2070.1	2242.5	2414.9
ρ/g cm^{-3}	0.497	0.503	0.508	0.519	0.533	0.543	0.553	0.562	0.571

Polymer (B):   **ethylene/1-butene copolymer**   **2000BY1**
Characterization: M_n/kg.mol^{-1} = 230, M_w/kg.mol^{-1} = 232, 20.2 mol% 1-butene,
                          Exxon Research and Engineering Co.
Solvent (A):   **n-pentane**   **C$_5$H$_{12}$**   **109-66-0**

T/K = 383.15 w_B 0.053

P/bar	70.7	139.6	208.6	277.5	346.5	518.8	691.2	1035.9	1380.7
ρ/g cm^{-3}	0.560	0.575	0.586	0.596	0.605	0.623	0.638	0.664	0.685

continued

continued

P/bar	2070.1	2414.9
ρ/g cm^{-3}	0.719	0.733

T/K = 403.15 w_B 0.053

P/bar	45.2	70.7	139.6	208.6	277.5	346.5	415.4	484.4	553.3
ρ/g cm^{-3}	0.528	0.536	0.555	0.569	0.581	0.591	0.599	0.607	0.614

P/bar	622.2	691.2	760.1	829.1	898.0	967.0	1035.9	1104.9	1173.8
ρ/g cm^{-3}	0.621	0.627	0.633	0.639	0.644	0.649	0.654	0.659	0.663

P/bar	1242.8	1311.7	1380.7	1449.6	1518.5	1587.5	1656.4	1725.4	1794.3
ρ/g cm^{-3}	0.668	0.672	0.676	0.680	0.684	0.688	0.691	0.695	0.698

P/bar	1863.3	1932.2	2001.2	2070.1
ρ/g cm^{-3}	0.702	0.705	0.708	0.712

T/K = 423.15 w_B 0.053

P/bar	105.1	139.6	208.6	277.5	346.5	518.8	691.2	1035.9	1380.7
ρ/g cm^{-3}	0.525	0.535	0.552	0.566	0.577	0.599	0.617	0.645	0.668

P/bar	2070.1	2414.9
ρ/g cm^{-3}	0.704	0.720

Polymer (B):	**ethylene/1-butene copolymer**		**2000BY1**
Characterization:	M_n/kg.mol^{-1} = 230, M_w/kg.mol^{-1} = 232, 20.2 mol% 1-butene, Exxon Research and Engineering Co.		
Solvent (A):	**n-pentane-d12** C_5D_{12}		**2031-90-5**

T/K = 383.15 w_B 0.051

P/bar	70.7	139.6	208.6	277.5	346.5	518.8	691.2	1035.9	1380.7
ρ/g cm^{-3}	0.644	0.662	0.676	0.688	0.699	0.722	0.740	0.771	0.796

P/bar	2070.1	2414.9
ρ/g cm^{-3}	0.837	0.855

T/K = 403.15 w_B 0.051

P/bar	70.7	77.6	84.5	91.3	105.1	125.8	139.6	174.1	208.6
ρ/g cm^{-3}	0.617	0.619	0.621	0.624	0.628	0.634	0.638	0.647	0.655

P/bar	277.5	346.5	518.8	689.5	1034.2	1380.7	2070.1	2414.9
ρ/g cm^{-3}	0.669	0.681	0.705	0.725	0.757	0.783	0.826	0.844

T/K = 423.15 w_B 0.051

P/bar	108.6	139.6	208.6	277.5	346.5	518.8	691.2	1035.9	1380.7
ρ/g cm^{-3}	0.605	0.616	0.635	0.651	0.664	0.692	0.714	0.747	0.774

P/bar	2070.1	2414.9
ρ/g cm^{-3}	0.818	0.837

Polymer (B):	**polybutadiene**							**1999INO**
Characterization:	M_n/kg.mol^{-1} = 159, M_w/kg.mol^{-1} = 165,							
	low *cis*-butadiene rubber							
Solvent (A):	**n-hexane**		**C$_6$H$_{14}$**					**110-54-3**

w_B = 0.92

P/MPa				T/K				
	303.15	328.15	353.15	378.15	403.15	428.15	453.15	478.15	503.15
				V_{spez}/cm^{3}g^{-1}				

P/MPa	303.15	328.15	353.15	378.15	403.15	428.15	453.15	478.15	503.15
0.1	1.23	1.24	1.26	1.28	1.30	1.32	1.34	1.36	1.38
20	1.22	1.23	1.25	1.26	1.28	1.30	1.31	1.33	1.35
40	1.21	1.22	1.24	1.25	1.26	1.28	1.30	1.31	1.32
60	1.20	1.21	1.22	1.24	1.25	1.27	1.28	1.29	1.31
80	1.19	1.20	1.21	1.23	1.24	1.25	1.27	1.28	1.29
100	1.18	1.19	1.21	1.22	1.23	1.24	1.26	1.27	1.28
120	1.17	1.19	1.20	1.21	1.22	1.23	1.25	1.26	1.27
150	1.17	1.18	1.19	1.20	1.21	1.22	1.23	1.24	1.25

Polymer (B):	**poly(dimethylsiloxane)**		**1994MER**
Characterization:	M_n/kg.mol^{-1} = 28, M_w/kg.mol^{-1} = 42,		
	Wacker GmbH, Munich, Germany		
Solvent (A):	**carbon dioxide** **CO$_2$**		**124-38-9**

T/K = 313.15 P/MPa = 30.0

φ_B	0.3	0.5	0.75	0.9
V^E/cm3 g^{-1}	−0.0055	−0.0089	−0.0082	−0.0043

Comments: The data were taken from Fig. 4 in the original source.

Polymer (B):	**poly(dimethylsiloxane)**		**2002DIN**
Characterization:	M_n/kg.mol^{-1} = 31.3, M_w/kg.mol^{-1} = 93.7		
Solvent (A):	**carbon dioxide** **CO$_2$**		**124-38-9**

w_B = 0.055 T/K = 328.15

P/MPa	48.70	44.93	41.74	39.29	34.72	33.19	33.19	30.78	28.04
ρ/g cm^{-3}	0.9801	0.9663	0.9534	0.9420	0.9189	0.9090	0.9090	0.8959	0.8765

w_B = 0.055 T/K = 343.15

P/MPa	48.45	46.38	44.90	43.38	41.54	40.07	38.17	35.86	35.16
ρ/g cm^{-3}	0.9349	0.9349	0.9222	0.9203	0.9064	0.9030	0.8887	0.8759	0.8723

continued

continued

$w_B = 0.055$ $T/K = 358.15$

P/MPa	48.28	44.83	41.59	39.41	37.75
ρ/g cm^{-3}	0.9054	0.8901	0.8734	0.8548	0.8476

$w_B = 0.055$ $T/K = 373.15$

P/MPa	48.21	46.55	44.83	43.79
ρ/g cm^{-3}	0.8610	0.8503	0.8396	0.8325

Polymer (B):   **poly(dimethylsiloxane)**   **1979REN**
Characterization: –
Solvent (A):   **cyclohexane**   **C₆H₁₂**   **110-82-7**

T-range/K = 298.15 to 338.15 P-range/bar = 0.1 to 1000

Coefficients of Tait equation (ϑ is the temperature in °C):

$$V_{spez} = \left(A_1 + A_2\vartheta + A_3\vartheta^2 \right) / \left\{ 1 - A_4 \ln\left[1 + P / A_5\, e^{-A_6\vartheta} \right] \right\}$$

φ_B	A_1/g cm^{-3}	$10^3 A_2$/g cm^{-3} °C^{-1}	$10^6 A_3$/g cm^{-3} °C^{-2}	$100 A_4$	$10^{-3} A_5$/bar	$10^3 B_6$/°C^{-1}
0.3530	1.07707	3.34760	−0.01903	9.136689	0.939672	7.512650
0.5017	1.10646	1.13189	1.24603	9.061028	0.962103	7.913653

Polymer (B):   **poly(dimethylsiloxane)**   **1979REN**
Characterization: –
Solvent (A):   **hexamethyldisiloxane**   **C₆H₁₈OSi₂**   **107-46-0**

T-range/K = 298.15 to 338.15 P-range/bar = 0.1 to 1000

Coefficients of Tait equation (ϑ is the temperature in °C):

$$V_{spez} = \left(A_1 + A_2\vartheta + A_3\vartheta^2 \right) / \left\{ 1 - A_4 \ln\left[1 + P / A_5\, e^{-A_6\vartheta} \right] \right\}$$

φ_B	A_1/g cm^{-3}	$10^3 A_2$/g cm^{-3} °C^{-1}	$10^6 A_3$/g cm^{-3} °C^{-2}	$100 A_4$	$10^{-3} A_5$/bar	$10^3 B_6$/°C^{-1}
0.4370	1.12518	1.38201	0.72986	9.227156	0.701099	7.974029
0.6497	1.04932	1.80918	−3.75208	8.568493	0.681781	7.013003

Polymer (B):   **polyethylene**   **1979KOB**
Characterization: M_n/kg.mol^{-1} = 0.74, M_w/kg.mol^{-1} = 1.60, LDPE-wax
Solvent (A):   **ethene**   **C$_2$H$_4$**   **74-85-1**

P/MPa	0.00	0.20	0.40	0.60	0.80	1.00
			w_B			
			ρ/g cm^{-3}			

T/K = 393.2

68.64	2.39	2.15	1.91	1.67	1.43	1.19
78.45	2.30	2.07	1.85	1.62	1.40	1.18
98.67	2.18	1.94	1.77	1.57	1.36	1.16
117.7	2.09	1.90	1.71	1.52	1.33	1.14
137.3	2.01	1.84	1.66	1.48	1.31	1.13
156.9	1.96	1.79	1.62	1.45	1.29	1.12
176.5	1.90	1.75	1.59	1.43	1.27	1.11
196.1	1.86	1.71	1.56	1.40	1.25	1.10

T/K = 433.2

68.64	2.58	2.30	2.03	1.76	1.49	1.22
78.45	2.46	2.21	1.96	1.71	1.46	1.21
98.67	2.31	2.08	1.86	1.64	1.41	1.19
117.7	2.19	1.99	1.78	1.58	1.38	1.17
137.3	2.11	1.92	1.73	1.54	1.35	1.16
156.9	2.04	1.86	1.68	1.50	1.32	1.14
176.5	1.98	1.81	1.64	1.47	1.30	1.13
196.1	1.94	1.77	1.61	1.44	1.28	1.12

T/K = 473.2

58.84	2.95	2.62	2.28	1.94	1.60	1.26
68.64	2.77	2.47	2.16	1.86	1.56	1.25
78.45	2.63	2.36	2.08	1.80	1.52	1.24
98.67	2.44	2.20	1.95	1.71	1.46	1.22
117.7	2.30	2.08	1.86	1.64	1.42	1.20
137.3	2.21	2.00	1.80	1.59	1.39	1.18
156.9	2.13	1.94	1.74	1.55	1.36	1.17
176.5	2.06	1.88	1.70	1.52	1.34	1.15
196.1	2.01	1.83	1.66	1.49	1.31	1.14

| Polymer (B): | polyethylene | | | | | **1979KOB** |

Polymer (B): polyethylene   **1979KOB**

Characterization: M_n/kg.mol^{-1} = 4.94, M_w/kg.mol^{-1} = 11.0, LDPE-wax

Solvent (A): ethene C_2H_4 **74-85-1**

P/MPa	w_B					
	0.00	0.20	0.40	0.60	0.80	1.00
			ρ/g cm^{-3}			

T/K = 393.2

137.3	2.01	1.83	1.65	1.47	1.29	1.11
156.9	1.96	1.78	1.61	1.44	1.27	1.10
176.5	1.90	1.74	1.58	1.42	1.26	1.09
196.1	1.86	1.71	1.55	1.40	1.24	1.08

T/K = 433.2

137.3	2.11	1.91	1.72	1.53	1.33	1.14
156.9	2.04	1.86	1.67	1.49	1.31	1.13
176.5	1.98	1.81	1.63	1.46	1.29	1.11
196.1	1.94	1.77	1.60	1.44	1.27	1.10

T/K = 473.2

137.3	2.21	2.00	1.79	1.58	1.37	1.16
156.9	2.13	1.93	1.73	1.54	1.34	1.15
176.5	2.06	1.87	1.69	1.50	1.32	1.13
196.1	2.01	1.83	1.65	1.47	1.30	1.12

Polymer (B): polyethylene   **2002DIN**

Characterization: M_n/kg.mol^{-1} = 28.1, M_w/kg.mol^{-1} = 121

Solvent (A): n-pentane C_5H_{12} **109-66-0**

w_B = 0.0574 T/K = 413.15

P/MPa	48.51	41.81	34.40	24.26	21.55	18.62	17.59		
ρ/g cm^{-3}	0.5751	0.5623	0.5511	0.5346	0.5296	0.5296	0.5255		

w_B = 0.0574 T/K = 423.15

P/MPa	48.29	44.86	41.65	34.88	31.34	27.76	20.83	15.52	12.97
ρ/g cm^{-3}	0.5819	0.5780	0.5737	0.5648	0.5597	0.5538	0.5411	0.5310	0.5255

Polymer (B):	**polyethylene**							**1998AAL**
Characterization:	$M_n/\text{kg.mol}^{-1} = 0.30$, $M_w/\text{kg.mol}^{-1} = 0.49$, LLDPE,							
	Borealis Pilot Plant, Finland							
Solvent (A):	**propane**			$\mathbf{C_3H_8}$				**74-98-6**

$w_B = 0.0094$

T/K	358.5	358.6	358.6	358.6	358.6	358.6	358.6	363.5	363.5
P/MPa	4.998	5.502	5.998	6.197	6.406	6.596	7.000	5.502	5.998
ρ/g cm^{-3}	0.3881	0.3951	0.3990	0.4008	0.4023	0.4045	0.4070	0.3782	0.3846

T/K	363.5	363.5	363.5	363.6	368.3	368.4	368.4	368.4	368.4
P/MPa	6.197	6.399	6.603	6.995	6.004	6.201	6.399	6.599	7.001
ρ/g cm^{-3}	0.3873	0.3892	0.3925	0.3956	0.3696	0.3727	0.3755	0.3789	0.3826

$w_B = 0.0097$

T/K	353.8	353.9	353.9	353.9	353.9	353.9	353.9	353.9
P/MPa	3.999	4.994	5.498	5.998	6.200	6.400	6.603	7.004
ρ/g cm^{-3}	0.3894	0.4018	0.4070	0.4115	0.4131	0.4149	0.4157	0.4199

T/K	373.5	373.5	373.5	373.5	378.2	378.2	378.2	378.2
P/MPa	6.198	6.399	6.603	6.998	6.202	6.400	6.600	6.999
ρ/g cm^{-3}	0.3560	0.3615	0.3647	0.3708	0.3375	0.3422	0.3469	0.3549

$w_B = 0.0101$

T/K	358.5	358.5	358.5	358.5	358.5	358.5	358.5	363.5	363.5
P/MPa	4.996	5.502	5.997	6.203	6.400	6.602	7.000	5.497	6.003
ρ/g cm^{-3}	0.3877	0.3937	0.3984	0.4006	0.4033	0.4050	0.4090	0.3792	0.3854

T/K	363.5	363.5	363.5	363.5	368.3	368.3	368.3	368.3	368.3
P/MPa	6.200	6.398	6.599	6.998	5.997	6.202	6.402	6.602	6.999
ρ/g cm^{-3}	0.3874	0.3901	0.3922	0.3970	0.3694	0.3728	0.3759	0.3781	0.3837

$w_B = 0.0232$

T/K	358.7	358.7	358.7	358.7	358.7	358.7	358.8	363.7	363.7
P/MPa	5.000	5.498	6.000	6.204	6.398	6.601	7.000	5.501	6.001
ρ/g cm^{-3}	0.3937	0.4005	0.4047	0.4072	0.4092	0.4112	0.4148	0.3853	0.3923

T/K	363.7	363.7	363.7	363.7	368.6	368.6	368.6	368.6	368.6
P/MPa	6.203	6.401	6.603	7.000	5.997	6.202	6.399	6.602	6.999
ρ/g cm^{-3}	0.3947	0.3967	0.3994	0.4033	0.3770	0.3799	0.3829	0.3856	0.3897

$w_B = 0.0355$

T/K	358.7	358.7	358.7	358.7	358.7	358.7	358.7	368.5	368.5
P/MPa	5.001	5.500	6.001	6.202	6.401	6.601	6.999	6.001	6.199
ρ/g cm^{-3}	0.3966	0.4027	0.4079	0.4101	0.4115	0.4131	0.4164	0.3811	

T/K	368.5	368.5	368.5	368.5	
P/MPa	6.399	6.600	6.997	7.097	
ρ/g cm^{-3}	0.3842	0.3871	0.3895	0.3937	0.3950

Polymer (B):	poly(ethylene glycol)							**2000LEE**
Characterization:	M_n/kg.mol^{-1} = 0.210, M_w/kg.mol^{-1} = 0.235,							
	Aldrich Chem. Co., Inc., Milwaukee, WI							
Solvent (A):	anisole			C_7H_8O				**100-66-3**

P/MPa				w_B					
	0.0000	0.1775	0.3268	0.4542	0.5642	0.6600	0.7444	0.8859	1.0000
				ρ/g cm^{-3}					

T/K = 298.15

0.1	0.9888	1.0061	1.0241	1.0405	1.0541	1.0665	1.0773	1.1008	1.1179
10	0.9951	1.0122	1.0298	1.0462	1.0596	1.0724	1.0832	1.1054	1.1235
20	1.0012	1.0177	1.0357	1.0512	1.0644	1.0770	1.0880	1.1101	1.1278
30	1.0068	1.0235	1.0406	1.0561	1.0697	1.0821	1.0926	1.1143	1.1316
40	1.0122	1.0287	1.0456	1.0611	1.0740	1.0863	1.0969	1.1186	1.1361
50	1.0172	1.0336	1.0504	1.0658	1.0789	1.0910	1.1017	1.1227	1.1401

T/K = 318.15

0.1	0.9689	0.9879	1.0061	1.0245	1.0367	1.0484	1.0611	1.0844	1.1108
10	0.9767	0.9943	1.0123	1.0306	1.0431	1.0554	1.0663	1.0894	1.1088
20	0.9836	1.0006	1.0185	1.0346	1.0485	1.0605	1.0713	1.0941	1.1131
30	0.9898	1.0065	1.0242	1.0397	1.0536	1.0657	1.0760	1.0988	1.1176
40	0.9954	1.0122	1.0293	1.0448	1.0585	1.0705	1.0805	1.1031	1.1215
50	1.0010	1.0176	1.0345	1.0499	1.0634	1.0750	1.0859	1.1075	1.1256

T/K = 338.15

0.1	0.9564	0.9692	0.9867	1.0062	1.0184	1.0327	1.0444	1.0659	1.0903
10	0.9581	0.9774	0.9945	1.0125	1.0258	1.0385	1.0502	1.0737	1.0930
20	0.9655	0.9832	1.0005	1.0169	1.0310	1.0434	1.0543	1.0779	1.0970
30	0.9724	0.9889	1.0068	1.0230	1.0369	1.0488	1.0594	1.0834	1.1019
40	0.9789	0.9961	1.0131	1.0290	1.0428	1.0544	1.0651	1.0884	1.1064
50	0.9851	1.0020	1.0187	1.0339	1.0481	1.0596	1.0705	1.0934	1.1120

Polymer (B):	poly(ethylene glycol)						1999LEE	
Characterization:	M_n/kg.mol^{-1} = 0.2, Aldrich Chem. Co., Inc., Milwaukee, WI							
Solvent (A):	1-octanol		$C_8H_{18}O$				111-87-5	

P/MPa				w_B					
	0.1458	0.2774	0.3969	0.5059	0.6056	0.6973	0.7818	0.8600	0.9325
				ρ/g cm^{-3}					

T/K = 298.15

0.1	0.8534	0.8849	0.9159	0.9466	0.9765	1.0061	1.0351	1.0634	1.0916
5	0.8566	0.8880	0.9189	0.9495	0.9793	1.0088	1.0376	1.0659	1.0940
10	0.8597	0.8910	0.9218	0.9523	0.9821	1.0114	1.0402	1.0683	1.0963
15	0.8627	0.8939	0.9246	0.9551	0.9847	1.0139	1.0426	1.0706	1.0986
20	0.8656	0.8967	0.9273	0.9577	0.9872	1.0164	1.0450	1.0729	1.1007
25	0.8683	0.8997	0.9299	0.9603	0.9897	1.0187	1.0472	1.0751	1.1028
30	0.8710	0.9020	0.9325	0.9628	0.9921	1.0210	1.0494	1.0772	1.1049

T/K = 318.15

0.1	0.8394	0.8707	0.9015	0.9321	0.9617	0.9910	1.0199	1.0480	1.0760
5	0.8428	0.8740	0.9047	0.9351	0.9648	0.9929	1.0226	1.0506	1.0785
10	0.8462	0.8773	0.9079	0.9382	0.9677	0.9967	1.0253	1.0532	1.0809
15	0.8495	0.8805	0.9109	0.9411	0.9705	0.9994	1.0279	1.0557	1.0833
20	0.8526	0.8835	0.9138	0.9440	0.9733	1.0021	1.0304	1.0581	1.0856
25	0.8556	0.8865	0.9167	0.9468	0.9761	1.0047	1.0329	1.0605	1.0879
30	0.8586	0.8894	0.9195	0.9495	0.9786	1.0072	1.0354	1.0629	1.0901

T/K = 328.15

0.1	0.8320	0.8630	0.8936	0.8241	0.9537	0.9830	1.0118	1.0399	1.0680
5	0.8356	0.8665	0.8970	0.8273	0.9569	0.9860	1.0146	1.0426	1.0706
10	0.8392	0.8700	0.9003	0.9305	0.9599	0.9890	1.0175	1.0453	1.0732
15	0.8427	0.8732	0.9035	0.9336	0.9629	0.9918	1.0202	1.0480	1.0757
20	0.8458	0.8764	0.9066	0.9366	0.9667	0.9945	1.0228	1.0505	1.0781
25	0.8489	0.8794	0.9095	0.9394	0.9685	0.9972	1.0254	1.0529	1.0804
30	0.8519	0.8824	0.9124	0.9422	0.9712	0.9998	1.0279	1.0553	1.0827

T/K = 338.15

0.1	0.8248	0.8558	0.8863	0.9167	0.9462	0.9756	1.0043	1.0322	1.0603
5	0.8286	0.8596	0.8899	0.9201	0.9495	0.9788	1.0072	1.0350	1.0630
10	0.8323	0.8632	0.8934	0.9235	0.9528	0.9818	1.0102	1.0378	1.0656
15	0.8358	0.8666	0.8967	0.9267	0.9558	0.9848	1.0130	1.0405	1.0682
20	0.8392	0.8699	0.8998	0.9297	0.9588	0.9876	1.0157	1.0431	1.0707
25	0.8425	0.8730	0.9029	0.9327	0.9616	0.9904	1.0184	1.0456	1.0731
30	0.8455	0.8760	0.9058	0.9355	0.9644	0.9931	1.0209	1.0481	1.0755

Polymer (B):   **poly(ethylene glycol)**   **1999LEE**
Characterization: M_n/kg.mol^{-1} = 0.6, Aldrich Chem. Co., Inc., Milwaukee, WI
Solvent (A):   **1-octanol**   **C$_8$H$_{18}$O**   **111-87-5**

P/MPa				w_B					
	0.3386	0.5353	0.6638	0.7544	0.8217	0.8736	0.9146	0.9485	0.9765
				ρ/g cm^{-3}					

T/K = 298.15

0.1	0.9019	0.9567	0.9968	1.0270	1.0509	1.0704	1.0861	1.0999	1.1113
5	0.9049	0.9596	0.9995	1.0297	1.0534	1.0729	1.0885	1.1023	1.1136
10	0.9079	0.9624	1.0022	1.0322	1.0559	1.0753	1.0909	1.1046	1.1159
15	0.9107	0.9651	1.0048	1.0348	1.0584	1.0777	1.0932	1.1069	1.1181
20	0.9135	0.9677	1.0073	1.0372	1.0607	1.0800	1.0954	1.1091	1.1203
25	0.9161	0.9702	1.0097	1.0395	1.0629	1.0822	1.0976	1.1112	1.1224
30	0.9187	0.9727	1.0120	1.0418	1.0652	1.0843	1.0997	1.1133	1.1244

T/K = 318.15

0.1	0.8875	0.9418	0.9813	1.0116	1.0351	1.0546	1.0702	1.0838	1.0949
5	0.8907	0.9449	0.9843	1.0144	1.0378	1.0573	1.0728	1.0863	1.0974
10	0.8939	0.9480	0.9872	1.0172	1.0406	1.0599	1.0754	1.0888	1.0999
15	0.8971	0.9509	0.9900	1.0199	1.0432	1.0624	1.0779	1.0912	1.1023
20	0.9001	0.9537	0.9928	1.0226	1.0457	1.0649	1.0803	1.0936	1.1045
25	0.9030	0.9565	0.9954	1.0251	1.0482	1.0673	1.0827	1.0960	1.1070
30	0.9058	0.9592	0.9980	1.0277	1.0507	1.0696	1.0850	1.0973	1.1092

T/K = 328.15

0.1	0.8797	0.9340	0.9734	1.0035	1.0268	1.0462	1.0617	1.0752	1.0865
5	0.8832	0.9372	0.9764	1.0065	1.0297	1.0490	1.0644	1.0779	1.0892
10	0.8866	0.9404	0.9795	1.0094	1.0326	1.0518	1.0671	1.0806	1.0918
15	0.8898	0.9435	0.9824	1.0122	1.0353	1.0544	1.0697	1.0832	1.0943
20	0.8929	0.9464	0.9852	1.0150	1.0380	1.0570	1.0723	1.0856	1.0967
25	0.8959	0.9493	0.9880	1.0176	1.0406	1.0595	1.0746	1.0880	1.0991
30	0.8988	0.9520	0.9907	1.0201	1.0431	1.0619	1.0771	1.0904	1.1014

T/K = 338.15

0.1	0.8724	0.9265	0.9657	0.9958	1.0193	1.0384	1.0539	1.0673	1.0786
5	0.8760	0.9299	0.9690	0.9989	1.0222	1.0413	1.0567	1.0701	1.0813
10	0.8795	0.9332	0.9721	1.0020	1.0252	1.0442	1.0595	1.0728	1.0840
15	0.8829	0.9364	0.9752	1.0049	1.0280	1.0470	1.0622	1.0755	1.0866
20	0.8861	0.9394	0.9781	1.0080	1.0308	1.0496	1.0648	1.0781	1.0891
25	0.8892	0.9424	0.9809	1.0104	1.0334	1.0522	1.0674	1.0805	1.0916
30	0.8922	0.9452	0.9836	1.0130	1.0360	1.0547	1.0698	1.0829	1.0940

Polymer (B):	poly(ethylene glycol) monomethyl ether						2000LEE

Polymer (B): poly(ethylene glycol) monomethyl ether   **2000LEE**
Characterization: M_n/kg.mol^{-1} = 0.36, M_w/kg.mol^{-1} = 0.39,
Aldrich Chem. Co., Inc., Milwaukee, WI
Solvent (A): anisole C_7H_8O **100-66-3**

P/MPa				w_B				
	0.0000	0.2689	0.4528	0.5866	0.6882	0.7680	0.8324	0.9298	1.0000

ρ/g cm^{-3}

T/K = 298.15

0.1	0.9888	1.0119	1.0289	1.0421	1.0509	1.0572	1.0644	1.0727	1.0827
10	0.9951	1.0177	1.0345	1.0479	1.0565	1.0627	1.0694	1.0784	1.0888
20	1.0012	1.0234	1.0400	1.0532	1.0618	1.0678	1.0745	1.0830	1.0934
30	1.0068	1.0290	1.0453	1.0580	1.0668	1.0727	1.0793	1.0882	1.0984
40	1.0122	1.0339	1.0502	1.0630	1.0715	1.0774	1.0839	1.0923	1.1025
50	1.0172	1.0389	1.0551	1.0678	1.0766	1.0820	1.0884	1.0971	1.1071

T/K = 318.15

0.1	0.9689	0.9941	1.0111	1.0243	1.0337	1.0399	1.0468	1.0549	1.0668
10	0.9767	1.0004	1.0171	1.0305	1.0395	1.0457	1.0526	1.0612	1.0721
20	0.9836	1.0060	1.0232	1.0363	1.0450	1.0510	1.0577	1.0669	1.0774
30	0.9898	1.0121	1.0287	1.0418	1.0503	1.0564	1.0630	1.0721	1.0825
40	0.9954	1.0176	1.0339	1.0470	1.0553	1.0613	1.0679	1.0767	1.0872
50	1.0010	1.0229	1.0393	1.0520	1.0608	1.0662	1.0728	1.0815	1.0917

T/K = 338.15

0.1	0.9564	0.9743	0.9926	1.0057	1.0147	1.0223	1.0294	1.0391	1.0491
10	0.9581	0.9819	0.9992	1.0131	1.0227	1.0288	1.0357	1.0447	1.0559
20	0.9655	0.9881	1.0057	1.0192	1.0279	1.0341	1.0409	1.0501	1.0613
30	0.9724	0.9946	1.0118	1.0252	1.0339	1.0401	1.0467	1.0557	1.0664
40	0.9789	1.0012	1.0181	1.0312	1.0398	1.0458	1.0525	1.0614	1.0721
50	0.9851	1.0066	1.0238	1.0360	1.0445	1.0512	1.0574	1.0655	1.0767

Polymer (B): poly(ethylene oxide)   **1987VE1, 1987VE2**
Characterization: M_w/kg.mol^{-1} = 25–30, M_w/M_n < 1.1
Solvent (A): water H_2O **7732-18-5**

T/K = 293.15   P/MPa = 0.1

z_B	0.0085	0.0246	0.0411	0.0701	0.1496	0.1866	0.2509	0.2914
V/cm^3 basemol^{-1}	18.229	18.530	18.841	19.381	20.861	21.552	22.770	23.561

continued

continued

T/K = 293.15 $\qquad$ P/MPa = 10

z_B	0.0085	0.0246	0.0411	0.0701	0.1496	0.1866	0.2509	0.2914
V/cm^3 basemol^{-1}	18.148	18.483	18.764	19.304	20.782	21.477	22.692	23.469

T/K = 293.15 $\qquad$ P/MPa = 20

z_B	0.0085	0.0246	0.0411	0.0701	0.1496	0.1866	0.2509	0.2914
V/cm^3 basemol^{-1}	18.071	18.375	18.688	19.229	20.710	21.404	22.617	23.393

T/K = 293.15 $\qquad$ P/MPa = 30

z_B	0.0085	0.0246	0.0411	0.0701	0.1496	0.1866	0.2509	0.2914
V/cm^3 basemol^{-1}	17.995	18.301	18.616	19.156	20.642	21.333	22.545	23.310

T/K = 293.15 $\qquad$ P/MPa = 40

z_B	0.0085	0.0246	0.0411	0.0701	0.1496	0.1866	0.2509	0.2914
V/cm^3 basemol^{-1}	17.992	18.228	18.542	19.085	20.572	21.264	22.474	23.246

Comments: V is given in cm^3/basemol mixture calculated with M_A = 18.015 g/mol and V_{0A} = 18.068 cm^3/mol for water and M_{0B} = 44.053 g/mol and V_{0B} = 39.382 cm^3/mol for the repeating unit of PEO.

Polymer (B): **poly(4-hydroxystyrene)** $\qquad$ **1995COM**
Characterization: M_n/kg.mol^{-1} = 15, M_w/kg.mol^{-1} = 30, Polysciences, Inc., Warrington, PA
Solvent (A): **ethanol** $\qquad$ **C$_2$H$_6$O** $\qquad$ **64-17-5**

T-range/K = 298.15 to 328.15 $\qquad$ P-range/MPa = 0.1 to 40

Coefficients of Tait equation for density ρ (standard deviation is 0.1 kg m^{-3}):

$$\rho = \left(B_1 + B_2 T + B_3 T^2\right)/\left\{1 - B_4 \ln\left[\left(P + B_5 e^{-B_6 T}\right)/\left(P_0 + B_5 e^{-B_6 T}\right)\right]\right\}$$

w_A	$10^{-3}B_1$/kg m^{-3}	B_2/kg m^{-3} K^{-1}	$10^3 B_3$/kg m^{-3} K^{-2}	$10B_4$	$10^{-3}B_5$/MPa	$10^2 B_6$/K^{-1}
0.70005	1.1009	−0.5824	−3.7116	0.7658	0.4740	0.5375
0.74083	1.1351	−0.8828	−0.6885	0.8149	0.5521	0.5803
0.79713	1.0924	−0.7348	−1.8545	0.7925	0.5667	0.6241
0.89815	1.0631	−0.7600	−1.7199	0.7920	0.5833	0.6747
0.97003	1.0106	−0.5710	−4.8404	0.7986	0.5681	0.6920
1.00000	1.0371	−0.7983	−1.2641	0.8428	0.6103	0.7028

Polymer (B): **poly(4-hydroxystyrene)** $\qquad$ **1994COM**
Characterization: M_n/kg.mol^{-1} = 15, M_w/kg.mol^{-1} = 30, Polysciences, Inc., Warrington, PA
Solvent (A): **2-propanone** $\qquad$ **C$_3$H$_6$O** $\qquad$ **67-64-1**

continued

continued

T-range/K = 298.15 to 328.15 P-range/MPa = 0.1 to 40

Coefficients of Tait equation for density ρ (standard deviation is about 0.0001 g/cm^3):

$$\rho = \left(B_1 + B_2 T + B_3 T^2 \right) / \left\{ 1 - B_4 \ln \left[\left(P + B_5 \, e^{-B_6 T} \right) / \left(P_0 + B_5 \, e^{-B_6 T} \right) \right] \right\}$$

w_A	B_1/g cm^{-3}	$10^3 B_2$/g cm^{-3} K^{-1}	$10^6 B_3$/g cm^{-3} K^{-2}	$10 B_4$	$10^{-3} B_5$/MPa	$10^2 B_6$/K^{-1}
0.96748	1.0956	−0.0893	−0.3587	0.9135	1.1777	0.9254
0.93684	1.1691	−0.1296	0.2703	0.9088	1.2046	0.9286
0.92248	1.4034	−0.2787	2.6940	0.9229	1.3020	0.9381
0.87947	1.1693	−0,1165	0.0589	0.9070	1.1966	0.8991
0.77489	1.3852	−0.2382	2.1434	0.8065	0.9405	0.8180
0.73891	1.2048	−0.1133	0.1551	0.8809	1.1722	0.8319
0.68443	1.4138	−0.2402	2.3193	0.8738	0.9886	0.7526
0.62253	1.3365	−0.174	1.2688	0.8034	0.8702	0.7081
0.49111	2.2944	−0.7797	11.6134	0.5368	0.5368	0.3779

Polymer (B):	**poly(4-hydroxystyrene)**	**1995COM**
Characterization:	M_n/kg.mol^{-1} = 15, M_w/kg.mol^{-1} = 30,	
	Polysciences, Inc., Warrington, PA	
Solvent (A):	**tetrahydrofuran C$_4$H$_8$O**	**109-99-9**

T/K = 298.15 P-range/MPa = 0.1 to 40

Coefficients of Tait equation for density ρ (standard deviation is 0.1 kg m^{-3}):

$$\rho = \left(B_1 + B_2 T + B_3 T^2 \right) / \left\{ 1 - B_4 \ln \left[\left(P + B_5 \, e^{-B_6 T} \right) / \left(P_0 + B_5 \, e^{-B_6 T} \right) \right] \right\}$$

w_A	$10^{-3} B_1$/kg m^{-3}	B_2/kg m^{-3} K^{-1}	$10^3 B_3$/kg m^{-3} K^{-2}	$10 B_4$	$10^{-3} B_5$/MPa	$10^2 B_6$/K^{-1}
0.69313	0.98051			0.6995	0.1032	
0.74866	0.96136			0.7627	0.1025	
0.77397	0.95322			0.8161	0.1099	
0.79769	0.94655			0.6901	0.0880	
0.85693	0.92908			0.8588	0.1070	
1.00000	1.21153	−1.0794	−0.6425	0.8176	1.0652	0.8540

Comments: The system had been studied at only one temperature due to the almost negligible temperature effect observed for the mixture.

Polymer (B):   **polyisobutylene**   **1979REN**
Characterization: –
Solvent (A):   **benzene** C_6H_6 **71-43-2**

T-range/K = 298.15 to 338.15   *P*-range/bar = 0.1 to 1000

Coefficients of Tait equation (ϑ is the temperature in °C):

$$V_{spez} = \left(A_1 + A_2\vartheta + A_3\vartheta^2\right)/\left\{1 - A_4 \ln\left[1 + P/A_5\, e^{-A_6\vartheta}\right]\right\}$$

φ_B	A_1/g cm^{-3}	$10^3 A_2$/g cm^{-3} °C^{-1}	$10^6 A_3$/g cm^{-3} °C^{-2}	$100 A_4$	$10^{-3} A_5$/bar	$10^3 B_6$/°C^{-1}
0.1989	1.10102	1.10965	1.56220	8.966583	1.251642	7.851594
0.4355	1.09412	1.05658	−0.60033	8.911763	1.376756	6.862239
0.5923	1.09088	0.90293	0.15281	8.211724	1.381603	6.473317

Polymer (B):   **polyisobutylene**   **1979REN**
Characterization: –
Solvent (A):   **cyclohexane** C_6H_{12} **110-82-7**

T-range/K = 298.15 to 338.15   *P*-range/bar = 0.1 to 1000

Coefficients of Tait equation (ϑ is the temperature in °C):

$$V_{spez} = \left(A_1 + A_2\vartheta + A_3\vartheta^2\right)/\left\{1 - A_4 \ln\left[1 + P/A_5\, e^{-A_6\vartheta}\right]\right\}$$

φ_B	A_1/g cm^{-3}	$10^3 A_2$/g cm^{-3} °C^{-1}	$10^6 A_3$/g cm^{-3} °C^{-2}	$100 A_4$	$10^{-3} A_5$/bar	$10^3 B_6$/°C^{-1}
0.1971	1.19508	1.21136	2.15655	9.125086	1.114994	7.853660
0.3731	1.17033	1.07970	0.92579	8.066102	1.134639	7.460571
0.5445	1.14672	0.91138	0.81216	8.122835	1.281160	6.457454

Polymer (B):   **poly(propylene glycol)**   **2001LIN**
Characterization: M_n/kg.mol^{-1} = 4.96, M_w/kg.mol^{-1} = 5.0,
     Aldrich Chem. Co., Inc., Milwaukee, WI
Solvent (A):   **acetophenone** C_8H_8O **98-86-2**

P/MPa				w_B					
	0.0000	0.7872	0.8927	0.9345	0.9569	0.9708	0.9804	0.9873	1.0000

ρ/g cm^{-3}

continued

continued

T/K = 298.15

0.1	1.0224	1.0052	1.0026	1.0011	1.0000	0.9994	0.9987	0.9983	0.9979
10	1.0281	1.0110	1.0086	1.0073	1.0060	1.0054	1.0049	1.0045	1.0042
15	1.0308	1.0140	1.0116	1.0103	1.0089	1.0084	1.0080	1.0075	1.0071
20	1.0335	1.0168	1.0145	1.0132	1.0119	1.0112	1.0109	1.0103	1.0100
25	1.0361	1.0197	1.0172	1.0160	1.0146	1.0141	1.0137	1.0131	1.0128
30	1.0387	1.0223	1.0199	1.0187	1.0173	1.0168	1.0164	1.0158	1.0156
35	1.0411	1.0249	1.0225	1.0214	1.0199	1.0194	1.0190	1.0185	1.0182
40	1.0436	1.0274	1.0250	1.0240	1.0226	1.0220	1.0216	1.0212	1.0208
45	1.0460	1.0300	1.0276	1.0266	1.0250	1.0246	1.0242	1.0237	1.0233
50	1.0483	1.0323	1.0301	1.0290	1.0275	1.0270	1.0267	1.0261	1.0258

T/K = 318.15

0.1	1.0053	0.9892	0.9867	0.9853	0.9844	0.9839	0.9832	0.9828	0.9828
10	1.0115	0.9958	0.9934	0.9920	0.9911	0.9904	0.9900	0.9897	0.9896
15	1.0145	0.9990	0.9966	0.9951	0.9944	0.9937	0.9933	0.9929	0.9928
20	1.0175	1.0020	0.9997	0.9983	0.9975	0.9968	0.9964	0.9961	0.9959
25	1.0204	1.0051	1.0027	1.0013	1.0005	0.9998	0.9994	0.9991	0.9990
30	1.0231	1.0079	1.0056	1.0042	1.0034	1.0028	1.0024	1.0021	1.0019
35	1.0258	1.0107	1.0085	1.0071	1.0063	1.0056	1.0053	1.0050	1.0048
40	1.0284	1.0135	1.0113	1.0099	1.0090	1.0084	1.0080	1.0078	1.0076
45	1.0310	1.0162	1.0140	1.0126	1.0118	1.0111	1.0107	1.0105	1.0103
50	1.0334	1.0188	1.0166	1.0152	1.0144	1.0137	1.0134	1.0132	1.0130

T/K = 348.15

0.1	0.9792	0.9651	0.9632	0.9620	0.9610	0.9606	0.9602	0.9599	0.9600
10	0.9864	0.9725	0.9708	0.9696	0.9685	0.9681	0.9676	0.9675	0.9675
15	0.9898	0.9760	0.9745	0.9732	0.9724	0.9718	0.9713	0.9712	0.9712
20	0.9932	0.9795	0.9779	0.9768	0.9758	0.9753	0.9749	0.9748	0.9747
25	0.9964	0.9828	0.9813	0.9802	0.9792	0.9788	0.9783	0.9782	0.9781
30	0.9995	0.9862	0.9846	0.9834	0.9825	0.9820	0.9815	0.9815	0.9814
35	1.0026	0.9893	0.9878	0.9866	0.9857	0.9852	0.9847	0.9847	0.9846
40	1.0055	0.9923	0.9908	0.9896	0.9887	0.9883	0.9878	0.9878	0.9877
45	1.0084	0.9952	0.9938	0.9927	0.9918	0.9913	0.9908	0.9908	0.9907
50	1.0112	0.9982	0.9967	0.9955	0.9947	0.9942	0.9937	0.9937	0.9936

Polymer (B):   **poly(propylene glycol)**   **2003LEE**

Characterization: M_n/kg.mol^{-1} = 0.485, M_w/kg.mol^{-1} = 0.500,
PPG-425, Aldrich Chem. Co., Inc., Milwaukee, WI

Solvent (A):   **anisole**   **C$_7$H$_8$O**   **100-66-3**

P/MPa				w_B					
	0.0000	0.3041	0.4956	0.6275	0.7238	0.7972	0.8550	0.9017	1.0000
				ρ/g cm^{-3}					

T/K = 298.15

P/MPa									
0.1	0.9887	0.9917	0.9953	0.9982	0.9997	1.0007	1.0014	1.0019	1.0031
10	0.9951	0.9981	1.0015	1.0044	1.0058	1.0069	1.0075	1.0080	1.0091
15	0.9981	1.0012	1.0046	1.0074	1.0088	1.0099	1.0105	1.0109	1.0120
20	1.0011	1.0042	1.0075	1.0103	1.0117	1.0127	1.0134	1.0138	1.0149
25	1.0040	1.0070	1.0102	1.0130	1.0145	1.0155	1.0161	1.0165	1.0176
30	1.0068	1.0099	1.0131	1.0158	1.0172	1.0182	1.0188	1.0193	1.0203
35	1.0095	1.0128	1.0158	1.0185	1.0199	1.0209	1.0215	1.0220	1.0230
40	1.0120	1.0153	1.0183	1.0210	1.0224	1.0234	1.0240	1.0245	1.0254
45	1.0147	1.0179	1.0208	1.0236	1.0250	1.0260	1.0265	1.0270	1.0280
50	1.0173	1.0205	1.0234	1.0261	1.0275	1.0284	1.0290	1.0294	1.0304

T/K = 318.15

P/MPa									
0.1	0.9700	0.9736	0.9779	0.9810	0.9829	0.9842	0.9850	0.9857	0.9872
10	0.9770	0.9805	0.9847	0.9877	0.9895	0.9908	0.9917	0.9923	0.9938
15	0.9803	0.9839	0.9880	0.9910	0.9929	0.9941	0.9950	0.9956	0.9970
20	0.9836	0.9873	0.9912	0.9942	0.9960	0.9972	0.9981	0.9987	1.0001
25	0.9869	0.9903	0.9943	0.9972	0.9991	1.0003	1.0011	1.0017	1.0031
30	0.9899	0.9934	0.9973	1.0002	1.0020	1.0032	1.0040	1.0046	1.0059
35	0.9929	0.9963	1.0002	1.0030	1.0049	1.0060	1.0069	1.0074	1.0088
40	0.9958	0.9992	1.0030	1.0059	1.0077	1.0089	1.0096	1.0102	1.0116
45	0.9987	1.0020	1.0058	1.0086	1.0103	1.0115	1.0122	1.0128	1.0141
50	1.0014	1.0046	1.0085	1.0113	1.0130	1.0142	1.0149	1.0155	1.0167

T/K = 338.15

P/MPa									
0.1	0.9409	0.9460	0.9511	0.9547	0.9571	0.9587	0.9600	0.9609	0.9630
10	0.9493	0.9542	0.9591	0.9627	0.9649	0.9666	0.9677	0.9687	0.9707
15	0.9533	0.9581	0.9630	0.9664	0.9686	0.9703	0.9714	0.9723	0.9743
20	0.9571	0.9618	0.9667	0.9701	0.9722	0.9738	0.9749	0.9759	0.9778
25	0.9607	0.9653	0.9701	0.9735	0.9757	0.9772	0.9784	0.9792	0.9811
30	0.9643	0.9688	0.9736	0.9769	0.9790	0.9806	0.9817	0.9825	0.9844
35	0.9677	0.9721	0.9769	0.9802	0.9822	0.9838	0.9848	0.9857	0.9875
40	0.9711	0.9754	0.9801	0.9834	0.9855	0.9870	0.9880	0.9889	0.9906
45	0.9743	0.9786	0.9832	0.9864	0.9885	0.9900	0.9910	0.9918	0.9936
50	0.9774	0.9816	0.9862	0.9894	0.9914	0.9930	0.9939	0.9947	0.9964

Polymer (B): **poly(propylene glycol)** 1997COL
Characterization: M_w/kg.mol^{-1} = 0.4, M_w/M_n < 1.5,
Polysciences, Inc., Warrington, PA
Solvent (A): **ethanol** **C_2H_6O** 64-17-5

T-range/K = 298.15 to 328.15   *P*-range/MPa = 0.1 to 40

Coefficients of Tait equation for density ρ (standard deviation is about 0.1 kg m^{-3}):

$$\rho = \left(B_1 + B_2T + B_3T^2\right)/\left\{1 - B_4 \ln\left[\left(P + B_5 e^{-B_6T}\right)/\left(P_0 + B_5 e^{-B_6T}\right)\right]\right\}$$

x_A	$10^{-3}B_1$/kg m^{-3}	B_2/kg m^{-3} K^{-1}	10^3B_3/kg m^{-3} K^{-2}	$10B_4$	$10^{-3}B_5$/MPa	10^2B_6/K^{-1}
0.00000	1.2567	−0.8796	0.1104	0.6849	0.4869	0.5079
0.13137	1.2685	−0.9746	0.2541	0.8076	0.6348	0.5292
0.22150	1.2604	−0.9412	0.1968	0.9170	0.6983	0.5157
0.41891	1.2397	−0.8740	0.0826	0.8069	0.6156	0.5330
0.51104	1.2254	−0.8274	0.0057	0.8176	0.6356	0.5429
0.60921	1.2124	−0.8069	−0.0363	0.7984	0.6242	0.5548
0.78808	1.2016	−0.9470	0.1687	0.8841	0.6622	0.5658
0.89978	1.1410	−0.8536	0.0056	0.8198	0.5837	0.5977
1.00000	1.0371	−0.7983	−1.2641	0.8428	0.6103	0.7028

Polymer (B): **poly(propylene glycol)** 1997COL
Characterization: M_w/kg.mol^{-1} = 0.4, M_w/M_n < 1.5,
Polysciences, Inc., Warrington, PA
Solvent (A): **n-hexane** **C_6H_{14}** 110-54-3

T-range/K = 298.15 to 328.15   *P*-range/MPa = 0.1 to 40

Coefficients of Tait equation for density ρ (standard deviation is about 0.1 kg m^{-3}):

$$\rho = \left(B_1 + B_2T + B_3T^2\right)/\left\{1 - B_4 \ln\left[\left(P + B_5 e^{-B_6T}\right)/\left(P_0 + B_5 e^{-B_6T}\right)\right]\right\}$$

x_A	$10^{-3}B_1$/kg m^{-3}	B_2/kg m^{-3} K^{-1}	10^3B_3/kg m^{-3} K^{-2}	$10B_4$	$10^{-3}B_5$/MPa	10^2B_6/K^{-1}
0.00000	1.2567	−0.8796	0.1104	0.6849	0.4869	0.5079
0.23651	1.2221	−0.8473	0.0392	0.9199	0.6877	0.5344
0.39687	1.2224	−1.0335	0.3294	0.8340	0.5956	0.5535
0.50500	1.1755	−0.8800	0.0596	0.8548	0.6248	0.5823
0.59756	1.1761	−1.0408	0.2920	0.8268	0.6595	0.6407
0.79920	1.1070	−1.0996	0.3329	0.8354	0.6682	0.7236
0.89964	1.0492	−1.1217	0.3522	0.8441	0.7779	0.8274
1.00000	1.0358	−1.5870	1.0805	0.8521	0.9049	0.9540

Polymer (B):	poly(propylene glycol)	**2001LIN**
Characterization:	M_n/kg.mol^{-1} = 4.96, M_w/kg.mol^{-1} = 5.0, Aldrich Chem. Co., Inc., Milwaukee, WI	
Solvent (A):	**1-octanol** $\quad$ C$_8$H$_{18}$O	**111-87-5**

P/MPa	w_B								
	0.0000	0.7334	0.8848	0.9294	0.9534	0.9788	0.9862	0.9919	1.0000
				ρ/g cm^{-3}					

T/K = 298.15

0.1	0.8216	0.9534	0.9745	0.9830	0.9876	0.9928	0.9945	0.9958	0.9979
10	0.8278	0.9596	0.9807	0.9893	0.9938	0.9990	1.0006	1.0021	1.0042
15	0.8308	0.9626	0.9837	0.9923	0.9969	1.0021	1.0036	1.0050	1.0071
20	0.8337	0.9654	0.9865	0.9952	0.9998	1.0050	1.0065	1.0079	1.0100
25	0.8364	0.9682	0.9893	0.9980	1.0025	1.0077	1.0093	1.0107	1.0128
30	0.8391	0.9709	0.9921	1.0007	1.0053	1.0104	1.0121	1.0134	1.0156
35	0.8417	0.9736	0.9947	1.0034	1.0080	1.0132	1.0147	1.0161	1.0182
40	0.8442	0.9762	0.9973	1.0060	1.0105	1.0158	1.0173	1.0187	1.0208
45	0.8467	0.9787	0.9998	1.0085	1.0131	1.0183	1.0198	1.0213	1.0233
50	0.8491	0.9812	1.0023	1.0110	1.0155	1.0208	1.0223	1.0237	1.0258

T/K = 318.15

0.1	0.8077	0.9381	0.9593	0.9680	0.9724	0.9776	0.9792	0.9805	0.9828
10	0.8149	0.9448	0.9659	0.9748	0.9792	0.9844	0.9861	0.9873	0.9896
15	0.8182	0.9481	0.9692	0.9780	0.9823	0.9877	0.9895	0.9906	0.9928
20	0.8213	0.9513	0.9724	0.9812	0.9856	0.9909	0.9925	0.9936	0.9959
25	0.8243	0.9543	0.9755	0.9842	0.9885	0.9940	0.9955	0.9967	0.9990
30	0.8272	0.9572	0.9785	0.9871	0.9914	0.9969	0.9985	0.9996	1.0019
35	0.8300	0.9601	0.9813	0.9900	0.9943	0.9998	1.0014	1.0025	1.0048
40	0.8327	0.9628	0.9841	0.9928	0.9973	1.0024	1.0042	1.0053	1.0076
45	0.8354	0.9655	0.9869	0.9955	0.9998	1.0053	1.0069	1.0080	1.0103
50	0.8380	0.9681	0.9895	0.9982	1.0025	1.0080	1.0095	1.0107	1.0130

T/K = 348.15

0.1	0.7855	0.9149	0.9362	0.9451	0.9496	0.9547	0.9563	0.9576	0.9600
10	0.7933	0.9226	0.9439	0.9529	0.9572	0.9624	0.9640	0.9653	0.9675
15	0.7969	0.9263	0.9476	0.9566	0.9609	0.9661	0.9677	0.9689	0.9712
20	0.8005	0.9298	0.9513	0.9601	0.9645	0.9697	0.9712	0.9724	0.9747
25	0.8039	0.9332	0.9548	0.9636	0.9679	0.9731	0.9746	0.9758	0.9781
30	0.8072	0.9365	0.9580	0.9668	0.9711	0.9764	0.9779	0.9791	0.9814
35	0.8104	0.9397	0.9612	0.9700	0.9743	0.9795	0.9811	0.9824	0.9846
40	0.8135	0.9428	0.9644	0.9731	0.9775	0.9826	0.9842	0.9854	0.9877
45	0.8164	0.9458	0.9673	0.9761	0.9804	0.9856	0.9872	0.9885	0.9907
50	0.8193	0.9486	0.9703	0.9790	0.9833	0.9886	0.9901	0.9914	0.9936

Polymer (B):	**poly(propylene glycol)**					**1999CRE**
Characterization:	M_w/kg.mol^{-1} = 0.4, M_w/M_n < 1.5,					
	Polysciences, Inc., Warrington, PA					
Solvent (A):	**water**		**H$_2$O**			**7732-18-5**

T-range/K = 298.15 to 328.15   *P*-range/MPa = 0.1 to 40

Coefficients of Tait equation for density ρ (standard deviation is 0.1 kg m^{-3}):

$$\rho = \left(B_1 + B_2 T + B_3 T^2 \right) / \left\{ 1 - B_4 \ln \left[\left(P + B_5\, e^{-B_6 T} \right) / \left(P_0 + B_5\, e^{-B_6 T} \right) \right] \right\}$$

x_A	$10^{-3}B_1$/kg m^{-3}	B_2/kg m^{-3} K^{-1}	10^3B_3/kg m^{-3} K^{-2}	$10B_4$	$10^{-3}B_5$/MPa	10^2B_6/K^{-1}
0.00000	1.2567	−0.8796	0.1104	0.6849	0.4869	0.5079
0.20105	1.2663	−0.9207	0.1560	0.8552	0.6609	0.5150
0.37701	1.2390	−0.7150	−0.1751	0.9205	0.7391	0.5148
0.61406	1.2227	−0.5678	−0.4161	0.7870	0.5934	0.4837
0.85537	1.2576	−0.7013	−0.2619	0.7695	0.7547	0.5296
0.95620	0.7619	2.6343	−5.8048	0.6260	3.4631	1.0627
0.97623	0.6962	2.8673	−5.9522	0.7450	4.3583	1.0479
0.98851	0.6397	2.9702	−5.7338	0.5937	0.6774	0.5301
1.00000	0.2871	4.6433	−7.5582	0.9362	0.1492	−0.1158

Polymer (B):	**tetra(ethylene glycol) dimethyl ether**	**2003ACE**
Characterization:	M/kg.mol^{-1} = 0.222, Fluka AG, Buchs, Switzerland	
Solvent (A):	**n-heptane** **C$_7$H$_{16}$**	**142-82-5**

P/MPa	x_B								
	0.00000	0.10504	0.29950	0.40213	0.50997	0.59095	0.69827	0.80114	1.00000
	ρ/g cm^{-3}								

T/K = 278.15

0.1	0.6963	0.7439	0.8225	0.8588	0.8941	0.9190	0.9499	0.9776	1.0251
1.0	0.6970	0.7446	0.8231	0.8594	0.8948	0.9196	0.9505	0.9782	1.0256
2.5	0.6983	0.7458	0.8243	0.8605	0.8958	0.9206	0.9515	0.9790	1.0265
5.0	0.7003	0.7478	0.8261	0.8623	0.8976	0.9223	0.9531	0.9806	1.0279
7.5	0.7023	0.7497	0.8279	0.8641	0.8993	0.9239	0.9546	0.9821	1.0293
10.0	0.7042	0.7516	0.8297	0.8658	0.9009	0.9255	0.9561	0.9836	1.0307
15.0	0.7079	0.7552	0.8332	0.8692	0.9041	0.9287	0.9592	0.9865	1.0334
20.0	0.7114	0.7586	0.8365	0.8724	0.9072	0.9317	0.9621	0.9893	1.0361
25.0	0.7147	0.7619		0.8755	0.9102	0.9346	0.9649	0.9921	1.0387

continued

continued

T/K = 283.15

0.1	0.6921	0.7396	0.8180	0.8544	0.8896	0.9144	0.9454	0.9728	1.0204
1.0	0.6929	0.7403	0.8187	0.8551	0.8902	0.9151	0.9460	0.9734	1.0209
2.5	0.6942	0.7416	0.8199	0.8562	0.8913	0.9161	0.9470	0.9744	1.0218
5.0	0.6963	0.7437	0.8218	0.8581	0.8931	0.9178	0.9486	0.9759	1.0233
7.5	0.6983	0.7456	0.8237	0.8599	0.8948	0.9195	0.9502	0.9775	1.0247
10.0	0.7002	0.7476	0.8255	0.8617	0.8965	0.9211	0.9518	0.9790	1.0261
15.0	0.7040	0.7513	0.8290	0.8651	0.8998	0.9243	0.9549	0.9820	1.0289
20.0	0.7076	0.7548	0.8324	0.8684	0.9029	0.9274	0.9579	0.9849	1.0316
25.0	0.7111	0.7582	0.8357	0.8716	0.9060	0.9304	0.9607	0.9877	1.0343

T/K = 293.15

0.1	0.6837	0.7313	0.8090	0.8453	0.8805	0.9053	0.9360	0.9634	1.0111
1.0	0.6846	0.7321	0.8097	0.8461	0.8811	0.9059	0.9366	0.9639	1.0116
2.5	0.6859	0.7335	0.8109	0.8473	0.8823	0.9070	0.9377	0.9649	1.0126
5.0	0.6882	0.7357	0.8129	0.8492	0.8841	0.9088	0.9394	0.9666	1.0141
7.5	0.6904	0.7378	0.8149	0.8511	0.8859	0.9106	0.9411	0.9682	1.0156
10.0	0.6925	0.7399	0.8169	0.8530	0.8877	0.9124	0.9428	0.9698	1.0171
15.0	0.6966	0.7439	0.8206	0.8566	0.8912	0.9158	0.9460	0.9729	1.0200
20.0	0.7004	0.7476	0.8242	0.8601	0.8945	0.9190	0.9491	0.9759	1.0229
25.0	0.7039	0.7512	0.8276	0.8634	0.8977	0.9221	0.9522	0.9789	1.0256

T/K = 303.15

0.1	0.6753	0.7226	0.8001	0.8366	0.8715	0.8962	0.9269	0.9543	1.0021
1.0	0.6762	0.7235	0.8009	0.8374	0.8722	0.8969	0.9276	0.9550	1.0027
2.5	0.6777	0.7249	0.8023	0.8386	0.8735	0.8980	0.9287	0.9561	1.0037
5.0	0.6801	0.7273	0.8044	0.8407	0.8755	0.9000	0.9305	0.9578	1.0053
7.5	0.6824	0.7295	0.8065	0.8427	0.8774	0.9018	0.9323	0.9595	1.0069
10.0	0.6847	0.7317	0.8085	0.8447	0.8793	0.9036	0.9341	0.9613	1.0085
15.0	0.6890	0.7359	0.8125	0.8485	0.8829	0.9072	0.9375	0.9645	1.0115
20.0	0.6930	0.7398	0.8162	0.8521	0.8865	0.9106	0.9408	0.9677	1.0145
25.0	0.6969	0.7436	0.8198	0.8556	0.8899	0.9139	0.9439	0.9708	1.0174

T/K = 313.15

0.1	0.6666	0.7137	0.7911	0.8275	0.8624	0.8869	0.9180	0.9453	0.9929
1.0	0.6676	0.7146	0.7920	0.8283	0.8632	0.8877	0.9187	0.9460	0.9935
5.0	0.6717	0.7187	0.7957	0.8318	0.8666	0.8908	0.9218	0.9490	0.9963
7.5	0.6742	0.7211	0.7980	0.8339	0.8687	0.8928	0.9237	0.9508	0.9979
10.0	0.6766	0.7235	0.8001	0.8360	0.8707	0.8947	0.9256	0.9525	0.9996
15.0	0.6812	0.7280	0.8043	0.8401	0.8746	0.8985	0.9292	0.9560	1.0028
20.0	0.6855	0.7322	0.8083	0.8439	0.8783	0.9021	0.9326	0.9593	1.0060
25.0	0.6895	0.7361	0.8121	0.8476	0.8818	0.9055	0.9360	0.9625	1.0090

continued

continued

T/K = 323.15

0.1	0.6576	0.7049	0.7820	0.8182	0.8533	0.8784	0.9089	0.9361	0.9836
1.0	0.6587	0.7060	0.7830	0.8191	0.8541	0.8792	0.9097	0.9367	0.9843
2.5	0.6604	0.7077	0.7845	0.8205	0.8554	0.8805	0.9109	0.9379	0.9853
5.0	0.6631	0.7103	0.7869	0.8228	0.8577	0.8826	0.9129	0.9398	0.9871
7.5	0.6658	0.7129	0.7893	0.8251	0.8598	0.8847	0.9149	0.9416	0.9889
10.0	0.6684	0.7153	0.7916	0.8273	0.8619	0.8867	0.9168	0.9434	0.9906
15.0	0.6733	0.7201	0.7960	0.8316	0.8660	0.8907	0.9206	0.9471	0.9940
20.0	0.6779	0.7246	0.8002	0.8357	0.8699	0.8944	0.9242	0.9506	0.9973
25.0	0.6821	0.7287	0.8042	0.8395	0.8736	0.8980	0.9277	0.9539	1.0005

T/K = 333.15

0.1	0.6489	0.6954	0.7725	0.8092	0.8439	0.8683	0.8990	0.9264	0.9741
1.0	0.6500	0.6965	0.7735	0.8101	0.8448	0.8691	0.8998	0.9272	0.9748
2.5	0.6518	0.6983	0.7751	0.8117	0.8462	0.8703	0.9011	0.9284	0.9760
5.0	0.6548	0.7011	0.7777	0.8141	0.8485	0.8725	0.9033	0.9305	0.9779
7.5	0.6576	0.7039	0.7802	0.8166	0.8508	0.8747	0.9053	0.9325	0.9798
10.0	0.6604	0.7065	0.7827	0.8189	0.8530	0.8768	0.9074	0.9344	0.9816
15.0	0.6656	0.7115	0.7873	0.8234	0.8573	0.8809	0.9114	0.9382	0.9851
20.0	0.6704	0.7162	0.7917	0.8277	0.8613	0.8849	0.9151	0.9418	0.9885
25.0	0.6749	0.7206	0.7958	0.8317	0.8652	0.8886	0.9187	0.9453	0.9918

T/K = 343.15

0.1	0.6398	0.6867	0.7638	0.8000	0.8351	0.8593	0.8904	0.9176	0.9650
1.0	0.6411	0.6879	0.7648	0.8010	0.8360	0.8602	0.8912	0.9184	0.9658
2.5	0.6431	0.6898	0.7666	0.8026	0.8375	0.8617	0.8926	0.9198	0.9670
5.0	0.6463	0.6929	0.7694	0.8052	0.8401	0.8641	0.8949	0.9220	0.9690
7.5	0.6494	0.6958	0.7721	0.8078	0.8425	0.8664	0.8971	0.9241	0.9710
10.0	0.6524	0.6987	0.7747	0.8102	0.8448	0.8687	0.8993	0.9262	0.9729
15.0	0.6580	0.7040	0.7796	0.8150	0.8494	0.8730	0.9034	0.9302	0.9766
20.0	0.6631	0.7090	0.7842	0.8194	0.8537	0.8772	0.9074	0.9340	0.9802
25.0	0.6679	0.7136	0.7886	0.8237	0.8578	0.8812	0.9112	0.9376	0.9837

T/K = 353.15

0.1	0.6306	0.6775	0.7545	0.7904	0.8257	0.8501	0.8811	0.9083	0.9557
1.0	0.6320	0.6788	0.7556	0.7915	0.8267	0.8511	0.8820	0.9091	0.9565
2.5	0.6342	0.6809	0.7575	0.7933	0.8284	0.8526	0.8835	0.9106	0.9578
5.0	0.6378	0.6842	0.7605	0.7961	0.8310	0.8552	0.8859	0.9128	0.9599
7.5	0.6411	0.6874	0.7634	0.7989	0.8336	0.8576	0.8883	0.9151	0.9620
10.0	0.6443	0.6904	0.7661	0.8015	0.8361	0.8600	0.8905	0.9172	0.9640
15.0	0.6503	0.6961	0.7713	0.8065	0.8409	0.8646	0.8949	0.9214	0.9679
20.0	0.6557	0.7013	0.7762	0.8112	0.8454	0.8689	0.8991	0.9254	0.9717
25.0	0.6608	0.7062	0.7809	0.8157	0.8497	0.8730	0.9031	0.9293	0.9753

Polymer (B):	tetra(ethylene glycol) dimethyl ether	**2002COM**
Characterization:	M/kg.mol^{-1} = 0.222, Aldrich Chem. Co., Inc., Milwaukee, WI	
Solvent (A):	**1,1,1,2-tetrafluoroethane** $\quad$ **C$_2$H$_2$F$_4$**	**811-97-2**

P/MPa				x_B					
	0.0710	0.1273	0.2712	0.3069	0.4298	0.6352	0.7104	0.8886	1.0000
				ρ/g cm^{-3}					

T/K = 293.15

10	1.2417	1.2182	1.1637	1.1528	1.1172	1.0701	1.0563	1.0315	1.0172
15	1.2545	1.2283	1.1703	1.1589	1.1221	1.0738	1.0598	1.0345	1.0201
20	1.2662	1.2376	1.1765	1.1646	1.1267	1.0774	1.0632	1.0375	1.0229
25	1.2770	1.2463	1.1824	1.1701	1.1313	1.0810	1.0665	1.0405	1.0257
30	1.2869	1.2546	1.1880	1.1754	1.1355	1.0845	1.0697	1.0433	1.0284
35	1.2962	1.2622	1.1934	1.1804	1.1397	1.0877	1.0728	1.0461	1.0310
40	1.3051	1.2695	1.1986	1.1853	1.1438	1.0910	1.0759	1.0488	1.0336
45	1.3134	1.2765	1.2036	1.1900	1.1477	1.0941	1.0788	1.0515	1.0361
50	1.3214	1.2832	1.2084	1.1946	1.1515	1.0972	1.0818	1.0541	1.0385
55	1.3288	1.2896	1.2130	1.1990	1.1552	1.1003	1.0846	1.0567	1.0410
60	1.3360	1.2957	1.2174	1.2032	1.1588	1.1032	1.0874	1.0593	1.0434

T/K = 303.15

10	1.2169	1.1972	1.1480	1.1377	1.1041	1.0593	1.0460	1.0220	1.0082
15	1.2311	1.2083	1.1551	1.1444	1.1094	1.0632	1.0497	1.0254	1.0113
20	1.2438	1.2184	1.1617	1.1505	1.1144	1.0671	1.0533	1.0285	1.0143
25	1.2555	1.2278	1.1680	1.1562	1.1191	1.0708	1.0568	1.0316	1.0172
30	1.2664	1.2366	1.1740	1.1619	1.1236	1.0744	1.0601	1.0346	1.0200
35	1.2765	1.2449	1.1797	1.1673	1.1280	1.0779	1.0634	1.0374	1.0227
40	1.2860	1.2528	1.1852	1.1724	1.1324	1.0813	1.0666	1.0403	1.0254
45	1.2949	1.2601	1.1905	1.1774	1.1364	1.0846	1.0697	1.0431	1.0281
50	1.3032	1.2671	1.1956	1.1822	1.1405	1.0878	1.0727	1.0457	1.0306
55	1.3114	1.2740	1.2004	1.1867	1.1443	1.0910	1.0757	1.0485	1.0332
60	1.3190	1.2803	1.2051	1.1912	1.1482	1.0939	1.0785	1.0510	1.0355

T/K = 313.15

10	1.1920	1.1761	1.1322	1.1227	1.0910	1.0483	1.0357	1.0124	0.9992
15	1.2078	1.1882	1.1398	1.1297	1.0966	1.0526	1.0396	1.0159	1.0025
20	1.2219	1.1992	1.1468	1.1362	1.1019	1.0566	1.0434	1.0192	1.0055
25	1.2343	1.2094	1.1535	1.1425	1.1069	1.0605	1.0470	1.0223	1.0086
30	1.2462	1.2188	1.1599	1.1485	1.1118	1.0643	1.0506	1.0255	1.0116
35	1.2569	1.2276	1.1660	1.1542	1.1164	1.0680	1.0540	1.0286	1.0144
40	1.2671	1.2359	1.1718	1.1595	1.1208	1.0715	1.0573	1.0316	1.0172
45	1.2766	1.2439	1.1773	1.1649	1.1252	1.0750	1.0606	1.0345	1.0201

continued

continued

50	1.2854	1.2514	1.1827	1.1698	1.1293	1.0783	1.0637	1.0374	1.0226
55	1.2939	1.2584	1.1878	1.1746	1.1334	1.0816	1.0668	1.0401	1.0252
60	1.3020	1.2654	1.1929	1.1795	1.1375	1.0849	1.0699	1.0430	1.0278

$T/K = 323.15$

10	1.1657	1.1543	1.1161	1.1074	1.0780	1.0376	1.0255	1.0031	0.9903
15	1.1833	1.1676	1.1242	1.1149	1.0838	1.0420	1.0296	1.0067	0.9937
20	1.1989	1.1797	1.1321	1.1222	1.0897	1.0464	1.0337	1.0103	0.9970
25	1.2128	1.1907	1.1392	1.1287	1.0949	1.0504	1.0375	1.0136	1.0001
30	1.2255	1.2009	1.1459	1.1351	1.1001	1.0544	1.0412	1.0169	1.0032
35	1.2371	1.2103	1.1524	1.1411	1.1050	1.0583	1.0447	1.0201	1.0063
40	1.2479	1.2192	1.1585	1.1469	1.1097	1.0620	1.0482	1.0232	1.0091
45	1.2581	1.2276	1.1644	1.1524	1.1143	1.0657	1.0517	1.0263	1.0121
50	1.2676	1.2355	1.1700	1.1576	1.1187	1.0691	1.0548	1.0291	1.0147
55	1.2766	1.2430	1.1754	1.1627	1.1229	1.0725	1.0581	1.0321	1.0175
60	1.2852	1.2503	1.1806	1.1677	1.1271	1.0759	1.0613	1.0349	1.0202

$T/K = 333.15$

10	1.1391	1.1321	1.0998	1.0920	1.0649	1.0267	1.0151	0.9936	0.9814
15	1.1589	1.1469	1.1087	1.1003	1.0712	1.0315	1.0196	0.9975	0.9850
20	1.1759	1.1599	1.1169	1.1078	1.0772	1.0360	1.0237	1.0011	0.9883
25	1.1912	1.1720	1.1247	1.1150	1.0829	1.0404	1.0278	1.0047	0.9917
30	1.2048	1.1830	1.1320	1.1218	1.0884	1.0446	1.0317	1.0081	0.9950
35	1.2175	1.1932	1.1388	1.1281	1.0935	1.0486	1.0355	1.0116	0.9982
40	1.2290	1.2026	1.1453	1.1342	1.0986	1.0526	1.0392	1.0148	1.0012
45	1.2397	1.2115	1.1514	1.1400	1.1034	1.0563	1.0428	1.0179	1.0042
50	1.2499	1.2198	1.1573	1.1455	1.1079	1.0600	1.0461	1.0210	1.0070
55	1.2593	1.2277	1.1629	1.1509	1.1123	1.0635	1.0495	1.0241	1.0098
60	1.2683	1.2352	1.1684	1.1560	1.1166	1.0670	1.0528	1.0269	1.0126

$T/K = 343.15$

10	1.1111	1.1092	1.0834	1.0766	1.0516	1.0160	1.0050	0.9844	0.9726
15	1.1336	1.1256	1.0931	1.0854	1.0585	1.0211	1.0097	0.9887	0.9764
20	1.1525	1.1401	1.1019	1.0936	1.0649	1.0259	1.0142	0.9925	0.9800
25	1.1692	1.1531	1.1102	1.1013	1.0710	1.0304	1.0185	0.9962	0.9835
30	1.1842	1.1649	1.1180	1.1084	1.0767	1.0349	1.0225	0.9998	0.9869
35	1.1976	1.1758	1.1253	1.1151	1.0822	1.0391	1.0265	1.0032	0.9901
40	1.2100	1.1859	1.1321	1.1216	1.0874	1.0431	1.0303	1.0065	0.9932
45	1.2214	1.1952	1.1385	1.1276	1.0924	1.0470	1.0339	1.0099	0.9963
50	1.2322	1.2040	1.1447	1.1335	1.0972	1.0508	1.0374	1.0129	0.9993

continued

continued

| 55 | 1.2422 | 1.2125 | 1.1507 | 1.1391 | 1.1021 | 1.0546 | 1.0410 | 1.0162 | 1.0023 |
| 60 | 1.2516 | 1.2202 | 1.1563 | 1.1444 | 1.1063 | 1.0581 | 1.0443 | 1.0192 | 1.0050 |

T/K = 353.15

10	1.0824	1.0858	1.0661	1.0609	1.0386	1.0052	0.9948	0.9752	0.9639
15	1.1079	1.1042	1.0768	1.0706	1.0458	1.0106	0.9997	0.9794	0.9679
20	1.1288	1.1200	1.0864	1.0793	1.0526	1.0156	1.0044	0.9834	0.9716
25	1.1473	1.1343	1.0952	1.0874	1.0590	1.0205	1.0089	0.9872	0.9752
30	1.1633	1.1470	1.1034	1.0949	1.0652	1.0251	1.0132	0.9911	0.9788
35	1.1778	1.1584	1.1111	1.1021	1.0707	1.0295	1.0173	0.9947	0.9821
40	1.1910	1.1692	1.1184	1.1089	1.0763	1.0338	1.0213	0.9983	0.9855
45	1.2032	1.1791	1.1253	1.1152	1.0816	1.0379	1.0250	1.0017	0.9886
50	1.2145	1.1884	1.1322	1.1214	1.0866	1.0419	1.0287	1.0050	0.9917
55	1.2252	1.1973	1.1384	1.1272	1.0915	1.0457	1.0325	1.0083	0.9947
60	1.2349	1.2055	1.1444	1.1329	1.0961	1.0495	1.0359	1.0114	0.9976

T/K = 363.15

10	1.0514	1.0613	1.0490	1.0450	1.0250	0.9945	0.9847	0.9659	0.9557
15	1.0806	1.0819	1.0606	1.0553	1.0331	1.0002	0.9899	0.9704	0.9598
20	1.1043	1.0993	1.0709	1.0649	1.0401	1.0055	0.9948	0.9746	0.9639
25	1.1245	1.1147	1.0805	1.0735	1.0471	1.0106	0.9995	0.9787	0.9675
30	1.1420	1.1284	1.0892	1.0816	1.0534	1.0153	1.0039	0.9828	0.9712
35	1.1577	1.1409	1.0975	1.0893	1.0595	1.0201	1.0083	0.9866	0.9748
40	1.1720	1.1523	1.1053	1.0964	1.0654	1.0245	1.0125	0.9903	0.9783
45	1.1848	1.1629	1.1124	1.1031	1.0708	1.0288	1.0164	0.9937	0.9815
50	1.1969	1.1728	1.1194	1.1096	1.0761	1.0330	1.0203	0.9972	0.9846
55	1.2080	1.1820	1.1259	1.1156	1.0813	1.0369	1.0241	1.0006	0.9880
60	1.2185	1.1908	1.1322	1.1216	1.0860	1.0408	1.0277	1.0038	0.9910

T/K = 373.15

10	1.0193	1.0360	1.0317	1.0289	1.0116	0.9835	0.9742	0.9564	0.9468
15	1.0532	1.0593	1.0443	1.0400	1.0200	0.9895	0.9798	0.9611	0.9511
20	1.0798	1.0786	1.0556	1.0502	1.0278	0.9951	0.9850	0.9657	0.9552
25	1.1019	1.0954	1.0658	1.0596	1.0350	1.0005	0.9899	0.9699	0.9593
30	1.1209	1.1102	1.0752	1.0682	1.0419	1.0057	0.9947	0.9742	0.9631
35	1.1377	1.1234	1.0839	1.0762	1.0482	1.0104	0.9991	0.9779	0.9669
40	1.1529	1.1356	1.0919	1.0837	1.0543	1.0151	1.0035	0.9817	0.9704
45	1.1667	1.1468	1.0997	1.0909	1.0601	1.0196	1.0076	0.9854	0.9739
50	1.1792	1.1571	1.1068	1.0976	1.0655	1.0239	1.0116	0.9890	0.9771
55	1.1911	1.1670	1.1137	1.1042	1.0709	1.0281	1.0157	0.9926	0.9805
60	1.2022	1.1762	1.1202	1.1102	1.0760	1.0321	1.0194	0.9955	0.9838

6.2.2. Ternary polymer solutions

Polymer (B):	ethylene/1-butene copolymer		2000BY1

Characterization: M_n/kg.mol^{-1} = 230, M_w/kg.mol^{-1} = 232, 20.2 mol% 1-butene, synthesized by complete hydrogenation of polybutadiene, Exxon Research and Engineering Co.

Solvent (A):	ethane	C$_2$H$_6$	74-84-0
Solvent (C):	n-pentane-d12	C$_5$D$_{12}$	2031-90-5

T/K = 383.15

w_A	0.231	0.231	0.231	0.231	0.231	0.231	0.231	0.231	0.231
w_B	0.049	0.049	0.049	0.049	0.049	0.049	0.049	0.049	0.049
P/bar	277.5	312.0	346.5	380.9	415.4	484.3	553.3	691.2	1035.9
ρ/g cm^{-3}	0.643	0.650	0.656	0.662	0.667	0.677	0.687	0.702	0.736

w_A	0.231	0.231	0.231	0.231
w_B	0.049	0.049	0.049	0.049
P/bar	1380.7	1725.4	2070.1	2414.9
ρ/g cm^{-3}	0.762	0.785	0.805	0.822

T/K = 403.15

w_A	0.231	0.231	0.231	0.231	0.231	0.231	0.231	0.231	0.231
w_B	0.049	0.049	0.049	0.049	0.049	0.049	0.049	0.049	0.049
P/bar	277.5	312.0	346.5	380.9	415.4	484.3	553.3	691.2	1035.9
ρ/g cm^{-3}	0.622	0.630	0.634	0.645	0.651	0.662	0.672	0.689	0.724

w_A	0.231	0.231	0.231	0.231
w_B	0.049	0.049	0.049	0.049
P/bar	1380.7	1725.4	2070.1	2414.9
ρ/g cm^{-3}	0.752	0.775	0.796	0.814

T/K = 423.15

w_A	0.231	0.231	0.231	0.231	0.231	0.231	0.231	0.231	0.231
w_B	0.049	0.049	0.049	0.049	0.049	0.049	0.049	0.049	0.049
P/bar	346.5	380.9	415.4	484.3	553.3	691.2	1035.9	1380.7	1725.4
ρ/g cm^{-3}	0.619	0.627	0.634	0.646	0.657	0.676	0.712	0.741	0.765

w_A	0.231	0.231
w_B	0.049	0.049
P/bar	2070.1	2414.9
ρ/g cm^{-3}	0.786	0.805

Polymer (B):	poly(ethylene glycol)	2003LEE

Characterization: M_n/kg.mol^{-1} = 0.26, M_w/kg.mol^{-1} = 0.28, PEG-200, Aldrich Chem. Co., Inc., Milwaukee, WI

Solvent (A):	anisole	C$_7$H$_8$O	100-66-3

continued

continued

Polymer (C):   **poly(propylene glycol)**
Characterization: M_n/kg.mol^{-1} = 0.485, M_w/kg.mol^{-1} = 0.50,
                 PPG-425, Aldrich Chem. Co., Inc., Milwaukee, WI

T/K = 298.15	P/MPa = 0.1			T/K = 298.15	P/MPa = 30		
w_A	0.0592	0.3031	0.0787	0.0797	0.0592	0.3031	0.0787	0.0797
w_B	0.1424	0.2429	0.5676	0.3203	0.1424	0.2429	0.5676	0.3203
w_C	0.7984	0.4540	0.3537	0.6000	0.7984	0.4540	0.3537	0.6000
ρ/g cm^{-3}	1.0168	1.0261	1.0667	1.0364	1.0335	1.0425	1.0809	1.0522

T/K = 298.15	P/MPa = 10			T/K = 298.15	P/MPa = 35		
w_A	0.0592	0.3031	0.0787	0.0797	0.0592	0.3031	0.0787	0.0797
w_B	0.1424	0.2429	0.5676	0.3203	0.1424	0.2429	0.5676	0.3203
w_C	0.7984	0.4540	0.3537	0.6000	0.7984	0.4540	0.3537	0.6000
ρ/g cm^{-3}	1.0226	1.0318	1.0716	1.0419	1.0360	1.0450	1.0831	1.0547

T/K = 298.15	P/MPa = 15			T/K = 298.15	P/MPa = 40		
w_A	0.0592	0.3031	0.0787	0.0797	0.0592	0.3031	0.0787	0.0797
w_B	0.1424	0.2429	0.5676	0.3203	0.1424	0.2429	0.5676	0.3203
w_C	0.7984	0.4540	0.3537	0.6000	0.7984	0.4540	0.3537	0.6000
ρ/g cm^{-3}	1.0254	1.0346	1.0739	1.0446	1.0384	1.0474	1.0852	1.0570

T/K = 298.15	P/MPa = 20			T/K = 298.15	P/MPa = 45		
w_A	0.0592	0.3031	0.0787	0.0797	0.0592	0.3031	0.0787	0.0797
w_B	0.1424	0.2429	0.5676	0.3203	0.1424	0.2429	0.5676	0.3203
w_C	0.7984	0.4540	0.3537	0.6000	0.7984	0.4540	0.3537	0.6000
ρ/g cm^{-3}	1.0281	1.0372	1.0763	1.0472	1.0409	1.0498	1.0873	1.0593

T/K = 298.15	P/MPa = 25			T/K = 298.15	P/MPa = 50		
w_A	0.0592	0.3031	0.0787	0.0797	0.0592	0.3031	0.0787	0.0797
w_B	0.1424	0.2429	0.5676	0.3203	0.1424	0.2429	0.5676	0.3203
w_C	0.7984	0.4540	0.3537	0.6000	0.7984	0.4540	0.3537	0.6000
ρ/g cm^{-3}	1.0308	1.0399	1.0785	1.0497	1.0432	1.0522	1.0893	1.0615

T/K = 318.15	P/MPa = 0.1			T/K = 318.15	P/MPa = 30		
w_A	0.0592	0.3031	0.0787	0.0797	0.0592	0.3031	0.0787	0.0797
w_B	0.1424	0.2429	0.5676	0.3203	0.1424	0.2429	0.5676	0.3203
w_C	0.7984	0.4540	0.3537	0.6000	0.7984	0.4540	0.3537	0.6000
ρ/g cm^{-3}	1.0005	1.0090	1.0502	1.0197	1.0187	1.0269	1.0661	1.0374

T/K = 318.15	P/MPa = 10			T/K = 318.15	P/MPa = 35		
w_A	0.0592	0.3031	0.0787	0.0797	0.0592	0.3031	0.0787	0.0797
w_B	0.1424	0.2429	0.5676	0.3203	0.1424	0.2429	0.5676	0.3203
w_C	0.7984	0.4540	0.3537	0.6000	0.7984	0.4540	0.3537	0.6000
ρ/g cm^{-3}	1.0069	1.0153	1.0559	1.0260	1.0215	1.0296	1.0685	1.0400

continued

continued

$T/K = 318.15$		$P/MPa = 15$			$T/K = 318.15$		$P/MPa = 40$	
w_A	0.0592	0.3031	0.0787	0.0797	0.0592	0.3031	0.0787	0.0797
w_B	0.1424	0.2429	0.5676	0.3203	0.1424	0.2429	0.5676	0.3203
w_C	0.7984	0.4540	0.3537	0.6000	0.7984	0.4540	0.3537	0.6000
$\rho/\text{g cm}^{-3}$	1.0100	1.0183	1.0585	1.0290	1.0242	1.0323	1.0709	1.0426

$T/K = 318.15$		$P/MPa = 20$			$T/K = 318.15$		$P/MPa = 45$	
w_A	0.0592	0.3031	0.0787	0.0797	0.0592	0.3031	0.0787	0.0797
w_B	0.1424	0.2429	0.5676	0.3203	0.1424	0.2429	0.5676	0.3203
w_C	0.7984	0.4540	0.3537	0.6000	0.7984	0.4540	0.3537	0.6000
$\rho/\text{g cm}^{-3}$	1.0130	1.0213	1.0611	1.0319	1.0267	1.0349	1.0731	1.0450

$T/K = 318.15$		$P/MPa = 25$			$T/K = 318.15$		$P/MPa = 50$	
w_A	0.0592	0.3031	0.0787	0.0797	0.0592	0.3031	0.0787	0.0797
w_B	0.1424	0.2429	0.5676	0.3203	0.1424	0.2429	0.5676	0.3203
w_C	0.7984	0.4540	0.3537	0.6000	0.7984	0.4540	0.3537	0.6000
$\rho/\text{g cm}^{-3}$	1.0159	1.0241	1.0636	1.0347	1.0293	1.0374	1.0754	1.0475

$T/K = 338.15$		$P/MPa = 0.1$			$T/K = 338.15$		$P/MPa = 30$	
w_A	0.0592	0.3031	0.0787	0.0797	0.0592	0.3031	0.0787	0.0797
w_B	0.1424	0.2429	0.5676	0.3203	0.1424	0.2429	0.5676	0.3203
w_C	0.7984	0.4540	0.3537	0.6000	0.7984	0.4540	0.3537	0.6000
$\rho/\text{g cm}^{-3}$	0.9757	0.9827	1.0243	0.9948	0.9963	1.0033	1.0418	1.0146

$T/K = 338.15$		$P/MPa = 10$			$T/K = 338.15$		$P/MPa = 35$	
w_A	0.0592	0.3031	0.0787	0.0797	0.0592	0.3031	0.0787	0.0797
w_B	0.1424	0.2429	0.5676	0.3203	0.1424	0.2429	0.5676	0.3203
w_C	0.7984	0.4540	0.3537	0.6000	0.7984	0.4540	0.3537	0.6000
$\rho/\text{g cm}^{-3}$	0.9831	0.9900	1.0305	1.0018	0.9993	1.0064	1.0444	1.0175

$T/K = 338.15$		$P/MPa = 15$			$T/K = 338.15$		$P/MPa = 40$	
w_A	0.0592	0.3031	0.0787	0.0797	0.0592	0.3031	0.0787	0.0797
w_B	0.1424	0.2429	0.5676	0.3203	0.1424	0.2429	0.5676	0.3203
w_C	0.7984	0.4540	0.3537	0.6000	0.7984	0.4540	0.3537	0.6000
$\rho/\text{g cm}^{-3}$	0.9866	0.9935	1.0334	1.0052	1.0023	1.0094	1.0471	1.0204

$T/K = 338.15$		$P/MPa = 20$			$T/K = 338.15$		$P/MPa = 45$	
w_A	0.0592	0.3031	0.0787	0.0797	0.0592	0.3031	0.0787	0.0797
w_B	0.1424	0.2429	0.5676	0.3203	0.1424	0.2429	0.5676	0.3203
w_C	0.7984	0.4540	0.3537	0.6000	0.7984	0.4540	0.3537	0.6000
$\rho/\text{g cm}^{-3}$	0.9899	0.9970	1.0364	1.0085	1.0052	1.0123	1.0495	1.0232

$T/K = 338.15$		$P/MPa = 25$			$T/K = 338.15$		$P/MPa = 50$	
w_A	0.0592	0.3031	0.0787	0.0797	0.0592	0.3031	0.0787	0.0797
w_B	0.1424	0.2429	0.5676	0.3203	0.1424	0.2429	0.5676	0.3203
w_C	0.7984	0.4540	0.3537	0.6000	0.7984	0.4540	0.3537	0.6000
$\rho/\text{g cm}^{-3}$	0.9932	1.0002	1.0391	1.0116	1.0080	1.0151	1.0520	1.0259

Polymer (B):	**poly(propylene glycol)**		**2003LEE**

Characterization: M_n/kg.mol^{-1} = 0.485, M_w/kg.mol^{-1} = 0.50,
PPG-425, Aldrich Chem. Co., Inc., Milwaukee, WI

Solvent (A): **anisole** C_7H_8O **100-66-3**

Polymer (C): **poly(ethylene glycol) monomethyl ether**

Characterization: M_n/kg.mol^{-1} = 0.366, M_w/kg.mol^{-1} = 0.373,
PEGME-350, Aldrich Chem. Co., Inc., Milwaukee, WI

T/K = 298.15	P/MPa = 0.1				T/K = 298.15		P/MPa = 30	
w_A	0.0639	0.2758	0.0560	0.1084	0.0639	0.2758	0.0560	0.1084
w_B	0.2872	0.4131	0.7546	0.5247	0.2872	0.4131	0.7546	0.5247
w_C	0.6489	0.3111	0.1894	0.3669	0.6489	0.3111	0.1894	0.3669
ρ/g cm^{-3}	1.0546	1.0285	1.0164	1.0334	1.0702	1.0449	1.0333	1.0496

T/K = 298.15	P/MPa = 10				T/K = 298.15		P/MPa = 35	
w_A	0.0639	0.2758	0.0560	0.1084	0.0639	0.2758	0.0560	0.1084
w_B	0.2872	0.4131	0.7546	0.5247	0.2872	0.4131	0.7546	0.5247
w_C	0.6489	0.3111	0.1894	0.3669	0.6489	0.3111	0.1894	0.3669
ρ/g cm^{-3}	1.0599	1.0341	1.0223	1.0390	1.0726	1.0474	1.0359	1.0521

T/K = 298.15	P/MPa = 15				T/K = 298.15		P/MPa = 40	
w_A	0.0639	0.2758	0.0560	0.1084	0.0639	0.2758	0.0560	0.1084
w_B	0.2872	0.4131	0.7546	0.5247	0.2872	0.4131	0.7546	0.5247
w_C	0.6489	0.3111	0.1894	0.3669	0.6489	0.3111	0.1894	0.3669
ρ/g cm^{-3}	1.0626	1.0369	1.0252	1.0413	1.0749	1.0498	1.0383	1.0545

T/K = 298.15	P/MPa = 20				T/K = 298.15		P/MPa = 45	
w_A	0.0639	0.2758	0.0560	0.1084	0.0639	0.2758	0.0560	0.1084
w_B	0.2872	0.4131	0.7546	0.5247	0.2872	0.4131	0.7546	0.5247
w_C	0.6489	0.3111	0.1894	0.3669	0.6489	0.3111	0.1894	0.3669
ρ/g cm^{-3}	1.0652	1.0396	1.0280	1.0444	1.0772	1.0522	1.0408	1.0568

T/K = 298.15	P/MPa = 25				T/K = 298.15		P/MPa = 50	
w_A	0.0639	0.2758	0.0560	0.1084	0.0639	0.2758	0.0560	0.1084
w_B	0.2872	0.4131	0.7546	0.5247	0.2872	0.4131	0.7546	0.5247
w_C	0.6489	0.3111	0.1894	0.3669	0.6489	0.3111	0.1894	0.3669
ρ/g cm^{-3}	1.0677	1.0422	1.0306	1.0470	1.0795	1.0546	1.0432	1.0591

T/K = 318.15	P/MPa = 0.1				T/K = 318.15		P/MPa = 30	
w_A	0.0639	0.2758	0.0560	0.1084	0.0639	0.2758	0.0560	0.1084
w_B	0.2872	0.4131	0.7546	0.5247	0.2872	0.4131	0.7546	0.5247
w_C	0.6489	0.3111	0.1894	0.3669	0.6489	0.3111	0.1894	0.3669
ρ/g cm^{-3}	1.0376	1.0112	0.9998	1.0168	1.0547	1.0293	1.0183	1.0345

T/K = 318.15	P/MPa = 10				T/K = 318.15		P/MPa = 35	
w_A	0.0639	0.2758	0.0560	0.1084	0.0639	0.2758	0.0560	0.1084
w_B	0.2872	0.4131	0.7546	0.5247	0.2872	0.4131	0.7546	0.5247
w_C	0.6489	0.3111	0.1894	0.3669	0.6489	0.3111	0.1894	0.3669
ρ/g cm^{-3}	1.0437	1.0176	1.0064	1.0231	1.0573	1.0320	1.0210	1.0372

continued

continued

$T/K = 318.15$		$P/MPa = 15$				$T/K = 318.15$		$P/MPa = 40$		
w_A	0.0639	0.2758	0.0560	0.1084		0.0639	0.2758	0.0560	0.1084
w_B	0.2872	0.4131	0.7546	0.5247		0.2872	0.4131	0.7546	0.5247
w_C	0.6489	0.3111	0.1894	0.3669		0.6489	0.3111	0.1894	0.3669
$\rho/\text{g cm}^{-3}$	1.0465	1.0207	1.0094	1.0261		1.0598	1.0347	1.0238	1.0399

$T/K = 318.15$		$P/MPa = 20$				$T/K = 318.15$		$P/MPa = 45$		
w_A	0.0639	0.2758	0.0560	0.1084		0.0639	0.2758	0.0560	0.1084
w_B	0.2872	0.4131	0.7546	0.5247		0.2872	0.4131	0.7546	0.5247
w_C	0.6489	0.3111	0.1894	0.3669		0.6489	0.3111	0.1894	0.3669
$\rho/\text{g cm}^{-3}$	1.0499	1.0236	1.0125	1.0290		1.0623	1.0373	1.0263	1.0423

$T/K = 318.15$		$P/MPa = 25$				$T/K = 318.15$		$P/MPa = 50$		
w_A	0.0639	0.2758	0.0560	0.1084		0.0639	0.2758	0.0560	0.1084
w_B	0.2872	0.4131	0.7546	0.5247		0.2872	0.4131	0.7546	0.5247
w_C	0.6489	0.3111	0.1894	0.3669		0.6489	0.3111	0.1894	0.3669
$\rho/\text{g cm}^{-3}$	1.0521	1.0265	1.0154	1.0318		1.0647	1.0398	1.0289	1.0448

$T/K = 338.15$		$P/MPa = 0.1$				$T/K = 338.15$		$P/MPa = 30$		
w_A	0.0639	0.2758	0.0560	0.1084		0.0639	0.2758	0.0560	0.1084
w_B	0.2872	0.4131	0.7546	0.5247		0.2872	0.4131	0.7546	0.5247
w_C	0.6489	0.3111	0.1894	0.3669		0.6489	0.3111	0.1894	0.3669
$\rho/\text{g cm}^{-3}$	1.0126	0.9852	0.9749	0.9916		1.0319	1.0060	0.9956	1.0120

$T/K = 338.15$		$P/MPa = 10$				$T/K = 338.15$		$P/MPa = 35$		
w_A	0.0639	0.2758	0.0560	0.1084		0.0639	0.2758	0.0560	0.1084
w_B	0.2872	0.4131	0.7546	0.5247		0.2872	0.4131	0.7546	0.5247
w_C	0.6489	0.3111	0.1894	0.3669		0.6489	0.3111	0.1894	0.3669
$\rho/\text{g cm}^{-3}$	1.0194	0.9926	0.9823	0.9989		1.0348	1.0091	0.9986	1.0149

$T/K = 338.15$		$P/MPa = 15$				$T/K = 338.15$		$P/MPa = 40$		
w_A	0.0639	0.2758	0.0560	0.1084		0.0639	0.2758	0.0560	0.1084
w_B	0.2872	0.4131	0.7546	0.5247		0.2872	0.4131	0.7546	0.5247
w_C	0.6489	0.3111	0.1894	0.3669		0.6489	0.3111	0.1894	0.3669
$\rho/\text{g cm}^{-3}$	1.0226	0.9961	0.9858	1.0023		1.0376	1.0121	1.0016	1.0179

$T/K = 338.15$		$P/MPa = 20$				$T/K = 338.15$		$P/MPa = 45$		
w_A	0.0639	0.2758	0.0560	0.1084		0.0639	0.2758	0.0560	0.1084
w_B	0.2872	0.4131	0.7546	0.5247		0.2872	0.4131	0.7546	0.5247
w_C	0.6489	0.3111	0.1894	0.3669		0.6489	0.3111	0.1894	0.3669
$\rho/\text{g cm}^{-3}$	1.0258	0.9995	0.9892	1.0056		1.0403	1.0150	1.0045	1.0208

$T/K = 338.15$		$P/MPa = 25$				$T/K = 338.15$		$P/MPa = 50$		
w_A	0.0639	0.2758	0.0560	0.1084		0.0639	0.2758	0.0560	0.1084
w_B	0.2872	0.4131	0.7546	0.5247		0.2872	0.4131	0.7546	0.5247
w_C	0.6489	0.3111	0.1894	0.3669		0.6489	0.3111	0.1894	0.3669
$\rho/\text{g cm}^{-3}$	1.0289	1.0028	0.9924	1.0088		1.0430	1.0178	1.0072	1.0235

Polymer (B):	**polystyrene**		**2001LI2**
Characterization:	M_n/kg.mol^{-1} = 71, M_w/kg.mol^{-1} = 78, State Key Laboratory of		
	Polymer Science, Chinese Academy of Sciences		
Solvent (A):	**carbon dioxide**	**CO_2**	**124-38-9**
Solvent (C):	**toluene**	**C_7H_8**	**108-88-3**

Type of data: volume expansion coefficients

Comments: Volume expansion coefficients, $(V - V_0)/V_0$, were determined for a given concentration of polystyrene in the CO_2-free toluene solution, c_{0B}, and the given equilibrium pressure of CO_2 in the ternary system.

T/K = 308.15

c_{0B}/g.ml^{-3}	0.00046	0.00046	0.00046	0.00046	0.00046	0.00046	0.00046	0.00046	0.00046
P/MPa	0.00	0.71	0.99	1.68	2.23	2.43	2.60	2.95	3.10
$(V-V_0)/V_0$	0.0000	0.0059	0.0103	0.0678	0.0997	0.1077	0.1237	0.1676	0.1835

c_{0B}/g.ml^{-3}	0.00046	0.00046	0.00046	0.00046	0.00046	0.00046	0.00046	0.00046	0.00046
P/MPa	3.26	3.55	3.63	3.76	3.86	3.96	4.06	4.15	4.20
$(V-V_0)/V_0$	0.1935	0.2414	0.2514	0.2654	0.2814	0.2913	0.3093	0.3312	0.3412

c_{0B}/g.ml^{-3}	0.00046	0.00046	0.00046	0.00046	0.00046	0.00096	0.00096	0.00096	0.00096
P/MPa	4.28	4.47	4.65	4.78	4.90	0.00	0.87	1.36	1.79
$(V-V_0)/V_0$	0.3632	0.4011	0.4670	0.5448	0.5648	0.0000	0.0106	0.0444	0.0824

c_{0B}/g.ml^{-3}	0.00096	0.00096	0.00096	0.00096	0.00096	0.00096	0.00096	0.00096	0.00096
P/MPa	2.29	2.54	2.87	3.14	3.60	3.80	3.98	4.08	4.25
$(V-V_0)/V_0$	0.1057	0.1290	0.1501	0.1755	0.2368	0.2939	0.3235	0.3425	0.3827

c_{0B}/g.ml^{-3}	0.00096	0.00096	0.00096	0.00096	0.00107	0.00107	0.00107	0.00107	0.00107
P/MPa	4.32	4.40	4.52	4.81	0.00	0.68	1.19	1.76	2.34
$(V-V_0)/V_0$	0.3911	0.4038	0.4419	0.5518	0.0000	0.0196	0.0414	0.0719	0.1220

c_{0B}/g.ml^{-3}	0.00107	0.00107	0.00107	0.00107	0.00107	0.00107	0.00107	0.00107	0.00107
P/MPa	2.98	3.16	3.35	3.65	3.74	3.82	3.88	4.04	4.16
$(V-V_0)/V_0$	0.1830	0.2004	0.2179	0.2658	0.2584	0.2941	0.2985	0.3355	0.3617

c_{0B}/g.ml^{-3}	0.00107	0.00107	0.00107	0.00107	0.00107	0.00107	0.00107	0.0032	0.0032
P/MPa	4.33	4.49	4.55	4.63	4.69	4.75	4.82	0.00	1.70
$(V-V_0)/V_0$	0.4074	0.4466	0.4575	0.4793	0.4880	0.5316	0.5599	0.0000	0.0466

c_{0B}/g.ml^{-3}	0.0032	0.0032	0.0032	0.0032	0.0032	0.0032	0.0032	0.0032	0.0032
P/MPa	2.07	2.48	3.10	3.47	3.80	3.98	4.12	4.33	4.63
$(V-V_0)/V_0$	0.0674	0.1036	0.1736	0.2073	0.2539	0.2850	0.3109	0.3523	0.4404

c_{0B}/g.ml^{-3}	0.0032	0.0092	0.0092	0.0092	0.0092	0.0092	0.0092	0.0092	0.0092
P/MPa	4.80	0.00	1.11	1.82	2.50	3.07	3.66	3.99	4.20
$(V-V_0)/V_0$	0.5130	0.0000	0.0483	0.0690	0.1402	0.2207	0.2713	0.2989	0.3494

c_{0B}/g.ml^{-3}	0.0092	0.0092
P/MPa	4.44	4.55
$(V-V_0)/V_0$	0.4207	0.4414

Comments: Cloud points of the ternary system are given in Chapter 4.

Polymer (B):	**polysulfone**		**2003ZH1**

Characterization: $M_n/\text{kg.mol}^{-1} = 36.1$, $M_w/\text{kg.mol}^{-1} = 66.5$, by GPC using PS standards, Scientific Polymer Products, Inc., Ontario, NY

Solvent (A):	**carbon dioxide**	**CO_2**	**124-38-9**
Solvent (C):	**tetrahydrofuran**	**C_4H_8O**	**109-99-9**

$w_A = 0.0984$ $w_B = 0.0150$ $w_C = 0.8866$

T/K	299.2		323.3		348.5		373.3		398.4		424.5	
$P/$ MPa	$\rho/$ g cm^{-3}	$P/$ MPa	$\rho/$ g cm^{-3}	$P/$ MPa	$\rho/$ g cm^{-3}	$P/$ MPa	$\rho/$ g cm^{-3}	$P/$ MPa	$\rho/$ g cm^{-3}	$P/$ MP a	$\rho/$ g cm^{-3}
---	---	---	---	---	---	---	---	---	---	---	---
45.58	0.8751	46.00	0.8551	49.64	0.8372	50.26	0.8177	54.63	0.8017	57.62	0.7845
40.83	0.8717	41.97	0.8522	45.77	0.8341	46.18	0.8139	51.14	0.7983	54.69	0.7813
35.17	0.8679	37.35	0.8485	42.28	0.8311	42.77	0.8105	47.04	0.7939	52.24	0.7783
29.78	0.8635	33.01	0.8451	38.72	0.8280	39.21	0.8070	44.02	0.7907	49.92	0.7755
25.72	0.8610	29.34	0.8419	35.66	0.8252	35.08	0.8026	40.04	0.7862	47.95	0.7731
23.93	0.8595	25.48	0.8386	34.60	0.8241	31.70	0.7989	35.01	0.7802	44.92	0.7694
20.45	0.8565	21.68	0.8352	30.85	0.8204	27.62	0.7944	30.01	0.7738	41.10	0.7643
16.39	0.8535	17.77	0.8316	26.53	0.8162	23.73	0.7898	25.74	0.7679	37.02	0.7587
12.20	0.8498	13.41	0.8272	22.77	0.8123					32.78	0.7523

$w_A = 0.0978$ $w_B = 0.0200$ $w_C = 0.8822$

T/K	297.1		323.5		347.2		372.7		399.4		425.4	
$P/$ MPa	$\rho/$ g cm^{-3}	$P/$ MPa	$\rho/$ g cm^{-3}	$P/$ MPa	$\rho/$ g cm^{-3}	$P/$ MPa	$\rho/$ g cm^{-3}	$P/$ MPa	$\rho/$ g cm^{-3}	$P/$ MP a	$\rho/$ g cm^{-3}
---	---	---	---	---	---	---	---	---	---	---	---
37.43	0.8784	42.87	0.8599	45.70	0.8423	49.63	0.8254	53.76	0.8083	57.85	0.7916
33.49	0.8759	38.61	0.8565	42.05	0.8394	46.14	0.8222	49.76	0.8043	54.16	0.7878
29.15	0.8726	34.67	0.8536	38.58	0.8362	42.31	0.8186	46.36	0.8009	50.97	0.7842
24.68	0.8694	30.59	0.8501	34.85	0.8330	38.49	0.8149	42.50	0.7965	47.27	0.7799
21.00	0.8666	26.88	0.8471	30.96	0.8294	34.90	0.8113	37.93	0.7912	43.56	0.7750
18.58	0.8647	23.28	0.8440	26.23	0.8248	32.25	0.8084	34.29	0.7867	42.66	0.7738
13.89	0.8608	20.64	0.8416	22.19	0.8207	28.77	0.8045	30.62	0.7820	38.62	0.7686
11.40	0.8588	17.34	0.8386	18.96	0.8173	25.60	0.8007	26.98	0.7772	34.86	0.7633
8.49	0.8562	13.64	0.8349	15.94	0.8140	22.29	0.7967			31.04	0.7577
		10.11	0.8312								

continued

continued

$w_A = 0.0974$ $w_B = 0.0246$ $w_C = 0.8780$

T/K	301.2		323.2		345.6		372.4		399.2		425.0	
P/ MPa	ρ/ g cm^{-3}	P/ MPa	ρ/ g cm^{-3}	P/ MPa	ρ/ g cm^{-3}	P/ MPa	ρ/ g cm^{-3}	P/ MPa	ρ/ g cm^{-3}	P/ MP a	ρ/ g cm^{-3}
53.48	0.9049	52.41	0.8850	52.86	0.8667	53.15	0.8455	54.55	0.8249	57.83	0.8076
49.54	0.9023	47.58	0.8816	47.94	0.8627	49.20	0.8420	51.38	0.8218	54.32	0.8035
45.34	0.8994	41.98	0.8772	43.48	0.8589	45.49	0.8383	48.71	0.8189	51.28	0.7999
42.01	0.8972	37.84	0.8740	38.99	0.8550	41.74	0.8346	46.01	0.8159	49.06	0.7971
37.95	0.8940	33.91	0.8709	35.73	0.8518	40.57	0.8332	45.07	0.8148	45.98	0.7933
34.29	0.8914	30.78	0.8683	31.95	0.8483	37.37	0.8298	42.32	0.8116	42.92	0.7892
30.31	0.8885	26.76	0.8648	28.09	0.8444	33.69	0.8258	38.43	0.8069	38.64	0.7832
27.11	0.8858	23.03	0.8614	24.38	0.8407	30.28	0.8219	34.60	0.8022	34.72	0.7775
26.39	0.8852	19.58	0.8582	21.20	0.8373	27.00	0.8179	30.97	0.7973	31.13	0.7719
23.32	0.8829	15.85	0.8546								
19.58	0.8797										
15.79	0.8765										

$w_A = 0.0967$ $w_B = 0.0301$ $w_C = 0.8732$

T/K	303.1		320.0		347.8		372.1		397.0		424.3	
P/ MPa	ρ/ g cm^{-3}	P/ MPa	ρ/ g cm^{-3}	P/ MPa	ρ/ g cm^{-3}	P/ MPa	ρ/ g cm^{-3}	P/ MPa	ρ/ g cm^{-3}	P/ MP a	ρ/ g cm^{-3}
53.29	0.8685	57.95	0.8558	57.06	0.8349	58.19	0.8174	59.53	0.7991	61.57	0.7800
50.21	0.8665	53.44	0.8530	52.95	0.8318	54.00	0.8138	56.62	0.7965	57.86	0.7761
47.69	0.8648	49.37	0.8500	49.80	0.8292	50.11	0.8103	52.72	0.7926	55.22	0.7732
44.06	0.8624	45.27	0.8471	46.07	0.8262	47.52	0.8078	50.90	0.7909	53.85	0.7716
40.99	0.8603	41.48	0.8444	43.79	0.8242	43.83	0.8042	47.42	0.7872	49.92	0.7670
38.63	0.8585	38.64	0.8423	40.01	0.8209	40.35	0.8007	42.96	0.7824	46.48	0.7628
35.01	0.8559	34.54	0.8392	36.64	0.8178	36.79	0.7971	39.09	0.7780	42.21	0.7574
30.89	0.8529	30.70	0.8362	32.74	0.8142	35.71	0.7740				
26.84	0.8497	26.61	0.8329	28.77	0.8104						
23.20	0.8470										

continued

continued

$w_A = 0.0969$ $w_B = 0.0321$ $w_C = 0.8710$

T/K	296.4		324.8		349.4		375.5		400.0		426.3	
P/ MPa	ρ/ g cm^{-3}	P/ MPa	ρ/ g cm^{-3}	P/ MPa	ρ/ g cm^{-3}	P/ MPa	ρ/ g cm^{-3}	P/ MPa	ρ/ g cm^{-3}	P/ MP a	ρ/ g cm^{-3}
56.87	0.9025	58.70	0.8808	56.98	0.8597	57.43	0.8387	57.65	0.8197	57.55	0.7985
52.53	0.8997	54.16	0.8777	53.40	0.8569	53.78	0.8356	54.91	0.8168	55.92	0.7966
45.15	0.8949	49.96	0.8746	49.86	0.8540	50.37	0.8325	53.02	0.8149	53.63	0.7941
42.03	0.8929	46.02	0.8715	46.74	0.8514	47.35	0.8295	50.68	0.8124	50.29	0.7901
37.85	0.8900	42.36	0.8686	43.71	0.8488	43.99	0.8262	46.63	0.8080	46.34	0.7850
36.29	0.8888	40.41	0.8670	40.31	0.8457	39.74	0.8218	42.39	0.8031	41.75	0.7791
31.87	0.8857	36.31	0.8638	36.08	0.8418	35.54	0.8174	37.98	0.7982	37.90	0.7739
26.35	0.8816	31.52	0.8598	31.57	0.8376	31.04	0.8125	33.39	0.7925	34.10	0.7683
22.19	0.8784	25.41	0.8545	26.99	0.8331						
17.67	0.8747										

$w_A = 0.0966$ $w_B = 0.0366$ $w_C = 0.8668$

T/K	298.6		323.6		348.6		373.8		398.8		423.7	
P/ MPa	ρ/ g cm^{-3}	P/ MPa	ρ/ g cm^{-3}	P/ MPa	ρ/ g cm^{-3}	P/ MPa	ρ/ g cm^{-3}	P/ MPa	ρ/ g cm^{-3}	P/ MP a	ρ/ g cm^{-3}
38.18	0.8852	41.90	0.8670	45.68	0.8489	46.02	0.8288	50.79	0.8128	53.95	0.7966
32.44	0.8813	37.51	0.8634	41.07	0.8451	41.89	0.8248	46.28	0.8080	51.02	0.7934
28.24	0.8782	33.59	0.8604	36.87	0.8413	37.67	0.8203	42.81	0.8042	48.04	0.7898
23.40	0.8747	29.16	0.8567	33.02	0.8378	34.26	0.8169	39.82	0.8008	44.76	0.7858
18.90	0.8712	24.89	0.8530	29.90	0.8348	33.14	0.8155	38.42	0.7991	42.95	0.7835
13.79	0.8670	20.94	0.8495	27.45	0.8325	29.10	0.8111	34.52	0.7946	39.55	0.7788
10.87	0.8644	17.24	0.8461	22.41	0.8271	25.10	0.8065	30.40	0.7893	35.73	0.7736
8.14	0.8620	14.17	0.8430	19.16	0.8238	20.67	0.8013	26.42	0.7841	31.42	0.7676
		10.43	0.8395	14.42	0.8186	16.84	0.7964	23.47	0.7799	28.09	0.7624

continued

continued

$w_A = 0.0948$ $w_B = 0.0444$ $w_C = 0.8608$

T/K	308.3		321.7		346.9		372.2		398.6		426.0	
$P/$ MPa	$\rho/$ g cm^{-3}		$P/$ MPa	$\rho/$ g cm^{-3}	$P/$ MPa	$\rho/$ g cm^{-3}	$P/$ MPa	$\rho/$ g cm^{-3}	$P/$ MPa	$\rho/$ g cm^{-3}	$P/$ MP a	$\rho/$ g cm^{-3}
52.72	0.8888	52.04	0.8783	53.25	0.8590	53.17	0.8378	54.16	0.8165	53.98	0.7951
48.62	0.8862	48.02	0.8756	48.45	0.8552	49.62	0.8348	51.05	0.8133	51.17	0.7919
44.32	0.8832	44.02	0.8726	44.66	0.8520	46.45	0.8319	47.19	0.8093	48.81	0.7890
39.86	0.8799	39.21	0.8691	40.73	0.8485	42.73	0.8284	43.98	0.8056	46.84	0.7865
35.16	0.8765	34.70	0.8656	37.81	0.8458	37.18	0.8226	42.43	0.8039	43.60	0.7823
30.89	0.8732	31.04	0.8626	31.99	0.8402	34.45	0.8199	38.31	0.7992	41.03	0.7789
27.38	0.8707	26.31	0.8585	28.13	0.8364	31.12	0.8161	35.07	0.7952	38.20	0.7751
22.97	0.8670	23.17	0.8559	24.21	0.8325	27.29	0.8119	31.39	0.7907	34.86	0.7705
19.47	0.8642	19.77	0.8528	20.20	0.8284						
15.94	0.8612	16.21	0.8494								
12.35	0.8581	12.24	0.8456								

$w_A = 0.0954$ $w_B = 0.0496$ $w_C = 0.8550$

T/K	299.2		323.2		348.9		374.8		400.8		425.3	
$P/$ MPa	$\rho/$ g cm^{-3}		$P/$ MPa	$\rho/$ g cm^{-3}	$P/$ MPa	$\rho/$ g cm^{-3}	$P/$ MPa	$\rho/$ g cm^{-3}	$P/$ MPa	$\rho/$ g cm^{-3}	$P/$ MP a	$\rho/$ g cm^{-3}
56.01	0.8837	57.15	0.8682	57.45	0.8480	58.09	0.8282	59.45	0.8089	60.14	0.7906
50.90	0.8810	53.62	0.8659	53.51	0.8450	54.01	0.8247	56.28	0.8061	58.32	0.7887
46.48	0.8783	49.44	0.8629	50.21	0.8423	51.24	0.8220	53.48	0.8031	56.31	0.7863
42.88	0.8761	45.68	0.8601	47.99	0.8405	49.24	0.8200	52.00	0.8015	54.91	0.7846
40.37	0.8745	42.96	0.8582	46.06	0.8388	45.52	0.8166	48.29	0.7976	51.51	0.7807
35.69	0.8713	37.43	0.8537	42.04	0.8354	41.33	0.8124	45.11	0.7941	47.95	0.7765
31.19	0.8682	33.08	0.8503	38.30	0.8320	37.09	0.8081	41.44	0.7902	43.64	0.7714
26.98	0.8648	28.88	0.8467	33.84	0.8280	32.02	0.8028	38.08	0.7863	39.59	0.7661

6.3. References

1976REN   Renuncio, J.A.R. and Prausnitz, J.M., Densities of polymer solutions to 1 kbar, *Macromolecules*, 9, 324, 1976.

1979KOB   Kobyakov, V.M., Kogan, V.B., Rätzsch, M., and Zernov, V.S., Sootnoshenie P-V-T-N dlya smesey etilena s niskomolekulyarnym polietilenom, *Plast. Massy*, 8, 21, 1979.

1987VE1   Vennemann, N., Lechner, M.D., and Oberthür, R.C., Thermodynamics and conformation of polyoxyethylene in aqueous solution under high pressure. 1. Small-angle neutron scattering and densitometric measurements at room temperature, *Polymer*, 28, 1738, 1987.

1987VE2   Vennemann, N., Der Einfluss von hohem Druck auf thermodynamische und strukturelle Eigenschaften von Polyoxiethylen in wäßriger Lösung, *Ph.D. Thesis*, Universität Osnabrück, 1987.

1989OUG   Ougizawa, T., Dee, G.T., and Walsh, D.J., PVT properties and equation of state of polystyrene. Molecular weight dependence of the characteristic parameters in equation-of-state theories (experimental data by D.J. Walsh), *Polymer*, 30, 1675, 1989.

1991FAK   Fakhreddine, Y.A. and Zoller, P., The equation of state of solid and molten poly(butylene terephthalate) to 300°C and 200 MPa (experimental data by P. Zoller), *J. Polym. Sci.: Part B: Polym. Phys.*, 29, 1141, 1991,

1993JAN   Janssen, S., Schwahn, D., Mortensen, K., and Springer, T., Pressure dependence of the Flory-Huggins interaction parameter in polymer blends. A SANS study and a comparison to the Flory-Orwoll-Vrij equation of state (experimental data by S. Janssen), *Macromolecules*, 26, 5587, 1993.

1992KIM   Kim, C.K. and Paul, D.R., Interaction parameters for blends containing polycarbonates. 1. Tetramethyl bisphenol-A polycarbonate/polystyrene, *Polymer*, 33, 1630, 1992.

1992WIN   Wind, R.W., Untersuchungen zum Phasenverhalten von Mischungen aus Ethylen, Acrylsäure und Ethylen-Acrylsäure-Copolymeren unter hohem Druck, *Ph.D. Thesis*, TH Darmstadt, 1992.

1993ROG   Rodgers, P.A., Pressure-volume-temperature relationships for polymeric liquids. A review of equations of state and their characteristic parameters for 56 polymers, *J. Appl. Polym. Sci.*, 48, 1061, 1993.

1994COM   Compostizo, A., Cancho, S.M., Colin, A.C., and Rubio, R.G., Polymer solutions with specific interactions: equation of state for poly(4-hydroxystyrene) + acetone, *Macromolecules*, 27, 3478, 1994.

1994MER   Mertsch, R. and Wolf, B.A., Solutions of poly(dimethylsiloxane) in supercritical CO_2: viscometric and volumetric behavior, *Macromolecules*, 27, 3289, 1994.

1995COM   Compostizo, A., Cancho, S.M., Rubio, R.G., and Colin, A.C., Equation of state of hydrogen-bonded polymer solutions: poly(4-hydroxystyrene) + ethanol and + tetrahydrofuran, *J. Phys. Chem.*, 99, 10261, 1995.

1995ZOL   Zoller, P. and Walsh, D.J., *Standard Pressure-Volume-Temperature Data for Polymers*, Technomic Publishing, Lancaster, 1995.

1996MAI   Maier, R.-D., Kressler, J., Rudolf, B., Reichert, P., Koopman, F., Frey, H., and Mühlhaupt, R., Miscibility of poly(sila-α-methylstyrene) with polystyrene (experimental data by J. Kressler), *Macromolecules*, 29, 1490, 1996.

1997COL   Colin, A.C., Cancho, S.M., Rubio, R.G., and Compostizo, A., Equation of state of hydrogen-bonded polymer solutions. Poly(propylene glycol) + n-hexane and poly(propylene glycol) + ethanol, *Macromolecules*, 30, 3389, 1997.

1997SAT   Sato, Y., Yamasaki, Y., Takishima, S., and Masuoka, H., Precise measurement of the PVT of polypropylene and polycarbonate up to 330°C and 200 MPa, *J. Appl. Polym. Sci.*, 66, 141, 1997.

1998AAL   Aalto, M.M. and Liukkonen, S.S., Liquid densities of propane + linear low-density poly-ethylene systems at (354-378) K and (4.00-7.00) MPa, *J. Chem. Eng. Data*, 43, 29, 1998.

1998LUN   Luna-Barcenas, G., Mawson, S., Takishima, S., DeSimone, J.M., Sanchez, I.C., and Johnston, K.P.: Phase behavior of poly(1,1-dihydroperfluorooctyl acrylate) in supercritical carbon dioxide, *Fluid Phase Equil.*, 146, 325, 1998.

1998SAC   Sachdev, V.K., Yashi, U., and Jain, R.K., Equation of state of poly(dimethylsiloxane) melts, *J. Polym. Sci.: Part B: Polym. Phys.*, 36, 841, 1998.

1999BRA   Brandrup, J., Immergut, E.H., and Grulke, E.A., Eds., *Polymer Handbook*, 4th ed., J. Wiley & Sons, New York, 1999.

1999CRE   Crespo Colin, A., Cancho, S.M., Rubio, R.G., and Compostizo, A., Equation of state of aqueous polymer systems: poly(propylene glycol) + water, *Phys. Chem. Chem. Phys.*, 1, 319, 1999.

1999CHU   Chu, J.H. and Paul, D.R., Interaction energies for blends of SAN with methyl methacrylate copolymers with ethyl acrylate and n-butyl acrylate, *Polymer*, 40, 2687, 1999.

1999INO   Inomata, H., Honma, Y., Imahori, M., and Arai, K., Fundamental study of desolventing polymer solutions with supercritical CO_2, *Fluid Phase Equil.*, 158-160, 857, 1999.

1999LEE   Lee, M.-J., Lo, C.-K., and Lin, H.-M., PVT measurements for mixtures of 1-octanol with oligomeric poly(ethylene glycol) from 298 K to 338 K and pressures up to 30 MPa, *J. Chem. Eng. Data*, 44, 1379, 1999.

1999MAI   Maier, R.-D., Thermodynamik und Kristallisation von Metallocen-Polyolefinen (experimental data by R.-D. Maier), *Ph.D. Thesis*, Albert-Ludwigs-Universität, Freiburg i. Br., 1999.

2000BY1   Byun, H.-S., DiNoia, T.P., and McHugh, M.A., High-pressure densities of ethane, pentane, pentane-d12, 25.5 wt% ethane in pentane-d12, 2.4 wt% deuterated poly(ethylene-*co*-butene) (PEB) in ethane, 5.3 wt% hydrogenated PEB in pentane, 5.1 wt % hydrogenated PEB in pentane-d12, and 4.9 wt % hydrogenated PEB in pentane-d12 + 23.1 wt % ethane, *J. Chem. Eng. Data*, 45, 810, 2000.

2000BY2   Byun, H.-S., Kim, K., and Lee, H.-S., High-pressure phase behavior and mixture density of binary poly(ethylene-*co*-butene)-dimethyl ether system, *Hwahak Konghak*, 38, 826, 2000.

2000LEE   Lee, M.-J., Tuan, Y.-C., and Lin, H.-M., Pressure, volume, and temperature for mixtures of poly(ethylene glycol methyl ether)-350 + anisole and poly(ethylene glycol)-200 + anisole from 298 K to 338 K and pressures up to 50 MPa, *J. Chem. Eng. Data*, 45, 1100, 2000.

2000SAT   Sato, Y., Inohara, K., Takishima, S., Masuoka, H., Imaizumi, M., Yamamoto, H., and Takasugi, M., Pressure-volume-temperature behavior of polylactide, poly(butylene succinate), and poly(butylene succinate-*co*-adipate), *Polym. Eng. Sci.*, 40, 2602, 2000.

2001LI2   Li, D., Han, B., Liu, Z., Liu, J., Zhang, X., Wang, S., and Zhang, X., Effect of gas antisolvent on conformation of polystyrene in toluene: Viscosity and small-angle X-ray scattering study (experimental data by B. Han), *Macromolecules*, 34, 2195, 2001.

2001LIN   Lin, H.-M., Hsu, T.-S., and Lee, M.-J., Pressure-volume-temperature properties for binary polymer solutions of poly(propylene glycol) with 1-octanol and acetophenone, *Macromolecules*, 34, 6297, 2001.

2001MEK   Mekhilef, N., Viscoelastic and pressure-volume-temperature properties of poly(vinylidene fluoride) and poly(vinylidene fluoride)-hexafluoropropylene copolymers (experimental PVT data by N. Mekhilef), *J. Appl. Polym. Sci.*, 80, 230, 2001.

2001WOH   Wohlfarth, C., *CRC Handbook of Thermodynamic Data of Copolymer Solutions*, CRC Press, Boca Raton, 2001.

2002COM   Comunas, M.J.P., Baylaucq, A., Boned, C., Canet, X., and Fernandez, J., High-pressure volumetric behavior of x 1,1,1,2-tetrafluoroethane + (1-x) 2,5,8,11,14-pentaoxapentadecane (TEGDME) mixtures, *J. Chem. Eng. Data*, 47, 233, 2002.

2002DIN   Dindar, C. and Kiran, E., High-pressure viscosity and density of polymer solutions at the critical polymer concentration in near-critical and supercritical fluids, *Ind. Eng. Chem. Res.*, 41, 6354, 2002.

2003ACE   Acevedo, I.L., Lugo, L., Comunas, M.J.P., Arancibia, E.L., and Fernandez, J., Volumetric properties of binary tetraethylene glycol dimethyl ether + heptane mixtures between (278.15 and 353.15) K and up to 25 MPa, *J. Chem. Eng. Data*, 48, 1271, 2003.

2003LEE   Lee, M.-J., Tuan, Y.-C., and Lin, H.-M., Pressure-volume-temperature properties for binary and ternary polymer solutions of poly(ethylene glycol), poly(propylene glycol), and poly(ethylene glycol methyl ether) with anisole, *Polymer*, 44, 3891, 2003.

2003WOH   Wohlfarth, C., Pressure-volume-temperature relationship for polymer melts, in *CRC Handbook of Chemistry and Physics*, 84th ed., pp. 13-16 to 13-20, 2003.

2003ZH1   Zhang, W. and Kiran, E., (*p, V, T*) Behaviour and miscibility of (polysulfone + THF + carbon dioxide) at high pressures, *J. Chem. Thermodyn.*, 35, 605, 2003.

2004BOL   Bolotnikov, M.F., Verveyko, V.N., and Verveyko, M.V., Speeds of sound, densities, and isentropic compressibilities of poly(propylene glycol)-425 at temperatures from (293.15 to 373.15) K and pressures up to 100 MPa, *J. Chem. Eng. Data*, 49, 631, 2004.

2004KIM   Kim, J.H., Whang, M.S., and Kim, C.K., Novel miscible blends composed of poly(ether sulfone) and poly(1-vinylpyrrolidone-*co*-styrene) copolymers and their interaction energies, *Macromolecules*, 37, 2287, 2004.

7. PRESSURE DEPENDENCE OF THE SECOND VIRIAL COEFFICIENTS (A_2) OF POLYMER SOLUTIONS

7.1. Experimental A_2 data

Polymer (B)	$M_n/$ kg/mol	$M_w/$ kg/mol	Solvent (A)	$T/$ K	$P/$ MPa	$10^4 A_2/$ cm^{3}mol/g^2	Ref.
Dextran							
		1000	water	303.15	0.1	0.85	1991STA
		1000	water	303.15	200.0	0.74	1991STA
		1000	water	343.15	0.1	2.1	1991STA
		1000	water	343.15	200.0	1.6	1991STA
		1000	water	373.15	0.1	1.8	1991STA
		1000	water	373.15	200.0	1.6	1991STA
		7400	water	303.15	0.1	0.099	1991STA
		7400	water	303.15	200.0	0.092	1991STA
		7400	water	343.15	0.1	0.064	1991STA
		7400	water	343.15	200.0	0.062	1991STA
		7400	water	373.15	0.1	0.078	1991STA
		7400	water	373.15	200.0	0.077	1991STA
		17500	water	303.15	0.1	0.013	1991STA
		17500	water	303.15	200.0	0.013	1991STA
		17500	water	323.15	0.1	0.018	1991STA
		17500	water	323.15	200.0	0.019	1991STA
		17500	water	343.15	0.1	0.014	1991STA
		17500	water	343.15	200.0	0.011	1991STA
		17500	water	373.15	0.1	0.011	1991STA
		17500	water	373.15	200.0	0.010	1991STA
		21100	water	303.15	0.1	0.017	1991STA
		21100	water	303.15	200.0	0.014	1991STA
		21100	water	323.15	0.1	0.015	1991STA
		21100	water	323.15	200.0	0.015	1991STA
		21100	water	343.15	0.1	0.015	1991STA
		21100	water	343.15	200.0	0.018	1991STA
		21100	water	373.15	0.1	0.0045	1991STA
		21100	water	373.15	200.0	0.0046	1991STA

Polymer (B)	$M_n/$ kg/mol	$M_w/$ kg/mol	Solvent (A)	$T/$ K	$P/$ MPa	$10^4 A_2/$ cm^{3}mol/g^2	Ref.
Polyacrylamide							
		1400	water	298.15	0.1	1.80	1981LEC
		1400	water	298.15	50.0	1.60	1981LEC
		1400	water	298.15	100.0	1.30	1981LEC
		1400	water	298.15	150.0	1.90	1981LEC
		1400	water	298.15	200.0	2.10	1981LEC
Poly(dimethyl siloxane)							
		578	benzene	300.15	0.1	2.95	1976KUB
		578	benzene	300.15	9.8	3.02	1976KUB
		578	benzene	300.15	19.6	3.02	1976KUB
		578	benzene	300.15	29.4	3.03	1976KUB
		578	benzene	300.15	39.2	3.00	1976KUB
		578	bromocyclohexane	300.15	0.1	−0.0158	1976KUB
		578	bromocyclohexane	300.15	9.8	0.0258	1976KUB
		578	bromocyclohexane	300.15	19.6	0.0538	1976KUB
		578	bromocyclohexane	300.15	29.4	0.0782	1976KUB
		578	bromocyclohexane	300.15	39.2	0.0995	1976KUB
		578	bromocyclohexane	309.65	0.1	0.362	1976KUB
		578	bromocyclohexane	309.65	9.8	0.437	1976KUB
		578	bromocyclohexane	309.65	19.6	0.504	1976KUB
		578	bromocyclohexane	309.65	29.4	0.548	1976KUB
		578	bromocyclohexane	309.65	39.2	0.576	1976KUB
		578	bromocyclohexane	320.35	0.1	0.657	1976KUB
		578	bromocyclohexane	320.35	9.8	0.739	1976KUB
		578	bromocyclohexane	320.35	19.6	0.794	1976KUB
		578	bromocyclohexane	320.35	29.4	0.820	1976KUB
		578	bromocyclohexane	320.35	39.2	0.852	1976KUB
		578	bromocyclohexane	329.35	0.1	0.954	1976KUB
		578	bromocyclohexane	329.35	9.8	1.00	1976KUB
		578	bromocyclohexane	329.35	19.6	1.04	1976KUB
		578	bromocyclohexane	329.35	29.4	1.07	1976KUB
		578	bromocyclohexane	329.35	39.2	1.11	1976KUB
		578	chlorobenzene	303.15	0.1	1.04	1976KUB
		578	chlorobenzene	303.15	9.8	1.10	1976KUB
		578	chlorobenzene	303.15	19.6	1.16	1976KUB
		578	chlorobenzene	303.15	29.4	1.21	1976KUB
		578	chlorobenzene	303.15	39.2	1.25	1976KUB
		578	toluene	300.15	0.1	4.50	1976KUB
		578	toluene	300.15	9.8	4.46	1976KUB
		578	toluene	300.15	19.6	4.63	1976KUB
		578	toluene	300.15	29.4	4.67	1976KUB
		578	toluene	300.15	39.2	4.67	1976KUB

Polymer (B)	M_n/ kg/mol	M_w/ kg/mol	Solvent (A)	T/ K	P/ MPa	$10^4 A_2$/ cm³mol/g²	Ref.
Poly(ethylene oxide)							
		20	deuterium oxide	298.15	0.1	24.0	1987VEN
		20	deuterium oxide	298.15	100.0	16.0	1987VEN
		20	deuterium oxide	298.15	200.0	13.0	1987VEN
Polyisobutylene							
	703	1020	2-methylbutane	297.15	0.1	0.70	1972GAE
	703	1020	2-methylbutane	297.15	3.0	1.00	1972GAE
	703	1020	2-methylbutane	297.15	6.0	1.25	1972GAE
	703	1020	2-methylbutane	297.15	9.0	1.50	1972GAE
	703	1020	2-methylbutane	297.15	12.0	1.60	1972GAE
	703	1020	2-methylbutane	330.15	0.1	−0.50	1972GAE
	703	1020	2-methylbutane	330.15	3.0	−0.05	1972GAE
	703	1020	2-methylbutane	330.15	6.0	0.45	1972GAE
	703	1020	2-methylbutane	330.15	9.0	0.85	1972GAE
	703	1020	2-methylbutane	330.15	12.0	1.20	1972GAE
	703	1020	2-methylbutane	337.15	0.1	−0.90	1972GAE
	703	1020	2-methylbutane	337.15	3.0	−0.40	1972GAE
	703	1020	2-methylbutane	337.15	6.0	0.10	1972GAE
	703	1020	2-methylbutane	337.15	9.0	0.60	1972GAE
		690	n-pentane	343.15	0.1	−0.55	1971WOL
		690	n-pentane	343.15	5.2	−0.10	1971WOL
		690	n-pentane	343.15	10.1	0.15	1971WOL
		690	n-pentane	343.15	20.3	0.50	1971WOL
		690	n-pentane	343.15	40.5	0.80	1971WOL
		690	n-pentane	343.15	60.8	1.00	1971WOL
		690	n-pentane	343.15	81.1	1.15	1971WOL
		690	2,2,4-trimethylpentane	298.15	0.1	2.30	1971WOL
		690	2,2,4-trimethylpentane	298.15	20.3	2.55	1971WOL
		690	2,2,4-trimethylpentane	298.15	40.5	2.75	1971WOL
		690	2,2,4-trimethylpentane	298.15	60.8	2.85	1971WOL
		690	2,2,4-trimethylpentane	298.15	81.1	3.00	1971WOL
Polystyrene							
		975	2-butanone	295.15	0.1	1.10	1972GAE
		975	2-butanone	295.15	5.0	1.20	1972GAE
		975	2-butanone	295.15	12.0	1.70	1972GAE
		100	cyclohexane	313.15	0.1	0.40	1970LEC
		100	cyclohexane	313.15	20.3	0.39	1970LEC
		100	cyclohexane	313.15	40.5	0.25	1970LEC
		100	cyclohexane	318.15	0.1	0.79	1970LEC
		100	cyclohexane	318.15	20.3	0.84	1970LEC
		100	cyclohexane	318.15	40.5	0.87	1970LEC
		100	cyclohexane	323.15	0.1	1.11	1970LEC
		100	cyclohexane	323.15	20.3	1.18	1970LEC
		100	cyclohexane	323.15	40.5	1.19	1970LEC

Polymer (B)	M_n/ kg/mol	M_w/ kg/mol	Solvent (A)	T/ K	P/ MPa	$10^4 A_2$/ cm³mol/g²	Ref.
Polystyrene (*continued*)							
	415	440	decalin	297.15	0.1	0.84	1976MCD
	415	440	decalin	297.15	137.2	−0.50	1976MCD
	739	850	decalin	297.15	0.1	0.55	1976MCD
	739	850	decalin	297.15	137.2	−0.20	1976MCD
	415	440	*cis*-decalin	297.15	0.1	1.12	1976MCD
	415	440	*cis*-decalin	297.15	137.2	−0.30	1976MCD
		100	*trans*-decalin	288.15	0.1	−0.25	1970LEC
		100	*trans*-decalin	288.15	20.3	−0.37	1970LEC
		100	*trans*-decalin	288.15	40.5	−0.54	1970LEC
		100	*trans*-decalin	288.15	60.8	−0.73	1970LEC
		100	*trans*-decalin	288.15	81.1	−0.95	1970LEC
		100	*trans*-decalin	293.15	0.1	+0.20	1970LEC
		100	*trans*-decalin	293.15	20.3	−0.04	1970LEC
		100	*trans*-decalin	293.15	40.5	−0.14	1970LEC
		100	*trans*-decalin	293.15	60.8	−0.26	1970LEC
		100	*trans*-decalin	293.15	81.1	−0.40	1970LEC
		100	*trans*-decalin	298.15	0.1	0.30	1970LEC
		100	*trans*-decalin	298.15	20.3	0.32	1970LEC
		100	*trans*-decalin	298.15	40.5	0.27	1970LEC
		100	*trans*-decalin	298.15	60.8	0.18	1970LEC
		100	*trans*-decalin	298.15	81.1	0.07	1970LEC
		100	*trans*-decalin	303.15	0.1	0.55	1970LEC
		100	*trans*-decalin	303.15	20.3	0.57	1970LEC
		100	*trans*-decalin	303.15	40.5	0.54	1970LEC
		100	*trans*-decalin	303.15	60.8	0.48	1970LEC
		100	*trans*-decalin	303.15	81.1	0.40	1970LEC
		100	*trans*-decalin	308.15	0.1	0.86	1970LEC
		100	*trans*-decalin	308.15	20.3	0.92	1970LEC
		100	*trans*-decalin	308.15	40.5	0.89	1970LEC
		100	*trans*-decalin	308.15	60.8	0.82	1970LEC
		100	*trans*-decalin	308.15	81.1	0.72	1970LEC
		100	*trans*-decalin	313.15	0.1	1.11	1970LEC
		100	*trans*-decalin	313.15	20.3	1.17	1970LEC
		100	*trans*-decalin	313.15	40.5	1.19	1970LEC
		100	*trans*-decalin	313.15	60.8	1.19	1970LEC
		100	*trans*-decalin	313.15	81.1	1.17	1970LEC
	415	440	*trans*-decalin	297.15	0.1	0.29	1976MCD
	415	440	*trans*-decalin	297.15	137.2	−0.30	1976MCD
		1800	*trans*-decalin	298.15	0.1	0.19	1970LEC
		1800	*trans*-decalin	298.15	19.0	0.10	1970LEC
		1800	*trans*-decalin	298.15	38.0	0.02	1970LEC
		1800	*trans*-decalin	298.15	57.0	−0.01	1970LEC
		1800	*trans*-decalin	303.15	0.1	0.40	1970LEC
		1800	*trans*-decalin	303.15	19.0	0.36	1970LEC
		1800	*trans*-decalin	303.15	38.0	0.29	1970LEC

Polymer (B)	M_n/ kg/mol	M_w/ kg/mol	Solvent (A)	T/ K	P/ MPa	10^4A_2/ cm^3mol/g^2	Ref.
Polystyrene (*continued*)							
		1800	*trans*-decalin	303.15	57.0	0.26	1970LEC
		1800	*trans*-decalin	303.15	76.0	0.20	1970LEC
		1800	*trans*-decalin	313.15	0.1	0.72	1970LEC
		1800	*trans*-decalin	313.15	19.0	0.65	1970LEC
		1800	*trans*-decalin	313.15	38.0	0.45	1970LEC
		1800	*trans*-decalin	313.15	57.0	0.42	1970LEC
		1800	*trans*-decalin	313.15	76.0	0.45	1970LEC
		1800	*trans*-decalin	298.15	0.1	0.19	1970SCH
		1800	*trans*-decalin	298.15	19.0	0.10	1970SCH
		1800	*trans*-decalin	298.15	38.0	0.02	1970SCH
		1800	*trans*-decalin	298.15	57.0	−0.01	1970SCH
		1800	*trans*-decalin	303.15	0.1	0.40	1970SCH
		1800	*trans*-decalin	303.15	19.0	0.36	1970SCH
		1800	*trans*-decalin	303.15	38.0	0.29	1970SCH
		1800	*trans*-decalin	303.15	57.0	0.26	1970SCH
		1800	*trans*-decalin	303.15	76.0	0.20	1970SCH
		1800	*trans*-decalin	313.15	0.1	0.72	1970SCH
		1800	*trans*-decalin	313.15	19.0	0.65	1970SCH
		1800	*trans*-decalin	313.15	38.0	0.45	1970SCH
		1800	*trans*-decalin	313.15	57.0	0.42	1970SCH
		1800	*trans*-decalin	313.15	76.0	0.45	1970SCH
		1800	*trans*-decalin	303.15	0.1	0.47	1971LEC
		1800	*trans*-decalin	303.15	50.7	0.44	1971LEC
	1563	2500	*trans*-decalin	299.15	0.1	0.26	1971LEC
	1563	2500	*trans*-decalin	299.15	60.8	0.00	1971LEC
	1563	2500	*trans*-decalin	299.15	121.6	−0.26	1971LEC
	1563	2500	*trans*-decalin	299.15	182.4	−0.50	1971LEC
	1563	2500	*trans*-decalin	315.65	0.1	0.77	1971LEC
	1563	2500	*trans*-decalin	315.65	60.8	0.61	1971LEC
	1563	2500	*trans*-decalin	315.65	121.6	0.61	1971LEC
	1563	2500	*trans*-decalin	315.65	182.4	0.41	1971LEC
	1563	2500	*trans*-decalin	332.65	0.1	1.06	1971LEC
	1563	2500	*trans*-decalin	332.65	60.8	1.11	1971LEC
	1563	2500	*trans*-decalin	332.65	121.6	1.12	1971LEC
	1563	2500	*trans*-decalin	332.65	182.4	1.06	1971LEC
	1563	2500	*trans*-decalin	356.15	0.1	1.37	1971LEC
	1563	2500	*trans*-decalin	356.15	60.8	1.25	1971LEC
	1563	2500	*trans*-decalin	356.15	121.6	1.32	1971LEC
	1563	2500	*trans*-decalin	356.15	182.4	1.31	1971LEC
	160	170	ethyl acetate	297.15	0.1	2.01	1976MCD
	160	170	ethyl acetate	297.15	392.2	2.98	1976MCD
	160	170	ethyl acetate	297.15	490.3	3.05	1976MCD
	415	440	ethyl acetate	297.15	0.1	1.29	1976MCD
	415	440	ethyl acetate	297.15	294.2	1.80	1976MCD
	415	440	ethyl acetate	297.15	392.2	1.75	1976MCD
	415	440	ethyl acetate	297.15	490.3	2.00	1976MCD

Polymer (B)	$M_n/$ kg/mol	$M_w/$ kg/mol	Solvent (A)	$T/$ K	$P/$ MPa	$10^4 A_2/$ cm^{3}mol/g^2	Ref.
Polystyrene (*continued*)							
	739	850	ethyl acetate	297.15	0.1	1.27	1976MCD
	739	850	ethyl acetate	297.15	392.2	1.96	1976MCD
		1700	ethyl acetate	297.15	0.1	1.18	1976MCD
		1700	ethyl acetate	297.15	392.2	1.76	1976MCD
		860	4-methyl-2-pentanone	297.15	0.1	0.50	1973MCD
		860	4-methyl-2-pentanone	297.15	49.0	0.85	1973MCD
		860	4-methyl-2-pentanone	297.15	98.0	1.25	1973MCD
		860	4-methyl-2-pentanone	297.15	147.1	1.50	1973MCD
		860	4-methyl-2-pentanone	297.15	245.2	1.75	1973MCD
		860	4-methyl-2-pentanone	297.15	343.2	1.70	1973MCD
		860	4-methyl-2-pentanone	297.15	400.0	1.75	1973MCD
		100	toluene	303.15	0.1	4.60	1970LEC
		100	toluene	303.15	20.3	4.60	1970LEC
		100	toluene	303.15	40.5	4.60	1970LEC
		100	toluene	303.15	60.8	4.60	1970LEC
		100	toluene	303.15	81.1	4.60	1970LEC
		100	toluene	303.15	0.1	4.60	1972SCH
		100	toluene	303.15	20.3	4.60	1972SCH
		100	toluene	303.15	40.5	4.60	1972SCH
		100	toluene	303.15	60.8	4.60	1972SCH
		100	toluene	303.15	81.1	4.60	1972SCH
	415	440	toluene	297.15	0.1	5.65	1976MCD
	415	440	toluene	297.15	392.2	6.50	1976MCD
		1700	toluene	297.15	0.1	4.43	1976MCD
		1700	toluene	297.15	392.2	5.70	1976MCD
		1800	toluene	288.15	0.1	2.24	1970LEC
		1800	toluene	288.15	19.0	2.40	1970LEC
		1800	toluene	288.15	38.0	2.35	1970LEC
		1800	toluene	288.15	57.0	2.42	1970LEC
		1800	toluene	288.15	76.0	2.39	1970LEC
		1800	toluene	298.15	0.1	2.27	1970LEC
		1800	toluene	298.15	19.0	2.38	1970LEC
		1800	toluene	298.15	38.0	2.40	1970LEC
		1800	toluene	298.15	57.0	2.46	1970LEC
		1800	toluene	298.15	76.0	2.41	1970LEC
		100	trichloromethane	303.15	0.1	5.40	1970LEC
		100	trichloromethane	303.15	20.3	5.30	1970LEC
		100	trichloromethane	303.15	40.5	5.20	1970LEC
		100	trichloromethane	303.15	60.8	5.15	1970LEC
		100	trichloromethane	303.15	81.1	5.10	1970LEC
		100	trichloromethane	303.15	0.1	5.40	1972SCH
		100	trichloromethane	303.15	20.3	5.30	1972SCH
		100	trichloromethane	303.15	40.5	5.20	1972SCH
		100	trichloromethane	303.15	60.8	5.15	1972SCH
		100	trichloromethane	303.15	81.1	5.10	1972SCH

Polymer (B)	$M_n/$ kg/mol	$M_w/$ kg/mol	Solvent (A)	$T/$ K	$P/$ MPa	$10^4 A_2/$ cm³mol/g²	Ref.
Polystyrene (*continued*)							
	415	440	trichloromethane	297.15	0.1	8.75	1976MCD
	415	440	trichloromethane	297.15	294.2	7.90	1976MCD
	415	440	trichloromethane	297.15	392.2	7.90	1976MCD

7.2. Pressure coefficient of the second virial coefficient $(\partial A_2/\partial P)_T$ at zero pressure

Polymer (B)	$M_n/$ kg/mol	$M_w/$ kg/mol	Solvent (A)	$T/$ K	$10^4 A_2/$ cm^3mol/g^2	$10^6(\partial A_2/\partial P)_T/$ cm^3mol/g^2MPa	Ref.
Poly(ethylene oxide)							
		20	deuterium oxide	298.15	24.0	−8.0	1987VEN
Polyisobutylene							
	703	1020	2-methylbutane	297.15	0.70	9.4	1972GAE
	703	1020	2-methylbutane	330.15	−0.50	15.5	1972GAE
	703	1020	2-methylbutane	337.15	−0.90	16.5	1972GAE
		690	n-pentane	343.15	−0.55	8.3	1971WOL
		690	2,2,4-trimethylpentane	298.15	2.30	1.5	1971WOL
Polystyrene							
		975	2-butanone	295.15	1.10	5.0	1972GAE
	24.6	25	cyclohexane	295.15	−2.10	−3.3	2003MOS
	24.6	25	cyclohexane	314.15	0.90	−1.7	2003MOS
		100	cyclohexane	313.15	0.40	−0.4	1970LEC
		100	cyclohexane	318.15	0.79	0.3	1970LEC
		100	cyclohexane	323.15	1.11	0.2	1970LEC
	377	400	cyclohexane	300.15	−2.90	1.3	2003MOS
	849	900	cyclohexane	302.15	−4.20	2.8	2003MOS
	86.5	90	decalin	298.15	2.00	−2.6	2003MOS
	415	440	decalin	297.15	0.84	−1.0	1976MCD
	739	850	decalin	297.15	0.55	−0.6	1976MCD
	415	440	cis-decalin	297.15	1.12	−1.0	1976MCD
		100	trans-decalin	288.15	−0.25	−0.7	1970LEC
		100	trans-decalin	293.15	0.00	−0.5	1970LEC
		100	trans-decalin	298.15	0.30	−0.3	1970LEC
		100	trans-decalin	303.15	0.55	−0.2	1970LEC
		100	trans-decalin	308.15	0.86	−0.1	1970LEC
		100	trans-decalin	313.15	1.11	0.1	1970LEC
	415	440	trans-decalin	297.15	0.29	−0.4	1976MCD
		1800	trans-decalin	298.15	0.19	−0.4	1970LEC
		1800	trans-decalin	303.15	0.40	−0.3	1970LEC
		1800	trans-decalin	313.15	0.72	−0.4	1970LEC

Polymer (B)	$M_n/$ kg/mol	$M_w/$ kg/mol	Solvent (A)	$T/$ K	$10^4 A_2/$ cm³mol/g²	$10^6(\partial A_2/\partial P)_T/$ cm³mol/g²MPa	Ref.
Polystyrene (*continued*)							
	1563	2500	*trans* -decalin	299.15	0.26	−0.4	1971LEC
	1563	2500	*trans* -decalin	315.65	0.77	−0.3	1971LEC
	1563	2500	*trans* -decalin	332.65	1.06	0.1	1971LEC
	1563	2500	*trans* -decalin	356.15	1.37	−0.1	1971LEC
	160	170	ethyl acetate	297.15	2.01	1.5	1976MCD
	415	440	ethyl acetate	297.15	1.29	0.5	1976MCD
	739	850	ethyl acetate	297.15	1.27	0.5	1976MCD
		1700	ethyl acetate	297.15	1.18	0.5	1976MCD
	24.6	25	methylcyclohexane	295.15	−6.00	5.5	2003MOS
		860	4-methyl-2-pentanone	297.15	0.50	0.75	1973MCD
	86.5	90	toluene	298.15	3.00	3.4	2003MOS
		100	toluene	303.15	4.60	0.0	1970LEC
	415	440	toluene	297.15	5.65	0.4	1976MCD
		1700	toluene	297.15	4.43	0.4	1976MCD
		1800	toluene	288.15	2.24	0.4	1970LEC
		1800	toluene	298.15	2.27	0.4	1970LEC
	86.5	90	trichloromethane	298.15	4.60	3.0	2003MOS
		100	trichloromethane	303.15	4.60	−3.8	1970LEC
	415	440	trichloromethane	297.15	8.75	−0.2	1976MCD

7.3. References

1970LEC   Lechner, M.D. and Schulz, G.V., Lichtstreuung von hochmolekularen Lösungen in Abhängigkeit von Temperature und Druck, *Eur. Polym. J.*, 6, 945, 1970.

1970SCH   Schulz, G.V. and Lechner, M.D., A light-scattering photometer for use at elevated pressures, *J. Polym. Sci.: Part A-2*, 8, 1885, 1970.

1971LEC   Lechner, M.D. and Schulz, G.V., Über die Beziehung zwischen der Winkelabhängigkeit und der Wellenlängenabhängigkeit bei Streulichtmessungen, *Makromol. Chem.*, 148, 325, 1971.

1971WOL   Wolf, Streulichtmessungen an exotherm-pseudoidealen Lösungen in Abhängigkeit von Druck und Temperatur, *Ber. Bunsenges. Phys. Chem.*, 75, 924, 1971.

1972LEC   Lechner, M.D., Schulz, G.V., and Wolf, B.A., Thermodynamics of polymer solutions as functions of pressure and temperature, *J. Coll. Interface Sci.*, 39, 462, 1972.

1972GAE   Gaeckle, D. and Patterson, D., Effect of pressure on the second virial coefficient and chain dimensions in polymer solutions, *Macromolecules*, 5, 136, 1972.

1972SCH   Schulz, G.V. and Lechner, M.D., Influence of pressure and temperature, in *Light Scattering from Polymer Solutions*, Huglin, M.B., Ed., Academic Press, New York, 1972, 503.

1973MCD   McDonald, C.J. and Claesson, S., Light scattering measurements under high pressure. I. Pure solvents and polystyrene in methyl isobutyl ketone, *Chem. Scripta*, 4, 155, 1973.

1976KUB   Kubota, K., Kubo, K., and Ogino, K., The pressure and temperature dependence of the second virial coefficients and the chain dimensions of polydimethylsiloxane solutions, *Bull. Chem. Soc. Japan*, 49, 2410, 1976.

1976MCD   McDonald, C.J. and Claesson, S., Light scattering measurements under high pressure. II. Dilute solution properties of polystyrene, *Chem. Scripta*, 9, 36, 1976.

1979HAM   Hammel, G.L., Schulz, G.V., and Lechner, M.D., Die Knäueldimensionen von Polystyrol in verschiedenen Lösungsmitteln unter hohen Drucken, *Eur. Polym. J.*, 15, 209, 1979.

1981LEC   Lechner, M.D. and Steinmeier, D.G., Das Verhalten von Polyacrylamidlösungen unter hohen Drücken, *Macromol. Chem., Rapid Commun.*, 2, 421, 1981.

1987VEN   Vennemann, N., Lechner, M.D., and Oberthür, R.C., Thermodynamics and conformation of polyoxyethylene in aqueous solution under high pressure. 1. Small-angle neutron scattering and densitometric measurements at room temperature, *Polymer*, 28, 1738, 1987.

1991STA   Stankovic, R.I., Jovanovic, S., Ilic, L., Nordmeier, E., and Lechner, M.D., Study of aqueous dextran solutions under high pressures and different temperatures by dynamic light scattering, *Polymer*, 32, 235, 1991.

2003MOS   Moses, C.L. and van Hook, W.A., Pressure dependence of the second virial coefficient of dilute polystyrene solutions, *J. Polym. Sci.: Polym. Phys.*, 41, 3070, 2003.

APPENDICES

Appendix 1   List of systems and properties in order of the polymers

Polymer(s)	Solvent(s)	Property	Page(s)
Butylene succinate/butylene adipate copolymer			
	carbon dioxide	gas solubility	17
Butyl methacrylate/N,N-dimethylaminoethyl methacrylate block copolymer			
	carbon dioxide	gas solubility	18
Butyl methacrylate/perfluoroalkylethyl acrylate block copolymer			
	carbon dioxide	gas solubility	19
Dextran			
	water	A_2	605
Ethylene/acrylic acid copolymer			
	–	PVT	551
	ethene	HPPE	227-232
	ethene + acrylic acid	HPPE	407-408
	ethene + n-decane	HPPE	409
	ethene + ethanol	HPPE	409-410
	ethene + ethyl acetate	HPPE	410-411
	ethene + n-heptane	HPPE	411
	ethene + octanoic acid	HPPE	411-412
	ethene + 2,2,4-trimethyl-pentane	HPPE	412
Ethylene/1-butene copolymer			
	–	PVT	551-552
	1-butene	HPPE	232-234
	dimethyl ether	density	567
	dimethyl ether-d6	density	567-568
	ethane	density	568
	ethane + n-pentane-d12	density	592

Polymer(s)	Solvent(s)	Property	Page(s)
Ethylene/1-butene copolymer (*continued*)			
	ethene	HPPE	234-235
	ethene + 1-butene	HPPE	412
	n-pentane	density	568-569
	n-pentane-d12	density	569
	propane	HPPE	235-237
Ethylene/butyl acrylate copolymer			
	ethene	HPPE	237-241
	ethene + butyl acrylate	HPPE	413-414
	ethene + methyl acrylate	HPPE	414-415
Ethylene/butyl methacrylate copolymer			
	ethene	HPPE	241
Ethylene/ethyl acrylate copolymer			
	ethene	HPPE	242
Ethylene/2-ethylhexyl acrylate copolymer			
	ethene	HPPE	243
	ethene + 2-ethylhexyl acrylate	HPPE	415
Ethylene/1-hexene copolymer			
	ethene	HPPE	243-244
	ethene + n-butane	HPPE	415
	ethene + carbon dioxide	HPPE	415-416
	ethene + ethane	HPPE	416
	ethene + helium	HPPE	416
	ethene + 1-hexene	HPPE	417-418
	ethene + 1-hexene + n-butane	HPPE	418-419
	ethene + 1-hexene + helium	HPPE	419-420
	ethene + 1-hexene + methane	HPPE	420-421
	ethene + 1-hexene + nitrogen	HPPE	421
	ethene + methane	HPPE	422
	ethene + nitrogen	HPPE	422
	ethene + propane	HPPE	422-423
	2-methylpropane	HPPE	244-246
	2-methylpropane + 1-hexene	HPPE	423
	propane	HPPE	246-249
Ethylene/methacrylic acid copolymer			
	ethene	HPPE	249-251

Polymer(s)	Solvent(s)	Property	Page(s)
Ethylene/methyl acrylate copolymer			
	ethene	HPPE	251-254
Ethylene/methyl methacrylate copolymer			
	ethene	HPPE	254-255
	ethene + butyl acrylate	HPPE	423
	ethene + 2-ethylhexyl acrylate	HPPE	424
	ethene + n-heptane	HPPE	424
	ethene + methyl acrylate	HPPE	424-425
Ethylene/norbornene copolymer			
	ethene + toluene + bicyclo[2,2,1]-2-heptene	VLE	84-86
Ethylene/1-octene copolymer			
	–	PVT	552
	ethene	HPPE	255-256
	propane	HPPE	257
Ethylene/propyl acrylate copolymer			
	ethene	HPPE	258-259
Ethylene/propylene copolymer			
	–	PVT	553
	1-butene	VLE	19
	1-butene	HPPE	259-260
	ethene	gas solubility	19-20
	ethene	HPPE	260-262
	ethene + 1-butene	HPPE	425
	ethene + carbon dioxide	HPPE	425-426
	ethene + 1-hexene	HPPE	426-427
	1-hexene	HPPE	262-263
	propane	HPPE	263
	propene	HPPE	264-266
	propene + ethylene/propylene copolymer	HPPE	427-428
	propene + ethylene/propylene copolymer + n-dodecane	HPPE	428
Ethylene/propylene block copolymer			
	ethene + n-hexane	VLE, K^∞	86
	propene	gas solubility	20
	propene + n-hexane	VLE, K^∞	87

Polymer(s)	Solvent(s)	Property	Page(s)
Ethylene/propylene/isoprene terpolymer			
	dimethyl ether	HPPE	267
	ethane	HPPE	267
	ethene	HPPE	267-268
	propane	HPPE	268
	propene	HPPE	268
Ethylene/vinyl acetate copolymer			
	–	PVT	553-554
	ethene	gas solubility	20-21
	ethene	HPPE	268-280
	ethene + carbon dioxide	HPPE	428
	ethene + ethane	HPPE	429
	ethene + helium	HPPE	429
	ethene + nitrogen	HPPE	430
	ethene + propane	HPPE	430
	ethene + vinyl acetate	HPPE	430-440
	ethene + vinyl acetate + carbon dioxide	HPPE	441
	ethene + vinyl acetate + helium	HPPE	442
	ethene + vinyl acetate + 2-methyl-2-propanol	HPPE	443-444
	ethene + vinyl acetate + nitrogen	HPPE	444-445
	cyclopentane	LLE	111
	cyclopentene	LLE	111
N-Isopropylacrylamide/1-deoxy-1-methacryl-amido-D-glucitol copolymer			
	water	LLE	111-112
DL-Lactide/glycolide copolymer			
	carbon dioxide	HPPE	281-282
	chlorodifluoromethane	HPPE	282-283
	dimethyl ether	HPPE	283-284
	trifluoromethane	HPPE	284
Nylon 6			
	carbon dioxide	gas solubility	21
	carbon dioxide + 2,2,2-trifluoroethanol	HPPE	445-446
	styrene	VLE	22

Polymer(s)	Solvent(s)	Property	Page(s)
Penta(ethylene glycol) monoheptyl ether			
	n-dodecane	LLE	112-113
	water	LLE	113-114
Poly(acrylamide)			
	water	A_2	606
Polybutadiene			
	carbon dioxide + cyclohexane	VLE	87-88
	carbon dioxide + cyclohexane	HPPE	446
	carbon dioxide + n-hexane	VLE, K^∞	88
	carbon dioxide + tetrahydrofuran	HPPE	446-448
	carbon dioxide + toluene	VLE	88-89
	carbon dioxide + toluene	HPPE	448-450
	cyclohexane	VLE	22
	1,2-dichlorobenzene + polystyrene	LLE	205-206
	ethene	HPPE	285-286
	n-hexane	density	570
Poly(1-butene)			
	–	PVT	554
	1-butene	HPPE	287-288
Poly(butyl acrylate)			
	–	PVT	556
	carbon dioxide	HPPE	288
Poly(butylene glycol)			
	carbon dioxide	HPPE	289
Poly(butylene oxide)-b-poly(ethylene oxide) diblock copolymer			
	carbon dioxide	HPPE	289
Poly(butyl methacrylate)			
	carbon dioxide	gas solubility	23-24
	carbon dioxide + butyl methacrylate	VLE	87
	carbon dioxide + butyl methacrylate	HPPE	450-451
	carbon dioxide	HPPE	288-289

Polymer(s)	Solvent(s)	Property	Page(s)
Poly(butylene succinate)			
	–	PVT	554-555
	carbon dioxide	gas solubility	22-23
Poly(butylene terephthalate)			
	–	PVT	555
Poly(butyl methacrylate)			
	ethanol	LLE	114
	ethene	HPPE	289
Poly(ε-caprolactone)			
	–	PVT	556-557
	chlorodifluoromethane	HPPE	289-290
Polycarbonate bisphenol-A			
	–	PVT	557
	carbon dioxide	gas solubility	24
Polycarbonate tetramethyl bisphenol-A			
	–	PVT	557-558
Poly(chlorotrifluoroethylene)			
	carbon dioxide	gas solubility	24
Poly(1,1-dihydroxyperfluorooctyl acrylate)			
	–	PVT	558
	carbon dioxide	HPPE	290
Poly(2,6-dimethyl-1,4-phenylene ether)			
	carbon dioxide	gas solubility	24-25
	carbon dioxide + polystyrene	gas solubility	89
Poly(dimethylsiloxane)			
	–	PVT	559
	benzene	A_2	606
	bromocyclohexane	A_2	606
	n-butane	LLE	114
	carbon dioxide	gas solubility	25-26
	carbon dioxide	swelling	25-26
	carbon dioxide + 1,1,1,3,3,3-hexadeutero-2-propanone	K_c	90

Polymer(s)	Solvent(s)	Property	Page(s)
Poly(dimethylsiloxane) (*continued*)			
	carbon dioxide + methanol	swelling	90-91
	carbon dioxide + 2-propanol	swelling	92-93
	carbon dioxide + 2-propanone	swelling	93
	carbon dioxide + toluene	VLE	93-94
	carbon dioxide	HPPE	290-291
	carbon dioxide	density	570-571
	chlorobenzene	A_2	606
	chlorodifluoromethane	gas solubility	26-27
	cyclohexane	PVT	571
	ethane	LLE	114-115
	hexamethyldisiloxane	PVT	571
	propane	LLE	115
	toluene	A_2	606
Poly(dimethylsiloxane) monomethacrylate			
	carbon dioxide	HPPE	291-292
Poly(dimethylsiloxane)-g-poly(ethylene oxide) graft copolymer			
	carbon dioxide	HPPE	292-293
Poly(dimethylsiloxane)-g-poly(ethylene oxide)-g-poly(propylene oxide) graft copolymer			
	carbon dioxide	HPPE	293-294
Poly(di-1H,1H,2H,2H-perfluorodecyl diitaconate)			
	carbon dioxide	HPPE	294
Poly(di-1H,1H,2H,2H-perfluorodecyl monoitaconate)			
	carbon dioxide	HPPE	295
Poly(di-1H,1H,2H,2H-perfluorododecyl diitaconate)			
	carbon dioxide	HPPE	294
Poly(di-1H,1H,2H,2H-perfluorododecyl monoitaconate)			
	carbon dioxide	HPPE	296

Polymer(s)	Solvent(s)	Property	Page(s)
Poly(di-1H,1H,2H,2H-perfluorohexyl diitaconate)			
	carbon dioxide	HPPE	295
Poly(di-1H,1H,2H,2H-perfluorohexyl monoitaconate)			
	carbon dioxide	HPPE	296
Poly(di-1H,1H,2H,2H-perfluorooctyl diitaconate)			
	carbon dioxide	HPPE	295
Poly(di-1H,1H,2H,2H-perfluorooctyl monoitaconate)			
	carbon dioxide	HPPE	296
Polyester (hyperbranched, aliphatic)			
	carbon dioxide + ethanol	HPPE	451-452
	carbon dioxide + water	HPPE	452-453
	ethanol	VLE	27
	water	VLE	27-28
Poly(ether sulfone)			
	–	PVT	559
Poly(ethyl acrylate)			
	ethene	HPPE	296
Polyethylene			
	–	PVT	560
	1-butene	VLE	28-30
	1-butene	HPPE	297
	carbon dioxide	gas solubility	30-31
	carbon dioxide + cyclohexane	HPPE	453-454
	carbon dioxide + n-heptane	HPPE	454
	cyclohexane	VLE	31-32
	cyclohexane	LLE	115-116
	cyclohexane + nitrogen	HPPE	454
	cyclohexane + polystyrene	LLE	204-205
	cyclopentane	LLE	116
	cyclopentene	LLE	116
	ethane	HPPE	298-299
	ethane + cyclohexane	HPPE	455
	ethene	gas solubility	32-37
	ethene	HPPE	299-349

Polymer(s)	Solvent(s)	Property	Page(s)
Polyethylene (*continued*)			
	ethene	density	32-33, 572-573
	ethene + butyl acrylate	HPPE	455
	ethene + n-hexane	HPPE	456-460
	ethene + n-hexane + 1-octene	HPPE	458-459
	ethene + 1-hexene	HPPE	460
	ethene + 4-methyl-1-pentene	HPPE	460-461
	ethene + polyethylene	HPPE	461-472
	ethene + vinyl acetate	HPPE	472-473
	n-heptane	VLE	37
	n-heptane	LLE	116
	n-hexane	VLE	37-38
	n-hexane	LLE	117-123
	n-hexane + nitrogen	VLE	94-95
	n-hexane + nitrogen	HPPE	474-476
	n-hexane + 1-octene	VLE	95
	n-hexane + 1-octene	LLE	202
	1-hexene	VLE	39-40
	methane	gas solubility	40
	2-methylpropane	gas solubility	40-41
	nitrogen	gas solubility	42-43
	1-octene	VLE	43
	1-octene	LLE	123-124
	n-pentane	VLE	43
	n-pentane	LLE	124-125
	n-pentane	density	573
	3-pentanol	VLE	43-44
	3-pentanone	VLE	44
	1-pentene	VLE	44
	propane	VLE	44-45
	propane	HPPE	349-350
	propane + cyclohexane	HPPE	476-477
	propane	density	574
	propyl acetate	VLE	45
	2-propylamine	VLE	45
Poly(ethylene glycol)			
	anisole	density	575
	anisole + poly(propylene glycol)	density	592-594
	carbon dioxide	gas solubility	45-47
	carbon dioxide	HPPE	351-357
	carbon dioxide + methanol	HPPE	477-478
	nitrogen	gas solubility	47-48
	1-octanol	density	576-577
	propane	gas solubility	49-50
	propane	HPPE	357-359

Polymer(s)	Solvent(s)	Property	Page(s)
Poly(ethylene glycol) dimethyl ether			
	carbon dioxide	gas solubility	51-52
	carbon dioxide	VLE	50-51
	carbon dioxide + water +		
	diisopropanolamine	gas solubility	95-96
	carbon dioxide	HPPE	359
	hydrogen	gas solubility	52
	hydrogen + water +		
	diisopropanolamine	gas solubility	96
	methanol	$\Delta_M H$	549
	methanol	VLE	52-53
	nitrogen	gas solubility	54
	nitrogen + water +		
	diisopropanolamine	gas solubility	96-97
Poly(ethylene glycol) monododecyl ether-b-poly(propylene glycol) diblock copolymer			
	carbon dioxide	gas solubility	54-55
Poly(ethylene glycol) monomethyl ether			
	anisole	density	578
	anisole + poly(propylene glycol)	density	595-596
Poly(ethylene oxide)			
	–	PVT	560
	deuterium oxide	A_2	607, 612
	water	LLE	125
	water	PVT	578-579
Poly(ethylene oxide)-b-poly(dimethyl-siloxane) diblock copolymer			
	toluene + poly(ethylene oxide)	LLE	203
Poly(ethylene oxide)-b-poly(propylene oxide) diblock copolymer			
	carbon dioxide	HPPE	359-360
Poly(ethylene oxide)-b-poly(propylene oxide)-b-poly(ethylene oxide) triblock copolymer			
	carbon dioxide	HPPE	360-362

Polymer(s)	Solvent(s)	Property	Page(s)
Polyisobutylene (*continued*)			
	n-pentane	A_2	607, 612
	propane	LLE	128
	propane	HPPE	371-372
	2,2,4-trimethylpentane	A_2	607, 612
Polyisoprene			
	carbon dioxide	gas solubility	55
	carbon dioxide	swelling	55
	dimethyl ether	HPPE	372-373
	ethane	HPPE	373
	ethene	HPPE	373
	propane	HPPE	373
	propene	HPPE	374
Poly(*N*-isopropylacrylamide)			
	deuterium oxide	LLE	128
	water	LLE	128-129
Poly(DL-lactide)			
	–	PVT	561
	carbon dioxide	HPPE	374-375
	carbon dioxide + dimethyl ether	HPPE	486-487
	carbon dioxide + trichloromethane	HPPE	488
	chlorodifluoromethane	HPPE	375-377
	difluoromethane	HPPE	378
	dimethyl ether	HPPE	378-379
	1,1,1,2-tetrafluoroethane	HPPE	379-380
	trifluoromethane	HPPE	380
Poly(L-lactide)			
	carbon dioxide	HPPE	375
	chlorodifluoromethane	HPPE	376-377
	carbon dioxide + chlorodifluoromethane	HPPE	480-484
	carbon dioxide + dichloromethane	HPPE	484-486
	carbon dioxide + 2-propanone	HPPE	487
	carbon dioxide + trichloromethane	HPPE	487

Polymer(s)	Solvent(s)	Property	Page(s)
Poly(methyl acrylate)			
	propene	HPPE	381
Poly(methyl methacrylate)			
	carbon dioxide	gas solubility	55-58
	carbon dioxide	swelling	55-56
	carbon dioxide	T_g	56
	carbon dioxide + 2-butanone	HPPE	488-490
	carbon dioxide + dichloromethane	HPPE	490
	carbon dioxide + ethanol	HPPE	490
	carbon dioxide + methyl methacrylate	HPPE	490-491
	carbon dioxide + 2-propanone	HPPE	491-497
	carbon dioxide + tetrahydrofuran	HPPE	497-500
	carbon dioxide + toluene	HPPE	500-505
	carbon dioxide + trichloromethane	HPPE	505-506
	chlorodifluoromethane	HPPE	381
	ethene	T_g	58
	methane	T_g	58
	propene	HPPE	381
Poly(α-methylstyrene)			
	–	PVT	561
Poly(octadecyl acrylate)			
	ethene	HPPE	382
	ethene + octadecyl acrylate	HPPE	506-507
Poly(octadecyl methacrylate)			
	ethene + octadecyl methacrylate	HPPE	508
Poly(1-octene)			
	–	PVT	562
Poly(phenylmethylsiloxane)			
	–	PVT	562-563
Poly(propyl acrylate)			
	ethene	HPPE	382

Polymer(s)	Solvent(s)	Property	Page(s)
Polypropylene			
	–	PVT	563-564
	n-butane	HPPE	382-383
	n-butane + toluene	VLE	98
	n-butane + toluene	LLE	203
	1-butene	HPPE	384-385
	1-butene + toluene	VLE	98
	1-butene + toluene	LLE	204
	carbon dioxide	gas solubility	59
	nitrogen	gas solubility	59
	propane	HPPE	386
	propene	gas solubility	60
	propene	HPPE	387
Poly(propylene glycol)			
	–	PVT	564
	acetophenone	density	581-582
	anisole	density	583
	anisole + poly(ethylene glycol)	density	592-594
	anisole + poly(ethylene glycol) monomethyl ether	density	595-596
	ethanol	PVT	584
	n-hexane	PVT	584
	1-octanol	PVT	585
	water	PVT	586
Poly(propylene oxide)			
	carbon dioxide	HPPE	388-389
Poly(propylene oxide)-b-poly(ethylene oxide)-b-poly(propylene oxide) triblock copolymer			
	carbon dioxide	HPPE	389-391
Polystyrene			
	–	PVT	564-565
	acetaldehyde	LLE	130-138
	tert-butyl acetate	LCST/UCST	217
	2-butanone	A_2	607, 612
	carbon dioxide	gas solubility	60-62
	carbon dioxide	swelling	60-62
	carbon dioxide	T_g	62
	carbon dioxide + 2-butanone	HPPE	508-509
	carbon dioxide + cyclohexane	HPPE	509-514

Polymer(s)	Solvent(s)	Property	Page(s)
Polystyrene (*continued*)			
	4-methyl-2-pentanone	A_2	610, 612
	nitroethane	LLE	159-170
	nitrogen	gas solubility	66-67
	nitrogen	swelling	66-67
	2-propanone	LLE	170-176
	2-propanone	LCST/UCST	220
	2-propanone + 1,1,1,3,3,3-hexadeutero-2-propanone	LLE	207-211
	propionitrile	LLE	176-194
	propionitrile + polystyrene	LLE	211-213
	sulfur dioxide	VLE	68
	sulfur dioxide	HPPE	392-393
	1,1,2,2-tetrafluoroethane	gas solubility	68
	toluene	A_2	610, 612
	trichlorofluoromethane	VLE	68
	trichloromethane	A_2	610-612
Polystyrene-b-polybutadiene-b-polystyrene triblock copolymer			
	carbon dioxide	gas solubility	69
Polystyrene-b-polyisoprene diblock copolymer			
	carbon dioxide	gas solubility	69
	carbon dioxide	swelling	69
Polystyrene-b-poly(methyl methacrylate) diblock copolymer			
	carbon dioxide	gas solubility	70-71
	carbon dioxide	swelling	70
Polystyrene-b-poly(vinyl pyridine) diblock copolymer			
	carbon dioxide	gas solubility	71
	carbon dioxide	swelling	71
Polysulfone			
	carbon dioxide + tetrahydrofuran	HPPE	528-529
	carbon dioxide + tetrahydrofuran	density	598-601

Polymer(s)	Solvent(s)	Property	Page(s)
Poly(1,1,2,2-tetrahydroperfluorodecyl acrylate)			
	carbon dioxide	HPPE	391-392
Poly[P-tris(trifluoroethoxy)-N-trimethylsilyl] phosphazene			
	carbon dioxide	HPPE	392
Poly(vinyl acetate)			
	carbon dioxide	gas solubility	71-73
	carbon dioxide	swelling	72
	carbon dioxide + benzene	VLE, K^∞	98-99
	carbon dioxide + toluene	VLE, K^∞	99
	cyclopentane	LLE	194
	cyclopentene	LLE	194
Poly(vinyl chloride)			
	–	PVT	565-566
	1,2-dimethylbenzene	LLE	195
	phenetole	LLE	195
	tetrahydrofuran + water	LLE	213-214
Poly(vinyl ethyl ether)			
	carbon dioxide	HPPE	393-394
	carbon dioxide + toluene	HPPE	529-530
Poly(vinylidene fluoride)			
	–	PVT	566
Poly(vinyl methyl ether)			
	–	PVT	566
Poly(vinyl pyridine)			
	carbon dioxide	gas solubility	73
	carbon dioxide	swelling	73
Starch			
	carbon dioxide + water	gas solubility	99

Polymer(s)	Solvent(s)	Property	Page(s)
Tetra(ethylene glycol) dimethyl ether			
	n-heptane	density	586-588
	methanol	$\Delta_M H$	550
	1,1,1,2-tetrafluoroethane	density	589-591
Tetra(ethylene glycol) monododecyl ether			
	carbon dioxide	HPPE	394

Appendix 2   List of solvents in alphabetical order

Name	Formula	CAS-RN	Page(s)
acetaldehyde	C_2H_4O	75-07-0	130-138
acetophenone	C_8H_8O	98-86-2	581-582
acrylic acid	$C_3H_4O_2$	79-10-7	407-408
anisole	C_7H_8O	100-66-3	575, 578, 583, 592-596
benzene	C_6H_6	71-43-2	98-99, 581, 606
bicyclo[2,2,1]-2-heptene	C_7H_{10}	498-66-8	84-86
bromocyclohexane	$C_6H_{11}Br$	108-85-0	606
n-butane	C_4H_{10}	106-97-8	98, 114, 126, 203, 382-383, 415, 418-419
2-butanone	C_4H_8O	78-93-3	488-490, 509, 607, 612
1-butcne	C_4H_8	106-98-9	19, 28-30, 98, 204, 232-233, 259-260, 287-288, 297, 384-385, 412, 425
tert-butyl acetate	$C_6H_{12}O_2$	540-88-5	217
butyl acrylate	$C_7H_{12}O_2$	141-32-2	413-414, 423, 455
butyl methacrylate	$C_8H_{14}O_2$	97-88-1	87, 450-451
carbon dioxide	CO_2	124-38-9	17-19, 21-26, 30-31, 45-47, 50-52, 54-62, 69-73, 87-94, 96-99, 281-282, 288-296, 351-357, 359-364, 374-375, 388-394, 415-416, 425-426, 428, 441, 445-454, 477-506, 508-525, 528-530, 570-571, 597-601
chlorobenzene	C_6H_5Cl	108-90-7	606
1-chloro-1,1-difluoroethane	$C_2H_3ClF_2$	75-68-3	62-63
chlorodifluoromethane	$CHClF_2$	75-45-6	26-27, 63, 282-283, 289-290, 365-366, 375-377, 381, 480-484
cyclohexane	C_6H_{12}	110-82-7	22, 87-88, 115-116, 138-139, 204-205, 217-218, 446, 453-454-455, 476-477, 509-514, 525-527, 571, 581, 607, 612
cyclopentane	C_5H_{10}	287-92-3	31-32, 111, 116, 194, 218-219
cyclopentene	C_5H_8	142-29-0	111, 116, 194
decalin	$C_{10}H_{18}$	91-17-8	608, 612
cis-decalin	$C_{10}H_{18}$	493-01-6	608, 612
trans-decalin	$C_{10}H_{18}$	493-02-7	139-140, 608-609, 612-613
n-decane	$C_{10}H_{22}$	124-18-5	409

Name	Formula	CAS-RN	Page(s)
deuterium oxide	D_2O	7789-20-0	128, 607, 612
1,2-dichlorobenzene	$C_6H_4Cl_2$	95-50-1	205-206
dichloromethane	CH_2Cl_2	75-09-2	484-486, 490
1,2-dichloro-1,1,2,2-tetrafluoroethane	$C_2Cl_2F_4$	76-14-2	63
diethyl ether	$C_4H_{10}O$	60-29-7	219
diethyl oxalate	$C_6H_{10}O_4$	95-92-1	140
1,1-difluoroethane	$C_2H_4F_2$	75-37-6	64
difluoromethane	CH_2F_2	75-10-5	378
diisopropanolamine	$C_6H_{15}NO_2$	110-97-4	96-97
1,2-dimethylbenzene	C_8H_{10}	95-47-6	195
dimethyl ether	C_2H_6O	115-10-6	267, 283-284, 366-368, 372-373, 378-379, 486-487, 567
dimethyl ether-d6	C_2D_6O	17222-37-6	567-568
n-dodecane	$C_{12}H_{26}$	112-40-3	112-113, 428
dodecadeuteromethylcyclopentane	C_6D_{12}	144120-51-4	206-207
ethane	C_2H_6	74-84-0	114-115, 267, 298-299, 368-370, 373, 416, 429, 455, 525-526, 528, 568, 592
ethanol	C_2H_6O	64-17-5	27, 114, 409-410, 451-452, 490, 578, 584
ethene	C_2H_4	74-85-1	19-21, 32-37, 58, 64, 84-86, 227-232, 234-235, 237-244, 249-256, 258-262, 267-280, 285-286, 289, 296, 299-349, 370, 373, 382, 407-445, 455-473, 507-508, 572-573
ethyl acetate	$C_4H_8O_2$	141-78-6	410-411, 609-610, 613
ethyl formate	$C_3H_6O_2$	109-94-4	219
ethyl methacrylate	$C_6H_{10}O_2$	97-63-2	97, 478
2-ethylhexyl acrylate	$C_{11}H_{20}O_2$	103-11-7	415, 424
ethyl phenyl ether	$C_8H_{10}O$	103-73-1	195
helium	He	7440-59-7	416, 419-420, 429, 442
n-heptane	C_7H_{16}	142-82-5	37, 55, 97, 116, 126, 411, 424, 454, 480, 586-588
1,1,1,3,3,3-hexadeutero-2-propanone	C_3D_6O	666-52-4	90, 141-145, 207-211
hexamethyldisiloxane	$C_6H_{18}OSi_2$	107-46-0	571
n-hexane	C_6H_{14}	110-54-3	37-38, 86-88, 94-95, 117-123, 126, 202, 456-460, 474-476, 570, 584
1-hexene	C_6H_{12}	592-41-6	39-40, 262-263, 417-421, 423, 426-427, 460
hexyl acrylate	$C_9H_{16}O_2$	2499-95-8	479
hexyl methacrylate	$C_{10}H_{18}O_2$	142-09-6	479-480
hydrogen	H_2	1333-74-0	52, 64, 94-96
isobutyl methyl ketone	$C_6H_{12}O$	108-10-1	610, 613
methane	CH_4	74-82-8	40, 59, 64-65, 420-422

Name	Formula	CAS-RN	Page(s)
methanol	CH$_4$O	67-56-1	52-53. 90-91, 477-478, 549-550
methanol-d4	CD$_4$O	811-98-3	91
methoxybenzene	C$_7$H$_8$O	100-66-3	575, 578, 583, 592-596
methyl acetate	C$_3$H$_6$O$_2$	79-20-9	145-146
methyl acrylate	C$_4$H$_6$O$_2$	96-33-3	414-415, 424-425
2-methylbutane	C$_6$H$_{12}$	78-78-4	127, 607, 612
methylcyclohexane	C$_7$H$_{14}$	108-87-2	65, 146-157, 219-220, 514-516, 613
methylcyclopentane	C$_6$H$_{12}$	96-37-7	158-159, 206-207
methyl methacrylate	C$_5$H$_8$O$_2$	82-62-6	490-491
4-methyl-2-pentanone	C$_6$H$_{12}$O	108-10-1	610, 613
4-methyl-1-pentene	C$_6$H$_{12}$	691-37-2	460-461
methyl phenyl ketone	C$_8$H$_8$O	98-86-2	581-582
2-methylpropane	C$_4$H$_{10}$	75-28-5	40-41, 244-246
2-methyl-2-propanol	C$_4$H$_{10}$O	75-65-0	443-444
nitroethane	C$_2$H$_5$NO$_2$	79-24-3	159-169
nitroethane-d5	C$_2$D$_5$NO$_2$	57817-88-6	165-166, 169-170
nitrogen	N$_2$	7727-37-9	42-43, 47-48, 54, 59, 66-67, 96-97, 421-422, 430, 444-445, 454, 474-476, 526-527
norbornene	C$_7$H$_{10}$	498-66-8	84-86
octadecyl acrylate	C$_{21}$H$_{40}$O$_2$	4813-57-4	507
octadecyl methacrylate	C$_{22}$H$_{42}$O$_2$	32360-05-7	508
octanoic acid	C$_8$H$_{16}$O$_2$	124-07-2	411-412
1-octanol	C$_8$H$_{18}$O	111-87-5	576-577, 585
1-octene	C$_8$H$_{16}$	111-66-0	43, 95, 123-124, 202, 458-459
n-pentane	C$_5$H$_{12}$	109-66-0	43, 124-125, 127, 568-569, 573, 607, 612
n-pentane-d12	C$_5$D$_{12}$	2031-90-5	569, 592
3-pentanol	C$_5$H$_{12}$O	584-02-1	43-44
3-pentanone	C$_5$H$_{10}$O	96-22-0	44
1-pentene	C$_5$H$_{10}$	109-67-1	44
phenetole	C$_8$H$_{10}$O	103-73-1	195
propane	C$_3$H$_8$	74-98-6	44-45, 48-50, 115, 128, 235-237, 246-249, 257, 263, 268, 349-351, 357-359, 371-373, 386, 422-423, 430, 476-477, 527, 574
2-propanol	C$_3$H$_8$O	67-63-0	92
2-propanol-d8	C$_3$D$_8$O	22739-76-0	92-93
2-propanone	C$_3$H$_6$O	67-64-1	93, 170-176, 207-211, 220, 487, 491-497, 579-580
propene	C$_3$H$_6$	115-07-1	20, 60, 264-266, 268, 374, 381, 387, 427-428
propionitrile	C$_3$H$_5$N	107-12-0	176-194, 211-213
propyl acetate	C$_5$H$_{10}$O$_2$	109-60-4	45

Name	Formula	CAS-RN	Page(s)
2-propylamine	C_3H_9N	75-31-0	45
styrene	C_8H_8	100-42-5	22
sulfur dioxide	SO_2	7446-09-5	68, 391-392
1,1,1,2-tetrafluoroethane	$C_2H_2F_4$	811-97-2	68, 379-380, 589-591
tetrahydrofuran	C_4H_8O	109-99-9	213-214, 446-448, 498-500, 516-519, 528-529, 580, 598-601
toluene	C_7H_8	108-88-3	84, 88-89, 93-94, 98-99, 203-204, 448-450, 500-505, 520-523, 528-530, 597, 606, 610, 613
trichlorofluoromethane	CCl_3F	75-69-4	68
trichloromethane	$CHCl_3$	67-66-3	487-488, 505-506, 524-525, 610-611, 613
2,2,2-trifluoroethanol	$C_2H_3F_3O$	75-89-8	445-446
trifluoromethane	CHF_3	75-46-7	284, 380
2,2,4-trimethylpentane	C_8H_{18}	540-84-1	412, 607, 612
vinyl acetate	$C_4H_6O_2$	108-05-4	21, 430-445, 472-473
water	H_2O	7732-18-5	27-28, 96-97, 99, 111-114, 125, 128-129, 213-214, 452-453, 578-579, 586, 605-606

Appendix 3   List of solvents in order of their molecular formulas

Formula	Name	CAS-RN	Page(s)
CCl_3F	trichlorofluoromethane	75-69-4	68
CD_4O	methanol-d4	811-98-3	91
$CHClF_2$	chlorodifluoromethane	75-45-6	26-27, 63, 282-283, 289-290, 365-366, 375-377, 381, 480-484
$CHCl_3$	trichloromethane	67-66-3	487-488, 505-506, 524-525, 610-611, 613
CHF_3	trifluoromethane	75-46-7	284, 380
CH_2Cl_2	dichloromethane	75-09-2	484-486, 490
CH_2F_2	difluoromethane	75-10-5	378
CH_4	methane	74-82-8	40, 59, 64-65, 420-422
CH_4O	methanol	67-56-1	52-53. 90-91, 477-478, 549-550
CO_2	carbon dioxide	124-38-9	17-19, 21-26, 30-31, 45-47, 50-52, 54-62, 69-73, 87-94, 96-99, 281-282, 288-296, 351-357, 359-364, 374-375, 388-394, 415-416, 425-426, 428, 441, 445-454, 477-506, 508-525, 528-530, 570-571, 597-601
$C_2Cl_2F_4$	1,2-dichloro-1,1,2,2-tetrafluoroethane	76-14-2	63
$C_2D_5NO_2$	nitroethane-d5	57817-88-6	165-166, 169-170
C_2D_6O	dimethyl ether-d6	17222-37-6	567-568
$C_2H_2F_4$	1,1,1,2-tetrafluoroethane	811-97-2	68, 379-380, 589-591
$C_2H_3ClF_2$	1-chloro-1,1-difluoroethane	75-68-3	62-63
$C_2H_3F_3O$	2,2,2-trifluoroethanol	75-89-8	445-446
C_2H_4	ethene	74-85-1	19-21, 32-37, 58, 64, 84-86, 227-232, 234-235, 237-244, 249-256, 258-262, 267-280, 285-286, 289, 296, 299-349, 370, 373, 382, 407-445, 455-473, 507-508, 572-573
$C_2H_4F_2$	1,1-difluoroethane	75-37-6	64
C_2H_4O	acetaldehyde	75-07-0	130-138
$C_2H_5NO_2$	nitroethane	79-24-3	159-169
C_2H_6	ethane	74-84-0	114-115, 267, 298-299, 368-370, 373, 416, 429, 455, 525-526, 528, 568, 592

Formula	Name	CAS-RN	Page(s)
C_2H_6O	dimethyl ether	115-10-6	267, 283-284, 366-368, 372-373, 378-379, 486-487, 567
C_2H_6O	ethanol	64-17-5	27, 114, 409-410, 451-452, 490, 578, 584
C_3D_6O	1,1,1,3,3,3-hexadeutero-2-propanone	666-52-4	90, 141-145, 207-211
C_3D_8O	2-propanol-d8	22739-76-0	92-93
$C_3H_4O_2$	acrylic acid	79-10-7	407-408
C_3H_5N	propionitrile	107-12-0	176-194, 211-213
C_3H_6	propene	115-07-1	20, 60, 264-266, 268, 374, 381, 387, 427-428
C_3H_6O	2-propanone	67-64-1	93, 170-176, 207-211, 220, 487, 491-497, 579-580
$C_3H_6O_2$	ethyl formate	109-94-4	219
$C_3H_6O_2$	methyl acetate	79-20-9	145-146
C_3H_8	propane	74-98-6	44-45, 48-50, 115, 128, 235-237, 246-249, 257, 263, 268, 349-351, 357-359, 371-373, 386, 422-423, 430, 476-477, 527, 574
C_3H_8O	2-propanol	67-63-0	92
C_3H_9N	2-propylamine	75-31-0	45
$C_4H_6O_2$	methyl acrylate	96-33-3	414-415, 424-425
$C_4H_6O_2$	vinyl acetate	108-05-4	21, 430-445, 472-473
C_4H_8	1-butene	106-98-9	19, 28-30, 98, 204, 232-233, 259-260, 287-288, 297, 384-385, 412, 425
C_4H_8O	2-butanone	78-93-3	488-490, 509, 607, 612
C_4H_8O	tetrahydrofuran	109-99-9	213-214, 446-448, 498-500, 516-519, 528-529, 580, 598-601
$C_4H_8O_2$	ethyl acetate	141-78-6	410-411, 609-610, 613
C_4H_{10}	n-butane	106-97-8	98, 114, 126, 203, 382-383, 415, 418-419
C_4H_{10}	2-methylpropane	75-28-5	40-41, 244-246
$C_4H_{10}O$	diethyl ether	60-29-7	219
$C_4H_{10}O$	2-methyl-2-propanol	75-65-0	443-444
C_5D_{12}	n-pentane-d12	2031-90-5	569, 592
C_5H_8	cyclopentene	142-29-0	111, 116, 194
$C_5H_8O_2$	methyl methacrylate	82-62-6	490-491
C_5H_{10}	cyclopentane	287-92-3	31-32, 111, 116, 194, 218-219
C_5H_{10}	1-pentene	109-67-1	44
$C_5H_{10}O$	3-pentanone	96-22-0	44
$C_5H_{10}O_2$	propyl acetate	109-60-4	45
C_5H_{12}	n-pentane	109-66-0	43, 124-125, 127, 568-569, 573, 607, 612
$C_5H_{12}O$	3-pentanol	584-02-1	43-44
C_6D_{12}	dodecadeuteromethylcyclopentane	144120-51-4	206-207

Formula	Name	CAS-RN	Page(s)
$C_9H_{16}O_2$	hexyl acrylate	2499-95-8	479
$C_{10}H_{18}$	decalin	91-17-8	608, 612
$C_{10}H_{18}$	*cis*-decalin	493-01-6	608, 612
$C_{10}H_{18}$	*trans*-decalin	493-02-7	139-140, 608-609, 612-613
$C_{10}H_{18}O_2$	hexyl methacrylate	142-09-6	479-480
$C_{10}H_{22}$	n-decane	124-18-5	409
$C_{11}H_{20}O_2$	2-ethylhexyl acrylate	103-11-7	415, 424
$C_{12}H_{26}$	n-dodecane	112-40-3	112-113, 428
$C_{21}H_{40}O_2$	octadecyl acrylate	4813-57-4	507
$C_{22}H_{42}O_2$	octadecyl methacrylate	32360-05-7	508
D_2O	deuterium oxide	7789-20-0	128, 607, 612
H_2	hydrogen	1333-74-0	52, 64, 94-96
H_2O	water	7732-18-5	27-28, 96-97, 99, 111-114, 125, 128-129, 213-214, 452-453, 578-579, 586, 605-606
He	helium	7440-59-7	416, 419-420, 429, 442
N_2	nitrogen	7727-37-9	42-43, 47-48, 54, 59, 66-67, 96-97, 421-422, 430, 444-445, 454, 474-476, 526-527
SO_2	sulfur dioxide	7446-09-5	68, 391-392

INDEX